FORENSIC BIOLOGY

SECOND EDITION

RICHARD LI

John Jay College of Criminal Justice
New York, New York, USA

CRC Press
Taylor & Francis Group
Boca Raton London New York

CRC Press is an imprint of the
Taylor & Francis Group, an **informa** business

CRC Press
Taylor & Francis Group
6000 Broken Sound Parkway NW, Suite 300
Boca Raton, FL 33487-2742

© 2015 by Taylor & Francis Group, LLC
CRC Press is an imprint of Taylor & Francis Group, an Informa business

No claim to original U.S. Government works

Printed on acid-free paper
Version Date: 20140818

International Standard Book Number-13: 978-1-4398-8970-1 (Hardback)

Library of Congress Cataloging-in-Publication Data

Li, Richard.
    Forensic biology / Richard Li. -- Second edition.
        pages cm
    Includes bibliographical references and index.
    ISBN 978-1-4398-8970-1
    1. Forensic biology. 2. Forensic genetics. 3. Forensic serology. I. Title.

QH313.5.F67L5 2015
363.25'62--dc23   2014028662

Visit the Taylor & Francis Web site at
http://www.taylorandfrancis.com

and the CRC Press Web site at
http://www.crcpress.com

Contents

Preface.. xv
Acknowledgments...xvii
Author.. xix
Introduction.. xxi

SECTION I   Biological Evidence

Chapter 1   Crime Scene Investigation of Biological Evidence...3

    1.1   Protection of Crime Scene...3
    1.2   Recognition of Biological Evidence...3
    1.3   Searches...7
    1.4   Documentation 16
    1.5   Chain of Custody...17
    1.6   Collection of Biological Evidence...18
    1.7   Marking Evidence... 20
    1.8   Packaging and Transportation 20
    1.9   Final Survey and the Release of the Crime Scene 25
    1.10 Crime Scene Reconstruction... 29
    Bibliography 30

Chapter 2   Crime Scene Bloodstain Pattern Analysis 35

    2.1   Basic Biological Properties of Human Blood................................35
    2.2   Formation of Bloodstains 35
    2.3   Chemical Enhancement and Documentation of Bloodstain
         Evidence 37
    2.4   Analyzing Spatter Stains...39
         2.4.1   Velocity of Blood Droplets..39
         2.4.2   Determining the Directionality of the Stains.................39
         2.4.3   Determining Angles of Impact.. 40
         2.4.4   Determining Area of Origin..41
    2.5   Types of Bloodstain Patterns.. 44
         2.5.1   Passive Bloodstains... 44
         2.5.2   Transfer Bloodstains 46
         2.5.3   Projected Bloodstains.. 46
    Bibliography 51

Contents

Chapter 3   Forensic Biology: A Subdiscipline of Forensic Science........................53

     3.1   Common Disciplines of Forensic Laboratory Services 53
     3.2   Laboratory Analysis of Biological Evidence.................................53
         3.2.1   Identification of Biological Evidence 53
         3.2.2   Comparison of Individual Characteristics of
                 Biological Evidence.. 54
         3.2.3   Reporting Results and Expert Testimony 54
     3.3   Forensic Science Services Related to Forensic Biology.....................57
         3.3.1   Forensic Pathology 58
         3.3.2   Forensic Anthropology... 58
         3.3.3   Forensic Entomology... 60
         3.3.4   Forensic Odontology 64
     3.4   Brief History of the Development of Forensic Biology..................... 65
         3.4.1   Antigen Polymorphism... 65
         3.4.2   Protein Polymorphism.. 66
         3.4.3   DNA Polymorphism 66
                 3.4.3.1   Genes and Related Sequences........................ 66
                 3.4.3.2   Intergenic Noncoding Sequences.................... 66
                 3.4.3.3   Human DNA Polymorphic Markers................. 68
                 3.4.3.4   Forensic DNA Polymorphism Profiling 69
     Bibliography 70

Chapter 4   Sources of Biological Evidence 77

     4.1   Bodily Fluids.. 77
         4.1.1   Extracellular Nucleic Acids.. 77
     4.2   Cells 81
         4.2.1   Cell Surface Markers..81
         4.2.2   Nucleated Cells..81
         4.2.3   Mitochondria and Other Organelles 83
         4.2.4   Cytosol 84
                 4.2.4.1   Messenger RNAs 84
                 4.2.4.2   MicroRNAs 85
     4.3   Tissues 87
         4.3.1   Skin 87
                 4.3.1.1   Biology of Skin.. 87
                 4.3.1.2   Skin as Source of DNA Evidence 87
         4.3.2   Hair 88
                 4.3.2.1   Biology of Hair 89
                 4.3.2.2   Hair as Source of DNA Evidence 91
         4.3.3   Bone 92
                 4.3.3.1   Biology of Bone... 92
                 4.3.3.2   Bone as Source of DNA Evidence 94
         4.3.4   Teeth 98
                 4.3.4.1   Biology of Teeth.. 98
                 4.3.4.2   Teeth as Source of DNA Evidence 100
     Bibliography 102

SECTION II   Basic Techniques in Forensic Biology

Chapter 5   Nucleic Acid Extraction...111

    5.1   Basic Principles of DNA Extraction...111
        5.1.1   Cell and Tissue Disruption.......................................111
        5.1.2   Lysis of Cellular and Organelle Membranes.........................114
        5.1.3   Removal of Proteins and Cytoplasmic Constituents...........114
        5.1.4   Storage of DNA Solutions..115
        5.1.5   Contamination...115
    5.2   Methods of DNA Extraction...115
        5.2.1   Extraction with Phenol–Chloroform.............................115
            5.2.1.1   Cell Lysis and Protein Digestion.................115
            5.2.1.2   Extraction with Organic Solvents........................115
            5.2.1.3   Concentrating DNA.....................................115
        5.2.2   Extraction by Boiling Lysis and Chelation....................117
            5.2.2.1   Washing...117
            5.2.2.2   Boiling..117
            5.2.2.3   Centrifugation 117
        5.2.3   Silica-Based Extraction...118
            5.2.3.1   Cell Lysis and Protein Digestion 118
            5.2.3.2   DNA Adsorption onto Silica 118
            5.2.3.3   Washing...119
            5.2.3.4   Elution of DNA...119
        5.2.4   Differential Extraction..121
    5.3   Essential Features of RNA... 122
    5.4   Methods of RNA Extraction 124
        5.4.1   RNA–DNA Coextraction 124
        5.4.2   miRNA Extraction 124
    Bibliography 127

Chapter 6   DNA Quantitation...133

    6.1   Slot Blot Assay 133
    6.2   Fluorescent Intercalating Dye Assay................................... 134
    6.3   Quantitative PCR Assay..137
        6.3.1   Real-Time Quantitative PCR....................................137
            6.3.1.1   TaqMan Method...................................... 138
    Bibliography 140

Chapter 7   Amplification by Polymerase Chain Reaction 143

    7.1   Denaturation and Renaturation of DNA..............................143
    7.2   Basic Principles of Polymerase Chain Reaction 143
    7.3   Essential PCR Components...145
        7.3.1   Thermostable DNA Polymerases.................................145
        7.3.2   PCR Primers...146
        7.3.3   Other Components...147

Contents

7.4   Cycle Parameters...147
7.5   Factors Affecting PCR..149
        7.5.1   Template Quality 149
        7.5.2   Inhibitors 149
        7.5.3   Contamination 149
7.6   Reverse Transcriptase PCR for RNA-Based Assays.......... 150
        7.6.1   Reverse Transcription 150
        7.6.2   Oligodeoxynucleotide Priming 151
        7.6.3   Reverse Transcriptase PCR 151
Bibliography 153

Chapter 8   DNA Electrophoresis 159

8.1   Basic Principles...159
8.2   Supporting Matrices..159
        8.2.1   Agarose..160
        8.2.2   Polyacrylamide..161
                8.2.2.1   Polymerization Reaction...................................161
                8.2.2.2   Cross-Linking Reaction 161
                8.2.2.3   Denaturing Polyacrylamide Electrophoresis 163
8.3   Apparatus and Forensic Applications......................................163
        8.3.1   Slab Gel Electrophoresis 163
                8.3.1.1   Agarose Gel Electrophoresis..............................163
                8.3.1.2   Polyacrylamide Gel Electrophoresis.................. 164
        8.3.2   Capillary Electrophoresis 164
        8.3.3   Microfluidic Devices 165
                8.3.3.1   Modular Microfluidic Devices 167
                8.3.3.2   Integrated Microfluidic Devices 168
8.4   Estimation of DNA Size..168
        8.4.1   Relative Mobility..168
        8.4.2   Local Southern Method 170
Bibliography 171

Chapter 9   Detection Methods..175

9.1   Direct Detection of DNA in Gels 175
        9.1.1   Fluorescent Intercalating Dye Staining 175
        9.1.2   Silver Staining 175
9.2   Detection of DNA Probes in Hybridization-Based Assays.............177
        9.2.1   Radioisotope Labeled Probes 177
        9.2.2   Enzyme-Conjugated Probe with
                Chemiluminescence Reporting System........................ 180
        9.2.3   Biotinylation of DNA with Colorimetric
                Reporting Systems 180
                9.2.3.1   Biotin..180
                9.2.3.2   Enzyme-Conjugated Avidin................................ 180
                9.2.3.3   Reporter Enzyme Assay 181
9.3   Detection Methods for PCR-Based Assays 182
        9.3.1   Fluorescence Labeling..182

                9.3.1.1 Fluorescent Dyes 182
                9.3.1.2 Labeling Methods...182
                9.3.1.3 Fluorophore Detection 185
        Bibliography 186

Chapter 10 Serology Concepts 189

        10.1 Serological Reagents 189
                10.1.1 Immunogens and Antigens...189
                10.1.2 Antibodies...189
                        10.1.2.1 Polyclonal Antibodies.................................191
                        10.1.2.2 Monoclonal Antibodies..............................191
                        10.1.2.3 Antiglobulins 192
        10.2 Strength of Antigen–Antibody Binding 193
        10.3 Antigen–Antibody Binding Reactions...................................193
                10.3.1 Primary Reactions 194
                10.3.2 Secondary Reactions 194
                        10.3.2.1 Precipitation...195
                        10.3.2.2 Agglutination 197
        Bibliography 198

Chapter 11 Serology Techniques: Past, Current, and Future........................... 199

        11.1 Introduction to Forensic Serology.. 199
                11.1.1 The Scope of Forensic Serology.............................. 199
                11.1.2 Class Characteristics and Individual Characteristics
                        of Biological Evidence 199
                11.1.3 Presumptive and Confirmatory Assays................... 200
                11.1.4 Primary and Secondary Binding Assays.................. 200
        11.2 Primary Binding Assays 201
                11.2.1 Enzyme-Linked Immunosorbent Assay (ELISA) 201
                11.2.2 Immunochromatographic Assays 202
        11.3 Secondary Binding Assays... 204
                11.3.1 Precipitation-Based Assays.................................... 204
                        11.3.1.1 Immunodiffusion 204
                        11.3.1.2 Immunoelectrophoretic Methods........................ 206
                11.3.2 Agglutination-Based Assays.................................... 208
        11.4 DNA Methylation Assays for Bodily Fluid Identification...............210
        11.5 Forensic Applications of RNA-Based Assays and RNA Profiling..... 212
                11.5.1 Messenger RNA-Based Assays................................212
                11.5.2 MicroRNA-Based Assays.......................................214
        11.6 Proteomic Approaches Using Mass Spectrometry for Bodily
                Fluid Identification 215
                11.6.1 Mass Spectrometric Instrumentation for
                        Protein Analysis 215
                11.6.2 Analysis Strategies for Protein Identification 216
        11.7 Microbial DNA Analysis for Bodily Fluid Identification.................217
        11.8 Nondestructive Assays for the Identification of Bodily Fluids....... 220
        Bibliography 221

Contents

SECTION III   Identification of Biological Evidence

Chapter 12 Identification of Blood 231
  12.1 Biological Properties 231
        12.1.1 Red Blood Cells..231
        12.1.2 White Blood Cells...231
        12.1.3 Platelets..231
  12.2 Presumptive Assays for Identification 232
        12.2.1 Mechanisms of Presumptive Assays 232
                12.2.1.1 Oxidation–Reduction Reactions..................... 234
        12.2.2 Colorimetric Assays 234
                12.2.2.1 Phenolphthalin Assay....................................... 234
                12.2.2.2 Leucomalachite Green (LMG) Assay 234
                12.2.2.3 Benzidine and Derivatives 234
        12.2.3 Chemiluminescence and Fluorescence Assays.................. 236
                12.2.3.1 Luminol (3-Aminophthalhydrazide)..................... 237
                12.2.3.2 Fluorescin 238
        12.2.4 Factors Affecting Presumptive Assay Results..................... 238
                12.2.4.1 Oxidants 238
                12.2.4.2 Plant Peroxidases.. 239
                12.2.4.3 Reductants... 239
  12.3 Confirmatory Assays for Identification 239
        12.3.1 Microcrystal Assays.. 239
                12.3.1.1 Hemochromagen Crystal Assay........................... 239
                12.3.1.2 Hematin Crystal Assay.................................... 240
        12.3.2 Other Assays.. 240
        Bibliography 241

Chapter 13 Species Identification 245
  13.1 General Considerations 245
        13.1.1 Types of Antibodies.. 245
        13.1.2 Titration of Antibodies 246
        13.1.3 Antibody Specificity 246
        13.1.4 Optimal Conditions for Antigen–Antibody Binding........ 247
  13.2 Assays 247
        13.2.1 Immunochromatographic Assays... 247
                13.2.1.1 Identification of Human
                          Hemoglobin Protein 247
                13.2.1.2 Identification of Human Glycophorin
                          A Protein 248
        13.2.2 Double Immunodiffusion Assays... 251
                13.2.2.1 Ring Assay.. 251
                13.2.2.2 Ouchterlony Assay 251
        13.2.3 Crossed-Over Electrophoresis 251
        Bibliography 255

Chapter 14 Identification of Semen 257

    14.1 Biological Characteristics 257
          14.1.1 Spermatozoa 257
          14.1.2 Acid Phosphatase 259
          14.1.3 Prostate-Specific Antigen 259
          14.1.4 Seminal Vesicle–Specific Antigen 259
    14.2 Analytical Techniques for Identifying Semen 260
          14.2.1 Presumptive Assays 260
                14.2.1.1 Lighting Techniques for Visual Examination
                        of Semen Stains 260
                14.2.1.2 Acid Phosphatase Techniques 261
          14.2.2 Confirmatory Assays 264
                14.2.2.1 Microscopic Examination of Spermatozoa 264
                14.2.2.2 Identification of Prostate-Specific Antigen 265
                14.2.2.3 Identification of Seminal
                        Vesicle–Specific Antigen 269
                14.2.2.4 RNA-Based Assays 270
    Bibliography 271

Chapter 15 Identification of Saliva 277

    15.1 Biological Characteristics of Saliva 277
          15.1.1 Amylases 277
    15.2 Analytical Techniques for Identification of Saliva 279
          15.2.1 Presumptive Assays 279
                15.2.1.1 Visual Examination 279
                15.2.1.2 Determination of Amylase Activity 279
           15.2.2 Confirmatory Assays 281
                15.2.2.1 Identification of Human Salivary α-Amylase 281
                15.2.2.2 RNA-Based Assays 285
    Bibliography 286

Chapter 16 Identification of Vaginal Secretions and Menstrual Blood 289

    16.1 Identification of Vaginal Stratified Squamous Epithelial Cells 289
          16.1.1 Lugol's Iodine Staining and Periodic
                Acid–Schiff Method 290
          16.1.2 Dane's Staining Method 293
    16.2 Identification of Vaginal Acid Phosphatase 293
    16.3 Identification of Vaginal Bacteria 294
    16.4 Outlook for Confirmatory Assays of Vaginal Secretions 295
    16.5 Menstruation 295
          16.5.1 Uterine Cycle 296
          16.5.2 Uterine Endometrial Hemostasis 298
    16.6 D-dimer Assay 299
    16.7 Lactate Dehydrogenase Assay 301
    16.8 RNA-Based Assays 302
    Bibliography 303

Contents

Chapter 17 Identification of Urine, Sweat, Fecal Matter, and Vomitus......................... 307
 17.1 Identification of Urine... 307
  17.1.1 Urine Formation 307
  17.1.2 Presumptive Assays.. 308
   17.1.2.1 The Identification of Urea 308
   17.1.2.2 Identification of Creatinine 314
  17.1.3 Confirmative Assays..315
   17.1.3.1 Identification of Tamm–Horsfall Protein.............315
   17.1.3.2 Identification of 17-Ketosteroids............................315
 17.2 Identification of Sweat...316
  17.2.1 Biology of Perspiration...316
  17.2.2 Sweat Identification Assays 317
 17.3 Identification of Fecal Matter...318
  17.3.1 Fecal Formation 319
  17.3.2 Fecal Matter Identification Assays 319
   17.3.2.1 Macroscopic and Microscopic Examination........ 320
   17.3.2.2 Urobilinoids Tests 322
   17.3.2.3 Fecal Bacterial Identification................................. 324
 17.4 Identification of Vomitus.. 324
  17.4.1 Biology of Gastric Fluid... 324
  17.4.2 Vomitus Identification Assays 325
 Bibliography 326

SECTION IV Individualization of Biological Evidence

Chapter 18 Blood Group Typing and Protein Profiling 331
 18.1 Blood Group Typing..331
  18.1.1 Blood Groups...331
  18.1.2 ABO Blood Group System..331
   18.1.2.1 Biosynthesis of Antigens 333
   18.1.2.2 Molecular Basis of the ABO System 333
   18.1.2.3 Secretors 335
   18.1.2.4 Inheritance of A and B Antigens 336
  18.1.3 Forensic Applications of Blood Group Typing................... 336
  18.1.4 Blood Group Typing Techniques.. 336
   18.1.4.1 Lattes Crust Assay.. 336
   18.1.4.2 Absorption–Elution Assay 338
 18.2 Forensic Protein Profiling... 338
  18.2.1 Methods 339
   18.2.1.1 Matrices Supporting Protein Electrophoresis...... 340
   18.2.1.2 Separation by Molecular Weight........................... 340
   18.2.1.3 Separation by Isoelectric Point.............................. 341
  18.2.2 Erythrocyte Protein Polymorphisms.. 342
   18.2.2.1 Erythrocyte Isoenzymes 342
   18.2.2.2 Hemoglobin 342

18.2.3 Serum Protein Polymorphisms 344
Bibliography 345

Chapter 19 Variable Number Tandem Repeat Profiling 353

19.1 Restriction Fragment Length Polymorphism....................................353
    19.1.1 Restriction Endonuclease Digestion 354
    19.1.2 Southern Transfer.. 356
    19.1.3 Hybridization with Probes...457
        19.1.3.1 Multilocus Probe Technique 357
        19.1.3.2 Single-Locus Probe Technique.......................... 358
    19.1.4 Detection..359
    19.1.5 Factors Affecting RFLP Results 360
        19.1.5.1 DNA Degradation 360
        19.1.5.2 Restriction Digestion–Related Artifacts..............361
        19.1.5.3 Electrophoresis and Blotting Artifacts................ 362
19.2 Amplified Fragment Length Polymorphism 363
Bibliography 364

Chapter 20 Autosomal Short Tandem Repeat Profiling.. 369

20.1 Characteristics of STR Loci.. 369
    20.1.1 Core Repeat and Flanking Regions.................................... 369
    20.1.2 Repeat Unit Length 370
    20.1.3 Repeat Unit Sequences 370
20.2 STR Loci Commonly Used for Forensic DNA Profiling.................371
20.3 Forensic STR Analysis..375
    20.3.1 Determining the Genotypes of STR Fragments...................376
    20.3.2 Interpretation of STR Profiling Results.............................376
        20.3.2.1 Inclusion (Match) 376
        20.3.2.2 Exclusion 376
        20.3.2.3 Inconclusive Result 376
20.4 Factors Affecting Genotyping Results..376
    20.4.1 Mutations.. 377
        20.4.1.1 Mutations at STR Core Repeat Regions................ 377
        20.4.1.2 Chromosomal and Gene Duplications 380
        20.4.1.3 Point Mutations...381
    20.4.2 Amplification Artifacts.. 382
        20.4.2.1 Stuttering.. 382
        20.4.2.2 Nontemplate Adenylation 383
        20.4.2.3 Heterozygote Imbalance 383
        20.4.2.4 Allelic Dropout... 383
    20.4.3 Electrophoretic Artifacts... 384
        20.4.3.1 Pull-Up Peaks 384
        20.4.3.2 Spikes 384
20.5 Genotyping of Challenging Forensic Samples.................................. 384
    20.5.1 Degraded DNA 384

20.5.2 Low Copy Number DNA Testing....................................386
20.5.3 Mixtures..386
Bibliography 389

Chapter 21 Sex Chromosome Haplotyping and Gender Identification....................407

21.1 Y Chromosome Haplotyping..407
21.1.1 Human Y Chromosome Genome 407
21.1.1.1 Pseudoautosomal Region 407
21.1.1.2 Male-Specific Y Region 407
21.1.1.3 Polymorphic Sequences.........................408
21.1.2 Y-STR...409
21.1.2.1 Core Y-STR Loci...................................410
21.1.2.2 Multilocal Y-STR Loci...........................411
21.1.2.3 Rapidly Mutating Y-STR 411
21.2 X Chromosome Haplotyping...412
21.3 Sex Typing for Gender Identification...............................416
21.3.1 Amelogenin Locus..416
21.3.2 Other Loci..418
Bibliography 419

Chapter 22 Single Nucleotide Polymorphism Profiling 437

22.1 Basic Characteristics of SNPs...437
22.2 Forensic Applications of SNP Profiling 438
22.2.1 HLA-DQA1 Locus...438
22.2.1.1 DQα AmpliType and Polymarker Assays.............438
22.2.1.2 Allele-Specific Oligonucleotide Hybridization 440
22.2.2 Current and Potential Applications of SNP Analysis.........443
22.2.2.1 Application of SNP Analysis for Forensic Identification 443
22.2.2.2 Potential Applications of SNPs for Phenotyping..444
22.3 SNP Techniques 444
22.3.1 Next-Generation Sequencing Technologies....................444
22.3.2 DNA Samples, Sequencing Library, and Template Preparation 446
22.3.3 NGS Chemistry..447
22.3.4 NGS Coverage 447
Bibliography 451

Chapter 23 Mitochondrial DNA Profiling 461

23.1 Human Mitochondrial Genome.......................................461
23.1.1 Genetic Contents of Mitochondrial Organelle Genomes.....461
23.1.2 Maternal Inheritance of mtDNA..............................463
23.2 mtDNA Polymorphic Regions..463
23.2.1 Hypervariable Regions...463
23.2.2 Heteroplasmy 463

23.3 Forensic mtDNA Testing 465
    23.3.1 General Considerations 465
    23.3.2 mtDNA Screen Assay 466
    23.3.3 mtDNA Sequencing 467
        23.3.3.1 PCR Amplification 467
        23.3.3.2 DNA Sequencing Reactions 468
        23.3.3.3 Electrophoresis, Sequence Analysis, and
                Mitotype Designations 471
    23.3.4 Interpretation of mtDNA Profiling Results 472
        23.3.4.1 Exclusion 474
        23.3.4.2 Cannot Exclude 474
        23.3.4.3 Inconclusive Result 474
Bibliography 474

SECTION V   Forensic Issues

Chapter 24 Forensic DNA Databases: Tools for Crime Investigations 485

    24.1 Brief History of Forensic DNA Databases 485
    24.2 Infrastructure of CODIS 485
    24.3 Indexes of CODIS 487
    24.4 Database Entries 489
    24.5 Database Expansion 490
    24.6 DNA Profiles 491
    24.7 Routine Database Searches for Forensic Investigations 492
        24.7.1 Case-to-Offender Searches 492
        24.7.2 Case-to-Case Searches 493
        24.7.3 Search Stringency and Partial Matches 493
    24.8 Familial Searches 495
        24.8.1 Legal and Ethical Issues of Familial Search 495
        24.8.2 Familial Search Strategies 496
        24.8.2.1 Identity-by-State and Kinship Index Method 496
        24.8.2.2 Focusing on Rare Alleles 496
        24.8.2.3 Excluding Candidates through Y-STR Screening 498
Bibliography 498

Chapter 25 Evaluation of the Strength of Forensic DNA Profiling Results 503

    25.1 A Review of Basic Principles of Genetics 503
        25.1.1 Mendelian Genetics 503
        25.1.2 Population Genetics 504
        25.1.2.1 Allele Frequency 504
        25.1.2.2 Genotype Frequency 504
        25.1.2.3 Heterozygosity 504
        25.1.2.4 Hardy–Weinberg Principle 505
        25.1.2.5 Testing for HW Proportions of Population
                Databases 505
        25.1.2.6 Probability of Match 511

Contents

25.2 Statistical Analysis of DNA Profiling Results.................................512
  25.2.1 Genotypes...513
    25.2.1.1 Profile Probability..513
    25.2.1.2 Likelihood Ratio...516
  25.2.2 Haplotypes...516
    25.2.2.1 Mitotypes Observed in Database............................518
    25.2.2.2 Mitotype Not Observed in Database.....................518
Bibliography 518

Chapter 26 Quality Assurance and Quality Control..523
26.1 US Quality Standards...523
26.2 International Quality Standards..524
26.3 Laboratory Accreditation...525
26.4 Laboratory Validation 525
26.5 Proficiency Testing 526
26.6 Certification...526
26.7 Forensic DNA Analyst Qualifications 527
26.8 Code of Ethics of Forensic Scientists...528
Bibliography 529

Index 533

Preface

Since the first edition of this book, new developments in forensic biology have led to a rapid expansion of the knowledge of the field. Therefore, it is necessary to create a new edition. This edition provides updates in most chapters of the original edition. Additionally, three new chapters (Chapters 2, 16, and 17) have been added and approximately 200 new figures have been created for this edition. Just like the first edition of this book, the new edition aims to inspire an undergraduate audience to tackle new challenges in the forensic biology field. It is written specifically to provide a general understanding of forensic biology and assist students in becoming more knowledgeable about the field of forensic biology and the wealth of available information. My readers should find that this edition of the book contains useful information, presented in a way that is more easily understood. Hopefully, it will be utilized by students, particularly those interested in forensic biology, to further enhance their education and training. I will continue to be open to suggestions in the future.

Richard C. Li
John Jay College of Criminal Justice
City University of New York
New York, NY

Acknowledgments

Many people contributed to this book. I would like to thank my family members for their support during the many hours I spent sitting in front of my computer. I would like to thank Jan Li for spending numerous hours assisting with the completion of this book by handling tasks such as image preparation and text editing. I would also like to thank Zully Santiago and Robert Greco for their proofreading of the manuscript. Additionally, I must acknowledge that my students at John Jay College of Criminal Justice provided excellent comments, suggestions, and even a few criticisms. The editorial and production staff at Taylor & Francis Group should also be acknowledged for their contributions. I would like to thank Mark Listewnik and Becky Masterman, senior acquisitions editors, for excellent suggestions and support, as well as project editor Richard Tressider for helping to keep the project on track. I would also like to thank Michelle van Kampen, project manager at Deanta Global Publishing Services, for overseeing the production of the book.

Author

Richard Li earned his MS in forensic science from the University of New Haven and his PhD in molecular biology from the University of Wisconsin–Madison. After completing his PhD, Dr. Li was awarded a postdoctoral fellowship at Weill Medical College of Cornell University and subsequently worked as a research faculty member at the School of Medicine of Yale University. Dr. Li has served as a criminalist at the Department of Forensic Biology, the Office of Chief Medical Examiner of New York City. For the past decade, he has held faculty positions in forensic science programs in the United States. Currently, Dr. Li is an associate professor of forensic biology at John Jay College of Criminal Justice. Additionally, he serves as a faculty member for the PhD program in forensic science for the college. Dr. Li's current research interests include the identification and analysis of biological specimens that are potentially useful for forensic investigations.

Introduction

This text defines forensic biology as analyses performed in the forensic biology sections of forensic laboratories and thus focuses on forensic serology and forensic DNA analysis. The aim of this book is to emphasize the basic science and its application to forensic science in an effort to make the principles more understandable. In addition, it introduces the language of forensic biology, thus enabling students to become comfortable with its use, and it provides clear explanations of the principles of forensic analysis.

To convey a general understanding of the concepts of forensic biology, it is necessary to include explanations of various techniques that are utilized in the field. The intent is to provide students with a scientific grounding in the area of forensic biology by offering an introduction to methods and techniques utilized by forensic biology laboratories. The techniques introduced in this text are accompanied by brief background descriptions and discussions of basic principles and techniques. Schematic illustrations are included where necessary. The text also acknowledges the benefits and limitations that apply to forensic biology techniques. Forensic techniques that were used in the past are also described. Learning past examples of forensic tests can help students to review historical forensic cases.

This text contains five modules, organized by section: Section I, Biological Evidence (Chapters 1–4); Section II, Basic Techniques in Forensic Biology (Chapters 5–11); Section III, Identification of Biological Evidence (Chapters 12–17); Section IV, Individualization of Biological Evidence (Chapters 18–23); and Section V, Forensic Issues (Chapters 24–26). The 26 chapters are designed to be covered in a single-semester course.

SECTION I

Biological Evidence

Crime Scene Investigation of Biological Evidence

A forensic investigation involving biological evidence usually begins at the crime scene. The crime scene investigation process includes maintaining scene security, preparing documentation, and collecting and preserving physical evidence. A crime scene investigation requires teamwork and effort. Each team member should be assigned specific tasks (Figure 1.1).

1.1 Protection of Crime Scene

A crime scene investigation begins with the initial response to a scene (Figure 1.2). Securing and protecting the scene are important steps in a crime scene investigation (Figures 1.3 through 1.6), and this task is usually carried out by the first responding officer arriving on the scene. The entry of authorized personnel admitted to the scene should be documented using a log sheet (Figure 1.7). Suspects, witnesses, and living victims should be evacuated from the scene. If a victim is wounded, medical attention should be sought.

Appropriate supplies and devices should be used to prevent the contamination of evidence by investigators. Protective wear and devices including a face mask or shield, safety eyeglasses, a disposable coverall bodysuit, gloves, shoe covers, and a hairnet should be used (Figures 1.8 and 1.9). Exposure to bodily fluids may occur during a crime scene investigation. An investigator can be exposed to bodily fluids through the mucous membranes, skin exposure, and needlestick injuries (especially when investigating a clandestine drug laboratory scene). Therefore, biosafety procedures must be followed for the protection of personnel from infectious blood-borne pathogens such as the human immunodeficiency virus (HIV), hepatitis B virus (HBV), and hepatitis C virus (HCV); infectious aerosol tuberculosis pathogens; and other biohazardous materials.

1.2 Recognition of Biological Evidence

A preliminary survey should be carried out to evaluate potential evidence. In particular, the recognition of evidence plays a critical role in solving or prosecuting crimes. The priority of the potential evidence at crime scenes should be assessed based on each item's relevance to the solution of the case. Higher priority should be assigned to evidence with probative value to the case. For example, the evidence related to a *corpus delicti* is considered to be of the highest priority. *Corpus delicti* is a Latin term meaning "body of crime." In Western law, it primarily refers to the principle that in order for an individual to be convicted, it is necessary to prove the occurrence

Figure 1.1 A crime scene investigation team. (© Richard C. Li.)

Figure 1.2 A crime scene unit vehicle that is used to respond to crime scenes. This type of vehicle is usually outfitted with devices and supplies that investigators need when processing a crime scene, as well as evidence packaging materials, fingerprint collection kits, and DNA collection kits. Additionally, it can be equipped with a workstation for computer access, a refrigerator for storing chemicals, and a compact fuming hood for processing latent fingerprints, as well as equipment cabinets and drawers. (© Richard C. Li.)

Figure 1.3 Crime scene barrier tape is used to ensure that only investigators are admitted to the scene. (Courtesy of H. Brewster.)

Figure 1.4   A police officer guards the crime scene. (© Richard C. Li.)

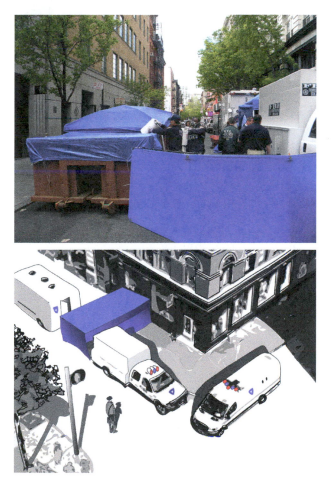

Figure 1.5   Crime scene privacy screen and tent. The screen (top) and the tent (bottom) are useful devices for shielding the evidence or body from viewing by unauthorized personnel. (© Richard C. Li.)

Figure 1.6   Barricades are set up to keep crowds at a distance from the scene. (© Richard C. Li.)

of the crime. In a forensic investigation, it also refers to the physical evidence proving that a crime was committed. For example, when an individual is missing, a missing persons investigation is usually initiated. If *corpus delicti*, such as a dead body or a victim's blood at a crime scene (Figure 1.10), is discovered during the investigation, a homicide case can be established and a suspect can be charged with homicide. Higher priority should also be attached to evidence that can establish connections such as *victim-to-perpetrator linkage*. For example, items found in a perpetrator's possession may be linked to a victim. This also applies to transfer evidence based on the principles of transfer theory, also known as the *Locard exchange principle*, which theorizes that the cross-transfer of evidence occurs when a perpetrator has any physical contact with an object or another person (Figures 1.11 through 1.14). Thus, trace evidence, such as hairs and fibers, may be transferred from a perpetrator to a victim or vice versa. This explains why it is important to ensure that perpetrators and their belongings are thoroughly searched for trace evidence. Likewise, victims and their belongings should be examined for the same reason.

Victim-to-scene and perpetrator-to-scene linkages can also be established. Blood belonging to a perpetrator or a victim found at a crime scene can establish such a linkage (Figure 1.15). Additionally, reciprocal transfers of trace evidence from crime scenes can be used to link a suspect or a victim to a crime scene. A perpetrator may present a unique *modus operandi* (MO). *Modus operandi*, a Latin term commonly used in criminal investigations, refers to a particular pattern of characteristics and the manner in which a crime is committed. For example, Richard Cottingham, a serial killer known as "the torso killer," dismembered his victims and took their limbs and heads with him but left their torsos at the scene. He then set the rooms on fire before

Crime Scene Processing

1: Date _____/_____/_____   2: Arrival Time: _____ HRS

3: Assigned Criminalist_____   4: Assisting Criminalist(s)_____

5: Supervisor_____ Present ____Yes ____No

6: _____ Case Number ____-_____

7: Individual/Agency Requesting Investigation _____

8: Location Address_____

    a. Type of Scene ____Indoor ____Outdoor
    b. Scene Secured ____Yes ____No
    c. Weather Conditions_____

9: Other Personnel	Shield #	Agency	Arrival	Departure
			HRS	HRS
			HRS	HRS
			HRS	HRS
			HRS	HRS
			HRS	HRS
			HRS	HRS
			HRS	HRS
			HRS	HRS
			HRS	HRS
			HRS	HRS
			HRS	HRS
			HRS	HRS
			HRS	HRS
			HRS	HRS
			HRS	HRS
			HRS	HRS
			HRS	HRS
			HRS	HRS
			HRS	HRS
			HRS	HRS

10: Departure Time _____ HRS

Figure 1.7 An example of a log sheet for documenting authorized personnel at a crime scene. (© Richard C. Li.)

fleeing the scenes. Evidence that provides information on the MO is also vital to an investigation. A distinct MO can establish a *case-to-case linkage* for serial offender cases.

1.3 Searches

Some investigations require a search for specific items of evidence such as biological stains, human remains, and all relevant evidence. A search usually has a specific purpose. Thus, the use of search patterns can be helpful, especially in cases involving large outdoor crime scenes.

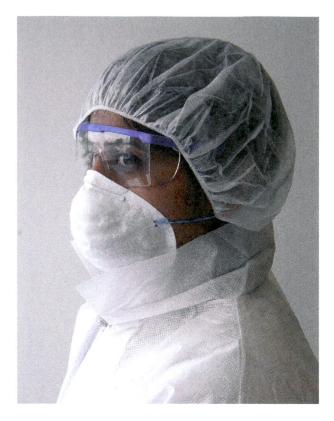

Figure 1.8   Personal protection wear and devices that are used at crime scenes. (© Richard C. Li.)

Figure 1.9   Disposable glove (left) and glove with extended cuff (right). (© Richard C. Li.)

Search patterns may include a grid, line, or zone (Figures 1.16 and 1.17). The method that is ultimately used depends on the type and size of the scene (Figures 1.18 through 1.20). Additionally, the points of entry and exit and the paths followed by a perpetrator should also be searched.

Searching for biological stains usually utilizes devices such as an alternate light source (ALS); see Figures 1.21 and 1.22. An ALS either produces a single specific wavelength of light or a desired wavelength by using specific filters. Biological materials such as blood, semen, and saliva emit fluorescent light under an ALS, which can facilitate the locating of biological materials. Additionally,

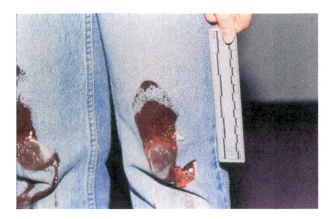

Figure 1.10 Photographic documentation of bloodstains on clothing. (© Richard C. Li.)

Figure 1.11 An electrostatic dust print lifting device can be utilized for processing impression evidence such as footprints and tire tracks. (© Richard C. Li.)

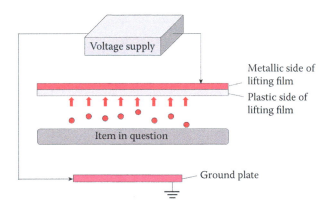

Figure 1.12 Basic components of an electrostatic dust print lifting device. The lifting film is placed on top of the item in question with the plastic side against the surface and the metallic side facing up. A ground plate is placed directly on the ground. The lifting film and ground plate are connected to the voltage supply apparatus. Once the charging voltage is turned on, the static charge transfers the dust particles from the surface to the plastic side of the lifting film. (© Richard C. Li.)

Figure 1.13   A high-intensity light-emitting diode (LED) device for locating evidence at a crime scene is particularly effective in highlighting trace evidence such as hairs, fibers, and shoe prints. (© Richard C. Li.)

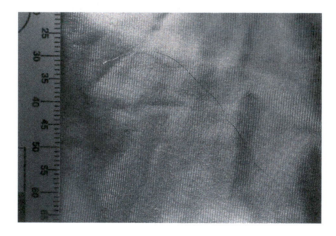

Figure 1.14   A hair found on a victim's clothing can be transferred evidence from a suspect. (© Richard C. Li.)

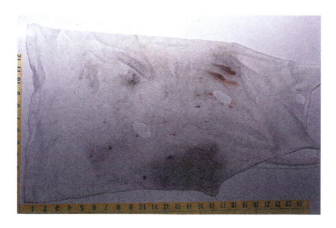

Figure 1.15   Finding a victim's blood on a suspect's clothing can establish a link between them. (© Richard C. Li.)

Figure 1.16   Grid search pattern for an outdoor scene. The investigators and anthropologists present are searching for human bone evidence within the grid. (Courtesy of H. Brewster.)

Figure 1.17   Line search pattern for an outdoor scene. (Courtesy of H. Brewster.)

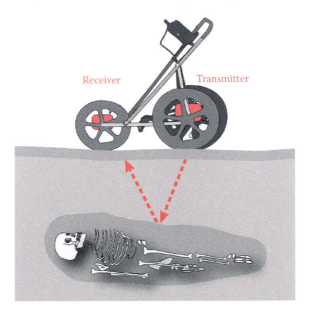

Figure 1.18   Using ground-penetrating radar (GPR) to locate clandestine graves of homicide victims. GPR uses electromagnetic waves emitted from a transmitter, which are detected by a receiver to locate clandestine burials and buried objects such as weapons embedded in soils. Images of the potential evidence are typically obtained by moving the antenna of the GPR device over the surface of the ground. (© Richard C. Li.)

Figure 1.19   Using cadaver-sniffing dogs to alert investigators to the presence of buried bodies. The odor produced by the decomposition of the human body may be sensed by cadaver-sniffing dogs. Odor-absorbing pads absorb and retain the scent of decomposed remains and can be placed at the scene for several days. Upon sniffing the pads, a cadaver dog may indicate the presence of buried human remains or may indicate that human remains were once buried in that location. However, it is not clear that this technique is reliable. (© Richard C. Li.)

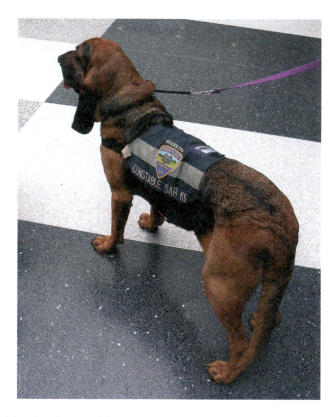

Figure 1.20   Tracking dog for searching suspects. In a situation of processing a recent crime scene when the suspect may be in close vicinity to the scene, a tracking dog can be used. A tracking dog, such as a bloodhound, can potentially follow the scent from items left at the scene to locate a suspect nearby. (© Richard C. Li.)

Figure 1.21   Compact alternate light source devices, which are intended specifically for use at crime scenes, reduce search time and improve the recovery of evidence such as biological evidence and chemically enhanced latent fingerprints. (© Richard C. Li.)

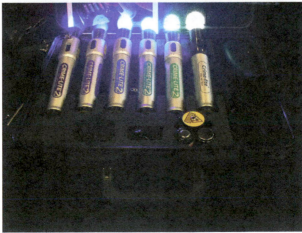

Figure 1.22   LED light sources provide illumination for locating evidence such as bodily fluid stains, hair, or fibers. (© Richard C. Li.)

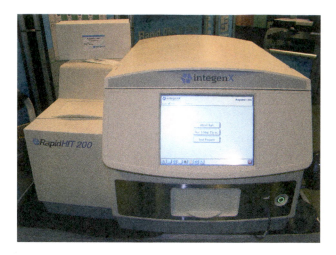

Figure 1.23 An example of a compact Rapid DNA device for processing DNA evidence in the field. (© Richard C. Li.)

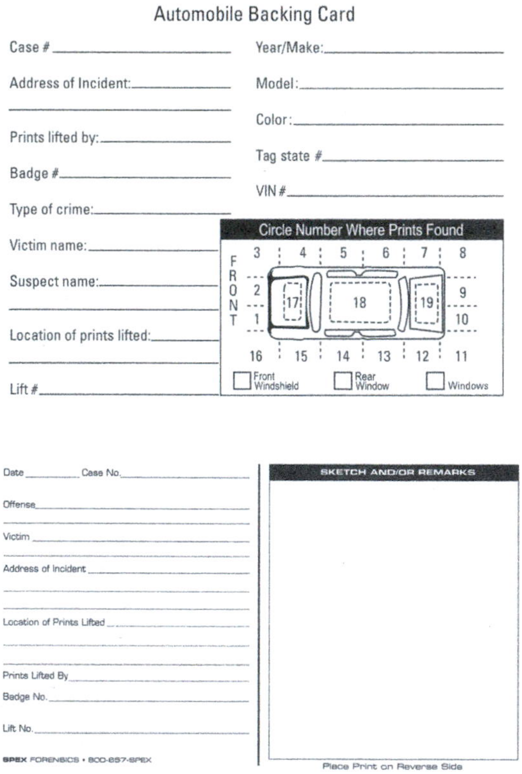

Automobile Backing Card

Case # _____   Year/Make: _____

Address of Incident: _____   Model: _____

_____   Color: _____

Prints lifted by: _____   Tag state # _____

Badge # _____   VIN # _____

Type of crime: _____

Victim name: _____

Suspect name: _____

_____

Location of prints lifted: _____

_____

Lift # _____

Circle Number Where Prints Found

F R O N T — 3 4 5 6 7 8 / 2 17 18 19 9 / 1 10 / 16 15 14 13 12 11

☐ Front Windshield   ☐ Rear Window   ☐ Windows

Date _____ Case No. _____

Offense _____

Victim _____

Address of Incident _____

_____

Location of Prints Lifted _____

_____

Prints Lifted By _____

Badge No. _____

Lift No. _____

SPEX FORENSICS • 800-897-SPEX

SKETCH AND/OR REMARKS

Place Print on Reverse Side

Figure 1.24 An example of sketch documentation. (© Richard C. Li.)

field tests and enhancement reagents can be used to facilitate crime scene searching (Chapter 12). These reagents can detect and identify biological evidence. The tests are very simple, rapid, and sensitive, and thus can be used at crime scenes. For example, phenolphthalin and leucomalachite green tests can be used for detecting blood evidence. Sometimes, minute amounts of blood may be

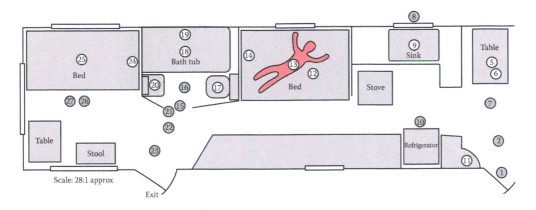

Figure 1.25　Sketches for documenting the sites of fingerprints. (© Richard C. Li.)

present at the scene as a result of attempts to clean up blood prior to the investigation. These stains may not be visible with the naked eye. Enhancement reagents such as luminol and fluorescein, which emit chemiluminant and fluorescent light upon reacting with certain biological materials, respectively, can be used. Additionally, the enhancement reagents can detect faint blood-containing pattern evidence such as faint bloody fingerprints, footprints, and other pattern evidence of physical contact such as drag marks in blood. However, precaution should be taken since these reagents are not usually very specific to blood. Certain substances such as bleach, various metals, and plants may also lead to chemical reactions with the field tests and the enhancement reagents. In these cases, the evidence collected is further tested with laboratory examination and analysis.

Recently, portable and field-deployable instruments have been developed that are capable of processing buccal swabs and potentially other evidence to produce a DNA profile on-site (Figure 1.23). It is a fully automated process, using the Rapid DNA technology (Chapter 8), that

Figure 1.26　A reflected ultraviolet imaging system (RUVIS) imager for documenting close-up views of evidence such as latent fingerprints. (© Richard C. Li.)

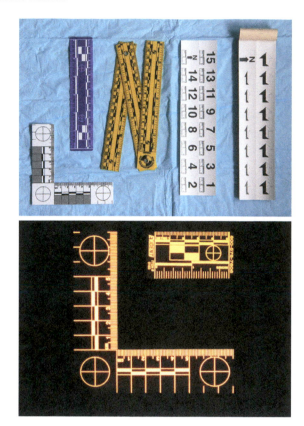

Figure 1.27 Scales for photographic documentation. Regular scales (top) can be used with visible light sources. Fluorescence scales (bottom) can be used for the documentation of fluorescing evidence after certain treatment, such as chemical enhancement, using alternate light sources. (© Richard C. Li.)

can be completed within 2 h by a trained crime scene investigator or police officer. These instruments may provide a new tool for expediting the identification of suspects and developing investigative leads at the scene. Additionally, this technology can enable law enforcement agents to rapidly determine whether the crimes were isolated incidents or part of serial crimes committed by the same offender, such as in serial burglary and arson cases. It can also be used in the identification of human remains in mass disasters.

1.4 Documentation

The conditions at a crime scene, including both the individual items of evidence and the overall scene, must be documented to provide vital information for investigators and for the courts. The most common documentation methods are drawing sketches and taking photographs and videographs. The sketch is to reflect the positions and the spatial relationships of items and persons with measurements using a scale. An investigator usually prepares a rough sketch first and later converts the rough sketch into a finished sketch (Figures 1.24 and 1.25). If bloodstains are present at the scene, the location of bloodstain patterns should be emphasized. Prior to handling and moving evidence, photographs should be taken with different views: an overall view of the entire scene, a medium-range view showing the positions and the relationships of items, and a close-up

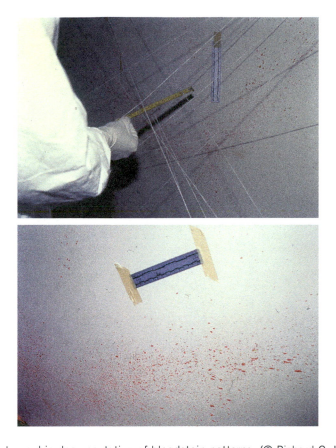

Figure 1.28 Photographic documentation of bloodstain patterns. (© Richard C. Li.)

view showing details of the evidence (Figure 1.26). Photographs should also include a measuring device such as a scale (Figure 1.27) to accurately depict the sizes of items such as bloodstains or bite marks. This can be achieved simply by placing a ruler adjacent to the evidence when it is photographed (see Figure 1.28). A photograph log sheet can be used to record the chronological order of crime scene photographs and to note filming conditions and any additional relevant information (Figure 1.29). Similar documentation should be prepared for videographs when appropriate. Additionally, written or audio-recorded notes can be used. Notes should include complete and accurate information of a crime scene investigation, such as the case identifier number, the identities of the investigators, and a description of the scene or items (e.g., location, size, and shape). Additionally, any disturbance of evidence occurring during crime scene processing should be noted.

1.5 Chain of Custody

Custody information should be recorded at each event when evidence is handled or transferred by authorized personnel. Usually, a custody form listing a specific evidence item is used to document the chain (Figure 1.30). Each individual who acquires custody of the evidence must sign a chain of custody document. An incomplete chain of custody may lead to an inference of possible tampering or contamination of evidence. As a result, the evidence may not be admissible in court.

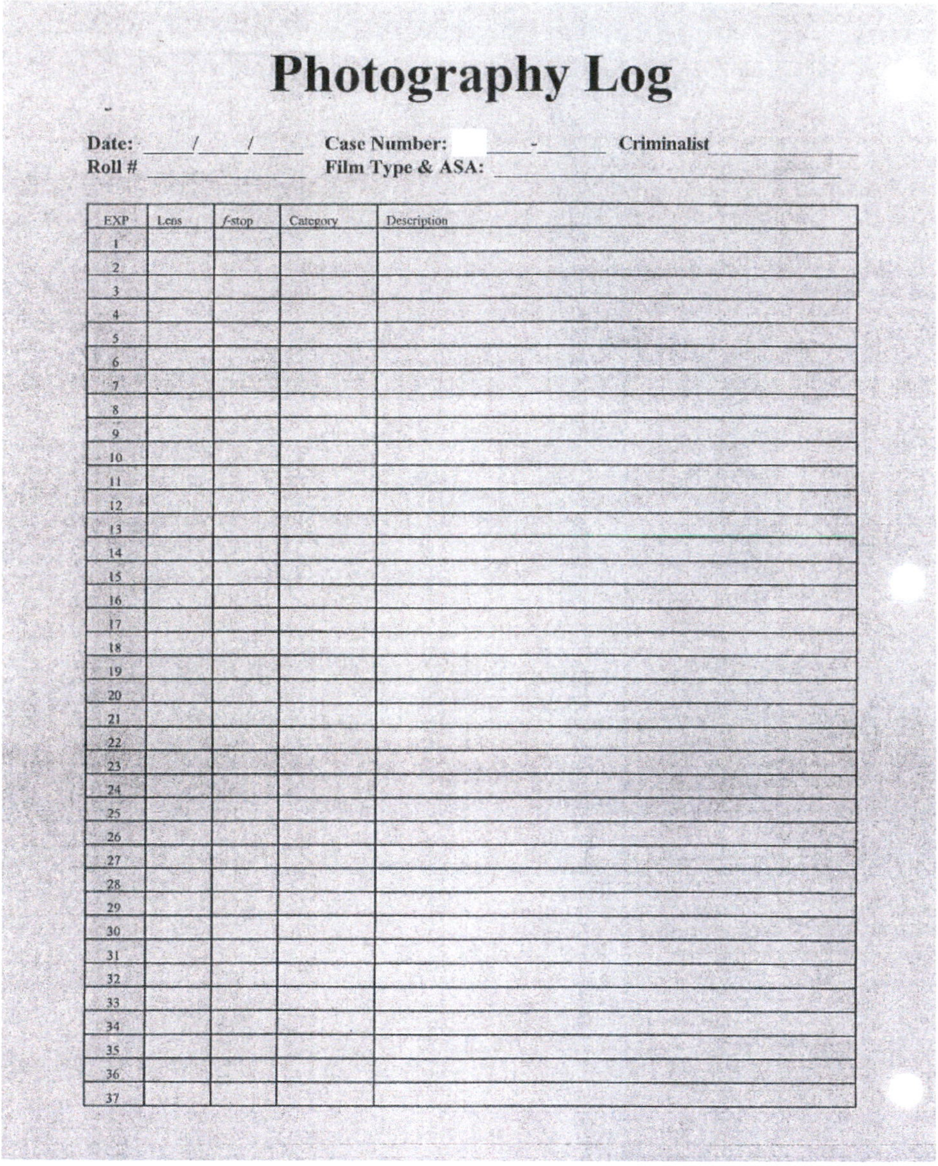

Photography Log

Date: _____ / _____ / _____   Case Number: [ ] - _____   Criminalist _____
Roll # _____   Film Type & ASA: _____

EXP	Lens	f-stop	Category	Description
1				
2				
3				
4				
5				
6				
7				
8				
9				
10				
11				
12				
13				
14				
15				
16				
17				
18				
19				
20				
21				
22				
23				
24				
25				
26				
27				
28				
29				
30				
31				
32				
33				
34				
35				
36				
37				

Figure 1.29   Crime scene photography log. (© Richard C. Li.)

1.6 Collection of Biological Evidence

After the crime scene documentation is completed, the collection of evidence can be initiated. Small or portable items, such as bloodstained knives, can be collected and submitted to a crime laboratory (Figure 1.31). Large or unmovable items of evidence (Figure 1.32) can be collected and submitted in sections, such as a section of wall where bloodstains are located. Table 1.1 and Figures 1.33 through 1.38 summarize and illustrate representative collection techniques. Specific care is required for the collection of biological evidence in the following situations:

Bloodstain pattern evidence: It is especially important to thoroughly document the bloodstain pattern evidence at a crime scene prior to collection. Bloodstain patterns can be especially useful in crime scene reconstruction.

Figure 1.30   Labels with the chain of custody that are used for marking the evidence contained in the packaging. (© Richard C. Li.)

Figure 1.31   Handling sharp objects. Bloodstained knives collected and submitted to laboratories (top) and a box for packing sharp objects (bottom). (© Richard C. Li.)

Figure 1.32   A section of bloodstained carpet is collected. (© Richard C. Li.)

- *Multiple analysis of evidence*: If multiple analyses are needed for a single item of evidence, nondestructive analyses should be carried out first. For example, a bloody fingerprint should be collected for ridge detail analysis prior to collecting blood for DNA analysis.

- *Trace evidence*: Trace evidence such as hairs and fibers can be present in bloodstained evidence and should be identified and properly collected.

- *Control samples*: Control (known or blank) samples should be collected from a control area (e.g., unstained area near a collected stain).

- *Size of stain*: Polymerase chain reaction (PCR)-based forensic DNA techniques are highly sensitive and allow for the successful analysis of very small bloodstains. All bloodstains, even if they are barely visible, should be collected at a crime scene.

- *Wet evidence*: Wet evidence should be air-dried (without heat) prior to packaging to prevent the degradation of proteins and nucleic acids, which are used for forensic serological and DNA analysis.

1.7 Marking Evidence

The marking of evidence is necessary for identification purposes so that it can be quickly recognized even years later (Figure 1.39). An investigator's initials, the item number, and the case number are usually included in marking. Information can be marked on a tag, a label attached to the item, or directly on garment evidence. The marking of evidence should not be proximal to bullet holes or biological stains to prevent the mark from interfering with analyses.

1.8 Packaging and Transportation

Packaging is intended to protect and preserve evidence. All evidence should be secured and protected from possible contamination. Fragile items should be protected to prevent any damage during transportation. Exposure to heat and humidity should be avoided to protect biological evidence from degradation during transport. Various packaging methods are available

Table 1.1 Methods for Collecting Biological Evidence

Type of Evidence	Condition	Method of Collection	Procedure
Blood	Dry	Swab	Best on nonabsorbent surfaces; lightly moisten sterile swab with distilled or sterile water, rub over stain while rotating; allow to air-dry; a combination of first a moistened swabbing followed by second a dry swabbing (both swabs submitted) is recommended
		Cutting	Cut stain from item
		Scraping	Scrape bloodstain into a clean piece of paper using a clean blade; wrap sample using druggist's fold
		Lifting	Works for nonabsorbent surfaces; use fingerprint lifting tape that does not interfere with DNA testing to lift stain; lifted stain should be covered with a piece of lifter's cover
		Collect entire item	Collect if item contains bloodstain pattern; difficult to swab; requires multiple exams
	Wet	Swab	Absorb blood sample onto sterile cotton swabs; stain should be concentrated on tip and allowed to air-dry
		FTA paper	Use a sterile disposable pipet to collect liquid blood; spot on FTA paper; allow to air-dry
	Reference liquid blood sample	Venous blood collection	Collect blood in a purple-topped vacutainer tube containing ethylenediaminetetraacetic acid (EDTA) anticoagulant; refrigerate but never freeze
Semen	Dry or wet	See "Blood" above	See "Blood" above
	Condom	Collect entire item	Secure condom with a tie and place in a container in a refrigerator; submit to laboratory as soon as possible
	Various conditions	Victim sexual assault kit	Standardized kit to collect biological evidence from the body of a victim includes swabs, microscope slides, and envelopes
	Various conditions	Suspect standard kit	Standardized kit to collect biological evidence from the body of a suspect includes swabs, microscope slides, and envelopes

(continued)

Table 1.1	(Continued) Methods for Collecting Biological Evidence		
Type of Evidence	Condition	Method of Collection	Procedure
Victim vaginal fluid	Dry or wet	See "Blood" above	Often collected from a suspect's pubic area or fingers
Saliva	Dry or wet	See "Blood" above	Often collected from a bite mark or a victim's pubic area
	Reference saliva (buccal) samples	Swab	Swab the inside of the cheek using two swabs, rotating them during collection; allow swabs to air-dry
		Filter paper	Place donor saliva sample on marked area of filter paper; allow to air-dry
Hairs	Head and pubic hairs	Lifting	Refer to dry bloodstain lifting method
		Transfer	Use forceps to transfer hair onto a piece of paper that can be folded
		Vacuum	Vacuum can be used if necessary; generally not recommended
Fingernails and scrapings	Various conditions	Clipping	Use a clean clipper to clip nails onto clean paper; wrap samples using druggist's fold
	Various conditions	Scraping	Scrape undersides of nails onto clean paper; wrap samples using druggist's fold
Bones	Various conditions	Freeze in container	Collect if blood is not available; collect bone with marrow if available; rib bone and vertebrae are preferred; place specimen in a container and freeze if wet
Teeth	Various conditions	Container	Collect teeth with dental pulp if possible into a container
Tissues and organs	Wet	Freeze in a container	Collect if blood is not available; place specimen in a container and freeze
DNA database samples	Various conditions	Various	Follow jurisdiction protocol for collecting samples from arrested individuals or convicted offenders or both

Source: Fisher, B., Techniques of Crime Scene Investigation, 7th edn., 2004, CRC Press, Boca Raton; Lee, H.C. et al., Henry Lee's Crime Scene Handbook, 2001, Academic Press, San Diego; National Institute of Justice. Using DNA to solve cold cases, Special Report, 2002, US Department of Justice, Office of Justice Programs, Washington, DC.

Note: Samples requiring refrigeration or freezing are noted; other samples may be stored at room temperature. Dry evidence should be packed in a porous material such as paper (envelope, bag, box) as described in the text.

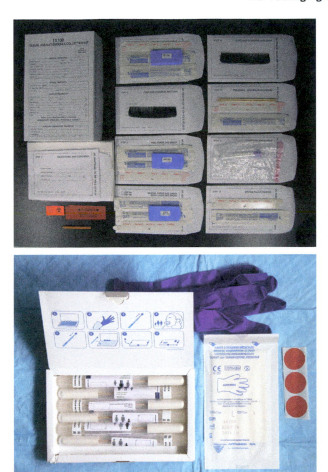

Figure 1.33 Evidence collection kits. Sexual assault evidence collection kit (top). Paternity evidence collection kit (bottom). (© Richard C. Li.)

Figure 1.34 Various types of swabs that are used for collecting biological evidence. (© Richard C. Li.)

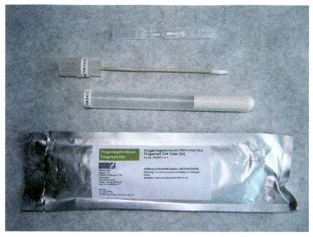

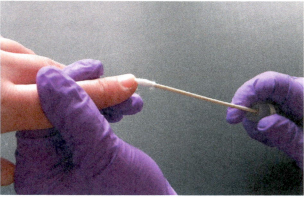

Figure 1.35 Fingernail swabbing for recovering evidence. Fingernail swabs are often collected from individuals who are involved in a struggle in violent crimes and in digital penetration in sexual assault cases. Fingernail swabs are sampled separately from both hands. (© Richard C. Li.)

depending on the type of evidence handled (Figures 1.40 through 1.42). The following are general considerations related to the packaging of evidence:

- *Evidence from different sources*: To prevent the transfer of evidence from different sources, items of evidence should not be grouped in a single package. However, evidence may be packed in a single container if the items were found together.

- *Folding of evidence*: Folding of clothing, especially items with wet bloodstains, can transfer evidence from one part of a garment to another. If a large, dry garment must be folded, a piece of clean paper should be placed between different parts of the garment to avoid direct contact between the different parts of the garment, thereby preventing the transfer of evidence.

- *Packing materials*: Envelopes, bags, and boxes that are made of porous materials such as paper are appropriate for packaging dry biological evidence. Dry, bloodstained evidence should not be sealed in plastic bags or containers that trap moisture.

- *Liquid evidence*: Tubes containing liquid such as blood should not be frozen because the volume of a liquid expands in freezing temperatures and this expansion may lead to cracking. Tubes should be placed in plastic bags to prevent leaks in case of

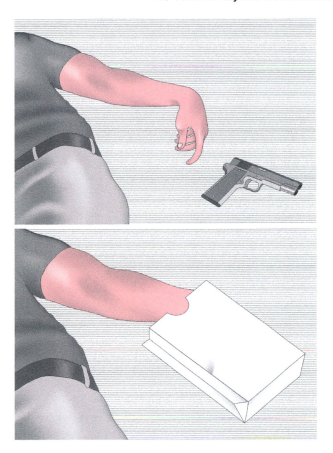

Figure 1.36 Hand bags for protecting the hands of a decedent in an alleged suicide. Gunshot residue can be found on hands after firing a weapon (top). The bagging of the hands, using paper bags and wide rubber bands, prevents the loss of gunshot residue during the transportation of the body (bottom). (© Richard C. Li.)

accidental breakage. Liquid evidence should be transported and submitted to a laboratory as soon as possible after the collection of evidence.

Trace evidence: All such evidence should be wrapped in paper with a druggist's fold (Figure 1.43). The wrapped trace evidence can be packed in an envelope.

Packaged evidence should be properly labeled with a description of the evidence and sealed with evidence tape. It is important for the person packaging the evidence to initial and date across the seal to show authenticity (Figure 1.44). A seal should not be cut when a sealed evidence bag is opened. Instead, an opening can be created by cutting at an area distal from the existing seal. After analysis is complete, the evidence packaging should be resealed. Table 1.1 summarizes additional steps for packaging evidence.

1.9 Final Survey and the Release of the Crime Scene

During a final survey, a discussion with all personnel in the crime scene investigation team should be carried out to thoroughly review all aspects of the search. It is important to ensure

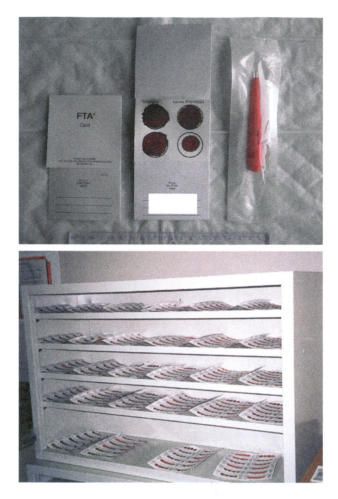

Figure 1.37 Blood cards are typically used for collecting blood evidence from a known source such as a suspect or a victim (top). A manual hole punch can be used to create a blood card punch for DNA extraction. Blood samples are air-dried on blood cards for storage (bottom). (© Richard C. Li.)

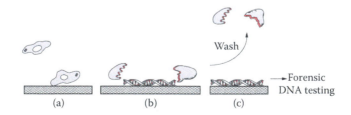

Figure 1.38 Application of Flinders Technology Associates (FTA) filter paper for the collection of biological evidence. (a) Biological fluid with cells is applied to FTA paper. (b) Cells are lysed and DNA is immobilized on FTA paper. (c) Cellular materials are washed away and DNA remaining on the FTA paper can be used for forensic testing. (© Richard C. Li.)

Figure 1.39   Photographic documentation of a knife. Note the evidence tag. (© Richard C. Li.)

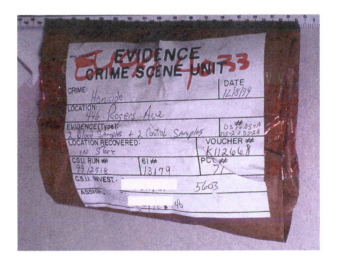

Figure 1.40   Evidence containing dried bodily fluid stains is packed in a paper bag. (© Richard C. Li.)

Figure 1.41   Alleged diluted blood collected from a pipe (left) and placed in plastic containers (right). (© Richard C. Li.)

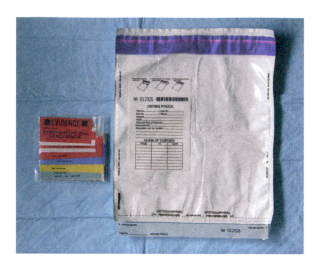

Figure 1.42 Evidence pouch. The front of the pouch is transparent for viewing the content. The back of the pouch is made of breathable materials allowing wet evidence, such as swabs, to dry inside the pouch. (© Richard C. Li.)

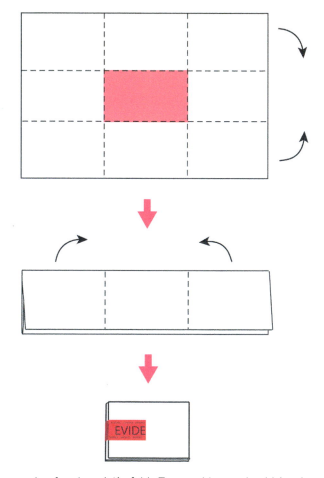

Figure 1.43 An example of a druggist's fold. Trace evidence should be deposited in the center (colored area) of the paper. (© Richard C. Li.)

Figure 1.44 Proper marking of sealed evidence. Note that the evidence packaging was cut a second time and resealed at a different location than that of the preexisting seal. (© Richard C. Li.)

that the scene has been searched correctly and completely, and that no area has been missed or overlooked. All documentation including the chain of custody document must be complete and all evidence should be collected, packed, documented, and marked. Photographs of the final condition of the scene should be taken. Once the final survey is completed, the crime scene can be released. Reentry into the crime scene may require a search warrant after the scene is released. Crime scene release documentation usually includes the time and date of release, to whom it is released, and by whom it is released.

1.10 Crime Scene Reconstruction

Crime scene reconstruction is the scientific process of determining the sequence of events and actions that occurred prior to, during, and after a crime. Reconstruction is carried out based on the information from the crime scene observations and the laboratory examination of physical evidence. The overall scientific process in reconstruction usually involves several steps. The process usually begins with the formulation of questions related to the problems that need to be solved. The questions can refer to the explanation of the specifics of the crime, for instance, "Where was the shooter's position when the shooting occurred?," "Where was the victim's position when shot?," and "What is the muzzle-to-target distance during the shooting?" In order to conduct a thorough crime scene reconstruction, all useful information is collected for review, such as photographs, videotapes, notes, sketches, autopsy reports, and analysis reports of the physical evidence. A hypothesis is then constructed based on the information obtained, which may explain the events and actions involved in a crime. The next step is making predictions that determine the logical consequences of the hypothesis. One or more predictions are selected for testing. The hypothesis is tested by conducting reconstruction experiments. One example of a reconstruction test is bloodstain pattern reconstruction in violent crimes (Chapter 2). Other examples of reconstruction may include trajectory and shooting, glass fracture, and accident reconstruction. The final step is to analyze experimental data and draw a conclusion. The experimental data are analyzed to see if the hypothesis is true or false. Additionally, the interpretation of physical evidence analysis, witness and confession statements, and investigative information should also be considered.

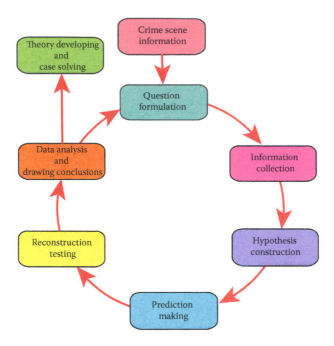

Figure 1.45 The scientific process of crime scene reconstruction. (© Richard C. Li.)

If the results of the experiment are consistent with the hypothesis, a theory can be developed that is intended to provide valuable information to the investigation and future prosecution of a case. Sometimes, forensic scientists may find that the hypothesis is inconsistent with the test results. In that case, an alternative hypothesis needs to be constructed to initiate another reconstruction process. Figure 1.45 illustrates the scientific process of crime scene reconstruction.

Bibliography

Ackermann, K., K.N. Ballantyne, and M. Kayser, Estimating trace deposition time with circadian biomarkers: A prospective and versatile tool for crime scene reconstruction. *Int J Legal Med*, 2010, **124**(5): 387–395.

Adderley, R. and J.W. Bond, The effects of deprivation on the time spent examining crime scenes and the recovery of DNA and fingerprints. *J Forensic Sci*, 2008, **53**(1): 178–182.

Barash, M., A. Reshef, and P. Brauner, The use of adhesive tape for recovery of DNA from crime scene items. *J Forensic Sci*, 2010, **55**(4): 1058–1064.

Benschop, C.C., et al., Post-coital vaginal sampling with nylon flocked swabs improves DNA typing. *Forensic Sci Int Genet*, 2010, **4**(2): 115–121.

Bond, J.W. and C. Hammond, The value of DNA material recovered from crime scenes. *J Forensic Sci*, 2008, **53**(4): 797–801.

Brownlow, R.J., K.E. Dagnall, and C.E. Ames, A comparison of DNA collection and retrieval from two swab types (cotton and nylon flocked swab) when processed using three QIAGEN extraction methods. *J Forensic Sci*, 2012, **57**(3): 713–717.

Buck, U., et al., Accident or homicide—Virtual crime scene reconstruction using 3D methods. *Forensic Sci Int*, 2013, **225**(1–3): 75–84.

Byard, R.W., et al., "Murder-suicide" or "murder-accident"? Difficulties with the analysis of cases. *J Forensic Sci*, 2010, **55**(5): 1375–1377.

Castello, A., F. Frances, and F. Verdu, Solving underwater crimes: Development of latent prints made on submerged objects. *Sci Justice*, 2013, **53**(3): 328–331.

Chen, T., et al., A rapid wire-based sampling method for DNA profiling. *J Forensic Sci*, 2012, **57**(2): 472–477.

Clarke, M., Alleged rape: An appeal case. *J Forensic Leg Med*, 2008, **15**(1): 32–36.

Courts, C., B. Madea, and C. Schyma, Persistence of biological traces in gun barrels—An approach to an experimental model. *Int J Legal Med*, 2011, **126**(3): 391–397.

Crispino, F., Nature and place of crime scene management within forensic sciences. *Sci Justice*, 2008, **48**(1): 24–28.

Durnal, E.W., Crime scene investigation (as seen on TV). *Forensic Sci Int*, 2010, **199**(1–3): 1–5.

Edelman, G., I. Alberink, and B. Hoogeboom, Comparison of the performance of two methods for height estimation. *J Forensic Sci*, 2010, **55**(2): 358–365.

Eldredge, K., E. Huggins, and L.C. Pugh, Alternate light sources in sexual assault examinations: An evidence-based practice project. *J Forensic Nurs*, 2012, **8**(1): 39–44.

Fisher, B., *Techniques of Crime Scene Investigation*, 7th edn., 2004. Boca Raton, FL: CRC Press.

Gokdogan, M.R. and J. Bafra, Development of a sexual assault evidence collection kit—The need for standardization in Turkey. *Nurse Educ Today*, 2010, **30**(4): 285–290.

Gosnell, J., Case three: Collection of evidence in a murder investigation. *Clin Lab Sci*, 2008, **21**(2): 124–125.

Gunasekera, R.S., A.B. Brown, and E.H. Costas, Tales from the grave: Opposing autopsy reports from a body exhumed. *J Forensic Leg Med*, 2012, **19**(5): 297–301.

Hammer, U. and A. Buttner, Distinction between forensic evidence and post-mortem changes of the skin. *Forensic Sci Med Pathol*, 2012, **8**(3): 330–333.

Hollmann, T., R.W. Byard, and M. Tsokos, The processing of skeletonized human remains found in Berlin, Germany. *J Forensic Leg Med*, 2008, **15**(7): 420–425.

Hornor, G., et al., Pediatric sexual assault nurse examiner care: Trace forensic evidence, ano-genital injury, and judicial outcomes. *J Forensic Nurs*, 2012, **8**(3): 105–111.

Hulse-Smith, L. and M. Illes, A blind trial evaluation of a crime scene methodology for deducing impact velocity and droplet size from circular bloodstains. *J Forensic Sci*, 2007, **52**(1): 65–69.

Ingemann-Hansen, O. and A.V. Charles, Forensic medical examination of adolescent and adult victims of sexual violence. *Best Pract Res Clin Obstet Gynaecol*, 2013, **27**(1): 91–102.

Ingemann-Hansen, O., et al., Legal aspects of sexual violence—Does forensic evidence make a difference? *Forensic Sci Int*, 2008, **180**(2–3): 98–104.

Jina, R., et al., Recovering of DNA evidence after rape. *S Afr Med J*, 2011, **101**(10): 758–759.

Kaliszan, M., et al., Striated abrasions from a knife with non-serrated blade—Identification of the instrument of crime on the basis of an experiment with material evidence. *Int J Legal Med*, 2011, **125**(5): 745–748.

Komar, D.A., S. Lathrop, and S. American, The use of material culture to establish the ethnic identity of victims in genocide investigations: A validation study from the American Southwest. *J Forensic Sci*, 2008, **53**(5): 1035–1039.

Larkin, B.A.J., et al., Crime scene investigation III: Exploring the effects of drugs of abuse and neurotransmitters on bloodstain pattern analysis. *Anal Methods*, 2012, **4**(3): 721–729.

Lee, H.C. and C. Ladd, Preservation and collection of biological evidence. *Croat Med J*, 2001, **42**(3): 225–228.

Lee, H.C. and E.M. Pagliaro, Forensic evidence and crime scene investigation. *J Forensic Invest*, 2013, **1**(1): 1–5.

Lee, H.C., T. Palmbach, and M.T. Miller, *Henry Lee's Crime Scene Handbook*, 2001. San Diego: Academic Press.

Lee, H.C., et al., Guidelines for the collection and preservation of DNA evidence. *J Forensic Ident*, 1991, **41**(5): 13.

Ma, M., H. Zheng, and H. Lallie, Virtual reality and 3D animation in forensic visualization. *J Forensic Sci*, 2010, **55**(5): 1227–1231.

Marchant, B. and C. Tague, Developing fingerprints in blood: A comparison of several chemical techniques. *J Forensic Ident*, 2007, **57**(1): 76–93.

Matte, M., et al., Prevalence and persistence of foreign DNA beneath fingernails. *Forensic Sci Int Genet*, 2012, **6**(2): 236–243.

Mennell, J., The future of forensic and crime scene science. Part II. A UK perspective on forensic science education. *Forensic Sci Int*, 2006, **157**(Suppl 1): S13–S20.

Morgan, J.A., Comparison of cervical os versus vaginal evidentiary findings during sexual assault exam. *J Emerg Nurs*, 2008, **34**(2): 102–105.

Mulligan, C.M., S.R. Kaufman, and L. Quarino, The utility of polyester and cotton as swabbing substrates for the removal of cellular material from surfaces. *J Forensic Sci*, 2011, **56**(2): 485–490.

National Institute of Justice. Using DNA to solve cold cases, Special Report, 2002. Washington, DC: US Department of Justice, Office of Justice Programs, Washington, DC, 2002.

Newton, M., The forensic aspects of sexual violence. *Best Pract Res Clin Obstet Gynaecol*, 2013, **27**(1): 77–90.

Nikolic, S. and V. Zivkovic, A healed bony puzzle: An old gunshot wound to the head. *Forensic Sci Med Pathol*, 2013, **9**(1): 112–116.

Nunn, S., Touch DNA collection versus firearm fingerprinting: Comparing evidence production and identification outcomes. *J Forensic Sci*, 2013, **58**(3): 601–608.

Oliva, A., et al., State of the art in forensic investigation of sudden cardiac death. *Am J Forensic Med Pathol*, 2011, **32**(1): 1–16.

Oliver, W.R. and L. Leone, Digital UV/IR photography for tattoo evaluation in mummified remains. *J Forensic Sci*, 2012, **57**(4): 1134–1136.

Osterkamp, T., K9 water searches: Scent and scent transport considerations. *J Forensic Sci*, 2011, **56**(4): 907–912.

Porta, D., et al., The importance of an anthropological scene of crime investigation in the case of burnt remains in vehicles: 3 case studies. *Am J Forensic Med Pathol*, 2013, **34**(3): 195–200.

Pringle, J.K., J.P. Cassella, and J.R. Jervis, Preliminary soilwater conductivity analysis to date clandestine burials of homicide victims. *Forensic Sci Int*, 2010, **198**(1–3): 126–133.

Ribaux, O., et al., Intelligence-led crime scene processing. Part I: Forensic intelligence. *Forensic Sci Int*, 2010, **195**(1–3): 10–16.

Ribaux, O., et al., Intelligence-led crime scene processing. Part II: Intelligence and crime scene examination. *Forensic Sci Int*, 2010, **199**(1–3): 63–71.

Rios, L., J.I. Ovejero, and J.P. Prieto, Identification process in mass graves from the Spanish Civil War I. *Forensic Sci Int*, 2010, **199**(1–3): e27–e36.

Rutty, G.N., A. Hopwood, and V. Tucker, The effectiveness of protective clothing in the reduction of potential DNA contamination of the scene of crime. *Int J Legal Med*, 2003, **117**(3): 170–174.

Sakelliadis, E.I., C.A. Spiliopoulou, and S.A. Papadodima, Forensic investigation of child victim with sexual abuse. *Indian Pediatr*, 2009, **46**(2): 144–151.

Sansoni, G., et al., Scene-of-crime analysis by a 3-dimensional optical digitizer: A useful perspective for forensic science. *Am J Forensic Med Pathol*, 2011, **32**(3): 280–286.

Schmidt, A., Crime scene investigation approach to sudden cardiac death. *J Am Coll Cardiol*, 2013, **62**(7): 630–631.

Schyma, C., B. Madea, and C. Courts, Persistence of biological traces in gun barrels after fatal contact shots. *Forensic Sci Int Genet*, 2013, **7**(1): 22–27.

Sharma, L., V.P. Khanagwal, and P.K. Paliwal, Homicidal hanging. *Legal Med*, 2011, **13**(5): 259–261.

Shaw, J. and R. Campbell, Predicting sexual assault kit submission among adolescent rape cases treated in forensic nurse examiner programs. *J Interpers Violence*, 2013, **28**(18): 3400–3417.

Shuttlewood, A.C., J.W. Bond, and L.L. Smith, The relationship between deprivation and forensic material recovered from stolen vehicles: Is it affected by vehicle condition and tidiness? *J Forensic Sci*, 2011, **56**(2): 510–513.

Smith, P.A., et al., Measuring team skills in crime scene investigation: Exploring ad hoc teams. *Ergonomics*, 2008, **51**(10): 1463–1488.

Solla, M., et al., Experimental forensic scenes for the characterization of ground-penetrating radar wave response. *Forensic Sci Int*, 2012, **220**(1–3): 50–58.

Taroni, F., et al., Whose DNA is this? How relevant a question? (a note for forensic scientists). *Forensic Sci Int Genet*, 2013, **7**(4): 467–470.

Thackeray, J.D., et al., Forensic evidence collection and DNA identification in acute child sexual assault. *Pediatrics*, 2011, **128**(2): 227–232.

Trombka, J.I., et al., Crime scene investigations using portable, non-destructive space exploration technology. *Forensic Sci Int*, 2002, **129**(1): 1–9.

Uzun, I., et al., Identification procedures as a part of death investigation in Turkey. *Am J Forensic Med Pathol*, 2012, **33**(1): 1–3.

Verdon, T.J., R.J. Mitchell, and R.A. van Oorschot, Swabs as DNA collection devices for sampling different biological materials from different substrates. *J Forensic Sci*, 2014, **59**(4): 1080–1089.

Wahlsten, P., V. Koiranen, and P. Saukko, Survey of medico-legal investigation of homicides in the city of Turku, Finland. *J Forensic Leg Med*, 2007, **14**(5): 243–252.

Walter, B.S. and J.J. Schultz, Mapping simulated scenes with skeletal remains using differential GPS in open environments: An assessment of accuracy and practicality. *Forensic Sci Int*, 2013, **228**(1–3): e33–e46.

Westen, A.A., R.R. Gerretsen, and G.J. Maat, Femur, rib, and tooth sample collection for DNA analysis in disaster victim identification (DVI): A method to minimize contamination risk. *Forensic Sci Med Pathol*, 2008, **4**(1): 15–21.

Wiegand, P., et al., Transfer of biological stains from different surfaces. *Int J Legal Med*, 2011, **125**(5): 727–731.

Winskog, C., Underwater disaster victim identification: The process and the problems. *Forensic Sci Med Pathol*, 2012, **8**(2): 174–178.

Yoon, D.Y., Newsmaker interview: Yoon Duk Yong. Crime scene investigation: The sinking of the Cheonan. Interview by Dennis Normile. *Science*, 2010, **328**(5984): 1336–1337.

Zeren, C., et al., A case of a serial sexual offender: The first ever report from Turkey. *J Forensic Leg Med*, 2012, **19**(7): 428–430.

Crime Scene Bloodstain Pattern Analysis

Bloodstain pattern analysis is the application of scientific knowledge to the examination and the interpretation of the morphology, the sequence, and the distribution of bloodstains associated with a crime. These analyses may determine the sequence of events; the approximate blood source locations; the positioning of the victim; and the position, the intensity, and the number of impacts applied to the blood source. They also can assist in the determination of the manner of death and can distinguish between accidents, homicides, and suicides. Bloodstain pattern analysis provides critical information for crime scene reconstructions in violent crime investigations (Chapter 1).

2.1 Basic Biological Properties of Human Blood

Blood is a bodily fluid circulating within the body. An average adult has a blood volume of approximately 8% of his or her body weight. Blood consists of a cellular portion as well as a liquid portion known as plasma (Chapter 12). The cellular portion consists of blood cells and platelets. The plasma is mostly composed of water and other substances such as proteins, inorganic salts, and other organic substances. The mass density of blood is only slightly greater than that of water. Blood can form clots (or *thrombi*) that are the result of blood coagulation (Chapter 16). Coagulation begins after an injury occurs, stopping blood loss from a damaged vessel. The normal coagulation time for 1 mL of venous blood in a glass tube is 5–15 min. The coagulation time can be affected by many factors such as blood volume and mechanical disturbance.

2.2 Formation of Bloodstains

The formation of a blood droplet is a complex event that is influenced by viscosity, surface tension, cohesion force, and gravity. Blood is viscous, and blood viscosity is a measure of the blood's resistance to flow. The viscosity of blood is approximately five times greater than that of water. During the formation of a drop of blood, blood leaks out from a blood source. The surface tension of the blood causes it to hang from the opening of a blood source and to form a pendant drop of blood (Figure 2.1). The molecules of a blood drop are held together by the cohesion force to maintain the shape of a blood drop (Figure 2.2). Surface tension causes liquids to minimize their surface. As a result, the formed blood drop is spherical. As the volume of the drop gradually increases and exceeds a certain size, it detaches itself and falls. The falling drop is also held

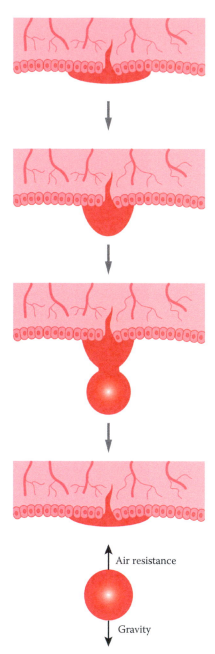

Figure 2.1   Forming a blood drop from a blood source. The blood that leaks out of the blood source forms a pendant drop of blood. As the volume of the pendant drop increases, the drop stretches in a downward direction. Eventually, the drop detaches and falls. The falling drop is largely influenced by the force of gravity and air resistance. (© Richard C. Li.)

together by surface tension. A falling blood drop is influenced by the downward force of grav-ity acting on the drop and the air resistance that acts in the opposite direction as the drop is in motion (Figure 2.1).

When a bloodstain lands on a surface, the shape and the size of the bloodstain is affected by the texture of the target surface. Bloodstains that land on porous or rough surfaces usually

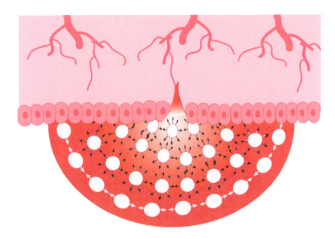

Figure 2.2   The particles of blood are attracted to each other by cohesive forces that are responsible for surface tension. As a result, a formed blood drop is spherical in order to minimize its surface area. Black arrow, cohesive force; white arrow, surface tension. (© Richard C. Li.)

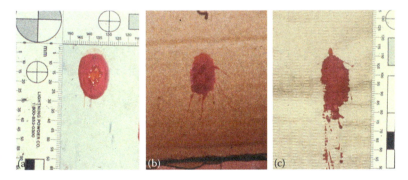

Figure 2.3   The respective morphologies of falling blood drops that land on surfaces with different textures at a 30° angle. (a) Tile, (b) cardboard, and (c) paper towel. (© Richard C. Li.)

have more distortion around the edges of the stains than those that land on smooth surfaces. A comparison of blood dropped onto different textures of target surfaces is shown in Figure 2.3.

2.3 Chemical Enhancement and Documentation of Bloodstain Evidence

Many chemical reagents react with blood to exhibit a color, a chemiluminescent light, or a fluorescent light (Chapter 12). These tests are extremely sensitive and thus are used as chemical enhancement reagents for detecting bloodstains. For bloodstain pattern analysis, the enhancement reagent is primarily used for detecting latent bloodstains such as diluted bloodstains that are visible on enhancement. A commonly used chemical reagent is luminol, which can be used for locating bloodstains at the scene. Other reagents such as phenolphthalein, leucomalachite green, and tetramethylbenzidine are not often used as enhancement reagents but rather as presumptive tests for blood. The positive reactions of all these reagents indicate the presence of blood.

Documenting bloodstain patterns at the scene is a major task of the investigation. Documenting bloodstain evidence can be done using a combination of photography, note-taking, and sketching. The general principle of crime scene documentation is described in Chapter 1. In bloodstain pattern analysis, special attention must obviously be given to bloodstains. The photographic documentation of bloodstains may be performed by multiple means, including film and digital photography, as well as videotaping. Photographs should be taken with an overall view followed by a medium-range and a close-up view of the bloodstain patterns. A scale of measurement must be included in the photograph, which is critical for bloodstain analysis. To avoid any distortion, the photographs should be taken with the camera lens parallel to the target surface where the bloodstains are located. An overall photograph provides an overall view of the scene including the bloodstain evidence (Figure 2.4a). A midrange photograph provides more details of the bloodstain pattern compared with that of the overall photograph (Figure 2.4b). Single bloodstains should be visible in midrange images. A close-up photograph, usually taken with a macro lens, provides a detailed image of single bloodstains, which is useful for spatter pattern analysis (Figure 2.4c).

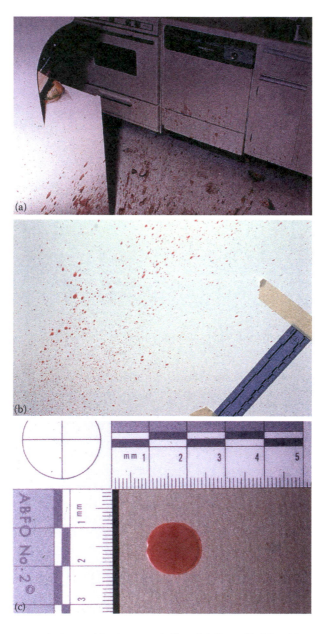

Figure 2.4   Crime scene photographic documentation. (a) Overall, (b) midrange, and (c) close-up photographs. (© Richard C. Li.)

2.4 Analyzing Spatter Stains

A *spatter stain*, based on the recommended terminology of the Scientific Working Group on Bloodstain Pattern Analysis (SWGSTAIN), is "a bloodstain resulting from a blood drop dispersed through the air due to an external force applied to a source of liquid blood." The patterns of spatter stains, including the shape and the size of the stains, are affected by the direction and the angle of impact (discussed in detail in Section 2.5.3) of the spatter stains that are projected. This information can be obtained from an analysis of the patterns of the spatter stains. Thus, it is possible to determine the area of origin (discussed in detail in Section 2.4.4) where an external force was directly applied to the blood source.

2.4.1 Velocity of Blood Droplets

The sizes of bloodstains are affected by the external force that is directly applied on a blood source. Increasing the energy of the external force will reduce the surface tension, thus decreasing the size of the droplets. Since these travelling blood droplets are driven by the energy derived from the external force, the higher the energy, the higher the velocities of the droplets. Bloodstains can be divided into three categories based on different travelling speeds. Low-velocity impact spatter is formed when a blood droplet is travelling at <1.5 m/s. The resulting stains are usually >4 mm in diameter (Figure 2.5a). As the travelling speed of blood droplets increases, the size of the spatter stain decreases. Medium-velocity impact spatter is formed when a blood source is subjected to a force associated with beatings or stabbings. The resulting stains range from 1 to 4 mm in diameter (Figure 2.5b). High-velocity impact spatter is formed when a blood source is subjected to a force associated with shooting using firearms. The resulting stains are usually <1 mm in diameter.

2.4.2 Determining the Directionality of the Stains

In this analysis, the effects of the directionality of the spatter stains projected are examined. SWGSTAIN defines the directionality to be "the characteristic of a bloodstain that indicates the direction blood was moving at the time of deposition." This analysis is applicable when the blood source is projected onto a surface at an angle of between 0° and 90°. Under this condition, the resulting spatter stain is an elongated ellipse (Figure 2.6), which is known as the *parent stain*. Additionally, *satellite stains* in the vicinity of the parent stain can be observed. As defined by SWGSTAIN, a satellite stain is "a smaller bloodstain that originated during the formation of the parent stain as a result of blood impacting a surface." More importantly, a spine is observed, which is the pointed edge away from the parent stain. When such a pattern is observed, the pointed end of the spine always points toward the direction of travel of the bloodstains.

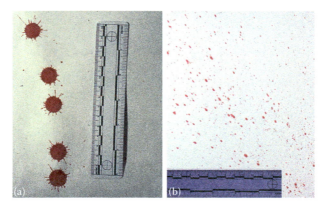

Figure 2.5 Bloodstains can be categorized based on their travelling velocities. (a) An example of a low-velocity impact spatter stain and (b) an example of a medium-velocity impact spatter stain. (© Richard C. Li.)

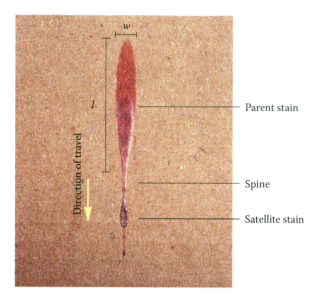

Figure 2.6 The morphology and directionality of a blood spatter stain. The arrow indicates the direction of travel. (© Richard C. Li.)

2.4.3 Determining Angles of Impact

SWGSTAIN defines the *angles of impact* to be "the acute angle (alpha), relative to the plane of a target at which a blood drop strikes the target." The shapes of the spatter stains are affected by the angle of impact. When a blood drop lands on a surface at a perpendicular angle (90°), a circular parent stain is formed (Figure 2.7), where the length and the width of the stain are equal. When a blood drop is projected onto a surface at an angle of between 0° and 90°, the stain is elongated. As the impact angle decreases, the shape of the spatter stain is more elongated (Figure 2.7), in which the length of the stain is greater than the width. It is observed that the ratio of the width and the length of the parent stain is proportional to the sine of the impact angle, which is summarized in the following trigonometric equation:

$$\sin \alpha = \frac{w}{l}$$

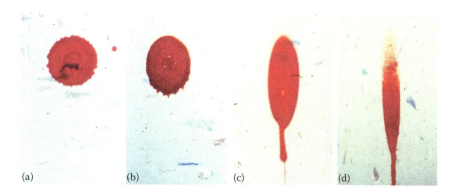

Figure 2.7 The effects of the impact angle on the shapes of blood spatter stains. Spatter stains are projected onto the surface of a ceramic tile at: (a) 90°, (b) 50°, (c) 20°, and (d) 10°. (© Richard C. Li.)

In this equation, α is the angle of impact, *l* is the length of the parent stain (major axis), and *w* is the width of the parent stain (minor axis). Thus, the angle of impact can be determined based on the relationship between the length and the width of the stain (Figure 2.8). Obviously, the measurement of the stain's axes is critical to the accuracy of the calculation of the angle of impact. To produce accurate and reproducible measurements, bloodstain pattern analysis software can be used, which superimposes a scaled close-up image of an individual bloodstain and calculates the angle of impact.

2.4.4 Determining Area of Origin

SWGSTAIN defines the *area of origin* to be "the three-dimensional location from which spatter originated." Using simple trigonometry, the area of origin can be determined based on the measurements from multiple elongated spatter stains (Figure 2.9). This can be accomplished by using the string method or the tangent method.

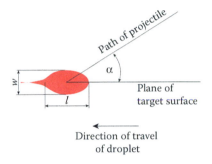

Figure 2.8 Impact angle. The angle between the path of a projectile and the plane of the target surface is shown. α, the impact angle; *l*, the length of the parent stain; and *w*, the width of the parent stain. (© Richard C. Li.)

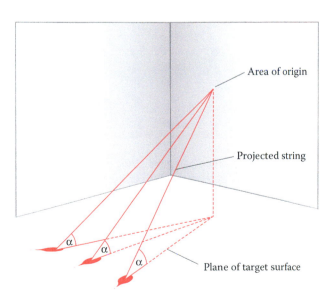

Figure 2.9 Area of origin. The area of origin is determined using the string method. Only three representative bloodstains are shown. α, impact angle. (© Richard C. Li.)

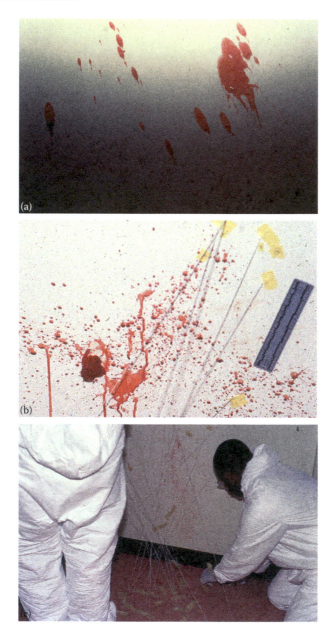

Figure 2.10 Determining the area of origin using the string method. (a) Selecting elongated spatter stains, (b) connecting strings, and (c) setting the path of the strings. (© Richard C. Li.)

In the string method, multiple (approximately two dozen) well-formed, elongated spatter stains are selected for analysis (Figure 2.10a). For each stain, the angle of impact is calculated. A piece of string is then connected between the stain and a surface with one end of the string precisely attached to the spatter stain (Figure 2.10b). The path of the string, indicating the trajectory of the stain, is set using a protractor based on the calculated angle of impact (Figures 2.10c and 2.11). This process is repeated until all the stains that have been selected are processed. For

Figure 2.11   Tools for finding the area of origin. (a) Laser trajectory pointer and (b) string and scales. (© Richard C. Li.)

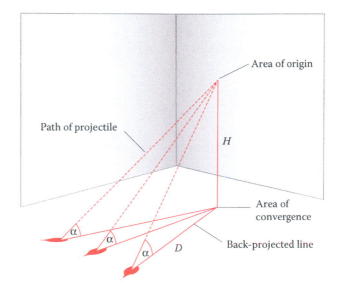

Figure 2.12   Determining the area of origin using the tangent method. Only three representative bloodstains are shown. α, the angle of impact; *H*, the height of the area of origin; and *D*, the distance from the spatter stain to the area of convergence. (© Richard C. Li.)

a spatter pattern generated from a single impact event, the strings converge. The area where the strings meet is the area of origin.

In the tangent method, the directionality of a single stain is determined first. A line is then back projected through the major axis of the bloodstain. For a single impact event, approximately two dozen stains are processed to determine the *area of convergence*. Based on SWGSTAIN's definition, the area of convergence is "the area containing the intersections generated by lines drawn through the long axes of individual stains that indicates in two dimensions the location of the blood source." Next, the angle of impact of each stain is calculated. The distance from the bloodstain to the area of convergence is measured (Figure 2.12). The height of the area of origin is calculated using the tangent function as shown:

$$H = D \cdot \tan \alpha$$

In the equation, α is the angle of impact, *H* is the height of the area of origin, and *D* is the distance from the spatter stain to the area of convergence.

2.5 Types of Bloodstain Patterns

Bloodstain patterns can be classified into three basic categories: passive, transfer, and projected bloodstains.

2.5.1 Passive Bloodstains

A *passive bloodstain* is formed due to bleeding from wounds, and the blood is deposited on a surface by the influence of the force of gravity alone. For example, a *drip stain* is formed when a falling drop of blood from an exposed wound or a blood-bearing object lands on a surface. If a blood source is moving, a *drip trail* is formed. A *drip pattern*, which is distinct from a drip stain, is formed when a liquid drips into another liquid, where one or both of the liquids are blood (Figure 2.13). As a result, secondary spatter stains are generated. As the dropping distance of the blood increases, the number of secondary spatter stains usually increases, and the size of these stains decreases. An approximate estimation of the dropping distance is possible. A *splash pattern* is formed when a volume of blood spills onto a surface (Figure 2.14). Splash patterns usually

Figure 2.13 A drip pattern. The secondary spatter stains are shown. (© Richard C. Li.)

Figure 2.14 A splash pattern. Peripheral, elongated bloodstains are shown. (© Richard C. Li.)

have large stains surrounded by numerous, peripheral, elongated bloodstains. A *flow pattern* is caused by the movement of a large volume of blood on a surface either due to gravity or to the movement of the target such as a victim or postmortem disturbance. A *pool* is a bloodstain resulting from the accumulation of liquid blood on a surface (Figure 2.15). Sometimes, air bubbles in the blood may cause a *bubble ring* pattern (Figure 2.15). If blood is coagulated, gelatinous *blood clots* can be observed. Additionally, a *serum stain*, which consists of the liquid portion of the blood after a clot is formed, may also be present.

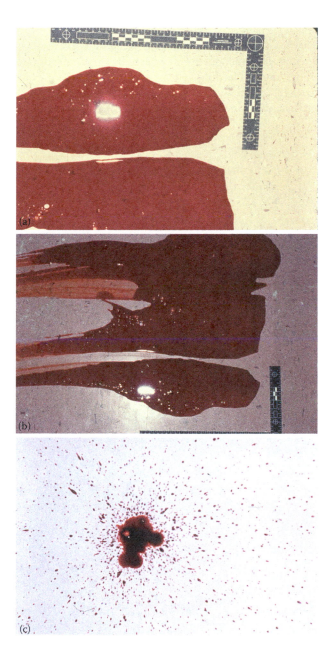

Figure 2.15   Pool and bubble ring patterns. (a) A pool pattern, (b) a disturbed pool pattern, and (c) a splash pattern with a bubble ring. Bubble rings are also present in (a) and (b). (© Richard C. Li.)

2.5.2 Transfer Bloodstains

A *transfer bloodstain*, based on SWGSTAIN, is "a bloodstain resulting from contact between a blood-bearing surface and another surface." For example, a *swipe pattern* is "a bloodstain pattern resulting from the transfer of blood from a blood-bearing surface onto another surface, with characteristics that indicate relative motion between the two surfaces." For example, bloody impressions can provide information about the shape, the size, and the pattern of the objects such as finger ridges, hands, and shoe soles. Examples of hand and shoe swipe patterns are shown in Figures 2.16 and 2.17. A *wipe pattern* is "an altered bloodstain pattern resulting from an object moving through a preexisting wet bloodstain." Examples of wipe patterns are shown in Figures 2.18 and 2.19. A *perimeter* stain, a type of wipe pattern, is a bloodstain that is disturbed before it is dried but it maintains the peripheral characteristics of the original stain (Figure 2.20). Perimeter stain patterns can be useful for the estimation of sequential events of acts. The pattern can also be used to estimate a time frame between the time of bleeding and the subsequent act. However, the drying time of a blood drop varies based on the surrounding conditions. Therefore, it is necessary to carry out a crime scene reconstruction under similar conditions to those of the scene to make such estimations.

2.5.3 Projected Bloodstains

A *projected bloodstain* is formed when a volume of blood is deposited on a surface under a pressure or a force that is greater than the force of gravity. For example, an *impact pattern* is

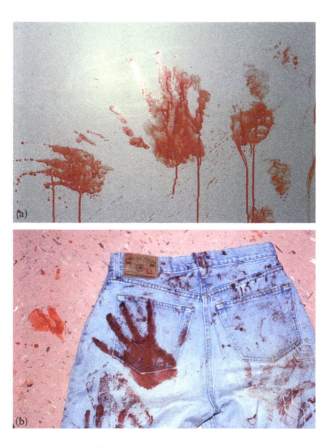

Figure 2.16 Bloody impressions. Bloody handprints are present on (a) a wall and (b) fabric. (© Richard C. Li.)

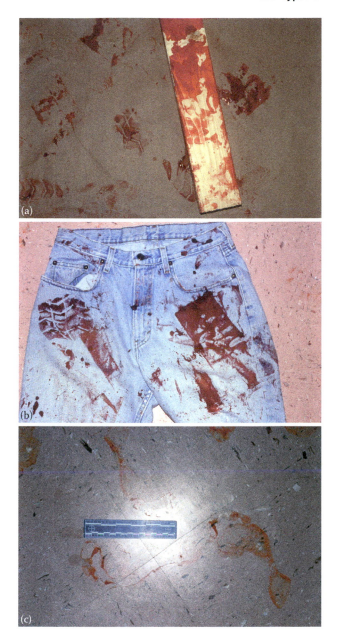

Figure 2.17 Bloody impressions. Bloody shoe prints on (a) paper and (b) fabric; and (c) bloody footprints on tile. (© Richard C. Li.)

formed when an object strikes liquid blood (Figure 2.21). A *cast-off pattern* is formed when blood drops are released from a moving blood-bearing object (Figure 2.22). Some spatter patterns are often associated with a wound penetrated by a projectile (Figure 2.23). A *forward spatter* is formed when blood drops travel from an exit wound in the same direction as a projectile, while a *back spatter* is formed when blood drops travel from an entry wound in the opposite direction of a projectile. Sometimes, internal bleeding caused by an injury may block the airway. An *expiration pattern* is formed when blood is forced by airflow through the

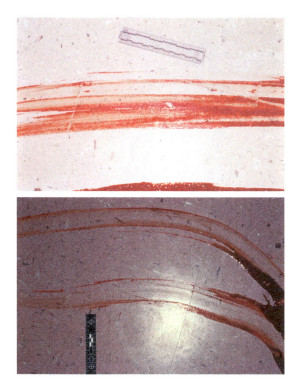

Figure 2.18   Wipe patterns. (© Richard C. Li.)

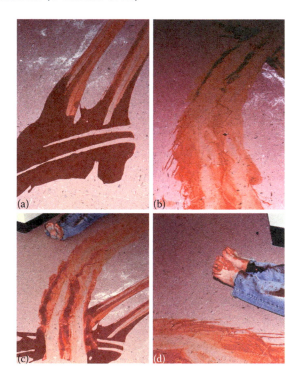

Figure 2.19   A wipe pattern caused by dragging a body through a pool of blood. (a) A pool of blood. Sections of the wipe pattern caused by dragging are shown in (b), (c), and (d). (© Richard C. Li.)

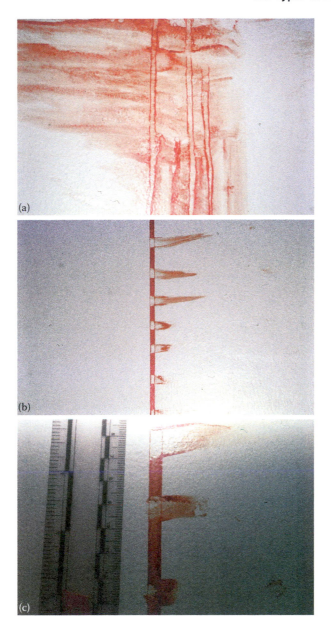

Figure 2.20 Perimeter stains. (a) Peripheral characteristics of the original stains are shown. Perimeter stains were created at different periods of time after the original stain was formed: (b) midrange view and (c) close-up view. (© Richard C. Li.)

trachea and out of the nose or mouth (Figure 2.24). An *arterial spurt pattern* is associated with wounds damaging arterial blood vessels where bloodstains are driven by arterial pressure. Although the shape of arterial patterns varies, these patterns usually have a series of large spurts with fluctuations corresponding to the systolic and the diastolic blood pressures. At a crime scene, if the projectile of bloodstains is blocked by an object, a *void pattern* is formed, which exhibits an area where there is an absence of blood surrounded by continuously distributed bloodstains.

Figure 2.21   An impact bloodstain pattern as a result of using blunt force. (© Richard C. Li.)

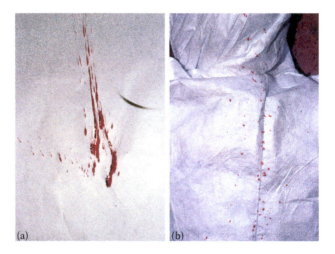

Figure 2.22   Cast-off patterns. Spatter stains are projected onto (a) a covered wall and (b) a lab coat. (© Richard C. Li.)

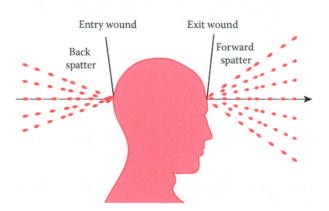

Figure 2.23   Forward and back spatter patterns. Arrow, the direction of a projectile. (© Richard C. Li.)

Figure 2.24   An expiration pattern. (© Richard C. Li.)

Bibliography

Adam, C.D., Fundamental studies of bloodstain formation and characteristics. *Forensic Sci Int*, 2012, **219**(1–3): 76–87.

Adelson, L., The coroner of Elsinore. *New Engl J Med*, 1960, **262**(5): 229–234.

Attinger, D., et al., Fluid dynamics topics in bloodstain pattern analysis: Comparative review and research opportunities. *Forensic Sci Int*, 2013, **231**(1–3): 375–396.

Behrooz, N., L. Hulse-Smith, and S. Chandra, An evaluation of the underlying mechanisms of bloodstain pattern analysis error. *J Forensic Sci*, 2011, **56**(5): 1136–1142.

Benecke, M. and L. Barksdale, Distinction of bloodstain patterns from fly artifacts. *Forensic Sci Int*, 2003, **137**(2–3): 152–159.

Buck, U., et al., 3D bloodstain pattern analysis: Ballistic reconstruction of the trajectories of blood drops and determination of the centres of origin of the bloodstains. *Forensic Sci Int*, 2011, **206**(1–3): 22–28.

Byard, R.W., et al., Blood stain pattern interpretation in cases of fatal haemorrhage from ruptured varicose veins. *J Forensic Leg Med*, 2007, **14**(3): 155–158.

Camana, F., Determining the area of convergence in bloodstain pattern analysis: A probabilistic approach. *Forensic Sci Int*, 2013, **231**(1–3): 131–136.

Connolly, C., M. Illes, and J. Fraser, Affect of impact angle variations on area of origin determination in bloodstain pattern analysis. *Forensic Sci Int*, 2012, **223**(1–3): 233–240.

Davidson, P.L., et al., Physical components of soft-tissue ballistic wounding and their involvement in the generation of blood backspatter. *J Forensic Sci*, 2012, **57**(5): 1339–1342.

de Bruin, K.G., R.D. Stoel, and J.C. Limborgh, Improving the point of origin determination in bloodstain pattern analysis. *J Forensic Sci*, 2011, **56**(6): 1476–1482.

de Castro, T., et al., Interpreting the formation of bloodstains on selected apparel fabrics. *Int J Legal Med*, 2013, **127**(1): 251–258.

Denison, D., et al., Forensic implications of respiratory derived blood spatter distributions. *Forensic Sci Int*, 2011, **204**(1–3): 144–155.

Donaldson, A.E., et al., Using oral microbial DNA analysis to identify expirated bloodspatter. *Int J Legal Med*, 2010, **124**(6): 569–576.

Donaldson, A.E., et al., Characterising the dynamics of expirated bloodstain pattern formation using high-speed digital video imaging. *Int J Legal Med*, 2011, **125**(6): 757–762.

Emes, A., Expirated blood—A review. *J Can Soc Forensic Sci*, 2001, **34**(4): 197–203.

Finnis, J., J. Lewis, and A. Davidson, Comparison of methods for visualizing blood on dark surfaces. *Sci Justice*, 2013, **53**(2): 178–186.

Fujikawa, A., et al., Changes in the morphology and presumptive chemistry of impact and pooled bloodstain patterns by *Lucilia sericata* (Meigen) (Diptera: Calliphoridae). *J Forensic Sci*, 2011, **56**(5): 1315–1318.

Illes, M. and M. Boue, Robust estimation for area of origin in bloodstain pattern analysis via directional analysis. *Forensic Sci Int*, 2013, **226**(1–3): 223–229.

Karger, B., et al., Bloodstain pattern analysis—Casework experience. *Forensic Sci Int*, 2008, **181**(1–3): 15–20.

Kettner, M., F. Ramsthaler, and A. Schnabel, "Bubbles"—A spot diagnosis. *J Forensic Sci*, 2010, **55**(3): 842–844.

Kunz, S.N., H. Brandtner, and H. Meyer, Unusual blood spatter patterns on the firearm and hand: A back-spatter analysis to reconstruct the position and orientation of a firearm. *Forensic Sci Int*, 2013, **228**(1–3): e54–e57.

Larkin, B.A. and C.E. Banks, Preliminary study on the effect of heated surfaces upon bloodstain pattern analysis. *J Forensic Sci*, 2013, **58**(5): 1289–1296.

Larkin, B.A.J., et al., Crime scene investigation III: Exploring the effects of drugs of abuse and neurotransmitters on bloodstain pattern analysis. *Anal Methods*, 2012, **4**(3): 721–729.

Lee, W.C., et al., Statistical evaluation of alternative light sources for bloodstain photography. *J Forensic Sci*, 2013, **58**(3): 658–663.

Liesegang, J., Bloodstain pattern analysis—Blood source location. *J Can Soc Forensic Sci*, 2004, **37**(4): 215–222.

Maclean, B., K. Powley, and D. Dahlstrom, A case study illustrating another logical explanation for high velocity impact spatter. *J Can Soc Forensic Sci*, 2001, **34**(4): 191–195.

Makovicky, P., et al., The use of trigonometry in bloodstain analysis. *Soud Lek*, 2013, **58**(2): 20–25.

Maloney, A., et al., One-sided impact spatter and area-of-origin calculations. *J Forensic Ident*, 2011, **61**(2): 123–135.

Nikolic, S. and V. Zivkovic, Bloodstain pattern in the form of gushing in a case of fatal exsanguination due to ruptured varicose vein. *Med Sci Law*, 2011, **51**(1): 61–62.

Peschel, O., et al., Blood stain pattern analysis. *Forensic Sci Med Pathol*, 2011, **7**(3): 257–270.

Pizzola, P.A., et al., Commentary on "3D bloodstain pattern analysis: Ballistic reconstruction of the trajectories of blood drops and determination of the centres of origin of the bloodstains" by Buck et al. [Forensic Sci. Int. 206 (2011) 22–28]. *Forensic Sci Int*, 2012, **220**(1–3): e39–e40; author reply e41.

Power, D.A., et al., PCR-based detection of salivary bacteria as a marker of expirated blood. *Sci Justice*, 2010, **50**(2): 59–63.

Ramsthaler, F., et al., Drying properties of bloodstains on common indoor surfaces. *Int J Legal Med*, 2012, **126**(5): 739–746.

Randall, B., Blood and tissue spatter associated with chainsaw dismemberment. *J Forensic Sci*, 2009, **54**(6): 1310–1314.

Raymond, M.A., E.R. Smith, and J. Liesegang, Oscillating blood droplets—Implications for crime scene reconstruction. *Sci Justice*, 1996, **36**(3): 161–171.

Raymond, M.A., E.R. Smith, and J. Liesegang, The physical properties of blood—Forensic considerations. *Sci Justice*, 1996, **36**(3): 153–160.

Sauvageau, A., et al., Bloodstain pattern analysis in a case of fatal varicose vein rupture. *Am J Forensic Med Pathol*, 2007, **28**(1): 35–37.

Stephens, B.G. and T.B. Allen, Back spatter of blood from gunshot wounds. Observations and experimental simulation. *J Forensic Sci*, 1983, **28**(2): 437–439.

Striman, B., et al., Alteration of expirated bloodstain patterns by *Calliphora vicina* and *Lucilia sericata* (Diptera: Calliphoridae) through ingestion and deposition of artifacts. *J Forensic Sci*, 2011, **56**(Suppl 1): S123–S127.

Taylor, M.C., et al., The effect of firearm muzzle gases on the backspatter of blood. *Int J Legal Med*, 2011, **125**(5): 617–628.

Thanakiatkrai, P., A. Yaodam, and T. Kitpipit, Age estimation of bloodstains using smartphones and digital image analysis. *Forensic Sci Int*, 2013, **233**(1–3): 288–297.

Trombka, J.I., et al., Crime scene investigations using portable, non-destructive space exploration technology. *Forensic Sci Int*, 2002, **129**(1): 1–9.

White, B., Bloodstain patterns on fabrics: The effect of drop volume, dropping height and impact angle. *J Can Soc Forensic Sci*, 1986, **19**(1): 3–36.

Yen, K., et al., Blood-spatter patterns: Hands hold clues for the forensic reconstruction of the sequence of events. *Am J Forensic Med Pathol*, 2003, **24**(2): 132–140.

Yi, R. and Y. Wang, The experimental observation of secondary spatter bloodstain in morphology. *Chin J Forensic Med*, 2011, **26**(1): 26–29.

Forensic Biology
A Subdiscipline of Forensic Science

Forensic laboratories provide scientific analysis, evidence evaluation, and consultations to various criminal justice agencies for the investigation of criminal cases. Additionally, forensic laboratories provide expert testimony related to the resolution of criminal cases to the courts.

3.1 Common Disciplines of Forensic Laboratory Services

Many of the disciplines of the forensic laboratory services are commonly practiced in various municipal, county, state, and federal forensic laboratories in the United States. Forensic biology is a subdiscipline of forensic science. A full range of forensic laboratory services, known as "full service," usually includes: crime scene investigation (Figure 3.1); latent print examination (Figure 3.2); forensic biology (Figure 3.3); controlled substance analysis (Figure 3.4); postmortem toxicology (Figure 3.5); questioned document examination (Figure 3.6); firearm, toolmark, and other impression evidence examination (Figures 3.7 through 3.10); explosive and fire debris examination (Figure 3.11); and transfer (trace) evidence examination (Figures 3.12 and 3.13). Table 3.1 describes the services that are normally provided by a forensic laboratory with their respective analyses.

3.2 Laboratory Analysis of Biological Evidence

Forensic biology uses scientifically accepted protocols to analyze biological evidence. Laboratory analysis (Figures 3.14 and 3.15) utilizes scientific techniques for the examination of evidence, the reconstruction of a crime scene, the identification of biological fluids, and the comparison of individual characteristics of biological evidence.

3.2.1 Identification of Biological Evidence

The identification of biological evidence is the first step that is performed before further analyses are carried out. This includes the identification of biological fluids such as blood, saliva, and semen; this process is discussed in more detail in subsequent chapters (Chapters 12 through 14). The *identification* is based on a comparison of *class characteristics*—a set of characteristics that allows a sample to be placed in a category with similar materials. By comparing the

Figure 3.1 Crime scene investigation. Recovery of fired casings at the scene can aid in determining the position of the shooter. (© Richard C. Li.)

class characteristics of a sample with known standards of its class, biological samples can be identified.

3.2.2 Comparison of Individual Characteristics of Biological Evidence

Individual characteristics refer to the unique characteristics of both the evidence and a reference sample such as fingerprints, which share a common origin to a high degree of certainty. An example of biological evidence possessing individual characteristics is DNA polymorphisms. In the case of biological evidence, current forensic DNA profiling can compare individual characteristics of DNA evidence with a known reference sample. It is possible to determine that a biological stain originated from a particular individual, which is useful for human identification. The examination of individual characteristics of evidence can also exclude the possibility of a common origin. The specific methods utilized for the individualization of evidence are also discussed in subsequent chapters (Chapters 19 through 23).

3.2.3 Reporting Results and Expert Testimony

After the analysis of evidence is completed, a report is prepared based on the results of the analysis, which may include sections discussing the specific evidence analyzed, the method of analysis used, the results obtained, and the conclusions drawn. In the case of DNA evidence, the strength of the conclusion is usually evaluated via statistical computations (Chapter 25). A forensic scientist often serves as an expert witness whose testimony

Figure 3.2 Developing latent fingerprints by dusting methods. Using magnetic fingerprint powder (top) that is held by a magnetic applicator (middle). Fingerprints (bottom) dusted by magnetic fingerprint powder, tape lifted and preserved on a fingerprint card. (© Richard C. Li.)

provides professional opinions about the evidence analyzed. Based on the federal rules of evidence, an expert witness is qualified based on his or her knowledge, skill, experience, training, or education, and may give an opinion to the court that is relevant to the analyses conducted. However, forensic scientists must also communicate their findings to attorneys, judges, and members of a jury. This requires the translation of technical information into layman's terms.

Figure 3.3 A section of a forensic biology laboratory showing an automated electrophoresis instrument used for forensic DNA profiling (left). Processing biological evidence in a biosafety cabinet (right). (© Richard C. Li.)

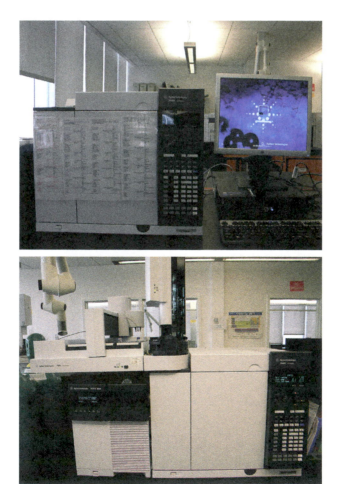

Figure 3.4 Gas chromatograph (top) and gas chromatograph-mass spectrometer (bottom) used for controlled substance analysis. (© Richard C. Li.)

Figure 3.5 Tissue samples (left) and gas chromatograph-mass spectrometer (right) for forensic postmortem toxicological analysis. (© Richard C. Li.)

Figure 3.6 A digital imaging system, using multiple illumination sources ranging from ultraviolet to infrared wavelengths, for examining altered and counterfeit documents (left). An electrostatic imaging system for detecting indented writing on questioned documents (right). This device generates an electrostatic image of indented writing, which is then visualized using charge-sensitive toners. (© Richard C. Li.)

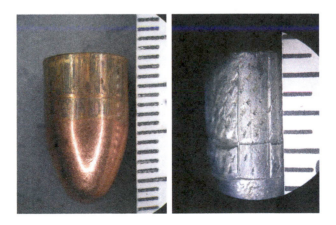

Figure 3.7 Striation marks on fired bullets can be analyzed to match a bullet to a gun. (Courtesy of P. Diaczuk.)

3.3 Forensic Science Services Related to Forensic Biology

A number of specialized forensic science services beyond those provided by forensic laboratories are routinely available to law enforcement agencies. For example, forensic services related to biological evidence and those involving more specialized analysis are available. These services are important aids to a criminal investigation and require the expertise of individuals who have highly specialized skills.

Figure 3.8   Fired hollowpoint bullets (.45 caliber ACP Winchester). Striation marks are visible on the side view of a bullet (right). (© Richard C. Li.)

Figure 3.9   A comparison microscope is used for the simultaneous comparison of two items of firearm evidence side by side. (© Richard C. Li.)

3.3.1 Forensic Pathology

When a death is deemed suspicious or unexplained, medical examiners frequently perform autopsies to determine the exact cause (Figures 3.16 and 3.17). The manner of death is classified into one of five categories based on the circumstances: natural, homicide, suicide, accident, or undetermined. Additionally, a medical examiner participating in a criminal investigation is often responsible for estimating the time of death.

3.3.2 Forensic Anthropology

Forensic anthropology is the identification and the examination of human skeletal remains (Figures 3.18 and 3.19). Skeletal remains can reveal a number of individual characteristics that can be useful in attempting to identify an individual. An examination of bones may reveal an individual's origin, sex, approximate age, race, and the presence of a skeletal injury. A forensic anthropologist may also assist in creating facial reconstructions to aid in the identification of skeletal remains or may be called on to help collect and organize bone fragments in the course of identifying victims of mass disasters such as plane crashes as well as victims in mass graves discovered after wars or genocides.

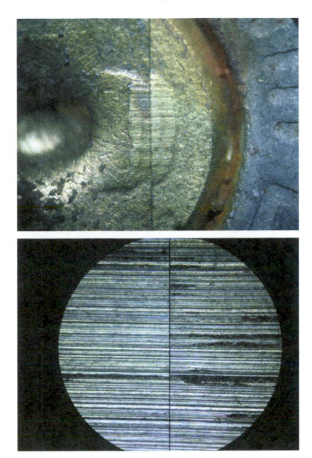

Figure 3.10 Comparing the striations between evidence and reference samples using a comparison microscope: casings (top) and bullets (bottom). (Courtesy of P. Diaczuk.)

Figure 3.11 Scanning electron microscope used in the analysis of gunshot residue and explosives. (© Richard C. Li.)

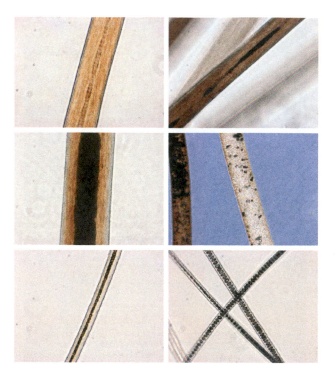

Figure 3.12 Trace evidence such as hairs can be transferred during the acts of a violent crime. The analysis and comparisons of these types of trace evidence can potentially establish a link between a suspect or a victim and a crime scene. Hairs from human (top), horse (middle left), deer (middle right), dog (bottom left), and cat (bottom right). (© Richard C. Li.)

Figure 3.13 Trace evidence: fibers. Cotton (left), nylon (middle), and polyester (right). (© Richard C. Li.)

3.3.3 Forensic Entomology

The study of insects in relation to a criminal investigation is known as forensic entomology. This forensic discipline is valuable for estimating the time of death when the circumstances surrounding the crime are otherwise unknown. The stages of development of certain insect species present in or on a body can be identified and allow a forensic entomologist to approximate how long the body was left exposed (Figure 3.20).

Table 3.1	Common Services Provided by US Forensic Laboratories	
Service	**Function**	**Method**
Crime scene investigation	Evidence recognition, documentation, collection, and preservation	Crime scene responses and related endeavors are diverse and vary with case and type of evidence
Latent print examination	Analysis of friction ridge detail in fingerprints Activities include visualization, recording, comparison, storage, and recovery of latent prints	Alternate light sources, physical (powder) and chemical enhancements Direct lifts, photography, and digital imaging Use of an Automated Fingerprint Identification System (AFIS) database
Forensic biology	Identification of biological fluids (blood, semen, and saliva) DNA profiling for individualization	Serological and biochemical methods Polymerase chain reaction (PCR)-based methods Automated electrophoresis platforms Use of Combined DNA Index System (CODIS)
Controlled substance analysis	Identification and quantification of drugs present in submitted evidence	Microscopic, chemical, chromatographic, and spectroscopic methodologies Gas chromatography–mass spectrometry or infrared spectrophotometry
Postmortem toxicology	Determination of concentrations of substances and their metabolites in biological fluids or tissues	Immunoassays and chemical methods Confirmatory techniques such as gas and liquid chromatography–mass spectrometry
Questioned document examination	Investigation of forgeries, tracings, disguised handwritings, computer manipulation of images, and recovery of altered documents Analysis of papers, inks, toners, word processors, typewriters, copiers, and printers	Macroscopic and microscopic comparisons Chromatographic and spectroscopic methods
Firearm and toolmark examination	Identification of firearms, tools, and other implements (expertise achieved predominantly through experience)	Microscopic comparisons of questioned and authenticated impressions Comparison of striae on recovered bullets Use of National Integrated Ballistics Information Network (NIBIN)
Explosive and fire debris examination	Identification, recovery, and detection of bulk explosives, residues, debris, and accelerants	Microscopic, spectroscopic, and chromatographic methods Gas chromatography–mass spectrometry may be needed to adequately characterize sample
Trace evidence examination	Analysis of transferred evidence such as hairs, fibers, soil, paints, and glass	Microscopic analysis of evidence with gas chromatograph-mass spectrometers, FTIR microscopes, scanning electron microscopes, basic and advanced microscopy, and capillary electrophoresis

Source: Adapted from US Department of Justice, Office of Justice Programs, National Institute of Justice (NIOJ), *Forensic Science: Review of Status and Needs*, 1999, US Department of Justice, Washington, DC.

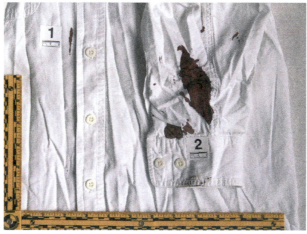

Figure 3.14   Photographic documentation of a bloodstained shirt after visual examination. (© Richard C. Li.)

Figure 3.15   A multiwavelength viewing and imaging device used for the examination of various types of evidence, including bodily fluid stains. (© Richard C. Li.)

Figure 3.16   View of a forensic pathology facility. (Courtesy of G. Ledwell.)

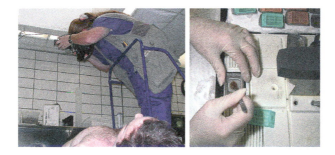

Figure 3.17   Photographic documentation prior to autopsy (left). Preparing specimens for histological sections for forensic pathological examination (right). (Courtesy of G. Ledwell.)

Figure 3.18   Buried human skeletons recovered by forensic anthropologists. (Courtesy of H. Brewster.)

Figure 3.19   Human skull recovered at the scene. (Courtesy of H. Brewster.)

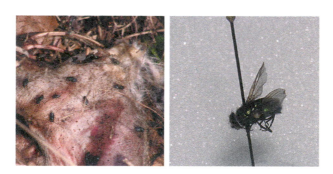

Figure 3.20   Insects found on a dead animal (left). Blow fly specimen. This insect is commonly encountered at crime scenes (right). (Courtesy of K. Wendler.)

3.3.4 Forensic Odontology

Practitioners of forensic odontology participate in the identification of victims whose bodies are left in an unrecognizable state. The characteristics of teeth, their alignment, and the overall structure of the mouth provide evidence that can identify a specific person. Dental records such as x-rays and dental casts allow a forensic odontologist to compare a set of dental remains with an alleged victim. Another application of forensic odontology in a criminal investigation is bite mark analysis (Figure 3.21). A forensic odontologist can analyze the marks left on a victim and compare them with the tooth structures of a suspect to make a comparison.

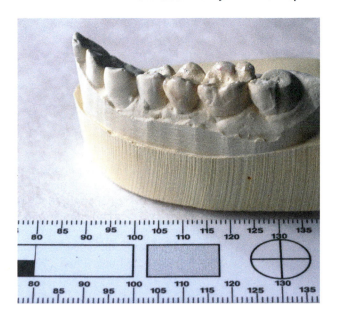

Figure 3.21   A dental cast can be utilized for identifying bite marks. (© Richard C. Li.)

3.4 Brief History of the Development of Forensic Biology

The developmental history of modern forensic biology spans three stages: (1) antigen polymorphism, (2) protein polymorphism, and (3) DNA polymorphism. Figure 3.22 illustrates this history.

3.4.1 Antigen Polymorphism

The human ABO blood groups were discovered in 1900 by Karl Landsteiner in a study of the causes of blood transfusion reactions. Landsteiner's discovery made blood transfusions feasible, and he received the Nobel Prize in 1930 when he revealed the four groups of human blood cells designated A, B, AB, and O. By the 1960s, a dozen more blood group systems had been characterized, and at least 29 systems are currently known (Chapter 18).

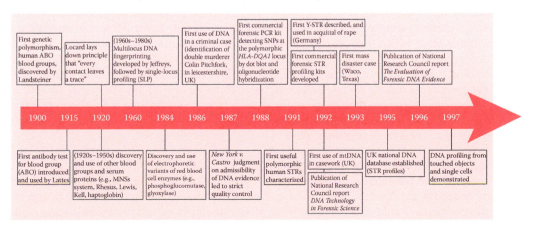

Figure 3.22   A brief history of the development of forensic biology. (From Jobling, M.A. and Gill, P., *Nat Rev Genet*, 5, 739–751, 2004. With permission.)

Subsequent studies found that the blood types in the ABO system were inherited, and the frequencies with which the four types appeared in specific human populations were found to differ. This led to the discovery of the first antigen polymorphic marker for use in human identification in forensic cases. In the past, forensic laboratories utilized blood group systems in a discipline known as forensic serology. While it is possible to exclude a suspect through the use of blood group typing, the evidence for the inclusion of a suspect is weak due to the high probability of a coincidental match between two unrelated persons.

3.4.2 Protein Polymorphism
Because of the limitations of antigen polymorphism, protein polymorphism was introduced for forensic identification (Chapter 18). Initially, a few polymorphisms in serum proteins and erythrocyte enzymes were reported. By the 1980s, however, approximately a hundred protein polymorphisms had been discovered. A few systems were commonly used in forensic laboratories, including the polymorphisms of erythrocyte enzymes, serum proteins, and hemoglobin. Blood groups and protein polymorphism analysis were combined in forensic investigations to lower the probability of a match between two unrelated individuals. However, more powerful methods were still sought.

3.4.3 DNA Polymorphism
The human genome contains all the necessary biological information for cellular and organ structure and function. It consists of the *nuclear genome* and the *mitochondrial genome* (Chapter 23 discusses the mitochondrial genome). The human nuclear genome, a set of 23 chromosomes, contains approximately 3 billion base pairs (bp). The Human Genome Project was initiated in 1990 to sequence the entire human nuclear genome. In 2003, 99% of the human genome, including the most important parts of the genome, was sequenced. Further analyses on the human genome sequences continue. The genome contains genes and intergenic noncoding sequences.

3.4.3.1 Genes and Related Sequences
Approximately 20,000–25,000 genes have been identified in the human genome, which encode the information for the synthesis of proteins. The functions of nearly half of these genes have been identified. Most encode the proteins that are responsible for the maintenance of the genome, the functioning of the cells, the immune response, and the structural proteins of cells.

Most human genes are discontinuous. The coding regions of genes are called *exons* and are separated by *introns*. During gene expression, the precursor messenger RNA transcript (pre-mRNA), consisting of both the exons and introns, is produced. The mRNA is a template for protein synthesis in which the sequence is based on a complementary strand of DNA. Through the process of splicing, the introns are removed and the exons are joined, producing the spliced mRNA form, which can be used for protein synthesis via the translation process. Other gene-related sequences include those responsible for gene transcription such as *promoter sequences*; those responsible for gene regulation such as *cis-regulatory sequences* (or enhancers); and *untranslated sequences*, which are transcribed but do not encode proteins. Figure 3.23 depicts the features of a representative human gene.

3.4.3.2 Intergenic Noncoding Sequences
More than 90% of the human genome sequence consists of intergenic noncoding sequences located between genes. The functions of these sequences are yet to be discovered. The intergenic noncoding sequences contain large quantities of various types of *repetitive DNA*, which falls into two categories: tandem repeats and genome-wide or interspersed repeats.

Tandem repeats are repeat units placed next to each other in an array. One type is called *satellite DNA* because of the observation of satellite bands containing DNA with tandem repeats

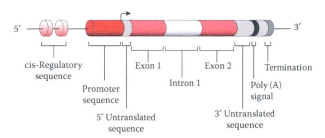

5′   3′

cis-Regulatory   Exon 1   Exon 2   Termination
sequence

Promoter   Intron 1   Poly (A)
sequence   signal

5′ Untranslated   3′ Untranslated
sequence   sequence

Figure 3.23   Gene structure. Transcription, which can be regulated by the cis-regulatory sequence, is initiated at the transcription start site (arrow) near the promoter. The exons, noncoding introns, and the untranslated sequences are also shown. (© Richard C. Li.)

during density gradient centrifugation. Satellite DNA can be found at centromeres and telomeres consisting of regions composed of long stretches of tandem repeats. Minisatellites and microsatellites are two other types of shorter tandem repeats. *Minisatellites*, also known as *variable number tandem repeats* (VNTRs), form arrays of tandem repeats with a *repeat unit length* from several to hundreds of base pairs. In a *microsatellite*, also known as a *short tandem repeat* (STR) or a *simple sequence repeat* (SSR), the repeat unit length can be 2–6 bp long.

Mobile elements (*interspersed repeats*) are randomly located throughout the human genome (Figure 3.24). Two human types have been characterized: *DNA transposons* and *retrotransposons*. The mobile elements change their locations, a process called *transposition*, by which these sequences are inserted into a new site in the genome. The transposition of DNA transposons is through a "cut-and-paste" mechanism. During transposition, DNA transposons are excised from one site and inserted at a new site in the genome. In contrast, retrotransposons duplicate themselves during transposition and propagate throughout the genome, which is a copy-and-paste mechanism: a copy of the original retrotransposons is generated at the new site and the original copy is retained. Additionally, the transposition of retrotransposons requires an RNA intermediate, a process called *retrotransposition*. Retrotransposons have two subtypes: *long terminal repeat* (LTR) and *non-LTR retroposons*. The non-LTR retroposons can be further divided into two subtypes: *long interspersed elements* (LINEs) and *short interspersed elements* (SINEs). Alu elements are the most abundant type of human SINE. There are more than one million copies of Alu elements in the human genome (Figure 3.25). Some members of the Alu elements are

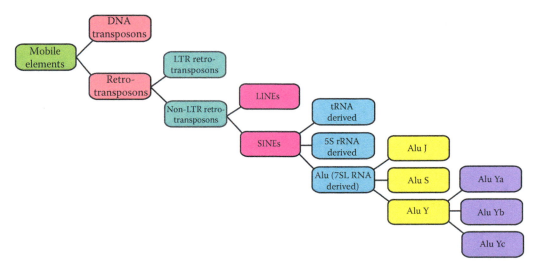

Figure 3.24   Mobile elements classification. (© Richard C. Li.)

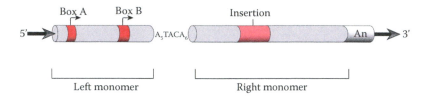

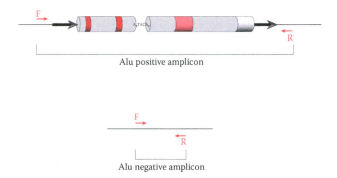

Figure 3.25   Structure of Alu elements. The name, Alu elements, was given since these elements usually contain DNA sequences cleaved by the restriction enzyme *Alu*I. The Alu element is a dimeric structure including two regions separated by an A-rich sequence. The 5′ region has boxes A and B containing RNA polymerase III promoter sequences. The 3′ region contains a 31-nucleotide insertion. There is a short polyadenylation tail located at the 3′ end of the Alu element (terminal A-stretch). Full-length Alu elements are approximately 300 bp long. (© Richard C. Li.)

Figure 3.26   Detecting Alu element insertion polymorphism for human identification. The Alu element insertion site is amplified using the polymerase chain reaction assay. The presence and absence of the Alu element at the site can be detected based on the length of the DNA fragment analyzed. F, forward primer; R, reverse primer. (© Richard C. Li.)

polymorphic for the presence or absence of insertion that can potentially be used for forensic human identification (Figure 3.26).

3.4.3.3 Human DNA Polymorphic Markers

Most individual human genome sequences are very similar. However, variations in sequences do occur. The differences between individual genomes that occur at the DNA level are called *DNA polymorphisms*. In particular, a DNA polymorphism with alternative forms of a chromosomal locus that differ in nucleotide sequence is known as a *sequence polymorphism*. A DNA polymorphism that differs in the numbers of tandem repeat units is known as a *length polymorphism*. A DNA polymorphism can occur anywhere in a genome including genes and other chromosomal locations. Many DNA polymorphisms are useful for genetic mapping studies and hence are called *DNA markers*. DNA polymorphisms form the basis of forensic DNA profiling. The focus of this text is on the human genome, but polymorphisms also occur in the genomes of other organisms.

Most DNA polymorphisms are single nucleotide polymorphisms (SNPs) involving a single-base-pair change or a point mutation. Over one million SNPs have been identified. SNPs arise by spontaneous mutation. Most SNPs occur in noncoding regions of the genome, although some appear in coding regions as well. Other forms of DNA polymorphisms are tandem repeats such as STRs and VNTRs. Although their biological functions are unknown, STRs and VNTRs are very useful for forensic DNA analysis. Many are highly polymorphic, and the number of repeat units varies greatly among different individuals of a population. It is unlikely that two unrelated individuals will have exactly the same combination of STR or VNTR polymorphisms if sufficient markers are examined. Thus, a resulting genetic profile can be used for human identification.

Alternative forms of DNA polymorphisms are called *alleles*. The same allele present in both homologous chromosomes is referred to as *homozygous*. Two different alleles present in homologous chromosomes are referred to as *heterozygous*. A combination of alleles at a given locus is a *genotype*. In forensic analysis, the genotype for a panel of analyzed loci is called the *DNA profile*.

3.4.3.4 Forensic DNA Polymorphism Profiling

In 1984, Sir Alec Jeffreys (Figure 3.27) developed a DNA profiling technique using a VNTR technique involving multilocus profiling and later followed by single-locus profiling (Chapter 19). This technique led to the solving of a double murder that had been committed in Leicestershire (United Kingdom) in the 1980s. The case was the first to apply DNA evidence to a criminal investigation. During the investigation, DNA profiling not only identified the true perpetrator but it also excluded an innocent suspect. This case demonstrated DNA profiling's great potential in forensic investigations.

DNA profiling offered a number of advantages compared with earlier systems. The most important is the ability of the technique to reveal far greater individual variability in DNA than can be revealed by antigen and protein polymorphic markers. The probability of two unrelated individuals having the same DNA profile is very low. The great variability of DNA polymorphisms has made it possible to offer strong support for concluding that DNA from a suspect and from a crime scene originated from the same source. This technique was subsequently implemented in forensic laboratories worldwide.

In the mid-1980s, Kary Mullis and his coworkers developed the polymerase chain reaction (PCR) technique, which amplifies a small quantity of DNA. Mullis's invention had a powerful impact on molecular biology and earned him a Nobel Prize in 1993. Since the introduction of PCR, new techniques have been developed for forensic DNA testing purposes.

Figure 3.27   Sir Alec Jeffreys. (© Richard C. Li.)

The application of PCR-based assays makes forensic DNA analysis possible when only minute quantities of DNA can be recovered from a crime scene, for example, from hairs and cigarette butts. These assays have greatly increased the sensitivity of forensic DNA testing. The first forensic application of a PCR-based assay utilizing SNPs at the *HLA-DQA1* locus (formerly called DQα) was developed in 1986 (Chapter 22). One major disadvantage of the assay was the high probability (approximately 1 in 4000) of a match between two unrelated persons. Amplified fragment length polymorphism (AFLP) at the D1S80 locus has also been implemented in forensic laboratories. The D1S80 locus is a small-size VNTR marker that can be amplified by PCR. The *HLA-DQA1* and AFLP assays were used for some years until the introduction of STR assays.

In the late 1990s, forensic laboratories started utilizing STR loci. STRs have a number of advantages compared with VNTRs. For example, STRs can be amplified by PCR because of their smaller size, which greatly increases the sensitivity of the assay. Furthermore, STR markers are as highly variable as VNTRs. With the application of multiple STR loci, the probability of a match between two unrelated persons becomes extremely low. As a result of DNA testing, perpetrators have been identified and wrongly convicted innocent people have been exonerated (Chapter 20).

In 1995, the United Kingdom established the first national DNA database for criminal investigations. By the end of 1998, several other countries, including the United States, had created their own national DNA databases. The United States has selected 13 STR loci for the Combined DNA Index System (CODIS). These national DNA databases play important roles in solving criminal cases.

Another technique known as mitochondrial DNA (mtDNA) profiling has also been used for forensic testing. mtDNA is maternally inherited genetic material and is therefore particularly useful for human identification. Each cell contains multiple copies of mtDNA. Thus, mtDNA typing can be useful for analysis when nuclear DNA is severely degraded or present in very limited amounts, such as in cases involving decomposed human remains. Alternatively, polymorphic markers at the Y chromosome have also been utilized for forensic DNA testing. Y chromosomal markers are paternally inherited so they can be used for paternity testing. These markers are also very useful in analyzing DNA from multiple contributors in sexual assault cases.

Bibliography

Aleksander, A., Forensic expertise and judicial practice: Evidence or proof? *J Eval Clin Pract*, 2012, **18**(6): 1147–1150.

Alvarez-Cubero, M.J., et al., Genetic identification of missing persons: DNA analysis of human remains and compromised samples. *Pathobiology*, 2012, **79**(5): 228–238.

Andelinovic, S., et al., Twelve-year experience in identification of skeletal remains from mass graves. *Croat Med J*, 2005, **46**(4): 530–539.

Apostolov, A., S. Hristov, and E. Angelova, DNA identification of biological traces and interpretation in a sexual assault case. *Am J Forensic Med Pathol*, 2009, **30**(1): 57–60.

Aquila, I., et al., The role of forensic botany in crime scene investigation: Case report and review of literature. *J Forensic Sci*, 2014, **59**(3): 820–824.

Asari, M., et al., Multiplex PCR-based Alu insertion polymorphisms genotyping for identifying individuals of Japanese ethnicity. *Genomics*, 2012, **99**(4): 227–232.

Auvdel, M.J., Comparison of laser and ultraviolet techniques used in the detection of body secretions. *J Forensic Sci*, 1987, **32**(2): 326–345.

Baraybar, J.P., When DNA is not available, can we still identify people? Recommendations for best practice. *J Forensic Sci*, 2008, **53**(3): 533–540.

Bauer, C.M., et al., Molecular genetic investigations on Austria's patron saint Leopold III. *Forensic Sci Int Genet*, 2013, **7**(2): 313–315.

Bogdanowicz, W., et al., Genetic identification of putative remains of the famous astronomer Nicolaus Copernicus. *Proc Natl Acad Sci U S A*, 2009, **106**(30): 12279–12282.

Boljuncic, J., DNA analysis of early mediaeval individuals from the Zvonimirovo burial site in Northern Croatia: Investigation of kinship relationships by using multiplex system amplification for short tandem repeat loci. *Croat Med J*, 2007, **48**(4): 536–546.

Bond, J.W., Value of DNA evidence in detecting crime. *J Forensic Sci*, 2007, **52**(1): 128–136.

Boric, I., J. Ljubkovic, and D. Sutlovic, Discovering the 60 years old secret: Identification of the World War II mass grave victims from the island of Daksa near Dubrovnik, Croatia. *Croat Med J*, 2011, **52**(3): 327–335.

Brettell, T.A., et al., Forensic science. *Anal Chem*, 1999, **71**(12): 235R–255R.

Bruce-Chwatt, R.M., A brief history of forensic odontology since 1775. *J Forensic Leg Med*, 2010, **17**(3): 127–130.

Budowle, B. and A. van Daal, Extracting evidence from forensic DNA analyses: Future molecular biology directions. *Biotechniques*, 2009, **46**(5): 339–340, 342–350.

Budowle, B., et al., Role of law enforcement response and microbial forensics in investigation of bioterrorism. *Croat Med J*, 2007, **48**(4): 437–449.

Burg, A., R. Kahn, and K. Welch, DNA testing of sexual assault evidence: The laboratory perspective. *J Forensic Nurs*, 2011, **7**(3): 145–152.

Byard, R.W., K. Both, and E. Simpson, The identification of submerged skeletonized remains. *Am J Forensic Med Pathol*, 2008, **29**(1): 69–71.

Caenazzo, L., P. Tozzo, and D. Rodriguez, Ethical issues in DNA identification of human biological material from mass disasters. *Prehosp Disaster Med*, 2013, **28**(4): 393–396.

Caramelli, D., et al., Genetic analysis of the skeletal remains attributed to Francesco Petrarca. *Forensic Sci Int*, 2007, **173**(1): 36–40.

Carey, L. and L. Mitnik, Trends in DNA forensic analysis. *Electrophoresis*, 2002, **23**(10): 1386–1397.

Cattaneo, C., et al., Unidentified bodies and human remains: An Italian glimpse through a European problem. *Forensic Sci Int*, 2010, **195**(1–3): 167, e1–e6.

Cavard, S., et al., Forensic and police identification of "X" bodies. A 6-years French experience. *Forensic Sci Int*, 2011, **204**(1–3): 139–143.

Charlier, P., et al., Genetic comparison of the head of Henri IV and the presumptive blood from Louis XVI (both Kings of France). *Forensic Sci Int*, 2013, **226**(1–3): 38–40.

Claes, P., et al., Computerized craniofacial reconstruction: Conceptual framework and review. *Forensic Sci Int*, 2010, **201**(1–3): 138–145.

Coble, M.D., et al., Mystery solved: The identification of the two missing Romanov children using DNA analysis. *PLoS One*, 2009, **4**(3): e4838.

Connery, S.A., Three decade old cold case murder solved with evidence from a sexual assault kit. *J Forensic Leg Med*, 2013, **20**(4): 355–356.

Cordner, S.M., N. Woodford, and R. Bassed, Forensic aspects of the 2009 Victorian Bushfires Disaster. *Forensic Sci Int*, 2011, **205**(1–3): 2–7.

Courts, C. and B. Madea, Genetics of the sudden infant death syndrome. *Forensic Sci Int*, 2010, **203**(1–3): 25–33.

Courts, C. and B. Madea, Full STR profile of a 67-year-old bone found in a fresh water lake. *J Forensic Sci*, 2011, **56**(Suppl 1): S172–S175.

Craft, K.J., J.D. Owens, and M.V. Ashley, Application of plant DNA markers in forensic botany: Genetic comparison of *Quercus* evidence leaves to crime scene trees using microsatellites. *Forensic Sci Int*, 2007, **165**(1): 64–70.

Davis, G.G., Forensic toxicology. *Clin Lab Sci*, 2012, **25**(2): 120–124.

Davoren, J., et al., Highly effective DNA extraction method for nuclear short tandem repeat testing of skeletal remains from mass graves. *Croat Med J*, 2007, **48**(4): 478–485.

De Ungria, M.C., et al., Forensic DNA evidence and the death penalty in the Philippines. *Forensic Sci Int Genet*, 2008, **2**(4): 329–332.

Definis Gojanovic, M. and D. Sutlovic, Skeletal remains from World War II mass grave: From discovery to identification. *Croat Med J*, 2007, **48**(4): 520–527.

Dekeirsschieter, J., et al., Forensic entomology investigations from Doctor Marcel Leclercq (1924–2008): A review of cases from 1969 to 2005. *J Med Entomol*, 2013, **50**(5): 935–954.

Deng, Y.J., et al., Preliminary DNA identification for the tsunami victims in Thailand. *Genomics Proteom Bioinform*, 2005, **3**(3): 143–157.

Dirkmaat, D.C., et al., New perspectives in forensic anthropology. *Am J Phys Anthropol*, 2008, **137**(Suppl 47): 33–52.

Dissing, J., et al., The last Viking king: A royal maternity case solved by ancient DNA analysis. *Forensic Sci Int*, 2007, **166**(1): 21–27.

Djuric, M., et al., Identification of victims from two mass-graves in Serbia: A critical evaluation of classical markers of identity. *Forensic Sci Int*, 2007, **172**(2–3): 125–129.

Dobberstein, R.C., et al., Degradation of biomolecules in artificially and naturally aged teeth: Implications for age estimation based on aspartic acid racemization and DNA analysis. *Forensic Sci Int*, 2008, **179**(2–3): 181–191.

Dolan, S.M., et al., The emerging role of genetics professionals in forensic kinship DNA identification after a mass fatality: Lessons learned from Hurricane Katrina volunteers. *Genet Med*, 2009, **11**(6): 414–417.

Dzijan, S., et al., Evaluation of the reliability of DNA typing in the process of identification of war victims in Croatia. *J Forensic Sci*, 2009, **54**(3): 608–609.

Enserink, M., Can this DNA sleuth help catch criminals? *Science*, 2011, **331**(6019): 838–840.

Ferri, G., et al., Forensic botany: Species identification of botanical trace evidence using a multigene barcoding approach. *Int J Legal Med*, 2009, **123**(5): 395–401.

Fondevila, M., et al., Case report: Identification of skeletal remains using short-amplicon marker analysis of severely degraded DNA extracted from a decomposed and charred femur. *Forensic Sci Int Genet*, 2008, **2**(3): 212–218.

Foran, D.R., et al., The conviction of Dr. Crippen: New forensic findings in a century-old murder. *J Forensic Sci*, 2011, **56**(1): 233–240.

Foster, E.A., et al., Jefferson fathered slave's last child. *Nature*, 1998, **396**(6706): 27–28.

Ge, J., B. Budowle, and R. Chakraborty, Choosing relatives for DNA identification of missing persons. *J Forensic Sci*, 2011, **56**(Suppl 1): S23–S28.

Geserick, G. and I. Wirth, Genetic kinship investigation from blood groups to DNA markers. *Transfus Med Hemother*, 2012, **39**(3): 163–175.

Gewin, V., Forensics: The call of the crime lab. *Nature*, 2011, **473**(7347): 409–411.

Gill, P., Role of short tandem repeat DNA in forensic casework in the UK—Past, present, and future perspectives. *Biotechniques*, 2002, **32**(2): 366–368, 370, 372, passim.

Gill, P., A.J. Jeffreys, and D.J. Werrett, Forensic application of DNA "fingerprints". *Nature*, 1985, **318**(6046): 577–579.

Gill, P., et al., Identification of the remains of the Romanov family by DNA analysis. *Nat Genet*, 1994, **6**(2): 130–135.

Gingerich, O., The Copernicus grave mystery. *Proc Natl Acad Sci U S A*, 2009, **106**(30): 12215–12216.

Gonzalez-Andrade, F., et al., Two fathers for the same child: A deficient paternity case of false inclusion with autosomic STRs. *Forensic Sci Int Genet*, 2009, **3**(2): 138–140.

Graham, E.A., DNA reviews: Low level DNA profiling. *Forensic Sci Med Pathol*, 2008, **4**(2): 129–131.

Grellner, W. and B. Madea, Demands on scientific studies: Vitality of wounds and wound age estimation. *Forensic Sci Int*, 2007, **165**(2–3): 150–154.

Hampikian, G., E. West, and O. Akselrod, The genetics of innocence: Analysis of 194 U.S. DNA exonerations. *Annu Rev Genomics Hum Genet*, 2011, **12**: 97–120.

Hartman, D., et al., The contribution of DNA to the disaster victim identification (DVI) effort. *Forensic Sci Int*, 2011, **205**(1–3): 52–58.

Hartman, D., et al., The importance of Guthrie cards and other medical samples for the direct matching of disaster victims using DNA profiling. *Forensic Sci Int*, 2011, **205**(1–3): 59–63.

Hatsch, D., et al., A rape case solved by mitochondrial DNA mixture analysis. *J Forensic Sci*, 2007, **52**(4): 891–894.

Hinchliffe, J., Forensic odontology, part 4. Human bite marks. *Br Dent J*, 2011, **210**(8): 363–368.

Holland, M.M., et al., Mitochondrial DNA sequence analysis of human skeletal remains: Identification of remains from the Vietnam War. *J Forensic Sci*, 1993, **38**(3): 542–553.

Holland, M.M., et al., Development of a quality, high throughput DNA analysis procedure for skeletal samples to assist with the identification of victims from the World Trade Center attacks. *Croat Med J*, 2003, **44**(3): 264–272.

Holobinko, A., Forensic human identification in the United States and Canada: A review of the law, admissible techniques, and the legal implications of their application in forensic cases. *Forensic Sci Int*, 2012, **222**(1–3): 394, e1–e13.

Honda, K., The Ashikaga case of Japan—Y-STR testing used as the exculpatory evidence to free a convicted felon after 17.5 years in prison. *Forensic Sci Int Genet*, 2013, **7**(1): e1–e2.

Horsman-Hall, K.M., et al., Development of STR profiles from firearms and fired cartridge cases. *Forensic Sci Int Genet*, 2009, **3**(4): 242–250.

Howard, C., et al., A *Cannabis sativa* STR genotype database for Australian seizures: Forensic applications and limitations. *J Forensic Sci*, 2009, **54**(3): 556–563.

Howlett, R., DNA forensics and the FBI. *Nature*, 1989, **341**(6239): 182–183.

Hughes, V.K. and N.E. Langlois, Use of reflectance spectrophotometry and colorimetry in a general linear model for the determination of the age of bruises. *Forensic Sci Med Pathol*, 2010, **6**(4): 275–281.

Irwin, J.A., et al., DNA identification of "Earthquake McGoon" 50 years postmortem. *J Forensic Sci*, 2007, **52**(5): 1115–1118.

Ishiko, A., et al., Experimental studies on identification of the driver based on STR analysis. *Leg Med (Tokyo)*, 2008, **10**(3): 115–118.

Janisch, S., et al., Analysis of clinical forensic examination reports on sexual assault. *Int J Legal Med*, 2010, **124**(3): 227–235.

Jeffreys, A.J., Genetic fingerprinting. *Nat Med*, 2005, **11**(10): 1035–1039.

Jewkes, R., et al., Medico-legal findings, legal case progression, and outcomes in South African rape cases: Retrospective review. *PLoS Med*, 2009, **6**(10): e1000164.

Jobling, M.A. and P. Gill, Encoded evidence: DNA in forensic analysis. *Nat Rev Genet*, 2004, **5**(10): 739–751.

Just, R.S., et al., Titanic's unknown child: The critical role of the mitochondrial DNA coding region in a re-identification effort. *Forensic Sci Int Genet*, 2011, **5**(3): 231–235.

Karlsson, A.O., et al., DNA-testing for immigration cases: The risk of erroneous conclusions. *Forensic Sci Int*, 2007, **172**(2–3): 144–149.

Kayser, M. and P.M. Schneider, DNA-based prediction of human externally visible characteristics in forensics: Motivations, scientific challenges, and ethical considerations. *Forensic Sci Int Genet*, 2009, **3**(3): 154–161.

Kim, N.Y., et al., A genetic investigation of Korean mummies from the Joseon Dynasty. *Mol Biol Rep*, 2011, **38**(1): 115–121.

Kjellstrom, A., et al., An analysis of the alleged skeletal remains of Carin Goring. *PLoS One*, 2012, **7**(12): e44366.

Krishan, K., Individualizing characteristics of footprints in Gujjars of North India—Forensic aspects. *Forensic Sci Int*, 2007, **169**(2–3): 137–144.

Kupiec, T. and W. Branicki, Genetic examination of the putative skull of Jan Kochanowski reveals its female sex. *Croat Med J*, 2011, **52**(3): 403–409.

Lalueza-Fox, C., et al., Genetic analysis of the presumptive blood from Louis XVI, King of France. *Forensic Sci Int Genet*, 2011, **5**(5): 459–463.

Leclair, B., et al., Bioinformatics and human identification in mass fatality incidents: The World Trade Center disaster. *J Forensic Sci*, 2007, **52**(4): 806–819.

Lee, H.C., et al., DNA typing in forensic science. I. Theory and background. *Am J Forensic Med Pathol*, 1994, **15**(4): 269–282.

Lee, H.Y., et al., Genetic characterization and assessment of authenticity of ancient Korean skeletal remains. *Hum Biol*, 2008, **80**(3): 239–250.

Lee, H.Y., et al., DNA typing for the identification of old skeletal remains from Korean War victims. *J Forensic Sci*, 2010, **55**(6): 1422–1429.

Lehman, D.C., Introduction to forensic science. *Clin Lab Sci*, 2012, **25**(2): 107–108.

Li, X., et al., Mitochondrial DNA and STR analyses for human DNA from maggots crop contents: A forensic entomology case from central-southern China. *Trop Biomed*, 2011, **28**(2): 333–338.

Loreille, O.M., et al., Integrated DNA and fingerprint analyses in the identification of 60-year-old mummified human remains discovered in an Alaskan glacier. *J Forensic Sci*, 2010, **55**(3): 813–818.

Luna, A., Is postmortem biochemistry really useful? Why is it not widely used in forensic pathology? *Leg Med (Tokyo)*, 2009, **11**(Suppl 1): S27–S30.

Madea, B., et al., Molecular pathology in forensic medicine—Introduction. *Forensic Sci Int*, 2010, **203**(1–3): 3–14.

Maiquilla, S.M., et al., Y-STR DNA analysis of 154 female child sexual assault cases in the Philippines. *Int J Legal Med*, 2011, **125**(6): 817–824.

Manasatienkij, C. and C. Ra-ngabpai, Clinical application of forensic DNA analysis: A literature review. *J Med Assoc Thai*, 2012, **95**(10): 1357–1363.

Manhart, J., A. Bittorf, and A. Buttner, Disaster victim identification-experiences of the "Autobahn A19" disaster. *Forensic Sci Med Pathol*, 2012, **8**(2): 118–124.

Manjunath, B.C., et al., DNA profiling and forensic dentistry—A review of the recent concepts and trends. *J Forensic Leg Med*, 2011, **18**(5): 191–197.

Marjanovic, D., et al., DNA identification of skeletal remains from the World War II mass graves uncovered in Slovenia. *Croat Med J*, 2007, **48**(4): 513–519.

Marjanovic, D., et al., Identification of skeletal remains of Communist Armed Forces victims during and after World War II: Combined Y-chromosome (STR) and MiniSTR approach. *Croat Med J*, 2009, **50**(3): 296–304.

Matheson, C.D., et al., Molecular exploration of the first-century Tomb of the Shroud in Akeldama, Jerusalem. *PLoS One*, 2009, **4**(12): e8319.

McDonald, J. and D.C. Lehman, Forensic DNA analysis. *Clin Lab Sci*, 2012, **25**(2): 109–113.

Meissner, C. and S. Ritz-Timme, Molecular pathology and age estimation. *Forensic Sci Int*, 2010, **203**(1–3): 34–43.

Melchior, L., et al., Evidence of authentic DNA from Danish Viking Age skeletons untouched by humans for 1000 years. *PLoS One*, 2008, **3**(5): e2214.

Milde-Kellers, A., et al., An illicit love affair during the Third Reich: Who is my grandfather? *J Forensic Sci*, 2008, **53**(2): 377–379.

Monckton, D.G. and A.J. Jeffreys, DNA profiling. *Curr Opin Biotechnol*, 1993, **4**(6): 660–664.

Montelius, K. and B. Lindblom, DNA analysis in disaster victim identification. *Forensic Sci Med Pathol*, 2012, **8**(2): 140–147.

Mundorff, A.Z., E.J. Bartelink, and E. Mar-Cash, DNA preservation in skeletal elements from the World Trade Center disaster: Recommendations for mass fatality management. *J Forensic Sci*, 2009, **54**(4): 739–745.

Nakamura, Y., DNA variations in human and medical genetics: 25 years of my experience. *J Hum Genet*, 2009, **54**(1): 1–8.

National Institute of Justice. Using DNA to solve cold cases, Special Report, 2002. Washington, DC: US Department of Justice, Office of Justice Programs.

Neuhuber, F., et al., An unusual case of identification by DNA analysis of siblings. *Forensic Sci Int Genet*, 2012, **6**(1): 121–123.

Njoroge, S.K., et al., Microchip electrophoresis of Alu elements for gender determination and inference of human ethnic origin. *Electrophoresis*, 2010, **31**(6): 981–990.

Nunez, C., et al., Genetic analysis of 7 medieval skeletons from the Aragonese Pyrenees. *Croat Med J*, 2011, **52**(3): 336–343.

Nuzzolese, E. and G. Di Vella, Future project concerning mass disaster management: A forensic odontology prospectus. *Int Dent J*, 2007, **57**(4): 261–266.

Onoja, A.M., Paternity testing. *Niger J Med*, 2011, **20**(4): 406–408.

Opdal, S.H., et al., Mitochondrial tRNA genes and flanking regions in sudden infant death syndrome. *Acta Paediatr*, 2007, **96**(2): 211–214.

Page, M., J. Taylor, and M. Blenkin, Expert interpretation of bitemark injuries—A contemporary qualitative study. *J Forensic Sci*, 2013, **58**(3): 664–672.

Palo, J.U., et al., Repatriation and identification of the Finnish World War II soldiers. *Croat Med J*, 2007, **48**(4): 528–535.

Park, D.K., et al., The role of forensic anthropology in the examination of the Daegu subway disaster (2003, Korea). *J Forensic Sci*, 2009, **54**(3): 513–518.

Parson, W., et al., Unravelling the mystery of Nanga Parbat. *Int J Legal Med*, 2007, **121**(4): 309–310.

Piccinini, A., et al., World War One Italian and Austrian soldier identification project: DNA results of the first case. *Forensic Sci Int Genet*, 2010, **4**(5): 329–333.

Pietrangeli, I., et al., Forensic DNA challenges: Replacing numbers with names of Fosse Ardeatine's victims. *J Forensic Sci*, 2009, **54**(4): 905–908.

Pinckard, J.K., Memorial Eckert paper for 2007 forensic DNA analysis for the medical examiner. *Am J Forensic Med Pathol*, 2008, **29**(4): 375–381.

Pollak, S., Medical criminalistics. *Forensic Sci Int*, 2007, **165**(2–3): 144–149.

Pomara, C., et al., A medieval murder. *Am J Forensic Med Pathol*, 2008, **29**(1): 72–74.

Prainsack, B. and M. Kitzberger, DNA behind bars: Other ways of knowing forensic DNA technologies. *Soc Stud Sci*, 2009, **39**(1): 51–79.

Pretty, I.A., Forensic dentistry: 1. Identification of human remains. *Dent Update*, 2007, **34**(10): 621–622, 624–626, 629–630, passim.

Pretty, I.A., Forensic dentistry: 2. Bitemarks and bite injuries. *Dent Update*, 2008, **35**(1): 48–50, 53–54, 57–58 passim.

Primorac, D. and M.S. Schanfield, Application of forensic DNA testing in the legal system. *Croat Med J*, 2000, **41**(1): 32–46.

Prottas, J.M. and A.A. Noble, Use of forensic DNA evidence in prosecutors' offices. *J Law Med Ethics*, 2007, **35**(2): 310–315.

Puentes, K., et al., Three-dimensional reconstitution of bullet trajectory in gunshot wounds: A case report. *J Forensic Leg Med*, 2009, **16**(7): 407–410.

Raymond, J.J., et al., Trace DNA analysis: Do you know what your neighbour is doing? A multi-jurisdictional survey. *Forensic Sci Int Genet*, 2008, **2**(1): 19–28.

Rios, L., et al., Identification process in mass graves from the Spanish Civil War II. *Forensic Sci Int*, 2012, **219**(1–3): e4–e9.

Roewer, L., Y chromosome STR typing in crime casework. *Forensic Sci Med Pathol*, 2009, **5**(2): 77–84.

Rogaev, E.I., et al., Genomic identification in the historical case of the Nicholas II royal family. *Proc Natl Acad Sci U S A*, 2009, **106**(13): 5258–5263.

Roper, S.M. and O.L. Tatum, Forensic aspects of DNA-based human identity testing. *J Forensic Nurs*, 2008, **4**(4): 150–156.

Ruffell, A. and A. Sandiford, Maximising trace soil evidence: An improved recovery method developed during investigation of a $26 million bank robbery. *Forensic Sci Int*, 2011, **209**(1–3): e1–e7.

Rutkowska, J., et al., Donor DNA is detected in recipient blood for years after kidney transplantation using sensitive forensic medicine methods. *Ann Transplant*, 2007, **12**(3): 12–14.

Saigusa, K., et al., Practical applications of molecular biological species identification of forensically important flies. *Leg Med (Tokyo)*, 2009, **11**(Suppl 1): S344–S347.

Sato, Y., et al., Multiplex STR typing of aortic tissues from unidentified cadavers. *Leg Med (Tokyo)*, 2009, **11**(Suppl 1): S455–S457.

Schou, M.P. and P.J. Knudsen, The Danish disaster victim identification effort in the Thai tsunami: Organisation and results. *Forensic Sci Med Pathol*, 2012, **8**(2): 125–130.

Schwark, T., A. Heinrich, and N. von Wurmb-Schwark, Genetic identification of highly putrefied bodies using DNA from soft tissues. *Int J Legal Med*, 2011, **125**(6): 891–894.

Schwark, T., et al., Reliable genetic identification of burnt human remains. *Forensic Sci Int Genet*, 2011, **5**(5): 393–399.

Scott, R. and C. Skellern, DNA evidence in jury trials: The "CSI effect". *J Law Med*, 2010, **18**(2): 239–262.

Seo, Y., et al., STR and mitochondrial DNA SNP typing of a bone marrow transplant recipient after death in a fire. *Leg Med (Tokyo)*, 2012, **14**(6): 331–335.

Skinner, M., D. Alempijevic, and A. Stanojevic, In the absence of dental records, do we need forensic odontologists at mass grave sites? *Forensic Sci Int*, 2010, **201**(1–3): 22–26.

Skopp, G., Postmortem toxicology. *Forensic Sci Med Pathol*, 2010, **6**(4): 314–325.

Speller, C.F., et al., Personal identification of cold case remains through combined contribution from anthropological, mtDNA, and bomb-pulse dating analyses. *J Forensic Sci*, 2012, **57**(5): 1354–1360.

Stadlbauer, C., et al., History of individuals of the 18th/19th centuries stored in bones, teeth, and hair analyzed by LA-ICP-MS—A step in attempts to confirm the authenticity of Mozart's skull. *Anal Bioanal Chem*, 2007, **388**(3): 593–602.

Stene, L.E., K. Ormstad, and B. Schei, Implementation of medical examination and forensic analyses in the investigation of sexual assaults against adult women: A retrospective study of police files and medical journals. *Forensic Sci Int*, 2010, **199**(1–3): 79–84.

Sudoyo, H., et al., DNA analysis in perpetrator identification of terrorism-related disaster: Suicide bombing of the Australian Embassy in Jakarta 2004. *Forensic Sci Int Genet*, 2008, **2**(3): 231–237.

Suzuki, S., et al., Guilty by his fibers: Suspect confession versus textile fibers reconstructed simulation. *Forensic Sci Int*, 2009, **189**(1–3): e27–e32.

Swift, A.M., Case three: Ethics of coercion. *Clin Lab Sci*, 2008, **21**(2): 122–123.

Thompson, R., S. Zoppis, and B. McCord, An overview of DNA typing methods for human identification: Past, present, and future. *Methods Mol Biol*, 2012, **830**: 3–16.

Tokdemir, M., et al., Forensic value of gunpowder tattooing in identification of multiple entrance wounds from one bullet. *Leg Med (Tokyo)*, 2007, **9**(3): 147–150.

Tokutomi, T., et al., Identification using DNA from skin contact: Case reports. *Leg Med (Tokyo)*, 2009, **11**(Suppl 1): S576–S577.

Tracey, M., Short tandem repeat-based identification of individuals and parents. *Croat Med J*, 2001, **42**(3): 6.

Travis, J., Forensic science. Scientists decry isotope, DNA testing of "nationality". *Science*, 2009, **326**(5949): 30–31.

Ubelaker, D.H., The forensic evaluation of burned skeletal remains: A synthesis. *Forensic Sci Int*, 2009, **183**(1–3): 1–5.

US Department of Justice, Office of Justice Programs, National Institute of Justice (NIOJ), *Forensic Science: Review of Status and Needs*, 1999. Washington, DC: US Department of Justice.

US Department of Justice, Office of Justice Programs, National Institute of Justice (NIOJ), *The Future of Forensic DNA Testing: Predictions of the Research and Development Working Group*, 2000. Washington, DC: US Department of Justice.

Vanek, D., L. Saskova, and H. Koch, Kinship and Y-chromosome analysis of 7th century human remains: Novel DNA extraction and typing procedure for ancient material. *Croat Med J*, 2009, **50**(3): 286–295.

Varsha, DNA fingerprinting in the criminal justice system: An overview. *DNA Cell Biol*, 2006, **25**(3): 181–188.

Vaz, M. and F.S. Benfica, The experience of the forensic anthropology service of the medical examiner's office in Porto Alegre, Brazil. *Forensic Sci Int*, 2008, **179**(2–3): e45–e49.

Verdiani, S., et al., An unusual observation of tetragametic chimerism: Forensic aspects. *Int J Legal Med*, 2009, **123**(5): 431–435.

von Wurmb-Schwark, N. and T. Schwark, Genetic determination of sibship and twin zygosity in a case of an alleged double infant homicide. *Leg Med (Tokyo)*, 2009, **11**(Suppl 1): S510–S511.

von Wurmb-Schwark, N., et al., A new multiplex-PCR comprising autosomal and y-specific STRs and mitochondrial DNA to analyze highly degraded material. *Forensic Sci Int Genet*, 2009, **3**(2): 96–103.

Walsh, S.J., Recent advances in forensic genetics. *Expert Rev Mol Diagn*, 2004, **4**(1): 31–40.

Weedn, V.W. and H.J. Baum, DNA identification in mass fatality incidents. *Am J Forensic Med Pathol*, 2011, **32**(4): 393–397.

Wenk, R.E., Testing for parentage and kinship. *Curr Opin Hematol*, 2004, **11**(5): 357–361.

Williamson, R. and R. Duncan, DNA testing for all. *Nature*, 2002, **418**(6898): 585–586.

Willott, G.M., The role of the forensic biologist in cases of sexual assault. *J Forensic Sci Soc*, 1975, **15**(4): 269–276.

Wilms, H.R., et al., Evaluation of autopsy and police reports in the investigation of sudden unexplained death in the young. *Forensic Sci Med Pathol*, 2012, **8**(4): 380–389.

Wyman, J.F., Principles and procedures in forensic toxicology. *Clin Lab Med*, 2012, **32**(3): 493–507.

Zaya, D.N. and M.V. Ashley, Plant genetics for forensic applications. *Methods Mol Biol*, 2012, **862**: 35–52.

Zietkiewicz, E., et al., Current genetic methodologies in the identification of disaster victims and in forensic analysis. *J Appl Genet*, 2012, **53**(1): 41–60.

Zupanic Pajnic, I., B. Gornjak Pogorelc, and J. Balazic, Molecular genetic identification of skeletal remains from the Second World War Konfin I mass grave in Slovenia. *Int J Legal Med*, 2010, **124**(4): 307–317.

Sources of Biological Evidence

Biological evidence analysis is one of the standard forensic examinations in the investigation of a wide variety of crimes. In particular, DNA evidence facilitates investigators' efforts to link offenders to crime scenes by matching DNA profiles. DNA evidence can also be used to eliminate suspects. The DNA evidence that is routinely encountered at crime scenes can often be categorized into several groups or types. Table 4.1 lists sources of DNA that are frequently found on personal items. Figures 4.1 and 4.2 illustrate representative types of evidence that are processed and their success rates.

4.1 Bodily Fluids

Bodily fluids and their stains are useful biological evidence for forensic, serological, and DNA analysis and may be useful in solving crimes. The most common bodily fluids in forensic analysis are blood, seminal fluid, and saliva. Blood evidence, such as peripheral blood that circulates through the heart, arteries, veins, and capillaries, is one of the most common types of biological evidence that is found at crime scenes (Chapter 12). The fluid portion of blood is called *plasma*, a subcompartment of extracellular fluid, which is the bodily fluid outside cells. Blood contains various suspended blood cells. The cellular portion of the blood consists of *erythrocytes* (also known as red blood cells), *leucocytes* (also known as white blood cells), and *platelets*. Because mature human erythrocytes and platelets do not have nuclei, they are not useful sources of nuclear DNA. For forensic DNA profiling, the nuclear DNA in blood samples (Figure 4.3) is primarily isolated from leucocytes, which are nucleated. Besides peripheral blood, menstrual blood can be analyzed to investigate the possibility of the occurrence of a sexual assault crime (Chapter 16).

Other bodily fluids are *transcellular fluids*. These fluids are considered to be a subcompartment of the extracellular fluid that is contained within epithelial-lined extracellular spaces. For example, seminal fluid (Chapter 14) and saliva (Chapter 15) stains as well as vaginal secretions are analyzed for the investigation of sexual assault crimes. Sometimes, urine stains and fecal materials (Chapter 16) are related to assault crimes as well. Sweat, which is secreted from the eccrine and apocrine sweat glands in the skin (Chapter 17), and cerumen, also known as earwax (a waxy substance secreted in the ear canal), can be potentially used for human identification. Fluids present in vomitus (Chapter 17) can be potentially important for forensic investigations of violent crimes.

4.1.1 Extracellular Nucleic Acids

Blood plasma and other various bodily fluids usually contain small amounts of nucleic acids (DNA or RNA) known as extracellular nucleic acids. The nucleic acids circulating in plasma

Table 4.1 Common Items of Evidence		
Evidence	Possible DNA Location	Source of DNA
Baseball bat	Handle	Skin cells, sweat, blood, tissue
Hat, bandana, mask	Inside surfaces	Sweat, hair, skin cells, dandruff, saliva
Eyeglasses	Nose, ear piece, lens	Sweat, skin cells
Facial tissue, cotton swab	Surface	Mucus, blood, sweat, semen, earwax
Dirty laundry	Surface	Blood, sweat, semen, saliva
Toothpick	Surface	Saliva
Used cigarette	Butt (filter area)	Saliva
Used stamp, envelope seal	Moistened area	Saliva
Tape or ligature	Inside or outside surface	Skin cells, sweat, saliva
Bottle, can, glass	Mouthpiece, rim, outer surface	Saliva, sweat, skin cells
Used condom	Inside surface, outside surface	Semen, vaginal cells, rectal cells
Bed linen	Surface	Sweat, hair, semen, saliva, blood
Through-and-through bullet	Outside surface	Blood, tissue
Bite mark	Skin surface	Saliva
Fingernail, partial fingernail	Scrapings	Blood, sweat, tissue, skin cells

Source: National Institute of Justice. Using DNA to solve cold cases, Special Report, 2002, US Department of Justice, Office of Justice Programs, Washington, DC.

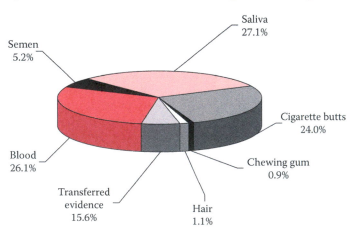

Figure 4.1 Representative types of evidence samples. Data compiled from the third quarter 2005 (July–September) results for all police forces in England and Wales. (Adapted from Bond, J.W., *J Forensic Sci*, 52, 128–136, 2007.)

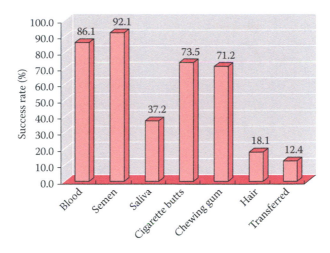

Figure 4.2 Success rate of obtaining suitable profiles from processed samples for submitting to DNA databases. Data are from the third-quarter 2005 (July–September) results for all police forces in England and Wales. (Adapted from Bond, J.W., *J Forensic Sci*, 52, 128–136, 2007.)

Figure 4.3 Biosafety cabinet for the extraction of DNA from biological evidence (left). Blood samples to be processed in a biosafety cabinet for DNA isolation (right). (© Richard C. Li.)

are referred to as *circulating* nucleic acids; the nucleic acids that are found in other body fluids, such as saliva and urine, are referred to as *cell-free* nucleic acids. Extracellular nucleic acids can remain soluble or form complexes with proteins and lipids.

The potential sources of extracellular nucleic acids are *extracellular vesicles* (EVs). EVs are endogenous vesicular structures, containing proteins and nucleic acids that are secreted by most eukaryotic cells (Figure 4.4). There are many types of EVs, including exosomes and microvesicles, which can be detected in bodily fluids. *Exosomes* are one potential source of extracellular nucleic acids. Exosomes are derived from multivesicular bodies (MVBs), which are intracellular organelles of the endocytic pathway (Figure 4.4a). MVBs fuse with the plasma membrane. As a result, the vesicles are released into the extracellular compartment as exosomes. *Microvesicles*, also called shed vesicles or ectosomes, are another possibility. Microvesicles shed from the plasma membrane and thus carry along membrane and cytosolic materials including nucleic acids (Figure 4.4b). *Apoptotic bodies* are a special type of microvesicle that is formed in apoptotic cells. During apoptosis, a process of programmed cell death, cells shrink and eventually form apoptotic bodies (Figure 4.5). Apoptotic bodies contain fragmented DNA by nucleolytic degradation that resembles similar characteristics to those observed in extracellular nucleic acids. However, messenger RNA (mRNA) within apoptotic bodies is protected from RNase degradation. These

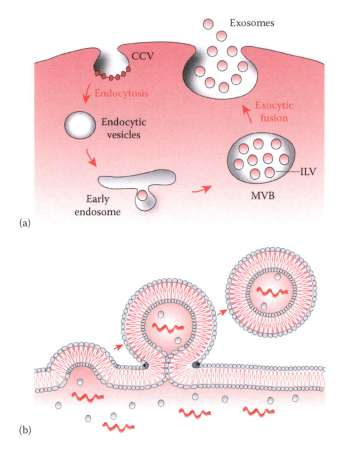

(a)

(b)

Figure 4.4 Exosomes and microvesicles released from healthy cells. (a) Exosomes are derived from the endocytic pathway. In the endocytic pathway, clathrin-coated vesicles (CCV) are formed at the plasma membrane. The endocytic vesicles are then formed through endocytosis and are transported to early endosomes. Multivesicular bodies (MVBs) containing intraluminal vesicles (ILVs) are then developed from early endosomes. Through the exocytic fusion of the MVB membrane with the plasma membrane of the cell, exosomes, containing cellular components including nucleic acids, are released to the extracellular space. (b) Microvesicles are derived from the budding of vesicles from the plasma membrane. Microvesicles are then shed from the plasma membrane carrying along cellular components including nucleic acids that are present in the cytoplasm. (© Richard C. Li.)

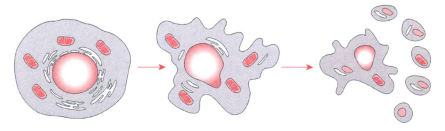

Figure 4.5 Apoptotic bodies released during apoptosis. During apoptosis, the preapoptotic cell (left) undergoes morphological changes. The early apoptotic cell (middle) shows deformation: the cell shrinks, the membrane blebs, and chromatin condense. The late apoptotic cells (right) are fragmented, releasing apoptotic bodies, which contain cytosol and organelles, and nuclear fragments. In addition, proteolytic enzymes are activated that cleave the genomic DNA into fragments. (© Richard C. Li.)

extracellular nucleic acids are potentially useful sources for the forensic identification of bodily fluids and for forensic DNA profiling.

4.2 Cells

4.2.1 Cell Surface Markers

Many of the forensic markers that are analyzed in forensic laboratories are from cells, the building blocks of the human body. All cells have membranes, also known as plasma membranes, which constitute their outer boundaries (Figure 4.6). The functions of cell membranes include exchanges with the environment, signal transduction, and structural support. The cell membrane is a phospholipid bilayer containing lipids, proteins, and carbohydrates. Membrane proteins can act as enzymes, receptors, or ion channels. Many cells also have carbohydrate-rich molecules, including proteoglycans, glycoproteins, and glycolipids, on their membrane surfaces (Figure 4.6). Many of these molecules act as cell surface antigens (Chapter 18).

4.2.2 Nucleated Cells

Probative biological evidence usually contains nucleated cells. The *nucleus* (Figure 4.7) is surrounded by a nuclear envelope and contains chromosomes and a nucleolus. A nucleolus is a dense, non-membrane-bound structure due to its high RNA content. The function of a nucleolus is to transcribe ribosomal RNA and to form ribosomes.

There are two types of cells in the body: sex cells (sperm and oocytes) and somatic cells (all other types). Spermatozoa and ova, which are formed by germ cells, are called *gametes*. In humans, each gamete is haploid, containing 22 *autosomes* (chromosomes other than sex chromosomes) plus one *sex chromosome*. In ova, the sex chromosome is always an X, while in spermatozoa it may be an X or a Y. After fertilization, the zygote, which is the fertilized egg cell formed when two gamete cells are joined, becomes diploid as a result of the fusion of the haploid spermatozoon and ovum. Most of the other cells of the body, known as *somatic cells*, are diploid. This means that they have two copies of each autosome plus two sex chromosomes, XX for females or XY for males. This results in a total of 46 chromosomes per diploid cell. The two chromosomes of a pair in a diploid cell are *homologous chromosomes*. One of the homologous chromosomes is inherited from the spermatozoon and the other from the ovum.

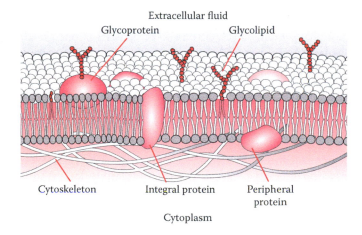

Figure 4.6   Cell membrane and carbohydrate-containing glycoproteins and glycolipids. (© Richard C. Li.)

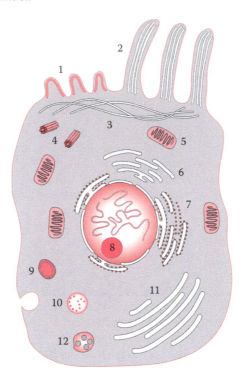

Figure 4.7   A cross-sectional view of a cell. 1, Microvilli; 2, cilia; 3, cytoskeleton; 4, centrioles; 5, mitochondrium; 6, smooth endoplasmic reticulum; 7, rough endoplasmic reticulum; 8, nucleolus; 9, peroxisome; 10, vesicle; 11, Golgi apparatus; and 12, lysosome. (© Richard C. Li.)

Although most somatic cells are diploid, exceptions exist. Some differentiated cells such as red blood cells and platelets have no nuclei and are designated *nulliploid*. A few other cells have more than two sets of chromosomes as a result of DNA replication without cell division and are referred to as *polyploid*. For example, the regenerating cells of the liver and other tissues are naturally tetraploid, while the giant megakaryocytes of the bone marrow may contain 8, 16, or even 32 copies of chromosomes.

The nuclear *chromosomes* of humans consist of complexes of DNA, histone proteins, and nonhistone chromosomal proteins. Each chromosome consists of one linear, double-stranded DNA molecule. The large amounts of DNA present in the human chromosome are compacted by their association with histones into nucleosomes and even further compacted by higher levels of folding of the nucleosomes into *chromatin* fibers. Each chromosome contains a large number of looped domains of chromatin fibers attached to a protein scaffold.

The degree of DNA packing varies throughout the cell cycle. During the metaphase of mitosis and meiosis of the cell cycle, chromatin is the most condensed. Two forms of chromatin have been defined on the basis of their chromosome-staining properties. *Euchromatin* regions are areas of chromosomes that undergo normal chromosome condensation and decondensation during the cell cycle. The intensity of staining of euchromatin is darkest in the metaphase and lightest in the synthesis (S) phase. Euchromatic regions account for most of the genome and lack repetitive DNA. Usually, the genes within the euchromatin can be expressed. *Heterochromatin* comprises the chromosomal regions that usually remain condensed throughout the cell cycle. It contains repetitive DNA and can be found at centromeres, much of the Y chromosome long arm, and the short arms of the acrocentric chromosomes (chromosomes with centromeres near one end). Genes within heterochromatic DNA are usually inactive.

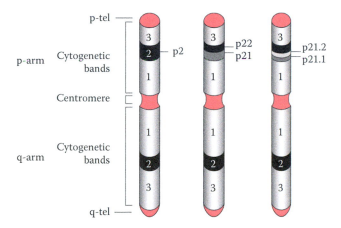

Figure 4.8 The ideogram of a human chromosome and cytogenetic banding nomenclature. The cytogenetic banding pattern is shown after chemical staining such as G-banding, which treats chromosomes with Giemsa dye. The short arm is designated p; the long arm is q. The centromere, p telomere (p-tel), and q telomere (q-tel) are also shown. G-bands at increasing resolution (from left to right) are shown. (© Richard C. Li.)

Each human chromosome has a short arm, designated p (for petit), and a long arm, designated q (for queue), separated by a centromere (Figure 4.8). *Centromeres* are the DNA sequences that are found near the points of attachment of mitotic or meiotic spindle fibers. The centromere region of each chromosome is responsible for accurately segregating the replicated chromosomes to daughter cells during cell divisions. The ends of the chromosome are called *telomeres* and they help stabilize the chromosome and play a role in the replication of DNA in the chromosome.

Chemical staining of metaphase chromosomes results in an alternating dark and light banding pattern (cytogenetic banding) that can be seen under a microscope. Each chromosome arm is divided into regions based on the cytogenic bands. This process is known as *cytogenetic mapping*. The cytogenetic bands are labeled p1, p2, p3, q1, q2, q3, and so on, counting from the centromere out toward the telomeres. At higher resolutions, subbands can be observed. For example, the cytogenetic map location of a gene termed *AMELY* (amelogenin, Y-linked; Chapter 21) is Yp11.2, which indicates its location on chromosome Y, p arm, band 11, subband 2. The visually distinct banding pattern gives each chromosome a unique appearance. Recently, the cytogenic map has been integrated with the human genome sequence to allow the determination of the positions of cytogenetic bands within the DNA sequence.

Chromosomes can be identified on the basis of the size and the positions of the centromeres and cytogenetic banding patterns. The chromosome constitution is described as a *karyotype* and can be displayed as a *karyogram*, which includes the total number of chromosomes and the sex chromosome composition (Figure 4.9). Chromosomes are numbered in order of their size, with chromosome 1 as the largest (except chromosome 21 is smaller than 22). In the cases of chromosomal abnormality, the karyotype can also reflect the type of abnormality and allow visualization of the affected chromosome bands.

4.2.3 Mitochondria and Other Organelles

The cytoplasm contains the cytosol fluid in which organelles are suspended (Figure 4.7). Multiple copies of *mitochondria* are located within the cytoplasm. Mitochondria are surrounded by phospholipid membranes that separate them from the cytosol. The mitochondria are responsible for energy production through aerobic metabolism by producing molecules containing high-energy bonds, such as adenosine triphosphate (ATP). Mitochondria have their own genome, which can be analyzed for human identification (Chapter 23).

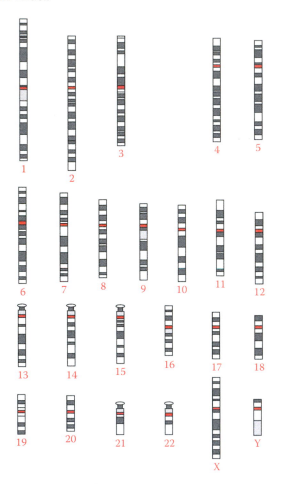

Figure 4.9 Human karyogram. The chromosomes are numbered. Cytogenic patterns show alternating dark and light bands (From www.genome.gov). Centromeres (in red) are shown.

Two types of *endoplasmic reticulum* (ER) can exist within the cytoplasm: the smooth ER (SER) is involved in lipid synthesis; the rough ER (RER) contains ribosomes on its outer surface and forms transport vesicles. The *Golgi apparatus* is responsible for the production of secretory vesicles and new membrane components, and also for the packaging of *lysosomes* (vesicles containing digestive enzymes for the degradation of injured cells). *Peroxisomes* carry enzymes that neutralize potentially harmful free radicals. Other organelles found within the cytoplasm of eukaryotic cells include the cytoskeleton, microvilli, centrioles, and cilia.

4.2.4 Cytosol
4.2.4.1 Messenger RNAs
The chromosomal DNA contains genes that encode for specific proteins. The genetic code is read as an array of triplet codes, a sequence of three bases that specifies the identity of a single amino acid. As gene expression is activated, transcription occurs in which precursor mRNA (pre-mRNA) is produced from a DNA template. After transcription, the pre-mRNA is capped, polyadenylated, and spliced to form matured mRNA. Only the matured mRNA is transported from the nucleus to the cytoplasm. Tissue-specific mRNA can be potentially used for the identification of biological evidence (Chapter 11). The proteins are synthesized in a process known as translation, in which amino acids are assembled based on the codons derived from the triplet

code of the DNA contained in the sequence of the mRNA strand. Various components including the ribosomal complex are involved in translation. The cytosol also contains many proteins that can be used for the identification of bodily fluids (Chapters 11 and 12).

4.2.4.2 MicroRNAs

MicroRNAs (miRNAs) are short RNA molecules that are 21–23 nucleotides in length. In eukaryotic organisms, miRNAs function as negative regulators of gene expression. They play roles in development and cell differentiation. Additionally, the altered expression of miRNAs can be detected in many human diseases. The biological function and the potential application of miRNAs in forensic biology are discussed in Chapter 11.

It is estimated that the human genome may encode approximately 1000 miRNAs. Based on their genomic location and structure, miRNAs can be characterized into three types: intergenic, intronic, and exonic (Figure 4.10). *Intergenic* miRNA genes are distinct transcriptional units that are found in genomic regions. In humans, most of the intergenic miRNA genes have a transcription start site and a polyadenylation site. *Intronic* miRNAs reside within the introns of protein-coding and noncoding genes. The orientation of the intronic miRNAs can be the same as that of the sense (coding) strand of a host gene or that of the antisense strand (complementary to the sense strand) of a host gene. Sense intronic miRNAs are transcribed from the same promoter as their host genes. The antisense intronic miRNAs are transcribed from their own promoters. *Exonic* miRNAs are rare in eukaryotic genomes and reside in genomic regions that overlap with an exon and an intron of a pseudogene, which is a noncoding gene or is no longer transcribed. These miRNAs are also transcribed from their host gene promoter. miRNAs can also be monocistronic or polycistronic. A *monocistronic* miRNA has a single transcriptional unit with its own promoter. In *polycistronic* miRNAs, several miRNAs reside as a cluster of transcriptional units with a shared promoter.

The biogenesis of miRNAs begins in the nucleus. miRNAs are transcribed by RNA polymerase II. Nascent transcripts, referred to as primary transcripts (pri-miRNAs), can be several hundreds to thousands of nucleotides in length. The pri-miRNAs have a hairpin secondary

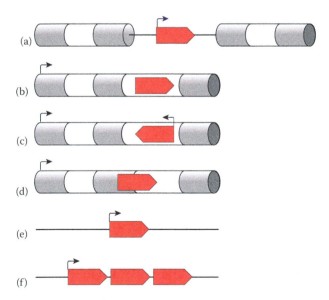

Figure 4.10 The genomic location and the structure of miRNAs. (a) An intergenic miRNA gene, (b) a sense intronic miRNA gene, (c) an antisense intronic miRNA gene, (d) an exonic miRNA gene, (e) a monocistronic miRNA gene, and (f) polycistronic miRNA genes. Exons (gray), introns (white), and miRNA genes (red) are shown. (© Richard C. Li.)

structure that is approximately 70 nucleotides long with imperfect base pairing in the stem. miRNA processing is initiated in the nucleus (Figure 4.11). The hairpin region of a pri-miRNA is then cleaved from the pri-miRNA by the nuclear RNase III endonuclease, Drosha. As a result, a pre-miRNA is formed. The pre-miRNA is transported from the nucleus to the cytoplasm by a nuclear transporter, Exportin 5. In the cytoplasm, a cytoplasmic RNAse III–like endonuclease, Dicer, cleaves the pre-miRNA to generate a double-stranded miRNA that is approximately 21–23 nucleotides in length. The mature miRNA strand is bound to the Argonaut protein to assemble the RNA-induced silencing complex (RISC). The complementary RNA strand is degraded.

A specific miRNA is designated with a prefix. The prefix for a mature miRNA is a capitalized "miR," while the prefix for a pre-miRNA is an uncapitalized "mir." The prefix "miR" is followed by a dash and a number (e.g., miR-135). Experimentally confirmed miRNAs are sequentially numbered and are deposited in the miRBase, which is a database that archives miRNA sequences and annotations.

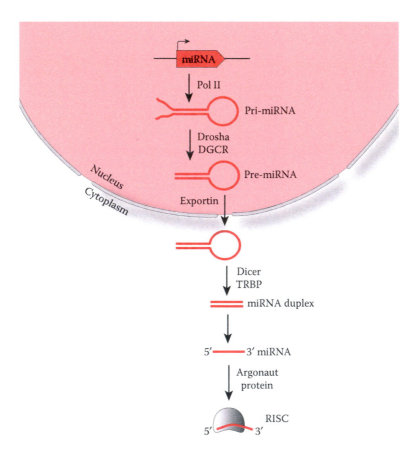

Figure 4.11   The biogenesis of miRNAs. In the nucleus, the pri-miRNA (primary transcript) is processed by Drosha (nuclear RNase for double-stranded RNA) and DGCR (nuclear double-stranded RNA-binding protein), forming the pre-miRNA. The pre-miRNA is transported from the nucleus to the cytoplasm by the exportin. In the cytoplasm, the pre-miRNA is further processed by Dicer (cytoplasmic RNase for double-stranded RNA) and TRBP (cytoplasmic double-stranded RNA-binding protein). The mature miRNA strand is bound to the Argonaut protein to assemble the RISC (RNA-induced silencing complex). (© Richard C. Li.)

4.3 Tissues

4.3.1 Skin

4.3.1.1 Biology of Skin

The biological evidence related to skin is important in forensic investigations. For example, fingerprints are ridge skin impressions that are usually collected at crime scenes, and shed skin tissue is a source of DNA for human identification. The skin covers the entire body surface. Additionally, the skin contains specialized structures that include sebaceous and sweat glands, hair follicles, and nails. The thickness of skin varies throughout the body. The skin of the dorsal area of the body is usually thicker than that of the ventral area of the body. The skin consists of different layers (Figure 4.12). The *epidermis* is the outer layer of the skin. The epidermis also contains melanocytes that produce the skin pigment melanin. The *dermis* is the middle layer of the skin. It is filled with fibrous collagen proteins secreted by fibroblasts and contains hair follicles and sweat glands. Additionally, it contains blood, lymph vessels, and nerves. The *subcutaneous layer* is the deepest layer of the skin. The subcutaneous layer consists of collagen networks and adipose tissue to prevent loss of heat.

The epidermis is a multilayered tissue that includes a number of morphologically distinct zones: the basal, the spinous, the granular, and the cornified layers (Figure 4.12). The epidermis renews continually through the proliferation and differentiation of keratinocytes. The *basal layer* contains newly formed keratinocytes that are proliferative. Epidermal differentiation begins with the migration of the keratinocytes from the basal layer toward the outer layer of skin. Once the migrating keratinocytes reach the *spinous* and *granular* layers, the keratinocytes become nonproliferating and partially differentiated. As the cells reach the *cornified* layer, these cells are filled with keratin filaments and are differentiated into corneocytes, which are dead, and terminally differentiated keratinocytes. In the course of differentiation, the cells are flattened, and all organelles including the nucleus are lost. The corneocytes are then shed from the skin surface.

4.3.1.2 Skin as Source of DNA Evidence

Evidence from skin contact, also referred to as *touched evidence*, can be collected and used for forensic DNA analysis. One example of this evidence is shed skin cells that are found on worn clothing, which is frequently encountered in crime scene investigations. For instance, a perpetrator's shed skin cells that are deposited on worn clothing are potential evidence to be collected. After collection, a DNA profile can be obtained from shed skin cells, providing the forensic evidence or "lead" that is required for the criminal investigation. Touched evidence becomes more

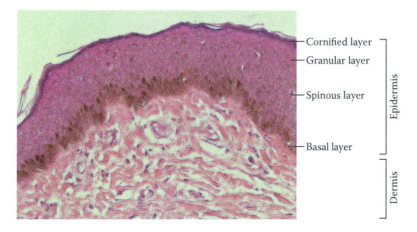

Figure 4.12   A sectional view of the skin. Multilayered epidermis tissues are shown. (© Richard C. Li.)

useful and important when no other type of evidence (e.g., a fingerprint, bloodstain, or seminal stain) can be collected at the crime scene, especially for the investigation of certain criminal cases such as cold cases and property crimes such as theft and burglary (Figures 4.13 through 4.15). Touched evidence usually contains minute quantities of nuclear DNA. DNA recovered from touched evidence is referred to as *transfer DNA*. However, the source of DNA transferred through physical contact is not well understood. Although the shed skin cells from touched evidence lack nuclei, these cells may contain DNA remnants from partial degradation during the differentiation process. These can be a possible DNA source. Additionally, the sweat glands in the skin produce sweat. When the skin makes contact with an item, a residue of sweat is left on the surface of the item. It is known that sweat contains cell-free DNA. Thus, cell-free DNA from sweat is another possible source of DNA. Furthermore, a small number of nucleated cells can be observed in touched evidence. These cells possibly originated from sweat glands and ducts.

4.3.2 Hair

Hairs, including scalp and pubic hairs, frequently constitute biological evidence that is found at crime scenes, and their identification can be of great forensic importance. Formerly, the

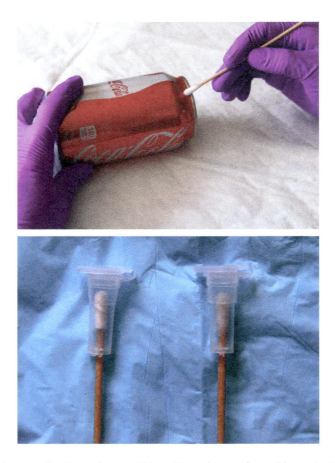

Figure 4.13 Evidence collection using swabbing. The evidence from skin contact for forensic DNA analysis is usually collected by swabbing. More DNA can be recovered when evidence is collected with a double swab method than with a single swab. The double swab method involves applying a moistened cotton swab followed by a second dry cotton swab onto the same target surface of evidence. The target surface is swabbed using a moistened swab first. The moisture left by the first swab is absorbed by the second dry swab. Both swabs can be pooled for DNA extraction. (© Richard C. Li.)

Figure 4.14   A worn glove as a source of DNA evidence. Shed skin cells on a worn glove can potentially generate DNA profiles. (© Richard C. Li.)

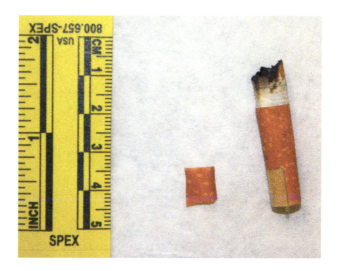

Figure 4.15   A cigarette butt as a source of DNA evidence. Shed cells left behind on a cigarette butt can be a source of DNA. A portion (~1 cm²) of the filter paper of a smoked cigarette butt can be cut for isolating DNA. (© Richard C. Li.)

principal methods that were utilized in forensic hair analysis were limited to morphological analysis and comparisons. Since then, protein polymorphisms provide some potential for identifying individuals from single hairs. However, human hairs contain DNA and as a DNA source they may be used for forensic analysis. The development of the polymerase chain reaction (PCR) amplification technique made it possible to analyze very small quantities of DNA in hair, and the use of hair as evidence of identification has become more significant.

4.3.2.1 Biology of Hair

The human hair shaft is a keratinized cylindrical structure (Figure 4.16). The center or core of the hair is called the *medulla*, which is present in the majority of hairs. The medulla is surrounded by a *cortex*, which is the outer layer of the hair shaft. The *cuticle* consists of overlapping layers of flattened keratinized cells that protect the hair. Hairs are produced in *hair follicles* (Figures 4.17 and 4.18). Each hair follicle is located deep in the dermis (a skin layer beneath the epidermis) and opens onto the surface of the epidermis (the outer layer of the skin). The hair follicle is composed

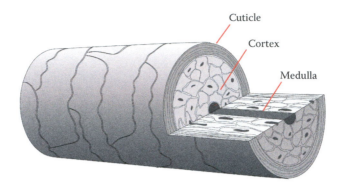

Figure 4.16   A sectional view of a hair shaft. (© Richard C. Li.)

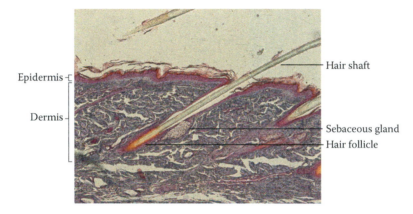

Figure 4.17   Longitudinal section view of a scalp hair follicle with accessory structures. (© Richard C. Li.)

of the *bulge* and the *bulb* regions. The bulge exports the stem cells that migrate down and give rise to bulb cells. The bulge also produces the stem cells that migrate up to form skin cells. The matrix cells, which generate the hair shaft cells, are located at the lower portion of the bulb. The *dermal papilla* is situated at the base of the bulb and contains cells that regulate hair growth, which are nourished by blood vessels and nerves. The growing hair shaft is surrounded by two concentric layers of cells, which are referred to as the inner root sheath and the outer root sheath, respectively. The entire hair follicle is surrounded by a connective tissue sheath.

Human scalp hairs grow for a few years and shed according to the hair follicle cycle (Figures 4.19 through 4.21). A human scalp hair can grow at its highest rate of approximately 1 mm per day. The growing phase of hair is called the *anagen* phase. The matrix cells undergo rapid proliferation and eventually become differentiated cells such as hair shaft cells. As a hair grows, it is pushed toward the surface of the skin and becomes longer. During the migration upward, keratinization occurs as cells are filled with fibril keratin proteins. In these keratinizing cells, nuclei are absent. However, mitochondrial remnants can be observed. By the time a hair approaches the skin surface, cell death occurs at the medulla, cortex, and cuticle. At the end of the anagen phase, the matrix cells enter the *catagen* phase and undergo cell death, thus leading to the regression of the bulb. Hair follicles then enter the *telogen* phase: the stage of rest. When another cycle begins, the follicle produces a new bulb and the telogenic hair, also known as the club hair, is pushed to the surface and shed. On average, an adult loses approximately dozens of hairs daily.

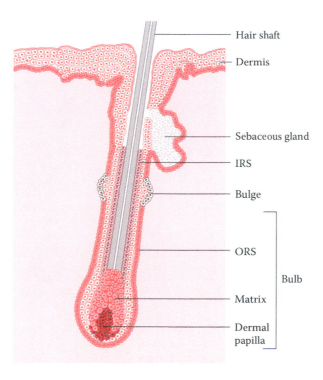

Figure 4.18 Structure of an anagen-phase hair follicle. The diagram shows the hair follicle structure including the dermal papilla, sebaceous gland, bulge, bulb, and hair shaft. Concentric layers of the outer root sheath (ORS), the inner root sheath (IRS), and the hair shaft are also shown. (© Richard C. Li)

4.3.2.2 Hair as Source of DNA Evidence

The isolation of DNA from intact hair roots is routinely used in nuclear DNA analysis. In comparison, the quantities of DNA in telogen hair roots are considerably less than the DNA that is found in the roots of anagen hairs. Nuclear DNA analysis is usually accomplished by using freshly plucked hair roots (Figures 4.22 and 4.23) because cells at the root region may contain nuclear DNA. Unfortunately, most human hairs recovered from crime scenes are shed naturally (in the telogen phase) and contain little nuclear DNA (Figure 4.24). Thus, multiple telogen hairs with roots are necessary to isolate enough nuclear DNA. However, shed hairs that are found at crime scenes may be derived from different individuals. Therefore, the ability to perform a forensic DNA analysis of a single shed hair would be highly desirable. Nuclear DNA isolation from hair shafts is still far less reliable because hair shafts contain very low amounts of nuclear DNA. In addition, variations in the amounts of DNA isolated from hair shafts are observed in a comparison between different hairs from the same head and hairs from different individuals.

A hair follicle cell contains multiple copies of mitochondria. As a result, mitochondrial DNA (mtDNA) can be successfully isolated from hair roots. Additionally, mtDNA is embedded in the keratin matrix of hair shaft cells, which protects the mtDNA molecules from degradation. Thus, mtDNA can also be isolated from hair shafts. A sequence polymorphism analysis of mtDNA from hair can be carried out. mtDNA is maternally inherited, which is useful to identify maternal relatives but cannot be used to perform paternity testing. Additionally, the mtDNA profiling results are not as discriminating as nuclear DNA profiling. Furthermore, mtDNA analysis is time-consuming. Therefore, the typing of nuclear DNA from hair would be preferable for forensic DNA analysis.

Sometimes, a mixture of more than one mtDNA sequence in the same individual is observed. This heterogeneous pool of mtDNA molecules is referred to as *heteroplasmy*. In hair, it is

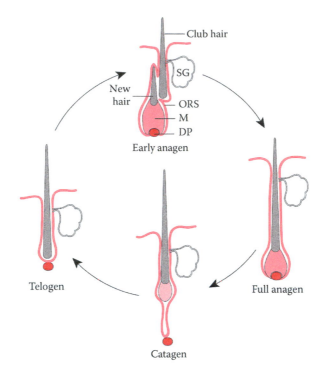

Figure 4.19   Diagrammatic representation of the scalp hair cycle. The morphological characteristics of a hair follicle are shown in three distinct phases encompassing the entire hair cycle: anagen (including early and full anagen), catagen, and telogen. DP, dermal papilla; M, matrix; ORS, outer root sheath; SG, sebaceous gland. (© Richard C. Li.)

believed that it is due to a mixture of mtDNA molecules from keratinocyte and melanocyte-derived mitochondria (Figure 4.25). Hair follicle melanocytes are formed at the beginning of each hair cycle and die at the end of the cycle. Melanocytes are located in the bulb hair follicle. Hair melanocytes play roles in hair pigmentation, which determine hair color. Melanin, which is produced by the melanocytes, is contained in an organelle called the *melanosome*. The melanosomes are transferred to neighboring keratinocytes though *dendritic processes*. In addition to melanosomes, the melanocyte mitochondria can also be transferred to keratinocytes. Thus, the keratinized cells in the hair shaft may carry more than one type of mitochondria, one from the keratinocytes and the other from the melanocytes. As a result, heteroplasmy of a mixture of different mtDNA molecules, with different DNA sequences, can occur in hairs.

4.3.3 Bone
4.3.3.1 Biology of Bone
The bodies of human remains begin to decompose shortly after death. The rate of decomposition of human remains varies greatly with environmental conditions such as climate, bacterial growth, and the presence of insects and other animal scavengers. However, soft tissues may be lost first while more stable bone tissues may remain. Identifying human skeletal remains can be applied in a variety of cases including mass fatality incidents, missing persons, fires, explosions, and violent crime cases involving skeletal remains.

An adult human skeleton consists of 206 bones (Figures 4.26 and 4.27). The shaft of a long bone, such as an arm or a leg bone, consists largely of an outer layer of *cortical* (or compact) bone, which is solid and strong. The shaft of a long bone forms a *marrow cavity*, which is filled with a specialized type of connective tissue called bone marrow. The portion at each end of a

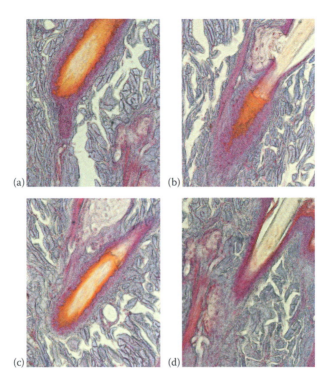

Figure 4.20 Longitudinal section view of a scalp hair follicle during the hair cycle. (a) Early anagen, (b) full anagen, (c) catagen, and (d) telogen. (© Richard C. Li.)

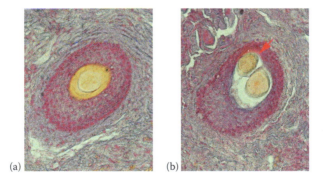

Figure 4.21 Cross-sectional view of hair follicles. (a) Early anagen and (b) full anagen. Note that the new hair (arrow) and the club hair are shown. (© Richard C. Li.)

long bone is called the *epiphysis*, which is composed largely of *cancellous* (or spongy) bone, and can bear the force of compression. A flat bone can have primarily either cortical or cancellous bone. For instance, a rib consists of primarily cancellous bone surrounded by a thin layer of cortical bone. A skull bone usually consists of largely cortical bone.

Bone, which is a connective tissue, contains a matrix and cells. The bone matrix consists of an inorganic and an organic matrix. Calcium and phosphorus are the major components of the inorganic matrix, which consists mainly of hydroxyapatite crystals, $Ca_{10}(PO_4)_6(OH)_2$. The organic matrix consists of collagens, primarily type I collagen, which are insoluble fibrous proteins. With the deposition of calcium hydroxyapatite crystals around the collagen fibrils, bone becomes a weight-bearing hard tissue.

Figure 4.22   Hair root of a pulled hair with visible soft tissue attached. (© Richard C. Li.)

Figure 4.23   Pulled dreadlocks recovered from a crime scene. (© Richard C. Li.)

Developing bones contain small numbers of *osteoprogenitor* cells. These cells can divide to produce cells that differentiate into *osteoblasts*. Osteoblast cells regulate the calcification of the bone matrix. Osteoblasts that are embedded in the bone matrix are termed *osteocytes* and are the most abundant cells in bone. Osteocytes play a role in maintaining the surrounding matrix and repairing damaged bone. Another type of cell that can be found in bone tissues is *osteoclasts*. These cells are giant cells containing 50 or more nuclei and are responsible for dissolving and recycling the bone matrix.

4.3.3.2 Bone as Source of DNA Evidence
A number of methods are used to identify human remains, for example: the identification of facial characteristics; the recognition of individualizing scars, marks, and other special body

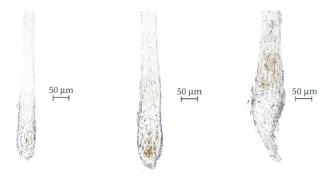

Figure 4.24   Telogen hair roots. From left to right: telogen club root with no visible soft tissue; telogen hair root with some visible soft tissue; telogen hair root with visible soft tissue. (From Bourguignon, L. et al., *Forensic Sci Int Genet*, 3, 27–31, 2008. With permission.)

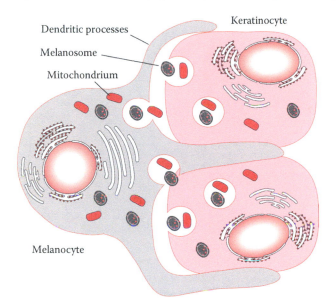

Figure 4.25   Scheme diagram of the melanosome transportation. Melanosomes are released into the extracellular space from the melanocyte dendrites through exocytosis and subsequent endocytosis by keratinocytes. During the process, mitochondria originating from the melanocytes can be potentially transported to the keratinocytes. (© Richard C. Li.)

features; the matching of dentition with premortem dental x-rays; and the comparison of fingerprints. In some situations, these methods cannot be used because of the extensive decomposition of the remains. The mass fatality terrorist attack on the World Trade Center in 2001 (Figure 4.28) serves as an example of a situation where common identification techniques may not be useful. Large quantities of compromised human skeletal fragments were recovered at the fatality site. In these cases, DNA typing is a powerful tool for identifying human remains.

Most DNA in cortical bone is located in the osteocytes. It has been estimated that there are approximately 20,000 osteocytes per cubic millimeter of calcified bone matrix. As a result, microgram quantities of DNA can potentially be obtained from a gram of bone. Thus, compact bone tissue should contain sufficient amounts of nuclear DNA for analysis. However, the skeletal fragments recovered from burial sites are often subjected to decomposition (Figures 4.29 and 4.30). During the decomposition process, both nuclear and mtDNA can be degraded.

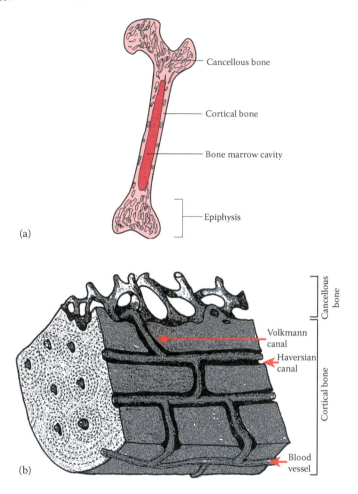

Figure 4.26 Bone structure. An adult human consists of 206 bones. A long bone, such as an arm or a leg bone, consists of an outer cylinder of cortical bone surrounding a marrow cavity. Each end of a long bone is called the epiphysis, which is composed largely of cancellous bone. Flat bones have variable structures; for example, the skull consists mainly of cortical bone whereas the spine consists mainly of cancellous bone. (a) Diagram of a long bone. (b) A bone fragment. Cortical and cancellous bones are shown. Blood vessels can be found in the Volkmann and Haversian canals. (© Richard C. Li.)

Additionally, burial conditions with high humidity and temperature promote the degradation of DNA. Thus, the identification of partial DNA profiles or the complete failure to obtain DNA profiles can occur after samples from decomposed remains are analyzed.

Processing bone samples for DNA extraction is a time-consuming task (Figure 4.31). Due to the potential for commingled remains and contaminants that interfere with forensic analysis, a bone sample initially must be cleaned prior to isolating DNA. The outer surfaces of bone fragments are usually removed by using a mechanical method such as sanding. However, to avoid cross-contamination of samples, the bone dust that is generated by sanding must be removed. Additionally, special protective equipment and safety procedures are necessary to protect analysts from exposure to blood-borne pathogens.

To obtain adequate quality and quantity of DNA from a bone sample, a high-yield DNA extraction method should be selected. The bone samples can be ground to powder to aid in DNA extraction (Chapter 5). The osteocytes containing DNA are embedded in a calcified bone matrix, which is a barrier for extracting DNA from the osteocytes. The bone matrix must be

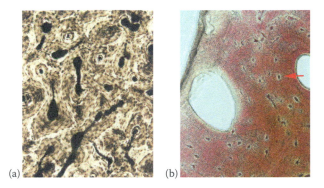

Figure 4.27 Cross-sectional view of cortical bone. (a) The functional unit of cortical bones is a cylindrical structure known as osteon. Haversian canals are shown in the center of the osteon. (b) Detailed view of an osteon. Osteocytes (arrows) are shown within osteons. (© Richard C. Li.)

Figure 4.28 Sections of the Fire Department of New York City (FDNY) Memorial Wall. Memorial to the Fallen Firefighters of 9/11, at FDNY Engine Co. 10 on Liberty Street, New York, by Rambusch Studios. (© Richard C. Li.)

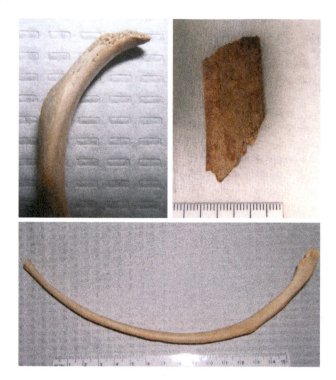

Figure 4.29   Human rib bone fragments recovered from burial site. (© Richard C. Li.)

Figure 4.30   Skeletal remains, exhumed from Frombork Cathedral in Poland, are thought to be those of astronomer Nicolaus Copernicus (1473–1543). (From Bogdanowicz, W., et al., *Proc Natl Acad Sci U S A*, 106, 12279–12282, 2009. With permission.)

removed to improve the yield of DNA. A decalcification method can be utilized to dissolve calcium ions to soften the bone tissue. Additionally, the application of proteinase can be used to digest the matrix proteins, thus increasing the yield of DNA that is harvested from osteocytes.

4.3.4 Teeth
4.3.4.1 Biology of Teeth
During embryonic development, two sets of teeth begin to form. The first to appear are the *deciduous teeth* or *primary teeth*. Most children have 20 deciduous teeth, which are later replaced by 32 teeth known as the *secondary dentition* or *permanent dentition* (Figure 4.32).

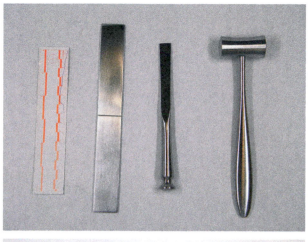

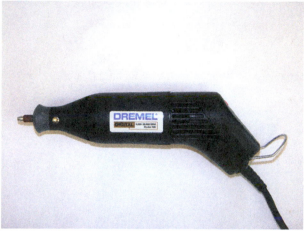

Figure 4.31 Tools for cutting bone samples. Osteotomes and a mallet (top), and a rotary device (bottom). (© Richard C. Li.)

The bulk of each tooth consists of a calcified connective tissue called *dentin* (Figure 4.33). The dentin of the crown is covered by a layer of *enamel*. The rest of the dentin is referred to as radicular dentin and is covered by a layer of *cementum*, which separates the tooth from the surrounding jawbone.

Similar to bone, tooth tissue contains a matrix. The inorganic matrix of tooth contains hydroxyapatite, a calcium phosphate in a crystalline form. The organic matrices of dentin and cementum are primarily collagens. In enamel, amelogenin is the major protein of the organic matrix. Other proteins include ameloblastin and enamelin, which are also the components of the enamel organic matrix.

The interior chamber within the tooth surrounded by dentin is known as the pulp cavity. The *dental pulp*, found within the pulp cavity, is the connective tissue made up of nerve fibers, blood vessels, and various cells. The blood vessels and nerves in the pulp cavity are innervated through the root canal, a narrow tunnel located at the root of the tooth. Incisor and cuspid teeth have single roots. Bicuspids have one or two roots. Molars typically have three or more roots.

The columnar cell bodies of *odontoblasts* are located along the peripheral dental pulp. A single *odontoblast process*, arising from each cell body of the odontoblasts, projects into the dentinal tubule (Figure 4.34). Odontoblasts play important roles in the formation of dentin. Odontoblasts secrete collagens and ground substances that are the components of the

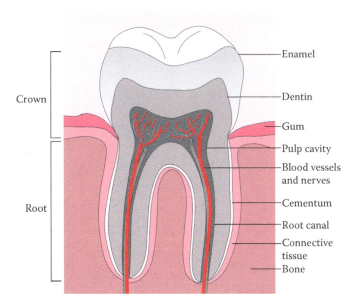

Figure 4.32 A sectional view of an adult tooth. (© Richard C. Li.)

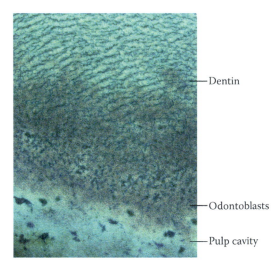

Figure 4.33 Cross-sectional view of a tooth. (© Richard C. Li.)

dentinal matrix. Additionally, odontoblasts regulate the calcification of the matrix of dentin. *Cementoblasts* are cells that play roles in forming the cementum. Cementoblasts secrete collagens and ground substances to form the extracellular matrix of the cementum. Through the process of forming cementum, cementoblasts become trapped in the extracellular matrix. The cementoblasts embedded in the cementum are referred to as *cementocytes. Ameloblasts* are cells that play a role in producing enamel and are subsequently lost during tooth eruption.

4.3.4.2 Teeth as Source of DNA Evidence

The characteristics of teeth, their alignment, and the overall structure of the mouth provide information for identifying a person. The use of dental records such as x-rays and dental casts can allow dental remains to be connected to a victim. Particularly in circumstances such as

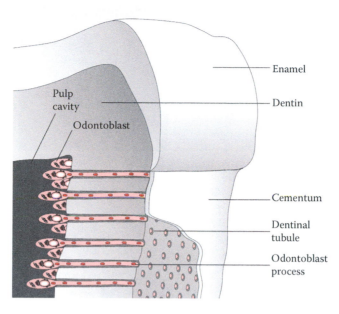

Figure 4.34 Odontoblast processes. See Section 4.3.4.1. (© Richard C. Li.)

Figure 4.35 Tools for dissecting teeth. (© Richard C. Li.) (a) A dental chisel and a mallet. (© Richard C. Li.) (b) An amalgam well. (Courtesy of Dr. Ken Hermsen.)

decomposition, an odontological comparison is possible since the dental evidence often remains intact. When no antemortem dental record is available for comparison, forensic DNA testing can be carried out for postmortem human identification. The mineralized dental structure protects DNA from degradation in cases where it may be degraded in other tissues of the body. Thus, teeth are an excellent source of DNA for forensic DNA analysis under such conditions.

Dental pulp tissue contains various cells and is a suitable source of DNA. However, when tooth evidence has been exposed to high temperature and humid environments, decomposition of the pulp tissue can occur. Thus, cementoblasts within the cementum, containing both nuclei and mitochondria, can then be utilized as a source of DNA. Additionally, odontoblast processes within dentin, containing mitochondria, can also be used. Several different methods are used to obtain dental tissues for DNA isolation (Figure 4.35). A vertical section is cut along the longitudinal axis of the tooth, which allows the dissection of the pulp, dentin, and cementum tissues

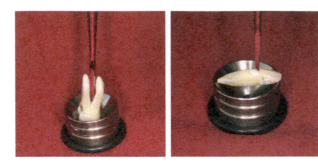

Figure 4.36   Dissecting teeth for DNA isolation. Vertical cutting and dissecting of the pulp, dentin, and cementum tissues (left) and horizontal cutting and dissecting of the root portion (right). (Courtesy of Dr. Ken Hermsen.)

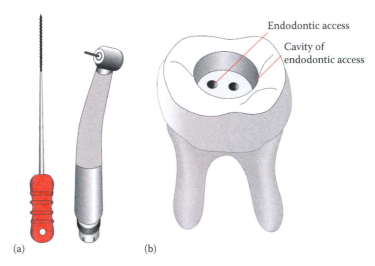

Figure 4.37   Extraction of pulp tissue using endodontic access procedures: trepanning the occlusal surface of a tooth using a dental bur mounted on a turbine (a), creating a cavity for endodontic access (b), and extracting pulp tissue using a nerve broach (a). (© Richard C. Li.)

(Figure 4.36). A horizontal section is cut through the cementum–enamel junction of the tooth (Figure 4.36). The root portion can be pulverized to a fine powder for DNA isolation. The crown can be preserved for forensic odontological comparisons if needed. Additionally, the extraction of pulp tissue can be carried out by standard endodontic access (Figure 4.37). For calcified tissues such as dentin and cementum, a decalcification step is needed to soften the dental matrix, which facilitates DNA isolation.

Bibliography

Abaz, J., et al., Comparison of the variables affecting the recovery of DNA from common drinking containers. *Forensic Sci Int*, 2002, **126**(3): 233–240.

Aditya, S., C.N. Bhattacharyya, and K. Chaudhuri, Generating STR profile from "Touch DNA". *J Forensic Leg Med*, 2011, **18**(7): 295–298.

Alakoc, Y.D. and P.S. Aka, "Orthograde entrance technique" to recover DNA from ancient teeth preserving the physical structure. *Forensic Sci Int*, 2009, **188**(1–3): 96–98.

Allouche, M., et al., Genetic identification of decomposed cadavers using nails as DNA source. *Forensic Sci Int Genet*, 2008, **3**(1): 46–49.

Alvarez Garcia, A., et al., Effect of environmental factors on PCR-DNA analysis from dental pulp. *Int J Legal Med*, 1996, **109**(3): 125–129.

Amory, S., et al., STR typing of ancient DNA extracted from hair shafts of Siberian mummies. *Forensic Sci Int*, 2007, **166**(2–3): 218–229.

Anderson, T.D., et al., A validation study for the extraction and analysis of DNA from human nail material and its application to forensic casework. *J Forensic Sci*, 1999, **44**(5): 1053–1056.

Anzai-Kanto, E., et al., DNA extraction from human saliva deposited on skin and its use in forensic identification procedures. *Pesqui Odontol Bras*, 2005, **19**(3): 216–222.

Barbaro, A., P. Cormaci, and A. Barbaro, Detection of STRs from body fluid collected on IsoCode paper-based devices. *Forensic Sci Int*, 2004, **146**(Suppl): S127–S128.

Barbaro, A., P. Cormaci, and A. Barbaro, DNA analysis from mixed biological materials. *Forensic Sci Int*, 2004, **146**(Suppl): S123–S125.

Barbaro, A., et al., Anonymous letters? DNA and fingerprints technologies combined to solve a case. *Forensic Sci Int*, 2004, **146**(Suppl): S133–S134.

Berger, B., et al., Chimerism in DNA of buccal swabs from recipients after allogeneic hematopoietic stem cell transplantations: Implications for forensic DNA testing. *Int J Legal Med*, 2013, **127**(1): 49–54.

Bogdanowicz, W., et al., Genetic identification of putative remains of the famous astronomer Nicolaus Copernicus. *Proc Natl Acad Sci U S A*, 2009, **106**(30), 12279–12282.

Bond, J.W., Value of DNA evidence in detecting crime. *J Forensic Sci*, 2007, **52**(1): 128–136.

Bourguignon, L., et al., A fluorescent microscopy-screening test for efficient STR-typing of telogen hair roots. *Forensic Sci Int Genet*, 2008, **3**(1): 27–31.

Bright, J.A., et al., The effect of cleaning agents on the ability to obtain DNA profiles using the Identifiler™ and PowerPlex® Y Multiplex Kits. *J Forensic Sci*, 2011, **56**(1): 181–185.

Bruck, S., et al., Single cells for forensic DNA analysis—From evidence material to test tube. *J Forensic Sci*, 2011, **56**(1): 176–180.

Burger, M.F., E.Y. Song, and J.W. Schumm, Buccal DNA samples for DNA typing: New collection and processing methods. *Biotechniques*, 2005, **39**(2): 257–261.

Castella, V., M. Lesta Mdel, and P. Mangin, One person with two DNA profiles: A(nother) case of mosaicism or chimerism. *Int J Legal Med*, 2009, **123**(5): 427–430.

Castella, V., et al., Successful DNA typing of ultrafiltered urines used to detect EPO doping. *Forensic Sci Int Genet*, 2007, **1**(3–4): 281–282.

Chiou, F.S., et al., Extraction of human DNA for PCR from chewed residues of betel quid using a novel "PVP/CTAB" method. *J Forensic Sci*, 2001, **46**(5): 1174–1179.

Cina, M.S., et al., Isolation and identification of male and female DNA on a postcoital condom. *Arch Pathol Lab Med*, 2000, **124**(7): 1083–1086.

Cook, O. and L. Dixon, The prevalence of mixed DNA profiles in fingernail samples taken from individuals in the general population. *Forensic Sci Int Genet*, 2007, **1**(1): 62–68.

Corte-Real, A., et al., The tooth for molecular analysis and identification: A forensic approach. *J Forensic Odontostomatol*, 2012, **30**(1): 22–28.

Courts, C. and B. Madea, Micro-RNA—A potential for forensic science? *Forensic Sci Int*, 2010, **203**(1–3): 106–111.

Courts, C., B. Madea, and C. Schyma, Persistence of biological traces in gun barrels—An approach to an experimental model. *Int J Legal Med*, 2012, **126**(3): 391–397.

Daly, D.J., C. Murphy, and S.D. McDermott, The transfer of touch DNA from hands to glass, fabric and wood. *Forensic Sci Int Genet*, 2012, **6**(1): 41–46.

de Lourdes Chavez-Briones, M., et al., Identification of human remains by DNA analysis of the gastrointestinal contents of fly larvae. *J Forensic Sci*, 2013, **58**(1): 248–250.

Delabarde, T., et al., The potential of forensic analysis on human bones found in riverine environment. *Forensic Sci Int*, 2013, **228**(1–3): e1–e5.

Dimo-Simonin, N., F. Grange, and C. Brandt-Casadevall, PCR-based forensic testing of DNA from stained cytological smears. *J Forensic Sci*, 1997, **42**(3): 506–509.

Dissing, J., A. Sondervang, and S. Lund, Exploring the limits for the survival of DNA in blood stains. *J Forensic Leg Med*, 2010, **17**(7): 392–396.

Dowlman, E.A., et al., The prevalence of mixed DNA profiles on fingernail swabs. *Sci Justice*, 2010, **50**(2): 64–71.

Durdle, A., R.J. Mitchell, and R.A. van Oorschot, The human DNA content in artifacts deposited by the blowfly *Lucilia cuprina* fed human blood, semen and saliva. *Forensic Sci Int*, 2013, **233**(1–3): 212–219.

Edson, S.M. and A.F. Christensen, Field contamination of skeletonized human remains with exogenous DNA. *J Forensic Sci*, 2013, **58**(1): 206–209.

Edson, J., et al., A quantitative assessment of a reliable screening technique for the STR analysis of telogen hair roots. *Forensic Sci Int Genet*, 2013, **7**(1): 180–188.

Ellis, M.A., et al., An evaluation of DNA yield, DNA quality and bite registration from a dental impression wafer. *J Am Dent Assoc*, 2007, **138**(9): 1234–1240; quiz 1267.

Farmen, R.K., et al., Assessing the presence of female DNA on post-coital penile swabs: Relevance to the investigation of sexual assault. *J Forensic Leg Med*, 2012, **19**(7): 386–389.

Flanagan, N. and C. McAlister, The transfer and persistence of DNA under the fingernails following digital penetration of the vagina. *Forensic Sci Int Genet*, 2011, **5**(5): 479–483.

Foran, D.R., M.E. Gehring, and S.E. Stallworth, The recovery and analysis of mitochondrial DNA from exploded pipe bombs. *J Forensic Sci*, 2009, **54**(1): 90–94.

Fridez, F. and R. Coquoz, PCR DNA typing of stamps: Evaluation of the DNA extraction. *Forensic Sci Int*, 1996, **78**(2): 103–110.

Goray, M., R.J. Mitchell, and R.A. van Oorschot, Investigation of secondary DNA transfer of skin cells under controlled test conditions. *Leg Med (Tokyo)*, 2010, **12**(3): 117–120.

Goray, M., J.R. Mitchell, and R.A. van Oorschot, Evaluation of multiple transfer of DNA using mock case scenarios. *Leg Med (Tokyo)*, 2012, **14**(1): 40–46.

Goray, M., R.A. van Oorschot, and J.R. Mitchell, DNA transfer within forensic exhibit packaging: Potential for DNA loss and relocation. *Forensic Sci Int Genet*, 2012, **6**(2): 158–166.

Goray, M., et al., Secondary DNA transfer of biological substances under varying test conditions. *Forensic Sci Int Genet*, 2010, **4**(2): 62–67.

Graham, E.A. and G.N. Rutty, Investigation into "normal" background DNA on adult necks: Implications for DNA profiling of manual strangulation victims. *J Forensic Sci*, 2008, **53**(5): 1074–1082.

Graham, E.A., E.E. Turk, and G.N. Rutty, Room temperature DNA preservation of soft tissue for rapid DNA extraction: An addition to the disaster victim identification investigators toolkit? *Forensic Sci Int Genet*, 2008, **2**(1): 29–34.

Grskovic, B., et al., Effect of ultraviolet C radiation on biological samples. *Croat Med J*, 2013, **54**(3): 263–271.

Grubwieser, P., et al., Systematic study on STR profiling on blood and saliva traces after visualization of fingerprint marks. *J Forensic Sci*, 2003, **48**(4): 733–741.

Gurvitz, A., L.Y. Lai, and B.A. Neilan, Exploiting biological materials in forensic science. *Australas Biotechnol*, 1994, **4**(2): 88–91.

Hellmann, A., et al., STR typing of human telogen hairs—A new approach. *Int J Legal Med*, 2001, **114**(4–5): 269–273.

Herber, B. and K. Herold, DNA typing of human dandruff. *J Forensic Sci*, 1998, **43**(3): 648–656.

Higgins, D. and J.J. Austin, Teeth as a source of DNA for forensic identification of human remains: A review. *Sci Justice*, 2013, **53**(4): 433–441.

Higuchi, R., et al., DNA typing from single hairs. *Nature*, 1988, **332**(6164): 543–546.

Hochmeister, M.N., PCR analysis of DNA from fresh and decomposed bodies and skeletal remains in medicolegal death investigations. *Methods Mol Biol*, 1998, **98**: 19–26.

Hochmeister, M.N., O. Rudin, and E. Ambach, PCR analysis from cigarette butts, postage stamps, envelope sealing flaps, and other saliva-stained material. *Methods Mol Biol*, 1998, **98**: 27–32.

Hochmeister, M.N., et al., PCR-based typing of DNA extracted from cigarette butts. *Int J Legal Med*, 1991, **104**(4): 229–233.

Hoile, R., et al., Gamma irradiation as a biological decontaminant and its effect on common fingermark detection techniques and DNA profiling. *J Forensic Sci*, 2010, **55**(1): 171–177.

Hopkins, B., et al., The use of minisatellite variant repeat-polymerase chain reaction (MVR-PCR) to determine the source of saliva on a used postage stamp. *J Forensic Sci*, 1994, **39**(2): 526–531.

Hopwood, A.J., A. Mannucci, and K.M. Sullivan, DNA typing from human faeces. *Int J Legal Med*, 1996, **108**(5): 237–243.

Hopwood, A., et al., Rapid quantification of DNA samples extracted from buccal scrapes prior to DNA profiling. *Biotechniques*, 1997, **23**(1): 18–20.

Johnson, D.J., A.C. Calderaro, and K.A. Roberts, Variation in nuclear DNA concentrations during urination. *J Forensic Sci*, 2007, **52**(1): 110–113.

Kaarstad, K., et al., The detection of female DNA from the penis in sexual assault cases. *J Forensic Leg Med*, 2007, **14**(3): 159–160.

Kaiser, C., et al., Molecular study of time dependent changes in DNA stability in soil buried skeletal residues. *Forensic Sci Int*, 2008, **177**(1): 32–36.

Kamodyova, N., et al., Prevalence and persistence of male DNA identified in mixed saliva samples after intense kissing. *Forensic Sci Int Genet*, 2013, **7**(1): 124–128.

Kamphausen, T., et al., Good shedder or bad shedder—The influence of skin diseases on forensic DNA analysis from epithelial abrasions. *Int J Legal Med*, 2012, **126**(1): 179–183.

Kenna, J., et al., The recovery and persistence of salivary DNA on human skin. *J Forensic Sci*, 2011, **56**(1): 170–175.

Kester, K.M., et al., Recovery of environmental human DNA by insects. *J Forensic Sci*, 2010, **55**(6): 1543–1551.

Kim, M., et al., Identification and long term stability of DNA captured on a dental impression wafer. *Pediatr Dent*, 2012, **34**(5): 373–377.

Kita, T., et al., Morphological study of fragmented DNA on touched objects. *Forensic Sci Int Genet*, 2008, **3**(1): 32–36.

Kitayama, T., et al., Evaluation of a new experimental kit for the extraction of DNA from bones and teeth using a non-powder method. *Leg Med (Tokyo)*, 2010, **12**(2): 84–89.

Kline, M.C., et al., Polymerase chain reaction amplification of DNA from aged blood stains: Quantitative evaluation of the "suitability for purpose" of four filter papers as archival media. *Anal Chem*, 2002, **74**(8): 1863–1869.

Kohnemann, S., et al., qPCR and mtDNA SNP analysis of experimentally degraded hair samples and its application in forensic casework. *Int J Legal Med*, 2010, **124**(4): 337–342.

Lee, E.J., et al., The effects of different maceration techniques on nuclear DNA amplification using human bone. *J Forensic Sci*, 2010, **55**(4): 1032–1038.

Li, R.C. and H.A. Harris, Using hydrophilic adhesive tape for collection of evidence for forensic DNA analysis. *J Forensic Sci*, 2003, **48**(6): 1318–1321.

Lijnen, I. and G. Willems, DNA research in forensic dentistry. *Methods Find Exp Clin Pharmacol*, 2001, **23**(9): 511–517.

Linacre, A., et al., Generation of DNA profiles from fabrics without DNA extraction. *Forensic Sci Int Genet*, 2010, **4**(2): 137–141.

Linch, C.A., The ultrastructure of tissue attached to telogen hair roots. *J Forensic Sci*, 2008, **53**(6): 1363–1366.

Linch, C.A., Degeneration of nuclei and mitochondria in human hairs. *J Forensic Sci*, 2009, **54**(2): 346–349.

Linch, C.A., et al., Specific melanin content in human hairs and mitochondrial DNA typing success. *Am J Forensic Med Pathol*, 2009, **30**(2): 162–166.

Lindahl, T., Instability and decay of the primary structure of DNA. *Nature*, 1993, **362**(6422): 709–715.

Lorente, M., et al., Dandruff as a potential source of DNA in forensic casework. *J Forensic Sci*, 1998, **43**(4): 901–902.

Maguire, S., et al., Retrieval of DNA from the faces of children aged 0–5 years: A technical note. *J Forensic Nurs*, 2008, **4**(1): 40–44.

Malsom, S., et al., The prevalence of mixed DNA profiles in fingernail samples taken from couples who co-habit using autosomal and Y-STRs. *Forensic Sci Int Genet*, 2009, **3**(2): 57–62.

Marrone, A. and J. Ballantyne, Changes in dry state hemoglobin over time do not increase the potential for oxidative DNA damage in dried blood. *PLoS One*, 2009, **4**(4): e5110.

Mayntz-Press, K.A., et al., Y-STR profiling in extended interval (> or = 3 days) postcoital cervicovaginal samples. *J Forensic Sci*, 2008, **53**(2): 342–348.

McCartney, C., LCN DNA: Proof beyond reasonable doubt? *Nat Rev Genet*, 2008, **9**(5): 325.

Medintz, I., et al., Restriction fragment length polymorphism and polymerase chain reaction-HLA DQ alpha analysis of casework urine specimens. *J Forensic Sci*, 1994, **39**(6): 1372–1380.

Misner, L.M., et al., The correlation between skeletal weathering and DNA quality and quantity. *J Forensic Sci*, 2009, **54**(4): 822–828.

National Institute of Justice. Using DNA to solve cold cases, Special Report, 2002, US Department of Justice, Office of Justice Programs, Washington, DC.

Ng, L.K., et al., Optimization of recovery of human DNA from envelope flaps using DNA IQ System for STR genotyping. *Forensic Sci Int Genet*, 2007, **1**(3–4): 283–286.

Niemcunowicz-Janica, A., et al., Typeability of PowerPlex Y (Promega) profiles in selected tissue samples incubated in various environments. *Arch Med Sadowej Kryminol*, 2007, **57**(4): 385–388.

Niemcunowicz-Janica, A., et al., Detectability of SGM Plus profiles in heart and lungs tissue samples incubated in different environments. *Leg Med (Tokyo)*, 2008, **10**(1): 35–38.

Nurit, B., et al., Evaluating the prevalence of DNA mixtures found in fingernail samples from victims and suspects in homicide cases. *Forensic Sci Int Genet*, 2011, **5**(5): 532–537.

O'Driscoll, L., Extracellular nucleic acids and their potential as diagnostic, prognostic and predictive biomarkers. *Anticancer Res*, 2007, **27**(3A): 1257–1265.

Offele, D., et al., Soft tissue removal by maceration and feeding of *Dermestes* sp.: Impact on morphological and biomolecular analyses of dental tissues in forensic medicine. *Int J Legal Med*, 2007, **121**(5): 341–348.

Ohira, H., et al., Effective appropriate use of dental remains and forensic DNA testing for personal identity confirmation. *Leg Med (Tokyo)*, 2009, **11**(Suppl 1): S560–S562.

Opel, K.L., et al., Evaluation and quantification of nuclear DNA from human telogen hairs. *J Forensic Sci*, 2008, **53**(4): 853–857.

Paneto, G.G., et al., Heteroplasmy in hair: Differences among hair and blood from the same individuals are still a matter of debate. *Forensic Sci Int*, 2007, **173**(2–3): 117–121.

Pelotti, S., et al., Cancerous tissues in forensic genetic analysis. *Genet Test*, 2007, **11**(4): 397–400.

Pfeiffer, H., et al., Influence of soil storage and exposure period on DNA recovery from teeth. *Int J Legal Med*, 1999, **112**(2): 142–144.

Phipps, M. and S. Petricevic, The tendency of individuals to transfer DNA to handled items. *Forensic Sci Int*, 2007, **168**(2–3): 162–168.

Pinchi, V., et al., Techniques of dental DNA extraction: Some operative experiences. *Forensic Sci Int*, 2011, **204**(1–3): 111–114.

Prado, V.F., et al., Extraction of DNA from human skeletal remains: Practical applications in forensic sciences. *Genet Anal*, 1997, **14**(2): 41–44.

Quinones, I. and B. Daniel, Cell free DNA as a component of forensic evidence recovered from touched surfaces. *Forensic Sci Int Genet*, 2012, **6**(1): 26–30.

Raimann, P.E., et al., Procedures to recover DNA from pre-molar and molar teeth of decomposed cadavers with different post-mortem intervals. *Arch Oral Biol*, 2012, **57**(11): 1459–1466.

Ramasamy, S., A. Houspian, and F. Knott, Recovery of DNA and fingermarks following deployment of render-safe tools for vehicle-borne improvised explosive devices (VBIED). *Forensic Sci Int*, 2011, **210**(1–3): 182–187.

Remualdo, V.R. and R.N. Oliveira, Analysis of mitochondrial DNA from the teeth of a cadaver maintained in formaldehyde. *Am J Forensic Med Pathol*, 2007, **28**(2): 145–146.

Riemer, L.B., D. Fairley, and D. Sweet, DNA collection from used toothbrushes as a means to decedent identification. *Am J Forensic Med Pathol*, 2012, **33**(4): 354–356.

Roberts, K.A. and C. Calloway, Mitochondrial DNA amplification success rate as a function of hair morphology. *J Forensic Sci*, 2007, **52**(1): 40–47.

Roeper, A., W. Reichert, and R. Mattern, The Achilles tendon as a DNA source for STR typing of highly decayed corpses. *Forensic Sci Int*, 2007, **173**(2–3): 103–106.

Rubio, L., et al., Study of short- and long-term storage of teeth and its influence on DNA. *J Forensic Sci*, 2009, **54**(6): 1411–1413.

Rubio, L., et al., Time-dependent changes in DNA stability in decomposing teeth over 18 months. *Acta Odontol Scand*, 2013, **71**(3–4): 638–643.

Salvador, J.M. and M.C. De Ungria, Isolation of DNA from saliva of betel quid chewers using treated cards. *J Forensic Sci*, 2003, **48**(4): 794–797.

Schneider, H., et al., Hot flakes in cold cases. *Int J Legal Med*, 2011, **125**(4): 543–548.

Schwark, T., et al., Phantoms in the mortuary—DNA transfer during autopsies. *Forensic Sci Int*, 2012, **216**(1–3): 121–126.

Sewell, J., et al., Recovery of DNA and fingerprints from touched documents. *Forensic Sci Int Genet*, 2008, **2**(4): 281–285.

Shaw, K., et al., Comparison of the effects of sterilisation techniques on subsequent DNA profiling. *Int J Legal Med*, 2007, **122**(1): 29–33.

Shintani-Ishida, K., et al., Usefulness of blood vessels as a DNA source for PCR-based genotyping based on two cases of corpse dismemberment. *Leg Med (Tokyo)*, 2010, **12**(1): 8–12.

Simons, J.L. and S.K. Vintiner, Effects of histological staining on the analysis of human DNA from archived slides. *J Forensic Sci*, 2011, **56**(Suppl 1): S223–S228.

Soares-Vieira, J.A., et al., Y-STRs in forensic medicine: DNA analysis in semen samples of azoospermic individuals. *J Forensic Sci*, 2007, **52**(3): 664–670.

Sosa, C., et al., Nuclear DNA typing from ancient teeth. *Am J Forensic Med Pathol*, 2012, **33**(3): 211–214.

Sosa, C., et al., Association between ancient bone preservation and DNA yield: A multidisciplinary approach. *Am J Phys Anthropol*, 2013, **151**(1): 102–109.

Springer, E.E., T.L. Laber, and B.B. Randall, The examination of vaginally inserted plastic tampon applicators for genetic markers and evidence of prior sexual intercourse. *J Forensic Sci*, 1988, **33**(5): 1139–1145.

Steadman, D.W., et al., The effects of chemical and heat maceration techniques on the recovery of nuclear and mitochondrial DNA from bone. *J Forensic Sci*, 2006, **51**(1): 11–17.

Szabo, S., et al., *In situ* labeling of DNA reveals interindividual variation in nuclear DNA breakdown in hair and may be useful to predict success of forensic genotyping of hair. *Int J Legal Med*, 2012, **126**(1): 63–70.

Taguchi, M., et al., DNA identification of formalin-fixed organs is affected by fixation time and type of fixatives: Using the AmpF l STR(R) Identifiler(R) PCR Amplification Kit. *Med Sci Law*, 2012, **52**(1): 12–16.

Tanaka, M., et al., Usefulness of a toothbrush as a source of evidential DNA for typing. *J Forensic Sci*, 2000, **45**(3): 674–676.

Tang, D.L., et al., Multiplex fluorescent PCR for noninvasive prenatal detection of fetal-derived paternally inherited diseases using circulatory fetal DNA in maternal plasma. *Eur J Obstet Gynecol Reprod Biol*, 2009, **144**(1): 35–39.

Tate, C.M., et al., Evaluation of circular DNA substrates for whole genome amplification prior to forensic analysis. *Forensic Sci Int Genet*, 2012, **6**(2): 185–190.

Theodore Harcke, H., et al., Forensic imaging-guided recovery of nuclear DNA from the spinal cord. *J Forensic Sci*, 2009, **54**(5): 1123–1126.

Tilotta, F., et al., A comparative study of two methods of dental pulp extraction for genetic fingerprinting. *Forensic Sci Int*, 2010, **202**(1–3): e39–e43.

Toothman, M.H., et al., Characterization of human DNA in environmental samples. *Forensic Sci Int*, 2008, **178**(1): 7–15.

Tsuchimochi, T., et al., Chelating resin-based extraction of DNA from dental pulp and sex determination from incinerated teeth with Y-chromosomal alphoid repeat and short tandem repeats. *Am J Forensic Med Pathol*, 2002, **23**(3): 268–271.

van Oorschot, R.A., et al., Beware of the possibility of fingerprinting techniques transferring DNA. *J Forensic Sci*, 2005, **50**(6): 1417–1422.

Vandewoestyne, M., et al., Presence and potential of cell free DNA in different types of forensic samples. *Forensic Sci Int Genet*, 2013, **7**(2): 316–320.

von Wurmb-Schwark, N., et al., Fast and simple DNA extraction from saliva and sperm cells obtained from the skin or isolated from swabs. *Leg Med (Tokyo)*, 2006, **8**(3): 177–181.

Warshauer, D.H., et al., An evaluation of the transfer of saliva-derived DNA. *Int J Legal Med*, 2012, **126**(6): 851–861.

Webb, L.G., S.E. Egan, and G.R. Turbett, Recovery of DNA for forensic analysis from lip cosmetics. *J Forensic Sci*, 2001, **46**(6): 1474–1479.

Wiegand, P., T. Bajanowski, and B. Brinkmann, DNA typing of debris from fingernails. *Int J Legal Med*, 1993, **106**(2): 81–83.

Yasuda, T., et al., A simple method of DNA extraction and STR typing from urine samples using a commercially available DNA/RNA extraction kit. *J Forensic Sci*, 2003, **48**(1): 108–110.

SECTION II

Basic Techniques in Forensic Biology

Nucleic Acid Extraction

5.1 Basic Principles of DNA Extraction

Deoxyribonucleic acid (DNA) is a linear polynucleotide consisting of four types of monomeric nucleotides. Each nucleotide contains three components: a deoxyribose, a nitrogenous base, and a phosphate group (Figure 5.1). The four bases for DNA are adenine (A), cytosine (C), guanine (G), and thymine (T) (Figure 5.2). The deoxyribose is attached to the nitrogen of a base. The phosphate group is attached to the deoxyribose. In a *polynucleotide*, individual nucleotides are linked by phosphodiester bonds (Figure 5.3).

Interestingly, the first attempt to isolate DNA from humans, which involved the use of human leucocytes, was accomplished by the Swiss physician Friedrich Miescher in 1869. This study led to the discovery of the DNA molecule, which he referred to as "nuclein." Over the years, various methods for isolating DNA from blood samples were developed. However, many protocols still use the basic principles Miescher developed more than 100 years ago.

This chapter introduces basic techniques for DNA extraction used in forensic laboratories (Table 5.1). The commonly used methods isolate total cellular DNA, which is suitable for most forensic DNA analyses. However, DNA extraction procedures may vary according to the type of biological evidence, which can include cell types, substrates, the quantity of biological evidence collected, and the type of test that is being performed. Specific DNA extraction procedures for a sample of interest can be found in the literature. The extraction method of choice yields an optimal quantity, quality, and purity of DNA to satisfy forensic DNA testing needs. A sufficient quantity of DNA ensures the generation of a complete DNA profile. Poor quality of DNA, such as fragmentation due to DNA degradation, may result in a partial DNA profile or a failure to obtain a profile. Poor purity of DNA may cause interference during subsequent DNA testing. For example, DNA polymerase inhibitors interfere with DNA amplification (Chapter 7). Additional criteria for selecting proper DNA extraction methods include adaptability to automation, throughput potential, simplicity, the reduction of contamination risks, and cost-effectiveness. The most common DNA extraction protocols include basic components discussed below.

5.1.1 Cell and Tissue Disruption

In most DNA extraction protocols, enzymatic digestions, such as those with proteinase K, are used for cell and tissue disruption. The disruption process can also be carried out by boiling and by using alkali treatment and mechanical methods (Figures 5.4 and 5.5). Materials such as bone and teeth can be frozen in liquid nitrogen and then ground to a fine powder with a mortar or cryogenic grinder such as the SPEX CertiPrep® freezer mill (Figure 5.6). In such specimens, cells containing DNA

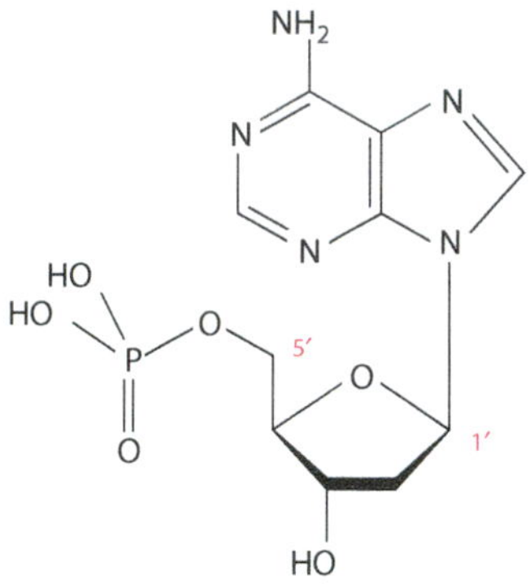

Figure 5.1   Structure of a nucleotide. Each nucleotide in a DNA polymer is made up of three components: a deoxyribose, a nitrogenous base (adenine is shown as an example), and a phosphate group. The 1′ carbon of the deoxyribose is attached to the nitrogen of a nitrogenous base. A phosphate group is attached to the 5′ carbon of the deoxyribose.

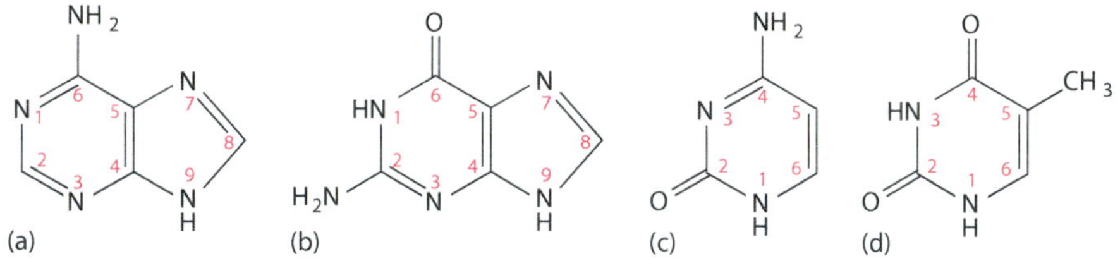

Figure 5.2   Chemical structures of nitrogenous bases: (a) adenine, (b) guanine, (c) cytosine, and (d) thymine.

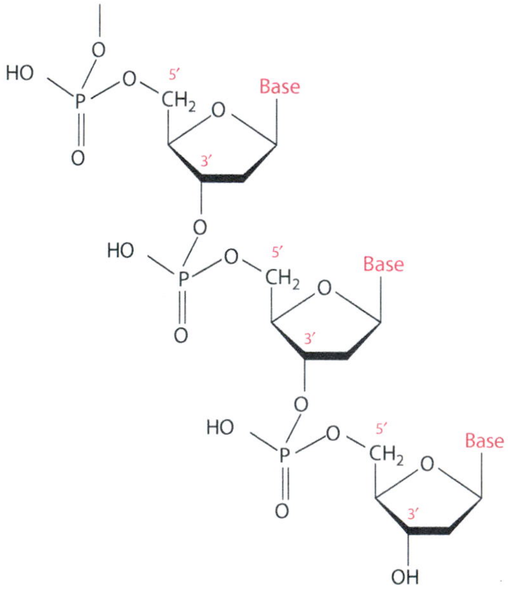

Figure 5.3   DNA polynucleotide chain. Individual nucleotides are linked by phosphodiester bonds between their 5′ and 3′ carbons.

Table 5.1 Three Basic DNA Extraction Methods

Method	Forensic Application	Purity	DNA Strand	DNA Size	Throughput	Note
Solvent-based	RFLP- and PCR-based assay	High	Double stranded	Large	Time-consuming; difficult to adapt for automation	Uses toxic solvent; multiple transfer steps between tubes
Boiling	PCR-based assay	Low	Single stranded	Small	~30 min per sample	No transfer required
Silica-based	RFLP- and PCR-based assay	High	Double stranded	Large	~1 h per sample; amenable to automation	Minimum transfer required

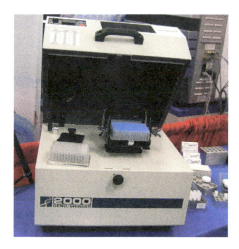

Figure 5.4 An automated mechanical disruption device for efficient disruption of tissues. It can be used for high-throughput applications involving sample preparation for the isolation of nucleic acids. (© Richard C. Li.)

Figure 5.5 Tissue disruption using a pressure-generating instrument also known as a barocycler. During the process, a barocycler is utilized to apply alternating cycles of high and low pressures onto specimens placed in single-use processing containers. The technique is known as pressure-cycling technology (PCT). (© Richard C. Li.)

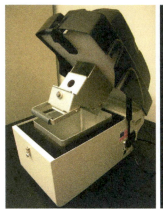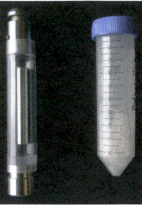

Figure 5.6   Tissue disruption using cryogenic grinding. A cryogenic mill (left), also known as freezer mill, can be used for the pulverization of hard tissue such as bone samples into fine powder prior to DNA extraction. During a grinding cycle, an impactor (right) moves back and forth, under a magnetic field, inside a grinding vial, grinding the sample to a fine powder. The grinding process is carried out at low temperatures (approximately −196°C), using liquid nitrogen to protect temperature-sensitive analytes such as DNA. (© Richard C. Li.)

are embedded in a calcified matrix that is responsible for the rigid structure of bone and teeth. In order to achieve a high yield of DNA extraction, it is necessary to remove the calcified matrix. A decalcification process removes calcium ions from the matrix, thus making the specimen suitable for DNA extraction. The most commonly used decalcifying agents for DNA extraction purposes are chelating agents such as ethylenediaminetetraacetic acid (EDTA), which sequesters the calcium ions. Decalcification is a lengthy procedure. Bone or teeth powder is usually treated with the decalcifying agent overnight or for a longer period of time, depending on the size and source of the sample.

5.1.2 Lysis of Cellular and Organelle Membranes

During or after tissue disruption, membranes—including those of cells, nuclei, and mitochondria—are lysed in order to release DNAs, including nuclear and mitochondrial DNA. The lysis can be carried out using salts and chaotropic agents (Section 5.2.3) such as guanidinium salts and detergents such as sarkosyl and sodium dodecyl sulfate (SDS). These substances can destroy membranes, denature proteins, and dissociate proteins such as histones from DNA. The lysis procedure is usually carried out in a buffer, such as Tris, in order to maintain a pH where endogenous *deoxyribonucleases* (DNases) remain inactive. DNases are a type of nuclease that catalyzes the cleavage of phosphodiester bonds of DNA. Endogenous DNases are located in cytoplasmic lysosomes and play a role in degrading the DNA of invading viruses. When cells are lysed, the DNases that are also released can degrade the extracted DNA. Chelating agents such as EDTA or Chelex® (Section 5.2.2) can therefore be used to chelate the divalent cations that are the cofactors of DNases, in order to inhibit DNase activities. Furthermore, reducing agents such as mercaptoethanol or dithiothreitol (DTT) can be used to inhibit the oxidization processes that can damage DNA.

5.1.3 Removal of Proteins and Cytoplasmic Constituents

After lysis, cytoplasmic constituents, such as proteins and liquids that interfere with DNA extraction, are removed. Proteins and lipids are usually removed by one or more rounds of extraction with organic solvents such as phenol–chloroform mixtures (Section 5.2.1). Another strategy to remove cytoplasmic constituents is to utilize the reversible binding of DNA to a solid material such as silica, which selectively binds DNA in chaotropic salt solutions. The proteins and cytoplasmic constituents can then be removed through washing steps.

5.1.4 Storage of DNA Solutions

Purified high-molecular-weight DNA is usually stored in TE buffer (10 mM Tris-HCl, 1 mM EDTA, pH 8.0). EDTA is usually included in storage solutions to chelate divalent cations and thereby inhibit DNases. Such a DNA solution may be stored at 4°C or −20°C. For long-term storage, −80°C is recommended. Frequent freezing or thawing cycles should be avoided because temperature fluctuations may cause breaks of single- and double-stranded DNA. Additionally, DNA containing impurities such as extracts generated via the Chelex method are much less stable. The presence of heavy metals, such as cadmium and cobalt, can cause breakage of phosphodiester bonds in the molecules. Additionally, unsuccessful removal of exogenous or intracellular free radicals, such as a hydroxyl radical (HO·), which have unpaired valence electrons, can cause DNA damage such as strand breaks. Furthermore, contamination by DNases may lead to subsequent degradation of DNA.

5.1.5 Contamination

Contamination is usually caused by the introduction of exogenous DNA to an evidence sample. It can occur between samples, between an individual and a sample, between other organisms and a sample, or between amplified DNA and a sample. Certain procedures can be utilized to prevent the occurrence of contamination. For example, evidence and reference samples should be processed separately in different rooms to avoid sample-to-sample contamination. In situations where space is limited, the evidence sample should be processed before the reference sample. Additionally, evidence samples should be processed and extracted for isolating DNA in separate areas (or at different times) from DNA amplification areas.

Solutions and test tubes used for extraction should be DNA-free, and aerosol-resistant pipet tips should be used during the extraction process. Additionally, the levels of contamination should be monitored by using *extraction reagent blanks* (having reagents but no samples), which monitor contamination from the extraction.

5.2 Methods of DNA Extraction

5.2.1 Extraction with Phenol–Chloroform

This method is also called organic extraction. Major steps include the following:

5.2.1.1 Cell Lysis and Protein Digestion

These steps can be achieved by digestion with proteolytic enzymes such as proteinase K before extraction with organic solvents.

5.2.1.2 Extraction with Organic Solvents

The removal of proteins is carried out by extracting aqueous solutions containing DNA with a mixture of phenol:chloroform:isoamyl alcohol (25:24:1). Phenol is used to extract the proteins from the aqueous solution. Although phenol has a slightly higher density than water, it is sometimes difficult to separate it from the aqueous phase. Therefore, chloroform is utilized as it has a higher density than phenol. As a result, the phenol–chloroform mixture forms the organic phase at the bottom of the tube and is easily separated from the aqueous phase. Isoamyl alcohol is often added to the phenol–chloroform mixture to reduce foaming. During partition, DNA is solubilized in the aqueous phase, while lipids are solubilized in the organic phase. Proteins are located at the interface between the two phases (Figure 5.7).

5.2.1.3 Concentrating DNA

Two common methods for concentrating DNA are ethanol precipitation and ultrafiltration. In the first method, the DNA is precipitated from the aqueous solution with ethanol and salts. Ethanol depletes the hydration shell of DNA, thus exposing its negatively charged phosphate groups. The

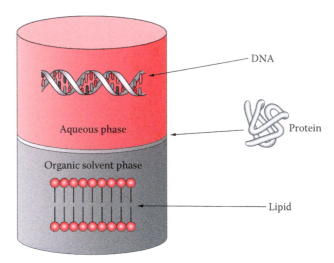

Figure 5.7 DNA extraction using organic solvent. The DNA is contained in the aqueous phase, while cellular materials such as lipids are contained in the organic-solvent phase. Proteins remain in the barrier between the two phases. (© Richard C. Li.)

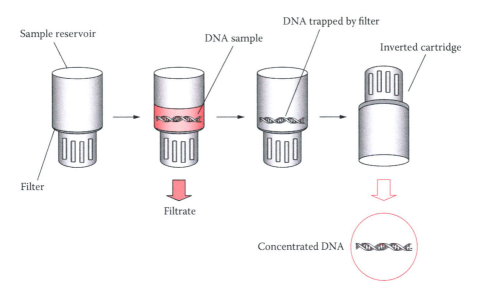

Figure 5.8 Concentrating DNA solutions using filtration devices. DNA samples are loaded into the reservoir. The liquid is filtered by centrifugation and the DNA becomes trapped in the membrane. The cartridge is then inverted to recover the trapped DNA by centrifugation. (© Richard C. Li.)

precipitation can only occur if sufficient quantities of cations are present in the solution. The most commonly used cations, such as ammonium, lithium, and sodium, neutralize the charges on the phosphate residues of DNA, forming a precipitate. Ultrafiltration is an alternative to ethanol precipitation for concentrating DNA solutions. The Microcon® and Amicon® are centrifugal ultrafiltration devices that can concentrate DNA samples (Figure 5.8). A proper Microcon® unit can be selected with a nucleotide cutoff equal to or smaller than the molecular weight of the DNA fragment of interest. Usually, the cutoff size is 100 kDa for forensic DNA samples.

Phenol–chloroform extraction yields large, double-stranded DNA and can be used for either restriction fragment length polymorphism (RFLP)-based or polymerase chain reaction

(PCR)-based analysis. However, the organic extraction method is time-consuming, involves the use of hazardous reagents, and requires the transfer of samples among tubes.

5.2.2 Extraction by Boiling Lysis and Chelation

This technique, also called the Chelex® extraction method, was introduced in the early 1990s to forensic DNA laboratories. It usually includes the following steps:

5.2.2.1 Washing

This step removes contaminants and inhibitors that may interfere with DNA amplification. For example, heme compounds found in blood samples should be removed from blood samples because it inhibits DNA amplification.

5.2.2.2 Boiling

Cells are suspended in a solution and incubated at 56°C, where DNases are not active, for 20 min. This preboiling step softens cell membranes and separates clumps of cells from each other. The cells are then lysed by heating to boiling temperature in order to break open the membranes and to release the DNA. Additionally, the lysis of cells releases all of their cellular constituents, including DNases. The DNase degradation of the extracted DNA can be blocked by applying a chelating resin (Chelex® 100) during the extraction process. Chelex® 100 is an ion-exchange resin composed of styrene divinylbenzene copolymers. The paired iminodiacetate ion groups in Chelex® 100 act as chelators by binding to divalent metal ions such as magnesium. Magnesium is a cofactor of endogenous DNases. Thus, sequestering magnesium in the solution using Chelex® 100 protects DNA from degradation by DNases (Figure 5.9).

5.2.2.3 Centrifugation

Brief centrifugation is performed to pull the Chelex® 100 resin and cellular debris to the bottom of the tube. The supernatant is used for DNA analysis. Carrying the Chelex® 100 resin over into the DNA amplification solutions should be avoided because the resin chelates magnesium, which is a necessary cofactor for DNA polymerases used for amplification.

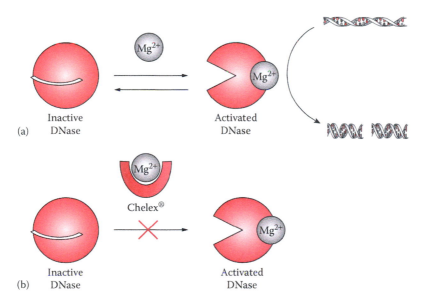

Figure 5.9 Utilizing Chelex® for preparing DNA samples. (a) Cations such as Mg^{2+} are required for the activity of endogenous DNase, which degrades DNA. (b) Chelex® prevents DNA degradation from endogenous DNase by sequestering the Mg^{2+}. (© Richard C. Li.)

This method is simple and rapid and uses only a single tube for extraction, thus reducing the risks of contamination and sample mix-ups. However, the heating step of this method disrupts and denatures proteins and also affects the chromosomal DNA. The resulting DNA extracted from the solution is fragmented single-stranded DNA. Thus, the DNA extracted is not suitable for RFLP analysis because RFLP requires double-stranded DNA samples. The DNA obtained by lysis and chelation can only be used for PCR-based DNA analysis.

5.2.3 Silica-Based Extraction

The method of adsorbing DNA molecules to solid silica surfaces is used for extraction (Figures 5.10 and 5.11). The method is based on the phenomenon that DNA is reversibly adsorbed to silica—silicon dioxide (SiO_2)—in the presence of high concentrations of chaotropic salts. Chaotropic salts can disrupt hydrogen bonding, affecting the three-dimensional structures of macromolecules. These salts are used to denature proteins. Additionally, chaotropic salts can facilitate the adsorption of DNA to silica.

In aqueous solutions, the hydration shells of nucleic acids shield the negative charges of phosphates at the phosphodiester backbone of nucleic acids. As a result, nucleic acids are usually hydrophilic in aqueous solutions. In the presence of chaotropic salts, nucleic acids become hydrophobic. The dehydrating effect caused by chaotropic salts allows the phosphate residues to become available for adsorption to the silica surface. Common chaotropic salts utilized for DNA extraction include guanidinium salts such as guanidinium thiocyanate (GuSCN) and guanidinium hydrochloride (GuHCl). GuSCN is a more potent chaotropic salt and also facilitates cell lysis and DNA adsorption. This technique usually includes the following steps:

5.2.3.1 Cell Lysis and Protein Digestion

This is carried out by proteinase K digestion. The cell membranes are broken open, and DNA is released.

5.2.3.2 DNA Adsorption onto Silica

This step utilizes silica as the stationary phase in a membrane configuration to which the DNA in the cell lysate binds. Adsorption of the DNA to the silica occurs in the presence of high concentrations of chaotropic agents (some protocols adjust pH conditions to enhance adsorption). Under these conditions, cellular materials and other contaminants that can inhibit DNA amplification reactions are not retained on the silica membrane. The adsorbed DNA is largely double stranded.

Figure 5.10   A silica-membrane spin column. (© Richard C. Li.)

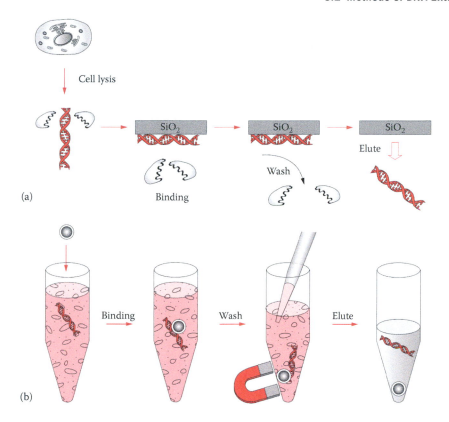

Figure 5.11 Silica-based DNA extractions. (a) Cells are lysed in the presence of proteinase. The DNA then binds to the silica matrix. A washing step removes unbound cellular materials and salts from the matrix. The purified DNA is then eluted for use in downstream applications. (b) Using silica-coated paramagnetic particles for DNA extraction. The particle is added into a lysate, and then DNA binds to the silica surface of the particle. The particle is then captured by a magnetic field. Afterwards, the supernatant is removed, and cellular materials are washed away. DNA elution follows. See Section 5.2.3. (© Richard C. Li.)

5.2.3.3 Washing

This step removes chaotropic agents and other contaminants. An ethanol-based wash solution is used. This wash solution does not remove DNA from the silica. The chaotropic agents and contaminants that are present in the solution can be removed using the ethanol-based wash solution.

5.2.3.4 Elution of DNA

The adsorbed DNA can be eluted by rehydration with aqueous low-salt solutions. The eluted DNA is double stranded and can be used for a wide variety of applications.

Silica-based extraction methods yield high-quality DNA. Silica membrane devices can also be adapted for automation; for example, this can be done by using 96-well silica membrane plates and a variety of robotic platforms. Another type of device utilizes silica-coated paramagnetic particles that adsorb DNA in the solution (Figure 5.11b). A magnet is used for particle capture instead of centrifugation or vacuum filtration. The magnetic particles can be resuspended during the wash steps, and the solution containing contaminants and cellular materials is then discarded. DNA is eluted after washing. This device can also be adapted to automated, high-throughput methods.

Over the years, high-throughput silica-based procedures have been developed to process large numbers of samples in parallel. Some of these methods are adapted for automated DNA extraction platforms (Figures 5.12 and 5.13).

Figure 5.12 Automated bench-top DNA purification systems that enable the isolation of genomic DNA from a wide variety of forensic samples. A low-throughput sample-preparation system that can process up to 12 samples (left) and a low-to-medium throughput sample-preparation system (right). (© Richard C. Li.)

Figure 5.13 Integrated platforms enabling automated DNA sample preparation and liquid handling for subsequent assay setup. These systems utilize silica-coated bead chemistries and can achieve moderate to high throughput of processing (1–96 samples) with bar coding for sample tracking. A Hamilton (top) and a Qiagen system (bottom). (© Richard C. Li.)

5.2.4 Differential Extraction

This method is very useful for the extraction of DNA from biological evidence derived from sexual assault cases, such as vaginal swabs and bodily fluid stains. These types of evidence often contain mixtures of spermatozoa from a male contributor and nonsperm cells such as epithelial cells from a female victim. Mixtures of individual DNA profiles can complicate data interpretation.

This method selectively lyses the nonsperm and spermatozoa in separate steps based on the differences in cell-membrane properties of spermatozoa and other types of cells. Thus, the DNA from spermatozoa and nonsperm cell fractions can be sequentially isolated.

First, the differential extraction procedure involves preferentially lysing the nonsperm cells with proteolytic degradation using proteinase. Sperm plasma membrane contains proteins cross-linked by disulfide bonds. The membrane exhibits a much higher mechanical stability than nonsperm cells and is thus resistant to proteolytic degradation. The nonsperm DNA is released into the supernatant and the liquid containing it (the nonsperm fraction) is extracted, yielding a fraction that predominantly contains DNA from nonsperm cells.

To lyse the sperm cells, it is necessary to cleave the disulfide bonds in addition to proteolytic digestion. The application of DTT, a reducing agent, is an approach that can be used for cleavage. In the presence of DTT and proteinase K, the sperm plasma membrane is then lysed (Figure 5.14). Subsequently, DNA from the sperm cells can be extracted (Figure 5.15).

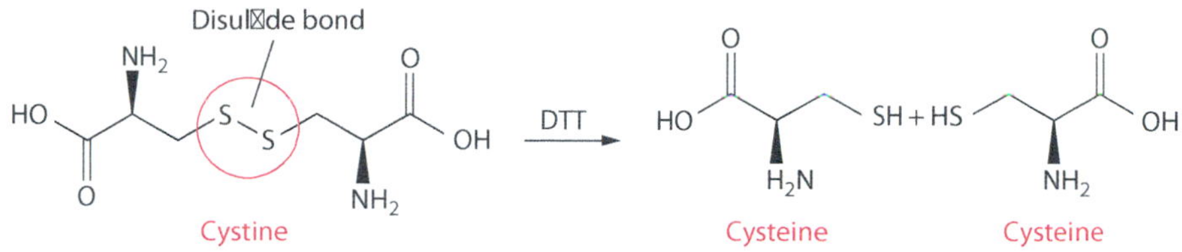

Figure 5.14 DTT reaction. The breaking of disulfide bonds in cystine residues is carried out by adding a reducing agent such as DTT.

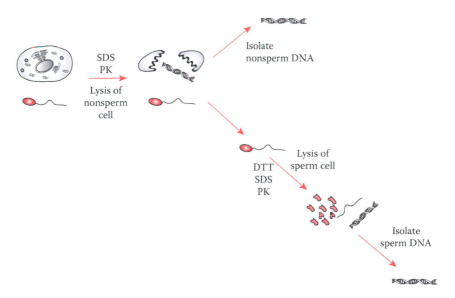

Figure 5.15 Differential extraction process used to separate sperm cells from nonsperm cells. Nonsperm cells are lysed in the presence of SDS and proteinase K (PK); sperm cells are resistant to such conditions. The nonsperm cell DNA is extracted. The sperm cells are then lysed separately in the presence of SDS and proteinase K plus DTT to extract the sperm DNA. (© Richard C. Li.)

This process isolates sperm and nonsperm cell DNA separately for obtaining DNA profiles from male and female contributors, respectively. However, the non-sperm-cell DNA and sperm-cell DNA may not be completely separated from one another; this can happen, for example, if the sperm cells have already lysed due to poor sample conditions. Some sperm DNA may be present in the non-sperm-cell fraction. Additionally, if a mixture has an abundance of non-sperm cells and fewer sperm cells, non-sperm-cell DNA may be detected in the sperm fraction. Thus, new methods that can overcome these problems are highly desired.

5.3 Essential Features of RNA

Like DNA, *ribonucleic acid* (RNA) is a linear molecule containing four types of nucleotides linked by phosphodiester bonds (Figures 5.16 and 5.17). However, certain properties of RNA differ from those of DNA. Unlike DNA containing deoxyribose (see Figure 5.18), the sugar residue

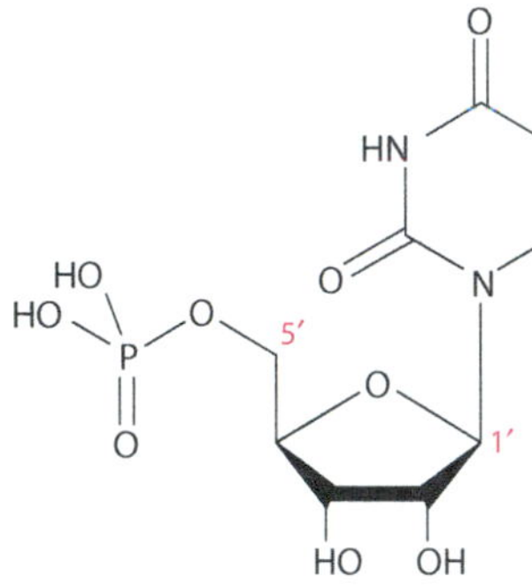

Figure 5.16 The structure of ribonucleotide. Each ribonucleotide in a RNA polymer consists of three components: a ribose, a nitrogenous base (uracil is shown as an example), and a phosphate group. The 1′ carbon of the ribose is attached to the nitrogen of the nitrogenous base. A phosphate group is attached to the 5′ carbon of the ribose.

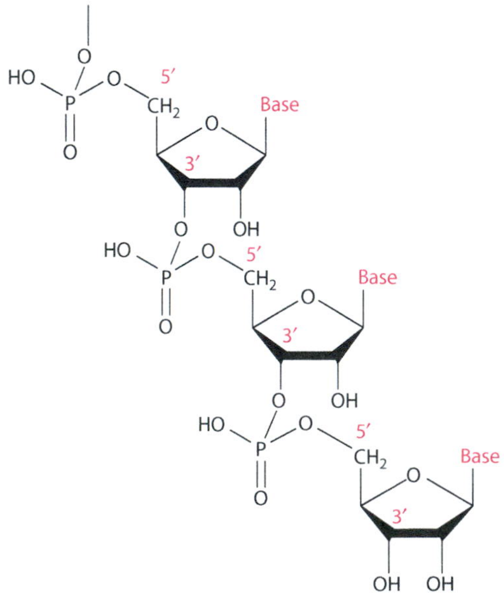

Figure 5.17 Chemical structure of RNA (polyribonucleotide) chain.

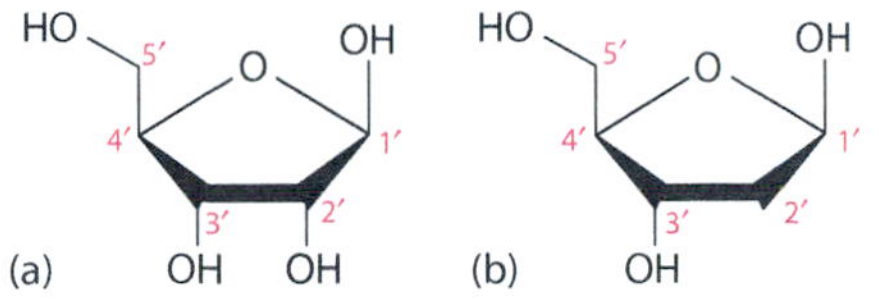

Figure 5.18 Comparison of (a) ribose and (b) deoxyribose. The hydroxyl group attached to the 2′ carbon of ribose is replaced by a hydrogen group in deoxyribose.

of RNA is a ribose, which has a hydroxyl (OH) group at the 2′ carbon position. Therefore, RNA has a lower pKa, which means that it is more acidic than DNA. RNA contains uracil (U) in the place of thymine in DNA (Figure 5.19). RNA is typically found in cells as a single-stranded molecule, while DNA is double stranded. Polyribonucleotides can form complementary helices with DNA strands by base pairing (with the exception of uracil pairs with adenine) (see Figure 5.20).

Total RNA contains all the RNA of the cells, including RNAs involved in protein synthesis and posttranscriptional modification, as well as regulatory RNAs. The RNA that carries codes from a DNA template is called messenger RNA (mRNA) and usually contains a cap and polyadenine tail at the 5′ and 3′ ends of the molecule, respectively (Figure 5.21).

The stability of RNA is an issue for mRNA-based forensic analysis. Single-stranded RNA is chemically less stable than DNA. The 2′ OH group of RNA can react with its phosphodiester backbone, potentially causing the nonenzymatic hydrolysis of an RNA molecule. Several factors, such as moisture, UV light, high temperature, and extreme pH, can facilitate the nonenzymatic

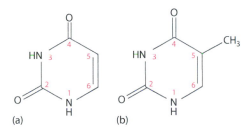

Figure 5.19 Comparison of (a) uracil and (b) thymine. Thymine has a methyl group attached at the 5-carbon position.

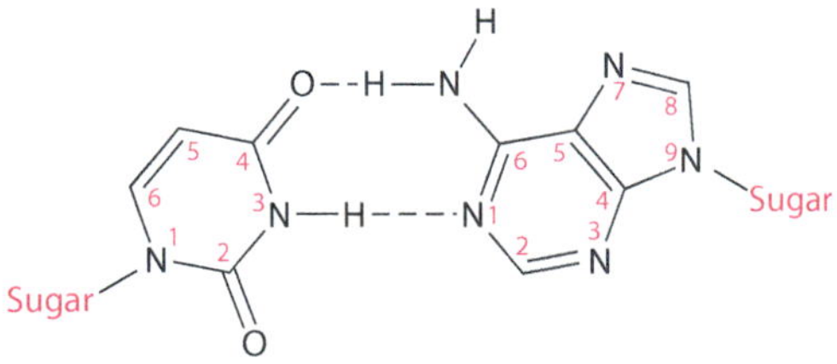

Figure 5.20 Base pairing of uracil (left) and adenine (right).

Figure 5.21 Structure of mRNA. Eukaryotic mRNA is modified at the 5′ end (cap) and 3′ end (polyadenine tail) needed for protein synthesis. Start and stop codons are required for the initiation and termination of protein synthesis. (© Richard C. Li.)

hydrolysis of RNA phosphodiesters, which leads to the degradation of RNA. Moreover, endogenous *ribonucleases* (RNases), present within cells, represent the major factor causing RNA degradation. Environmental microorganisms are also a common source of RNase contamination. RNases are very stable and usually do not require cofactors to carry out enzymatic reactions. Small amounts of RNases are sufficient to degrade RNA. RNA can be protected by adding RNase inhibitors such as diethylpyrocarbonate. Additionally, using RNase-free laboratory supplies and reagents and wearing disposable gloves while handling samples and reagents can reduce the risks of RNase contamination.

5.4 Methods of RNA Extraction

5.4.1 RNA–DNA Coextraction

For most forensic applications, total RNA is usually isolated. The procedure for isolating total RNA is simpler than that for mRNA. The total RNA isolated usually contains a sufficient amount of mRNA for subsequent reverse transcriptase PCR (RT-PCR; see Chapter 7). However, if the quantity of a target mRNA is very low, the procedures specifically for isolating mRNA should be used. There are a variety of protocols for isolating total RNA. However, RNA–DNA coextraction methods allow for the simultaneous extraction of high-quality DNA and RNA for forensic DNA analysis and bodily-fluid identification (Chapter 11), respectively, without the consumption of additional samples when extracted separately.

Biological samples are first lysed in lysis buffer. The lysis buffer usually contains chaotropic salts that can facilitate the lysis of cells and denature proteins, thus inactivating endogenous RNases to protect RNA. The lysate is then passed through a silica membrane column. A high-salt environment in the lysate allows selective binding of silica to genomic DNA over RNA. The column is washed, and purified DNA is then eluted.

The extraction of RNA is achieved by utilizing a high concentration of chaotropic salts that are already present in the lysate, along with ethanol, in order to decrease the hydrophilic property of RNA and increase its affinity for silica. After ethanol is added to the flow-through that passed the first column, the sample is then applied to a second silica column, where total RNA molecules, longer than 200 nucleotides, bind to the silica membrane. RNA including mRNA is then eluted (Figure 5.22). This method does not isolate small RNAs such as miRNA, rRNA, and tRNA, which comprise 15%–20% of total RNA.

5.4.2 miRNA Extraction

miRNAs (Chapter 4) are low-molecular-weight RNAs ranging between 15 and 30 nucleotides in length. Conventional methods routinely used for extracting total RNA do not effectively recover small RNAs; thus, they are not suitable for isolating miRNA. One approach to extracting small RNA molecules including miRNA involves two steps: organic-solvent extraction to isolate total RNA and solid-phase extraction to enrich small RNA.

To begin the process, tissues are disrupted and cells are lysed. The lysis reagents also inactivate RNases to protect RNA. The lysate is then extracted with an organic solvent, such as phenol and chloroform, with a high concentration of chaotropic salts such as GuSCN. The partitioning of DNA and RNA between the organic phase and the aqueous phase is determined by the pH during the organic extraction. At an acidic pH, DNA partitions to the organic phase. The pKa of RNA is usually lower than DNA; thus, RNA remains hydrophilic and is retained in the aqueous phase under this condition. This step removes most of the DNA, proteins, and other cellular components from the lysate into the organic phase.

The second step, silica-based extraction, further purifies and enriches small RNAs. It is achieved by utilizing the high concentration of chaotropic salt that is already present in the lysate, along with ethanol. The size-fractioning of RNAs is achieved using different ethanol

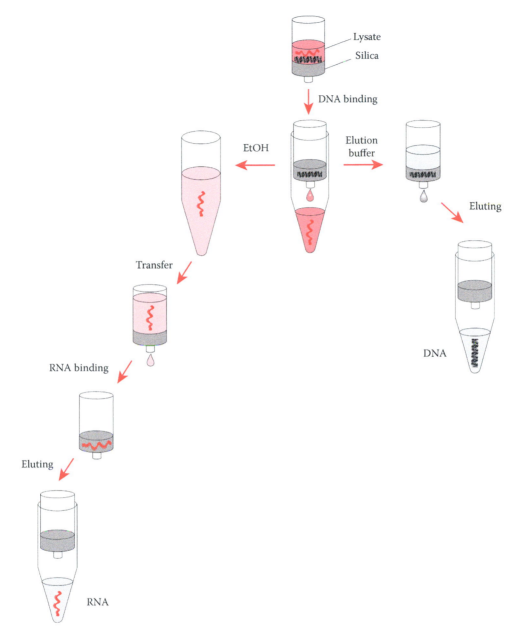

Figure 5.22   RNA and DNA coextraction using a silica-based method. See Section 5.4.1. (© Richard C. Li.)

concentrations. At a low ethanol concentration, the solvent-extracted lysate is passed through a silica membrane filter. Large RNAs are retained on the silica membrane, while small RNAs are not and are collected in the filtrate. The filtrate is passed through a second silica membrane filter. At a high ethanol concentration, the small RNAs are retained on the silica membrane. The small RNAs are then eluted in a low ionic strength solution (Figure 5.23). RNA less than 200 nucleotides, including miRNA, can be obtained.

After RNA is extracted, the small amounts of RNA can be quantified using a quantitative RT-PCR (Chapter 7) or fluorescent intercalating dye assay (Chapter 6). Purified RNA can be

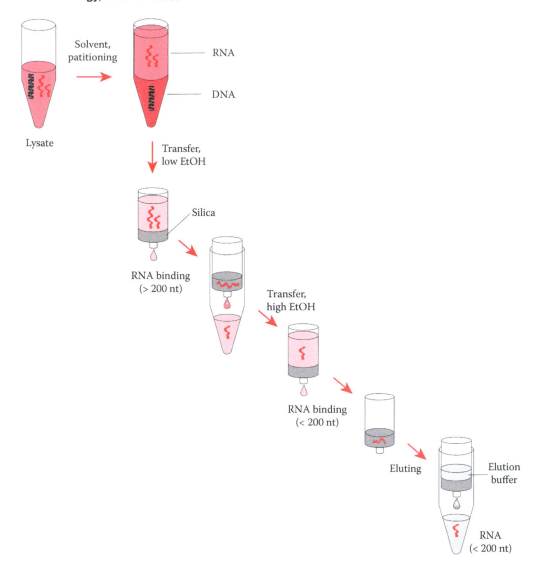

Figure 5.23 Schematic illustration of an miRNA extraction method. See Section 5.4.2. (© Richard C. Li.)

stored at −80°C in RNase-free water or Tris-EDTA buffer (pH 7). Single-use aliquots are preferred when possible to avoid multiple freeze–thaw cycles.

The integrity of extracted RNA is a great concern for RNA-based forensic analysis. RNA quality is traditionally assessed using the 28S:18S rRNA ratio. There are some conditions where this method has been shown to be inconsistent; that is, in which the 28S:18S rRNA ratio does not reflect the true level of RNA degradation. Recently, the *RNA integrity number* (RIN) has provided an effective method for determining RNA quality. RIN is a software algorithm that takes the electrophoretic RNA measurements into account in order to assign integrity values to RNA samples. RIN ranges from 1 to 10, with 10 being the most intact. Thus, the RIN method facilitates the assessment of the integrity of RNA samples. As reflected by RIN, RNA degradation has a negative influence on the reproducibility of the results of RNA-based forensic analysis.

Bibliography

Alvarez, M., J. Juusola, and J. Ballantyne, An mRNA and DNA co-isolation method for forensic casework samples. *Anal Biochem*, 2004, **335**(2): 289–298.

Amory, S., et al., Automatable full demineralization DNA extraction procedure from degraded skeletal remains. *Forensic Sci Int Genet*, 2012, **6**(3): 398–406.

Ananian, V., et al., Tumoural specimens for forensic purposes: Comparison of genetic alterations in frozen and formalin-fixed paraffin-embedded tissues. *Int J Legal Med*, 2011, **125**(3): 327–332.

Anzai-Kanto, E., et al., DNA extraction from human saliva deposited on skin and its use in forensic identification procedures. *Pesqui Odontol Bras*, 2005, **19**(3): 216–222.

Archer, E., et al., Validation of a dual cycle ethylene oxide treatment technique to remove DNA from consumables used in forensic laboratories. *Forensic Sci Int Genet*, 2010, **4**(4): 239–243.

Baker, L.E., W.F. McCormick, and K.J. Matteson, A silica-based mitochondrial DNA extraction method applied to forensic hair shafts and teeth. *J Forensic Sci*, 2001, **46**(1): 126–130.

Benoit, J.N., et al., An alternative procedure for extraction of DNA from ancient and weathered bone fragments. *Med Sci Law*, 2013, **53**(2): 100–106.

Bowden, A., R. Fleming, and S. Harbison, A method for DNA and RNA co-extraction for use on forensic samples using the Promega DNA IQ system. *Forensic Sci Int Genet*, 2011, 5(1): 64–68.

Brevnov, M.G., et al., Developmental validation of the PrepFiler Forensic DNA Extraction Kit for extraction of genomic DNA from biological samples. *J Forensic Sci*, 2009, **54**(3): 599–607.

Castella, V., et al., Forensic evaluation of the QIAshredder/QIAamp DNA extraction procedure. *Forensic Sci Int*, 2006, **156**(1): 70–73.

Cattaneo, C., et al., A simple method for extracting DNA from old skeletal material. *Forensic Sci Int*, 1995, **74**(3): 167–174.

Crouse, C.A., J.D. Ban, and J.K. D'Alessio, Extraction of DNA from forensic-type sexual assault specimens using simple, rapid sonication procedures. *Biotechniques*, 1993, **15**(4): 641–644, 646, 648.

Davis, C.P., et al., Extraction platform evaluations: A comparison of AutoMate Express, EZ1(R) Advanced XL, and Maxwell(R) 16 Bench-top DNA extraction systems. *Leg Med (Tokyo)*, 2012, **14**(1): 36–39.

Di Nunno, N., et al., DNA extraction: An anthropologic aspect of bone remains from sixth to seventh-century ad bone remains. *Am J Forensic Med Pathol*, 2007, **28**(4): 333–341.

Diegoli, T.M., et al., An optimized protocol for forensic application of the PreCR Repair Mix to multiplex STR amplification of UV-damaged DNA. *Forensic Sci Int Genet*, 2012, **6**(4): 498–503.

Dieltjes, P., et al., A sensitive method to extract DNA from biological traces present on ammunition for the purpose of genetic profiling. *Int J Legal Med*, 2011, **125**(4): 597–602.

Duarte, G.R., et al., Characterization of dynamic solid phase DNA extraction from blood with magnetically controlled silica beads. *Analyst*, 2010, **135**(3): 531–537.

Duval, K., et al., Optimized manual and automated recovery of amplifiable DNA from tissues preserved in buffered formalin and alcohol-based fixative. *Forensic Sci Int Genet*, 2010, **4**(2): 80–88.

Edson, S.M. and A.F. Christensen, Field contamination of skeletonized human remains with exogenous DNA. *J Forensic Sci*, 2013, **58**(1): 206–209.

Elliott, K., et al., Use of laser microdissection greatly improves the recovery of DNA from sperm on microscope slides. *Forensic Sci Int*, 2003, **137**(1): 28–36.

Eminovic, I., et al., A simple method of DNA extraction in solving difficult criminal cases. *Med Arh*, 2005, **59**(1): 57–58.

Farrugia, A., C. Keyser, and B. Ludes, Efficiency evaluation of a DNA extraction and purification protocol on archival formalin-fixed and paraffin-embedded tissue. *Forensic Sci Int*, 2010, **194**(1–3): e25–e28.

Fattorini, P., et al., CE analysis and molecular characterisation of depurinated DNA samples. *Electrophoresis*, 2011, **32**(21): 3042–3052.

Fattorini, P., et al., Assessment of DNA damage by micellar electrokinetic chromatography. *Methods Mol Biol*, 2013, **984**: 341–351.

Fracasso, T., et al., Ultrasound-accelerated formalin fixation improves the preservation of nucleic acids extraction in histological sections. *Int J Legal Med*, 2009, **123**(6): 521–525.

Fregeau, C.J. and A. De Moors, Competition for DNA binding sites using Promega DNA IQ paramagnetic beads. *Forensic Sci Int Genet*, 2012, **6**(5): 511–522.

Fregeau, C.J., C.M. Lett, and R.M. Fourney, Validation of a DNA IQ-based extraction method for TECAN robotic liquid handling workstations for processing casework. *Forensic Sci Int Genet*, 2010, **4**(5): 292–304.

Fregeau, C.J., et al., Automated processing of forensic casework samples using robotic workstations equipped with nondisposable tips: Contamination prevention. *J Forensic Sci*, 2008, **53**(3): 632–651.

Fridez, F. and R. Coquoz, PCR DNA typing of stamps: Evaluation of the DNA extraction. *Forensic Sci Int*, 1996, **78**(2): 103–110.

Frippiat, C., et al., Evaluation of novel forensic DNA storage methodologies. *Forensic Sci Int Genet*, 2011, **5**(5): 386–392.

Fujita, Y. and S. Kubo, Application of FTA technology to extraction of sperm DNA from mixed body fluids containing semen. *Leg Med (Tokyo)*, 2006, **8**(1): 43–47.

Garvin, A.M. and A. Fritsch, Purifying and concentrating genomic DNA from mock forensic samples using Millipore Amicon filters. *J Forensic Sci*, 2013, **58**(Suppl 1): S173–S175.

Garvin, A.M., et al., DNA preparation from sexual assault cases by selective degradation of contaminating DNA from the victim. *J Forensic Sci*, 2009, **54**(6): 1297–1303.

Ginestra, E., et al., DNA extraction from blood determination membrane card test. *Forensic Sci Int*, 2004, **146**(Suppl): S145–S146.

Greenspoon, S.A., et al., QIAamp spin columns as a method of DNA isolation for forensic casework. *J Forensic Sci*, 1998, **43**(5): 1024–1030.

Greenspoon, S.A., et al., Application of the BioMek 2000 Laboratory Automation Workstation and the DNA IQ System to the extraction of forensic casework samples. *J Forensic Sci*, 2004, **49**(1): 29–39.

Grubb, J.C., et al., Implementation and validation of the Teleshake unit for DNA IQ robotic extraction and development of a large volume DNA IQ method. *J Forensic Sci*, 2010, **55**(3): 706–714.

Hagan, K.A., et al., Chitosan-coated silica as a solid phase for RNA purification in a microfluidic device. *Anal Chem*, 2009, **81**(13): 5249–5256.

Hanselle, T., et al., Isolation of genomic DNA from buccal swabs for forensic analysis, using fully automated silica-membrane purification technology. *Leg Med (Tokyo)*, 2003, **5**(Suppl 1): S145–S149.

Hedman, J., et al., A fast analysis system for forensic DNA reference samples. *Forensic Sci Int Genet*, 2008, **2**(3): 184–189.

Heinrich, M., et al., Successful RNA extraction from various human postmortem tissues. *Int J Legal Med*, 2007, **121**(2): 136–142.

Hennekens, C.M., et al., The effects of differential extraction conditions on the premature lysis of spermatozoa. *J Forensic Sci*, 2013, **58**(3): 744–752.

Ho, T.T. and R. Roy, Generating DNA profiles from immunochromatographic cards using LCN methodology. *Forensic Sci Int Genet*, 2011, **5**(3): 210–215.

Hochmeister, M.N., et al., PCR-based typing of DNA extracted from cigarette butts. *Int J Legal Med*, 1991, **104**(4): 229–233.

Hoff-Olsen, P., et al., Extraction of DNA from decomposed human tissue. An evaluation of five extraction methods for short tandem repeat typing. *Forensic Sci Int*, 1999, **105**(3): 171–183.

Howlett, S.E., et al., Evaluation of DNAstable for DNA storage at ambient temperature. *Forensic Sci Int Genet*, 2014, **8**(1): 170–178.

Hudlow, W.R. and M.R. Buoncristiani, Development of a rapid, 96-well alkaline based differential DNA extraction method for sexual assault evidence. *Forensic Sci Int Genet*, 2012, **6**(1): 1–16.

Hudlow, W.R., et al., The NucleoSpin(R) DNA Clean-up XS kit for the concentration and purification of genomic DNA extracts: An alternative to microdialysis filtration. *Forensic Sci Int Genet*, 2011, **5**(3): 226–230.

Hulme, P., J. Lewis, and G. Davidson, Sperm elution: An improved two phase recovery method for sexual assault samples. *Sci Justice*, 2013, **53**(1): 28–33.

Iwasa, M., et al., Y-chromosomal short tandem repeats haplotyping from vaginal swabs using a chelating resin-based DNA extraction method and a dual-round polymerase chain reaction. *Am J Forensic Med Pathol*, 2003, **24**(3): 303–305.

Jakubowska, J., A. Maciejewska, and R. Pawlowski, Comparison of three methods of DNA extraction from human bones with different degrees of degradation. *Int J Legal Med*, 2012, **126**(1): 173–178.

Karija Vlahovic, M. and M. Kubat, DNA extraction method from bones using Maxwell(R) 16. *Leg Med (Tokyo)*, 2012, **14**(5): 272–275.

Kim, K., et al., Technical note: Improved ancient DNA purification for PCR using ion-exchange columns. *Am J Phys Anthropol*, 2008, **136**(1): 114–121.

Kochl, S., H. Niederstatter, and W. Parson, DNA extraction and quantitation of forensic samples using the phenol–chloroform method and real-time PCR. *Methods Mol Biol*, 2005, **297**: 13–30.

Kopka, J., et al., New optimized DNA extraction protocol for fingerprints deposited on a special self-adhesive security seal and other latent samples used for human identification. *J Forensic Sci*, 2011, **56**(5): 1235–1240.

Lee, E.J., et al., The effects of different maceration techniques on nuclear DNA amplification using human bone. *J Forensic Sci*, 2010, **55**(4): 1032–1038.

Lee, H.Y., et al., Simple and highly effective DNA extraction methods from old skeletal remains using silica columns. *Forensic Sci Int Genet*, 2010, **4**(5): 275–280.

Lee, S.B., et al., Assessing a novel room temperature DNA storage medium for forensic biological samples. *Forensic Sci Int Genet*, 2012, **6**(1): 31–40.

Li, R. and L. Liriano, A bone sample cleaning method using trypsin for the isolation of DNA. *Leg Med (Tokyo)*, 2011, **13**(6): 304–308.

Li, R. and S. Klempner, The effect of an enzymatic bone processing method on short tandem repeat profiling of challenged bone specimens. *Leg Med (Tokyo)*, 2013, **15**(4): 171–176.

Li, R., et al., Developing a simple method to process bone samples prior to DNA isolation. *Leg Med (Tokyo)*, 2009, **11**(2): 76–79.

Li, C.X., et al., New cell separation technique for the isolation and analysis of cells from biological mixtures in forensic caseworks. *Croat Med J*, 2011, **52**(3): 293–298.

Liu, J.Y., et al., AutoMate Express forensic DNA extraction system for the extraction of genomic DNA from biological samples. *J Forensic Sci*, 2012, **57**(4): 1022–1030.

Loreille, O.M., et al., High efficiency DNA extraction from bone by total demineralization. *Forensic Sci Int Genet*, 2007, **1**(2): 191–195.

Lounsbury, J.A., et al., An enzyme-based DNA preparation method for application to forensic biological samples and degraded stains. *Forensic Sci Int Genet*, 2012, **6**(5): 607–615.

Ma, H.W., J. Cheng, and B. Caddy, Extraction of high quality genomic DNA from microsamples of human blood. *J Forensic Sci Soc*, 1994, **34**(4): 231–235.

McMichael, G.L., et al., Comparison of DNA extraction methods from small samples of newborn screening cards suitable for retrospective perinatal viral research. *J Biomol Tech*, 2011, **22**(1): 5–9.

Meredith, M., et al., Development of a one-tube extraction and amplification method for DNA analysis of sperm and epithelial cells recovered from forensic samples by laser microdissection. *Forensic Sci Int Genet*, 2012, **6**(1): 91–96.

Montpetit, S.A., I.T. Fitch, and P.T. O'Donnell, A simple automated instrument for DNA extraction in forensic casework. *J Forensic Sci*, 2005, **50**(3): 555–563.

Nagy, M., et al., Optimization and validation of a fully automated silica-coated magnetic beads purification technology in forensics. *Forensic Sci Int*, 2005, **152**(1): 13–22.

Norris, J.V., et al., Acoustic differential extraction for forensic analysis of sexual assault evidence. *Anal Chem*, 2009, **81**(15): 6089–6095.

Parker, C., E. Hanson, and J. Ballantyne, Optimization of dried stain co-extraction methods for efficient recovery of high quality DNA and RNA for forensic analysis. *Forensic Sci Int: Genet Suppl Series*, 2011, **3**(1): e309–e310.

Phillips, K., N. McCallum, and L. Welch, A comparison of methods for forensic DNA extraction: Chelex-100(R) and the QIAGEN DNA Investigator Kit (manual and automated). *Forensic Sci Int Genet*, 2012, **6**(2): 282–285.

Piglionica, M., et al., Extraction of DNA from bones in cases where expectations for success are low. *Am J Forensic Med Pathol*, 2012, **33**(4): 322–327.

Raimann, P.E., et al., Procedures to recover DNA from pre-molar and molar teeth of decomposed cadavers with different post-mortem intervals. *Arch Oral Biol*, 2012, **57**(11): 1459–1466.

Reedy, C.R., et al., Volume reduction solid phase extraction of DNA from dilute, large-volume biological samples. *Forensic Sci Int Genet*, 2010, **4**(3): 206–212.

Rerkamnuaychoke, B., et al., Comparison of DNA extraction from blood stain and decomposed muscle in STR polymorphism analysis. *J Med Assoc Thai*, 2000, **83**(Suppl 1): S82–S88.

Reshef, A., et al., STR typing of formalin-fixed paraffin embedded (FFPE) aborted foetal tissue in criminal paternity cases. *Sci Justice*, 2011, **51**(1): 19–23.

Rothe, J., L. Roewer, and M. Nagy, Individual specific extraction of DNA from male mixtures—First evaluation studies. *Forensic Sci Int Genet*, 2011, **5**(2): 117–121.

Rucinski, C., et al., Comparison of two methods for isolating DNA from human skeletal remains for STR analysis. *J Forensic Sci*, 2012, **57**(3): 706–712.

Scherczinger, C.A., et al., DNA extraction from liquid blood using QIAamp. *J Forensic Sci*, 1997, **42**(5): 893–896.

Seo, S.B., et al., Technical note: Efficiency of total demineralization and ion-exchange column for DNA extraction from bone. *Am J Phys Anthropol*, 2010, **141**(1): 158–162.

Shalhoub, R., et al., The recovery of latent fingermarks and DNA using a silicone-based casting material. *Forensic Sci Int*, 2008, **178**(2–3): 199–203.

Shan, Z., et al., PCR-ready human DNA extraction from urine samples using magnetic nanoparticles. *J Chromatogr B Analyt Technol Biomed Life Sci*, 2012, **881–882**: 63–68.

Shaw, K., et al., Comparison of the effects of sterilisation techniques on subsequent DNA profiling. *Int J Legal Med*, 2008, **122**(1): 29–33.

Sinclair, K. and V.M. McKechnie, DNA extraction from stamps and envelope flaps using QIAamp and QIAshredder. *J Forensic Sci*, 2000, **45**(1): 229–230.

Stangegaard, M., et al., Automated extraction of DNA from blood and PCR setup using a Tecan Freedom EVO liquid handler for forensic genetic STR typing of reference samples. *J Lab Autom*, 2011, **16**(2): 134–140.

Stangegaard, M., et al., Biomek 3000: The workhorse in an automated accredited forensic genetic laboratory. *J Lab Autom*, 2012, **17**(5): 378–386.

Stangegaard, M., et al., Automated extraction of DNA from biological stains on fabric from crime cases. A comparison of a manual and three automated methods. *Forensic Sci Int Genet*, 2013, **7**(3): 384–388.

Stangegaard, M., et al., Evaluation of four automated protocols for extraction of DNA from FTA cards. *J Lab Autom*, 2013, **18**(5): 404–410.

Startari, L., et al., Comparison of extractable DNA from bone following six-month exposure to outdoor conditions, garden loam, mold contamination or room storage. *Med Sci Law*, 2013, **53**(1): 29–32.

Sweet, D., et al., Increasing DNA extraction yield from saliva stains with a modified Chelex method. *Forensic Sci Int*, 1996, **83**(3): 167–177.

Tack, L.C., M. Thomas, and K. Reich, Automated forensic DNA purification optimized for FTA card punches and identifiler STR-based PCR analysis. *Clin Lab Med*, 2007, **27**(1): 183–191.

Tsuchimochi, T., et al., Chelating resin-based extraction of DNA from dental pulp and sex determination from incinerated teeth with Y-chromosomal alphoid repeat and short tandem repeats. *Am J Forensic Med Pathol*, 2002, **23**(3): 268–271.

Vandenberg, N., R.A. van Oorschot, and R.J. Mitchell, An evaluation of selected DNA extraction strategies for short tandem repeat typing. *Electrophoresis*, 1997, **18**(9): 1624–1626.

Vandewoestyne, M., et al., Evaluation of three DNA extraction protocols for forensic STR typing after laser capture microdissection. *Forensic Sci Int Genet*, 2012, **6**(2): 258–262.

von Wurmb-Schwark, N., et al., The impact of DNA contamination of bone samples in forensic case analysis and anthropological research. *Leg Med (Tokyo)*, 2008, **10**(3): 125–130.

Walsh, P.S., D.A. Metzger, and R. Higuchi, Chelex 100 as a medium for simple extraction of DNA for PCR-based typing from forensic material. *Biotechniques*, 1991, **10**(4): 506–513.

Wilkinson, D.A., et al., The fate of the chemical warfare agent during DNA extraction. *J Forensic Sci*, 2007, **52**(6): 1272–1283.

Winskog, C., et al., The use of commercial alcohol products to sterilize bones prior to DNA sampling. *Forensic Sci Med Pathol*, 2010, **6**(2): 127–129.

Witt, S., et al., Establishing a novel automated magnetic bead-based method for the extraction of DNA from a variety of forensic samples. *Forensic Sci Int Genet*, 2012, **6**(5): 539–547.

Wolfe, K.A., et al., Toward a microchip-based solid-phase extraction method for isolation of nucleic acids. *Electrophoresis*, 2002, **23**(5): 727–733.

Wolfgramm Ede, V., et al., Simplified buccal DNA extraction with FTA Elute Cards. *Forensic Sci Int Genet*, 2009, **3**(2): 125–127.

Ye, J., et al., A simple and efficient method for extracting DNA from old and burned bone. *J Forensic Sci*, 2004, **49**(4): 754–759.

Yoshida-Yamamoto, S., et al., Efficient DNA extraction from nail clippings using the protease solution from Cucumis melo. *Mol Biotechnol*, 2010, **46**(1): 41–48.

Zamir, A., Y. Cohen, and M. Azoury, DNA profiling from heroin street dose packages. *J Forensic Sci*, 2007, **52**(2): 389–392.

Zehner, R., "Foreign" DNA in tissue adherent to compact bone from tsunami victims. *Forensic Sci Int Genet*, 2007, **1**(2): 218–222.

Zhang, H., et al., Solid-phase based on-chip DNA purification through a valve-free stepwise injection of multiple reagents employing centrifugal force combined with a hydrophobic capillary barrier pressure. *Analyst*, 2013, **138**(6): 1750–1757.

Zhang, S.H., et al., Genotyping of urinary samples stored with EDTA for forensic applications. *Genet Mol Res*, 2012, **11**(3): 3007–3012.

6

DNA Quantitation

Determining the amount of DNA in a sample is essential for polymerase chain reaction (PCR)-based DNA testing (Chapter 7). Because PCR-based DNA testing is very sensitive, a narrow concentration range is required for amplification to be successful. If too much DNA template is used in PCR, the resulting artifacts may interfere with data analysis and interpretation. On the other hand, very low amounts of DNA template may result in a partial DNA profile or failure to attain a profile. Samples with poor purity may contain PCR inhibitors that lead to a failure of DNA amplification. Thus, a test that can measure the quality and quantity of the DNA template of a sample is desirable.

Additionally, forensic samples containing DNA are often mixtures that may include nonhuman DNA. For instance, microbial DNA may be present. Thus, it is necessary to use human-specific DNA quantitation methods to selectively determine the amount of human DNA present. In the United States, quality-assurance guidelines require the use of a quantitation method that estimates the amount of human nuclear DNA in crime-scene evidence samples.

In this chapter, the slot blot, intercalating dye, and quantitative PCR methods are introduced. The quantitative PCR method is the most sensitive of the three methods. It is the only method that can detect PCR inhibitors. Both the slot blot and the quantitative PCR methods can detect human and nonhuman primate DNA. Current technology cannot distinguish between human and nonhuman primate DNA.

6.1 Slot Blot Assay

Historically, the slot blot assay was used to detect human genomic DNA in a sample. Methodically, the slot blot assay works based on the following principles. Prior to the quantitation, an alkaline solution is added to the genomic DNA sample, which denatures DNA. Generating single-stranded DNA is necessary for DNA to be cross-linked onto a nitrocellulose membrane. The DNA sample is then spotted, using a slot blot device, onto a nitrocellulose membrane (Figure 6.1). The single-stranded DNA is then immobilized onto a nylon membrane. The targeted sequence is revealed by hybridization with a labeled 40-nucleotide probe complementary to a primate-specific α-satellite DNA sequence at the D17Z1 locus (Figure 6.2). In humans, the α-satellite DNA sequences are highly repetitive sequences located near the centromeres of chromosomes. These sequences are usually distinct for each chromosome.

Three detecting schemes of the slot blot assay have been developed. Initially, the D17Z1 probe was labeled with radioisotopes that could be visualized by exposing the slot blot membrane to x-ray film. The hazardous radioisotope detection method was then replaced by alkaline phosphatase-labeled and biotinylated probes. The alkaline phosphatase-labeled probe can be coupled with chemiluminescent detection (Lumi-Phos Plus kit, Lumigen, Inc.). The biotinylated probe

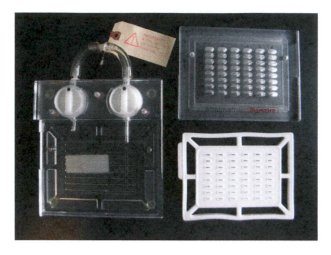

Figure 6.1 A slot blot device. (© Richard C. Li.)

(QuantiBlot Human DNA Quantitation Kit, Applied Biosystems) can be coupled with either colorimetric or chemiluminescent detection.

In colorimetric detection, the biotin moiety of the probe is bound to streptavidin. The streptavidin is conjugated with horseradish peroxidase, which catalyzes the oxidation reaction of tetramethylbenzidine (TMB), a substrate, forming a blue precipitate. With the chemiluminescent detection method, the horseradish peroxidase catalyzes the oxidation reaction of a substrate such as luminol, emitting photons that can be detected by exposure to x-ray film. The sensitivity of chemiluminescent detection is slightly higher than that of the colorimetric detection. The detection mechanisms will be discussed in Chapter 9.

The detected signal intensity is proportional to the concentration of the DNA sample in question. Quantitative measurements can be made by comparing an unknown sample to a set of standards with known DNA concentrations (Figure 6.3).

The assay typically quantifies DNA over the range of 150 pg–10 ng. However, the quantitation results are manually read, and conclusions are based on subjective judgments. Additionally, the D17Z1 sequences of humans and other primates share homology. The probe cannot distinguish between human and other primate genomic DNA. This is not a great concern because cases involving nonhuman primates are rare. Nevertheless, the probe does not cross-react with all other species. This assay has been replaced by recently developed quantitative PCR assays (see Section 6.3).

6.2 Fluorescent Intercalating Dye Assay

Small quantities of DNA can also be quantified by using a fluorescent intercalating dye method. Intercalating dyes, usually planar molecules, can slide themselves in between base pairs of DNA without breaking the DNA double helix. The Quant-iT™ PicoGreen® dsDNA reagent (Invitrogen) is a fluorescent intercalating dye that stains double-stranded DNA (dsDNA) for quantitation in a sample (Figure 6.4a). The detection limit of this method is approximately 250 pg. Intercalating dyes, not specific to human DNA, bind to all DNA molecules. Therefore, fluorescent intercalating dye assay can be utilized for the quantitation of known reference samples. For instance, DNA database samples from known sources can be quantified using this method. The assay has also been adapted for automation and is, thus, a high-throughput method.

DNA samples are simply added to a solution containing the fluorescent intercalating dye. The fluorescence, proportional to the quantities of DNA (Figure 6.4b), is measured using a standard spectrofluorometer with excitation and emission wavelengths of the light source. A standard

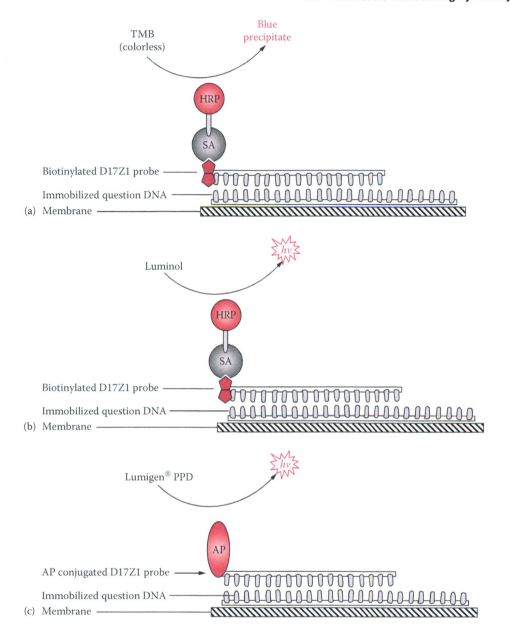

Figure 6.2   Slot blot assay. Questioned DNA is immobilized onto a solid-phase membrane then hybridized with a biotinylated D17Z1 probe. The detection of the hybridization is carried out by (a) streptavidin (SA) and horseradish peroxidase (HRP) conjugate, and a colorimetric reaction is catalyzed by HRP using tetramethylbenzidine (TMB) as a substrate; (b) SA and HRP conjugate, and a chemiluminescent reaction is catalyzed by HRP using Luminol as a substrate; (c) immobilized DNA is hybridized with an alkaline phosphatase (AP)-labeled D17Z1 probe. The detection of the hybridization is carried out by a chemiluminescent reaction catalyzed by AP using Lumigen® PPD as a substrate. (© Richard C. Li.)

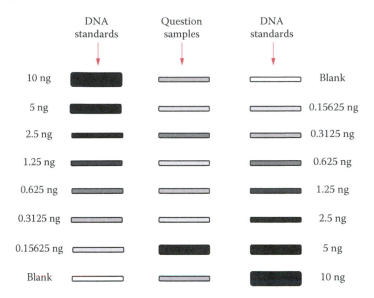

Figure 6.3   Human DNA quantitation using the slot blot assay. Standards with known amounts of human DNA are applied, and unknown samples and a set of standards are compared. The quantities in the unknown samples are estimated by visual comparison to the standards. (© Richard C. Li.)

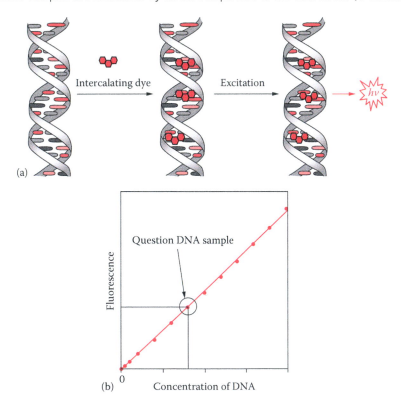

Figure 6.4   DNA quantitation using intercalating dye. (a) Fluorescent dye intercalates into DNA. The fluorescence can be measured upon applying an excitation light source. (b) A standard curve can be constructed using known amounts of DNA standards. The amount of questioned DNA can be determined by comparing the standard curve. (© Richard C. Li.)

curve is first created using samples containing known amounts of DNA. The assay is then performed for the unknown samples and the quantities of DNA in the samples are determined by comparing the results to the standard curve.

6.3 Quantitative PCR Assay

Based on the principle of PCR amplification (Chapter 7), the amount of PCR product amplified correlates with the initial concentration of DNA templates. Thus, the DNA concentration of a sample can be determined. There are two types of quantitative PCR methods. *End-point PCR* methods measure the quantity of amplified product at the end of PCR. Usually, the fluorescence is emitted by the dyes that intercalate into the double-stranded DNA. The quantity of amplified DNA is measured from the amount of fluorescence emitted from dyes such as SYBR (Figures 6.5 and 9.1b).

Real-time PCR methods can quantify the amplified DNA during the exponential phase of PCR (Chapter 7). The quantitation result is not affected to a significant extent by slight variations in PCR conditions. Thus, the precision of the quantitation of target sequences is improved with this method.

6.3.1 Real-Time Quantitative PCR

Real-time quantitative PCR (qPCR) was developed in the early 1990s, and it analyzes the amplification of a target sequence at each cycle of PCR. A fluorescent reporter is used to monitor the accumulation of amplified products during PCR. The fluorescence signals of the reporter molecule increase as amplified products accumulate with each cycle of PCR. qPCR is commonly used because of the following advantages:

- Better objectivity than the QuantiBlot method
- Increased sensitivity with a large dynamic range (30 pg–100 ng)
- More accurate measurements of small quantities of DNA in samples
- Fewer laboratory manipulations; amenable to automation
- Ability to detect PCR inhibitors (Section 6.3.1.1)

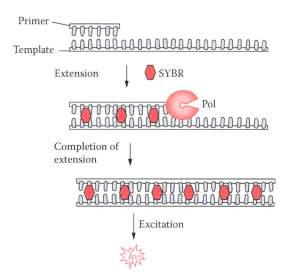

Figure 6.5 End-point PCR using SYBR Green detection. During the extension phase of PCR, in which DNA synthesis occurs, the dye binds to the double-stranded amplicons. Upon excitation, the emission intensity of the dye can be measured. Pol represents Taq polymerase. (© Richard C. Li.)

The technique is amenable to multiplexing to detect more than one type of DNA target sequence in a single reaction. Commercial qPCR kits for human DNA and Y-chromosome DNA quantitation are available. Additionally, the qPCR method for mtDNA quantitation is possible. qPCR uses commercially available fluorescence-detecting thermocyclers to amplify specific DNA sequences and measure their concentrations simultaneously. The fluorescent reporter can be a nonspecific intercalating double-stranded DNA-binding dye or a sequence-specific fluorescently labeled oligonucleotide probe (Chapter 9). The target sequences are amplified and detected by the same instrument, and the reporter fluorescence is monitored externally. Thus the reaction tubes do not need to be opened. This minimizes aerosol contamination and reduces the risk of false-positive results. A widely used qPCR probe technique is the TaqMan method (Applied Biosystems).

6.3.1.1 TaqMan Method

This method utilizes the 5′ exonuclease activity of *Taq* polymerase to cleave the probe during PCR (also known as the 5′ exonuclease assay). The probe is designed to anneal to the target sequence between the upstream and downstream primers and is added to the PCR mixture together with primers (Figure 6.6). The probe T_m (melting temperature; see Chapter 7) should be higher than the amplification primer T_m. A minor groove binder (MGB), such as dihydrocyclopyrroloindole tripeptide, is often linked at the 3′ end of the probe (Figure 6.6). A conjugated MGB binds to the minor groove of a B-form DNA helix (Figure 6.7), which is stabilized by van

Figure 6.6 TaqMan probe. Each probe is labeled with a reporter dye (R) on the 5′ end and a fluorescence quencher (Q) on the 3′ end. MGB represents a minor groove binder. (© Richard C. Li.)

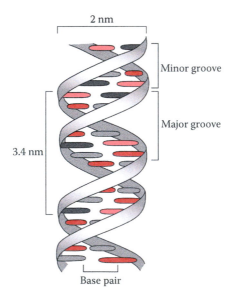

Figure 6.7 DNA double helix. The double-helical structure of B-DNA, the most common form of DNA, is shown. With a helical diameter of 2 nm, each turn of the helix takes 3.4 nm, which corresponds to 10 base pairs per turn. The major and minor grooves are shown. (© Richard C. Li.)

der Waals forces. As a result, conjugating an MGB into a probe increases the T_m values, allowing for the use of shorter probes. As a result, having the probe T_m higher than that of the primers ensures that the probe is fully hybridized during primer extension. The oligonucleotide probe is labeled with both a *reporter* fluorescent dye, usually 6-carboxyfluorescein (6-FAM) or tetrachlorofluorescein (TET) at the 5′ end; and a nonfluorescent *quencher* moiety, such as tetramethylrhodamine (TAMRA), usually at the 3′ end or any thymine position. While the probe is intact, the quencher greatly reduces the fluorescence emitted by the reporter via fluorescent resonance energy transfers (FRET). FRET is a distance-dependent interaction between two molecules in which the excitation energy is transferred from a photon donor molecule (reporter) to an acceptor molecule (quencher) without emission of a photon. During the extension phase of the PCR cycle, the 5′ exonuclease activity of Taq polymerase cleaves the reporter dye from the probe. Because the reporter dye is no longer in close proximity to the quencher, the FRET is disrupted and the probe begins to fluoresce (Figure 6.8). The intensity of fluorescence can be measured (Chapter 9) and is proportional to the amount of target DNA synthesized during the PCR.

In this assay, the rate of accumulation of amplified DNA over the entire course of a PCR is generated. The greater the initial concentration of target templates in a sample, the fewer cycles required to reach a particular quantity of amplified product. The initial concentration of target templates can be expressed using the *cycle threshold* (C_T). C_T is defined as the number of PCR cycles required for the fluorescent signal to cross a threshold of amplification where the signal exceeds background level or baseline noise. A plot of C_T against the $\log_{10}$ of the initial concentration of a set of DNA standards yields a straight line as a standard curve (Figure 6.9). The target sequences in an unknown sample can be quantified by comparing to the standard curve. Additionally, qPCR has the ability to detect PCR inhibitors (Chapter 7) that interact with DNA or with DNA polymerases. The presence of PCR inhibitors in the DNA extracts can be measured by monitoring the amplification of the internal positive control (IPC). Most human DNA quantitation kits contain a known amount of exogenous DNA as IPC that can be fortified to the sample and amplified. Monitoring the amplification of IPC enables the detection of PCR failure due to inhibition when the IPC's C_T value is higher than that of an uninhibited PCR reaction.

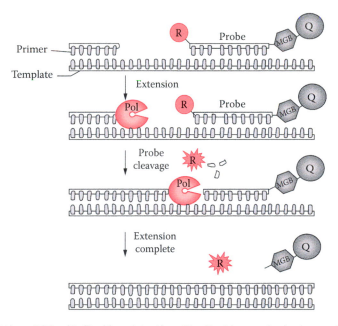

Figure 6.8 Real-time PCR with TaqMan detection. The TaqMan probe is shown. During extension, the 5′ nuclease activity of Taq polymerase (Pol) cleaves the probe. Reporter dye is released during each cycle of PCR. (© Richard C. Li.)

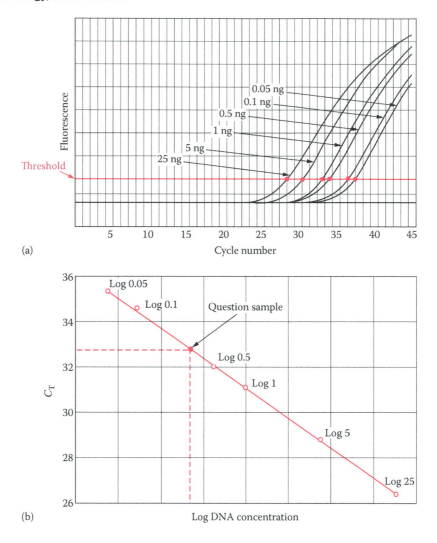

(a)

(b)

Figure 6.9 Real-time quantitative PCR. (a) Amplification curves for a dilution series of standard**s** with known quantities of DNA. C_T is the cycle threshold at which the amplification curve crosses the threshold, as indicated by the red line. (b) A standard curve based on data obtained from the amplification curves. The quantity of DNA in a questioned sample can be determined from the standard curve. (© Richard C. Li.)

Bibliography

Ahn, S.J., J. Costa, and J.R. Emanuel, PicoGreen quantitation of DNA: Effective evaluation of samples pre- or post-PCR. *Nucleic Acids Res*, 1996, **24**(13): 2623–2625.

Alonso, A. and P. Martin, A real-time PCR protocol to determine the number of amelogenin (X-Y) gene copies from forensic DNA samples. *Methods Mol Biol*, 2005, **297**: 31–44.

Alonso, A., et al., Specific quantification of human genomes from low copy number DNA samples in forensic and ancient DNA studies. *Croat Med J*, 2003, **44**(3): 273–280.

Alonso, A., et al., Real-time PCR designs to estimate nuclear and mitochondrial DNA copy number in forensic and ancient DNA studies. *Forensic Sci Int*, 2004, **139**(2–3): 141–149.

Andreasson, H. and M. Allen, Rapid quantification and sex determination of forensic evidence materials. *J Forensic Sci*, 2003, **48**(6): 1280–1287.

Andreasson, H., U. Gyllensten, and M. Allen, Real-time DNA quantification of nuclear and mitochondrial DNA in forensic analysis. *Biotechniques*, 2002, **33**(2): 402–404, 407–411.

Andreasson, H., et al., Nuclear and mitochondrial DNA quantification of various forensic materials. *Forensic Sci Int*, 2006, **164**(1): 56–64.

Budowle, B., et al., Simple protocols for typing forensic biological evidence: Chemiluminescent detection for human DNA quantitation and restriction fragment length polymorphism (RFLP) analyses and manual typing of polymerase chain reaction (PCR) amplified polymorphisms. *Electrophoresis*, 1995, **16**(9): 1559–1567.

Budowle, B., et al., The presumptive reagent fluorescein for detection of dilute bloodstains and subsequent STR typing of recovered DNA. *J Forensic Sci*, 2000, **45**(5): 1090–1092.

Budowle, B., et al., Using a CCD camera imaging system as a recording device to quantify human DNA by slot blot hybridization. *Biotechniques*, 2001, **30**(3): 680–685.

Del Rio, S.A., M.A. Marino, and P. Belgrader, Reusing the same blood-stained punch for sequential DNA amplifications and typing. *Biotechniques*, 1996, **20**(6): 970–972, 974.

Duewer, D.L., et al., NIST mixed stain studies #1 and #2: Interlaboratory comparison of DNA quantification practice and short tandem repeat multiplex performance with multiple-source samples. *J Forensic Sci*, 2001, **46**(5): 1199–1210.

Fox, J.C., C.A. Cave, and J.W. Schumm, Development, characterization, and validation of a sensitive primate-specific quantification assay for forensic analysis. *Biotechniques*, 2003, **34**(2): 314–318, 320, 322.

Green, R.L., et al., Developmental validation of the quantifiler real-time PCR kits for the quantification of human nuclear DNA samples. *J Forensic Sci*, 2005, **50**(4): 809–825.

Hayn, S., et al., Evaluation of an automated liquid hybridization method for DNA quantitation. *J Forensic Sci*, 2004, **49**(1): 87–91.

Hewakapuge, S., et al., Investigation of telomere lengths measurement by quantitative real-time PCR to predict age. *Leg Med* (Tokyo), 2008, **10**(5): 236–242.

Hudlow, W.R., et al., A quadruplex real-time qPCR assay for the simultaneous assessment of total human DNA, human male DNA, DNA degradation and the presence of PCR inhibitors in forensic samples: A diagnostic tool for STR typing. *Forensic Sci Int Genet*, 2008, **2**(2): 108–125.

Kihlgren, A., A. Beckman, and S. Holgersson, Using D3S1358 for quantification of DNA amenable to PCR and for genotype screening. *Prog Forensic Genet*, 1998, **7**: 3.

Kline, M.C., et al., NIST Mixed Stain Study 3: DNA quantitation accuracy and its influence on short tandem repeat multiplex signal intensity. *Anal Chem*, 2003, **75**(10): 2463–2469.

Koukoulas, I., et al., Quantifiler observations of relevance to forensic casework. *J Forensic Sci*, 2008, **53**(1): 135–141.

Krenke, B.E., et al., Developmental validation of a real-time PCR assay for the simultaneous quantification of total human and male DNA. *Forensic Sci Int Genet*, 2008, **3**(1): 14–21.

LaSalle, H.E., G. Duncan, and B. McCord, An analysis of single and multi-copy methods for DNA quantitation by real-time polymerase chain reaction. *Forensic Sci Int Genet*, 2011, **5**(3): 185–193.

Mandrekar, M.N., et al., Development of a human DNA quantitation system. *Croat Med J*, 2001, **42**(3): 336–339.

Morrison, T.B., J.J. Weis, and C.T. Wittwer, Quantification of low-copy transcripts by continuous SYBR Green I monitoring during amplification. *Biotechniques*, 1998, **24**(6): 954–958, 960, 962.

Myers, J.R., Validation of a DNA quantitation method on the Biomek® 3000. *J Forensic Sci*, 2010, **55**(6): 1570–1575.

Nicklas, J.A. and E. Buel, Development of an Alu-based, real-time PCR method for quantitation of human DNA in forensic samples. *J Forensic Sci*, 2003, **48**(5): 936–944.

Nicklas, J.A. and E. Buel, Quantification of DNA in forensic samples. *Anal Bioanal Chem*, 2003, **376**(8): 1160–1167.

Nicklas, J.A., T. Noreault-Conti, and E. Buel, Development of a real-time method to detect DNA degradation in forensic samples. *J Forensic Sci*, 2012, **57**(2): 466–471.

Niederstatter, H., et al., Characterization of mtDNA SNP typing and mixture ratio assessment with simultaneous real-time PCR quantification of both allelic states. *Int J Legal Med*, 2006, **120**(1): 18–23.

Niederstatter, H., et al., A modular real-time PCR concept for determining the quantity and quality of human nuclear and mitochondrial DNA. *Forensic Sci Int Genet*, 2007, **1**(1): 29–34.

Nielsen, K., et al., Comparison of five DNA quantification methods. *Forensic Sci Int Genet*, 2008, **2**(3): 226–230.

Omelia, E.J., M.L. Uchimoto, and G. Williams, Quantitative PCR analysis of blood- and saliva-specific microRNA markers following solid-phase DNA extraction. *Anal Biochem*, 2013, **435**(2): 120–122.

Ong, Y.L. and A. Irvine, Quantitative real-time PCR: A critique of method and practical considerations. *Hematology*, 2002, **7**(1): 59–67.

Puch-Solis, R., et al., Practical determination of the low template DNA threshold. *Forensic Sci Int Genet*, 2011, **5**(5): 422–427.

Richard, M.L., R.H. Frappier, and J.C. Newman, Developmental validation of a real-time quantitative PCR assay for automated quantification of human DNA. *J Forensic Sci*, 2003, **48**(5): 1041–1046.

Shewale, J.G., et al., Human genomic DNA quantitation system, H-Quant: Development and validation for use in forensic casework. *J Forensic Sci*, 2007, **52**(2): 364–370.

Sifis, M.E., K. Both, and L.A. Burgoyne, A more sensitive method for the quantitation of genomic DNA by Alu amplification. *J Forensic Sci*, 2002, **47**(3): 589–592.

Swango, K.L., et al., A quantitative PCR assay for the assessment of DNA degradation in forensic samples. *Forensic Sci Int*, 2006, **158**(1): 14–26.

Swango, K.L., et al., Developmental validation of a multiplex qPCR assay for assessing the quantity and quality of nuclear DNA in forensic samples. *Forensic Sci Int*, 2007, **170**(1): 35–45.

Tak, Y.K., et al., Highly sensitive polymerase chain reaction-free quantum dot-based quantification of forensic genomic DNA. *Anal Chim Acta*, 2012, **721**: 85–91.

Thomas, J.T., et al., Qiagen's Investigator Quantiplex Kit as a predictor of STR amplification success from low-yield DNA samples. *J Forensic Sci*, 2013, **58**(5): 1306–1309.

Timken, M.D., et al., A duplex real-time qPCR assay for the quantification of human nuclear and mitochondrial DNA in forensic samples: Implications for quantifying DNA in degraded samples. *J Forensic Sci*, 2005, **50**(5): 1044–1060.

Tozzo, P., et al., Discrimination between human and animal DNA: Application of a duplex polymerase chain reaction to forensic identification. *Am J Forensic Med Pathol*, 2011, **32**(2): 180–182.

Walker, J.A., et al., Multiplex polymerase chain reaction for simultaneous quantitation of human nuclear, mitochondrial, and male Y-chromosome DNA: Application in human identification. *Anal Biochem*, 2005, **337**(1): 89–97.

Walsh, P.S., J. Varlaro, and R. Reynolds, A rapid chemiluminescent method for quantitation of human DNA. *Nucleic Acids Res*, 1992, **20**(19): 5061–5065.

Waye, J.S., et al., A simple and sensitive method for quantifying human genomic DNA in forensic specimen extracts. *Biotechniques*, 1989, **7**(8): 852–855.

Westring, C.G., et al., Validation of reduced-scale reactions for the Quantifiler Human DNA kit. *J Forensic Sci*, 2007, **52**(5): 1035–1043.

Wilson, J., et al., Molecular assay for screening and quantifying DNA in biological evidence: The modified Q-TAT assay. *J Forensic Sci*, 2010, **55**(4): 1050–1057.

Amplification by Polymerase Chain Reaction

7.1 Denaturation and Renaturation of DNA

The DNA double helix is stabilized by chemical interactions. Base pairing of the two strands involves the formation of hydrogen bonds that provide weak electrostatic attractions between electronegative atoms. An adenine always pairs with a thymine (two hydrogen bonds), and a cytosine pairs with a guanine (three hydrogen bonds) (see Figure 7.1). Base stacking involves hydrophobic interactions between adjacent base pairs and provides stability to the double helix.

Double-stranded DNA is maintained by hydrogen bonding between the bases of complementary pairs. *Denaturation* occurs when the hydrogen bonds of DNA are disrupted, and the strands are separated. A *melting curve* can be obtained from measuring DNA denaturation by slowly heating a solution of DNA. As shown in Figure 7.2, an increasing temperature increases the percentage of the DNA that is denatured. The temperature at which 50% of DNA strands are denatured is defined as the *melting temperature* (T_m). The value of T_m is affected by the salt concentration of the solution, but can also be affected by nucleotide content, high pH, and length of the molecule.

Nucleotide content affects the value of T_m because GC pairs are joined by three hydrogen bonds while the AT pairs are joined by only two. Increasing the GC content of a DNA molecule increases the T_m. Excessively high pH causes the hydrogen bonds to break and the paired strands to separate. Finally, the length of the molecule also affects the T_m simply because a longer molecule of DNA requires more energy to break more bonds than a shorter molecule.

The single strands in a solution of denatured DNA can, under certain conditions, reanneal into double-stranded DNA. The process is called *renaturation* and two requirements must be met for it to occur. First, sufficient amounts of charged molecules, such as salts, must be present in the solution to neutralize the negative charges of the phosphate groups in DNA. This prevents the complementary strands from repelling each other. Additionally, the temperature must be high enough to disrupt hydrogen bonds that formed randomly between the bases of DNA strands. However, excessively elevated temperatures can disrupt the base pairs between the complementary DNA strands.

7.2 Basic Principles of Polymerase Chain Reaction

The *polymerase chain reaction* (*PCR*) allows the exponential amplification of specific sequences of DNA to yield sufficient amplified products, also known as *amplicons*, for various downstream

143

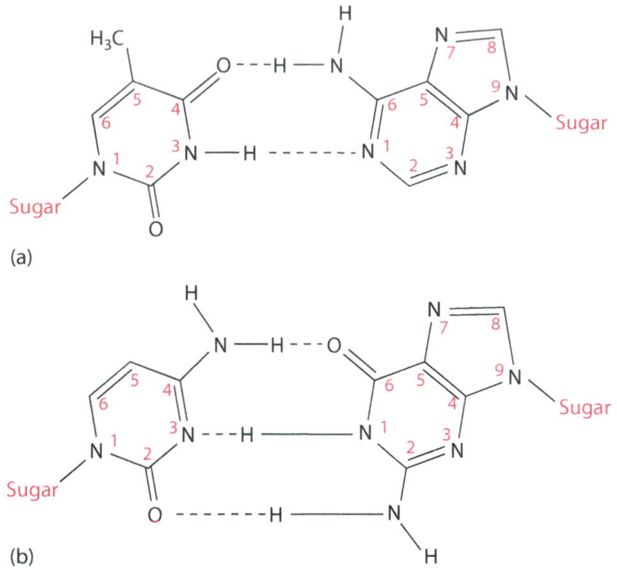

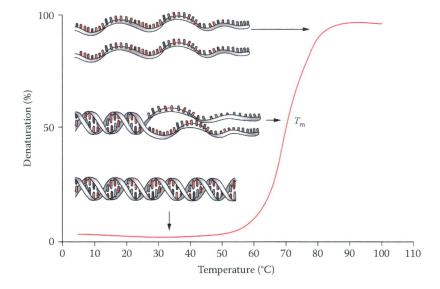

Figure 7.1 Base pairing between two DNA strands. (a) An adenine (right) pairs with a thymine (left). (b) A cytosine (left) pairs with a guanine (right).

Figure 7.2 Melting curve of DNA. The degree of DNA denaturation is increased by increasing the temperature. T_m and possible shapes of a DNA molecule are shown. (© Richard C. Li.)

applications. The technique is highly sensitive and can amplify very small quantities of DNA. Therefore, it can be utilized for the analysis of samples of limited quantity. PCR-based assays are rapid and robust. Thus, PCR forms the basis of many forensic DNA assays such as DNA quantitation (Chapter 6), short tandem repeat (STR) profiling (Chapter 20), and mitochondrial DNA (mtDNA) sequencing (Chapter 23).

The concept of synthesizing DNA by a cycling process was first proposed in the early 1970s. In the mid-1980s, PCR technology was finally developed by Kary Mullis and his coworkers (Cetus Corporation) to amplify the β-globin gene for the diagnosis of sickle-cell anemia. In the late 1980s, a

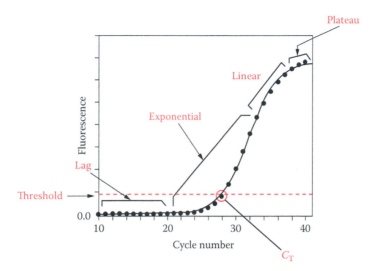

Figure 7.3 PCR amplification curve. A fluorescence signal results from the accumulation of ampli-con. The location of the signal threshold is indicated as a dashed line. The C_T value for the curve is the cycle at which the amplification curve crosses the threshold. (© Richard C. Li.)

thermostable polymerase from *Thermus aquaticus* was utilized for PCR. This step greatly increased efficiency and allowed the process to be automated. The result was a powerful impact on molecular biology. In 1993, Mullis was awarded the Nobel Prize for the invention of PCR technology.

In the early 1990s, the technique for the simultaneous amplification and detection of the accumulation of amplicons at each PCR cycle was developed, and the concept of *real-time PCR* was born. This process allows the monitoring of amplicon production at each cycle of the PCR process. The fundamental processes were studied by characterizing the amplification kinetics of PCR using a graph that plotted the amount of amplicon yield at each cycle versus the cycle number (Figure 7.3).

An S-shaped amplification curve is obtained and divided into an exponential phase, a linear phase, and a plateau. During the exponential phase, the amplicon accumulates exponentially. It was revealed that the amplicon accumulation during PCR was correlated to the starting copy number of DNA template. Thus, the amount of amplicon produced during the exponential amplification phase can be used to determine the amount of starting material. This relationship can be further examined using a plot of cycle numbers versus a log scale of the serial dilution of the starting concentration of DNA template, which results in a linear relationship (see Chapter 6). It demonstrates that fewer cycles are needed if larger quantities of starting DNA template are present. The slope of this linear curve (Figure 6.9b) is known as the *amplification efficiency*. The exponential phase continues until one or more of the components (Section 7.3) in the reaction become limited. At this point, the amplification efficiency decreases, the amplicon no longer accumulates exponentially, and PCR enters the linear phase of the curve. At the plateau phase, no more amplicon is accumulated due to the exhaustion of reagents and polymerase.

7.3 Essential PCR Components

A PCR reaction requires thermostable DNA polymerases, primers, and other components, as described below (Figure 7.4).

7.3.1 Thermostable DNA Polymerases

A wide variety of DNA polymerases are available. They vary in fidelity, efficiency, and ability to synthesize longer DNA fragments. Nonetheless, *Taq* polymerase is the most commonly used

Figure 7.4   A liquid handling workstation for automated PCR assay setup. (© Richard C. Li.)

enzyme for routine PCR applications (0.5–5 units per reaction). Currently, AmpliTaq Gold™ DNA polymerase (Applied Biosystems) is the most common DNA polymerase for forensic applications.

PCR reactions are usually set up at room temperature. Nonspecific annealing between primers and template DNA can occur, resulting in the formation of nonspecific amplicons. Additionally, annealing between the primers can occur to form primer dimers. Nonspecific annealing interferes with PCR amplification by reducing the amplification efficiency of the specific sequences of interest.

Such interference can be minimized by a *hot-start PCR* approach. The AmpliTaq Gold™ DNA polymerase, a modified enzyme, remains in an inactive form until activated with a pH below 7 prior to the PCR cycling in which the inhibitory motif is inactivated. The pH of the buffer system used in the PCR reaction is temperature sensitive; increasing the temperature decreases the pH of the solution. Thus, the activation of the enzyme can actually be carried out by a heating step at 95°C prior to the start of the cycling. During the heating process, the DNA strands also denature, which can prevent the formation of nonspecific PCR products.

7.3.2 PCR Primers

PCR primers are the oligonucleotides that are complementary to the sequences that flank the target region of the template. A pair (forward and reverse) of primers (typically 0.1–1 μM) is required. Properly designed primers are critical to the success of a PCR reaction. Computer software such as Primer3 is available to assist and optimize the designing of primers.

A primer must be specific to the target sequence; otherwise, nonspecific products that might interfere with the proper interpretation of a DNA profile might be produced. The primers within a pair should have similar T_m values. The estimated T_m values of a primer pair should not differ by more than 5°C. The T_m of an oligonucleotide primer can be predicted and calculated using the following equation:

$$T_m = 81.5°C + 16.6\left(\log_{10}[K^+]\right) + 0.41\left(\% \ [G+C]\right) - \left(\frac{675}{n}\right)$$

where:
$[K^+]$ = concentration of the potassium ion
$[G+C]$ = GC content (%) of the oligonucleotide
n = number of bases in the oligonucleotide

This equation shows that the T_m can be affected by primer base composition (GC content) and primer length. The GC content of an oligonucleotide primer should be 40%–60%, and the length of an oligonucleotide primer should be 15–25 base pairs, although longer primers can be used.

A primer should not contain self-complementary sequences that may form hairpin structures interfering with the annealing of primers and the template. Additionally, the primers in a pair should not share similar sequences, to avoid the primers from annealing to each other. These annealed primers may then be amplified during PCR, creating products known as *primer dimers*, which compete with the target DNA template for PCR components.

Multiplex PCR is the simultaneous amplification of more than one region of a DNA template in a single reaction to achieve high-throughput analysis. Multiplex PCR consists of multiple primer sets in a single reaction to produce amplicons of multiple target DNA regions. The primers should be designed to yield proper sizes of amplicons to be resolvable in downstream separation and detection procedures such as electrophoresis. To prevent preferential amplification of one target sequence over another, the annealing temperatures should be similar among multiplex PCR primer pairs. Additionally, the primers should lead to similar amplification efficiencies among the loci to be tested. Forensic applications of multiplex PCR, such as autosomal and Y-chromosomal short tandem repeat analysis, are discussed in Chapters 20 and 21.

7.3.3 Other Components

Essential components include template DNA with target sequences in either linear form (nuclear genomic DNA) or circular form (mitochondrial DNA). Both single- and double-stranded DNA can be used as a template for PCR. Typically, 1–2.5 ng of template DNA is utilized for forensic applications using PCR.

Deoxynucleoside triphosphates (dNTPs) are the substrates for DNA synthesis. A PCR assay usually contains equal molar amounts (typically, 200 µM) of dATP, dCTP, dGTP, and dTTP.

Divalent cations, such as Mg^{2+}, are required for the enzymatic activity of DNA polymerases. A PCR assay usually contains 1.5–2.5 mM Mg^{2+}. Monovalent cations, such as K^+ (50 mM), are usually recommended, and a buffer is often utilized to maintain pH between 8.3 and 8.8 at room temperature.

Controls should be used to monitor the effectiveness of PCR amplification. A *positive control* shows that PCR components such as reagents and PCR cycle parameters are working properly during a PCR. A standard DNA template should be used as a positive control and amplified with the same PCR components used on the rest of the samples. The amplification negative control and extraction reagent blank are discussed in Section 7.5.4.

7.4 Cycle Parameters

PCR cycling protocols may vary according to the type of analysis. Figure 7.5 shows representative PCR cycling conditions commonly used by forensic DNA laboratories. PCR utilizes a number of cycles for the replication of a specific region of a DNA template. During each cycle, a copy of the target DNA sequence is synthesized. A PCR cycle consists of three elements: denaturation, annealing, and extension. Precise and accurate temperatures at denaturation, annealing, and extension are critical to achieve a successful amplification. At the beginning of each cycle, the two complementary DNA template strands are separated at high temperatures (94°C–95°C) in a process called *denaturation*. The temperature is then decreased to allow *annealing* between the oligonucleotide primers and the template. The temperature for annealing is usually 3°C–5°C lower than the T_m of the oligonucleotide primer.

The annealing temperature is critical. If it is too high, a very low quantity of amplicon is yielded because of the failure of annealing between the primer and the template. If the annealing temperature is too low, nonspecific amplification can occur. Next, optimal temperature for DNA polymerase is reached, thus allowing for DNA replication (*extension*). By the end of each cycle, the copy number of the amplicon is nearly doubled (Figure 7.6).

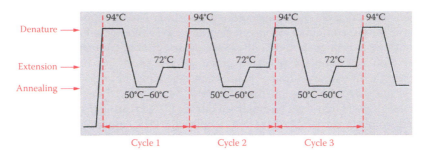

Figure 7.5 Temperature parameters during thermal cycling of the PCR process. The first three cycles are shown. (© Richard C. Li.)

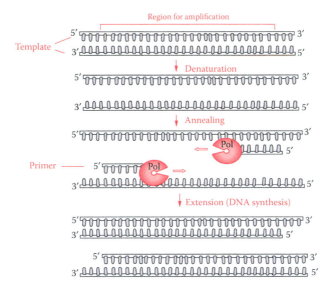

Figure 7.6 First cycle of PCR. The amplified region is determined by the positions of the primers. The direction (5′ to 3′) of DNA synthesis is indicated by arrows. Pol represents *Taq* polymerase. (© Richard C. Li.)

The number of cycles needed for PCR depends primarily on the number of copies of starting DNA template. The relationship can be expressed as the following equation:

$$N_x = N_0 (1 + E)^x$$

where:

X = number of PCR cycles
N_x = copy number of the amplicon after x cycles of PCR
N_0 = initial copy number of the template
E = amplification efficiency of the *Taq* polymerase

For example, in a 28-cycle PCR amplification, the DNA template can theoretically be amplified by a factor of approximately 10^8. If the cycle number is increased to 34, a factor of 10^{10} can theoretically be reached.

PCR amplification can be carried out using an instrument known as a *thermal cycler* (Figure 7.7). Various types of thermal cyclers differ in the number of samples they can process, the sizes of the sample tubes, and temperature control features.

Figure 7.7   Thermal cyclers. A thermal cycler can provide rapid temperature changes as desired to carry out PCR. A PCR thermal cycler (left) and a real-time PCR instrument (right). (© Richard C. Li.)

7.5 Factors Affecting PCR

7.5.1 Template Quality

It is important to prevent degradation of the DNA during the collection and processing of evidence. Degradation causes DNA to break into smaller fragments. If the damage occurs at a region to be amplified, the result can be failure in PCR amplification. In a degraded sample, the longer the amplicon length, the higher the risk of failure in PCR. Low copy number (LCN) of DNA template is often encountered in forensic samples. When amplifying very low levels (approximately 100 pg of DNA) of a template, the following phenomenon is often observed: one of the two alleles fails to be detected from a heterozygote and can falsely be identified as a homozygote. This phenomenon is also known as the *stochastic effect* in which the two alleles in a heterozygous individual are unequally detected at a low level of starting DNA template. Approaches such as increasing the cycle number, from 28 to 34 (Section 7.4), have been introduced to address the LCN problem.

7.5.2 Inhibitors

Inhibitors, if present, can interact with the DNA template or polymerase, causing PCR amplification failure. The presence of PCR inhibitors can be detected using an internal positive control (Section 6.3.1.1). A number of PCR inhibitors commonly encountered in evidence samples include heme molecules from blood, indigo dyes from fabrics, and melanin from hair samples. Thus, it is important to remove PCR inhibitors during DNA extraction. If PCR inhibitors are not eliminated during the extraction process, additional procedures such as the use of centrifugal filtration devices can be used. Centrifugal filtration devices can separate molecules by size. After the centrifugation step, small molecular weight inhibitors are filtered by passing through the membrane and are discarded. Alternatively, increasing the amount of DNA polymerase or adding bovine serum albumin (BSA) in the reaction can overcome the inhibition effects.

7.5.3 Contamination

PCR is a highly sensitive method; therefore, procedures that minimize the risk of contamination are necessary. To prevent contamination, pre- and post-PCR samples should be processed in separate areas or at different times. Additionally, reagents, supplies, and equipment used for pre- and post-PCR steps should be separated as well. Protective gear should include laboratory coats and disposable gloves. Facial masks and hair caps may be used if necessary. Aerosol-resistant pipet tips and DNA-free solutions and test tubes should also be used.

The levels of contamination must be monitored using controls. Extraction *reagent blanks*, which contain all extraction reagents but no sample, monitor contamination from extraction to PCR. Contamination detected in an extraction reagent blank but not in an amplification-negative control indicates that the reagents used for extraction are contaminated. Amplification-*negative controls*, which contain all PCR reagents and no DNA template, monitor contamination of the amplification. Contamination observed in an amplification-negative control but not in an extraction reagent blank indicates that the contamination occurred during the amplification step. Contamination observed in both an amplification reagent blank and an extraction-negative control indicates that amplification reagents are contaminated. A collection of DNA profiles of each member of a laboratory should be readily available for comparisons. Sources of laboratory contamination can be identified by comparing results with an analyst's DNA profile.

7.6 Reverse Transcriptase PCR for RNA-Based Assays

The pathway for the flow of genetic information is called the *central dogma*—a term coined by Francis Crick in 1956. According to the central dogma, parental DNA serves as the template for *DNA replication*. With RNA synthesis or *transcription*, the process is carried out using the DNA as a template. Conversely, RNA chains can be used as templates for the synthesis of a DNA strand of complementary sequence, in which the end product is referred to as *complementary DNA (cDNA)*. Protein synthesis, also known as *translation*, is directed by an RNA template (Figure 7.8).

The flow of genetic information from RNA to DNA is referred to as *reverse transcription*. It was discovered independently by David Baltimore and Howard Temin in 1970, and they shared the Nobel Prize for their work. Reverse transcription is carried out by a *reverse transcriptase* that forms the basis of reverse transcriptase PCR as described below.

Reverse transcriptase PCR (RT-PCR) is highly sensitive and can be used to detect very small quantities of mRNA. It can be utilized to measure levels of gene expression even when the RNA of interest is expressed at very low levels. Detecting mRNAs of tissue-specific genes can be utilized for bodily fluid identification in forensic investigations (Chapter 11). During an RT-PCR process, a single-stranded cDNA is synthesized from a template mRNA using reverse transcription. The cDNA is then amplified by PCR for detection and analysis.

7.6.1 Reverse Transcription

The synthesis of single-stranded cDNA from an mRNA template is catalyzed by reverse transcriptase. Reverse transcriptases share many features in structure and function with DNA polymerases. The catalytic function of cDNA synthesis requires a primer that anneals to a complementary mRNA template. The primer can be either RNA or DNA; however, DNA primers are more efficient than RNA primers. During the elongation of the primer, reverse transcriptase incorporates the corresponding deoxyribonucleotide triphosphate according to the rules of base pairing with the RNA template. The RNA template is then degraded by an intrinsic RNase H activity of reverse transcriptase during the reverse transcription reaction. The retroviral RNase H is a domain of the viral reverse transcriptase enzyme. It is a nonspecific endonuclease that cleaves RNA.

Several reverse transcriptases derived from retroviruses can be used to generate cDNA from an RNA template. The most common reverse transcriptases used for cDNA synthesis are

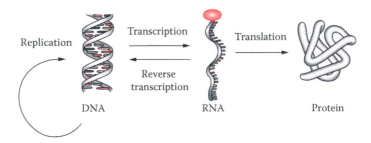

Figure 7.8 Pathway for the flow of genetic information. (© Richard C. Li.)

encoded by the *pol* gene from the avian myeloblastosis virus (AMV) and the Moloney strain of the murine leukemia virus (MMLV). These enzymes are genetically engineered reverse transcriptases but lack RNase H activity. Genetically engineered reverse transcriptases that are stable at higher temperatures (up to 60°C) are also produced. Increased thermal stability of reverse transcriptase allows the reverse transcription to be carried out at a higher than ambient temperature, which eliminates RNA secondary structures and improves the specificity of the reaction and the yields of the synthesis of full-length cDNA.

7.6.2 Oligodeoxynucleotide Priming

Oligodeoxynucleotide primers are an essential component for a reverse transcriptase reaction. Different priming strategies can be utilized to synthesize cDNA from a particular target mRNA or from all mRNA in a sample. *Gene-specific primers* are designed to hybridize to a particular mRNA sequence for the conversion of a specific gene sequence into cDNA. *Universal primers* can hybridize to any mRNA sequence in a sample to convert all mRNAs to cDNA. Two types of universal primers can be used: oligo (dT) and random hexamer primers. An *oligo (dT) primer* can hybridize to the 3′ termini poly (A) tails of eukaryotic mRNAs. Reverse transcriptases with oligo (dT) only synthesize cDNA from transcribed genes. *Random hexamer primers* are another type of universal primer. Random hexamers are nonspecific primers that can hybridize, at multiple sites, to any RNA sequence including non-mRNA templates such as ribosomal RNA.

7.6.3 Reverse Transcriptase PCR

During an RT-PCR process, a single-stranded cDNA is synthesized from a template mRNA using reverse transcription. The cDNA is then amplified by PCR with a pair of oligonucleotide primers corresponding to a specific sequence in the cDNA (Figure 7.9).

Two strategies of RT-PCR exist: a one-step and a two-step RT-PCR (Figure 7.10). One-step RT-PCR combines the reverse transcription reaction and PCR in a single tube. Only gene-specific

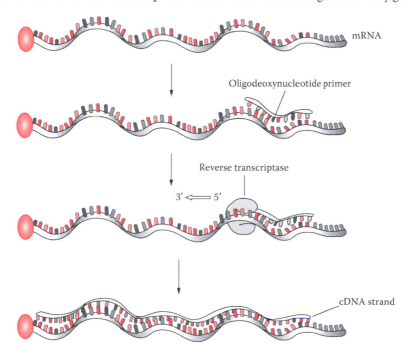

Figure 7.9 cDNA synthesis. The synthesis of a DNA strand transcribed from mRNA can be carried out using a primer and reverse transcriptase. A gene-specific primer that amplifies a specific target sequence is shown. (© Richard C. Li.)

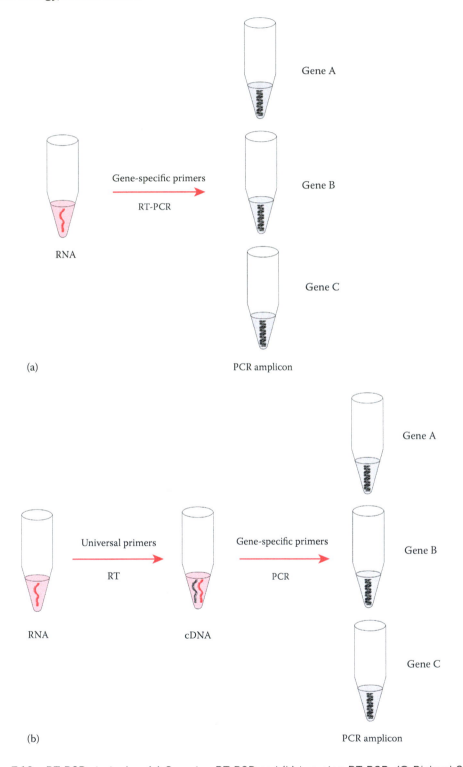

Figure 7.10   RT-PCR strategies. (a) One-step RT-PCR and (b) two-step RT-PCR. (© Richard C. Li.)

primers can be used for the reverse transcription reaction and PCR. RNAs from either total RNA or mRNA can be used. This approach simplifies the reaction setup as it is useful for processing large numbers of samples. The tubes are not opened between reverse transcription and PCR, thus minimizing the risk of pipetting errors and carryover contamination.

During a two-step PCR, the reverse transcription reaction and PCR are carried out in separate tubes. Following the reverse transcription reaction, the cDNA is transferred to a separate tube for the PCR amplification. Either oligo (dT) or random hexamer primers can be used for the reverse transcription reaction. Total RNA or polyadenylated RNA can be used for the reverse transcriptase reaction. This allows for the ability to convert all the messages in an RNA sample into cDNA. The two-step RT-PCR is useful for analyzing multiple mRNAs from a single sample.

Two types of PCR methods can be utilized for analyzing amplified products. The end-point PCR method measures the amount of amplified product synthesized during PCR at the end of the PCR amplification. The detection of the amplified product indicates the presence of the mRNA of interest. With the real-time PCR method, the amplified product is quantified during the exponential phase of PCR. The hot-start PCR strategy is typically used to increase sensitivity, specificity, and yield. Usually, the hot-start strategy is carried out at a high temperature, for example 94°C, prior to PCR cycling. Under this condition, *Taq* DNA polymerase is activated. Additionally, the RNA-cDNA hybrid is denatured, and reverse transcriptase is inactivated at 94°C. Multiplex RT-PCR assays have been developed that can detect multiple bodily fluid mRNAs from single or mixed stains (Chapter 11). The commonly used RT-PCR for forensic identification of bodily fluids is the two-step RT-PCR. The end-point PCR is usually utilized with the forward primers labeled with florescent dyes at the 5′ end. The amplified products are separated and detected in a standard capillary electrophoresis instrument.

Bibliography

Aboud, M., H.H. Oh, and B. McCord, Rapid direct PCR for forensic genotyping in under 25 min. *Electrophoresis*, 2013, **34**(11): 1539–1547.

Afonina, I.A., et al., Minor groove binder-conjugated DNA probes for quantitative DNA detection by hybridization-triggered fluorescence. *Biotechniques*, 2002, **32**(4): 940–944, 946–949.

Alaeddini, R., Forensic implications of PCR inhibition—A review. *Forensic Sci Int Genet*, 2012, **6**(3): 297–305.

Ballantyne, K.N., R.A. van Oorschot, and R.J. Mitchell, Comparison of two whole genome amplification methods for STR genotyping of LCN and degraded DNA samples. *Forensic Sci Int*, 2007, **166**(1): 35–41.

Ballantyne, K.N., R.A. van Oorschot, and R.J. Mitchell, Increased amplification success from forensic samples with locked nucleic acids. *Forensic Sci Int Genet*, 2011, **5**(4): 276–280.

Ballantyne, K.N., R.A. van Oorschot, and R.J. Mitchell, Locked nucleic acids in PCR primers increase sensitivity and performance. *Genomics*, 2008, **91**(3): 301–305.

Ballantyne, K.N., et al., Decreasing amplification bias associated with multiple displacement amplification and short tandem repeat genotyping. *Anal Biochem*, 2007, **368**(2): 222–229.

Bassam, B.J., et al., Nucleic acid sequence detection systems: Revolutionary automation for monitoring and reporting PCR products. *Aust Biotechnol*, 1996, **6**: 285–294.

Bekaert, B., et al., Automating a combined composite-consensus method to generate DNA profiles from low and high template mixture samples. *Forensic Sci Int Genet*, 2012, **6**(5): 588–593.

Birch, D.E., et al., Simplified hot start PCR. *Nature*, 1996, **318**: 2.

Bloch, W., A biochemical perspective of the polymerase chain reaction. *Biochemistry*, 1991, **30**(11): 2735–2747.

Bright, J.A., et al., The effect of cleaning agents on the ability to obtain DNA profiles using the Identifiler and PowerPlex® Y multiplex kits. *J Forensic Sci*, 2011, **56**(1): 181–185.

Brotherton, P., et al., Preferential access to genetic information from endogenous hominin ancient DNA and accurate quantitative SNP-typing via SPEX. *Nucleic Acids Res*, 2010, **38**(2): e7.

Budowle, B., A.J. Eisenberg, and A. van Daal, Validity of low copy number typing and applications to forensic science. *Croat Med J*, 2009, **50**(3): 207–217.

Chung, D.T., et al., A study on the effects of degradation and template concentration on the amplification efficiency of the STR Miniplex primer sets. *J Forensic Sci*, 2004, **49**(4): 733–740.

Date-Chong, M., et al., An examination of the utility of a nuclear DNA/mitochondrial DNA duplex qPCR assay to assess surface decontamination of hair. *Forensic Sci Int Genet*, 2013, **7**(3): 392–396.

Davis, C.P., et al., Multiplex short tandem repeat amplification of low template DNA samples with the addition of proofreading enzymes. *J Forensic Sci*, 2011, **56**(3): 726–732.

Deguilloux, M.F., et al., Analysis of ancient human DNA and primer contamination: One step backward one step forward. *Forensic Sci Int*, 2011, **210**(1–3): 102–109.

Dieffenbach, C.W., T.M. Lowe, and G.S. Dveksler, General concepts for PCR primer design. *PCR Methods Appl*, 1993, **3**(3): S30–S37.

Eckert, K.A. and T.A. Kunkel, DNA polymerase fidelity and the polymerase chain reaction. *PCR Methods Appl*, 1991, **1**(1): 17–24.

Eckert, K.A. and T.A. Kunkel, High fidelity DNA synthesis by the *Thermus aquaticus* DNA polymerase. *Nucleic Acids Res*, 1990, **18**(13): 3739–3744.

Edwards, M.C. and R.A. Gibbs, Multiplex PCR: Advantages, development, and applications. *PCR Methods Appl*, 1994, **3**(4): S65–S75.

Eilert, K.D. and D.R. Foran, Polymerase resistance to polymerase chain reaction inhibitors in bone*. *J Forensic Sci*, 2009, **54**(5): 1001–1007.

Endicott, P., et al., Genotyping human ancient mtDNA control and coding region polymorphisms with a multiplexed Single-Base-Extension assay: The singular maternal history of the Tyrolean Iceman. *BMC Genet*, 2009, **10**: 29.

Erlich, H.A., D. Gelfand, and J.J. Sninsky, Recent advances in the polymerase chain reaction. *Science*, 1991, **252**(5013): 1643–1651.

Estes, M.D., et al., Optimization of multiplexed PCR on an integrated microfluidic forensic platform for rapid DNA analysis. *Analyst*, 2012, **137**(23): 5510–5519.

Forster, L., J. Thomson, and S. Kutranov, Direct comparison of post-28-cycle PCR purification and modified capillary electrophoresis methods with the 34-cycle "low copy number" (LCN) method for analysis of trace forensic DNA samples. *Forensic Sci Int Genet*, 2008, **2**(4): 318–328.

Funes-Huacca, M.E., et al., A comparison of the effects of PCR inhibition in quantitative PCR and forensic STR analysis. *Electrophoresis*, 2011, **32**(9): 1084–1089.

Gasparini, P., et al., Amplification of DNA from epithelial cells in urine. *N Engl J Med*, 1989, **320**(12): 809.

Gavazaj, F.Q., et al., Optimization of DNA concentration to amplify short tandem repeats of human genomic DNA. *Bosn J Basic Med Sci*, 2012, **12**(4): 236–239.

Gefrides, L.A., et al., UV irradiation and autoclave treatment for elimination of contaminating DNA from laboratory consumables. *Forensic Sci Int Genet*, 2010, **4**(2): 89–94.

Ghasemi, A., et al., Design of a biological method for rapid detection of presence of PCR inhibitors in aged bone DNA. *Clin Lab*, 2012, **58**(7–8): 681–686.

Giardina, E., et al., Whole genome amplification and real-time PCR in forensic casework. *BMC Genom*, 2009, **10**: 159.

Giese, H., et al., Fast multiplexed polymerase chain reaction for conventional and microfluidic short tandem repeat analysis. *J Forensic Sci*, 2009, **54**(6): 1287–1296.

Gill, P., et al., An investigation of the rigor of interpretation rules for STRs derived from less than 100 pg of DNA. *Forensic Sci Int*, 2000, **112**(1): 17–40.

Grgicak, C.M., Z.M. Urban, and R.W. Cotton, Investigation of reproducibility and error associated with qPCR methods using Quantifiler® Duo DNA quantification kit. *J Forensic Sci*, 2010, **55**(5): 1331–1339.

Gyllensten, U.B. and H.A. Erlich, Generation of single-stranded DNA by the polymerase chain reaction and its application to direct sequencing of the HLA-DQA locus. *Proc Natl Acad Sci U S A*, 1988, **85**(20): 7652–7656.

Haas, C., E. Hanson, and J. Ballantyne, Capillary electrophoresis of a multiplex reverse transcription-polymerase chain reaction to target messenger RNA markers for body fluid identification. *Methods Mol Biol*, 2012, **830**: 169–183.

Haff, L., et al., A high-performance system for automation of the polymerase chain reaction. *Biotechniques*, 1991, **10**(1): 102–103, 106–112.

Hedman, J., et al., Improved forensic DNA analysis through the use of alternative DNA polymerases and statistical modeling of DNA profiles. *Biotechniques*, 2009, **47**(5): 951–958.

Henegariu, O., et al., Multiplex PCR: Critical parameters and step-by-step protocol. *Biotechniques*, 1997, **23**(3): 504–511.

Herrmann, D., et al., Microarray-based STR genotyping using RecA-mediated ligation. *Nucleic Acids Res*, 2010, **38**(17): e172.

Higuchi, R., et al., Kinetic PCR analysis: Real-time monitoring of DNA amplification reactions. *Biotechnology (N Y)*, 1993, **11**(9): 1026–1030.

Higuchi, R., et al., Simultaneous amplification and detection of specific DNA sequences. *Biotechnology (N Y)*, 1992, **10**(4): 413–417.

Hochmeister, M.N., et al., Confirmation of the identity of human skeletal remains using multiplex PCR amplification and typing kits. *J Forensic Sci*, 1995, **40**(4): 701–705.

Holland, P.M., et al., Detection of specific polymerase chain reaction product by utilizing the 5′–3′ exonuclease activity of *Thermus aquaticus* DNA polymerase. *Proc Natl Acad Sci U S A*, 1991, **88**(16): 7276–7280.

Hughes-Stamm, S.R., K.J. Ashton, and A. van Daal, Assessment of DNA degradation and the genotyping success of highly degraded samples. *Int J Legal Med*, 2011, **125**(3): 341–348.

Irwin, J.A., et al., Characterization of a modified amplification approach for improved STR recovery from severely degraded skeletal elements. *Forensic Sci Int Genet*, 2012, **6**(5): 578–587.

Isacsson, J., et al., Rapid and specific detection of PCR products using light-up probes. *Mol Cell Probes*, 2000, **14**(5): 321–328.

Ju, J., et al., Fluorescence energy transfer dye-labeled primers for DNA sequencing and analysis. *Proc Natl Acad Sci U S A*, 1995, **92**(10): 4347–4351.

Kaminiwa, J., et al., Vanadium accelerates polymerase chain reaction and expands the applicability of forensic DNA testing. *J Forensic Leg Med*, 2013, **20**(4): 326–333.

Kavlick, M.F., et al., Quantification of human mitochondrial DNA using synthesized DNA standards. *J Forensic Sci*, 2011, **56**(6): 1457–1463.

Kermekchiev, M.B., et al., Mutants of *Taq* DNA polymerase resistant to PCR inhibitors allow DNA amplification from whole blood and crude soil samples. *Nucleic Acids Res*, 2009, **37**(5): e40.

Kimpton, C., et al., Evaluation of an automated DNA profiling system employing multiplex amplification of four tetrameric STR loci. *Int J Legal Med*, 1994, **106**(6): 302–311.

Kitayama, T., et al., Estimation of the detection rate in STR analysis by determining the DNA degradation ratio using quantitative PCR. *Leg Med (Tokyo)*, 2013, **15**(1): 1–6.

Koon, H.E., et al., Diagnosing post-mortem treatments which inhibit DNA amplification from US MIAs buried at the Punchbowl. *Forensic Sci Int*, 2008, **178**(2–3): 171–177.

Koppelkamm, A., et al., Validation of adequate endogenous reference genes for the normalization of qPCR gene expression data in human post mortem tissue. *Int J Legal Med*, 2010, **124**(5): 371–380.

Krawczak, M., et al., Polymerase chain reaction: Replication errors and reliability of gene diagnosis. *Nucleic Acids Res*, 1989, **17**(6): 2197–2201.

Kunkel, T.A., DNA replication fidelity. *J Biol Chem*, 1992, **267**(26): 18251–18254.

Kutyavin, I.V., et al., 3′-minor groove binder-DNA probes increase sequence specificity at PCR extension temperatures. *Nucleic Acids Res*, 2000, **28**(2): 655–661.

Kwok, S. and R. Higuchi, Avoiding false positives with PCR. *Nature*, 1989, **339**(6221): 237–238.

Laurin, N. and C. Fregeau, Optimization and validation of a fast amplification protocol for AmpFlSTR® Profiler Plus® for rapid forensic human identification. *Forensic Sci Int Genet*, 2012, **6**(1): 47–57.

Lee, J.C., et al., Evaluating the performance of whole genome amplification for use in low template DNA typing. *Med Sci Law*, 2012, **52**(4): 223–228.

Maciejewska, A., J. Jakubowska, and R. Pawlowski, Whole genome amplification of degraded and nondegraded DNA for forensic purposes. *Int J Legal Med*, 2013, **127**(2): 309–319.

Markoulatos, P., N. Siafakas, and M. Moncany, Multiplex polymerase chain reaction: A practical approach. *J Clin Lab Anal*, 2002, **16**(1): 47–51.

Marshall, P.L., et al., Pressure cycling technology (PCT) reduces effects of inhibitors of the PCR. *Int J Legal Med*, 2013, **127**(2): 321–333.

McCloskey, M.L., et al., Encoding PCR products with batch-stamps and barcodes. *Biochem Genet*, 2007, **45**(11–12): 761–767.

Mitsuhashi, M., Technical report: Part 2. Basic requirements for designing optimal PCR primers. *J Clin Lab Anal*, 1996, **10**(5): 285–293.

Moretti, T., B. Koons, and B. Budowle, Enhancement of PCR amplification yield and specificity using AmpliTaq Gold DNA polymerase. *Biotechniques*, 1998, **25**(4): 716–722.

Morling, N., PCR in forensic genetics. *Biochem Soc Trans*, 2009, **37**(Pt 2): 438–440.

Myers, B.A., J.L. King, and B. Budowle, Evaluation and comparative analysis of direct amplification of STRs using PowerPlex® 18D and Identifiler® Direct systems. *Forensic Sci Int Genet*, 2012, **6**(5): 640–645.

Nazarenko, I.A., S.K. Bhatnagar, and R.J. Hohman, A closed tube format for amplification and detection of DNA based on energy transfer. *Nucleic Acids Res*, 1997, **25**(12): 2516–2521.

Nguyen, Q., et al., STR melting curve analysis as a genetic screening tool for crime scene samples. *J Forensic Sci*, 2012, **57**(4): 887–899.

Nicklas, J.A., T. Noreault-Conti, and E. Buel, Development of a fast, simple profiling method for sample screening using high resolution melting (HRM) of STRs. *J Forensic Sci*, 2012, **57**(2): 478–488.

Nogami, H., et al., Rapid and simple sex determination method from dental pulp by loop-mediated isothermal amplification. *Forensic Sci Int Genet*, 2008, **2**(4): 349–353.

Opel, K.L., D. Chung, and B.R. McCord, A study of PCR inhibition mechanisms using real time PCR. *J Forensic Sci*, 2010, **55**(1): 25–33.

Park, S.J., et al., Direct STR amplification from whole blood and blood- or saliva-spotted FTA without DNA purification. *J Forensic Sci*, 2008, **53**(2): 335–341.

Reynolds, R., G. Sensabaugh, and E. Blake, Analysis of genetic markers in forensic DNA samples using the polymerase chain reaction. *Anal Chem*, 1991, **63**(1): 2–15.

Roeder, A.D., et al., Maximizing DNA profiling success from sub-optimal quantities of DNA: A staged approach. *Forensic Sci Int Genet*, 2009, **3**(2): 128–137.

Rothe, J., N.E. Watkins Jr., and M. Nagy, New prediction model for probe specificity in an allele-specific extension reaction for haplotype-specific extraction (HSE) of Y chromosome mixtures. *PLoS One*, 2012, **7**(9): e45955.

Saiki, R.K., et al., Enzymatic amplification of beta-globin genomic sequences and restriction site analysis for diagnosis of sickle cell anemia. *Science*, 1985, **230**(4732): 1350–1354.

Saiki, R.K., et al., Primer-directed enzymatic amplification of DNA with a thermostable DNA polymerase. *Science*, 1988, **239**(4839): 487–491.

Sanchez, J.J., et al., Multiplex PCR and minisequencing of SNPs—A model with 35 Y chromosome SNPs. *Forensic Sci Int*, 2003, **137**(1): 74–84.

Schoske, R., et al., Multiplex PCR design strategy used for the simultaneous amplification of 10 Y chromosome short tandem repeat (STR) loci. *Anal Bioanal Chem*, 2003, **375**(3): 333–343.

Seo, S.B., et al., Effects of humic acid on DNA quantification with Quantifiler® Human DNA Quantification kit and short tandem repeat amplification efficiency. *Int J Legal Med*, 2012, **126**(6): 961–968.

Shuber, A.P., V.J. Grondin, and K.W. Klinger, A simplified procedure for developing multiplex PCRs. *Genome Res*, 1995, **5**(5): 488–493.

Smith, P.J. and J. Ballantyne, Simplified low-copy-number DNA analysis by post-PCR purification. *J Forensic Sci*, 2007, **52**(4): 820–829.

Solinas, A., N. Thelwell, and T. Brown, Intramolecular TaqMan probes for genetic analysis. *Chem Commun (Camb)*, 2002, (19): 2272–2273.

Steadman, S.A., et al., Recovery and STR amplification of DNA from RFLP membranes. *J Forensic Sci*, 2008, **53**(2): 349–358.

Sutlovic, D., et al., Interaction of humic acids with human DNA: Proposed mechanisms and kinetics. *Electrophoresis*, 2008, **29**(7): 1467–1472.

Swaran, Y.C. and L. Welch, A comparison between direct PCR and extraction to generate DNA profiles from samples retrieved from various substrates. *Forensic Sci Int Genet*, 2012, **6**(3): 407–412.

Tindall, K.R. and T.A. Kunkel, Fidelity of DNA synthesis by the *Thermus aquaticus* DNA polymerase. *Biochemistry*, 1988, **27**(16): 6008–6013.

Tokutomi, T., et al., Real-time PCR method for identification of Asian populations in forensic casework. *Leg Med (Tokyo)*, 2009, **11**(Suppl 1): S106–S108.

Turrina, S., et al., STR typing of archival Bouin's fluid-fixed paraffin-embedded tissue using new sensitive redesigned primers for three STR loci (CSF1PO, D8S1179 and D13S317). *J Forensic Leg Med*, 2008, **15**(1): 27–31.

Verheij, S., J. Harteveld, and T. Sijen, A protocol for direct and rapid multiplex PCR amplification on forensically relevant samples. *Forensic Sci Int Genet*, 2012, **6**(2): 167–175.

Wallin, J.M., et al., Constructing universal multiplex PCR systems for comparative genotyping. *J Forensic Sci*, 2002, **47**(1): 52–65.

Walsh, P.S., H.A. Erlich, and R. Higuchi, Preferential PCR amplification of alleles: Mechanisms and solutions. *PCR Methods Appl*, 1992, **1**(4): 241–250.

Wang, J. and B. McCord, The application of magnetic bead hybridization for the recovery and STR amplification of degraded and inhibited forensic DNA. *Electrophoresis*, 2011, **32**(13): 1631–1638.

Weiler, N.E., A.S. Matai, and T. Sijen, Extended PCR conditions to reduce drop-out frequencies in low template STR typing including unequal mixtures. *Forensic Sci Int Genet*, 2012, **6**(1): 102–107.

Westwood, S.A. and D.J. Werrett, An evaluation of the polymerase chain reaction method for forensic applications. *Forensic Sci Int*, 1990, **45**(3): 201–215.

Weusten, J. and J. Herbergs, A stochastic model of the processes in PCR based amplification of STR DNA in forensic applications. *Forensic Sci Int Genet*, 2012, **6**(1): 17–25.

Whitcombe, D., et al., A homogeneous fluorescence assay for PCR amplicons: Its application to real-time, single-tube genotyping. *Clin Chem*, 1998, **44**(5): 918–923.

Wong, H.Y., E.S. Lim, and W.F. Tan-Siew, Amplification volume reduction on DNA database samples using FTA Classic Cards. *Forensic Sci Int Genet*, 2012, **6**(2): 176–179.

Yu, P.H. and M.M. Wallace, Effect of 1,2-indanedione on PCR-STR typing of fingerprints deposited on thermal and carbonless paper. *Forensic Sci Int*, 2007, **168**(2–3): 112–118.

8

DNA Electrophoresis

It is necessary to separate various sizes of DNA fragments so that the DNA fragment in question can be identified and analyzed. This can be achieved by electrophoresis, a process in which fragments are separated based on the migration of charged macromolecules in an electric field.

8.1 Basic Principles

DNA is a negatively charged molecule in an aqueous environment, with the phosphate groups of DNA nucleotides carrying negative charges. DNA molecules migrate from the negative electrode (cathode) toward the positive electrode (anode) in an electric field during electrophoresis. The *electrical potential*, a measure of the work required to move a charged molecule in an electric field, is the force responsible for moving the charged macromolecules during electrophoresis.

The *electrophoretic mobility* of macromolecules is primarily determined by their *charge-to-mass ratios* and their shapes. However, because the phosphate group of every DNA nucleotide carries a negative charge, the charge-to-mass ratio of DNA molecules is almost the same even if the length of the DNA fragment varies. Additionally, forensic DNA testing usually analyzes linear DNA molecules such as double-stranded linear DNA for variable number tandem repeat (VNTR) analysis (Chapter 19) and single-stranded DNA for short tandem repeat (STR) analysis (Chapter 20). The shapes of linear DNA molecules are similar. Therefore, the electrophoretic separation of different-sized DNA fragments, through a series of pores of the supporting matrix in which the fragments travel, is based more on their sizes than their shapes.

It is obviously easier for smaller molecules to migrate through the pores of a supporting matrix than larger molecules; this is why smaller molecules migrate faster through the matrix. Hence, the electrophoretic mobility increases as the size of the DNA molecule decreases. Conversely, larger DNA molecules migrate much more slowly because they experience more friction and collisions as they travel through the net of pores in a matrix. Therefore, the separation of DNA molecules with different sizes can be accomplished. Figure 8.1 depicts the models for DNA electrophoresis.

8.2 Supporting Matrices

Although the actual separation occurs in an aqueous phase, most variants of electrophoresis use a physical support material also called a *matrix*. As discussed in Section 8.1, the matrix can be used as a molecular sieve for the separation of DNA molecules, and it also reduces diffusion and convection during electrophoresis. Agarose and polyacrylamide are commonly used electrophoresis matrices because of their good reproducibility, reliability, and versatility.

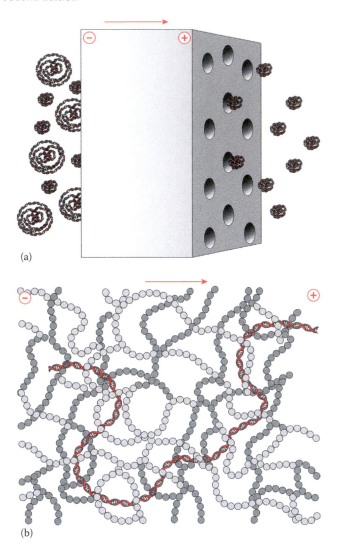

(a)

(b)

Figure 8.1 Schematic illustrations showing some of the nucleic acid separation mechanisms. (a) Ogston sieving model for the separation of DNA fragments through electrophoresis. Based on this model, the matrix used for gel electrophoresis consists of randomly distributed pores. DNA molecules are considered to behave as globular objects. During the migration, smaller DNA molecules migrate faster than their larger counterparts through the matrix. However, this model does not apply to DNA molecules with a radius much larger than the pore size of the matrix. (b) A reptation model presents an alternative mechanism of gel electrophoretic separation of DNA molecules. Based on this model, DNA molecules are flexible and can migrate through the pores of the gel matrix through a reptation process that is driven by an applied electric field. This model can also apply to capillary electrophoresis separation of DNA molecules in entangled polymer solutions. +, anode; −, cathode; arrow, direction of migration. (© Richard C. Li.)

8.2.1 Agarose

Agarose is a linear polymer composed of alternating residues of D- and L-galactose (Figure 8.2). A gelatinized agarose forms a three-dimensional sieve with pores from 50 to 200 nm in diameter (Figure 8.3). DNA fragments ranging from 50 to 20,000 base pairs in size are best resolved in agarose gels.

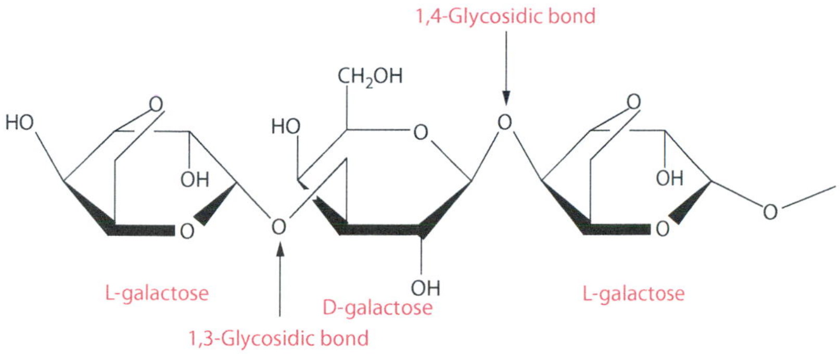

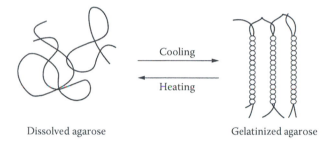

1,4-Glycosidic bond

CH₂OH

L-galactose   D-galactose   L-galactose

1,3-Glycosidic bond

Figure 8.2   Chemical components of agarose, a linear polysaccharide composed of alternating units of D- and L-galactose. Often, the L-galactose residue has an anhydro bridge between the 3 and 6 positions and is called 3,6-anhydro-L-galactose.

Cooling

Heating

Dissolved agarose   Gelatinized agarose

Figure 8.3   Agarose chains. Gelatinized agarose consists of a three-dimensional network in which chains of agarose form helical fibers that aggregate into supercoiled structures. (© Richard C. Li.)

The electrophoretic mobility of double-stranded DNA through agarose gel matrices is inversely proportional to the $\log_{10}$ of the size of DNA fragments ranging from 50 to 20,000 base pairs. The electrophoretic mobility of a linear DNA fragment is also inversely proportional to the concentration of agarose in the gel. A low concentration of agarose in a gel forms larger pores, allowing the separation of larger DNA fragments.

8.2.2 Polyacrylamide

A polyacrylamide gel matrix is very effective for the separation of smaller fragments of DNA (5–500 base pairs). Additionally, a single nucleotide difference in the length of a DNA fragment can be resolved with this type of gel. Polyacrylamide produces much smaller pore sizes than agarose gels and thus has a much higher resolving power than agarose gels for low-molecular-weight DNA molecules. A polyacrylamide gel matrix is formed by polymerization and cross-linking reactions.

8.2.2.1 Polymerization Reaction

Long linear chains of polyacrylamide are polymerized from acrylamide monomers (Figure 8.4). This polymerization reaction is initiated in the presence of free radicals that are generated from the reduction of ammonium persulfate (APS) by *N,N,N′,N′*-tetramethylethylene diamine (TEMED). See Figure 8.5.

8.2.2.2 Cross-Linking Reaction

As shown in Figure 8.5, three-dimensionally cross-linked polyacrylamide chains can be formed with the use of cross-linking agents, such as *N,N′*-methylenebisacrylamide (BIS; Figure 8.4).

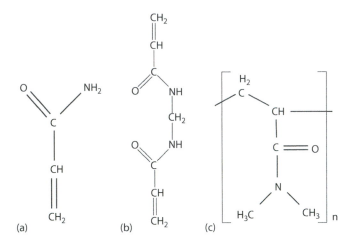

Figure 8.4 Chemical structures of (a) acrylamide, (b) N,N'-methylenebisacrylamide (BIS), and (c) polydimethylacrylamide (a linear polymer for capillary electrophoresis).

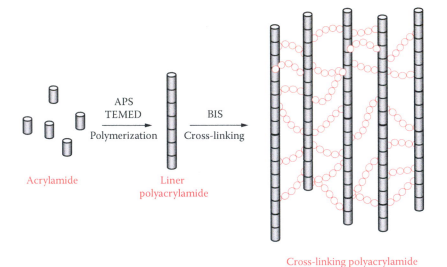

Figure 8.5 Formation of polyacrylamide gel. Polymerization is followed by cross-linking in which the pore size of the gel is determined by its degree of polymerization and cross-linking. (© Richard C. Li.)

The porosity of the resulting gel is determined by the lengths of these chains and the degree of cross-linking between them. Therefore, the sizes of the pores that are formed can be adjusted by altering the concentrations of the acrylamide monomer and the cross-linking reagent.

Polyacrylamide can also be used in a capillary electrophoresis matrix. It is very difficult to insert a cross-linked polyacrylamide matrix into a capillary. Therefore, a solution of a linear polymer (non-cross-linked) of polydimethylacrylamide (Figure 8.4) is used as a matrix for capillary electrophoresis. For instance, POP-4 (4% un-cross-linked polydimethylacrylamide polymer) is used for fragment analysis, such as forensic STR analysis (Chapter 18), in which fluorescently labeled DNA amplicons are separated using electrophoresis and are identified based on their sizes by comparing them with a size standard. POP-6 (6% un-cross-linked polydimethylacrylamide polymer) is used for sequencing, such as mitochondrial DNA (mtDNA) sequencing (Chapter 21). These un-cross-linked polymers are commercially available (Applied Biosystems).

8.2.2.3 Denaturing Polyacrylamide Electrophoresis

Polyacrylamide-based electrophoresis can be carried out with both double- and single-stranded DNA. Electrophoresis that is performed under the conditions of single-stranded DNA is called *denaturing electrophoresis*. Denatured DNA migrates through the gel as linear molecules at a rate independent of the base composition and sequence. With short single-stranded fragments, molecules differing in size by a single nucleotide can be separated. This extraordinary sensitivity to size is extremely useful for the separation of small DNA fragments that are used for DNA sequencing and STR analysis in forensic applications. The denaturing condition can be best achieved by adding chemicals such as urea or formamide to the matrix or by increasing the temperature or pH during electrophoresis.

8.3 Apparatus and Forensic Applications

Agarose electrophoresis is carried out in a slab gel. Agarose gel can be prepared in a variety of shapes and sizes and can be run in different configurations. The most common gel for forensic use is the horizontal slab. Polyacrylamide can be used either in a slab gel or in a capillary electrophoretic apparatus.

8.3.1 Slab Gel Electrophoresis

8.3.1.1 Agarose Gel Electrophoresis

One forensic application of agarose gel electrophoresis is restriction fragment length polymorphism (RFLP) analysis of VNTR loci (see Chapter 19). An agarose gel is used to separate the DNA fragments by size ranging from 500 to 20,000 base pairs of commonly used VNTR loci for forensic testing. This type of electrophoresis is done under nondenatured conditions (double-stranded DNA).

Figure 8.6 shows an example of the apparatus used for gel electrophoresis of DNA. A slab (3–4 mm) of agarose gel is prepared by allowing liquid agarose to gelatinize in a cast. The gel is submerged in a buffer tank filled with an electrophoretic buffer with proper ionic strength, which is necessary to achieve efficient electrical conductance. The gel contains small wells for loading samples. The DNA samples are mixed with a gel loading buffer prior to loading them into the gel. The gel loading buffer contains dyes that add color to the sample to facilitate the process of loading. Dyes such as bromophenol blue and xylene cyanol FF migrate toward the anode during electrophoresis. In most forensic applications, these dyes migrate faster than DNA fragments. The dyes are visible and can thus be used to track the progression of electrophoresis. The electrophoresis can be stopped as the dye front reaches the bottom of the gel (anode side).

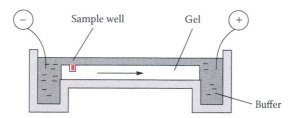

Figure 8.6 Horizontal agarose slab gel apparatus. The gel can be prepared by heating an agarose suspension to dissolve it. The agarose solution can then be poured into a gel cast and allowed to cool until gelatinized. An electrophoresis buffer is poured over the gel to submerge it. The samples, usually mixed with loading buffer containing dye to allow visualization, are loaded into the wells of the submerged gel using a pipet. The electrodes are indicated. (© Richard C. Li.)

The samples are loaded into the sample wells situated proximal to the cathode. An electric field is applied and the negatively charged DNA molecules migrate through the agarose toward the anode. Slab gel electrophoresis is capable of running multiple samples simultaneously. The sample is made visible by using a fluorescent intercalating reagent that can make DNA fluoresce under ultraviolet (UV) light (Chapter 9), either using the fluorescent intercalating reagent incorporated into the gel or staining the gel after electrophoresis. The separated DNA samples appear as bands and the sizes of the DNA fragments can be estimated by comparing them with the sizes of standards run concurrently.

8.3.1.2 Polyacrylamide Gel Electrophoresis

The forensic application of this apparatus is the separation of STR fragments and DNA sequencing reaction products of mtDNA. The sizes of DNA fragments that can be separated range from 100 to 500 base pairs—much smaller than what can be separated efficiently with agarose gels. Single-nucleotide resolution to distinguish similarly sized fragments can be achieved with this technique under denatured conditions with only single-stranded DNA. Polyacrylamide thin slab gels (0.75–1.5 mm) are usually used. Samples are loaded into wells flanked by two pieces of glass. Therefore, polyacrylamide gels are usually run in a vertical configuration (Figure 8.7). The detection of DNA bands in polyacrylamide gels will be described in Chapter 9. The sizes of DNA fragments can be calculated by including an internal size standard. As with agarose gel, polyacrylamide gel electrophoresis is capable of running multiple samples simultaneously and high throughput (a measurement of the rate that a sample is processed by a given analysis) can be achieved. However, caution should be taken to ensure that cross-contamination does not occur from the spilling of samples from adjacent wells. Additionally, polyacrylamide gels are more difficult to prepare than agarose gels.

8.3.2 Capillary Electrophoresis

Capillary electrophoresis (Figures 8.8 through 8.11) is a newer method than the slab gel method, and can be utilized to separate charged macromolecules, such as DNA, RNA, polysaccharides, and proteins.

One essential component of the capillary electrophoresis instrument is the capillary, a thin hollow tube made of fused silica (which typically has a diameter between 50 and 100 μm and a length between 10 and 50 cm). The capillary contains a translucent detection window for the instrument to detect signals from the labeled DNA fragments during electrophoresis.

Linear polydimethylacrylamide is used as the matrix. Capillary electrophoresis is conducted under denatured conditions for forensic applications such as STR analysis and mtDNA sequencing analysis. The denatured condition is achieved by including urea in the electrophoresis matrix and formamide during sample preparation. The injection of samples into the capillary is performed by an autosampler using an *electrokinetic* mechanism (an injection based on the charge of molecules). Only small quantities of sample are required for each injection, and any remaining sample can be saved in case an analysis needs to be repeated.

During electrophoresis, the capillary is connected to buffer reservoirs that are connected to electrodes. The efficient heat dissipation property of thin capillaries allows the separation to be performed at higher voltages, as the electric field (typically 200–300 V/cm) is much higher than that of a slab gel platform. Thus, the separation in capillary electrophoresis is rapid. During electrophoresis, DNA fragments migrate through the capillaries toward the anode. Similar to gel electrophoresis, DNA fragments are separated according to their sizes, with the shorter fragments moving faster than the longer fragments. During capillary electrophoresis, linear polymers in the solution can act as obstacles to the migrating DNA fragments. Thus, electrophoretic separation models (Figure 8.1) developed for gels are applicable for capillary electrophoresis. Currently, capillary electrophoresis can separate DNA fragments of up to 1000 nucleotides with single-nucleotide resolution.

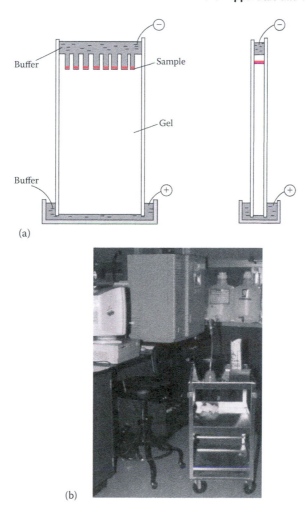

(a)

(b)

Figure 8.7   Vertical slab polyacrylamide gel. (a) Front and side view of vertical slab polyacrylamide gel apparatus. (b) Automated gel electrophoresis instrument, ABI PRISM 377® Genetic Analyzer, which was used in forensic laboratories but is now discontinued. (© Richard C. Li.)

Capillary electrophoresis instruments are equipped with detection systems utilizing laser excitation sources that excite the fluorescently labeled DNA fragments. They also include fluorescence detectors that record the signals emitted from the labeled DNA fragments (Chapter 9). In capillary electrophoresis, samples can only be analyzed sequentially so the throughput is much more limited than with slab gels. This disadvantage can be overcome by utilizing capillary array systems that can run up to 16 capillaries at one time (Figure 8.10).

8.3.3 Microfluidic Devices

Forensic DNA analysis includes the extraction of DNA from a sample, DNA quantification, polymerase chain reaction (PCR) amplification, capillary electrophoresis, data collection, and genotyping. Various microfluidic devices have been developed for forensic DNA analysis. Microfluidic devices control the movement of samples and reagents in a small, geometrically constrained environment and carry out biochemical reactions and analysis. These devices are made using microfabrication technology. This technology fabricates miniature structures historically used for integrated circuit fabrication in semiconductor manufacturing. The microfluidic

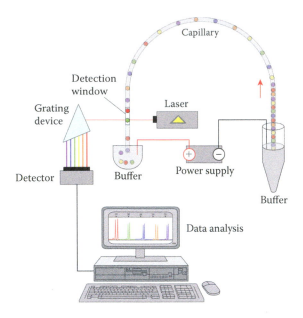

Figure 8.8 Essential components of a capillary electrophoresis system. The system includes a capillary, buffer reservoirs, two electrodes, a laser excitation source, a fluorescence detector, and an autosampler that holds the sample. Sample injection, electrophoresis, and data collection are automated and controlled by a computer. (© Richard C. Li.)

Figure 8.9 A single-capillary electrophoresis instrument, ABI PRISM 310® Genetic Analyzer. This was used in forensic laboratories but is now discontinued. (© Richard C. Li.)

devices can be classified as modular and integrated devices. Overall, the microfluidic device has many advantages compared with conventional techniques. The reaction volume required in microfluidic devices is usually in the nanoliter range, which decreases reagent and sample consumption. Additionally, due to the high surface area-to-volume ratio of the system, the efficiency of the system is greatly improved. Lastly, it can be automated, which makes the device a potential new platform for forensic DNA analysis.

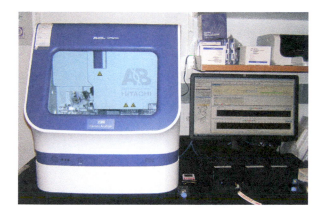

Figure 8.10 Photo of a multicapillary electrophoresis instrument. ABI PRISM 3500® Genetic Analyzer. (© Richard C. Li.)

Figure 8.11 Components of the ABI PRISM 3500® Genetic Analyzer. A capillary array (left), electrophoretic buffer chambers (middle), and an autosampler (right). (© Richard C. Li.)

8.3.3.1 Modular Microfluidic Devices

The modular devices are separate devices for DNA extraction, quantitation, amplification, electrophoresis, and so on. Each processing step is carried out on different devices. The modular design is flexible, allowing laboratories to choose the best device suited to the procedure for each step. For example, an electrophoresis microfluidic device has been developed (Figure 8.12). The electrophoretic assay is carried out on a modular microfluidic chip format using the same basic principle of capillary electrophoresis. The single-use chip contains wells for samples, a sieving polymer matrix for electrophoresis, and size standards. The wells are connected through microchannels. When the interconnected wells and channels are filled, the chip becomes an integrated electrical circuit. Each well is then connected to an electrode with an independent power supply. DNA molecules, carrying negative charges, are

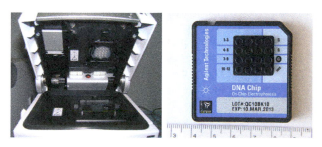

Figure 8.12 A modular instrument for gel electrophoresis. Left: Agilent 2100 Bioanalyzers contain 16-pin electrodes that fit into the wells of a chip. Right: The chip can accommodate sample wells, gel wells, and a well for a size standard. (© Richard C. Li.)

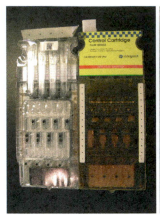

Figure 8.13 Integrated instruments for rapid DNA analysis. Cartridges developed by IntegeneX (left) and Zygem (right). (© Richard C. Li.)

electrophoretically separated based on their size. Smaller molecular weight fragments migrate faster than larger fragments. The polymer matrix contains fluorescent dye molecules that intercalate with DNA in a sample, which can be detected by laser-induced fluorescence. The duration of electrophoresis is approximately 30 min. The sensitivity is approximately 0.1 ng/μL; thus, sample consumption is small. The device provides the sizing and quantitation information of the nucleic acid fragment. In addition to DNA, the device can also be used for the analysis of RNA and protein. In the forensic laboratory, it has been used for the quantitation of amplified mtDNA product to monitor successful amplification and to normalize the quantities of DNA template for cycle sequencing.

8.3.3.2 Integrated Microfluidic Devices

It is possible to integrate several different modular devices into a single device (Figure 8.13). These devices are also known as "micro-total analysis" systems. An integrated system is a fully automated process. It also reduces the risk of contamination from laboratory sources. Full integration of all the constituent steps required for forensic STR DNA analysis has been achieved by a *rapid DNA* instrument. This instrument integrates various microfluidic devices including DNA extraction, PCR amplification, electrophoresis, detection, and genotyping steps into a single process on a cartridge. The cartridge, made from a variety of materials such as polycarbonate, incorporates pumps and valves controlling fluidic movement for the delivery of samples and prefilled reagents to the chambers for processing. DNA extraction usually utilizes silica-coated magnetic particles that can be collected by a magnet for DNA extraction. Microquantitative PCR utilizes the Taqman real-time quantitative PCR technique. During amplification, multiplex PCR is accomplished using commercial STR kits. Additionally, the cartridge also integrates a microcapillary electrophoresis chip for the separation of amplified products. The fluorescently labeled DNA fragments are detected based on laser-induced fluorescence. Data analysis and genotyping are then carried out to produce a DNA profile, which is compatible with the CODIS DNA database (Chapter 24). Rapid DNA instruments are fully automated platforms that are designed to process a single-source sample such as a buccal swab to generate a DNA profile and database search in less than 2 h. The goal is for these instruments to be deployed in field tests at the scene or at the booking stations operated by trained law enforcement agents.

8.4 Estimation of DNA Size

8.4.1 Relative Mobility

The *relative mobility* (R_f) of a DNA molecule during electrophoresis can be calculated as the distance of band migration divided by the distance of tracking dye migration. The DNA

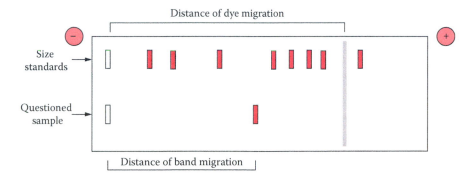

Figure 8.14 Measurement of the electromobility of DNA. Questioned DNA samples are analyzed concurrently with a set of standard DNA fragments of known size. The distances of DNA migration can thus be measured. The samples are usually mixed with a gel loading buffer prior to loading samples to the gel. The buffer contains dyes that add color to the sample to facilitate the process of loading. Dyes such as bromophenol blue and xylene cyanol FF migrate toward the positive electrode during electrophoresis. They can be used for tracking purposes as well. (© Richard C. Li.)

band migration is the distance from the sample origin to the center of the band. The tracking dye migration distance extends from the sample origin to the center of the dye band (Figure 8.14). To estimate the size of the DNA, standards containing DNA of known size and questioned samples are run on the same gel at the same time. The standards can be used to estimate the size of an unknown DNA molecule. A plot of $\log_{10}$ base pair of the standards versus R_f for a given gel can be constructed. A linear relationship, over a size range, between $\log_{10}$ base pair of the DNA molecule and the R_f can be observed. The R_f of the test sample is interpolated on the plot from which the size of an unknown DNA molecule can be determined (Figure 8.15).

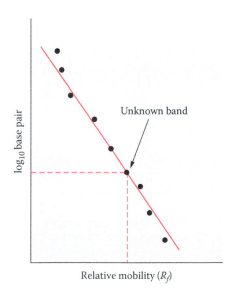

Figure 8.15 Estimation of the size of a DNA fragment. A plot of size standards versus relative migration during electrophoresis is shown. The size of the questioned DNA fragment can be estimated using this plot. (© Richard C. Li.)

8.4.2 Local Southern Method

The size of a DNA fragment is determined by an internal size standard that is a set of synthetic fragments with known molecular weight. The standard is labeled with a different colored dye so that it can be spectrally distinguished from DNA fragments of an unknown size (Figure 8.16). The sample including the internal size standard is then mixed in with DNA samples and is analyzed by electrophoresis. To determine the sizes of DNA fragments, a standard curve using the internal size standards must be established based on the reciprocal relationship between the electrophoretic mobility and the sizes of DNA fragments. However, this relationship is not exactly linear; instead, it appears to be sigmoidal (Figure 8.17). Therefore, the *Local Southern method*, described by Sir Edwin Southern, is used to generate standard curves for determining the sizes of DNA fragments. The equation is as follows:

$$L = \left[\frac{C}{M - M_0}\right] + L_0$$

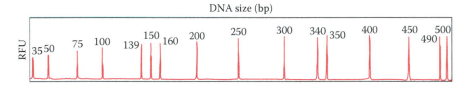

Figure 8.16   Electropherogram of a synthetic molecular weight size standard, GeneScan™ 500 size standard (Applied Biosystems). RFU, relative fluorescence units. (© Richard C. Li.)

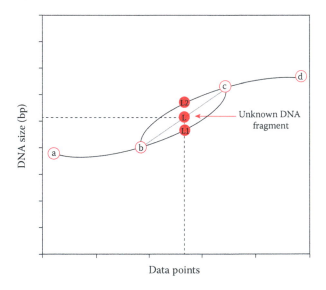

Figure 8.17   Estimation of unknown DNA fragment sizes using the Local Southern method. The migration of ssDNA fragments (20–1200 base pairs) is not linear over the entire fragment size range and is somewhat sigmoidal. To determine the size of a DNA fragment in question, two curves are generated. The first curve is generated using three standard points: two points (a and b) that are smaller and one point (c) that is larger than the questioned fragment. As a result, a fragment size (L_1) is determined. The second curve is then generated using an additional three standard points: one point (b) that is smaller and two points (c and d) that are larger than the questioned fragment. Thus, a fragment size, L_2, is determined. The two values (L_1 and L_2) are averaged to determine the size (L) of the questioned fragment. (Adapted from Southern, E.M., *Anal Biochem*, 100, 319–323, 1979.)

In the equation, L is the size of the unknown fragment; M is the mobility of the fragment; and C, M_0, and L_0 are constants that are obtained from the fragments of known size. To determine the size of an unknown fragment, two curves are generated based on the equation. Each curve utilizes three data points. The first curve utilizes two size standards that are smaller and one size standard that is greater than that of the unknown fragment. As a result, size L_1 is obtained. The second curve utilizes a size standard that is smaller and two size standards that are greater than that of the unknown fragment. Thus, size L_2 is obtained. Finally, the average of L_1 and L_2 is defined as the size of the unknown fragment. In this method, only the size standards that are in the proximity of the unknown fragment are analyzed. However, the accuracy of the standard curve relies on the proper separation of the standards during electrophoresis.

Bibliography

Aboud, M.J., M. Gassmann, and B.R. McCord, The development of mini pentameric STR loci for rapid analysis of forensic DNA samples on a microfluidic system. *Electrophoresis*, 2010, **31**(15): 2672–2679.

Biega, L.A. and B.W. Duceman, Substitution of H_2O for formamide in the sample preparation protocol for STR analysis using the capillary electrophoresis system: The effects on precision, resolution, and capillary life. *J Forensic Sci*, 1999, **44**: 3.

Bienvenue, J.M., et al., An integrated microfluidic device for DNA purification and PCR amplification of STR fragments. *Forensic Sci Int Genet*, 2010, **4**(3): 178–186.

Buel, E., M.B. Schwartz, and M.J. LaFountain, Capillary electrophoresis STR analysis: Comparison to gel-based systems. *J Forensic Sci*, 1998, **43**(1): 164–170.

Butler, J.M., C.M. Ruitberg, and P.M. Vallone, Capillary electrophoresis as a tool for optimization of multiplex PCR reactions. *Fresenius J Anal Chem*, 2001, **369**(3–4): 200–205.

Butler, J.M., et al., Rapid analysis of the short tandem repeat HUMTH01 by capillary electrophoresis. *Biotechniques*, 1994, **17**(6): 1062–1064, 1066, 1068, passim.

Butler, J.M., et al., Forensic DNA typing by capillary electrophoresis using the ABI Prism 310 and 3100 genetic analyzers for STR analysis. *Electrophoresis*, 2004, **25**(10–11): 1397–1412.

Chen, Y., et al., Sample stacking capillary electrophoretic microdevice for highly sensitive mini Y short tandem repeat genotyping. *Electrophoresis*, 2010, **31**(17): 2974–2980.

Choi, J.Y. and T.S. Seo, An integrated microdevice for high-performance short tandem repeat genotyping. *Biotechnol J*, 2009, **4**(11): 1530–1541.

Fernandez, C. and A. Alonso, Microchip capillary electrophoresis protocol to evaluate quality and quantity of mtDNA amplified fragments for DNA sequencing in forensic genetics. *Methods Mol Biol*, 2012, **830**: 367–379.

Frazier, R.R., et al., Validation of the applied biosystems Prism 377 automated sequencer for the forensic short tandem repeat analysis. *Electrophoresis*, 1996, **17**(10): 1550–1552.

French, D.J., et al., Interrogation of short tandem repeats using fluorescent probes and melting curve analysis: A step towards rapid DNA identity screening. *Forensic Sci Int Genet*, 2008, **2**(4): 333–339.

Gale, N., et al., Rapid typing of STRs in the human genome by HyBeacon melting. *Org Biomol Chem*, 2008, **6**(24): 4553–4559.

Gill, P., P. Koumi, and H. Allen, Sizing short tandem repeat alleles in capillary array gel electrophoresis instruments. *Electrophoresis*, 2001, **22**(13): 2670–2678.

Greenspoon, S.A., et al., A forensic laboratory tests the Berkeley microfabricated capillary array electrophoresis device. *J Forensic Sci*, 2008, **53**(4): 828–837.

Hagan, K.A., et al., A valveless microfluidic device for integrated solid phase extraction and polymerase chain reaction for short tandem repeat (STR) analysis. *Analyst*, 2011, **136**(9): 1928–1937.

Horsman, K.M., et al., Forensic DNA analysis on microfluidic devices: A review. *J Forensic Sci*, 2007, **52**(4): 784–799.

Hurth, C., et al., An automated instrument for human STR identification: Design, characterization, and experimental validation. *Electrophoresis*, 2010, **31**(21): 3510–3517.

Isenberg, A.R., et al., Analysis of two multiplexed short tandem repeat systems using capillary electrophoresis with multiwavelength fluorescence detection. *Electrophoresis*, 1998, **19**(1): 94–100.

Issaq, H.J., K.C. Chan, and G.M. Muschik, The effect of column length, applied voltage, gel type, and concentration on the capillary electrophoresis separation of DNA fragments and polymerase chain reaction products. *Electrophoresis*, 1997, **18**(7): 1153–1158.

Koumi, P., et al., Evaluation and validation of the ABI 3700, ABI 3100, and the MegaBACE 1000 capillary array electrophoresis instruments for use with short tandem repeat microsatellite typing in a forensic environment. *Electrophoresis*, 2004, **25**(14): 2227–2241.

Lazaruk, K., et al., Genotyping of forensic short tandem repeat (STR) systems based on sizing precision in a capillary electrophoresis instrument. *Electrophoresis*, 1998, **19**(1): 86–93.

Liu, P., et al., Integrated portable polymerase chain reaction-capillary electrophoresis microsystem for rapid forensic short tandem repeat typing. *Anal Chem*, 2007, **79**(5): 1881–1889.

Liu, P., et al., Real-time forensic DNA analysis at a crime scene using a portable microchip analyzer. *Forensic Sci Int Genet*, 2008, **2**(4): 301–309.

Liu, P. and R.A. Mathies, Integrated microfluidic systems for high-performance genetic analysis. *Trends Biotechnol*, 2009, **27**(10): 572–581.

Liu, P., et al., Integrated DNA purification, PCR, sample cleanup, and capillary electrophoresis microchip for forensic human identification. *Lab Chip*, 2011, **11**(6): 1041–1048.

Liu, P., et al., Integrated sample cleanup and capillary array electrophoresis microchip for forensic short tandem repeat analysis. *Forensic Sci Int Genet*, 2011, **5**(5): 484–492.

Liu, P., et al., Integrated sample cleanup and microchip capillary array electrophoresis for high-performance forensic STR profiling. *Methods Mol Biol*, 2012, **830**: 351–365.

Lounsbury, J.A., et al., From sample to PCR product in under 45 minutes: A polymeric integrated microdevice for clinical and forensic DNA analysis. *Lab Chip*, 2013, **13**(7): 1384–1393.

Madabhushi, R.S., Separation of 4-color DNA sequencing extension products in noncovalently coated capillaries using low viscosity polymer solutions. *Electrophoresis*, 1998, **19**(2): 224–230.

Mansfield, E.S., et al., Sensitivity, reproducibility, and accuracy in short tandem repeat genotyping using capillary array electrophoresis. *Genome Res*, 1996, **6**(9): 893–903.

Mansfield, E.S., et al., Analysis of multiplexed short tandem repeat (STR) systems using capillary array electrophoresis. *Electrophoresis*, 1998, **19**(1): 101–107.

McLaren, R.S., et al., Post-injection hybridization of complementary DNA strands on capillary electrophoresis platforms: A novel solution for dsDNA artifacts. *Forensic Sci Int Genet*, 2008, **2**(4): 257–273.

Montesino, M. and L. Prieto, Capillary electrophoresis of big-dye terminator sequencing reactions for human mtDNA control region haplotyping in the identification of human remains. *Methods Mol Biol*, 2012, **830**: 267–281.

Moretti, T.R., et al., Validation of STR typing by capillary electrophoresis. *J Forensic Sci*, 2001, **46**(3): 661–676.

Nai, Y.H., S.M. Powell, and M.C. Breadmore, Capillary electrophoretic system of ribonucleic acid molecules. *J Chromatogr A*, 2012, **1267**: 2–9.

Njoroge, S.K., et al., Integrated microfluidic systems for DNA analysis. *Top Curr Chem*, 2011, **304**: 203–260.

Pascali, J.P., F. Bortolotti, and F. Tagliaro, Recent advances in the application of CE to forensic sciences, an update over years 2009–2011. *Electrophoresis*, 2012, **33**(1): 117–126.

Pereira, J., et al., MtDNA typing of single-sperm cells isolated by micromanipulation. *Forensic Sci Int Genet*, 2012, **6**(2): 228–235.

Rosenblum, B.B., et al., Improved single-strand DNA sizing accuracy in capillary electrophoresis. *Nucleic Acids Res*, 1997, **25**(19): 3925–3929.

Roy, R., et al., Producing STR locus patterns from bloodstains and other forensic samples using an infrared fluorescent automated DNA sequencer. *J Forensic Sci*, 1996, **41**(3): 418–424.

Schneider, C., et al., Low copy number DNA profiling from isolated sperm using the aureka® micromanipulation system. *Forensic Sci Int Genet*, 2012, **6**(4): 461–465.

Sgueglia, J.B., S. Geiger, and J. Davis, Precision studies using the ABI prism 3100 genetic analyzer for forensic DNA analysis. *Anal Bioanal Chem*, 2003, **376**(8): 1247–1254.

Siles, B.A., et al., The use of a new gel matrix for the separation of DNA fragments: A comparison study between slab gel electrophoresis and capillary electrophoresis. *Appl Theor Electrophor*, 1996, **6**(1): 15–22.

Sinville, R. and S.A. Soper, High resolution DNA separations using microchip electrophoresis. *J Sep Sci*, 2007, **30**(11): 1714–1728.

Southern, E.M., Measurement of DNA length by gel electrophoresis. *Anal Biochem*, 1979, **100**, 319–323.

Tagliaro, F. and F. Bortolotti, Recent advances in the applications of CE to forensic sciences (2005–2007). *Electrophoresis*, 2008, **29**(1): 260–268.

Tereba, A., K.A. Micka, and J.W. Schumm, Reuse of denaturing polyacrylamide gels for short tandem repeat analysis. *Biotechniques*, 1998, **25**(5): 892–897.

Villablanca, A., et al., Suspension bead array branch migration displacement assay for rapid STR analysis. *Electrophoresis*, 2008, **29**(19): 4109–4114.

Wang, D.Y., et al., Identification and secondary structure analysis of a region affecting electrophoretic mobility of the STR locus SE33. *Forensic Sci Int Genet*, 2012, **6**(3): 310–316.

Wang, Y., et al., Rapid sizing of short tandem repeat alleles using capillary array electrophoresis and energy-transfer fluorescent primers. *Anal Chem*, 1995, **67**(7): 1197–1203.

Wenz, H., et al., High-precision genotyping by denaturing capillary electrophoresis. *Genome Res*, 1998, **8**(1): 69–80.

Westen, A.A., et al., Higher capillary electrophoresis injection settings as an efficient approach to increase the sensitivity of STR typing. *J Forensic Sci*, 2009, **54**(3): 591–598.

Yang, J., et al., An integratable microfluidic cartridge for forensic swab samples lysis. *Forensic Sci Int Genet*, 2014, **8**(1): 147–158.

Detection Methods

A variety of techniques are available for the detection of DNA fragments in forensic DNA analysis. For example, the direct detection of DNA fragments in a gel can be achieved via staining. The detection of DNA probes in a hybridization-based assay can be performed with radioisotopes, colorimetric assays, and chemiluminescence labeling. For polymerase chain reaction (PCR)-based assays, DNA primers can be labeled directly with fluorescent dyes.

9.1 Direct Detection of DNA in Gels

This section describes two simple and rapid detection methods used for detecting DNA: staining agarose gels with fluorescent intercalating dyes and staining denatured polyacrylamide gels with silver.

9.1.1 Fluorescent Intercalating Dye Staining

The location of DNA in an agarose gel can be determined directly by staining with low concentrations of fluorescent intercalating dyes (Chapter 6) such as ethidium bromide (Figures 9.1 through 9.3). These techniques allow the detection of DNA bands as small as 10 ng in agarose gels. Staining of DNA in agarose gels can be achieved by including an intercalating dye in the gel or staining the gel after electrophoresis in a dye-containing solution, followed by a washing step known as de-staining to reduce nonspecific staining. Historically, ethidium bromide-stained gels were photographed with a standard ultraviolet transilluminator at approximately 300 nm using a camera with an orange filter, whereas gels can currently be documented using digital techniques. When examining a DNA-containing gel under a UV lamp, the eyes and the skin should be protected from UV exposure. Ethidium bromide is a mutagen and a potential carcinogen. Ethidium bromide should be handled according to the Material Safety Data Sheet and safety protection wear should be used while handling the chemical. Ethidium bromide waste is usually disposed of as hazardous waste and is handled in accordance with laboratory guidelines. Alternatives to ethidium bromide are available, for example, fluorescent dyes such as SYBR® stains (Molecular Probes; Figures 9.2 and 9.3). Some of these alternatives are less mutagenic and have better performance than ethidium bromide.

9.1.2 Silver Staining

Electrophoretically separated DNA fragments can also be detected with silver nitrate staining. Silver staining of polyacrylamide gels has been used for the amplified fragment length polymorphism (AFLP; see Chapter 19) method of variable number tandem repeat (VNTR; see

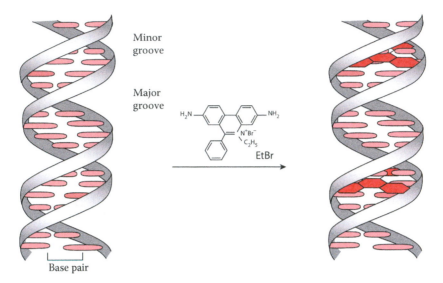

Figure 9.1   Schematic illustration showing binding of an intercalating agent, ethidium bromide (EtBr) to DNA. It intercalates into the minor groove of DNA. (© Richard C. Li.)

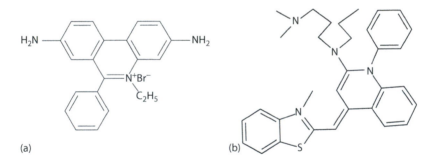

Figure 9.2   Intercalating agents used as nucleic acid stains. (a) Chemical structure of ethidium bromide. Ethidium bromide has UV absorbance. The emission maximum of the DNA–dye complex in aqueous solution is approximately 590 nm. (b) Chemical structure of SYBR Green 1. The stain is a cyanine dye that binds to DNA and is used as a nucleic acid stain. The excitation and emission maxima of the DNA–dye complex are 494 and 521 nm, respectively. The stain preferentially binds to double-stranded DNA rather than single-stranded DNA and RNA.

Chapter 19) profiling. The sensitivity of silver staining is approximately 100 times higher than that obtained with ethidium bromide, and silver staining is less hazardous than ethidium bromide detection. Also, the developing chemicals are readily available at low cost.

Silver staining involves processing a gel followed by exposure to a series of chemicals. First, the gel is submerged in a silver nitrate solution. Silver ions are positively charged, and DNA is negatively charged. Therefore, silver ions bind to the DNA and are subsequently reduced using formaldehyde to form a deposit of metallic silver on the DNA in the gel (Figure 9.4). A photograph of the gel with images of the silver-stained DNA strands is kept as a permanent record. Alternatively, the gels may be sealed and preserved for record purposes. One disadvantage of this method is that silver stains RNA and proteins along with DNA. The presence of restriction enzymes and polymerase should therefore be minimized. Additionally, bands from both complementary DNA strands may be detected in a denatured polyacrylamide gel, which leads to a two-band pattern.

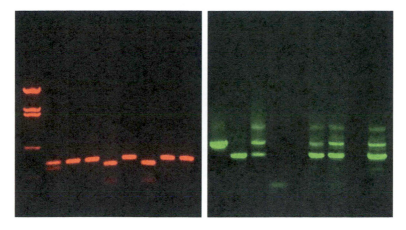

Figure 9.3   A DNA–dye complex emitting fluorescence on UV-light exposure. Ethidium bromide–containing agarose gel (left). SYBR Green 1–containing agarose gel (right). (© Richard C. Li.)

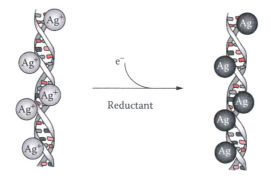

Figure 9.4   Silver staining of DNA. Silver (Ag+) ions bind to DNA and are reduced to metallic silver (Ag) to form dark particles. (© Richard C. Li.)

9.2 Detection of DNA Probes in Hybridization-Based Assays

9.2.1 Radioisotope Labeled Probes

Radioisotope probe labeling was used for early versions of VNTR testing and DNA quantitation. Labeling can be accomplished in several ways (Figure 9.5). For example, nick translation incorporates labeled deoxyribonucleotides (dNTPs) into double-stranded DNA. DNase I is used to introduce single-strand nicks within the DNA fragment to be labeled. Next, DNA Polymerase I recognizes the nicks and replaces the preexisting nucleotides with new strands containing labeled dNTPs, resulting in the generation of ^{32}P-labeled double-stranded DNA molecules. ^{32}P, with a half-life of approximately 14 days, is the most common radioisotope used in this technique. Nick translation can utilize any dNTP labeled with ^{32}P.

Prior to hybridization, the probe is denatured into single-stranded fragments by boiling for a few minutes followed by rapid cooling on ice. After the hybridization process, these probes can be visualized by exposing the DNA-containing membrane to an x-ray film. The radioactive object is commonly placed in an x-ray cassette. The energy released from the decay of the radioisotopes is absorbed by the silver halide grains in the film emulsion and forms a latent image. A chemical development process amplifies the latent image and renders it visible on film (Figure 9.6). Because most ^{32}P emissions pass through the thin film emulsion without

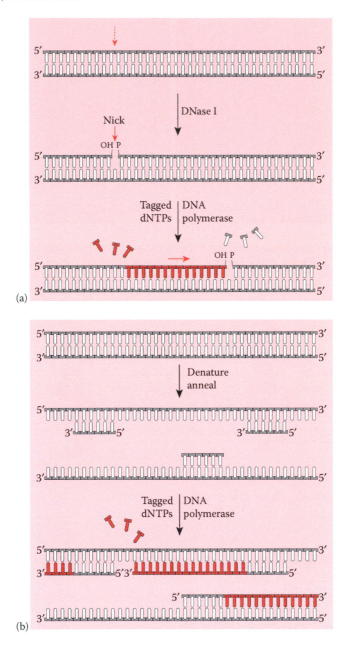

Figure 9.5   DNA probe labeling using dNTPs. (a) Nick translation method. A DNase is utilized to create nicks along a DNA strand. Tagged nucleotides (one or more than one of the dNTPs) are then incorporated into the DNA strand using a DNA polymerase. DNA synthesis extends from the 5' to 3' end and the original strand is degraded. (b) Random primer labeling method. A template DNA is denatured and separated into single strands. A random mixture of hexameric deoxynucleotide fragments, as primers, is then annealed to the template strand. The tagged nucleotides are then incorporated by the Klenow fragment of the DNA polymerase. (© Richard C. Li.)

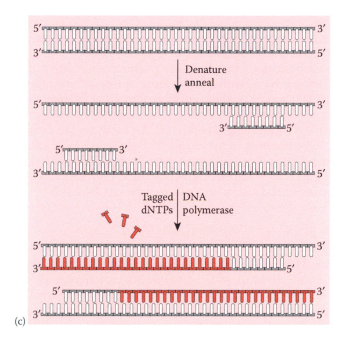

(c)

Figure 9.5 (Continued) (c) PCR method. DNA templates are denatured and then annealed to primers flanking the region of interest. During the DNA synthesis phase of a PCR cycle (see Chapter 7), tagged nucleotides are incorporated into the DNA strand using a Taq DNA polymerase, producing new double-stranded DNA. The process is repeated for many cycles. Only the first cycle is shown. (© Richard C. Li.)

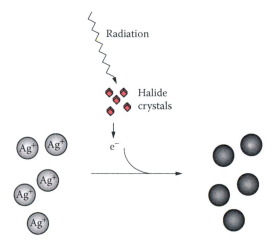

Figure 9.6 Autoradiography. Exposure to radiation causes halide crystals to release electrons, thus reducing silver (Ag+) ions to metallic silver (Ag). (© Richard C. Li.)

contributing to the final image, the detection process may require long exposure times. Signal intensity can be enhanced, however, by using intensifying screens at low temperatures. The screens emit photons upon receiving radioactive β-particles, thus further enhancing signals. Additionally, highly sensitive x-ray film can be used. One disadvantage of using ^{32}P is that it is a safety hazard. Additionally, autoradiography is a lengthy process. Therefore, nonradioisotopic detection methods have become popular alternatives.

9.2.2 Enzyme-Conjugated Probe with Chemiluminescence Reporting System

The use of alkaline phosphatase (AP)-conjugated probes with chemiluminescent substrates comprises a highly sensitive nonradioisotopic detection system.

Alkaline phosphatase can cleave the phosphate groups from a variety of substrate molecules. Its enzymatic activity can be measured using dioxetane-based chemiluminescent substrates such as Lumigen® PPD (Figure 9.7). The Lumi-Phos Plus kit of Lumigen Inc. contains this substrate and can serve as a detection system for slot blot assays for DNA quantitation (Chapter 6) and RFLP assays for VNTR profiling (Chapter 19). AP catalyzes the cleavage of the phosphate ester of Lumigen® PPD, resulting in the release of a photon (Figure 9.8). The Lumigen® PPD substrate yields a long-lasting light emission that can be detected by exposure to x-ray film. This system provides a highly sensitive chemiluminescent detection method for AP-conjugated DNA probes in solution or on a solid matrix such as a membrane (Table 9.1).

9.2.3 Biotinylation of DNA with Colorimetric Reporting Systems

9.2.3.1 Biotin

Biotin, also known as vitamin H, is a water-soluble molecule found in egg yolk (Figure 9.9). It can be incorporated onto oligonucleotide probes without interfering with the ability of probes to hybridize because of its small size (molecular weight: 244.31 u). Signals from a biotinylated probe can be detected with an enzyme-conjugated avidin system. Two steps are required to detect biotin-labeled probes. First, an avidin conjugate consisting of a reporter enzyme is added. Then, the reporter enzyme is assayed with substrates.

9.2.3.2 Enzyme-Conjugated Avidin

Avidin is a glycoprotein found in egg white; it binds to biotin with extremely high affinity. Thus, a biotin–avidin complex is very stable. However, avidin detection has a high background due to nonspecific binding. The nonspecific binding can be reduced by replacing avidin with its

Figure 9.7 Detection system using AP-conjugated probe. Chemiluminescence is generated by using the Lumigen® PPD as an AP substrate. (© Richard C. Li.)

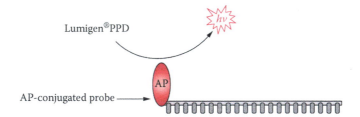

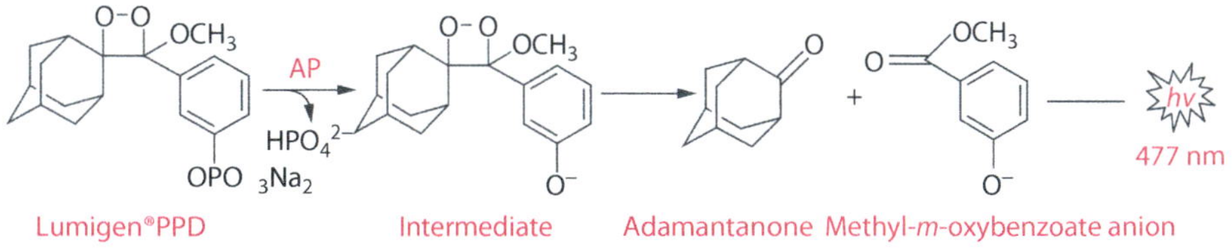

Lumigen®PPD   Intermediate   Adamantanone Methyl-*m*-oxybenzoate anion

Figure 9.8 Lumi-Phos Plus contains Lumigen® PPD (4-methoxy-4-[3-phosphatephenyl]spiro[1,2-dioxetane-3,2′-adamantane], disodium salt). AP catalyzes the removal of the phosphate group of Lumigen® PPD and generates a chemiluminescent intermediate that is subsequently broken down to the excited state of methyl-*m*-oxybenzoate. The decay of the excited state end-product releases a photon.

Table 9.1   Forensic Applications of Enzyme Reporting Systems for Detecting DNA

Reporter Enzyme	Labeling	Detection Mechanism	Substrate	Forensic Application
AP	AP-conjugated probe	Chemiluminescent	Lumigen®PPD	RFLP; blot assay for DNA quantitation
Horseradish peroxidase (HRP)	Biotinylated probe, recognized by streptavidin conjugated with HRP	Colorimetric/ chemiluminescent	TMB/luminol	Blot assay for DNA quantitation
	Biotinylated primer, recognized by streptavidin conjugated with HRP	Colorimetric	TMB	Reverse blot assay for DQA1 typing and for mtDNA typing

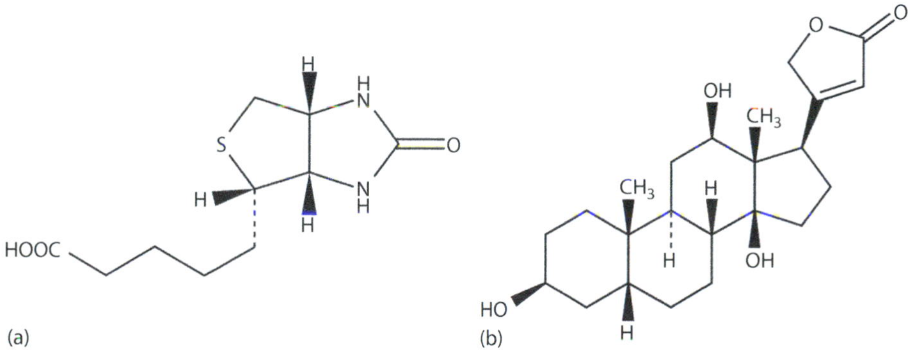

Figure 9.9   Nonradioactive tags for labeling and detecting nucleic acids. (a) Chemical structure of biotin, also known as vitamin H. Molecular formula: $C_{10}H_{16}N_2O_3S$. Molecular weight: 244.31. (b) Chemical structure of digoxigenin. Molecular formula: $C_{23}H_{34}O_5$. Molecular weight: 390.51.

streptavidin counterpart from *Streptomyces avidinii*. To detect binding, an enzyme-conjugated streptavidin such as horseradish peroxidase (HRP)-conjugated streptavidin can be used. HRP isolated from horseradish roots contains heme residues that catalyze the oxidation reactions of substrates (Chapter 12).

9.2.3.3 Reporter Enzyme Assay

HRP can be assayed with colorimetric, chemiluminescent, or fluorogenic substrates. One common colorimetric substrate for forensic DNA testing is 3,3′,5,5′-tetramethylbenzidine (TMB), which is oxidized by the peroxidase to form an insoluble precipitate of intense blue at an acidic pH (Figure 9.10). Because the colored precipitate is difficult to remove from membranes, TMB is not suitable if reprobing for RFLP analysis is required. This technique has been used for forensic DNA testing such as slot blot assays for DNA quantitation (Table 9.1). A chemiluminescent substrate such as a luminol-based reagent can also be utilized with

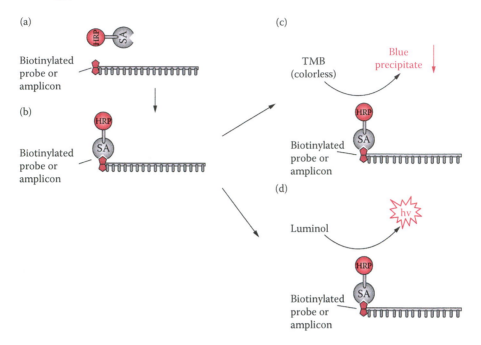

Figure 9.10 Detection system using biotinylated DNA probes with colorimetric and chemiluminescent reactions. (a) Biotinylated probe is incubated with a streptavidin (SA) and horseradish peroxidase (HRP) conjugate complex. (b) Biotin is recognized by the complex. Reporter enzyme assays can be carried out using either a colorimetric reaction with a TMB substrate (c) or a chemiluminescent reaction using luminol analogs (d). (© Richard C. Li.)

HRP. The peroxidase catalyzes the oxidation of luminol to form a chemiluminescent product (Figure 9.10).

9.3 Detection Methods for PCR-Based Assays

9.3.1 Fluorescence Labeling

9.3.1.1 Fluorescent Dyes

The advantages of fluorescence detection methods include a higher sensitivity and broader dynamic range than comparable colorimetric detection methods. Furthermore, they have the capacity for simultaneous analysis of complex samples such as multiplex PCR products with different fluorescent labels (Figure 9.11), allowing the distinction of various amplicons. Commonly used fluorescent dyes in DNA labeling emit fluorescence in the range of 400–600 nm (Figures 9.12 and 9.13).

9.3.1.2 Labeling Methods

Fluorescent dye labeling can be incorporated into a DNA fragment using a 5′-end fluorescently labeled oligonucleotide primer (Figures 9.14 and 9.15). The dye-labeled primer method is usually used for STR profiling (see Chapter 20) in which only one primer from each primer pair is labeled; therefore, only one strand can be detected. The two-band pattern observed with silver staining does not appear with this method. Additionally, dye-labeled primers allow multiplex PCR amplifications in the same tube.

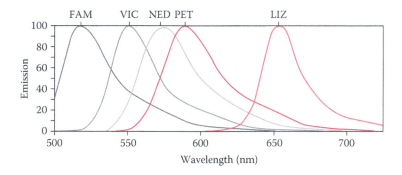

Figure 9.11   Emission spectra of common fluorescent dyes used for forensic DNA analysis. (© Richard C. Li.)

Color	Fluorescent dye	Globalfiler	AmpFlSTR Yfiler	PowerPlex fusion system	PowerPlex Y23 system
Purple dye	SID	☑			
Blue dyes	FAM	☑	☑		
	Fluorescein			☑	☑
Green dyes	JOE			☑	☑
	VIC	☑	☑		
Yellow dyes	NED	☑	☑		
	TMR (TAMRA)			☑	☑
Orange dyes	CC5			☑	☑
	LIZ	☑	☑		
Red dyes	PET		☑		
	ROX (CXR)			☑	☑
	TAZ	☑			

Figure 9.12   Examples of fluorescent dyes and their applications in STR kits. *Note*: Globalfiler and AmpFlSTR Yfiler are manufactured by Applied Biosystems. PowerPlex Fusion system and PowerPlex® Y23 System are manufactured by Promega.

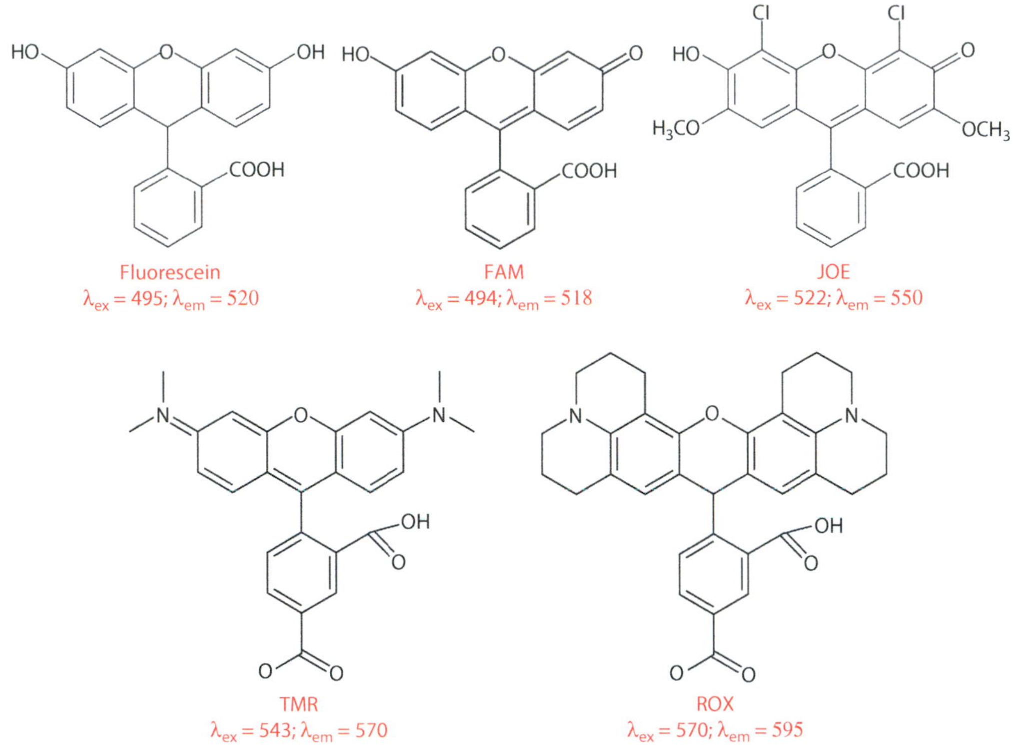

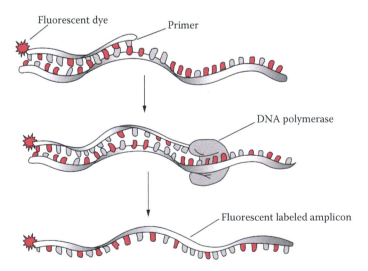

Fluorescein
$\lambda_{ex} = 495; \lambda_{em} = 520$

FAM
$\lambda_{ex} = 494; \lambda_{em} = 518$

JOE
$\lambda_{ex} = 522; \lambda_{em} = 550$

TMR
$\lambda_{ex} = 543; \lambda_{em} = 570$

ROX
$\lambda_{ex} = 570; \lambda_{em} = 595$

Figure 9.13 Chemical structures of representative fluorescent dyes. Fluorescein, 3′,6′-dihydroxyspiro[2-benzofuran-3,9′-xanthene]-1-one; FAM, 5(6)-carboxyfluorescein; JOE, 6-carboxy-4′,5′-dichloro-2′,7′-dimethoxy-fluorescein; TMR, 5-carboxytetramethylrhodamine; ROX, 5(6)-carboxy-X-rhodamine; λ_{ex}, peak excitation wavelength (nm); λ_{em}, peak emission wavelength (nm).

Fluorescent dye

Primer

DNA polymerase

Fluorescent labeled amplicon

Figure 9.14 A fluorescent dye–labeled primer can be used for the amplification of DNA. The dye is conjugated at the 5′ end of the primer. The amplified product is fluorescently labeled. (© Richard C. Li.)

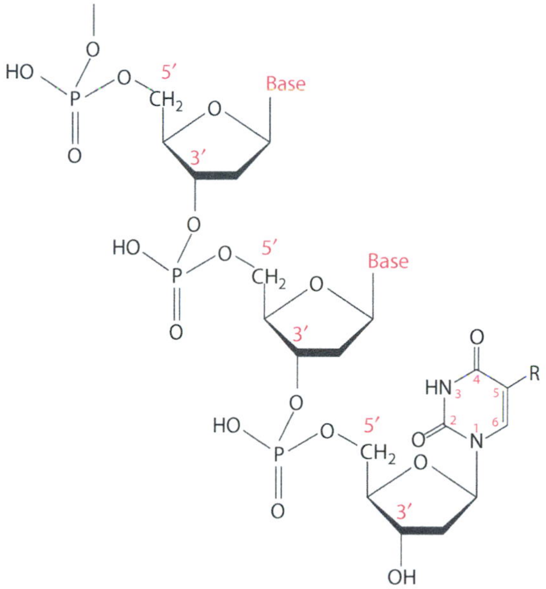

Figure 9.15 Direct labeling of nucleic acids with fluorescent dye. Labeled cytosine is shown as an example. R, fluorescent dye.

Alternatively, fluorescent dye labeling of DNA fragments can be carried out by incorporating fluorescently labeled dideoxynucleotides (ddNTPs) in the PCR product. This labeling method is usually used in DNA sequencing such as mtDNA sequence profiling (Chapter 23).

9.3.1.3 Fluorophore Detection

The *fluorophore* is a component of a florescent dye molecule that causes the molecule to be fluorescent. First, a laser strikes a fluorophore covalently linked to the end of a DNA fragment. An electron of the fluorophore is then excited, rising to an excited state from a ground state. The excited electron then descends to the ground state and releases a photon (Figure 9.16). The emitted photon has a longer wavelength (emission spectrum) than that of the excitation photon (excitation spectrum). The wavelengths of excitation and emission spectra (Figure 9.17) are largely dependent on the chemical structure of the fluorophore.

Lasers are commonly used as *excitation sources* because laser light emissions have high intensity and are monochromatic (single wavelength). The argon ion gas laser is frequently used in applications such as fluorescence-labeled STR and mtDNA sequence analysis because the excitation wavelength of commonly used fluorescent dyes matches the wavelength of the argon laser.

Optical filters are used to filter out undesired light and to allow only one particular wavelength to pass through. An essential optical filter consists of three components: an excitation filter, a dichroic beam splitter, and an emission filter. The excitation filter selectively transmits light from an excitation source. The light is then directed by the dichroic beam splitter to DNA molecules labeled with fluorescent dyes. The light emitted from a fluorophore is also transmitted by the dichroic beam splitter toward the detector. The emission filter selectively blocks undesired light, thus transmitting a specific wavelength of the emitted fluorescence. The light intensity emitted from a fluorophore is detected using a photosensitive device such as a charge-coupled device (CCD). The signal from the fluorophore is collected and converted to an electronic signal expressed in an arbitrary unit such as a *relative fluorescence unit* (RFU). Signals from multiple fluorophores in the same sample can be recorded separately using optical filters and a mathematical *matrix* (fluorophore separation algorithm). The function of a matrix is to subtract the

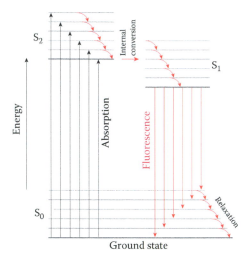

Figure 9.16 A Jablonski diagram illustrating the absorption of a single photon and subsequent emission of fluorescence. During an excitation stage, a single photon with the appropriate energy excites a molecule to the first excited energy level (S_2). In fluorescence reactions, the energy is supplied by an external light source such as a laser, while in chemiluminescence reactions, the energy is derived from a chemical reaction. The excited molecule is not stable and relaxes to its lowest vibrational level in the excited state (S_1). The molecule then returns to the ground state (S_0), emitting a photon. (© Richard C. Li.)

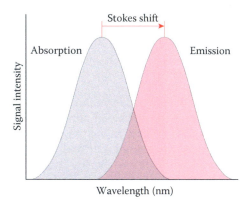

Figure 9.17 A fluorescence excitation spectrum. In fluorescence reactions, the energy of the photon emitted is lower than that of the excitation photon. As a result, the wavelength of an emission photon is longer than that of an excitation photon. The difference in wavelength is known as the Stokes shift. (© Richard C. Li.)

color overlaps of the various fluorescent dyes. The matrix can be established by the calibration of fluorescently labeled standards.

Bibliography

Bassam, B.J., G. Caetano-Anolles, and P.M. Gresshoff, Fast and sensitive silver staining of DNA in polyacrylamide gels. *Anal Biochem*, 1991, **196**(1): 80–83.

Bregu, J., et al., Analytical thresholds and sensitivity: Establishing RFU thresholds for forensic DNA analysis. *J Forensic Sci*, 2013, **58**(1): 120–129.

Bronstein, I. and P. McGrath, Chemiluminescence lights up. *Nature*, 1989, **338**(6216): 599–600.

Bronstein, I., et al., Rapid and sensitive detection of DNA in Southern blots with chemiluminescence. *Biotechniques*, 1990, **8**(3): 310–314.

Brunk, C.F. and L. Simpson, Comparison of various ultraviolet sources for fluorescent detection of ethidium bromide-DNA complexes in polyacrylamide gels. *Anal Biochem*, 1977, **82**(2): 455–462.

Budowle, B., et al., Simple protocols for typing forensic biological evidence: Chemiluminescent detection for human DNA quantitation and restriction fragment length polymorphism (RFLP) analyses and manual typing of polymerase chain reaction (PCR) amplified polymorphisms. *Electrophoresis*, 1995, **16**(9): 1559–1567.

Buel, E. and M. Schwartz, The use of DAPI as a replacement for ethidium bromide in forensic DNA analysis. *J Forensic Sci*, 1995, **40**(2): 275–278.

Cardullo, R.A., et al., Detection of nucleic acid hybridization by nonradiative fluorescence resonance energy transfer. *Proc Natl Acad Sci U S A*, 1988, **85**(23): 8790–8794.

Decorte, R. and J.J. Cassiman, Evaluation of the ALF DNA sequencer for high-speed sizing of short tandem repeat alleles. *Electrophoresis*, 1996, **17**(10): 1542–1549.

Feinberg, A.P. and B. Vogelstein, A technique for radiolabeling DNA restriction endonuclease fragments to high specific activity. *Anal Biochem*, 1983, **132**(1): 6–13.

French, B.T., H.M. Maul, and G.G. Maul, Screening cDNA expression libraries with monoclonal and poly-clonal antibodies using an amplified biotin-avidin-peroxidase technique. *Anal Biochem*, 1986, **156**(2): 417–423.

Gillespie, P.G. and A.J. Hudspeth, Chemiluminescence detection of proteins from single cells. *Proc Natl Acad Sci U S A*, 1991, **88**(6): 2563–2567.

Giusti, A.M. and B. Budowle, A chemiluminescence-based detection system for human DNA quantitation and restriction fragment length polymorphism (RFLP) analysis. *Appl Theor Electrophor*, 1995, **5**(2): 89–98.

Green, N.M., Avidin. *Adv Protein Chem*, 1975, **29**: 85–133.

Green, N.M. and E.J. Toms, The properties of subunits of avidin coupled to sepharose. *Biochem J*, 1973, **133**(4): 687–700.

Halpern, M.D. and J. Ballantyne, An STR melt curve genotyping assay for forensic analysis employing an intercalating dye probe FRET. *J Forensic Sci*, 2011, **56**(1): 36–45.

Johnson, E.D. and T.M. Kotowski, Chemiluminescent detection of RFLP patterns in forensic DNA analysis. *J Forensic Sci*, 1996, **41**(4): 569–578.

Laskey, R.A. and A.D. Mills, Enhanced autoradiographic detection of 32P and 125I using intensifying screens and hypersensitized film. *FEBS Lett*, 1977, **82**(2): 314–316.

Lee, L.G., et al., Seven-color, homogeneous detection of six PCR products. *Biotechniques*, 1999, **27**(2): 342–349.

LePecq, J.B. and C. Paoletti, A fluorescent complex between ethidium bromide and nucleic acids. Physical-chemical characterization. *J Mol Biol*, 1967, **27**(1): 87–106.

Mansfield, E.S. and M.N. Kronick, Alternative labeling techniques for automated fluorescence based analysis of PCR products. *Biotechniques*, 1993, **15**(2): 274–279.

Mayrand, P.E., et al., The use of fluorescence detection and internal lane standards to size PCR products automatically. *Appl Theor Electrophor*, 1992, **3**(1): 1–11.

McKimm-Breschkin, J.L., The use of tetramethylbenzidine for solid phase immunoassays. *J Immunol Methods*, 1990, **135**(1–2): 277–280.

Merril, C.R., Silver staining of proteins and DNA. *Nature*, 1990, **343**(6260): 779–780.

Mitchell, L.G., A. Bodenteich, and C.R. Merril, Use of silver staining to detect nucleic acids. *Methods Mol Biol*, 1994, **31**: 197–203.

Morin, P.A. and D.G. Smith, Nonradioactive detection of hypervariable simple sequence repeats in short polyacrylamide gels. *Biotechniques*, 1995, **19**(2): 223–228.

Renz, M. and C. Kurz, A colorimetric method for DNA hybridization. *Nucleic Acids Res*, 1984, **12**(8): 3435–3444.

Roberts, I.M., et al., A comparison of the sensitivity and specificity of enzyme immunoassays and time-resolved fluoroimmunoassay. *J Immunol Methods*, 1991, **143**(1): 49–56.

Selvin, P.R., Fluorescence resonance energy transfer. *Methods Enzymol*, 1995, **246**: 300–334.

Southern, E.M., Detection of specific sequences among DNA fragments separated by gel electrophoresis. *J Mol Biol*, 1975, **98**(3): 503–517.

Stone, T. and I. Durrant, Enhanced chemiluminescence for the detection of membrane-bound nucleic acid sequences: Advantages of the Amersham system. *Genet Anal Tech Appl*, 1991, **8**(8): 230–237.

Tizard, R., et al., Imaging of DNA sequences with chemiluminescence. *Proc Natl Acad Sci U S A*, 1990, **87**(12): 4514–4518.

Tuma, R.S., et al., Characterization of SYBR Gold nucleic acid gel stain: A dye optimized for use with 300-nm ultraviolet transilluminators. *Anal Biochem*, 1999, **268**(2): 278–288.

Waring, M.J., Complex formation between ethidium bromide and nucleic acids. *J Mol Biol*, 1965, **13**(1): 269–282.

Watkins, T.I. and G. Woolfe, Effect of changing the quaternizing group on the trypanocidal activity of dimidium bromide. *Nature*, 1952, **169**(4299): 506–507.

Yeung, S.H., et al., Fluorescence energy transfer-labeled primers for high-performance forensic DNA profiling. *Electrophoresis*, 2008, **29**(11): 2251–2259.

Serology Concepts

10.1 Serological Reagents

10.1.1 Immunogens and Antigens

A foreign substance that is capable of eliciting antibody formation when introduced into a host is called an *immunogen*. Natural immunogens are usually macromolecules such as proteins and polysaccharides. Other molecular structures can also act as immunogens, for example, glyco-lipids (such as A, B, and O blood group antigens) and glycoproteins (such as Rh and Lewis antigens). However, they must be foreign to their hosts. The molecular structure of an immu-nogen, usually a small portion recognized by an antibody, is called an *epitope* or determinant site. An immunogen usually consists of multiple epitopes and is thus considered *multivalent* (Figure 10.1). Each epitope can elicit the production of its own corresponding antibody.

An *antigen* is a foreign substance that is capable of reacting with an antibody. All immu-nogens can be considered antigens, but not all antigens can elicit antibody formation. *Hapten* is one example of a substance that is antigenic but not immunogenic. Haptens are chemical compounds that are too small to elicit antibody production when they are introduced to a host animal. However, a hapten can be coupled to a *carrier*, usually a macromolecule, to produce antibodies. A hapten-conjugated carrier can become immunogenic to elicit the formation of an antibody specific to the hapten. The resulting antibody can bind to free haptens. Certain con-trolled substances such as cocaine and amphetamines are haptens and can be detected through corresponding antibodies for forensic toxicological analysis.

10.1.2 Antibodies

Antibodies, also known as *immunoglobulins*, are capable of binding specifically to antigens and are designated with an Ig prefix. The five major classes of immunoglobulins are designated IgG, IgA, IgM, IgD, and IgE. Additionally, IgG immunoglobulins can be further divided into four subclasses (IgG1–IgG4) and IgA immunoglobulins can be further divided into two subclasses (IgA1 and IgA2). Thus, there are a total of nine immunoglobulin isotypes in humans. *Isotypes* are the immunoglobulins that differ based on the molecular variations in the constant domains of the heavy and light chains (Figure 10.2). IgD, IgE, and IgG are usually monomers. IgM can be a membrane-bound monomer or a cross-linked pentamer (secreted form). IgA can be a monomer, dimer, or trimer. In an immune response, an initial exposure to an immunogen elicits a primary response, producing IgM immunoglobulins. Further exposure to the immunogen can elicit a sec-ondary response, producing IgG, IgA, IgE, and IgD immunoglobulins. IgG is the most abundant immunoglobulin in serum. A majority of serology tests are based on the IgG immunoglobulins.

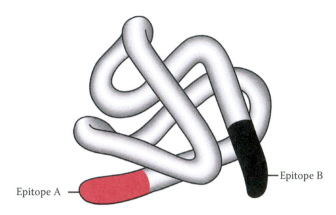

Figure 10.1   Multivalent immunogen. A protein with two different epitopes is shown. (© Richard C. Li)

Immunoglobulins have many similarities in their molecular structures. Figure 10.2 illustrates a diagram of the IgG molecule. The structures of immunoglobulins were first revealed by Gerald Edelman and Rodney Porter who shared a Nobel Prize in 1972. Immunoglobulins are composed of four polypeptide chains: two heavy (H) chains and two light (L) chains. The polypeptide chains are linked by disulfide bonds into a Y-shaped complex. The H chain can be divided into fragment antigen-binding (*Fab*) and fragment crystallizable (*Fc*) fragments. The L chain consists of a Fab fragment only. A typical antibody has two identical antigen-binding sites and is thus considered *bivalent*. The antigen-binding activity is located within the Fab fragments. In particular, the N-terminal ends of the L and H chains together form antigen-binding sites. At the amino acid sequence level, both H and L chains have *variable and constant domains* (Figure 10.2). The variable domains are located at the N-terminal ends of the immunoglobulins. Additionally, three small *hypervariable regions* are located within the variable domain of each chain.

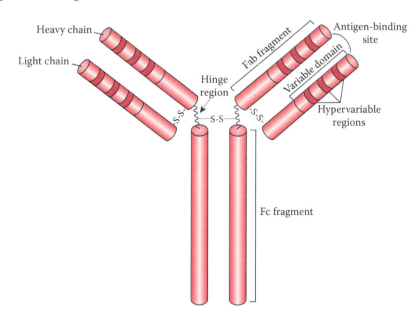

Figure 10.2   Immunoglobulin IgG structure. Immunoglobulin IgG is composed of light and heavy chains that contain variable domains. The remaining portions of chains form the constant domains. The variable domains of both light and heavy chains contain three hypervariable regions that form the antigen-binding site of the immunoglobulin. (© Richard C. Li.)

The diversity in the amino acid sequences of these hypervariable regions determines the specificity of the antigen-binding sites. The hinge regions provide flexibility to the antibody molecule and are important for the efficiency of the binding and cross-linking reactions. The basal portion of the H chain consists of *Fc* fragment.

The binding affinity and specificity of antibodies make them useful reagents for serological testing. Two types of antibodies are commonly used: polyclonal and monoclonal antibodies.

10.1.2.1 Polyclonal Antibodies

To produce an antibody, an immunogen is usually introduced into a host animal. A multivalent immunogen is capable of eliciting a mixture of antibodies with diverse specificities for the immunogen. As a result, a polyclonal antibody is produced by different B lymphocyte clones in response to the different epitopes of the immunogen. Antibodies can be circulating (in blood or other bodily fluids) or tissue bound (in cell surface antibodies). Circulating immunoglobulins are referred to as humoral antibodies.

The blood from an immune host is drawn and allowed to clot, resulting in the formation of a solid consisting largely of blood cells and a liquid portion known as *serum* containing antibodies (Figure 10.3). Such a preparation of humoral antibodies is also called a *polyclonal antiserum*. Depending on the type of animal used, the antibodies produced are classified as avian (B), rabbit (R), or horse (H) type. The characteristics of polyclonal antibodies may vary if they are produced from different individual host animals of the same species. Variations in reactions among different sources of antibodies should be monitored by quality-control procedures.

10.1.2.2 Monoclonal Antibodies

To produce a monoclonal antibody, spleen cells are harvested from a host animal, such as a mouse, inoculated with an immunogen (Figure 10.4). Next, the plasma cells of the spleen, which produce antibodies, are fused with myeloma cells. Since only a small population of cells fuse, a selection step is needed to allow only fused cells to grow (Figure 10.4). The fused cells, called *hybridoma* cells, are immortal (proliferate indefinitely) in cell cultures. Pools of hybridoma cells are diluted into single clones and are allowed to proliferate. The clones are then screened for the specific antibody of interest.

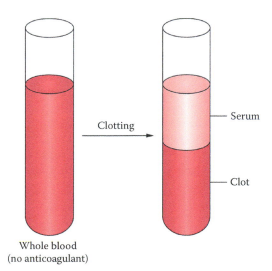

Whole blood
(no anticoagulant)

Clotting

Serum

Clot

Figure 10.3 Serum component of blood. The blood of an immunized animal is collected in the absence of an anticoagulant and is allowed to clot. The resulting liquid portion of the blood is serum. (© Richard C. Li.)

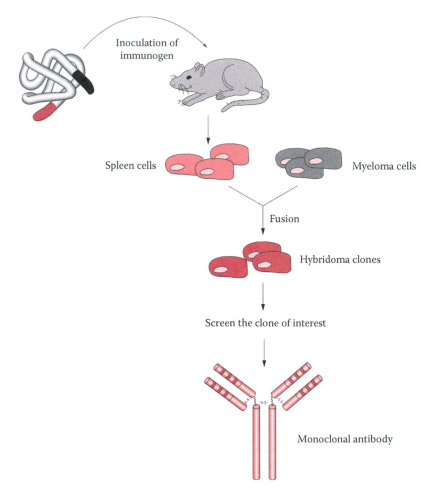

Figure 10.4 Preparation of a monoclonal antibody. A mouse is immunized with an immunogen, and its spleen cells are fused with myeloma cells to generate hybridoma cells. The clone that synthesizes and secretes the monoclonal antibody of interest can then be identified. (© Richard C. Li.)

The desired hybridoma clone can be maintained indefinitely, and it produces a monoclonal antibody that reacts with a single epitope. The hybridoma-derived monoclonal antibodies are specific and homogenous, and can be obtained in unlimited quantities. Monoclonal antibodies have been utilized in many serological assays, as discussed in Chapter 11. However, they have certain limitations in serology assays. For instance, monoclonal antibodies react with only a single epitope of a multivalent antigen and, therefore, cannot form cross-linked networks in precipitation assays (see Section 10.3.2.1).

10.1.2.3 Antiglobulins

Immunoglobulins are proteins that can also function as immunogens. If a purified foreign immunoglobulin or a fragment of a foreign immunoglobulin is introduced into a host, the antibodies produced are known as *antiglobulins*. Antiglobulins that are specific to a particular isotype can be produced in laboratories. In addition to specific antiglobulins, it is possible to produce nonspecific antiglobulins, which recognize an epitope that is common to all isotypes of an immunoglobulin class, such as the Fc portion of the heavy chain of all subclasses of human IgG. Antiglobulins are important reagents in many serological tests.

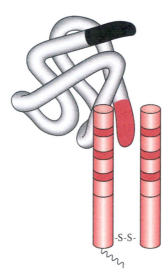

Figure 10.5   Affinity is a measure of the interaction between a single epitope on an antigen and a single binding site on an antibody. (© Richard C. Li.)

10.2 Strength of Antigen–Antibody Binding

The binding of an antigen to its specific antibody is mediated by the interaction between the epitope of an antigen and the binding site of its antibody. Noncovalent bonds can be formed during antigen–antibody binding. Various forces act cooperatively during antigen–antibody binding. These include hydrogen bonding, hydrophobic interactions that exclude water molecules from the area of contact, and van der Waals forces arising from the asymmetric distribution of the charges of electrons. The binding process occurs rapidly and the formation of the antigen–antibody complex is reversible. Such binding occurs at short distances when the antigen and antibody are in close proximity. Additionally, the strongest binding occurs only if the shape of the epitope fits the binding site of the antibody. The strength of the interaction between the antigen and the antibody depends on two characteristics, designated affinity and avidity.

Affinity is the energy of the interaction of a single epitope on an antigen and a single binding site on a corresponding antibody (Figure 10.5). The strength of the interaction depends on the specificity of the antibody for the antigen. Nevertheless, antibodies can bind with lower strength to antigens that are structurally similar to the immunogen. Such binding is known as *cross-reaction*.

Avidity is the overall strength of the binding of an antibody and an antigen (Figure 10.6). Since an antigen is usually considered multivalent and an antibody is bivalent, the avidity reflects the combined synergistic strength of the binding of all the binding sites of antigens and antibodies rather than the sum of individual affinities. It also reflects the overall stability of an antigen–antibody complex that is essential for many serological assays.

10.3 Antigen–Antibody Binding Reactions

The binding of an antigen to an antibody is an equilibrium reaction consisting of three types of reactions. The primary and secondary reactions form the basis for many forensic serological assays and will be discussed in the following subsections. The third type is called the tertiary reaction. It is used to measure *in vivo* immune responses such as inflammation and phagocytosis. Because most forensic serology tests are *in vitro* assays, the tertiary reaction is not commonly utilized in forensic serology testing and will not be discussed here.

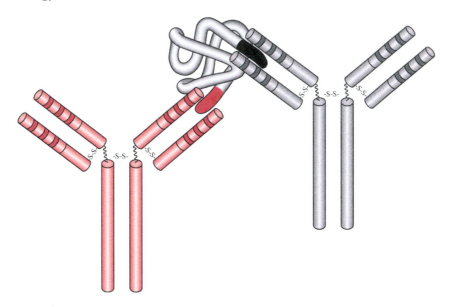

Figure 10.6 Avidity is a measure of the overall strength of the binding between antigens and antibodies. (© Richard C. Li.)

10.3.1 Primary Reactions

A primary reaction is the initial binding of a single epitope of an antigen (Ag) and a single binding site of an antibody (Ab) to form an antigen–antibody complex (Figure 10.7). This rapid and reversible binding reaction can be expressed as

$$Ag + Ab \rightleftarrows AgAb$$

At equilibrium, the strength of the interaction can be expressed as the affinity constant (K_a) that reflects the affinity of binding, where:

$$K_a = \frac{[AgAb]}{[Ag][Ab]}$$

The square brackets indicate the concentration of each component at equilibrium. K_a is the reciprocal of the concentration of free epitopes when half the antibody-binding sites are occupied. Thus, a higher K_a corresponds to a stronger binding interaction.

Techniques such as enzyme immunoassays, immunofluorescence assays, radioimmunoassays, and dye-labeled immunochromatography can measure the concentrations of antigen–antibody complexes formed by primary reactions (Chapter 11). These techniques are the most sensitive for detecting the presence of an antigen and an antibody in a sample. Additionally, many forensic serology assays are based on the detection of primary reactions and will be discussed in Chapters 12 through 17.

10.3.2 Secondary Reactions

The primary reaction between an antigen and an antibody is often followed by a secondary reaction. The three types of secondary reactions are precipitation, agglutination, and complement fixation. The techniques that detect secondary reactions are usually less sensitive but easier to perform than primary reaction assays. The precipitation and agglutination reactions form the basis for many serologic assays performed in forensic laboratories. These reactions will be discussed in the following subsections in detail. The third type of reaction is called *complement fixation*. If an antigen is located on a cell surface, the binding of the antigen and the antibody

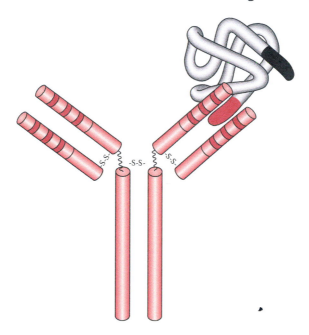

Figure 10.7   Primary reaction. Initial binding forms an antigen–antibody complex. (© Richard C. Li.)

may activate the classical complement pathway and lead to cell lysis, also known as a complement fixation reaction. The detection of this type of reaction is not commonly used in forensic serology.

10.3.2.1 Precipitation

If a soluble antigen is mixed and incubated with its antibody, the antigen–antibody complexes can form cross-linked complexes at the optimal ratio of antigen-to-antibody concentration. The cross-linked complex is insoluble and eventually forms a precipitate that settles to the bottom of a test tube. Antibodies that produce such precipitation are also called *precipitins*.

This precipitation reaction can be characterized by examining the effect of varying the relative ratio of antigen and antibody. If an increasing amount of soluble antigen is mixed with a constant amount of antibody, the amount of precipitate formed can be plotted. A precipitin curve (Figure 10.8)

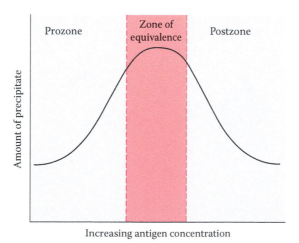

Figure 10.8   Precipitin curve. (© Richard C. Li.)

illustrates the results observed when antigens and antibodies are mixed in various concentration ratios. The curve can be divided into three zones known as the prozone, the zone of equivalence, and the postzone.

10.3.2.1.1 Prozone

At this zone, the ratio of antigen–antibody concentration is low. In other words, the antibody is in excess. Each antigen molecule is rapidly saturated with antibody, thus preventing cross-linking (Figure 10.9a). No precipitate is formed at the prozone stage.

10.3.2.1.2 Zone of Equivalence

As the concentrations of antigen increase, the amount of precipitate increases until it reaches a maximum. The amount of precipitation depends on the relative proportions of antigens and antibodies present. The maximum precipitation occurs in what is called the zone of equivalence. In the zone of equivalence, the ratio of antibody to antigen concentration is optimal and

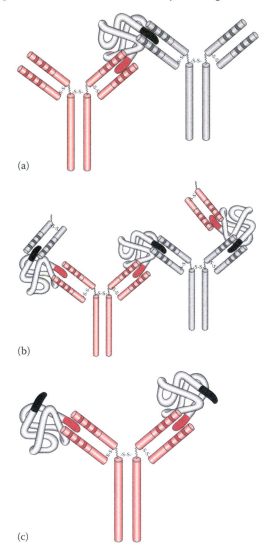

(a)

(b)

(c)

Figure 10.9 Antigen–antibody binding in (a) prozone, (b) zone of equivalence, and (c) postzone. (© Richard C. Li.)

precipitation occurs as a result of forming cross-linked networks (Figure 10.9b). Precipitation assays are usually carried out under the condition of the zone of equivalence, forming a sufficient quantity of precipitation to be detected.

10.3.2.1.3 Postzone

With the addition of more antigens, the ratio of antigen–antibody concentrations is high. In other words, the antigen is in excess. The amount of precipitate decreases and eventually diminishes. Each antibody molecule is saturated with antigen molecules (Figure 10.9c). Cross-linkage cannot form, and precipitation does not occur.

10.3.2.2 Agglutination

As discussed in the previous subsection, precipitation reactions involve soluble antigens. If the antigens are located on the surfaces of cells or carriers (carrier cells such as sheep erythrocytes,

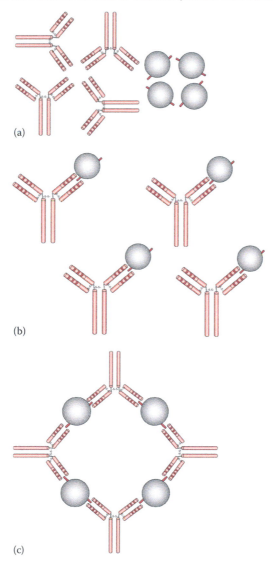

Figure 10.10 Agglutination reaction: (a) antigens are mixed with antibodies, (b) an antigen–antibody complex is formed during initial binding, and (c) the lattice is formed. (© Richard C. Li.)

bacteria, or latex particles), antibodies can bind to the surface antigens and can form cross-links among cells or carriers, causing them to aggregate. This aggregation is referred to as *agglutination*. In agglutination, a visible clumping can be observed as an indicator of the reaction of the antigen and the antibody. If the antigen is located on an erythrocyte, the agglutination reaction is designated *hemagglutination*. Agglutination is a two-step process that includes initial binding and lattice formation (Figure 10.10a).

10.3.2.2.1 Initial Binding

The first step of the reaction involves antigen–antibody binding at a single epitope on the cell surface (Figure 10.10b). This initial binding is rapid and reversible.

10.3.2.2.2 Lattice Formation

The second step involves the formation of a cross-linked network resulting in visible aggregates that constitute a lattice (Figure 10.10c). This involves an antibody binding to multiple epitopes because each antibody has two binding sites and antigens are multivalent. Lattice formation is a much slower process than the initial binding step. The cross-linking of cells requires physical contact. Additionally, an antibody must bind to epitopes on two different cells. The ability to cross-link cells depends on the nature of the antibody. Antibodies that produce such reactions are often called *agglutinins*.

Additionally, a *complete antibody* is capable of carrying out both primary and secondary interactions that result in agglutination. An antibody that can carry out initial binding but fails to form agglutination is called an *incomplete antibody*. This type of antibody is believed to have only one active antigen-binding site and is thus not capable of agglutination. It is potentially caused by the presence of steric obstruction due to the conformation of the incomplete antibody molecule, preventing the binding of antigens at the second binding site. However, other incomplete antibodies have two active sites but cannot bridge the distance between cells, thus failing to form lattices. Certain antibodies such as IgG are small and lack flexibility at the hinge region, and this may prevent agglutination. In contrast, the large IgM antibodies produce agglutination much more easily than IgG. Agglutination reactions have a wide variety of applications in the detection of antigens and antibodies. Such assays have high degrees of sensitivity and have been used for many years in forensic serology.

Bibliography

Baxter, S.J. and B. Rees, The use of anti-human haemoglobin in forensic serology. *Med Sci Law*, 1974, **14**(3): 159–162.

Dodd, B.E., Some recent advances in forensic serology. *Med Sci Law*, 1972, **12**(3): 195–199.

Fletcher, S. and M.J. Davie, Monoclonal antibody: A major new development in immunology. *J Forensic Sci Soc*, 1980, **20**(3): 163–167.

Giusti, G.V., Leone Lattes: Italy's pioneer in forensic serology. *Am J Forensic Med Pathol*, 1982, **3**(1): 79–81.

Goldsby, R., Kindt, T, and Osborne, B., *Immunology*, 6th edn. 2007, New York: W.H. Freeman and Company.

Lee, H.C., et al., Bits and pieces: Serology, criminalistics, anthropology and odontology. *J Forensic Sci Soc*, 1991, **31**(2): 293–296.

Patzelt, D., History of forensic serology and molecular genetics in the sphere of activity of the German Society for Forensic Medicine. *Forensic Sci Int*, 2004, **144**(2–3): 185–191.

Tietz, N. (ed.), *Fundamentals of Clinical Chemistry*, 1987. Philadelphia, PA: W.B. Saunders Company.

Serology Techniques
Past, Current, and Future

11.1 Introduction to Forensic Serology

11.1.1 The Scope of Forensic Serology

Forensic serology is the component of forensic biology that deals with the examination and identification of biological evidence. In particular, it focuses on determining the presence and identification of various bodily fluids such as blood, semen, and saliva in a questioned sample.

Bodily fluid stains are commonly associated with violent criminal cases. For instance, the identification of blood evidence (Chapter 12) is often necessary for investigating cases involving homicide, aggravated assault, sexual assault, and burglary. Proving the presence of blood has probative value or may corroborate allegations of violent acts. This evidence can also be used for investigative purposes. For example, the forensic DNA analysis of evidence such as a victim's blood on a suspect's weapon can establish a link between a victim and a suspect. The identification of semen (Chapter 14) and saliva (Chapter 15) is important for the investigation of a sexual assault case. For instance, a stain from a victim's clothing would be processed with forensic DNA analysis in a sexual assault case so that the DNA profile of the stain could be compared with that of an alleged suspect. Matching DNA profiles prove that the suspect's DNA was found on the victim's clothing, establishing a link between the suspect and the victim. Additionally, the identification of a biological stain as a semen stain through forensic serology testing proves that semen was found on the clothing taken from the victim and that the suspect is the source of the DNA from the semen stain. This evidence can then be used in court to support allegations that a sexual act occurred. Likewise, the presence of a suspect's saliva stains on a victim's genital area may corroborate an alleged oral copulation. The identification of bodily fluids other than blood, semen, and saliva will be discussed in Chapters 16 and 17.

11.1.2 Class Characteristics and Individual Characteristics of Biological Evidence

The identification of an unknown fluid sample is based on a comparison of the class characteristics (Section 3.2.1) of a sample with known standards of its class. Forensic identification typically involves bodily fluids, such as blood, semen, and saliva. If the presence of the bodily fluid is confirmed, the individual characteristics (Section 3.2.2) of the biological evidence are then determined to find out whether or not a bodily fluid sample has come from a particular individual. Today, this analysis can be achieved through forensic DNA analysis. The most commonly

utilized forensic DNA analysis is the short tandem repeat (STR) analysis used for human identification. Additionally, Y-chromosomal STR analysis is often utilized for the investigation of
sexual assault crimes. Furthermore, mitochondrial DNA analysis is used for the identification
of human remains. Therefore, establishing the probative value of a sample requires both the
identification of its class characteristics and the individualization of its contributor. Therefore,
the identification of bodily fluids cannot be replaced by forensic DNA analysis.

11.1.3 Presumptive and Confirmatory Assays

The identification of bodily fluids can be carried out using presumptive and confirmatory assays
to identify the type of bodily fluid in question. The advantages of *presumptive assays* are that
these assays are sensitive, rapid, and simple. A positive reaction of a presumptive assay indicates
the possibility of the presence of the bodily fluid in question. However, presumptive assays are
not very specific. Therefore, they should not be considered conclusive for the presence of a type
of bodily fluid. In contrast, a negative assay indicates that the questioned bodily fluid is absent.
Thus, presumptive assays can be used as a screening method and for narrowing down bodily
fluid stains prior to other types of analyses, such as forensic DNA testing. Moreover, these assays
can be used as a search method to locate bodily fluid stains at the crime scene. Additional assays,
such as confirmatory assays, should be conducted afterward if necessary.

Confirmatory assays are more specific for the bodily fluid in question. These assays are utilized to identify bodily fluids with higher certainty than presumptive assays. For example,
bloodstains are commonly associated with criminal investigations. A reddish-brown stain that
has been identified through visual examination is usually tested by using presumptive assays. If
the result of the presumptive assay on the alleged bloodstain is positive, the stain is then further
analyzed by forensic DNA analysis. This approach *indicates* the presence of blood. Confirmatory
assays are performed when a sample has to be *identified* as blood (Figure 11.1). Additionally, the
human or animal origin of blood evidence can be determined if necessary and the techniques to
do so will be discussed in Chapter 13.

11.1.4 Primary and Secondary Binding Assays

Traditionally, the detection and measurement of the antigen–antibody binding reactions serve
as the bases of forensic serology. These assays fall into two categories: primary and secondary binding assays. Recall that the primary binding assays involve the initial binding between
a single epitope of an antigen and a single binding site of an antibody (see Section 10.3.1).
Therefore, they are very sensitive. Although the secondary binding assays are less sensitive
than the primary binding assays, they are easier to perform than primary assays. The secondary assays consist of precipitation-based and agglutination-based assays. Precipitation-based
assays have been used for species identification. Agglutination-based assays are more sensitive

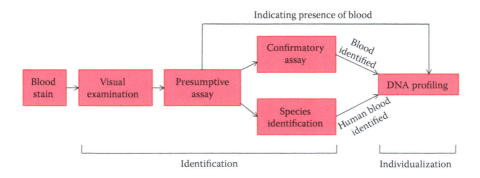

Figure 11.1 An example of work flow for processing blood samples. (© Richard C. Li.)

than precipitation-based assays. Agglutination-based assays, which detect antigens located on the surface of cells or carriers, are normally applied to blood group typing. Recently, emerging techniques such as RNA, proteomic, and DNA methylation assays, and fluorescence as well as Raman spectroscopy can potentially be used for the identification of bodily fluids.

11.2 Primary Binding Assays

11.2.1 Enzyme-Linked Immunosorbent Assay (ELISA)

An enzyme-linked immunosorbent assay (ELISA) is an immunoenzyme assay that can be used to detect and measure the antibody or antigen in question. The most common ELISA that is used in forensic serology is the antibody-sandwich ELISA (Figure 11.2). It is utilized to detect the prostate-specific antigen (PSA) to identify seminal stains and amylase for the identification of saliva (Chapters 13 and 14).

An antibody (usually monoclonal) coating is formed by nonspecific adsorption onto a solid phase such as the wells of a polystyrene plate. A sample containing the antigen to be tested is then added and binds to the solid-phase antibody. Subsequently, a second antibody is added to form an antibody–antigen–antibody sandwich complex. The second antibody binds to different epitopes of the antigen. Next, an enzyme-labeled antiglobulin (Chapter 10) is added to bind the sandwich. Subsequently, the excess enzyme-labeled antiglobulin is removed by washing and the bound antiglobulin can be detected.

A number of enzymes such as alkaline phosphatase and horseradish peroxidase have been used as reporting enzymes (Chapter 9) to label the antiglobulin for ELISA. The enzyme catalyzes the substrate and produces colorimetric or fluorometric signals. The intensity of the signals can

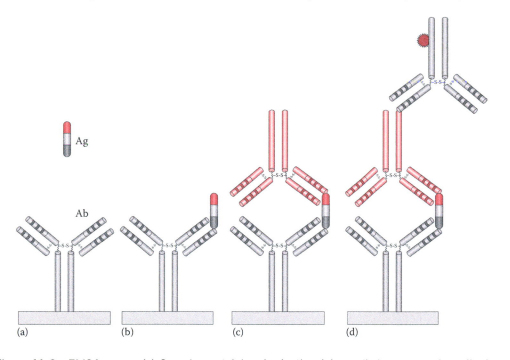

Figure 11.2 ELISA assay. (a) Sample containing Ag (antigen) is applied to a sample well where Ab (antibody) is immobilized. (b) Ag binds to an immobilized Ab to form an Ag–Ab complex. (c) A second Ab to a different epitope is added to form an Ab–Ag–Ab sandwich. (d) Labeled antiglobulin binds to the sandwich. The bound antiglobulin can be detected by various reporting schemes. (© Richard C. Li.)

be detected spectrophotometrically and is proportional to the amount of bound antigen. The amount of antigen can be quantified by comparing the standard with known concentrations. Recall that antiglobulin can recognize an epitope that is common to all isotypes within an antibody class (Section 10.1.2.3); the enzyme-labeled antiglobulin is a universal reagent regardless of the antibody used, as long as they are of the same class.

Alternatively, antibodies in a sample can also be detected and may be quantified by an ELISA system in which the antigen is bound to a solid phase instead of the antibody. After an antibody in a sample binds to the solid-phase antigen, an enzyme-labeled antiglobulin is added to bind to the bound antibody. The bound antiglobulin can be detected and measured by the addition of an enzyme substrate. The enzymatic catalytic reaction is similar to the antibody sandwich ELISA procedure previously described.

11.2.2 Immunochromatographic Assays

Figure 11.3 depicts a test using an immunochromatographic membrane device. A dye-labeled monoclonal antibody is contained in a sample well and a polyclonal antibody for the antigen (or a second monoclonal antibody to a different epitope of the antigen) is immobilized onto a test zone of a nitrocellulose membrane. An antiglobulin that recognizes the antibody is immobilized onto a control zone.

The assay is carried out by loading a sample into the sample well. The antigen in the sample binds to the dye-labeled antibody already in the sample well to form an antigen–antibody complex. The complex then diffuses across the nitrocellulose membrane until it reaches the test zone. The antibody immobilized at the test zone traps the antigen–antibody complex to form an antibody–antigen–antibody sandwich. The presence of the antigen in the sample results in a colored vertical line at the test zone. (Figure 11.4)

The immunochromatographic device also utilizes a control zone to ensure that the device works properly and that the sample has diffused completely along the test strip. Unbound

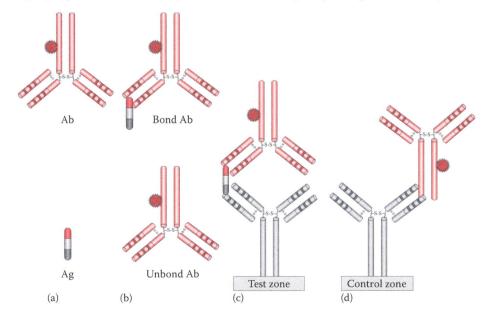

Figure 11.3 Immunochromatographic assay. (a) Sample containing Ag (antigen) is loaded in a sample well. (b) Ag binds to a labeled Ab (antibody) to form a labeled Ab–Ag complex. (c) At the test zone, the labeled Ab–Ag complex binds to an immobilized Ab to form a labeled Ab–Ag–Ab sandwich. (d) At the control zone, a labeled Ab binds to an immobilized antiglobulin and is captured at the control zone. (© Richard C. Li.)

monoclonal antibodies diffuse across the membrane until they reach the control zone where they are trapped by the immobilized antiglobulin. This antibody–antiglobulin complex at the control zone also results in a colored vertical line (Figure 11.4). The test is considered valid only if the line in the control zone is observed. The presence of an antigen results in a line at both the test zone and the control zone, while the absence of an antigen results in a line in the control zone only (Figure 11.4). This method is rapid and simple and thus can be used as a screening test in laboratories and as a field test at crime scenes to identify semen, saliva, and species (Table 11.1).

However, a false-negative result may be obtained if a sample contains a very high concentration of an antigen (Figure 11.5). Under this condition, the dye-labeled antibody will become saturated with the antigen to form antigen–antibody complexes. The unbound antigen will diffuse along with the

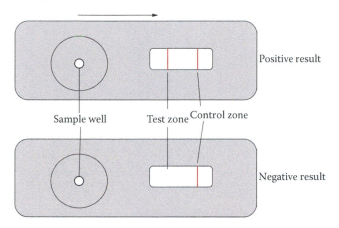

Figure 11.4   Immunochromatographic device. Positive and negative results are shown. (© Richard C. Li.)

Table 11.1 Common Immunochromatographic Assays for Forensic Applications				
Assay	**Antigen**	**Labeled Antibody**	**Immobilized Antibody**	**Forensic Application**
ABAcard® HemaTrace® (Abacus Diagnostics)	Hemoglobin (Hb)	Monoclonal antihuman Hb antibody	Polyclonal antihuman Hb antibody	Blood and species identification
RSID™-Blood (Independent Forensics)	Glycophorin A (GPA)	Monoclonal antihuman GPA antibody	Monoclonal antihuman GPA antibody	Blood and species identification
RSID™-Saliva (Independent Forensics)	Human salivary α-amylase (HAS)	Monoclonal antihuman HAS antibody	Monoclonal antihuman HAS antibody[a]	Saliva identification
One-Step ABAcard PSA® (Abacus Diagnostics)	Prostate-specific antigen (PSA)	Monoclonal antihuman PSA antibody	Polyclonal antihuman PSA antibody	Semen identification
RSID™-Semen (Independent Forensics)	Semenogelin (Sg)	Monoclonal antihuman Sg antibody	Monoclonal antihuman Sg antibody[a]	Semen identification

[a] The epitope recognized by the immobilized antibody is different from that of the labeled antibody.

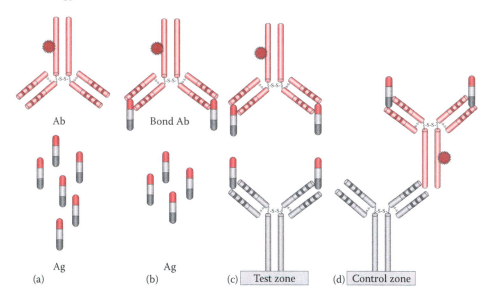

Figure 11.5 High-dose hook effect of immunochromatographic assay. (a) Ag (antigen) in sample is loaded in the sample well. (b) Ag binds to a labeled Ab (antibody) to form a labeled Ag–Ab complex. (c) At the test zone, free Ag binds to an immobilized Ab to form an unlabeled Ab–Ag complex and prevents the formation of a labeled Ab–Ag–Ab sandwich. (d) At the control zone, a labeled Ab binds to immobilized antiglobulin and is captured at the control zone. (© Richard C. Li.)

antigen–antibody complex toward the test zone. At the test zone, the unbound antigen will compete with the antigen–antibody complexes for the antibody immobilized at the test zone. Since the antigen is in excess, the unbound antigen binds to the immobilized antibody, preventing the formation of the antibody–antigen–antibody sandwich. The resulting reading appears as a negative result. This artifact is known as the *high-dose hook* effect. To prevent the high-dose hook effect, a smaller volume of sample can be applied, or the sample can be diluted to reduce the amount of antigen applied.

11.3 Secondary Binding Assays

11.3.1 Precipitation-Based Assays

These techniques are based on the precipitation reaction and are used primarily for species identification in forensic laboratories.

11.3.1.1 Immunodiffusion

Immunodiffusion is a passive method in which an antigen or an antibody or both are allowed to diffuse and therefore a gradient, from low to high concentration, is established for an antigen or an antibody or both. As a result, precipitation occurs due to the interaction between an antigen and an antibody. The assay can be carried out in a liquid or a semisolid medium such as agarose gel. The semisolid medium can stabilize the diffusion process and reduce interference such as convection, which is the movement of molecules within liquids. The two types are single immunodiffusion and double immunodiffusion.

11.3.1.1.1 Single Immunodiffusion

A concentration gradient is established for either an antigen or an antibody. One serology technique based on this principle is radial immunodiffusion—a single diffusion method in which a concentration gradient is established for an antigen. The antibody is uniformly distributed in the gel matrix (Figure 11.6). The antigen is loaded into a sample well and allowed to diffuse from the well into the gel until a precipitation reaction occurs. The precipitate ring around the well is observed; the area within the ring of precipitate is proportional to the amount of antigen loaded in the well. Standards using known concentrations of antigen can be included in the same assay along with the samples, and a standard curve can be plotted. The amounts of antigen in samples can then be quantified by comparing the results with the standard curve.

11.3.1.1.2 Double Immunodiffusion

The second type of assay is double diffusion in which a concentration gradient is established for both an antigen and an antibody. The most common examples are the ring assay and the Ouchterlony assay. The *ring assay* can be performed in a test tube or capillary tube (Figure 11.7). An antiserum, a denser phase, is placed in a small tube. An antigen solution is carefully layered on top of an antibody solution without mixing. Both the antigen and the antibody will diffuse toward each other. In a positive reaction, a ring of precipitate can be observed at the interface of the two solutions. A negative reaction is indicated by a lack of precipitation. The assay requires positive and negative controls along with questioned samples.

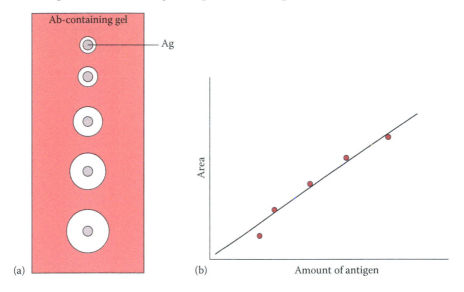

Figure 11.6 Radial immunodiffusion assay. (a) Immunodiffusion of antigens from a well into an antibody-containing gel. (b) Standard curve based on results of standards with known amounts of antigens. (© Richard C. Li.)

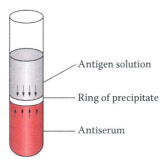

Figure 11.7 Ring assay. (© Richard C. Li.)

The *Ouchterlony assay* is named after the Swedish immunologist, Örjan Ouchterlony, who developed it. The assay can be performed in an agarose gel supported by a glass slide or polyester film (Figure 11.8). Wells are created by punching holes in the gel layer at desired locations. Often, a pattern with six wells surrounding a center well is used. The antibody is loaded in the central well while the questioned samples and the controls are loaded in the surrounding wells. The double diffusion of the antigen and the antibody from the wells is allowed to occur during incubation. If the reaction is positive, a precipitate line between wells can be observed at the end of incubation. The precipitate can be stained, enhancing visibility to aid observation. A single assay can compare more than one antigen to determine whether the antigens in question react the same way or differently with the antibody (see Chapter 7). This method is sometimes used to determine whether samples have come from the same or different origins.

11.3.1.2 Immunoelectrophoretic Methods

Diffusion techniques can be combined with electrophoresis to enhance test results. Electrophoresis separates molecules according to the differences in their electrophoretic mobility.

11.3.1.2.1 Immunoelectrophoresis

Immunoelectrophoresis (IEP) is a two-step procedure that can analyze a wide range of antigens. This technique uses electrophoresis to separate the antigen mixture prior to immunodiffusion (Figure 11.9). In the first step, the antigens in a sample are separated using agarose gel electrophoresis.

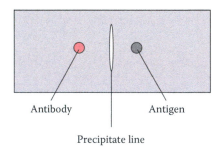

Antibody | Antigen

Precipitate line

Figure 11.8   Ouchterlony assay. (© Richard C. Li.)

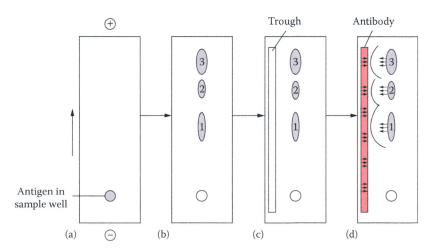

Figure 11.9   IEP assay. (a) Electrophoresis of antigens is carried out. (b) Various antigens are separated after electrophoresis. (c) Trough is cut. (d) Antibody is applied to the trough allowing diffusion to occur and forming precipitate lines. +, anode; −, cathode. (© Richard C. Li.)

Then, a trough is cut in the gel parallel to the array of antigens separated by electrophoresis. In the second step, an antibody is loaded in the trough and the gel is incubated for double diffusion. An arc-shaped precipitate line can be observed for a positive reaction. Multiple precipitate lines can occur if more than one antigen reacts with the antibody. The shapes, intensities, and locations of the precipitate lines of a known control and a questioned sample can be compared.

11.3.1.2.2 Crossed Immunoelectrophoresis

Crossed immunoelectrophoresis (CRIE), also known as two-dimensional IEP, is a modification of IEP. The first step utilizes electrophoresis to separate antigens contained in a sample (Figure 11.10). A strip of gel containing separated antigens is cut for the second round of electrophoresis. The gel including the gel strip is turned at a 90° angle and is further separated by a second-dimension electrophoresis. This drives the antigens from the gel strip into an agarose gel that contains uniformly distributed antibodies. Following the second-dimension electrophoresis, an arc-shaped precipitate line is formed. The area of the arc can be measured, and a sample can be identified by comparison with a known standard. This technique is more sensitive than IEP.

11.3.1.2.3 Rocket Immunoelectrophoresis

An antibody-containing agarose gel is used. The antigen is loaded into the well (Figure 11.11). Electrophoresis then drives the antigen from the well into the agarose gel. In a positive reaction,

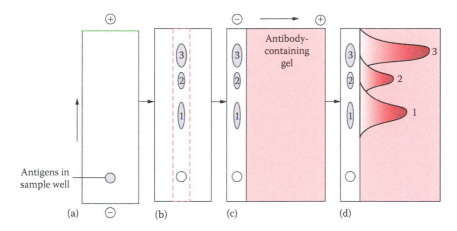

Figure 11.10 CRIE assay. (a) Electrophoresis of antigens is carried out. (b) Various antigens are separated after electrophoresis. A strip of gel containing separated antigens is excised and then used for second-dimension electrophoresis. (c) As shown, second-dimension electrophoresis is carried out to drive the antigens into an antibody-containing gel; (d) arc-shaped precipitate lines are formed as a result. (© Richard C. Li.)

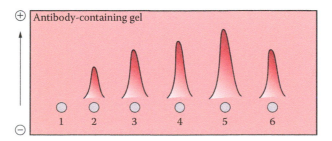

Figure 11.11 Rocket immunoelectrophoresis assay. Antigens are driven from the wells into an antibody-containing gel forming precipitate lines. (© Richard C. Li.)

a rocket-shaped precipitate line can be observed. The height of the rocket is in proportion to the amount of antigen in the sample. Quantitation can be achieved by comparing standards and the sample in the same gel.

11.3.1.2.4 Crossed-Over Immunoelectrophoresis

This technique is also known as counterimmunoelectrophoresis (CIE). Two arrays of opposing wells are created by punching holes in the agarose gel (Figure 11.12). The antibody and samples are loaded in opposing wells arranged by pairs. Electrophoresis is used to drive the antigen and the antibody toward each other. The wells containing the antibody should be proximal to the anode and the wells containing the samples should be proximal to the cathode. During gel electrophoresis, the antigen, which is usually negatively charged, migrates toward the anode. The antibody migrates in the opposite direction as a result of electroendosmosis—a phenomenon in which the movement of molecules is caused by fluid flow. A precipitate line is formed between the opposing wells if the antigen reacts with its specific antibody.

11.3.2 Agglutination-Based Assays

Agglutination reactions can be used as forensic serological assays such as for blood group typing (Chapter 18) and menstrual blood identification (Chapter 16). Agglutination assays are qualitative, indicating the absence or presence of antigens or antibodies. Semiquantitative results can be obtained by titration (diluting the antigen or antibody). Many types of agglutination reactions are available; only those used in forensic serology are discussed next.

Direct agglutination assays involve reactions in which an antibody interacts with antigens originally located on cell surfaces. In a *hemagglutination* reaction (Figure 11.13a), an antibody binds to the antigens located on erythrocytes. This method is used for the identification of blood types, for example, the testing of erythrocytes for ABO blood group typing. The assay can be carried out on a glass slide and the agglutination, as indicated by forming cell clumps, can be observed under a microscope (Chapter 18). The assay can also be carried out in a test tube. In a positive reaction, agglutinated cells are formed on the bottom of the test tube. Test tubes can also be centrifuged and then swirled to determine whether the cell clumps can be resuspended. Agglutinated cells cannot be resuspended. In a negative reaction, unagglutinated cells can be resuspended.

In *agglutination inhibition assays*, the presence of an antigen in question is indirectly detected (Figure 11.13b). If a known antigen is added to a mixture consisting of antigen-containing cells and antibodies, the added antigen will compete with the antigen located on the cell surfaces for antibody binding and inhibit the agglutination reaction. Another indirect agglutination assay is the absorption–elution assay (Chapter 18), which can be used for ABO blood group typing.

Passive agglutination assays are different from direct agglutination assays, in which the antigen is coated on the surface of carrier cells such as tannic acid–treated sheep erythrocytes. Tannic acid is considered to be a fixative. Treating with tannic acid can stabilize the carrier cells

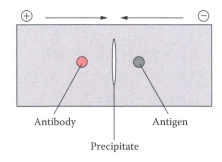

Figure 11.12 Crossed-over immunoelectrophoresis assay. (© Richard C. Li.)

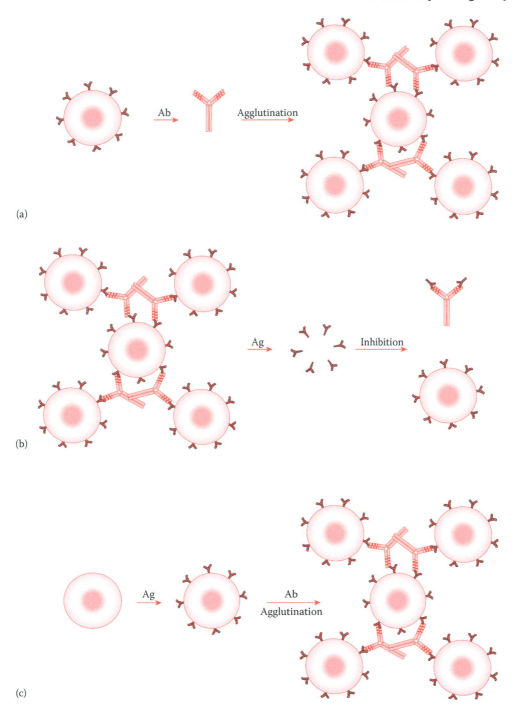

Figure 11.13 Agglutination-based assays. (a) Direct agglutination assay and (b) agglutination inhibition assay. Hemagglutination reactions are shown; antigens are located on erythrocytes. (c) Passive agglutination assay. (© Richard C. Li.)

for a subsequent coating of antigens. The carrier cells are then incubated with samples containing antigens. A coating of antigens can be formed on the surfaces of carriers. Such antigen-coated carrier cells can then be agglutinated using an antibody that is specific to the absorbed antigens (Figure 11.13c).

11.4 DNA Methylation Assays for Bodily Fluid Identification

In the eukaryotic genome, methylation occurs at the cytosine residues commonly in the CpG dinucleotide sequences of both DNA strands. The "p" in CpG refers to the phosphodiester bond between cytosine and guanine. Cytosine methylation is carried out *in vivo* by methyltransferase (DNMT). DNMT catalyzes the transfer of a methyl group (CH_3) from S-adenosylmethionine (SAM) to the C^5 position of cytosine (Figure 11.14). As a result, cytosine is converted to 5-methylcytosine. Cytosine methylation is observed throughout the eukaryotic genomes in both the coding and intergenic regions located between genes. However, cytosine methylation is rare at CpG islands, which are stretches of DNA containing a high frequency of CpG dinucleotides in the regulatory regions of genes (Figure 11.15). It is implicated that cytosine methylation alters the chromatin structure and causes chromatin condensation. As a result, gene expression is usually inhibited by cytosine methylation.

Genomic loci, known as *tissue-specific differentially methylated regions* (TDMRs), have been identified. TDMRs show differential DNA methylation patterns between different tissues. For example, semen-specific TDMRs are consistently hypomethylated in spermatozoa, which can be used to identify semen samples. Other TDMRs display varying degrees of methylation in different types of bodily fluids. This suggests that differential DNA methylation assays may be potentially useful for bodily fluid identification.

The detection of DNA methylation can be carried out by a number of methods. *Methylation-sensitive restriction enzyme digestion polymerase chain reaction* (MSRE-PCR) can be used for the

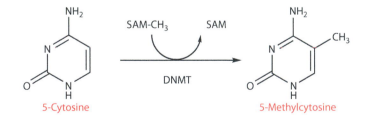

Figure 11.14 Eukaryotic cytosine methylation catalyzed by methyltransferase.

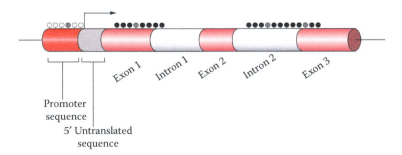

Figure 11.15 Methylation at the cytosine residues in CpG dinucleotide sequences. Note that the CpG dinucleotide sequences are usually unmethylated in the promoter regions. Black circle: methylated. Clear circle: unmethylated. Gray circle: partial methylated. (© Richard C. Li.)

rapid detection of DNA methylation. This method utilizes the *methylation-sensitive restriction enzyme* (MSRE). MSREs cleave DNA at unmethylated restriction sites, but they are not able to cleave once a cytosine residue is methylated. Thus, the methylated DNA remains intact. After PCR amplification, the methylated DNA is amplified and the amplicon is detected. In contrast, the unmethylated DNA is cleaved by the MSRE and cannot be amplified (Figure 11.16). *Bisulfite sequencing* is another widely used method. The genomic DNA is treated with sodium bisulfite, which catalyzes the hydrolytic deamination of unmethylated cytosines (Figure 11.17). As a result, unmethylated cytosines are converted to uracils, while methylated cytosines are resistant to conversion and remain unchanged. Next, uracils are replaced by thymines during PCR. After sequencing, the methylation sites can be deduced at a single-nucleotide resolution by comparing them with the reference sequence (Figure 11.18). *Methyl-DNA immunoprecipitation* (MeDIP) is a useful method for isolating methylated DNA. In this method, DNA is first sheared into fragments. The fragmented DNA is then treated with an antimethylcytosine antibody that binds to methylated cytosines. Methylated DNA fragments can then be isolated by immunoprecipitation for further analysis (Figure 11.19).

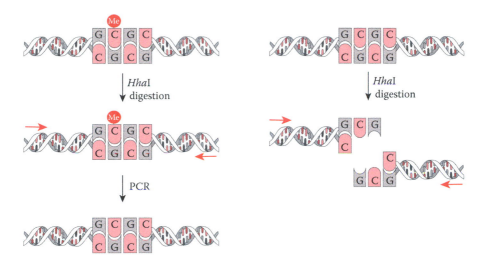

Figure 11.16 Methylation-sensitive restriction enzyme digestion PCR (MSRE-PCR) for the detection of DNA methylation. This method utilizes the methylation-sensitive restriction enzyme (MSRE) such as *Hha*I. *Hha*I cleaves DNA at the unmethylated sequence GCGC but is not able to cleave once a cytosine residue is methylated (GCmGC). The *Hha*I-treated DNA is then PCR amplified. The *Hha*I uncleaved DNA results in an amplicon, indicating the presence of a methylated cytosine in the sequence. The *Hha*I cleaved DNA cannot be amplified, indicating the absence of methylated cytosine in the sequence. Me, 5-methylcytosine site; red arrow, PCR primers. (Adapted from Frumkin, D. et al., *Forensic Sci Int Genet*, 5, 517–524, 2011.)

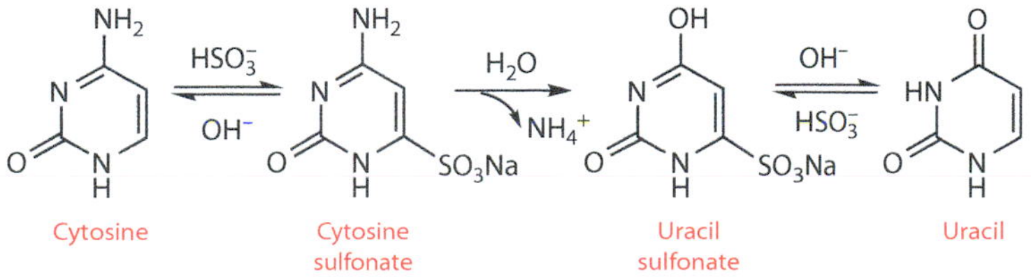

Figure 11.17   Bisulfite conversion of 5-methylcytosine.

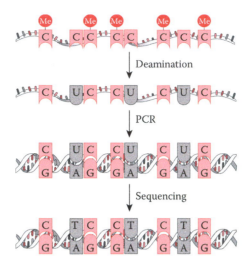

Figure 11.18   Schematic illustration of bisulfite sequencing of cytosine methylation. In a deamination reaction, unmethylated cytosines are converted to uracil. PCR and DNA sequencing are then carried out. The retention of C indicates that the site was methylated; the conversion of C to T indicates that the site was unmethylated in the DNA sample. Me, 5-methylcytosines site. (© Richard C. Li)

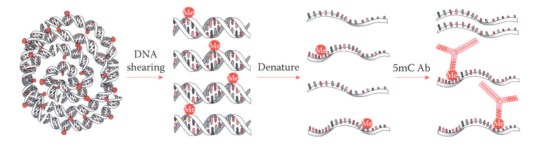

Figure 11.19   Methylated DNA immunoprecipitation (MeDIP) for the detection of methylated DNA. Genomic DNA is randomly and mechanically sheared into fragments. Fragments are then denatured to single-stranded DNA. The fragments containing the methylated region can be immunoprecipitated using an antibody that specifically recognizes 5-methylcytosine. The methylated DNA sequence in the immunoprecipitated fragments can then be determined by DNA sequencing and methylation analysis. Me, 5-methylcytosine site; 5mC Ab, 5-methylcytosine antibody. (© Richard C. Li.)

11.5 Forensic Applications of RNA-Based Assays and RNA Profiling

11.5.1 Messenger RNA-Based Assays

mRNA-based assays have been developed to identify bodily fluids for forensic investigation. In forensic pathological investigation, they can potentially be used for wound age estimation and the age of biological stains. The assays are based on the expression of certain genes in certain cell or tissue types. Candidate tissue-specific genes that may be useful for the identification of bodily fluids have been identified. Thus, the techniques that are used in the identification of bodily fluids are based on the detection of specific types of mRNA that are expressed exclusively in certain cells. The *tissue-specific genes* that are utilized for bodily fluid identification are summarized in Table 11.2. Additionally, *reference genes* that are constitutively expressed housekeeping genes are utilized as internal controls. The amount of mRNA can be assessed by normalizing the target gene to the expression level of the reference genes. Compared with conventional assays that are used for bodily fluid identification, the mRNA-based assay has higher specificity and is amenable to automation. However, one limitation is that the mRNA is unstable due to degradation by

Table 11.2   Representative Markers of mRNA-Based Assays for Bodily Fluid Identification

Bodily Fluid	Gene Symbol	Description
Blood	ALAS2	Aminolevulinatesynthase 2
	AMICA1	Adhesion molecule, interacts with CXADR antigen 1
	ANK1	Ankyrin1
	CD3G	CD3 gammamolecule
	CD93	CD93 molecule
	HBA1	Alpha 1 hemoglobin
	HBB	Beta hemoglobin
	PBGD	Porphobilinogen deaminase
	SPTB	Beta spectrin
Saliva	HTN3	Histatin 3
	MUC7	Mucin 7
	STATH	Statherin
Semen	KLK3	Kallikrein 3 (prostate-specific antigen)
	PRM1	Protamine 1
	PRM2	Protamine 2
	SEMG1	Semenogelin 1
	TGM4	Transglutaminase 4
Vaginal secretions	CYP2B7P1	Cytochrome P450, family 2, subfamily B, polypeptide 7 pseudogene 1
	DKK4	Dickkopf homolog 4
	FUT6	Fucosyltransferase 6
	HBD1	Beta defensin 1
	IL19	Interleukin 19
	MUC4	Mucin 4
	MYOZ1	Myozenin 1
	SFTA2	Surfactant associated 2
Menstrual blood	MMP7	Matrix metalloproteinase 7
	MMP11	Matrix metalloproteinase 11
Skin	CDSN	Corneodesmosin
	LOR	Loricrin

(continued)

Table 11.2 (Continued) Representative Markers of mRNA-Based Assays for Bodily Fluid Identification

Bodily Fluid	Gene Symbol	Description
Reference gene	*18S rRNA*	18S ribosomal RNA
	ACTB	Beta actin
	G6PDH	Hexose-6-phosphate dehydrogenase
	RPS15	Ribosomal protein S15

Sources: Adapted from Bauer, M. and Patzelt, D., *J Forensic Sci*, 47, 1278–1282, 2002; Hanson, E.K. and Ballantyne, J., *Sci Justice*, 53, 14–22, 2013; Lindenbergh, A., Maaskant, P., and Sijen, T., *Forensic Sci Int Genet*, 7, 159–166, 2013; Nussbaumer, C., Gharehbaghi-Schnell, E., and Korschineck, I., *Forensic Sci Int*, 157, 181–186, 2006.

endogenous ribonucleases. Additionally, bodily fluid stains collected from crime scenes are often exposed to ultraviolet (UV) light, moisture, and high temperature, which can promote mRNA degradation. Nevertheless, the successful detection of mRNA from aged samples is possible.

11.5.2 MicroRNA-Based Assays

The biological function of microRNAs (miRNAs) is to regulate gene expression. The mature miRNA strand associated with the RNA-induced silencing complex (RISC) binds to its target mRNA. In animal cells, the mature miRNA forms a base pairing with its complementary sequence of the *miRNA responsive element* (MRE) within the 3′ untranslated regions (3′UTRs) of the target mRNA (Figure 11.20). A single miRNA can bind different mRNA transcripts encoded by multiple genes. Animal miRNAs usually form a base pairing with their target mRNAs through partial complementary sequences, which lead to translation repression to inhibit protein synthesis. If the base pairing is exactly complementary to its target mRNA sequence, then the cleavage and the degradation of the target mRNA occurs. As a result, the synthesis of the protein encoded by the mRNA is reduced. Thus, miRNAs play important roles in cellular function.

miRNAs have potential applications in forensic identification. Recent studies have revealed that miRNAs can be detected from various bodily fluids such as blood, saliva, semen, vaginal secretions, and menstrual blood. The type of miRNA expressed is not exclusively tissue- or cell-type specific. However, the expression patterns of specific miRNAs are unique for various bodily fluids. Thus, a bodily fluid in question can potentially be identified using multiple differentially expressed miRNAs as markers (Table 11.3). miRNA markers for bodily fluid identification have their advantages compared with those of mRNA. Mature miRNAs are much

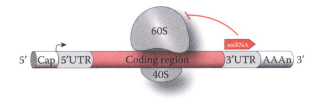

Figure 11.20 Schematic illustration of the roles of miRNA in translational inhibition. In animal cells, miRNAs bind to 3′UTR of their target mRNA with partially complementary Watson–Crick base pairing. As a result, the binding of miRNA to the target mRNA inhibits translation of the target mRNA. Eukaryotic 40S and 60S ribosomal subunits are shown. (© Richard C. Li.)

Table 11.3 Differentially Expressed miRNA Markers for Bodily Fluid Identification	
Biological Fluid	**miRNA Set**
Blood	miR16; miR451
	miR16; miR486
	miR20a; miR106a; miR144; miR185
	miR126; miR150; miR451
Semen	miR10b; miR135b
	miR10a; miR135a; miR943; miR507; miR891a
Saliva	miR205; miR658
	miR200c; miR203; miR205
Vaginal secretions	miR124a; miR372
Menstrual blood	miR214
	miR412; miR451
Reference gene	*RNU6-2* (U6 small nuclear 2 RNA)

Sources: Adapted from Courts, C. and Madea, B., *J Forensic Sci*, 56, 1464–1470, 2011; Hanson, E.K., Lubenow, H., and Ballantyne, J., *Anal Biochem*, 387, 303–314, 2009; Wang, Z. et al., *Forensic Sci Int Genet*, 7, 116–123, 2013; Zubakov, D. et al., *Int J Legal Med*, 124, 217–226, 2010.

smaller, approximately 21–23 nucleotides in length, than mRNA. As a result, miRNAs are less susceptible to degradation and more stable than mRNAs. Additionally, the sensitivity of the miRNA-based method is much higher than that of the mRNA-based methods.

11.6 Proteomic Approaches Using Mass Spectrometry for Bodily Fluid Identification

The forensic application of proteomic approaches is to identify biomarker proteins derived from bodily fluids to determine the type of bodily fluid in question for forensic investigations. Mass spectrometry (MS) is a highly sensitive and rapid technique for protein identification from complex biological samples.

11.6.1 Mass Spectrometric Instrumentation for Protein Analysis

A mass spectrometer is an analytical technique to identify a molecule by measuring the ratio of the mass (m) to the charge (z) of a charged molecule (Figure 11.21). A typical mass spectrometer instrument analyzes ionized molecules while in its gas phase. A mass spectrometer consists of an *ion source* that converts analytes into gaseous phase ions, through a process known as *ionization*, by gaining a positive or a negative charge from a neutral species. The ions are then introduced into the *mass analyzer* of a mass spectrometer, where ions are accelerated under electric fields and separated based on their *m/z* ratio. The separated ions are detected by an *ion detector* that records the number of ions at each *m/z* value. The result of molecular ionization,

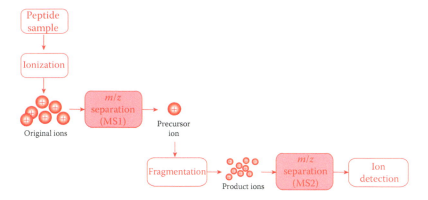

Figure 11.21   Diagram of tandem mass spectrometry. For the initial mass analysis of thermolabile analytes including peptides and proteins, soft ionization techniques such as matrix-assisted laser desorption ionization (MALDI) and electrospray ionization (ESI) can be used to produce original ions. MALDI utilizes a laser beam to trigger ionization and facilitate the vaporization of the analyte. In the ESI method, a fine mist of droplets is formed in the presence of a high electric field, producing charged ions. In the first round of mass analysis, ions are separated based on their *m/z* ratio. A particular ion, called a precursor ion, is then selected for fragmentation, generating product ions. The fragmentation can be carried out using collision techniques such as collision-induced dissociation (CID) by multiple collisions with rare gas atoms. Additionally, a new fragmentation technique, electron-capture dissociation (ECD), can be used in which the fragmentation is induced through the capture of a thermal electron by a protonated peptide cation. The product ions are analyzed by the second round of mass analysis and detected by an ion detector. MS, mass analyzer. (© Richard C. Li.)

ion separation, and ion detection is a spectrum that can be used to determine the mass (or molecular weight) and potentially the structure of the molecule.

Furthermore, biomarker proteins can be identified by *peptide sequencing* using the tandem mass spectrometry (MS/MS) mode of operation in which a mass spectrometer uses two or more mass analyzers. After an initial mass analysis, individual peptide ions in the mass spectrum can be selected. The selected ions are known as *precursor ions*. A precursor ion is then subjected to a second round of fragmentation through collision and is broken into smaller ions known as *product ions*. The resulting product ions are further analyzed by MS. Thus, the corresponding protein can be identified through searching a protein database, based on the mass of a specific peptide obtained.

11.6.2 Analysis Strategies for Protein Identification

To date, various MS-based protein identification strategies have been developed, which can be divided into two categories: "top-down" and "bottom-up" strategies (Figure 11.22). In the *top-down* strategy, intact proteins in a complex mixture are fractionated and separated into less complex protein mixtures or single proteins. Intact proteins are then analyzed using MS. A liquid chromatography–mass spectrometry (LC–MS) system can be used for such an application. The advantage of this approach is that an entire protein is analyzed for identification purposes. However, the throughput and efficiency of this approach is still a major challenge for large-scale sample analysis in forensic applications.

The *bottom-up* strategy is commonly used for high-throughput analysis of small peptides derived from highly complex samples. Current MS usually cannot resolve the mass measurement of large proteins; rather, MS is ideal for the analysis of small peptides. One approach is the *sort-then-break* approach in which proteins in a complex mixture are first isolated and then enzymatically cleaved. The target proteins are isolated from their complex biological source using 1-D or 2-D gel electrophoresis (Chapter 18). The isolated proteins are then enzymatically

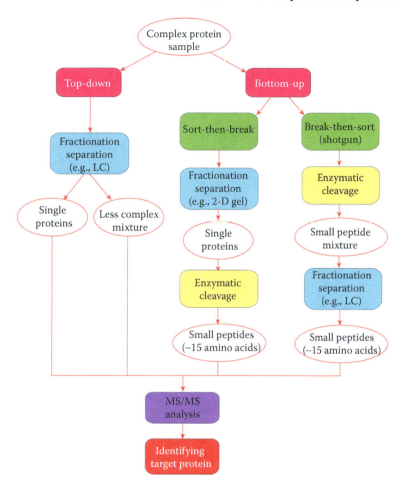

Figure 11.22 Top-down and bottom-up strategies for MS-based protein identification. LC, liquid chromatography; MS/MS, tandem mass spectrometry. (© Richard C. Li.)

cleaved into small peptides, approximately 15 amino acids in length, using proteases such as trypsin. These peptides are then analyzed by peptide sequencing using tandem MS to identify the target protein.

Break-then-sort, also known as the *shotgun* approach, is another approach of the bottom-up strategy. In this approach, proteins are cleaved first and the resulting peptide mixture is then separated by LC and analyzed by tandem MS (usually coupled to the LC). In a complex mixture, a peptide mass measurement alone is not sufficient for the identification of a target protein. Database searches are carried out utilizing software that compares the observed MS peptide spectra against all candidate spectra of the proteins that are present in a protein sequence (or translated from genomic DNA sequences) database for possible matches. Thus, a biomarker protein can be identified.

11.7 Microbial DNA Analysis for Bodily Fluid Identification

Human bodies harbor a large number of microbes. It is estimated that there are over 10 times more microbial cells than human cells in and on the human body, in the nose, mouth, throat, intestine, vagina, and skin. This microbial community is referred to as the human *microbiota*.

217

It is known that the microbiota has an important influence on human health. A decade ago, the concept of the *microbiome* was proposed by Nobel laureate Joshua Lederberg to identify and characterize the microbiota, their genomes, and their environmental interactions in a defined community. Traditionally, the studies of microbial genomes rely on cultivated individual species. The disadvantage of cultivation-based methods, however, is that large numbers of microbial species cannot be successfully isolated because many microbial species require specific growth conditions. Based on the advances in DNA sequencing technologies, a new approach, known as *metagenomics*, has been developed. Instead of studying the genome of an individual cultivated microbial species, metagenomics allows the study of the microbial genome of complete microbial communities harvested directly from natural environments. Metagenomics has been a powerful tool to analyze the human microbiome. Recently, human microbiota have been surveyed, revealing that microbial diversity is greatest in the teeth and the intestines, intermediate on the skin and the inside surface of the cheek, and lowest in the vagina (Table 11.4). More interestingly, each habitat is characterized by a small number of highly predominant bacterial taxa. The predominant taxa of each habitat can potentially be utilized for the forensic analysis of biological evidence. Bodily fluids identification is an important task in the forensic analysis of biological evidence. The detection of predominant taxa of bacteria can aid in the forensic identification of a particular type of bodily fluid (Table 11.4).

Table 11.4 Representative Microbiota that are Potentially Can Be Used for Forensic Identification

Body Habitats	Estimated Number of Microbial Species	Predominent Taxa	Potential Forensic Applications	DNA Marker	Further Reading
Cheek	800	*Streptococcus* spp.	Saliva identification	*gtf*	Nakanishi et al. (2011)
			Expirated blood	*gtf*	Donaldson et al. (2010)
				16S rRNA	Power et al. (2010)
Vagina	300	*Lactobacillus* spp.	Vaginal secretion identification	16S rRNA	Akutsu et al. (2012)
				16S–23S ISR	Fleming and Harbison (2010)
					Giampaoli et al. (2012)
Gastrointestinal tract	4000	*Bacteroides* spp.	Fecal matter identification	*rpoB*	Nakanishi et al. (2013)
Skin	1000	*Staphylococcus* spp.	Touched evidence identification	16S rRNA	Fierer et al. (2010)

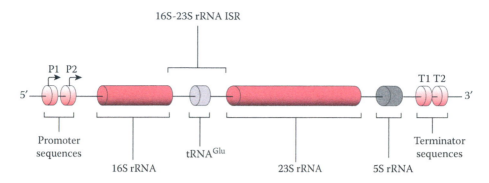

Figure 11.23 Schematic diagram of an *Escherichia coli* rRNA operon. *E. coli* has seven rRNA operons. A typical rRNA operon is shown. The rRNA operon encodes for three ribosomal RNA (rRNA) genes: the 5S and 23S genes code for the large ribosomal subunit, while the 16S gene encodes for the small ribosomal subunit. *E. coli* encodes one tRNA gene in the rRNA operon. The intergenic spacer regions (ISRs) and the promoter and terminator sequences are also shown. The rRNA operon is transcribed into a single precursor RNA transcript that is cleaved into separate rRNA and tRNA transcripts. The diagram is not to scale to the physical map of rRNA operons. (© Richard C. Li.)

The detection of bacteria can be achieved by utilizing DNA markers. The most commonly used DNA markers are the essential genes present in all bacterial species. These markers have highly variable regions that can be used to distinguish various bacterial species. The DNA sequences within these variable regions can also suggest taxonomic relationships between different taxa. Additionally, DNA markers contain conserved regions that are necessary for use as primer-binding sites for PCR-based detection.

For example, the rRNA genes (Figure 11.23), encoding for the ribosomal RNA, are conserved in all living organisms including bacteria. The bacterial rRNA operon contains three rRNA genes: the 5S and 23S rRNA genes encode the RNA components of the large subunit of the ribosome, while the 16S rRNA gene encodes the RNA component of the small subunit of the ribosome. The 16S rRNA gene is the most commonly used marker for the taxonomic identification of bacteria. The bacterial 16S rRNA genes are similar in length, approximately 1.5 kb. They contain highly conserved regions that can be used as primer-binding sites for the amplification of the adjacent DNA regions. Additionally, they contain variable regions that allow for taxonomic identification. In forensic applications, they are used as a marker for the identification of vaginal bacteria as an indicator of vaginal secretions (Chapter 16). However, the number of variable regions in the 16S rRNA genes of some bacteria species is limited. Furthermore, closely related species have high levels of sequence similarity to the 16S rRNA genes. Thus, the 16S rRNA gene marker is not adequate for distinguishing closely related species.

DNA markers of bacteria identification can also be located in noncoding regions of DNA. The *intergenic spacer region* (ISR) between the 16S rRNA and 23S rRNA genes in the rRNA operon is another commonly used marker. The 16S–23S rRNA ISR contains both length and sequence variations between species. These variations are partially caused by the number and type of tRNA genes that this region of operon contains. For example, most gram-negative bacteria contain both a copy of the tRNA^{ala} and the tRNA^{ile} genes, while some contain only a copy of the tRNA^{glu} gene. In contrast, most gram-positive bacteria have no tRNA gene; some contain either a copy of the tRNA^{ala} gene or the tRNA^{ile} gene or both. Thus, heterogeneity in the length and in the sequence of the 16S–23S rRNA ISR allows bacteria to be identified at the species level.

In addition to rRNA genes, many other genes are utilized as DNA markers for bacterial species identification. For example, the *rpoB* gene encodes the β subunit of bacterial RNA polymerase and is used as a marker for the forensic identification of fecal matter (Chapter 17). The sequence variation of the *rpoB* gene allows the distinguishing of species when they are not distinguishable, using 16S rRNA gene sequences. Several oral streptococcal species that produce extracellular polysaccharides contain the *gtf* gene, which encodes glucosyltransferase. Thus, the *gtf* gene is a useful marker for oral streptococcal species identification as a potential indicator of saliva in forensic investigations. However, this marker cannot be used to identify bacteria other than *streptococcus*. Nevertheless, combining several DNA markers can be used for the identification of bacteria species.

11.8 Nondestructive Assays for the Identification of Bodily Fluids

Most of the bodily fluid identification techniques mentioned previously consume a portion of the evidence. When the amount of sample evidence is very limited, such destructive assays may consume the evidence that is also needed for subsequent forensic DNA analysis. Moreover, most of these tests are used for a single type of bodily fluid. Tests for multiple bodily fluids in question will consume additional amounts of the evidence. Therefore, nondestructive identification techniques are highly desired. Additionally, it is important to develop an assay that can be utilized to test multiple types of bodily fluids. Furthermore, a portable device that can identify bodily fluids at a crime scene is very useful. It provides investigators with test results immediately to aid the crime scene investigations. Two techniques have been developed that can potentially be used for nondestructive bodily fluid identification: fluorescence spectroscopy and Raman spectroscopy.

Fluorescence spectroscopy is one example of a nondestructive technique. Fluorescence is the emission of light by a fluorophore. A fluorophore is a moiety in a molecule that fluoresces on absorbing the energy from an excitation light source or radiation. The emitted light usually has a longer wavelength and lower energy than the absorbed radiation. Constituents present in bodily fluids, such as nucleic acids, proteins, lipids, and metabolites, can exhibit fluorescence. The unique composition of a bodily fluid emits characteristic emission spectra, thus making it identifiable. Using multiple wavelengths of an excitation light source, fluorescent emissions at a wide range of wavelengths can be detected, which allows for the identification of various bodily fluids. This technique is sensitive and rapid. It does not utilize chemical reagents and the detection does not require physical contact with a sample. After identification, the same biological sample can be used for forensic DNA analysis. Additionally, portable fluorescence instruments can potentially be used at a crime scene for bodily fluid identification. However, it is not clear if exposure to an excitation light source can damage DNA evidence.

Raman spectroscopy is another example of a nondestructive technique. Raman spectroscopy utilizes a near-infrared excitation light source and measures the scattering of laser light caused by the vibrating molecules of a sample. A typical Raman spectrum provides information about the molecular structure and the "signature" based on the properties of the constituents of a sample. Thus, it is possible to obtain unique Raman spectra for a particular type of bodily fluid. Raman spectroscopy is highly sensitive. The measurements can be carried out on very small amounts of sample, ranging from a few picoliters to femtoliters. This technique does not require any chemical reagent. Additionally, it does not consume the sample, which can be used for subsequent DNA analysis. Raman spectroscopy has been utilized for various forensic purposes including the identification of drugs, trace evidence such as fibers, and questioned document evidence such as inks and paints. It can potentially be used for the nondestructive identification of various bodily fluids. A portable Raman spectrometer will allow bodily fluids to be identified at crime scenes.

Bibliography

Aaspõllu, A., et al., Can microbes on skin help linking persons and crimes? *Forensic Sci Int Genet Suppl Series*, 2011, **3**(1): e269–e270.

Akutsu, T., et al., Detection of bacterial 16S ribosomal RNA genes for forensic identification of vaginal fluid. *Leg Med*, 2012, **14**(3): 160–162.

Allen, S.M., An enzyme linked immunosorbent assay (ELISA) for detection of seminal fluid using a monoclonal antibody to prostatic acid phosphatase. *J Immunoassay*, 1995, **16**(3): 297–308.

Alvarez, M., J. Juusola, and J. Ballantyne, An mRNA and DNA co-isolation method for forensic casework samples. *Anal Biochem*, 2004, **335**(2): 289–298.

Amin, K.M., et al., The exon-intron organization of the human erythroid beta-spectrin gene. *Genomics*, 1993, **18**(1): 118–125.

An, J.H., et al., Body fluid identification in forensics. *BMB Rep*, 2012, **45**(10): 545–553.

An, J.H., et al., DNA methylation-specific multiplex assays for body fluid identification. *Int J Legal Med*, 2013, **127**(1): 35–43.

Anderson, J.R., The agglutination of sensitised red cells by antibody to serum, with special reference to non-specific reactions. *Br J Exp Pathol*, 1952, **33**(5): 468–483.

Bauer, M., RNA in forensic science. *Forensic Sci Int Genet*, 2007, **1**(1): 69–74.

Bauer, M. and D. Patzelt, Evaluation of mRNA markers for the identification of menstrual blood. *J Forensic Sci*, 2002, **47**(6): 1278–1282.

Bauer, M. and D. Patzelt, Protamine mRNA as molecular marker for spermatozoa in semen stains. *Int J Legal Med*, 2003, **117**(3): 175–179.

Bauer, M., A. Kraus, and D. Patzelt, Detection of epithelial cells in dried blood stains by reverse transcriptase-polymerase chain reaction. *J Forensic Sci*, 1999, **44**(6): 1232–1236.

Benschop, C.C.G., et al., Vaginal microbial flora analysis by next generation sequencing and microarrays; can microbes indicate vaginal origin in a forensic context? *Int J Legal Med*, 2012, **126**(2): 303–310.

Berti, A., et al., Expression of seminal vesicle-specific antigen in serum of lung tumor patients. *J Forensic Sci*, 2005, **50**(5): 1114–1115.

Black, C.M., et al., Multicenter study of nucleic acid amplification tests for detection of *Chlamydia trachomatis* and *Neisseria gonorrhoeae* in children being evaluated for sexual abuse. *Pediatr Infect Dis J*, 2009, **28**(7): 608–613.

Bocklandt, S., et al., Epigenetic predictor of age. *PLoS One*, 2011, **6**(6): e14821.

Bowden, A., R. Fleming, and S. Harbison, A method for DNA and RNA co-extraction for use on forensic samples using the Promega DNA IQ™ system. *Forensic Sci Int Genet*, 2011, **5**(1): 64–68.

Brenig, B., J. Beck, and E. Schutz, Shotgun metagenomics of biological stains using ultra-deep DNA sequencing. *Forensic Sci Int Genet*, 2010, **4**(4): 228–231.

Brzezinski, J.L. and D.L. Craft, Characterization of microorganisms isolated from counterfeit toothpaste. *J Forensic Sci*, 2012, **57**(5): 1365–1367.

Chu, Z.L., et al., Erythroid-specific processing of human beta spectrin I pre-mRNA. *Blood*, 1994, **84**(6): 1992–1999.

Counsil, T.I. and J.L. McKillip, Forensic blood evidence analysis using RNA targets and novel molecular tools. *Biologia*, 2010, **65**(2): 175–182.

Courts, C. and B. Madea, Micro-RNA—A potential for forensic science? *Forensic Sci Int*, 2010, **203**(1–3): 106–111.

Courts, C. and B. Madea, Specific micro-RNA signatures for the detection of saliva and blood in forensic body-fluid identification. *J Forensic Sci*, 2011, **56**(6): 1464–1470.

Courts, C. and B. Madea, [Ribonucleic acid—Importance in forensic molecular biology]. *Rechtsmedizin*, 2012, **22**(2): 135–144.

Courts, C., M. Grabmuller, and B. Madea, Dysregulation of heart and brain specific micro-RNA in sudden infant death syndrome. *Forensic Sci Int*, 2013, **228**(1–3): 70–74.

Culliford, B.J., Precipitin reactions in forensic problems. A new method for precipitin reactions on forensic blood, semen and saliva stains. *Nature*, 1964, **201**: 1092–1093.

Cummings, C.A. and D.A. Relman, Genomics and microbiology. Microbial forensics— "cross-examining pathogens". *Science*, 2002, **296**(5575): 1976–1979.

Donaldson, A.E., et al., Using oral microbial DNA analysis to identify expirated bloodspatter. *Int J Legal Med*, 2010, **124**(6): 569–576.

Elkins, K.M., Rapid presumptive "fingerprinting" of body fluids and materials by ATR FT-IR spectroscopy. *J Forensic Sci*, 2011, **56**(6): 1580–1587.

Ferguson, L.S., et al., Direct detection of peptides and small proteins in fingermarks and determination of sex by MALDI mass spectrometry profiling. *Analyst*, 2012, **137**(20): 4686–4692.

Fierer, N., et al., Forensic identification using skin bacterial communities. *Proc Natl Acad Sci U S A*, 2010, **107**(14): 6477–6481.

Fleming, R.I. and S. Harbison, The development of a mRNA multiplex RT-PCR assay for the definitive identification of body fluids. *Forensic Sci Int Genet*, 2010, **4**(4): 244–256.

Fleming, R.I. and S. Harbison, The use of bacteria for the identification of vaginal secretions. *Forensic Sci Int Genet*, 2010, **4**(5): 311–315.

Fordyce, S.L., et al., Long-term RNA persistence in postmortem contexts. *Investig Genet*, 2013, **4**(1): 7.

Frascione, N., et al., Detection and identification of body fluid stains using antibody-nanoparticle conjugates. *Analyst*, 2012, **137**(2): 508–512.

Frumkin, D., et al., Authentication of forensic DNA samples. *Forensic Sci Int Genet*, 2010, **4**(2): 95–103.

Frumkin, D., et al., DNA methylation-based forensic tissue identification. *Forensic Sci Int Genet*, 2011, **5**(5): 517–524.

Gao, L.L., et al., Application and progress of RNA in forensic science. *J Forensic Med*, 2011, **27**(6): 455–459.

Gauvin, J., et al., Forensic pregnancy diagnostics with placental mRNA markers. *Int J Legal Med*, 2010, **124**(1): 13–17.

Gelmini, S., et al., Real-time quantitative reverse transcriptase-polymerase chain reaction (RT-PCR) for the measurement of prostate-specific antigen mRNA in the peripheral blood of patients with prostate carcinoma using the taqman detection system. *Clin Chem Lab Med*, 2001, **39**(5): 385–391.

Giampaoli, S., et al., Molecular identification of vaginal fluid by microbial signature. *Forensic Sci Int Genet*, 2012, **6**(5): 559–564.

Gipson, I.K., et al., Mucin genes expressed by human female reproductive tract epithelia. *Biol Reprod*, 1997, **56**(4): 999–1011.

Gipson, I.K., et al., MUC4 and MUC5B transcripts are the prevalent mucin messenger ribonucleic acids of the human endocervix. *Biol Reprod*, 1999, **60**(1): 58–64.

Goga, H., Comparison of bacterial DNA profiles of footwear insoles and soles of feet for the forensic discrimination of footwear owners. *Int J Legal Med*, 2012, **126**(5): 815–823.

Grskovic, B., et al., DNA methylation: The future of crime scene investigation? *Mol Biol Rep*, 2013, **40**(7): 4349–4360.

Gubin, A.N. and J.L. Miller, Human erythroid porphobilinogen deaminase exists in 2 splice variants. *Blood*, 2001, **97**(3): 815–817.

Gunn, A. and S.J. Pitt, Microbes as forensic indicators. *Trop Biomed*, 2012, **29**(3): 311–330.

Haas, C., E. Hanson, and J. Ballantyne, Capillary electrophoresis of a multiplex reverse transcription-polymerase chain reaction to target messenger RNA markers for body fluid identification. *Methods Mol Biol*, 2012, **830**: 169–183.

Haas, C., et al., mRNA profiling for body fluid identification by reverse transcription endpoint PCR and realtime PCR. *Forensic Sci Int Genet*, 2009, **3**(2): 80–88.

Haas, C., et al., mRNA profiling for the identification of sperm and seminal plasma. *Forensic Sci Int Genet Suppl Series*, 2009, **2**(1): 534–535.

Haas, C., et al., Collaborative EDNAP exercises on messenger RNA/DNA co-analysis for body fluid identification (blood, saliva, semen) and STR profiling. *Forensic Sci Int Genet Suppl Series*, 2011, **3**(1): e5–e6.

Haas, C., et al., MRNA profiling for the identification of blood—Results of a collaborative EDNAP exercise. *Forensic Sci Int Genet*, 2011, **5**(1): 21–26.

Haas, C., et al., Selection of highly specific and sensitive mRNA biomarkers for the identification of blood. *Forensic Sci Int Genet*, 2011, **5**(5): 449–458.

Haas, C., et al., RNA/DNA co-analysis from blood stains—Results of a second collaborative EDNAP exercise. *Forensic Sci Int Genet*, 2012, **6**(1): 70–80.

Haas, C., et al., RNA/DNA co-analysis from human menstrual blood and vaginal secretion stains: Results of a fourth and fifth collaborative EDNAP exercise. *Forensic Sci Int Genet*, 2014, **8**(1): 203–212.

Haas, C., et al., RNA/DNA co-analysis from human saliva and semen stains—Results of a third collaborative EDNAP exercise. *Forensic Sci Int Genet*, 2013, **7**(2): 230–239.

Hampson, C., J. Louhelainen, and S. McColl, An RNA expression method for aging forensic hair samples. *J Forensic Sci*, 2011, **56**(2): 359–365.

Hanson, E.K. and J. Ballantyne, Circulating microRNA for the identification of forensically relevant body fluids. *Methods Mol Biol*, 2013, **1024**: 221–234.

Hanson, E.K. and J. Ballantyne, Highly specific mRNA biomarkers for the identification of vaginal secretions in sexual assault investigations. *Sci Justice*, 2013, **53**(1): 14–22.

Hanson, E.K., H. Lubenow, and J. Ballantyne, Identification of forensically relevant body fluids using a panel of differentially expressed microRNAs. *Anal Biochem*, 2009, **387**(2): 303–314.

Hanson, E., et al., Identification of skin in touch/contact forensic samples by messenger RNA profiling. *Forensic Sci Int Genet Suppl Series*, 2011, **3**(1): e305–e306.

Hanson, E., et al., Specific and sensitive mRNA biomarkers for the identification of skin in 'touch DNA' evidence. *Forensic Sci Int Genet*, 2012, **6**(5): 548–558.

Harteveld, J., A. Lindenbergh, and T. Sijen, RNA cell typing and DNA profiling of mixed samples: Can cell types and donors be associated? *Sci Justice*, 2013, **53**(3): 261–269.

Heinrich, M., et al., Real-time PCR detection of five different "endogenous control gene" transcripts in forensic autopsy material. *Forensic Sci Int Genet*, 2007, **1**(2): 163–169.

Hochmeister, M.N., et al., Evaluation of prostate-specific antigen (PSA) membrane test assays for the forensic identification of seminal fluid. *J Forensic Sci*, 1999, **44**(5): 1057–1060.

Hochmeister, M.N., et al., Validation studies of an immunochromatographic 1-step test for the forensic identification of human blood. *J Forensic Sci*, 1999, **44**(3): 597–602.

Hsu, L., et al., Amplification of oral streptococcal DNA from human incisors and bite marks. *Curr Microbiol*, 2012, **65**(2): 207–211.

Iida, R., T. Yasuda, and K. Kishi, Identification of novel fibronectin fragments detected specifically in juvenile urine. *FEBS J*, 2007, **274**(15): 3939–3947.

Jakubowska, J., A. Maciejewska, and R. Pawłowski, mRNA profiling in identification of biological fluids in forensic genetics. *Z Zagadnien Nauk Sadowych*, 2011, **87**: 204–215.

Jakubowska, J., et al., mRNA profiling for vaginal fluid and menstrual blood identification. *Forensic Sci Int Genet*, 2013, **7**(2): 272–278.

Johnston, S., J. Newman, and R. Frappier, Validation study of the abacus diagnostics ABAcard HemaTrace Membrane Test for the forensic identification of human blood. *Can Soc Forensic Sci J*, 2003, **36**(3): 173–183.

Juusola, J. and J. Ballantyne, Messenger RNA profiling: A prototype method to supplant conventional methods for body fluid identification. *Forensic Sci Int*, 2003, **135**(2): 85–96.

Juusola, J. and J. Ballantyne, Multiplex mRNA profiling for the identification of body fluids. *Forensic Sci Int*, 2005, **152**(1): 1–12.

Juusola, J. and J. Ballantyne, mRNA profiling for body fluid identification by multiplex quantitative RT-PCR. *J Forensic Sci*, 2007, **52**(6): 1252–1262.

Kakizaki, E., et al., Bioluminescent bacteria have potential as a marker of drowning in seawater: Two immersed cadavers retrieved near estuaries. *Leg Med*, 2009, **11**(2): 91–96.

Kakizaki, E., et al., Detection of marine and freshwater bacterioplankton in immersed victims: Post-mortem bacterial invasion does not readily occur. *Forensic Sci Int*, 2011, **211**(1–3): 9–18.

Kakizaki, E., et al., *In vitro* study of possible microbial indicators for drowning: Salinity and types of bacterioplankton proliferating in blood. *Forensic Sci Int*, 2011, **204**(1–3): 80–87.

Kakizaki, E., et al., Detection of diverse aquatic microbes in blood and organs of drowning victims: First metagenomic approach using high-throughput 454-pyrosequencing. *Forensic Sci Int*, 2012, **220**(1–3): 135–146.

Kohlmeier, F. and P.M. Schneider, Successful mRNA profiling of 23 years old blood stains. *Forensic Sci Int Genet*, 2012, **6**(2): 274–276.

Koppelkamm, A., et al., Validation of adequate endogenous reference genes for the normalisation of qPCR gene expression data in human post mortem tissue. *Int J Legal Med*, 2010, **124**(5): 371–380.

Koppelkamm, A., et al., RNA integrity in post-mortem samples: Influencing parameters and implications on RT-qPCR assays. *Int J Legal Med*, 2011, **125**(4): 573–580.

Kumagai, A., N. Nakayashiki, and Y. Aoki, Analysis of age-related carbonylation of human vitreous humor proteins as a tool for forensic diagnosis. *Leg Med (Tokyo)*, 2007, **9**(4): 175–180.

Lee, H.Y., et al., Potential forensic application of DNA methylation profiling to body fluid identification. *Int J Legal Med*, 2012, **126**(1): 55–62.

Lee, H.Y., et al., Potential forensic application of DNA methylation profiling to body fluid identification. *Int J Legal Med*, 2012, **126**(1): 55–62.

Lehman, D.C., Forensic microbiology. *Clin Lab Sci*, 2012, **25**(2): 114–119.

Li, W.C., et al., Estimation of postmortem interval using microRNA and 18S rRNA degradation in rat cardiac muscle. *J Forensic Med*, 2010, **26**(6): 413–417.

Li, C., et al., Differences of DNA methylation profiles between monozygotic twins' blood samples. *Mol Biol Rep*, 2013, **40**(9): 5275–5280.

Lincoln, P.J. and B.E. Dodd, The use of low ionic strength solution (LISS) in elution experiments and in combination with papain-treated cells for the titration of various antibodies, including eluted antibody. *Vox Sang*, 1978, **34**(4): 221–226.

Lindenbergh, A., et al., A multiplex (m)RNA-profiling system for the forensic identification of body fluids and contact traces. *Forensic Sci Int Genet*, 2012, **6**(5): 565–577.

Lindenbergh, A., P. Maaskant, and T. Sijen, Implementation of RNA profiling in forensic casework. *Forensic Sci Int Genet*, 2013, **7**(1): 159–166.

Lucci, A., et al., A promising microbiological test for the diagnosis of drowning. *Forensic Sci Int*, 2008, **182**(1–3): 20–26.

Madi, T., et al., The determination of tissue-specific DNA methylation patterns in forensic biofluids using bisulfite modification and pyrosequencing. *Electrophoresis*, 2012, **33**(12): 1736–1745.

Moreno, L.I., et al., Determination of an effective housekeeping gene for the quantification of mRNA for forensic applications. *J Forensic Sci*, 2012, **57**(4): 1051–1058.

Nagasawa, S., et al., Detection of *Helicobacter pylori* (*H. pylori*) DNA in digestive systems from cadavers by real-time PCR. *Leg Med (Tokyo)*, 2009, **11**(Suppl 1): S458–S459.

Nakanishi, H., et al., A novel method for the identification of saliva by detecting oral streptococci using PCR. *Forensic Sci Int*, 2009, **183**(1–3): 20–23.

Nakanishi, H., et al., A simple identification method of saliva by detecting *Streptococcus salivarius* using loop-mediated isothermal amplification. *J Forensic Sci*, 2011, **56**(Suppl. 1): S158–S167.

Nakanishi, H., et al., Identification of feces by detection of *Bacteroides* genes. *Forensic Sci Int Genet*, 2013, **7**(1): 176–179.

Nakatome, M., et al., Methylation analysis of circadian clock gene promoters in forensic autopsy specimens. *Leg Med (Tokyo)*, 2011, **13**(4): 205–209.

Nakayashiki, N., et al., Analysis of the methylation profiles in imprinted genes applicable to parental allele discrimination. *Leg Med (Tokyo)*, 2009, **11**(Suppl 1): S471–S472.

Nakayashiki, N., et al., Investigation of the methylation status around parent-of-origin detectable SNPs in imprinted genes. *Forensic Sci Int Genet*, 2009, **3**(4): 227–232.

Nussbaumer, C., E. Gharehbaghi-Schnell, and I. Korschineck, Messenger RNA profiling: A novel method for body fluid identification by real-time PCR. *Forensic Sci Int*, 2006, **157**(2–3): 181–186.

Oakley, C.L. and A.J. Fulthorpe, Antigenic analysis by diffusion. *J Pathol Bacteriol*, 1953, **65**(1): 49–60.

Odriozola, A., et al., miRNA analysis in vitreous humor to determine the time of death: A proof-of-concept pilot study. *Int J Legal Med*, 2013, **127**(3): 573–578.

Omelia, E.J., M.L. Uchimoto, and G. Williams, Quantitative PCR analysis of blood- and saliva-specific microRNA markers following solid-phase DNA extraction. *Anal Biochem*, 2013, **435**(2): 120–122.

Ota, M., et al., Restriction enzyme analysis of PCR products. *Methods Mol Biol*, 2009, **578**: 405–414.

Ouchterlony, O., Antigen-antibody reactions in gels. *Acta Pathol Microbiol Scand*, 1949, **26**: 507.

Pang, B.C. and B.K. Cheung, Identification of human semenogelin in membrane strip test as an alternative method for the detection of semen. *Forensic Sci Int*, 2007, **169**(1): 27–31.

Park, J.L., et al., Forensic body fluid identification by analysis of multiple RNA markers using nanoString technology. *Genomics Inform*, 2013, **11**(4): 277–281.

Park, S.M., et al., Genome-wide mRNA profiling and multiplex quantitative RT-PCR for forensic body fluid identification. *Forensic Sci Int Genet*, 2013, **7**(1): 143–150.

Parker, C., E. Hanson, and J. Ballantyne, Optimization of dried stain co-extraction methods for efficient recovery of high quality DNA and RNA for forensic analysis. *Forensic Sci Int Genet Suppl Series*, 2011, **3**(1): e309–e310.

Partemi, S., et al., Analysis of mRNA from human heart tissue and putative applications in forensic molecular pathology. *Forensic Sci Int*, 2010, **203**(1–3): 99–105.

Petrie, G.F., A specific precipitin reaction associated with with growth on agar plates on *meningococcus*, *pneumococcus*, and *B. dysenteriae* (shiga). *Br J Exp Pathol*, 1932, **13**: 380.

Phang, T.W., et al., Amplification of cDNA via RT-PCR using RNA extracted from postmortem tissues. *J Forensic Sci*, 1994, **39**(5): 1275–1279.

Power, D.A., et al., PCR-based detection of salivary bacteria as a marker of expired blood. *Sci Justice*, 2010, **50**(2): 59–63.

Qi, B., L. Kong, and Y. Lu, Gender-related difference in bloodstain RNA ratio stored under uncontrolled room conditions for 28 days. *J Forensic Legal Med*, 2013, **20**(4): 321–325.

Qin, J., et al., Application situation and prospect of mRNA in forensic medicine. *Chin J Forensic Med*, 2013, **28**(1): 30–33.

Quarino, L., et al., An ELISA method for the identification of salivary amylase. *J Forensic Sci*, 2005, **50**(4): 873–876.

Richard, M.L., et al., Evaluation of mRNA marker specificity for the identification of five human body fluids by capillary electrophoresis. *Forensic Sci Int Genet*, 2012, **6**(4): 452–460.

Roeder, A.D. and C. Haas, mRNA profiling using a minimum of five mRNA markers per body fluid and a novel scoring method for body fluid identification. *Int J Legal Med*, 2013, **127**(4): 707–721.

Sabatini, L.M., Y.Z. He, and E.A. Azen, Structure and sequence determination of the gene encoding human salivary statherin. *Gene*, 1990, **89**(2): 245–251.

Saeed, M., R.M. Berlin, and T.D. Cruz, Exploring the utility of genetic markers for predicting biological age. *Leg Med (Tokyo)*, 2012, **14**(6): 279–285.

Sakurada, K., et al., Evaluation of mRNA-based approach for identification of saliva and semen. *Leg Med*, 2009, **11**(3): 125–128.

Sakurada, K., et al., Identification of nasal blood by real-time RT-PCR. *Leg Med (Tokyo)*, 2012, **14**(4): 201–204.

Salamonsen, L.A. and D.E. Woolley, Matrix metalloproteinases in normal menstruation. *Hum Reprod*, 1996, **11**(Suppl 2): 124–133.

Salamonsen, L.A. and D.E. Woolley, Menstruation: Induction by matrix metalloproteinases and inflammatory cells. *J Reprod Immunol*, 1999, **44**(1–2): 1–27.

Sampaio-Silva, F., et al., Profiling of RNA degradation for estimation of post mortem interval. *PLoS One*, 2013, **8**(2): e56507.

Sato, I., et al., Rapid detection of semenogelin by one-step immunochromatographic assay for semen identification. *J Immunol Methods*, 2004, **287**(1–2): 137–145.

Schulz, M.M., et al., A new approach to the investigation of sexual offenses—Cytoskeleton analysis reveals the origin of cells found on forensic swabs. *J Forensic Sci*, 2010, **55**(2): 492–498.

Seidl, S., R. Hausmann, and P. Betz, Comparison of laser and mercury-arc lamp for the detection of body fluids on different substrates. *Int J Legal Med*, 2008, **122**(3): 241–244.

Setzer, M., J. Juusola, and J. Ballantyne, Recovery and stability of RNA in vaginal swabs and blood, semen, and saliva stains. *J Forensic Sci*, 2008, **53**(2): 296–305.

Sikirzhytski, V., A. Sikirzhytskaya, and I.K. Lednev, Multidimensional Raman spectroscopic signatures as a tool for forensic identification of body fluid traces: A review. *Appl Spectrosc*, 2011, **65**(11): 1223–1232.

Sikirzhytski, V., K. Virkler, and I.K. Lednev, Discriminant analysis of Raman spectra for body fluid identification for forensic purposes. *Sensors (Basel)*, 2010, **10**(4): 2869–2884.

Sikirzhytskaya, A., V. Sikirzhytski, and I.K. Lednev, Raman spectroscopic signature of vaginal fluid and its potential application in forensic body fluid identification. *Forensic Sci Int*, 2012, **216**(1–3): 44–48.

Sikirzhytski, V., A. Sikirzhytskaya, and I.K. Lednev, Advanced statistical analysis of Raman spectroscopic data for the identification of body fluid traces: Semen and blood mixtures. *Forensic Sci Int*, 2012, **222**(1–3): 259–265.

Sikirzhytski, V., A. Sikirzhytskaya, and I.K. Lednev, Multidimensional Raman spectroscopic signature of sweat and its potential application to forensic body fluid identification. *Anal Chim Acta*, 2012, **718**: 78–83.

Spencer, S.E., et al., Matrix-assisted laser desorption/ionization-time of flight-mass spectrometry profiling of trace constituents of condom lubricants in the presence of biological fluids. *Forensic Sci Int*, 2011, **207**(1–3): 19–26.

Steger, K., et al., Expression of protamine-1 and -2 mRNA during human spermiogenesis. *Mol Hum Reprod*, 2000, **6**(3): 219–225.

Suto, M., et al., PCR detection of bacterial genes provides evidence of death by drowning. *Leg Med*, 2009, **11**(Suppl. 1): S354–S356.

Sweet, G.H. and J.W. Elvins, Studies by crossed electroimmunodiffusion on the individuality and sexual origin of bloodstains. *J Forensic Sci*, 1976, **21**(3): 498–509.

Tims, S., et al., Microbial DNA fingerprinting of human fingerprints: Dynamic colonization of fingertip microflora challenges human host inferences for forensic purposes. *Int J Legal Med*, 2010, **124**(5): 477–481.

Tims, S., et al., Microbial DNA fingerprinting of human fingerprints: Dynamic colonization of fingertip microflora challenges human host inferences for forensic purposes. *Int J Legal Med*, 2010, **124**(5): 477–481.

Tran, T.N., et al., Classification of ancient mammal individuals using dental pulp MALDI-TOF MS peptide profiling. *PLoS One*, 2011, **6**(2): e17319.

Tran-Hung, L., et al., A new method to extract dental pulp DNA: Application to universal detection of bacteria. *PLoS One*, 2007, **2**(10): e1062.

Uchimoto, M.L., et al., Considering the effect of stem-loop reverse transcription and real-time PCR analysis of blood and saliva specific microRNA markers upon mixed body fluid stains. *Forensic Sci Int Genet*, 2013, **7**(4): 418–421.

Uchiyama, T., et al., A new molecular approach to help conclude drowning as a cause of death: Simultaneous detection of eight bacterioplankton species using real-time PCR assays with TaqMan probes. *Forensic Sci Int*, 2012, **222**(1–3): 11–26.

Valore, E.V., et al., Human beta-defensin-1: An antimicrobial peptide of urogenital tissues. *J Clin Invest*, 1998, **101**(8): 1633–1642.

van der Meer, D., M.L. Uchimoto, and G. Williams, Simultaneous analysis of micro-RNA and DNA for determining the body fluid origin of DNA profiles. *J Forensic Sci*, 2013, **58**(4): 967–971.

Van Steendam, K., et al., Mass spectrometry-based proteomics as a tool to identify biological matrices in forensic science. *Int J Legal Med*, 2013, **127**(2): 287–298.

Vennemann, M. and A. Koppelkamm, mRNA profiling in forensic genetics I: Possibilities and limitations. *Forensic Sci Int*, 2010, **203**(1–3): 71–75.

Vennemann, M. and A. Koppelkamm, Postmortem mRNA profiling II: Practical considerations. *Forensic Sci Int*, 2010, **203**(1–3): 76–82.

Virkler, K. and I.K. Lednev, Analysis of body fluids for forensic purposes: From laboratory testing to non-destructive rapid confirmatory identification at a crime scene. *Forensic Sci Int*, 2009, **188**(1–3): 1–17.

Virkler, K. and I.K. Lednev, Blood species identification for forensic purposes using Raman spectroscopy combined with advanced statistical analysis. *Anal Chem*, 2009, **81**(18): 7773–7777.

Virkler, K. and I.K. Lednev, Raman spectroscopic signature of semen and its potential application to forensic body fluid identification. *Forensic Sci Int*, 2009, **193**(1–3): 56–62.

Virkler, K. and I.K. Lednev, Raman spectroscopy offers great potential for the nondestructive confirmatory identification of body fluids. *Forensic Sci Int*, 2008, **181**(1–3): e1–e5.

Virkler, K. and I.K. Lednev, Forensic body fluid identification: The Raman spectroscopic signature of saliva. *Analyst*, 2010, **135**(3): 512–517.

Virkler, K. and I.K. Lednev, Raman spectroscopic signature of blood and its potential application to forensic body fluid identification. *Anal Bioanal Chem*, 2010, **396**(1): 525–534.

Visser, M., et al., mRNA-based skin identification for forensic applications. *Int J Legal Med*, 2011, **125**(2): 253–263.

Wang, Z., et al., A model for data analysis of microRNA expression in forensic body fluid identification. *Forensic Sci Int Genet*, 2012, **6**(3): 419–423.

Wang, Z., et al., Screening and confirmation of microRNA markers for forensic body fluid identification. *Forensic Sci Int Genet*, 2013, **7**(1): 116–123.

Wasserstrom, A., et al., Demonstration of DSI-semen—A novel DNA methylation-based forensic semen identification assay. *Forensic Sci Int Genet*, 2013, **7**(1): 136–142.

Wiener, A.S., M.A. Hyman, and L. Handman, A new serological test (inhibition test) for human serum globulin. *Proc Soc Exp Biol Med*, 1949, **71**: 96.

Wilkinson, D.A., et al., The effects of aerosolized bacteria on fingerprint impression evidence. *J Forensic Identif*, 2009, **59**(1): 65–79.

Xu, H., et al., Bisulfite genomic sequencing of DNA from dried blood spot microvolume samples. *Forensic Sci Int Genet*, 2012, **6**(3): 306–309.

Yoshida, K., et al., Quantification of seminal plasma motility inhibitor/semenogelin in human seminal plasma. *J Androl*, 2003, **24**(6): 878–884.

Young, S.T., et al., Estimating postmortem interval using RNA degradation and morphological changes in tooth pulp. *Forensic Sci Int*, 2013, **229**(1–3): 163.e1–e6.

Zhao, D., et al., Postmortem mRNA quantification for investigation of infantile death: A comparison with adult cases. *Leg Med (Tokyo)*, 2009, **11**(Suppl 1): S286–S289.

Zhao, D., et al., Postmortem quantitative mRNA analyses of death investigation in forensic pathology: An overview and prospects. *Leg Med (Tokyo)*, 2009, **11**(Suppl 1): S43–S45.

Zhao, G., et al., Study on the application of parent-of-origin specific DNA methylation markers to forensic genetics. *Forensic Sci Int*, 2005, **154**(2–3): 122–127.

Zhao, G.S. and Q.E. Yang, Applications of DNA methylation markers in forensic medicine. *Fa Yi Xue Za Zhi*, 2005, **21**(1): 61–64.

Zhao, S.M. and C.T. Li, Perspective of DNA methylation in forensic genetics and new progress of its detection methods. *J Forensic Med*, 2009, **25**(4): 290–295.

Zhao, S.M., et al., Differences of DNA methylation profiles in monozygotic twins' blood samples. *J Forensic Med*, 2011, **27**(4): 260–264.

Zubakov, D., et al., MicroRNA markers for forensic body fluid identification obtained from microarray screening and quantitative RT-PCR confirmation. *Int J Legal Med*, 2010, **124**(3): 217–226.

Zubakov, D., et al., New markers for old stains: Stable mRNA markers for blood and saliva identification from up to 16-year-old stains. *Int J Legal Med*, 2009, **123**(1): 71–74.

Zubakov, D., et al., Stable RNA markers for identification of blood and saliva stains revealed from whole genome expression analysis of time-wise degraded samples. *Int J Legal Med*, 2008, **122**(2): 135–142.

SECTION III

Identification of
Biological Evidence

12

Identification of Blood

12.1 Biological Properties

Blood constitutes about 8% of the human body weight of a healthy individual. *Plasma* is the fluid portion of the blood. The cellular portion of the blood consists of *red blood cells*, *white blood cells*, and *platelets*, all of which are suspended in the plasma (Figures 12.1 and 12.2).

12.1.1 Red Blood Cells

These cells are also called *erythrocytes*. Their life span in humans is approximately 3–4 months. Additionally, mature human erythrocytes do not have nuclei, and therefore lack nuclear DNA. Erythrocytes consist of hemoglobin—proteins that are responsible for the transportation of oxygen. Most adult human hemoglobin consists of four polypeptide chains, two α chains and two β chains. Thus, adult hemoglobin is designated as $\alpha_2\beta_2$. Other forms of hemoglobin will be discussed in Chapter 18. Under normal physiological conditions, each hemoglobin subunit contains a heme moiety that binds to oxygen (Figure 12.3). A heme molecule consists of an organic component known as protoporphyrin IX and a *ferrous* (Fe^{2+}) iron ion (Figure 12.4). A heme molecule is also known as *ferroprotoporphyrin*. The ferrous ion of heme forms four bonds with the nitrogens of protoporphyrin IX, along with a fifth bond with a hemoglobin chain and a sixth bond with a molecule of O_2. Other chemicals such as carbon monoxide and cyanide also bind to the ferrous iron of the heme molecule and can cause chemical asphyxia. Heme groups are also present in the blood of various animals and in other proteins such as myoglobin in muscles and neuroglobin in the brain.

12.1.2 White Blood Cells

Also called leucocytes, white blood cells are subdivided into three types: granulocytes, lymphocytes, and monocytes. White blood cells are involved in defending the body against infection. They have nuclei and thus represent the main sources of nuclear DNA from the blood.

12.1.3 Platelets

These cells are also known as thrombocytes, and they play a role in blood clotting (Chapter 16). Platelets aggregate at sites of vascular and blood vessel injury. Like erythrocytes, they lack nuclei.

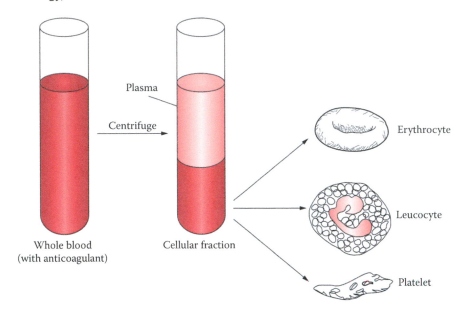

Figure 12.1 Basic composition of blood. Blood can be separated into two phases in the presence of an anticoagulant. The liquid portion called plasma accounts for approximately 55% of blood volume. The cellular elements include erythrocytes, leucocytes, and platelets. (© Richard C. Li.)

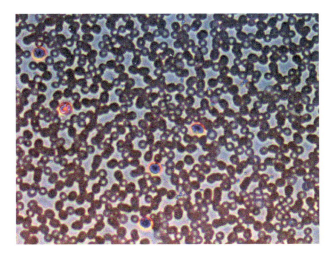

Figure 12.2 Blood cells. (© Richard C. Li.)

12.2 Presumptive Assays for Identification
12.2.1 Mechanisms of Presumptive Assays

Presumptive blood assays are designed to detect traces of blood. These assays are based on the basic principle of the oxidation–reduction reaction catalyzed by the heme moiety of the hemoglobin. As a result, colorless substrates catalyzed by heme undergo an oxidation reaction, causing either chemiluminescence, fluorescence, or a change of color. These assays are very sensitive and can detect blood in samples with 10^{-5}–10^{-6}-fold dilutions. A positive reaction indicates the possible presence of blood. Additionally, most of these assays do not interfere with forensic DNA analysis.

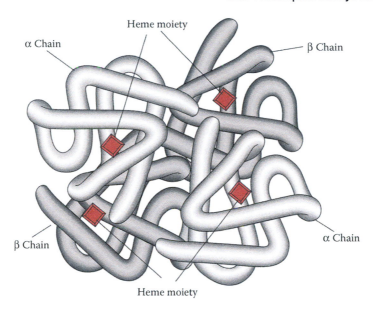

Figure 12.3   Human adult hemoglobin. Four subunits, two α and two β chains, of human adult hemoglobin are shown. Each hemoglobin subunit contains a heme moiety. (© Richard C. Li.)

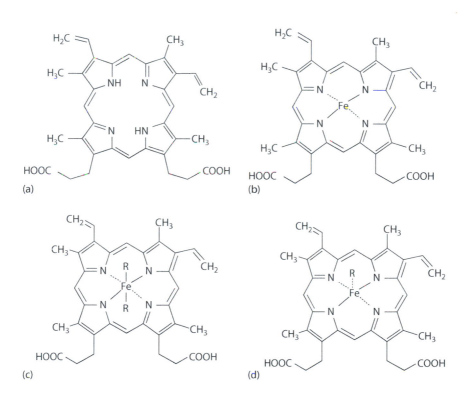

Figure 12.4   Chemical structures of heme and its precursor and derivatives. (a) Protoporphyrin IX. (b) Heme (ferroprotoporphyrin). (c) Hemochromagen, R = pyridine (pyridineferroprotoporphyrin). (d) Hematin hydroxide, R = OH (ferriprotoporphyrin hydroxide); hematin chloride, R = Cl (ferriprotoporphyrin chloride).

12.2.1.1 Oxidation–Reduction Reactions

An oxidation–reduction reaction involves a change in the oxidation state of a molecule. Specifically, the *oxidation* of a molecule means that the molecule has lost electrons, and the *reduction* of a molecule means that the molecule has gained electrons. Chemicals that can be reduced and therefore gain electrons from other molecules are called *oxidants*. In contrast, *reductants* are chemicals that can be oxidized and therefore lose electrons to other molecules. In biochemical reactions, oxidation often coincides with a loss of hydrogen. Figure 12.5 depicts an example of an oxidation–reduction reaction for blood identification. In presumptive assays, heme is utilized as the catalyst, and hydrogen peroxide is utilized as the oxidant for the reaction. In the presence of heme, a colorless substrate is oxidized, yielding a product with color, chemiluminescence, or fluorescence.

12.2.2 Colorimetric Assays

Many procedures are available for detecting heme in blood through color reactions. The most common agents are phenolphthalin, leucomalachite green, and benzidine derivatives. The color reactions produced by these assays can be observed immediately with the naked eye.

12.2.2.1 Phenolphthalin Assay

Phenolphthalein, a member of a class of indicators and dyes, is used in titrations of mineral and organic acids as well as most alkalis. The phenolphthalin assay for blood identification is also known as the Kastle–Meyer test. Kastle published a study in 1901, presenting the results of a reaction in which phenolphthalin, a colorless compound, is catalyzed by heme with hydrogen peroxide as the oxidant (Figures 12.6 and 12.7). The oxidized derivative is phenolphthalein, which appears pink under alkaline conditions.

12.2.2.2 Leucomalachite Green (LMG) Assay

Malachite green is a triphenylmethane dye. The leuco base form of malachite green is colorless and can be oxidized by the catalysis of heme to produce a green color. The reaction is carried out under acid conditions with hydrogen peroxide as the oxidant (Figures 12.8 and 12.9).

12.2.2.3 Benzidine and Derivatives

Historically, *benzidine* was used as an intermediate in dye manufacturing (Figure 12.10). Subsequently, it was used as a presumptive assay for the presence of blood after the discovery that the oxidation of benzidine can be catalyzed by heme to produce a blue to dark blue color (carried out in an acid solution). Since the blue color may eventually turn brown, the reaction must be read

$$AH_2 \;+\; H_2O_2 \;\xrightarrow{\text{Heme}}\; A \;+\; 2H_2O$$

(Colorless)   (Color)

Figure 12.5 Oxidation–reduction reaction as the basis for presumptive assays for blood identification. AH_2, substrate; A, oxidized substrate.

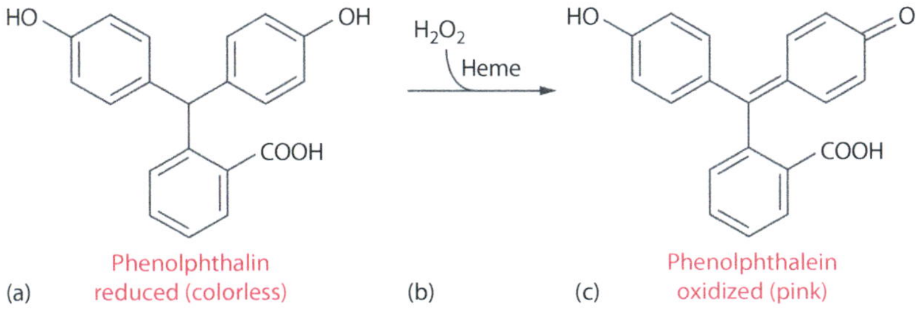

Figure 12.6 Chemical reaction of phenolphthalin assay.

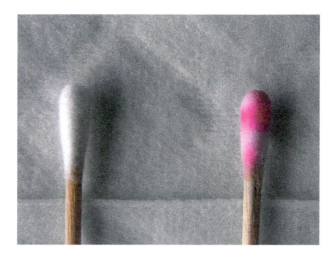

Figure 12.7 Photograph of phenolphthalin assay results. Negative (left) and positive (right) reaction. (© Richard C. Li.)

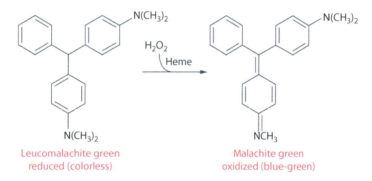

Leucomalachite green
reduced (colorless)

Malachite green
oxidized (blue-green)

Figure 12.8 Chemical reaction of leucomalachite green assay.

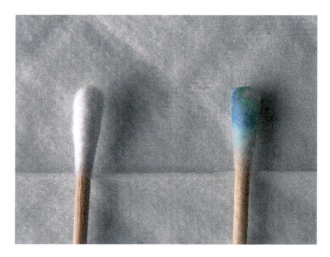

Figure 12.9 Photograph of leucomalachite green assay results. Negative (left) and positive (right) reaction. (© Richard C. Li.)

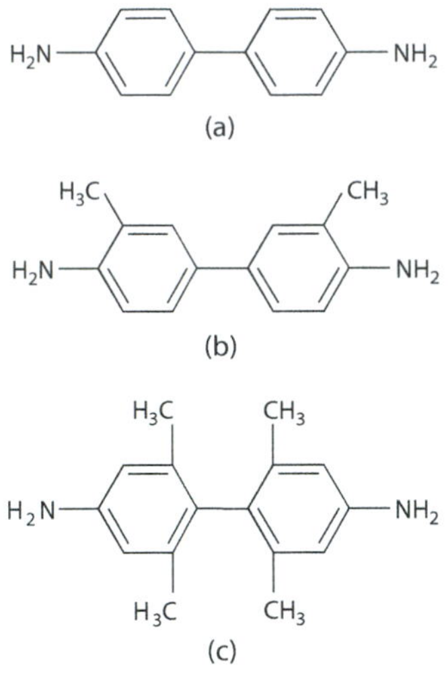

Figure 12.10 Chemical structures of benzidine and derivatives: (a) benzidine; (b) orthotolidine; (c) tetramethylbenzidine.

immediately. Benzidine was found to be a carcinogen and is therefore no longer used for forensic testing. *Orthotolidine* is a dimethyl derivative of benzidine. Its oxidation reaction can be catalyzed by heme to produce a blue color reaction under acidic conditions (Figure 12.11). Orthotolidine is also considered a potential carcinogen based on animal studies, and for this reason it has been replaced by tetramethylbenzidine. *Tetramethylbenzidine* (TMB) is a tetramethyl derivative of benzidine. The oxidation of TMB can be catalyzed by heme to produce a green to blue-green color under acidic conditions. TMB continues to be used. The Hemastix® assay kit (Miles Laboratories) is a TMB-based assay that utilizes a TMB-containing strip device. A test is carried out by applying a moistened sample to a Hemastix® strip. The appearance of a green or a blue-green color indicates the presence of blood.

12.2.3 Chemiluminescence and Fluorescence Assays

Other organic compounds whose oxidation products have chemiluminescent or fluorescent properties are utilized for testing. In the *chemiluminescence assay*, light is emitted as a product of a chemical reaction. In this category, luminol produces chemiluminescence when blood is present. In contrast, a *fluorescence assay* requires the exposure of an oxidized product, such as fluorescin, to a particular wavelength of an excitation light source. The fluorescence is then emitted at longer wavelengths than that of the excitation light source.

One advantage of chemiluminescent and fluorescent reagents is that they can be sprayed over large areas where latent bloodstains are potentially located. A positive reaction identifies blood and also reveals the patterns of bloody impressions such as footprints and fingerprints. These methods are very sensitive and can pinpoint the locations of even small traces of blood. Additionally, they are useful for detecting blood at crime scenes that have been cleaned and show no visible staining. One disadvantage of chemiluminescent and fluorescent reagents is that precautions must be taken if stains are very small or have been washed. The spraying of

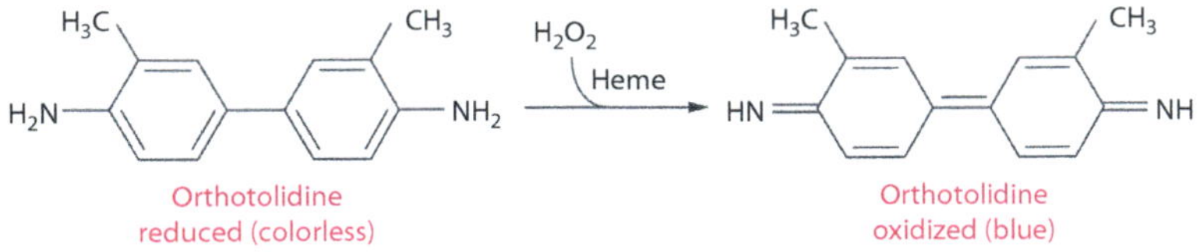

Figure 12.11 Chemical reaction of orthotolidine assay.

presumptive assay reagents may further dilute a sample and thus lead to difficulty in isolating sufficient amounts of DNA for forensic DNA analysis.

12.2.3.1 Luminol (3-Aminophthalhydrazide)

Luminol is usually utilized as a chemiluminescent reagent. The oxidation reaction of luminol catalyzed by heme produces light in the presence of an oxidant (Figures 12.12 through 12.14). The light emitted from a positive reaction can only be observed in the dark, which limits the

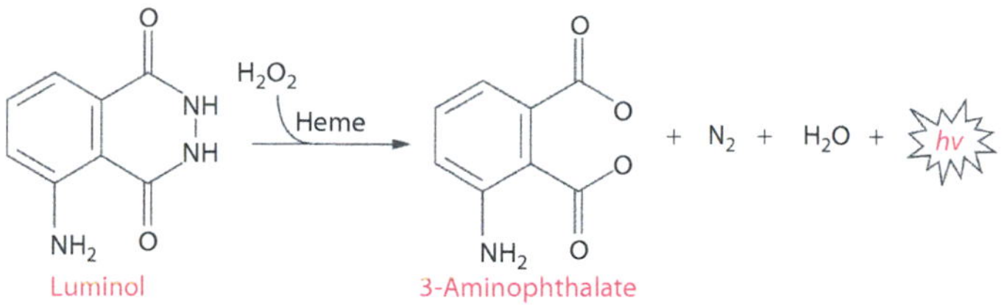

Figure 12.12 Chemical reaction of luminol assay.

Figure 12.13 Results obtained from testing a drop of blood on a glass plate with luminol. From left to right: drop of blood; drop of blood after spraying with luminol; drop of blood after spraying with luminol and viewing in the dark. A blue chemiluminescence indicates the presence of blood. (From Bergervoet, P.W., et al., *J Hosp Infect*, 68, 329–333, 2008. With permission.)

Figure 12.14 Detecting latent bloodstains on a floor with luminol. Left: with the light on; right: immediately after the light is switched off. A blue chemiluminescence indicates the presence of blood traces. (From Bergervoet, P.W., et al., *J Hosp Infect*, 68, 329–333, 2008. With permission.)

applications of luminol. The photodocumentation of a luminol-enhanced pattern should be done immediately before it fades away.

12.2.3.2 Fluorescin

Fluorescin is another reagent that is used to test for the presence of bloodstains at a crime scene (Figures 12.15 and 12.16). When oxidized and catalyzed by heme, fluorescin demonstrates fluorescent properties. Usually, fluorescin-sprayed stains are exposed to light in the range of 425–485 nm using an alternate light source device. In a positive reaction, the oxidized fluorescin emits an intense yellowish-green fluorescent light, which indicates the presence of a bloodstain. The light emitted from fluorescin-sprayed stains lasts longer than that of luminol.

12.2.4 Factors Affecting Presumptive Assay Results

The catalytic assays discussed in the previous sections are not specific to blood only, which can possibly lead to the observation of false-positive or false-negative results (Figure 12.17).

12.2.4.1 Oxidants

Chemicals that are strong oxidants may cause a false-positive reaction. Such chemicals can catalyze the oxidation reaction even in the absence of heme and result in a false-positive reaction. Certain metal salts, such as copper and nickel salts, household bleaches and cleaners that contain hypochlorite ions, and hair-coloring products that contain hydrogen peroxide, work as oxidants. To address this problem, a two-step catalytic assay should be performed. The substrate

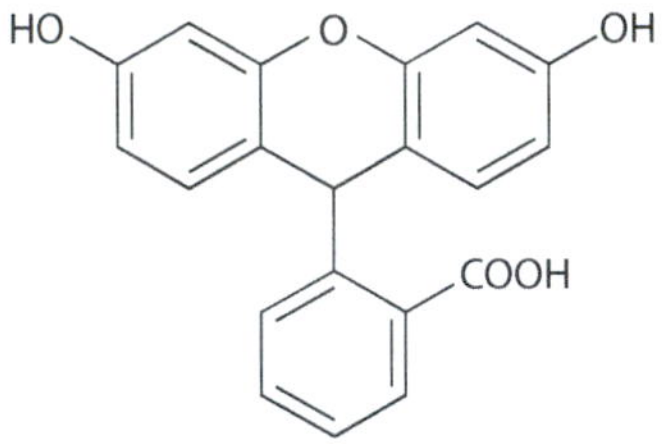

Figure 12.15 Chemical structure of fluorescein.

Figure 12.16 Detecting diluted bloodstains on black cotton fabric with fluorescin (left) and Bluestar (right). (From Finnis, J., Lewis, J., and Davidson, A., *Sci Justice*, 53, 178–186, 2013. With permission.)

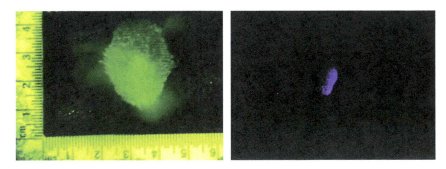

$$AH_2 \;+\; H_2O_2 \;\underset{\text{(3) Reductant (color)}}{\overset{\substack{\text{(1) Oxidant}\\\text{(2) Peroxidase}}}{\rightleftharpoons}}\; A \;+\; 2H_2O$$

(colorless)

Figure 12.17 Factors affecting presumptive assay results. Strong oxidant and peroxidase may cause false-positive results; reductant may cause false-negative results.

is applied first to the sample in question. A color change occurring before the addition of hydrogen peroxide indicates a false-positive result due to a possible oxidant in the sample. If a color change is observed after the addition of hydrogen peroxide, the result is a true positive.

12.2.4.2 Plant Peroxidases
Many types of plants such as horseradish contain peroxidases. Plant peroxidases may also catalyze oxidation reactions and lead to false-positive results. However, plant peroxidases are usually heat sensitive and may be inactivated by high temperatures. Because the heme molecule is relatively stable at high temperatures, samples can be retested after heating. This will inactivate any plant peroxidases in a sample.

12.2.4.3 Reductants
Although not common, a false-negative result can occur when a strong reductant is present in a sample. Strong reductants such as certain metal ions including lithium and zinc may inhibit the oxidation reaction.

12.3 Confirmatory Assays for Identification
12.3.1 Microcrystal Assays
Microcrystal assays apply chemicals to treat bloodstains, forming crystals of heme molecules. The morphologies of the resulting crystals are distinctive for heme and can be compared with a known standard using a microscope. A positive microcrystal assay strongly indicates the presence of blood. However, confirmatory assays are usually not as sensitive as presumptive assays. Additionally, these assays cannot distinguish between human and animal blood.

12.3.1.1 Hemochromagen Crystal Assay
Hemochromagens are heme derivatives in which the ferrous iron of the heme forms two bonds with nitrogenous bases (Figure 12.4). The method for forming hemochromagen crystals was documented in 1864. Since then, various modifications have been reported. The *Takayama crystal assay*, published in 1912, has been the method preferred by many forensic laboratories. A bloodstain is treated with pyridine and glucose (a reducing sugar that is capable of reducing ferric ion) under alkaline conditions to form crystals of pyridine ferroprotoporphyrin (Figure 12.18).

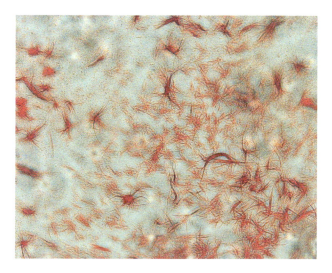

Figure 12.18   Microcrystal assays using the Takayama method. (© Richard C. Li.)

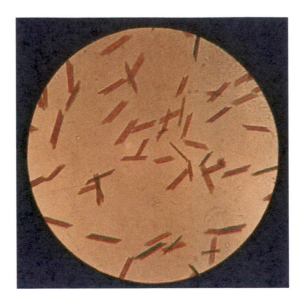

Figure 12.19   Microcrystal assays using the Teichmann method. (From James, S.H., Nordby, J.J., and Bell, S., *Forensic Science: An Introduction to Scientific and Investigative Techniques*, 4th edn., CRC Press, Boca Raton, 2014. With permission.)

12.3.1.2 Hematin Crystal Assay

This assay is also known as the *Teichmann crystal assay*. In 1853, Teichmann documented a method of forming crystals of blood specimens. When blood specimens are treated with glacial acetic acid and salts, and subsequently heated, hematin chloride (ferriprotoporphyrin chloride), a prismatic brown-colored crystal, is formed (Figure 12.19). Hematin (Figure 12.4) is a heme derivative; its iron is in the *ferric* (Fe^{+3}) state. This hematin assay has a similar sensitivity and specificity as hemochromagen assays. The hematin assay has the advantage of being more reliable than hemochromagen assays for aged blood samples.

12.3.2 Other Assays

Additional techniques may be used to confirm the presence of hemoglobin. For example, *chromatographic* and *electrophoretic methods* can identify hemoglobin by its mobility characteristics. *Spectrophotometric methods* for identifying hemoglobin are based on measurements of the

Table 12.1	Application of RT-PCR Assay for Blood Identification		
Gene Symbol	**Gene Product**	**Description**	**Further Reading**
HBA1	Hemoglobin α1	Hemoglobin α1 chain (abundant in erythrocytes)	Waye and Chui (2001)
PBGD[a]	Porphobilinogen deaminase	Erythrocyte-specific isoenzyme of heme biosynthesis pathway	Gubin and Miller (2001)
SPTB	β-Spectrin	Subunit of major protein of erythrocyte membrane skeleton	Amin et al. (1993)

Source: Adapted from Juusola, J. and Ballantyne, J., *Forensic Sci Int*, 152, 1–12, 2005; Nussbaumer, C., Gharehbaghi-Schnell, E., and Korschineck, I., *Forensic Sci Int*, 157, 181–186, 2006.

[a] Also known as hydroxymethylbilane synthase (HMBS).

characteristic light spectra, with peak absorbance at 400–425 nm, absorbed by hemoglobin and its derivatives. Finally, *immunological methods* utilize antihuman hemoglobin antibodies. This antibody can be used to detect human hemoglobin and thus indicate the presence of human blood (see Chapter 13).

Recently, *RNA-based assays* have been developed to identify blood. These assays are based on the fact that certain genes are specifically expressed in certain cell types (Chapter 11). Thus, the techniques used in the identification of blood are based on the detection of specific types of messenger RNA (mRNA) that are expressed exclusively in erythrocytes. These assays utilize reverse transcriptase polymerase chain reaction (RT-PCR; see Chapter 7) methods to detect the gene expression levels of mRNAs for blood identification. Table 12.1 summarizes the tissue-specific genes utilized for blood identification. Compared with conventional assays that are used for blood identification, the RNA-based assays have higher specificity and are amenable to automation. However, one limitation is that RNA is unstable due to degradation by endogenous and environmentally born ribonucleases.

Bibliography

Amin, K.M., et al., The exon-intron organization of the human erythroid beta-spectrin gene. *Genomics*, 1993, **18**(1): 118–125.

Andersen, J. and S. Bramble, The effects of fingermark enhancement light sources on subsequent PCR-STR DNA analysis of fresh bloodstains. *J Forensic Sci*, 1997, **42**(2): 303–306.

Anderson, S.E., G.R. Hobbs, and C.P. Bishop, Multivariate analysis for estimating the age of a bloodstain. *J Forensic Sci*, 2011, **56**(1): 186–193.

Anderson, S., et al., A method for determining the age of a bloodstain. *Forensic Sci Int*, 2005, **148**(1): 37–45.

Andrasko, J., The estimation of age of bloodstains by HPLC analysis. *J Forensic Sci*, 1997, **42**(4): 601–607.

Anslinger, K., et al., Ninhydrin treatment as a screening method for the suitability of swabs taken from contact stains for DNA analysis. *Int J Legal Med*, 2004, **118**(2): 122–124.

Arany, S. and S. Ohtani, Age estimation of bloodstains: A preliminary report based on aspartic acid racemization rate. *Forensic Sci Int*, 2011, **212**(1–3): e36–e39.

Barnett, P.D., et al., Discussion of "Effects of presumptive test reagents on the ability to obtain restriction fragment length polymorphism (RFLP) patterns from human blood and semen stains". *J Forensic Sci*, 1992, **37**(2): 369–370.

Bauer, M., S. Polzin, and D. Patzelt, Quantification of RNA degradation by semi-quantitative duplex and competitive RT-PCR: A possible indicator of the age of bloodstains? *Forensic Sci Int*, 2003, **138**(1–3): 94–103.

Bergervoet, P.W., et al., Application of the forensic Luminol for blood in infection control. *J Hosp Infect*, 2008, **68**: 329–333.

Botoniic-Sehic, E., et al., Forensic application of near-infrared spectroscopy: Aging of bloodstains. *Spectroscopy* (Santa Monica), 2009, **24**(2): 42–48.

Boyd, S., et al., Highly sensitive detection of blood by surface enhanced Raman scattering. *J Forensic Sci*, 2013, **58**(3): 753–756.

Bremmer, R.H., et al., Remote spectroscopic identification of bloodstains. *J Forensic Sci*, 2011, **56**(6): 1471–1475.

Bremmer, R.H., et al., Forensic quest for age determination of bloodstains. *Forensic Sci Int*, 2012, **216**(1–3): 1–11.

Budowle, B., et al., The presumptive reagent fluorescein for detection of dilute bloodstains and subsequent STR typing of recovered DNA. *J Forensic Sci*, 2000, **45**(5): 1090–1092.

Caldwell, J.P., W. Henderson, and N.D. Kim, ABTS: A safe alternative to DAB for the enhancement of blood fingerprints. *J Forensic Sci*, 2000, **45**(4): 785–794.

Cheeseman, R., Direct sensitivity comparison to the fluorescein and luminol bloodstain enhancement techniques. *J Forensic Ident*, 1999, **49**(3): 8.

Cheeseman, R. and L.A. DiMeo, Fluorescein as a field-worthy latent bloodstain detection system. *J Forensic Ident*, 1995, **45**(6): 1.

Cheeseman, R. and R. Tomboc, Fluorescein technique performance study on bloody foot trails. *J Forensic Ident*, 2001, **51**(1): 12.

Courts, C. and B. Madea, Specific micro-RNA signatures for the detection of saliva and blood in forensic body-fluid identification. *J Forensic Sci*, 2011, **56**(6): 1464–1470.

Cox, M., Effect of fabric washing on the presumptive identification of bloodstains. *J Forensic Sci*, 1990, **35**(6): 1335–1341.

Cox, M., A study of the sensitivity and specificity of four presumptive tests for blood. *J Forensic Sci*, 1991, **36**(5): 1503–1511.

Creamer, J.I., et al., A comprehensive experimental study of industrial, domestic and environmental interferences with the forensic luminol test for blood. *Luminescence*, 2003, **18**(4): 193–198.

Creamer, J.I., et al., Attempted cleaning of bloodstains and its effect on the forensic luminol test. *Luminescence*, 2005, **20**(6): 411–413.

DeForest, P., R. Gaensslen, and H.C. Lee, *Forensic Science: An Introduction to Criminalistics*, 1983. New York: McGraw-Hill Book Company.

Della Manna, A. and S. Montpetit, A novel approach to obtaining reliable PCR results from luminol treated bloodstains. *J Forensic Sci*, 2000, **45**(4): 886–890.

Dixon, T.R., et al., A scanning electron microscope study of dried blood. *J Forensic Sci*, 1976, **21**(4): 797–803.

Dorward, D.W., Detection and quantitation of heme-containing proteins by chemiluminescence. *Anal Biochem*, 1993, **209**(2): 219–223.

Edelman, G., et al., Identification and age estimation of blood stains on colored backgrounds by near infrared spectroscopy. *Forensic Sci Int*, 2012, **220**(1–3): 239–244.

Farrugia, K.J., et al., Chemical enhancement of footwear impressions in blood on fabric—Part 3: Amino acid staining. *Sci Justice*, 2013, **53**(1): 8–13.

Finnis, J., J. Lewis, and A. Davidson, Comparison of methods for visualizing blood on dark surfaces. *Sci Justice*, 2013, **53**(2): 178–186.

Frascione, N., V. Pinto, and B. Daniel, Development of a biosensor for human blood: New routes to body fluid identification. *Anal Bioanal Chem*, 2012, **404**(1): 23–28.

Fregeau, C.J., O. Germain, and R.M. Fourney, Fingerprint enhancement revisited and the effects of blood enhancement chemicals on subsequent profiler plus fluorescent short tandem repeat DNA analysis of fresh and aged bloody fingerprints. *J Forensic Sci*, 2000, **45**(2): 354–380.

Fujita, Y., et al., Estimation of the age of human bloodstains by electron paramagnetic resonance spectroscopy: Long-term controlled experiment on the effects of environmental factors. *Forensic Sci Int*, 2005, **152**(1): 39–43.

Gaensslen, R.E., *Sourcebook in Forensic Serology, Immunology, and Biochemistry*, 1983. Washington, DC: US Government Printing Office.

Garner, D.D., et al., An evaluation of tetramethylbenzidine as a presumptive test for blood. *J Forensic Sci*, 1976, **21**(4): 816–821.

Gimeno, F.E., Fill flash color photography to photograph luminol blood stain patterns. *J Forensic Ident*, 1989, **39**(5): 305–306.

Gimeno, F.E. and G.E. Rini, Fill flash luminescence to photograph luminol blood stain patterns. *J Forensic Ident*, 1989, **39**(3): 1.

Grafit, A., A. Cohen, and Y. Cohen, Estimation of original volume of dry bloodstains using spectrophotometric method. *J Forensic Ident*, 2012, **62**(4): 305–314.

Greenfield, A., M. Sloan, and R. Spaulding, Identification of blood and body fluids, in S. James, J. Nordby and S. Bell (eds), *Forensic Science: An Introduction to Scientific and Investigative Techniques*, pp. 205–228, 2014. Boca Raton, FL: CRC Press.

Grispino, R.R.J., The effect of luminol on the serological analysis of dried human bloodstains. *Crime Lab Digest*, 1990, **17**(1): 13–23.

Gross, A.M., K.A. Harris, and G.L. Kaldun, The effect of luminol on presumptive tests and DNA analysis using the polymerase chain reaction. *J Forensic Sci*, 1999, **44**(4): 837–840.

Gubin, A.N. and J.L. Miller, Human erythroid porphobilinogen deaminase exists in 2 splice variants. *Blood*, 2001, **97**(3): 815–817.

Guo, K., S. Achilefu, and M.Y. Berezin, Dating bloodstains with fluorescence lifetime measurements. *Chemistry*, 2012, **18**(5): 1303–1305.

Haas, C., et al., MRNA profiling for the identification of blood—Results of a collaborative EDNAP exercise. *Forensic Sci Int Genet*, 2011, **5**(1): 21–26.

Haas, C., et al., Selection of highly specific and sensitive mRNA biomarkers for the identification of blood. *Forensic Sci Int Genet*, 2011, **5**(5): 449–458.

Haas, C., et al., RNA/DNA co-analysis from blood stains—Results of a second collaborative EDNAP exercise. *Forensic Sci Int Genet*, 2012, **6**(1): 70–80.

Hanson, E., A. Albornoz, and J. Ballantyne, Validation of the hemoglobin (Hb) hypsochromic shift assay for determination of the time since deposition (TSD) of dried bloodstains. *Forensic Sci Int Genet Suppl Series*, 2011, **3**(1): e307–e308.

Hanson, E.K. and J. Ballantyne, A blue spectral shift of the hemoglobin soret band correlates with the age (time since deposition) of dried bloodstains. *PloS ONE*, 2010, **5**(9): e12830.

Hatch, A.L., A modified reagent for the confirmation of blood. *J Forensic Sci*, 1993, **38**(6): 1502–1506.

Higaki, R.S. and W.M.S. Philp, A study for the sensitivity, stability, and specificity of phenolphthalein as an indicator test for blood. *Can Soc Forensic Sci*, 1976, **9**(3): 97–102.

Hochmeister, M.N., B. Budowle, and F.S. Baechtel, Effects of presumptive test reagents on the ability to obtain restriction fragment length polymorphism (RFLP) patterns from human blood and semen stains. *J Forensic Sci*, 1991, **36**(3): 656–661.

Inoue, H., et al., Identification of fetal hemoglobin and simultaneous estimation of bloodstain age by high-performance liquid chromatography. *Int J Legal Med*, 1991, **104**(3): 127–131.

Inoue, H., et al., A new marker for estimation of bloodstain age by high performance liquid chromatography. *Forensic Sci Int*, 1992, **57**(1): 17–27.

Introna, F. Jr., G. Di Vella, and C.P. Campobasso, Determination of postmortem interval from old skeletal remains by image analysis of luminol test results. *J Forensic Sci*, 1999, **44**(3): 535–538.

James, S.H. and Nordby, J.J., *Forensic Science: An Introduction to Scientific and Investigative Techniques*, 2nd edn., 2005. Boca Raton, FL: CRC Press.

Juusola, J. and Ballantyne, J., Multiplex mRNA profiling for the identification of body fluids. *Forensic Sci Int*, 2005, **152**: 1–12.

Kashyap, V.K., A simple immunosorbent assay for detection of human blood. *J Immunoassay*, 1989, **10**(4): 315–324.

Kent, E.J., D.A. Elliot, and G.M. Miskelly, Inhibition of bleach-induced luminol chemiluminescence. *J Forensic Sci*, 2003, **48**(1): 64–67.

Kohlmeier, F. and P.M. Schneider, Successful mRNA profiling of 23 years old blood stains. *Forensic Sci Int Genet*, 2012, **6**(2): 274–276.

Laux, D.L., Effects of luminol on the subsequent analysis of bloodstains. *J Forensic Sci*, 1991, **36**(5): 1512–1520.

Laux, D.L., The detection of blood using luminol, in S.H. James, P.E. Kish, and T.P. Sutton (eds), *Principles of Bloodstain Pattern Analysis: Theory and Practice*, pp. 369–390, 2005. Boca Raton, FL: CRC Press.

Lee, H.C., Identification and grouping of bloodstains, in R. Saferstein (ed.), *Forensic Science Handbook*, pp. 267–337, 1982. Englewood Cliffs, NJ: Prentice Hall.

Lee, H.C., T. Palmbach, and M.T. Miller, *Henry Lee's Crime Scene Handbook*, 2001. San Diego: Academic Press.

Lytle, L.T. and D.G. Hedgecock, Chemiluminescence in the visualization of forensic bloodstains. *J Forensic Sci*, 1978, **23**(3): 550–562.

Matsuoka, T., T. Taguchi, and J. Okuda, Estimation of bloodstain age by rapid determinations of oxyhemoglobin by use of oxygen electrode and total hemoglobin. *Biol Pharm Bull*, 1995, **18**(8): 1031–1035.

Miyaishi, S., et al., Discrimination between postmortem and antemortem blood by a dot-ELISA for human myoglobin. *Nihon Hoigaku Zasshi*, 1994, **48**(6): 433–438.

Nussbaumer, C., Gharehbaghi-Schnell, E., and Korschineck, I., Messenger RNA profiling: a novel method for body fluid identification by real-time PCR. *Forensic Sci Int*, 2006, **157**: 181–186.

Omelia, E.J., M.L. Uchimoto, and G. Williams, Quantitative PCR analysis of blood- and saliva-specific microRNA markers following solid-phase DNA extraction. *Anal Biochem*, 2013, **435**(2): 120–122.

Ponce, A.C. and F.A. Verdu Pascual, Critical revision of presumptive tests for bloodstains. *Forensic Sci Commun*, 1999, **1**(2): 1–15.

Quickenden, T.I. and J.I. Creamer, A study of common interferences with the forensic luminol test for blood. *Luminescence*, 2001, **16**(4): 295–298.

Quickenden, T.I. and P.D. Cooper, Increasing the specificity of the forensic luminol test for blood. *Luminescence*, 2001, **16**(3): 251–253.

Quickenden, T.I., C.P. Ennis, and J.I. Creamer, The forensic use of luminol chemiluminescence to detect traces of blood inside motor vehicles. *Luminescence*, 2004, **19**(5): 271–277.

Richards, A. and R. Leintz, Forensic reflected ultraviolet imaging. *J Forensic Ident*, 2013, **63**(1): 46–69.

Santos, V.R.D., W.X. Paula, and E. Kalapothakis, Influence of the luminol chemiluminescence reaction on the confirmatory tests for the detection and characterization of bloodstains in forensic analysis. *Forensic Sci Int Genet Suppl Series*, 2009, **2**(1): 196–197.

Shaler, R.C., Modern forensic biology, in R. Saferstein (ed.), *Forensic Science Handbook*, pp. 525–614, 2002. Upper Saddle River, NJ: Pearson Education.

Shipp, E., et al., Effects of argon laser light, alternate source light, and cyanoacrylate fuming on DNA typing of human bloodstains. *J Forensic Sci*, 1993, **38**(1): 184–191.

Spalding, R., The identification and characterization of blood and bloodstains, in S. James and J. Nordby (eds), *Forensic Science: An Introduction to Scientific and Investigative Techniques*, pp. 237–260, 2005. Boca Raton, FL: CRC Press.

Stoilovic, M., Detection of semen and blood stains using polilight as a light source. *Forensic Sci Int*, 1991, **51**(2): 289–296.

Strasser, S., et al., Age determination of blood spots in forensic medicine by force spectroscopy. *Forensic Sci Int*, 2007, **170**(1): 8–14.

Sutton, T.P., Presumptive blood testing, in S.H. James (ed.), *Scientific and Legal Applications of Bloodstain Pattern Interpretation*, pp. 48–70, 1999. Boca Raton, FL: CRC Press.

Suwa, N., et al., Human blood identification using the genome profiling method. *Leg Med*, 2012, **14**(3): 121–125.

Thomas, P. and K. Farrugia, An investigation into the enhancement of fingermarks in blood on paper with genipin and lawsone. *Sci Justice*, 2013, **53**(3): 315–320.

Tobe, S.S., N. Watson, and N.N. Daeid, Evaluation of six presumptive tests for blood, their specificity, sensitivity, and effect on high molecular-weight DNA. *J Forensic Sci*, 2007, **52**(1): 102–109.

Tsutsumi, H. and Y. Katsumata, Forensic study on stains of blood and saliva in a chimpanzee bite case. *Forensic Sci Int*, 1993, **61**(2–3): 101–110.

Tsutsumi, A., Y. Yamamoto, and H. Ishizu, Determination of the age of bloodstains by enzyme activities in blood cells. *Nihon Hoigaku Zasshi*, 1983, **37**(6): 770–776.

Vandenberg, N. and R.A. van Oorschot, The use of Polilight in the detection of seminal fluid, saliva, and bloodstains and comparison with conventional chemical-based screening tests. *J Forensic Sci*, 2006, **51**(2): 361–370.

Virkler, K. and I.K. Lednev, Raman spectroscopic signature of blood and its potential application to forensic body fluid identification. *Anal Bioanal Chem*, 2010, **396**(1): 525–534.

Waye, J.S. and D.H. Chui, The alpha-globin gene cluster: Genetics and disorders. *Clin Invest Med*, 2001, **24**(2): 103–109.

Webb, J.L., J.I. Creamer, and T.I. Quickenden, A comparison of the presumptive luminol test for blood with four non-chemiluminescent forensic techniques. *Luminescence*, 2006, **21**(4): 214–220.

Yanagida, J., M. Hara, and H. Nakamura, Estimation of age of dried bloodstains by a surface absorption spectrophotometric method. *J Saitama Med School*, 1978, **5**(3): 221–225.

Yanagida, J., et al., Colorimetric study on the color changes of ageing bloodstains. *J Saitama Med School*, 1981, **8**(1–2): 15–20.

Zubakov, D., et al., Stable RNA markers for identification of blood and saliva stains revealed from whole genome expression analysis of time-wise degraded samples. *Int J Legal Med*, 2008, **122**(2): 135–142.

Zubakov, D., et al., New markers for old stains: Stable mRNA markers for blood and saliva identification from up to 16-year-old stains. *Int J Legal Med*, 2009, **123**(1): 71–74.

Zweidinger, R.A., L.T. Lytle, and C.G. Pitt, Photography of bloodstains visualized by luminol. *J Forensic Sci*, 1973, **18**(4): 296–302.

13

Species Identification

Chapter 12 discussed the principles of the identification of blood. If a stain is identified as blood, the evidence can be tested to determine whether the blood is of human origin. If the bloodstain is nonhuman, further analysis is usually not necessary.

Before forensic DNA techniques were implemented, species identification was largely determined by serological methods. Currently, most forensic laboratories perform DNA quantitation prior to DNA profile analysis. The quantitation method specifically detects higher-primate DNA. The presence of DNA measured by the quantitation assays concurrently identifies a sample as being of human origin (since crimes involving primate blood are extremely rare). Thus, species identification is usually not performed in forensic laboratories.

Nevertheless, species identification assays can be useful for screening to exclude or eliminate nonhuman samples unrelated to an investigation. Thus, it is practical for small laboratories to eliminate unnecessary analyses due to time and budget constraints. Additionally, species identification kits such as immunochromatographic devices allow field testing by crime scene investigators.

In cases involving the killing, trading, and possession of products derived from species that are protected from illegal hunting, it may be necessary to identify the animal species prior to further analysis. Species identification using DNA analysis can be performed using commonly used loci at the mitochondrial cytochrome b gene (*Cytb*), the cytochrome c oxidase I gene (*COI*), and the D-loop region. This type of identification is usually within the scope of wildlife forensic science and is thus not discussed here.

13.1 General Considerations

Most assays for species identification are based on serological techniques, including primary and secondary binding assays. The most common primary binding assays are immunochromatographic assays. The most commonly used secondary binding assays are precipitation-based assays that rely on the binding of an antigen to an antibody, causing the formation of visible precipitation. These precipitation-based assays include ring assays, Ouchterlony assays, and crossed-over immunoelectrophoresis. These assays utilize antihuman and antianimal antibodies to identify human and animal species, respectively.

13.1.1 Types of Antibodies

An antihuman antibody that is used in the identification of human samples can be made by introducing human serum into a host animal, which then produces specific antibodies against the human serum proteins. Antibodies produced from different species of host animals may produce variations in the characteristics of reactions. Since albumin is the most abundant protein in human serum, the antihuman antibody that is produced reacts strongly with human albumin. Albumin is a protein that plays important roles in the maintenance of the vascular circulating fluid and the transportation of various substances such as nutrients, hormones, and metabolic products. Blood is drawn from the host animal and the serum portion is collected. The collected serum is a polyclonal antihuman antiserum containing a mixture of antibodies against various human serum proteins. Likewise, an antibody against animal serum proteins can also be made to identify animal species of interest.

Other antibodies such as antihuman hemoglobin (Hb) antibodies can also be used to identify the human origin of a sample. Hb is an oxygen-transport protein that is found in erythrocytes (Chapter 12). Purified Hb can be used to generate monoclonal and polyclonal antihuman Hb antibodies. Likewise, antibodies recognizing glycophorin A (GPA), a human erythrocyte membrane antigen (see Section 13.2.1.2), can also be produced in a similar manner.

13.1.2 Titration of Antibodies

Recall that the ratio of antigen to antibody is critical for the success of a secondary reaction (Chapter 10). An extreme excess of antigen or antibody concentrations can inhibit secondary reactions. The prozone and postzone phenomena must be considered, and the concentrations of antigen and antibody must be carefully determined for forensic serology assays. For instance, in the prozone situation, a false-negative reaction may occur due to the presence of a high concentration of antibody.

Quality-control procedures can be used to estimate the amount of a specific antibody that is present, often via titration (Figure 13.1). To titrate an antiserum, a series of dilutions are made and each dilution is then tested for activity using precipitation or agglutination methods. The reciprocal of the highest dilution giving a positive reaction is known as the *titer*. This reflects the amount of antibody in the antiserum. Additionally, the polyclonal antiserum is a mixture of antibodies; thus, the reaction of the antiserum may vary from animal to animal (of the same species). Each lot or batch of antiserum must be validated by titration.

13.1.3 Antibody Specificity

In addition to the titer, the specificity of the antihuman antibody must be tested. Most antihuman antibodies usually have cross-reactivity with higher primates. This is not a great

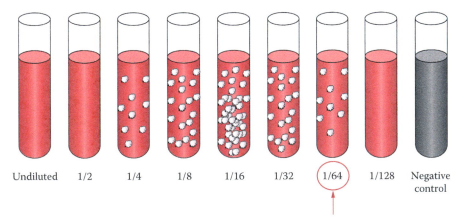

| Undiluted | 1/2 | 1/4 | 1/8 | 1/16 | 1/32 | 1/64 | 1/128 | Negative control |

Figure 13.1 Titration of antibodies. Serum is serially diluted and a constant amount of antigen is applied to each tube. The mixture is incubated, allowing agglutination to occur. The reciprocal of the highest dilution giving a positive agglutination reaction is 64 (the titer). (© Richard C. Li.)

concern because crimes involving nonhuman primates are very rare. Nevertheless, the antihuman antibody must not cross-react with other commonly encountered animals. Antisera and positive control samples must be validated for cross-reactivity. Tissue specificity must also be validated. The antiserum against human serum is usually reactive with other human biological fluids such as semen and saliva.

13.1.4 Optimal Conditions for Antigen–Antibody Binding

A number of factors can affect antigen–antibody binding. For example, increasing ionic strength can inhibit the binding of an antigen and an antibody. Stronger inhibition is usually observed for ions with large ionic radii and small radii of hydration. It is believed that the lower degree of hydration permits interactions of ions and the antibody-binding site, leading to inhibition. A proper buffer system must be selected in serological assays to ensure reliable results. The introduction of polymers can facilitate precipitation in secondary binding reactions because the presence of a polymer in a solution decreases the solubility of proteins. Linear hydrophilic polymers with high molecular weights (e.g., polyethylene glycol) are preferred. Additional factors such as temperature and pH can also affect antigen–antibody binding.

13.2 Assays

Samples can be prepared by cutting out a portion of a stain or scraping stains from a surface. A sample is usually extracted with a small volume of saline or buffer. The extracted sample can be tested using the assays described in the following subsections. Controls should be included, for example, by using a known human serum as a positive control and an extraction blank as a negative control.

13.2.1 Immunochromatographic Assays

Immunochromatographic assays are rapid, specific, and sensitive and can be used in both laboratory and field tests for species identification. Two types of assays are discussed, including those based on the detection of human erythrocyte proteins. Chapter 11 discusses the principle of immunochromatographic assays in more detail.

13.2.1.1 Identification of Human Hemoglobin Protein

Commercially produced immunochromatographic kits such as the Hexagon OBTI (Human Gesellschaft für Biochemica und Diagnostica mbH, Wiesbaden) and the ABAcard HemaTrace® (Abacus Diagnostics, California) are available. They utilize the antibody–antigen–antibody sandwich method by using antibodies that recognize human Hb. The ABAcard HemaTrace assay utilizes a labeled monoclonal antihuman Hb antibody contained in a sample well, and a polyclonal antihuman Hb antibody immobilized at a test zone of a nitrocellulose membrane. Additionally, an antiglobulin that recognizes the antibody is immobilized onto a control zone (Figure 13.2).

A sample can be prepared by cutting a small portion (2 mm diameter) of a stain or a swab. Each sample is extracted for 5 min in 2 mL of extraction buffer. A longer extraction time may be used for older stains. The samples are loaded into the sample well, and the antigen in the sample binds to the labeled antibody in the well to form an antigen–antibody complex, which then diffuses across the nitrocellulose membrane. At the test zone, the solid-phase antihuman Hb antibody binds to the antigen–antibody complex to form a labeled antibody–antigen–antibody sandwich.

The ABAcard HemaTrace® uses a pink dye that is visualized in a positive result as a pink horizontal line at the test zone (Figure 13.3). In the control zone, unbound labeled antihuman

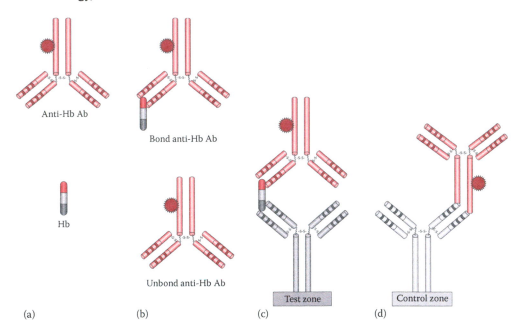

Figure 13.2 Immunochromatographic assays for the identification of Hb in human blood. (a) In a sample well, Hb in a blood sample is mixed with a labeled anti-Hb Ab. (b) The Hb binds to the labeled anti-Hb Ab to form a labeled Ab–Hb complex. The complex diffuses toward the test zone. (c) At the test zone, the labeled Ab–Hb complex binds to an immobilized anti-Hb Ab to form a labeled Ab–Hb–Ab sandwich. (d) At the control zone, the labeled anti-Hb Ab binds to an immobilized antiglobulin and is captured. Ab and Hb represent the antibody and hemoglobin, respectively. (© Richard C. Li.)

Hb antibody binds to the solid-phase antiglobulin. This antibody–antiglobulin complex at the control zone also produces a pink horizontal line. The test is considered valid only if the line in the control zone is observed. The presence of human Hb results in a pink line at both the test and control zones. The absence of human Hb results in a pink line in the control zone only. A positive result can appear in less than a minute.

Validation studies have revealed that the sensitivity of the ABAcard HemaTrace® can be as low as 0.07 μg/mL of Hb. The normal blood Hb concentration is 14–18 and 12–16 g/dL among males and females, respectively. This assay is more sensitive than the Kastle–Meyer assay (Chapter 12). Additionally, the assay is responsive to aged stains and degraded materials. Specificity studies have shown that it is specific for blood of higher primates, including humans. However, it is also responsive to seminal stains, and oral, vaginal, anal, and rectal swabs. It is believed that these biological fluids contain very low amounts of Hb, which can still be detected by highly sensitive assays. However, if the concentration of blood is too high, a false negative can result due to the high-dose hook effect described in Chapter 11.

13.2.1.2 Identification of Human Glycophorin A Protein

Commercially produced immunochromatographic kits such as RSID™-Blood (Independent Forensics, Hillside, IL) use antibodies that recognize human GPA (Figure 13.4). A labeled monoclonal antihuman GPA antibody is contained in a sample well, and a second monoclonal antihuman GPA antibody, to a different epitope of GPA, is immobilized onto a test zone of the membrane. An antiglobulin that recognizes the antibody is immobilized onto a control zone (Figure 13.5).

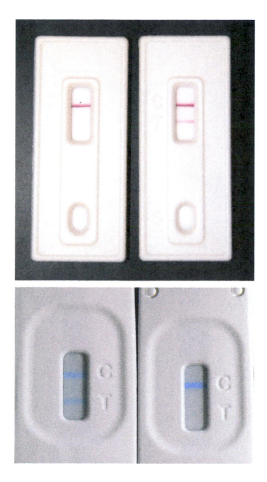

Figure 13.3 Human blood identification using immunochromatic devices. Top: In an assay using an ABAcard HemaTrace device, the negative (left) and positive (right) results are shown. The "C" band indicates that the test is valid. The "T" band indicates the presence of human blood. The sample well is labeled "S." (© Richard C. Li.) Bottom: The positive (left) and negative (right) results are shown using a Hexagon-OBTI device. The "C" band indicates that the test is valid. The "T" band indicates the presence of human blood. (From Ramsthaler, F., et al., Postmortem interval of skeletal remains through the detection of intraosseal hemin traces. A comparison of UV-fluorescence, luminol, Hexagon-OBTI(R), and Combur(R) tests. *Forensic Sci Int*, 209, 59–63, 2011. With permission.)

The sample can be collected by cutting out a small portion of a stain or a swab. The sample is then extracted overnight in an extraction buffer. The extract is removed and mixed with a running buffer. The assay is carried out by loading the extracted sample into the sample well. Again, the presence of GPA results in a pink line at both the test zone and the control zone, while the absence of GPA results in a pink line in the control zone only. The test is considered valid only if the line in the control zone is observed. A result can be read after 10 min.

Validation studies revealed that the sensitivity of the RSID kit can be as low as 100 nL of human blood. Species specificity studies showed no cross-reactivity with various animal species, including nonhuman primates. Biological fluid specificity studies revealed that the kit is not responsive to other human biological fluids such as semen, saliva, urine, milk, and amniotic and vaginal fluid. No high-dose hook effects were observed in samples containing up to 5 μL of blood.

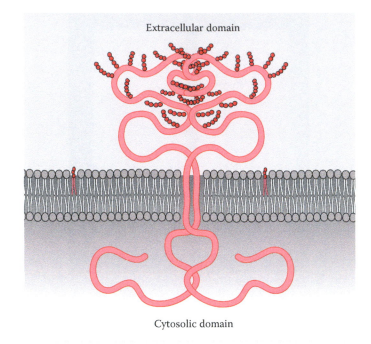

Figure 13.4 Diagram of the structure of glycophorin (GPA) protein. GPA is a transmembrane protein on the human erythrocyte membrane. A GPA dimer is shown. The extracellular domain of the GPA is glycosylated with carbohydrate side chains (red). Various GPA epitopes are antigenic determinants of several blood group systems. (© Richard C. Li.)

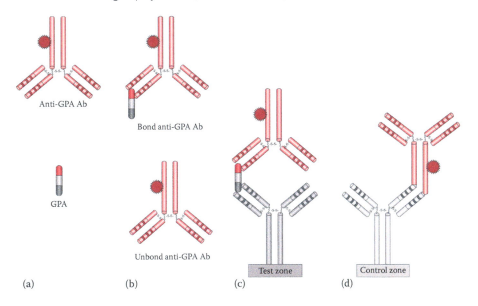

Figure 13.5 Immunochromatographic assays for the identification of GPA in human blood. (a) In a sample well, GPA in a blood sample is mixed with a labeled anti-GPA Ab. (b) The GPA binds to the labeled anti-GPA Ab to form a labeled Ab–GPA complex. (c) At the test zone, the labeled Ab–GPA complex binds to an immobilized anti-GPA Ab to form a labeled Ab–GPA–Ab sandwich. (d) At the control zone, the labeled anti-GPA Ab binds to an immobilized antiglobulin and is captured. Ab and GPA represent the antibody and GPA protein, respectively. (© Richard C. Li.)

13.2.2 Double Immunodiffusion Assays

13.2.2.1 Ring Assay

Chapter 11 discussed the basic principle of ring assay in detail. In this double immunodiffusion assay, an antihuman antibody reagent is placed at the bottom of a test tube and a bloodstain extract is placed on top of the bottom layer, as illustrated in Figure 13.6. The procedure is described in Box 13.1. In a positive reaction, a white precipitate between the two layers observed after several minutes indicates that a sample is of human origin. If the bloodstain extract is not human, no precipitation should appear.

13.2.2.2 Ouchterlony Assay

The basic principle of this double immunodiffusion assay has also been discussed in Chapter 11. The procedure of the Ouchterlony assay is described in Box 13.2. In a positive reaction, a line of precipitate will form between each antigen well and antibody well. This assay can also determine the similarity of the antigens (Figure 13.7). During the diffusion process, different antigen–antibody complexes migrate at different rates. Consequently, a separate line of precipitate will appear in the gel for each antigen–antibody complex. In an assay in which two antigens are loaded in adjacent wells and an antibody in the third well, the following results can be observed:

- If the two antigens are identical, the two lines will become fused. This phenomenon is referred to as *identity*.

- If the two antigens are totally unrelated, the lines will cross each other but not fuse; this is known as *nonidentity*.

- If the two antigens are related (share a common epitope) but are not identical, the lines will merge with spur formation. The spurs are continuations of the line formed by the antigen due to its unique epitope. This phenomenon is known as *partial identity*.

Thus, in a species test, a positive result is noted when the precipitate lines for the positive controls and the samples fuse. No spur formation should be observed.

13.2.3 Crossed-Over Electrophoresis

This method is a combination of immunodiffusion and electrophoresis (also see Chapter 11). The procedure for the assay is described in Box 13.3. With this technique, a sharp precipitate band is visualized in a positive reaction (Figure 13.8). However, false-negative results can occur due to the postzone phenomenon, in which excess antigen may inhibit precipitation. In this

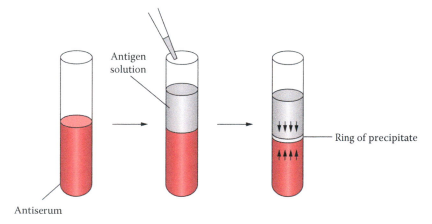

Figure 13.6 Ring assay. The antigen solution is carefully applied over the antiserum solution. Incubation may be necessary in order to form a ring of precipitate. (© Richard C. Li.)

BOX 13.1   RING ASSAY PROCEDURE

Sample preparation and extraction

- Extract a portion of a stain with saline at 4°C overnight.

Controls

- Include a positive control (known human serum sample) and a negative control (extraction blank).

Loading of antibody and samples

- Spin the antihuman antibody in a microfuge and transfer the supernatant into test tubes or capillary tubes (depending on the volume of the stain and the antiserum extracted).
- Place the sample carefully over the top of the antiserum solution, which is usually denser than the sample.

Immunodiffusion reaction

- Carry out the reaction at room temperature.
- In a positive reaction, white precipitate between the two layers can be observed after several minutes. This indicates that the sample is of human origin. No precipitate is formed if a bloodstain extract is from a nonhuman origin.

BOX 13.2   OUCHTERLONY ASSAY PROCEDURE

Sample preparation and extraction

- Cut out a small portion (approximately 5 × 5 mm) of a stain or a portion of a swab.
- Extract at room temperature in 100 µL of water for 30 min. The extract can be diluted if necessary. Alternatively, a very small piece of the stain or swab can be inserted directly into the well.

Controls

- Positive (known serum)
- Negative (extraction blank)
- Substrate controls (extraction of substrate from unstained area) if applicable

Agarose gel preparation

Heat a suspension of agarose (4%) until liquefied. Cool the solution in a water bath at 55°C. Pour the agarose onto a piece of glass slide and let the gel solidify to a thickness of about 2–3 mm. Alternatively, a polyester support film such as GelBond (Cambrex, New Zealand) can be used as a gel support. The agarose should be poured onto the hydrophilic side of a piece of GelBond film (6 × 9 cm). Punch wells consisting of a central well surrounded by four wells using a template.

Loading antibodies and samples

- Apply antihuman antibody to the central well. Apply the positive control to one of the surrounding wells.
- Apply the sample(s) in question next to a positive control.
- Apply negative and substrate controls to the remaining wells; only one negative control is needed per gel.

Immunodiffusion reaction

Incubate the plate overnight in a moisture chamber at 37°C.

Staining

Soak the gel overnight in saline solution and then soak it in deionized water for 10 min. Repeat once. Dry the gel between paper towels with a weight on top for 30 min. Dry in an oven for 30 min. Stain the gel with Coomassie blue. Stained precipitate bands appear blue.

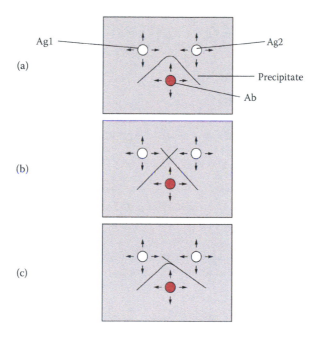

Figure 13.7   Results of Ouchterlony assay. Ab indicates an antibody. Ag1 and Ag2 are the antigen samples in question. (a) Identity. The two antigens are identical (fused line). (b) Nonidentity. The two antigens are unrelated (spur). (c) Partial identity. The two antigens are related but not identical (fused line with spur). (© Richard C. Li.)

situation, the sample can be diluted and the assay can be repeated. False-negative results can also occur due to simple mistakes made during electrophoresis:

- Electrophoresis is carried out in the opposite direction, which results in samples running off the gel.
- Electrophoresis is carried out using an incorrect buffer system, affecting antigen–antibody binding. The amount of current applied during the electrophoresis is too strong and generates heat and denatures proteins.

BOX 13.3 CROSSED-OVER ELECTROPHORESIS PROCEDURE

Sample preparation and extraction

- Cut out a small portion (~5 × 5 mm) of a stain or a portion of a swab.
- Extract at room temperature in 100 μL of water for 30 min. The extract can be diluted if necessary. Alternatively, a very small piece of the stain or swab can be inserted directly into the well.

Controls

- Positive (known serum)
- Negative (extraction blank)
- Substrate (extraction of substrate from unstained area) if applicable

Agarose gel preparation

Heat a suspension of agarose (4%) until liquefied. Cool the solution in a water bath at 55°C. Pour the agarose onto a piece of glass slide and let it solidify. Alternatively, a polyester support film such as GelBond can be used as a gel support. The agarose should be poured onto the hydrophilic side of a piece of GelBond (6 × 9 cm). Punch small wells (about 1–2 mm) in rows using a template.

Loading antibodies and samples

- Apply antihuman antibody in one row of wells.
- Apply samples in the other row of wells. Apply the positive, negative, and substrate controls.

Electrophoresis

Submerge the agarose gel in an electrophoresis tank in proper orientation. The wells containing antihuman antibody should be closest to the anode (positive electrode) and the wells containing samples should be closest to the cathode (negative electrode). During electrophoresis, the antibody in the antiserum should migrate toward the cathode while the antigen migrates toward the anode. Electrophoresis is carried out at 10 V/cm for 20 min.

Staining

Soak the gel overnight in a saline solution and then soak it in deionized water for 10 min. Repeat once. Dry the gel between paper towels with a weight on top for 30 min. Dry in an oven for 30 min. Stain the gel with Coomassie blue. Stained precipitate bands appear blue.

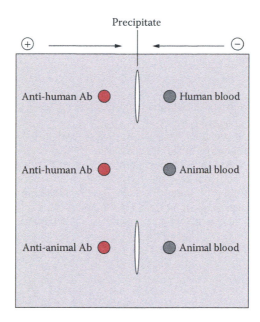

Figure 13.8 Results of crossed-over electrophoresis. A precipitate line is formed between a human blood sample and an antihuman antibody. No precipitate line is formed when the antihuman antibody is tested for an animal blood sample. A precipitate line is formed between the animal blood sample and the antianimal antibody. (© Richard C. Li.)

Bibliography

Allison, A.C. and J.A. Morton, Species specificity in the inhibition of antiglobulin sera: A technique for the identification of human and animal bloods. *J Clin Pathol*, 1953, **6**(4): 314–319.

Berger, B., et al., Validation of two canine STR multiplex-assays following the ISFG recommendations for non-human DNA analysis. *Forensic Sci Int Genet*, 2014, **8**(1): 90–100.

Bhatia, R.Y., Specificity of some plant lectins in the differentiation of animal blood. *Forensic Sci*, 1974, **4**(1): 47–52.

Coomber, N., et al., Validation of a short tandem repeat multiplex typing system for genetic individualization of domestic cat samples. *Croat Med J*, 2007, **48**(4): 547–555.

Dawnay, N., et al., Validation of the barcoding gene *COI* for use in forensic genetic species identification. *Forensic Sci Int*, 2007, **173**(1): 1–6.

DeForest, P., R. Gaensslen, and H.C. Lee, *Forensic Science: An Introduction to Criminalistics*, 1983. New York: McGraw-Hill.

Dorrill, M. and P.H. Whitehead, The species identification of very old human blood-stains. *Forensic Sci Int*, 1979, **13**(2): 111–116.

El-Sayed, Y.S., et al., Using species-specific repeat and PCR-RFLP in typing of DNA derived from blood of human and animal species. *Forensic Sci Med Pathol*, 2010, **6**(3): 158–164.

Grobbelaar, B.G., D. Skinner, and H.N. van de Gertenbach, The anti-human globulin inhibition test. In the identification of human blood stains. *J Forensic Med*, 1970, **17**(3): 103–111.

Hochmeister, M.N., et al., Validation studies of an immunochromatographic 1-step test for the forensic identification of human blood. *J Forensic Sci*, 1999, **44**(3): 597–602.

Imaizumi, K., et al., Development of species identification tests targeting the 16S ribosomal RNA coding region in mitochondrial DNA. *Int J Legal Med*, 2007, **121**(3): 184–191.

Johnston, S., J. Newman, and R. Frappier, Validation study of the Abacus Diagnostics ABAcard® HemaTrace® membrane test for the forensic identification of human blood. *Can Soc Forensic Sci*, 2003, **36**(3): 173–183.

Juusola, J. and J. Ballantyne, Messenger RNA profiling: A prototype method to supplant conventional methods for body fluid identification. *Forensic Sci Int*, 2003, **135**(2): 85–96.

Kanthaswamy, S., et al., Quantitative real-time PCR (qPCR) assay for human-dog-cat species identification and nuclear DNA quantification. *Forensic Sci Int Genet*, 2012, **6**(2): 290–295.

Karlsson, A.O. and G. Holmlund, Identification of mammal species using species-specific DNA pyrosequencing. *Forensic Sci Int*, 2007, **173**(1): 16–20.

Kun, T., et al., Developmental validation of Mini-DogFiler for degraded canine DNA. *Forensic Sci Int Genet*, 2013, **7**(1): 151–158.

Lawton, M.E. and J.G. Sutton, Species identification of deer blood by isoelectric focusing. *J Forensic Sci Soc*, 1982, **22**(4): 361–366.

Lee, H.C., Identification and grouping of bloodstains, in R. Saferstein (ed.), *Forensic Science Handbook*, pp. 267–337, 1982. Englewood Cliffs, NJ: Prentice Hall.

Lee, H.C. and P.R. De Forest, A precipitin-inhibition test on denatured bloodstains for the determination of human origin. *J Forensic Sci*, 1976, **21**(4): 804–810.

Lowenstein, J.M., et al., Identification of animal species by protein radioimmunoassay of bone fragments and bloodstained stone tools. *Forensic Sci Int*, 2006, **159**(2–3): 182–188.

Nakaki, S., et al., Study of animal species (human, dog and cat) identification using a multiplex single-base primer extension reaction in the cytochrome b gene. *Forensic Sci Int*, 2007, **173**(2–3): 97–102.

Ogden, R., Forensic science, genetics and wildlife biology: Getting the right mix for a wildlife DNA forensics lab. *Forensic Sci Med Pathol*, 2010, **6**(3): 172–179.

Parson, W., et al., Species identification by means of the cytochrome b gene. *Int J Legal Med*, 2000, **114**(1–2): 23–28.

Ramsthaler, F., et al., Postmortem interval of skeletal remains through the detection of intraosseal hemin traces. A comparison of UV-fluorescence, luminol, Hexagon-OBTI(R), and Combur(R) tests. *Forensic Sci Int*, 2011, **209**: 59–63.

Sato, I., et al., Forensic hair analysis to identify animal species on a case of pet animal abuse. *Int J Legal Med*, 2010, **124**(3): 249–256.

Shaler, R.C., Modern forensic biology, in R. Saferstein (ed.), *Forensic Science Handbook*, pp. 525–614, 2002. Upper Saddle River, NJ: Pearson Education.

Sivaram, S., et al., Differentiation between stains of human blood and blood of monkey. *Forensic Sci*, 1975, **6**(3): 145–152.

Spear, T.F. and S.A. Binkley, The HemeSelect test: A simple and sensitive forensic species test. *J Forensic Sci Soc*, 1994, **34**(1): 41–46.

Tabata, N. and M. Morita, Immunohistochemical demonstration of bleeding in decomposed bodies by using anti-glycophorin A monoclonal antibody. *Forensic Sci Int*, 1997, **87**(1): 1–8.

Tanaka, Y., et al., Enzyme-linked immunospot assay for detecting cells secreting antibodies against human blood group A epitopes. *Transplant Proc*, 2003, **35**(1): 555–556.

Tobe, S. and A. Linacre, Species identification of human and deer from mixed biological material. *Forensic Sci Int*, 2007, **169**(2–3): 278–279.

Tobe, S.S., et al., Recovery of human DNA profiles from poached deer remains part 2: Improved recovery protocol without the need for LCN analysis. *Sci Justice*, 2013, **53**(1): 23–27.

Tsutsumi, H. and Y. Katsumata, Forensic study on stains of blood and saliva in a chimpanzee bite case. *Forensic Sci Int*, 1993, **61**(2–3): 101–110.

van Asch, B. and A. Amorim, Capillary electrophoresis analysis of a 9-plex STR assay for canine genotyping. *Methods Mol Biol*, 2012, **830**: 231–240.

van Asch, B., et al., A new autosomal STR nineplex for canine identification and parentage testing. *Electrophoresis*, 2009, **30**(2): 417–423.

van Asch, B., et al., Forensic analysis of dog (*Canis lupus familiaris*) mitochondrial DNA sequences: An interlaboratory study of the GEP-ISFG working group. *Forensic Sci Int Genet*, 2009, **4**(1): 49–54.

Virkler, K. and I.K. Lednev, Blood species identification for forensic purposes using Raman spectroscopy combined with advanced statistical analysis. *Anal Chem*, 2009, **81**(18): 7773–7777.

Whitehead, P.H. and A. Brech, A micro-technique involving species identification and ABO grouping on the same fragment of blood. *J Forensic Sci Soc*, 1974, **14**(2): 109–110.

Wictum, E., et al., Developmental validation of DogFiler, a novel multiplex for canine DNA profiling in forensic casework. *Forensic Sci Int Genet*, 2013, **7**(1): 82–91.

Wilson-Wilde, L., et al., Current issues in species identification for forensic science and the validity of using the cytochrome oxidase I (COI) gene. *Forensic Sci Med Pathol*, 2010, **6**(3): 233–241.

Identification of Semen

14.1 Biological Characteristics

A typical ejaculation releases 2–5 mL of semen, which contains seminal fluid and sperm cells (*spermatozoa*). A normal sperm count ranges from 10^7 to 10^8 spermatozoa per milliliter of semen. The spermatozoa are formed from spermatogonia in the seminiferous tubules of the testes. This process of generating spermatozoa is referred to as spermatogenesis (Figure 14.1). The spermatozoa are then transported and stored in the tubular network of the epididymis where they undergo functional maturation (spermatogenesis and maturation take approximately 3 months). The epididymis joins the ductus deferens, which transports matured sperm from the epididymis to the ejaculatory duct. From there, spermatozoa follow the ejaculatory ducts into the prostatic urethra where they are joined with secretions from the prostate. Figure 14.2 illustrates the anatomy of the male reproductive system.

Seminal fluid is a complex mixture of glandular secretions. A typical sample of seminal fluid contains the combined secretions of several accessory glands. Seminal vesicle fluid accounts for approximately 60% of the ejaculate. Various proteins secreted by the seminal vesicles play a role in the coagulation of the ejaculate. Additionally, seminal vesicle fluid contains flavin, which causes semen to fluoresce under ultraviolet light, often utilized when searching for semen-stain evidence.

Prostatic fluid secretions account for approximately 30% of the ejaculate. The components of this fluid are complex as well. This portion of semen contains high concentrations of *acid phosphatase* (AP) and *prostate-specific antigen* (PSA). Both are useful markers for the identification of semen in forensic laboratories. The epididymis and the bulbourethral secretions each account for approximately 5% of the ejaculate.

A *vasectomy* is the surgical removal of a bilateral segment of the ductus deferens. The surgery prevents spermatozoa from reaching the distal portions of the male reproductive tract. However, a vasectomized male can still produce ejaculate that contains only seminal vesicle fluid, prostatic fluid, and bulbourethral fluid. The condition by which males have abnormally low sperm counts is known as *oligospermia*. *Azoospermia* is a condition that causes males to produce no spermatozoa. However, the secretion of seminal fluid is not affected in males who have these conditions. DNA derived from epithelial cells can be isolated from the seminal fluids of these individuals.

14.1.1 Spermatozoa

A human spermatozoon has three morphologically distinct structures: the head, the middle piece, and the tail (Figure 14.3). The head contains a nucleus with densely packed chromosomes. At the tip of the head is the *acrosomal cap*, which is a membranous compartment

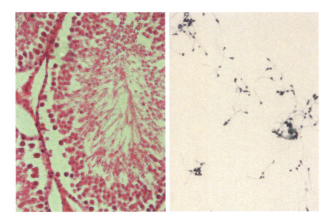

Figure 14.1 Spermatogenesis. In the seminiferous tubules (left) of the testes, spermatogonia (located at peripheral area of the seminiferous tubules) are differentiated to spermatids (located at the center of the seminiferous tubules). The spermatids are eventually differentiated to matured spermatozoa (right). (© Richard C. Li.)

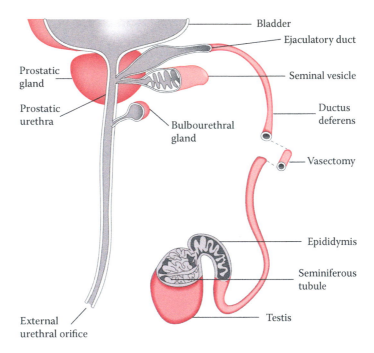

Figure 14.2 Male reproductive system and accessory glands (unilateral view). (© Richard C. Li.)

containing enzymes essential for fertilization. The head is attached to the middle piece through a short neck where the mitochondria that provide the energy for moving the tail are located. The tail or flagellum is responsible for spermatozoon motility. In contrast to other cell types, a mature spermatozoon lacks various intracellular organelles such as an endoplasmic reticulum, Golgi apparatus, lysosomes, and peroxisomes. In a normal male, at least 60% of spermatozoa have normal morphology, so morphological abnormalities can often be observed.

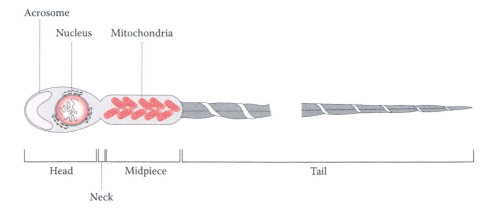

Figure 14.3   Structure of spermatozoon. (© Richard C. Li.)

14.1.2 Acid Phosphatase

Acid phosphatase (AP) consists of a group of phosphatases with optimal activity in an acidic pH environment. The greatest forensic importance of AP is that the prostate-derived AP contributes most of the AP activity present in semen. AP levels in semen are not affected by vasectomies. AP isoenzymes are also found in other tissues (Section 16.2).

The half-life of AP activity at 37°C is 6 months. However, the half-life is decreased if a sample is stored in a wet environment. AP activity can be detected from dry seminal stains stored at −20°C up to 1 year. Low levels of prostatic AP are present in the sera of healthy males. Elevated levels of prostatic AP found in serum are useful in diagnosing and monitoring prostate carcinoma. Many AP tests utilized in clinical testing may be used to identify semen for forensic applications.

14.1.3 Prostate-Specific Antigen

Prostate-specific antigen (PSA) is a major protein present in seminal fluid at concentrations of 0.5–2.0 mg/mL. PSA is produced in the prostate epithelium and secreted into the semen. PSA can also be found in the paraurethral glands, perianal glands, apocrine sweat glands, and mammary glands. Thus small quantities can be detected in urine, fecal material, sweat, and milk. PSA can also be found at much lower levels in the bloodstream. An elevated plasma PSA is present in prostate cancer patients, and it is widely used as a screening test for this disease. PSA is also elevated in cases of benign prostatic hyperplasia and prostatitis. The synthesis of PSA is stimulated by androgen, a steroid hormone.

PSA is a protein that has a molecular weight of 30 kDa and is thus also known as *P30*. It is responsible for hydrolyzing *semenogelin* (Sg), which mediates gel formation in semen (Section 14.1.4). PSA is a member of the tissue *kallikrein* (serine protease) family and is encoded by the *KLK3* locus located on chromosome 19. In addition to PSA, other tissue kallikreins encoded by *KLK2* and *KLK4* loci are expressed in the prostate. The half-life for PSA in a dried semen stain is about 3 years at room temperature. The half-life is greatly reduced when a sample is stored in wet conditions.

14.1.4 Seminal Vesicle–Specific Antigen

Human *seminal vesicle–specific antigen* (SVSA) includes two major types, semenogelin I (SgI) and semenogelin II (SgII), and constitutes the major seminal vesicle–secreted protein in semen. On ejaculation, SVSA forms a coagulum that is liquefied after a few minutes due to the degradation of SVSA by PSA. In humans, both SgI and SgII are present in a number of tissues of the male reproductive system, including the seminal vesicles, ductus deferens, prostate, and epididymis. They are also present in several other tissues such as skeletal muscle, kidney, colon, and trachea. They have also been found in the sera of lung cancer patients.

The use of Sg as a marker for semen identification instead of PSA presents certain advantages. The concentration of Sg in seminal fluid is much higher than that of PSA, and this is beneficial for the sensitivity of detection. Sg is present in seminal fluid and absent in urine, milk, and sweat, where PSA can be found. Although Sg compounds are present in skeletal muscle, kidney, and colon, this is not a great concern because these tissue samples are not routinely collected for semen detection in sexual assault cases.

14.2 Analytical Techniques for Identifying Semen

The location of semen stains is usually carried out through visual examination. Particularly, the application of alternate light sources (ALSs) can facilitate searches for semen stains. The presumptive identification of semen is largely based on the detection of the presence of prostatic AP activity in a sample. However, most presumptive assays cannot completely distinguish prostatic AP from nonprostatic AP. Confirmatory assays for the identification of semen are available, including the microscopic examination of spermatozoa, the identification of PSA and SVSA, and the RNA-based assay.

14.2.1 Presumptive Assays

14.2.1.1 Lighting Techniques for Visual Examination of Semen Stains

Lighting techniques can be used to aid in searching for semen stains. A dried semen stain fluoresces under certain light sources such as ALSs or argon lasers. ALSs are most commonly utilized for the visual examination of semen stains (Chapter 1; Figures 14.4 and 14.5). Excitation wavelengths between 450 and 495 nm can be used, allowing for the visualization of fluorescence with orange goggles. However, this approach is not specific for semen. Other bodily fluid stains, such as saliva and urine stains, can also fluoresce with less intensity. Additionally, the intensity of the fluorescence can be affected by different colors of substrates, and the material, such as clothing, where semen stains have been deposited.

Figure 14.4   A tabletop ALS device (left) for the detection of semen stains (right). (© Richard C. Li.)

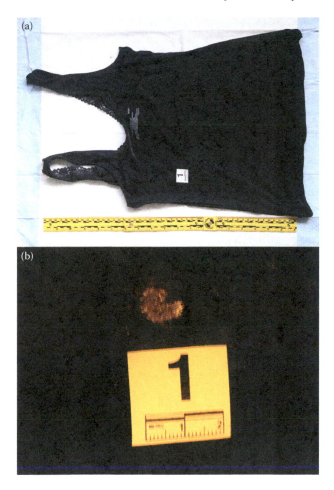

Figure 14.5 Examining a garment for semen stains. (a) A potential semen stain (labeled) is found on a garment, and (b) the stain fluoresces when irradiated with an ALS device. (© Richard C. Li.)

14.2.1.2 Acid Phosphatase Techniques

14.2.1.2.1 Colorimetric Assays

Colorimetric assays can be used for the presumptive identification of semen. The AP contained in semen can hydrolyze a variety of phosphate esters. It catalyzes the removal of the phosphate group from a substrate (Figure 14.6). Subsequently, an insoluble colored precipitate at sites of acid phosphatase activity is formed with a stabilized diazonium salt (usually in the form of zinc double salts).

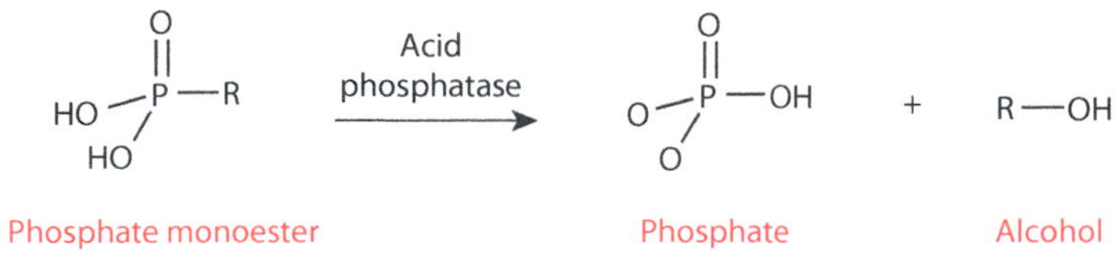

Phosphate monoester   Phosphate   Alcohol

Figure 14.6 Reaction catalyzed by acid phosphatases (EC 3.1.3.2). The optimal pH of the reaction is usually under pH 7.

However, interference during a test by nonprostatic AP isoenzymes (multiple forms of AP), such as contamination by AP commonly present in vaginal secretions (Chapter 16), can create problems in specimens collected from victims. Thus, it is desirable to be able to increase the specificity of the assay for prostatic AP. One solution is the application of substrates that are hydrolyzed rapidly by the prostatic enzyme and at a slower rate by the other forms of AP isoenzymes. For example, α-naphthylphosphate and thymolphthalein monophosphate are more specific to prostatic AP than phenyl phosphate and 4-nitrophenyl phosphate (Figure 14.7). The most common method for forensic applications is the use of α-naphthyl phosphate as a substrate. In the presence of AP, α-naphthylphosphate is hydrolyzed to phosphate and α-naphthol. Subsequently, the Fast Blue B, a stabilized diazonium salt, is added to carry out an azo coupling reaction, producing a purple azo dye (Figures 14.8 through 14.10).

Prostatic AP is water soluble. Thus, a moistened cotton swab or piece of filter paper can be used to transfer a small amount of sample from a stain by briefly pressing onto the questioned stain area. The α-naphthylphosphate reagent is added to the swab or filter paper followed by the addition of Fast Blue B reagent. If a purple coloration develops within 1 min, the test is considered a positive indication for semen. Color that develops after more than 1 min may arise from the activity of nonprostatic AP.

Additionally, the prostatic enzyme is strongly inhibited by dextrorotatory tartrate ions. Thus, these inhibitors, particularly tartrate, allow a distinction to be made between prostatic AP and other AP isoenzymes. Prostate and vaginal acid phosphatase can also be distinguished by using gel electrophoresis (Chapter 16).

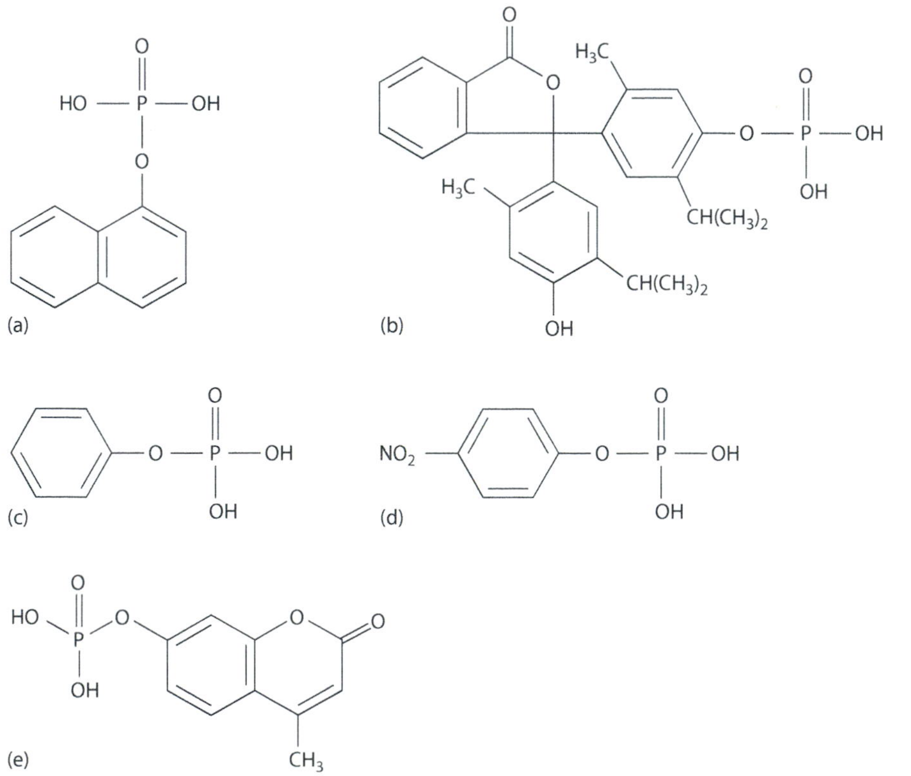

Figure 14.7 Chemical structures of acid phosphatase substrates. (a) α-Naphthyl phosphate, (b) thymolphthalein monophosphate, (c) phenyl phosphate, (d) 4-nitrophenyl phosphate, and (e) MUP.

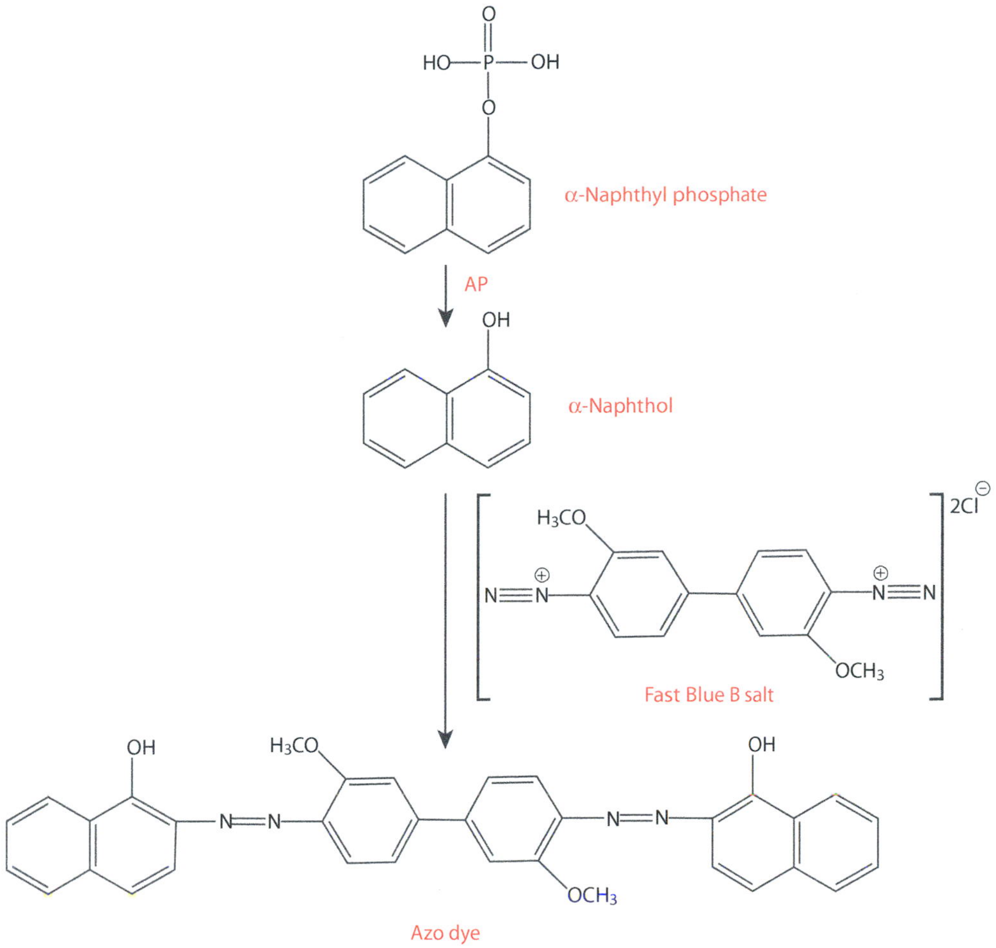

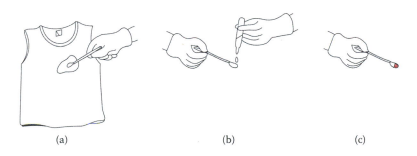

Figure 14.8 A colorimetric acid-phosphatase assay. In this assay, α-naphthylphosphate is hydrolyzed by acid phosphatase to phosphate and α-naphthol. The α-naphthol is subsequently converted into a purple azo dye with a diazonium salt such as Fast Blue B salt. AP, acid phosphatase.

Figure 14.9 AP colorimetric assay. (a) A small amount of sample from the stain is transferred using a moistened cotton swab. (b) The substrate reagent is applied, followed by adding Fast Blue B reagent. (c) Purple coloration indicates a positive reaction. (© Richard C. Li.)

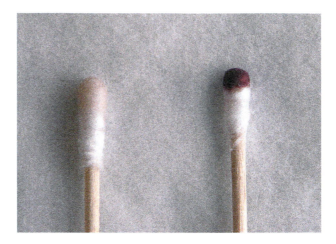

Figure 14.10 Photo of a colorimetric acid-phosphatase assay using α-naphthylphosphate as a substrate. (© Richard C. Li.)

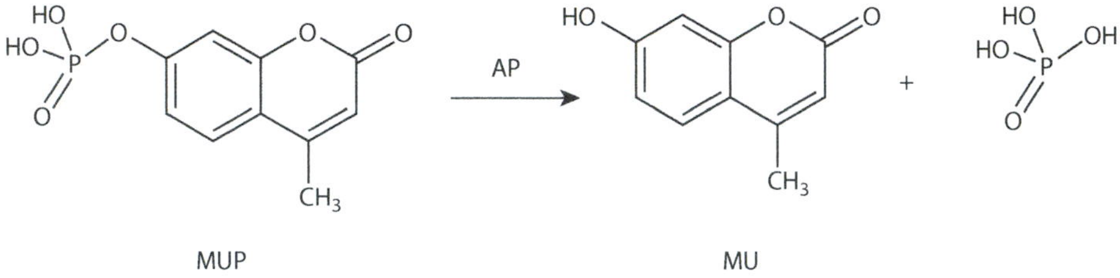

MUP → MU

Figure 14.11 The principle of MUP assay for detecting acid phosphatase activity. In the presence of acid phosphatase, 4-methylumbelliferone phosphate (MUP) is hydrolyzed, forming phosphate and 4-methylumbelliferone (MU), which fluoresces. AP, acid phosphatase.

14.2.1.2.2 Fluorometric Assays

Fluorometric methods are more sensitive than the colorimetric detection of AP and are used for semen stain mapping. AP catalyzes the removal of the phosphate residue on a 4-methylumbelliferone phosphate (MUP) substrate (Figure 14.11), a reaction that generates fluorescence under ultraviolet light. A piece of moistened filter paper, marked for proper orientation and identification, is used for transferring the prostatic AP. The evidence to be tested, a garment for example, is covered by the filter paper. Gloved hands are used to press the filter paper onto the stained area, ensuring that the evidence is in close contact with the paper. The filter paper is lifted from the evidence and examined in a dark room using long-wave ultraviolet light to detect any background fluorescence, which is then marked on the paper. The paper can then be sprayed with MUP reagent in a fume hood. The AP reaction on the paper can be visualized immediately. Areas where semen is present can be visualized as fluorescent areas on the filter paper (Figure 14.12).

14.2.2 Confirmatory Assays

14.2.2.1 Microscopic Examination of Spermatozoa

The cells from a questioned stain on an absorbent material can be transferred to a microscope slide by extracting a small portion of a stain with water, followed by gentle vortexing. The suspension is then transferred to a slide and evaporated at room temperature or fixed with low heat. Alternatively, it can be transferred by dampening the stain with water and rubbing or rolling it onto a microscope slide.

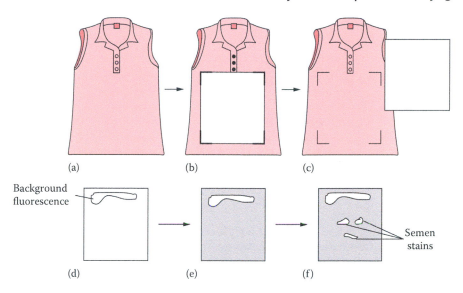

Figure 14.12 Fluorometric assay of acid phosphatase for locating semen stains. Evidence item (a) is closely covered by a piece of moistened filter paper (b) to allow transfer of a small amount of stain. The orientation of the paper is marked (c). The paper is lifted. The background fluorescence is marked under UV light (d). The filter paper is treated with MUP (e). The presence of fluorescence under ultraviolet light indicates semen stains (f). (© Richard C. Li.)

Microscopic identification of spermatozoa provides the proof of a seminal stain. Histological staining can facilitate microscopic examination. The most common staining technique is the *Christmas tree stain* (Figure 14.13). The red component known as Nuclear Fast Red (NFR) is a dye used for staining the nuclei of spermatozoa in the presence of aluminum ions. The green component, picroindigocarmine (PIC), stains the neck and tail portions of the sperm. The acrosomal cap and the nucleus stain pink-red, and the sperm tails and the midpiece stain blue-green. Epithelial cells, if present in the sample, appear blue-green and have red nuclei. Additionally, fluorescent detection utilizing SPERM HY-LITER Fluorescent Staining Kit can facilitate the identification of spermatozoa.

Laser capture microdissection (LCM) has been shown to be an effective technique for separating spermatozoa from nonsperm cells (i.e., epithelial cells from the victim) on a glass slide (Figure 14.14). This technique involves using a thin layer of a thermosensitive polymer that is placed on the surface of an LCM cap. Once spermatozoa are identified on the slide under a microscope, a polymer-containing LCM cap is placed over the spermatozoa on the slide. An infrared laser melts the polymer and causes it to adhere only to the targeted spermatozoa. The spermatozoa are then lifted off the slide. This allows spermatozoa to be separated and placed into snap-cap tubes for forensic DNA analysis.

14.2.2.2 Identification of Prostate-Specific Antigen

Over the years, a number of methods have been utilized to detect PSA: immunodiffusion, immunoelectrophoresis, enzyme-linked immunosorbent assay (ELISA), and immunochromatographic assays. ELISA and immunochromatographic assays have been found to be the most sensitive methods (Chapter 11).

14.2.2.2.1 Immunochromatographic Assays

Commercially produced immunochromatographic kits such as the PSA-check-1 (VED-LAB, Alencon), Seratec® PSA Semiquant (Seratec Diagnostica, Göttingen), and One Step ABAcard

Figure 14.13   Human spermatozoa stained with Christmas tree stain. (a) Staining spermatozoa on a microscope slide and (b) stained spermatozoa. (© Richard C. Li.)

Figure 14.14   A microscopic device for fluorescent detection utilizing SPERM HY-LITER Fluorescent Staining Kit (left) and a laser-capture microdissection device for separating sperm cells from other types of cells (right). (© Richard C. Li.)

PSA® (Abacus Diagnostics, California) are available. These devices utilize antihuman PSA antibodies. In the ABAcard PSA® assay, a labeled monoclonal antihuman PSA antibody is contained in a sample well, a polyclonal antihuman PSA antibody is immobilized on a test zone of a nitrocellulose membrane, and an antiglobulin that recognizes the antibody is immobilized on a control zone (Figures 14.15 and 14.16).

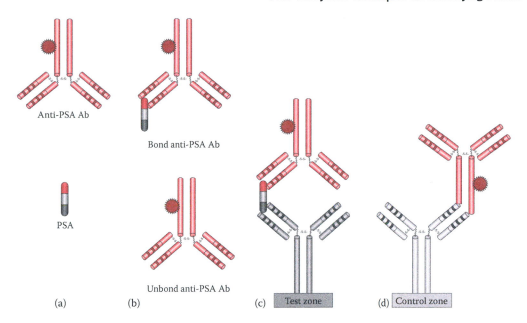

Figure 14.15 Immunochromatographic assays for identification of PSA in semen. (a) In a sample well, PSA in a semen sample is mixed with labeled anti-PSA Ab. (b) The PSA binds to the labeled anti-PSA Ab to form a labeled Ab–PSA complex. (c) At the test zone, the labeled complex binds to an immobilized anti-PSA Ab to form a labeled Ab–PSA–Ab sandwich. (d) At the control zone, the labeled anti-PSA Ab binds to an immobilized antiglobulin and is captured at the control zone. Ab and PSA represent antibody and prostate-specific antigen, respectively. (© Richard C. Li.)

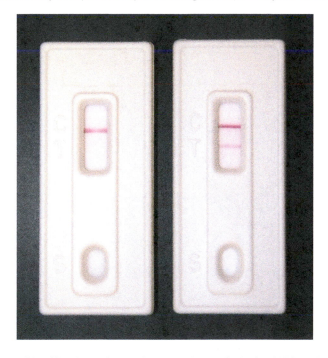

Figure 14.16 Semen identification using an immunochromatic device (ABAcard PSA). The negative (left) and positive (right) results are shown. The "C" band indicates that the test is valid. The "T" band indicates the presence of human blood. The sample well is labeled as "S". (© Richard C. Li.)

The assay is carried out by loading an extracted sample into the sample well. The antigen in the sample binds to the labeled antibody in the sample well to form an antigen–antibody complex. The complex then diffuses across the nitrocellulose membrane. At the test zone, the immobilized antihuman PSA antibody binds with the antigen–antibody complex to form an antibody–antigen–antibody sandwich. The ABAcard PSA® uses a pink dye that allows for the visualization of a positive test with a pink line at the test zone. In the control zone, unbound labeled antihuman PSA antibody binds to the immobilized antiglobulin. This antibody–antiglobulin complex at the control zone also results in a pink line. The test is considered valid only if the line in the control zone is observed.

The presence of human PSA results in a pink line at both the test and control zones. The absence of human PSA produces a pink line in the control zone only. A positive result can appear within 1 min; a negative result is read after 10 min. However, the high-dose hook effect, an artifact that may cause false-negative results (Chapter 10), occurs when high quantities of seminal fluid are tested.

14.2.2.2.2 ELISA

The ELISA method can be used to detect PSA with anti-PSA antibodies. The most common method used in forensic serology is antibody sandwich ELISA, in which an antibody–antigen–antibody sandwich complex is formed (Figure 14.17). The intensity of the signal can be detected and is proportional to the amount of bound antigen. The amount of PSA can also be quantified by comparing a standard with known concentrations. Although this method is specific and highly sensitive, it is time-consuming. Chapter 11 discusses the principle of ELISA in further detail.

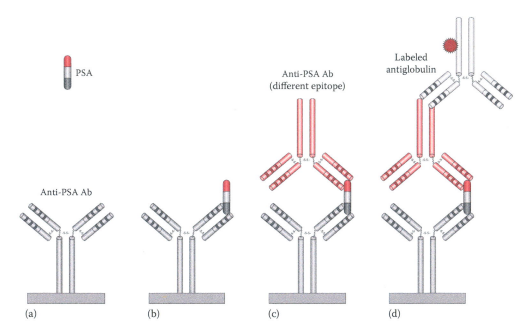

Figure 14.17 Use of ELISA for identification of PSA in semen. (a) Sample containing PSA is applied to polystyrene tubes in which anti-PSA Ab is immobilized. (b) The PSA binds to immobilized Ab to form a PSA–Ab complex. (c) A second anti-PSA Ab, specific for a different epitope of PSA, is added to form an Ab–PSA–Ab sandwich. (d) A labeled antiglobulin then binds to the Ab–PSA–Ab sandwich. The bound antiglobulin can be detected by various reporting schemes. Ab and PSA represent antibody- and prostate-specific antigen, respectively. (© Richard C. Li.)

14.2.2.3 Identification of Seminal Vesicle–Specific Antigen

14.2.2.3.1 *Immunochromatographic Assays*

Commercially produced immunochromatographic kits include the RSID®-Semen test (Independent Forensics, Hillside, IL) and the Nanotrap Sg. In the RSID®-Semen assay, a labeled monoclonal anti-Sg antibody is contained in a sample well, and a second monoclonal anti-Sg antibody, to a different epitope of Sg, is immobilized on the test zone of the membrane. An anti-globulin that recognizes the antibody is immobilized on a control zone (Figure 14.18).

The sample can be prepared by cutting a small portion of a stain or a swab and is extracted for 1–2 h in an extraction buffer (200–300 μL). Approximately 10% of the extract is removed and mixed with the running buffer. The assay is carried out by loading an extracted sample into the sample well. The antigen in the sample binds to the labeled anti-Sg antibody in the sample well to form a labeled antibody–antigen complex that then diffuses across the membrane. At the test zone, the solid-phase anti-Sg antibody binds with the labeled complex to form a labeled antibody–antigen–antibody sandwich. The antigen in the sample produces a pink line at the test zone. In the control zone, unbound labeled anti-Sg antibody binds to the solid-phase antiglobulin. This labeled antibody–antiglobulin complex at the control zone also results in a pink line. The presence of Sg generates a pink line at both the test and control zones. The absence of Sg results in a pink line in the control zone only. Results may be read after 10 min.

Validation studies have revealed that the sensitivity of the RSID-Semen kit for detecting seminal fluid can be as low as a 5×10^4-fold dilution. Species specificity studies have shown no cross-reactivity with various animal species including ruminants and small mammals. Bodily fluid specificity studies have also shown that the assay is not responsive to human blood, saliva,

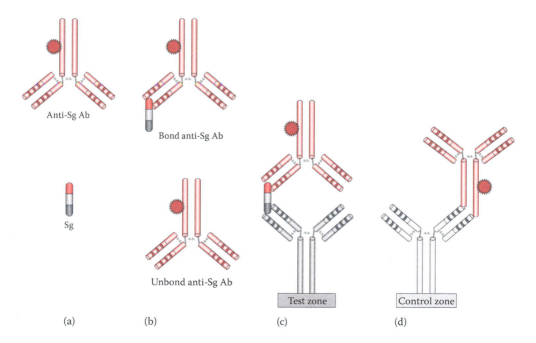

Figure 14.18 Immunochromatographic assays for identification of semenogelin (Sg) in semen. (a) In a sample well, Sg in a semen sample is mixed with labeled anti-Sg Ab. (b) Sg binds to the labeled anti-Sg Ab to form a labeled Ab–Sg complex. (c) At the test zone, the labeled complex binds to an immobilized anti-Sg Ab to form a labeled Ab–Sg–Ab sandwich. (d) At the control zone, the labeled anti-Sg Ab binds to an immobilized antiglobulin and is captured at the control zone. Ab represents antibody. (© Richard C. Li.)

urine, sweat, fecal matter, milk, or vaginal secretions. The assay results are not affected by condom lubricants or spermicides such as nonoxynol-9 and menfegol. However, the high-dose hook effect occurs when more than 3 μL of seminal fluid is tested.

14.2.2.3.2 ELISA

Identification of Sg for semen detection has also been carried out with ELISA. Anti-Sg antibodies are utilized. An antibody–antigen–antibody sandwich complex is formed (Figure 14.19). The intensity of the colorimetric or fluorometric signals can be detected spectrophotometrically and is proportional to the amount of bound antigen. The amount of Sg can be quantified by comparing a standard with known concentrations.

14.2.2.4 RNA-Based Assays

RNA-based assays (Chapter 11) have been developed to identify semen. The assays are based on the expression of certain genes in certain cell or tissue types. Thus, the techniques used in the identification of semen are based on the detection of specific types of mRNA expressed exclusively in spermatozoa and in certain cells of male accessory glands. These assays utilize reverse transcriptase polymerase chain reaction (RT-PCR; see Chapter 7) methods to detect gene expression levels of mRNAs for semen identification. Table 14.1 lists the tissue-specific genes utilized for semen identification. Compared to conventional assays used for semen identification, the RNA-based assay has higher specificity and is amenable to automation. However, one limitation is that the RNA is unstable due to degradation by endogenous ribonucleases.

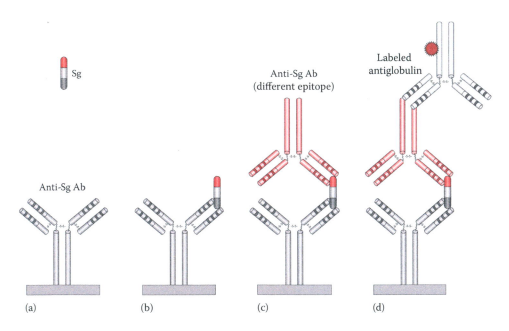

Figure 14.19   Use of ELISA for identification of semenogelin (Sg) in semen. (a) Sample containing Sg is applied to polystyrene tubes in which anti-Sg Ab is immobilized. (b) Sg binds to immobilized Ab to form a Sg–Ab complex. (c) A second anti-Sg Ab, specific for a different epitope of Sg, is added to form an Ab–Sg–Ab sandwich. (d) A labeled antiglobulin then binds to the Ab–Sg–Ab sandwich. The bound antiglobulin can be detected by various reporting schemes. Ab represents antibody. (© Richard C. Li.)

Table 14.1	Application of RT-PCR for Semen Identification		
Gene Symbol	**Gene Product**	**Description**	**Further Reading**
KLK3	Kallikrein 3	Also called prostate-specific antigen (PSA)	Gelmini et al. (2001)
PRM1	Protamine 1	DNA-binding proteins involved in condensation of sperm chromatin	Steger et al. (2000)
PRM2	Protamine 2	DNA-binding proteins involved in condensation of sperm chromatin	Steger et al. (2000)

Source: Adapted from Juusola, J. and Ballantyne, J., *Forensic Sci Int*, 152, 1–12, 2005; Nussbaumer, C., Gharehbaghi-Schnell, E., and Korschineck, I., *Forensic Sci Int*, 157, 181–186, 2006.

Bibliography

Ablett, P.J., The identification of the precise conditions for seminal acid phosphatase (SAP) and vaginal acid phosphatase (VAP) separation by isoelectric focusing patterns. *J Forensic Sci Soc*, 1983, **23**(3): 255–256.

Allard, J.E., The collection of data from findings in cases of sexual assault and the significance of spermatozoa on vaginal, anal and oral swabs. *Sci Justice*, 1997, **37**(2): 99–108.

Allard, J.E., et al., A comparison of methods used in the UK and Ireland for the extraction and detection of semen on swabs and cloth samples. *Sci Justice*, 2007, **47**(4): 160–167.

Allen, S.M., An enzyme linked immunosorbent assay (ELISA) for detection of seminal fluid using a monoclonal antibody to prostatic acid phosphatase. *J Immunoassay*, 1995, **16**(3): 297–308.

Allery, J.P., et al., Cytological detection of spermatozoa: Comparison of three staining methods. *J Forensic Sci*, 2001, **46**(2): 349–351.

Anoruo, B., et al., Isolating cells from non-sperm cellular mixtures using the PALM microlaser micro dissection system. *Forensic Sci Int*, 2007, **173**(2–3): 93–96.

Anslinger, K., et al., Sex-specific fluorescent labelling of cells for laser microdissection and DNA profiling. *Int J Legal Med*, 2007, **121**(1): 54–56.

Astrup, B.S., et al., Detection of spermatozoa following consensual sexual intercourse. *Forensic Sci Int*, 2012, **221**(1–3): 137–141.

Auvdel, M.J., Comparison of laser and high-intensity quartz arc tubes in the detection of body secretions. *J Forensic Sci*, 1988, **33**(4): 929–945.

Baechtel, S., The identification and individualization of semen stains, in R. Saferstein (ed.), *Forensic Science Handbook*, pp. 369–374, 1988. Englewood Cliffs, NJ: Prentice Hall.

Bauer, M. and D. Patzelt, Protamine mRNA as molecular marker for spermatozoa in semen stains. *Int J Legal Med*, 2003, **117**(3): 175–179.

Baxter, S.J., Immunological identification of human semen. *Med Sci Law*, 1973, **13**(3): 155–165.

Berti, A., et al., Expression of seminal vesicle–specific antigen in serum of lung tumor patients. *J Forensic Sci*, 2005, **50**(5): 1114–1145.

Bitner, S.E., False positives observed on the Seratec(R) PSA SemiQuant cassette test with condom lubricants. *J Forensic Sci*, 2012, **57**(6): 1545–1548.

Blake, E.T. and G.F. Sensabaugh, Genetic markers in human semen: A review. *J Forensic Sci*, 1976, **21**(4): 785–796.

Chapman, R.L., N.M. Brown, and S.M. Keating, The isolation of spermatozoa from sexual assault swabs using proteinase K. *J Forensic Sci Soc*, 1989, **29**(3): 207–212.

Chen, J.T. and G.L. Hortin, Interferences with semen detection by an immunoassay for a seminal vesicle-specific antigen. *J Forensic Sci*, 2000, **45**(1): 234–235.

Christoffersen, S., Immunohistochemical staining of human sperm cells in smears from sexual assault cases. *Int J Legal Med*, 2011, **125**(6): 887–890.

Clery, J.M., Stability of prostate specific antigen (PSA), and subsequent Y-STR typing, of *Lucilia* (*Phaenicia*) *sericata* (Meigen) (Diptera: Calliphoridae) maggots reared from a simulated postmortem sexual assault. *Forensic Sci Int*, 2001, **120**(1–2): 72–76.

Collins, K.A., et al., Identification of sperm and non-sperm male cells in cervicovaginal smears using fluorescence *in situ* hybridization: Applications in alleged sexual assault cases. *J Forensic Sci*, 1994, **39**(6): 1347–1355.

Cortner, G.V. and A.J. Boudreau, Phase contrast microscopy versus differential interference contrast microscopy as applicable to the observation of spermatozoa. *J Forensic Sci*, 1978, **23**(4): 830–832.

Dahlke, M.B., et al., Identification of semen in 500 patients seen because of rape. *Am J Clin Pathol*, 1977, **68**(6): 740–746.

Davidson, G. and T.B. Jalowiecki, Acid phosphatase screening—Wetting test paper or wetting fabric and test paper? *Sci Justice*, 2012, **52**(2): 106–111.

Davies, A., Evaluation of results from tests performed on vaginal, anal and oral swabs received in casework. *J Forensic Sci Soc*, 1978, **17**(2–3): 127–133.

Davies, A. and E. Wilson, The persistence of seminal constituents in the human vagina. *Forensic Sci*, 1974, **3**(1): 45–55.

De Moors, A., et al., Sperm Hy-Liter: An effective tool for the detection of spermatozoa in sexual assault exhibits. *Forensic Sci Int Genet*, 2013, **7**(3): 367–379.

Duenhoelter, J.H., et al., Detection of seminal fluid constituents after alleged sexual assault. *J Forensic Sci*, 1978, **23**(4): 824–829.

Elliott, K., et al., Use of laser microdissection greatly improves the recovery of DNA from sperm on microscope slides. *Forensic Sci Int*, 2003, **137**(1): 28–36.

Engelbertz, F., et al., Longevity of spermatozoa in the post-ejaculatory urine of fertile men. *Forensic Sci Int*, 2010, **194**(1–3): 15–19.

Enos, W.F. and J.C. Beyer, Prostatic acid phosphatase, aspermia, and alcoholism in rape cases. *J Forensic Sci*, 1980, **25**(2): 353–356.

Enos, W.F. and J.C. Beyer, Spermatozoa in the anal canal and rectum and in the oral cavity of female rape victims. *J Forensic Sci*, 1978, **23**(1): 231–233.

Eungprabhanth, V., Finding of the spermatozoa in the vagina related to elapsed time of coitus. *Z Rechtsmed*, 1974, **74**(4): 301–304.

Evers, H., et al., Investigative strategy for the forensic detection of sperm traces. *Forensic Sci Med Pathol*, 2009, **5**(3): 182–188.

Fraysier, H.D., A rapid screening technique for the detection of spermatozoa. *J Forensic Sci*, 1987, **32**(2): 527–530.

Gaensslen, R.E., *Sourcebook in Forensic Serology, Immunology, and Biochemistry*, 1983. Washington, DC: US Government Printing Office.

Gelmini, S., et al., Real-Time quantitative reverse transcriptase-polymerase chain reaction (RT-PCR) for the measurement of prostate-specific antigen mRNA in the peripheral blood of patients with prostate carcinoma using the taqman detection system. *Clin Chem Lab Med*, 2001, **39**(5): 385–391.

Greenfield, A. and Sloan, M., Identification of biological fluids and stains, in James, S.H., Nordby, J.J., and Bell, S. (eds), *Forensic Science: An Introduction to Scientific and Investigative Techniques*, 2009. Boca Raton, FL: Taylor & Francis.

Haas, C., et al., mRNA profiling for the identification of sperm and seminal plasma. *Forensic Sci Int Genet Suppl*, 2009, **2**(1): 534–535.

Haas, C., et al., RNA/DNA co-analysis from human saliva and semen stains—Results of a third collaborative EDNAP exercise. *Forensic Sci Int Genet*, 2013, **7**(2): 230–239.

Hardinge, P., et al., Optimisation of choline testing using Florence Iodine reagent, including comparative sensitivity and specificity with PSA and AP tests. *Sci Justice*, 2013, **53**(1): 34–40.

Healy, D.A., et al., Biosensor developments: Application to prostate-specific antigen detection. *Trends Biotechnol*, 2007, **25**(3): 125–131.

Herr, J.C. and M.P. Woodward, An enzyme-linked immunosorbent assay (ELISA) for human semen identification based on a biotinylated monoclonal antibody to a seminal vesicle–specific antigen. *J Forensic Sci*, 1987, **32**(2): 346–356.

Herr, J.C., et al., Characterization of a monoclonal antibody to a conserved epitope on human seminal vesicle–specific peptides: A novel probe/marker system for semen identification. *Biol Reprod*, 1986, **35**(3): 773–784.

Hochmeister, M.N., et al., High levels of alpha-amylase in seminal fluid may represent a simple artifact in the collection process. *J Forensic Sci*, 1997, **42**(3): 535–536.

Hochmeister, M.N., et al., Evaluation of prostate-specific antigen (PSA) membrane test assays for the forensic identification of seminal fluid. *J Forensic Sci*, 1999, **44**(5): 1057–1060.

Hueske, E.E., Techniques for extraction of spermatozoa from stained clothing: A critical review. *J Forensic Sci*, 1977, **22**(3): 596–598.

Ishiyama, I., Rapid histological examination of trace evidence by means of cellophane tape. *J Forensic Sci*, 1981, **26**(3): 570–575.

Iwasaki, M., et al., A demonstration of spermatozoa on vaginal swabs after complete destruction of the vaginal cell deposits. *J Forensic Sci*, 1989, **34**(3): 659–664.

Jones, E., The identification of semen and other body fluids, in R. Saferstein (ed.), *Forensic Science Handbook*, 2005. New Jersey: Pearson Prentice Hall.

Juusola, J. and J. Ballantyne, Messenger RNA profiling: A prototype method to supplant conventional methods for body fluid identification. *Forensic Sci Int*, 2003, **135**(2): 85–96.

Juusola, J. and J. Ballantyne, Multiplex mRNA profiling for the identification of body fluids. *Forensic Sci Int*, 2005, **152**(1): 1–12.

Keil, W., J. Bachus, and H.D. Troger, Evaluation of MHS-5 in detecting seminal fluid in vaginal swabs. *Int J Legal Med*, 1996, **108**(4): 186–190.

Khaldi, N., et al., Evaluation of three rapid detection methods for the forensic identification of seminal fluid in rape cases. *J Forensic Sci*, 2004, **49**(4): 749–753.

Laffan, A., et al., Evaluation of semen presumptive tests for use at crime scenes. *Med Sci Law*, 2011, **51**(1): 11–17.

LaRue, B.L., J.L. King, and B. Budowle, A validation study of the Nucleix DSI-Semen kit—A methylation-based assay for semen identification. *Int J Legal Med*, 2013, **127**(2): 299–308.

Laux, D.L. and J.P. Barnhart, Validation of the Seratec(R) SeraQuant for the quantitation of prostate-specific antigen levels on immunochromatographic membranes. *J Forensic Sci*, 2011, **56**(6): 1574–1579.

Levine, B., et al., Use of prostate specific antigen in the identification of semen in postmortem cases. *Am J Forensic Med Pathol*, 2004, **25**(4): 288–290.

Lewis, J., et al., The fallacy of the two-minute acid phosphatase cut off. *Sci Justice*, 2012, **52**(2): 76–80.

Liedtke, R.J. and J.D. Batjer, Measurement of prostate-specific antigen by radioimmunoassay. *Clin Chem*, 1984, **30**(5): 649–652.

Lilja, H., A kallikrein-like serine protease in prostatic fluid cleaves the predominant seminal vesicle protein. *J Clin Invest*, 1985, **76**(5): 1899–1903.

Lin, M.F., et al., Fundamental biochemical and immunological aspects of prostatic acid phosphatase. *Prostate*, 1980, **1**(4): 415–425.

Linde, H.G. and K.E. Molnar, The simultaneous identification of seminal acid phosphatase and phosphoglucomutase by starch gel electrophoresis. *J Forensic Sci*, 1980, **25**(1): 113–117.

Lunetta, P. and H. Sippel, Positive prostate-specific antigen (PSA) reaction in post-mortem rectal swabs: A cautionary note. *J Forensic Leg Med*, 2009, **16**(7): 397–399.

Maher, J., et al., Evaluation of the BioSign PSA membrane test for the identification of semen stains in forensic casework. *N Z Med J*, 2002, **115**(1147): 48–49.

Masibay, A.S. and N.T. Lappas, The detection of protein p30 in seminal stains by means of thin-layer immunoassay. *J Forensic Sci*, 1984, **29**(4): 1173–1177.

Mauck, C.K., G.F. Doncel, and G. Biomarkers of semen exposure clinical working, biomarkers of semen in the vagina: Applications in clinical trials of contraception and prevention of sexually transmitted pathogens including HIV. *Contraception*, 2007, **75**(6): 407–419.

McAlister, C., The use of fluorescence *in situ* hybridisation and laser microdissection to identify and isolate male cells in an azoospermic sexual assault case. *Forensic Sci Int Genet*, 2011, **5**(1): 69–73.

McGee, R.S. and J.C. Herr, Human seminal vesicle–specific antigen is a substrate for prostate-specific antigen (or P-30). *Biol Reprod*, 1988, **39**(2): 499–510.

McWilliams, S. and B. Gartside, Identification of prostate-specific antigen and spermatozoa from a mixture of semen and simulated gastric juice. *J Forensic Sci*, 2009, **54**(3): 610–611.

Miller, K.W., et al., Developmental validation of the SPERM HY-LITER kit for the identification of human spermatozoa in forensic samples. *J Forensic Sci*, 2011, **56**(4): 853–865.

Mitchell, A., Validation study of KPICS SpermFinder™ by NicheVision Forensics, LLC for the identification of human spermatozoa. *J Forensic Sci*, 2012, **57**(4): 1042–1050.

Montagna, C.P., The recovery of seminal components and DNA from the vagina of a homicide victim 34 days postmortem. *J Forensic Sci*, 1996, **41**(4): 700–702.

Murray, C., C. McAlister, and K. Elliott, Identification and isolation of male cells using fluorescence *in situ* hybridisation and laser microdissection, for use in the investigation of sexual assault. *Forensic Sci Int Genet*, 2007, **1**(3–4): 247–252.

Nettikadan, S., et al., Detection and quantification of protein biomarkers from fewer than 10 cells. *Mol Cell Proteomics*, 2006, **5**(5): 895–901.

Nishi, K., et al., Utilization of lectin-histochemistry in forensic neuropathology: Lectin staining provides useful information for postmortem diagnosis in forensic neuropathology. *Leg Med* (Tokyo), 2003, **5**(3): 117–131.

Norris, J.V., et al., Expedited, chemically enhanced sperm cell recovery from cotton swabs for rape kit analysis. *J Forensic Sci*, 2007, **52**(4): 800–805.

Nussbaumer, C., E. Gharehbaghi-Schnell, and I. Korschineck, Messenger RNA profiling: A novel method for body fluid identification by real-time PCR. *Forensic Sci Int*, 2006, **157**(2–3): 181–186.

Old, J., et al., Developmental validation of RSID-Semen: A lateral flow immunochromatographic strip test for the forensic detection of human semen. *J Forensic Sci*, 2012, **57**(2): 489–499.

Pang, B.C. and B.K. Cheung, Identification of human semenogelin in membrane strip test as an alternative method for the detection of semen. *Forensic Sci Int*, 2007, **169**(1): 27–31.

Peonim, V., et al., Comparable between rapid one step immunochromatographic assay and ELISA in the detection of prostate specific antigen in vaginal specimens of raped women. *J Med Assoc Thai*, 2007, **90**(12): 2624–2629.

Poyntz, F.M. and P.D. Martin, Comparison of p30 and acid phosphatase levels in post-coital vaginal swabs from donor and casework studies. *Forensic Sci Int*, 1984, **24**(1): 17–25.

Randall, B., Glycogenated squamous epithelial cells as a marker of foreign body penetration in sexual assault. *J Forensic Sci*, 1988, **33**(2): 511–514.

Rao, A.R., H.G. Motiwala, and O.M. Karim, The discovery of prostate-specific antigen. *BJU Int*, 2008, **101**(1): 5–10.

Redhead, P. and M.K. Brown, The acid phosphatase test two minute cut-off: An insufficient time to detect some semen stains. *Sci Justice*, 2013, **53**(2): 187–191.

Rodrigues, R.G., et al., Semenogelins are ectopically expressed in small cell lung carcinoma. *Clin Cancer Res*, 2001, **7**(4): 854–860.

Romero-Montoya, L., et al., Relationship of spermatoscopy, prostatic acid phosphatase activity and prostate-specific antigen (p30) assays with further DNA typing in forensic samples from rape cases. *Forensic Sci Int*, 2011, **206**(1–3): 111–118.

Roy, A.V., M.E. Brower, and J.E. Hayden, Sodium thymolphthalein monophosphate: A new acid phosphatase substrate with greater specificity for the prostatic enzyme in serum. *Clin Chem*, 1971, **17**(11): 1093–1102.

Sakurada, K., et al., Evaluation of mRNA-based approach for identification of saliva and semen. *Leg Med* (Tokyo), 2009, **11**(3): 125–128.

Sato, I., et al., A dot-blot-immunoassay for semen identification using a polyclonal antibody against semenogelin, a powerful seminal marker. *Forensic Sci Int*, 2001, **122**(1): 27–34.

Sato, I., et al., Use of the "SMITEST" PSA card to identify the presence of prostate-specific antigen in semen and male urine. *Forensic Sci Int*, 2002, **127**(1–2): 71–74.

Sato, I., et al., Rapid detection of semenogelin by one-step immunochromatographic assay for semen identification. *J Immunol Methods*, 2004, **287**(1–2): 137–145.

Sato, I., et al., Applicability of Nanotrap Sg as a semen detection kit before male-specific DNA profiling in sexual assaults. *Int J Legal Med*, 2007, **121**(4): 315–319.

Sato, I., et al., Urinary prostate-specific antigen is a noninvasive indicator of sexual development in male children. *J Androl*, 2007, **28**(1): 150–154; discussion 155–157.

Schiff, A.F., Reliability of the acid phosphatase test for the identification of seminal fluid. *J Forensic Sci*, 1978, **23**(4): 833–844.

Sensabaugh, G.F., Isolation and characterization of a semen-specific protein from human seminal plasma: A potential new marker for semen identification. *J Forensic Sci*, 1978, **23**(1): 106–115.

Sensabaugh, G.F., The quantitative acid phosphatase test. A statistical analysis of endogenous and postcoital acid phosphatase levels in the vagina. *J Forensic Sci*, 1979, **24**(2): 19.

Shaler, R.C. and P. Ryan, High acid phosphatase levels as a possible false indicator of the presence of seminal fluid. *Am J Forensic Med Pathol*, 1982, **3**(2): 161–163.

Simich, J.P., et al., Validation of the use of a commercially available kit for the identification of prostate specific antigen (PSA) in semen stains. *J Forensic Sci*, 1999, **44**(6): 1229–1231.

Sokoll, L.J. and D.W. Chan, Prostate-specific antigen. Its discovery and biochemical characteristics. *Urol Clin North Am*, 1997, **24**(2): 253–259.

Standefer, J.C. and E.W. Street, Postmortem stability of prostatic acid phosphatase. *J Forensic Sci*, 1977, **22**(1): 165–172.

Stefanidou, M., G. Alevisopoulos, and C. Spiliopoulou, Fundamental issues in forensic semen detection. *West Indian Med J*, 2010, **59**(3): 280–283.

Steger, K., et al., Expression of protamine-1 and -2 mRNA during human spermiogenesis. *Mol Hum Reprod*, 2000, **6**(3): 219–225.

Stoilovic, M., Detection of semen and blood stains using polilight as a light source. *Forensic Sci Int*, 1991, **51**(2): 289–296.

Stowell, L.I., L.E. Sharman, and K. Hamel, An enzyme-linked immunosorbent assay (ELISA) for prostate-specific antigen. *Forensic Sci Int*, 1991, **50**(1): 125–138.

Stubbings, N.A. and P.J. Newall, An evaluation of gamma-glutamyl transpeptidase (GGT) and p30 determinations for the identification of semen on postcoital vaginal swabs. *J Forensic Sci*, 1985, **30**(3): 604–614.

Talthip, J., et al., An autopsy report case of rape victim by the application of PSA test kit as a new innovation for sexual assault investigation in Thailand. *J Med Assoc Thai*, 2007, **90**(2): 348–351.

Vandewoestyne, M. and D. Deforce, Laser capture microdissection in forensic research: A review. *Int J Legal Med*, 2010, **124**(6): 513–521.

Vandewoestyne, M., et al., Automatic detection of spermatozoa for laser capture microdissection. *Int J Legal Med*, 2009, **123**(2): 169–175.

Vandewoestyne, M., et al., Suspension fluorescence *in situ* hybridization (S-FISH) combined with automatic detection and laser microdissection for STR profiling of male cells in male/female mixtures. *Int J Legal Med*, 2009, **123**(5): 441–447.

Vennemann, M., et al., Sensitivity and specificity of presumptive tests for blood, saliva and semen. *Forensic Sci Med Pathol*, 2014, **10**(1): 69–75.

Virkler, K. and I.K. Lednev, Raman spectroscopic signature of semen and its potential application to forensic body fluid identification. *Forensic Sci Int*, 2009, **193**(1–3): 56–62.

Ward, A.M., J.W. Catto, and F.C. Hamdy, Prostate specific antigen: Biology, biochemistry and available commercial assays. *Ann Clin Biochem*, 2001, **38**(Pt 6): 633–651.

Wasserstrom, A., et al., Demonstration of DSI-semen—A novel DNA methylation-based forensic semen identification assay. *Forensic Sci Int Genet*, 2013, **7**(1): 136–142.

Willott, G.M. and J.E. Allard, Spermatozoa—Their persistence after sexual intercourse. *Forensic Sci Int*, 1982, **19**(2): 135–154.

Willott, G.M. and M.A. Crosse, The detection of spermatozoa in the mouth. *J Forensic Sci Soc*, 1986, **26**(2): 125–128.

Yokota, M., et al., Evaluation of prostate-specific antigen (PSA) membrane test for forensic examination of semen. *Leg Med* (Tokyo), 2001, **3**(3): 171–176.

Yoshida, K., et al., Quantification of seminal plasma motility inhibitor/semenogelin in human seminal plasma. *J Androl*, 2003, **24**(6): 878–884.

Yoshida, M., et al., Examination of seminal stain by HPLC assay of phenolphthalein. *Leg Med* (Tokyo), 2009, **11**(Suppl 1): S357–S359.

Yu, H. and E.P. Diamandis, Prostate-specific antigen in milk of lactating women. *Clin Chem*, 1995, **41**(1): 54–58.

Zarghami, N., et al., Prostate-specific antigen in serum during the menstrual cycle. *Clin Chem*, 1997, **43**(10): 1862–1867.

Identification of Saliva

15.1 Biological Characteristics of Saliva

The human salivary glands produce 1.0–1.5 L of saliva daily. About 70% of saliva is produced from the submandibular salivary glands, 25% from the parotids, and 5% from the sublingual salivary glands (Figure 15.1). Although a continuous basal level of saliva secretion is maintained, a large amount of saliva is produced during eating. Saliva is largely water containing small quantities of electrolytes, proteins, antibodies, and enzymes. The digestion of carbohydrates in the diet begins in the oral cavity, where amylase in the saliva breaks down carbohydrates such as starch. Thus, detecting amylase indicates the presence of saliva.

15.1.1 Amylases

Amylases are enzymes that cleave polysaccharides such as starches, which are composed of D-glucose units connected by α1→4 linkages. Starches contain two types of glucose polymers: *amylose* and *amylopectin* (Figure 15.2). Amylose consists of long, linear chains of glucose residues connected by α1→4 linkages. Amylopectin is highly branched and consists of linear chains of glucose residues connected by α1→4 linkages with the branch points connected by α1→6 linkages. Both linear amylose and amylopectin can be hydrolyzed by amylase by cleaving the chains at alternate α1→4 linkages. Amylase cleaves off one maltose (two glucose units) at a time. However, the α1→6 linkages at the branch points are not cleaved by the amylase.

Two types of amylases are characterized. β-*Amylases* found in plant and bacterial sources cleave only at the terminal reducing end of a polysaccharide chain. The end of a chain with a free anomeric carbon (not involved in a glycosidic bond) is called the reducing end. Human α-*amylases* cleave at α1→4 linkages randomly along the polysaccharide chain.

Human α-amylases have two major isoenzymes (multiple forms that differ in their amino acid sequences). Human *salivary* α-*amylase* (HSA) is encoded by the *Amy1* locus, synthesized at the salivary glands and secreted into the oral cavity. Human *pancreatic* α-amylase (HPA), encoded by the *Amy2* locus, is synthesized by the pancreas and secreted into the duodenum through the pancreatic duct. The amino acid sequences of the HSA and HPA are highly homologous. Therefore, monoclonal antibodies against HSA also cross-react with HPA. However, HSA is inactivated by acids in the stomach, while most HPA is inactivated in the lower portions of the intestine, and some amylase activity remains in the feces.

Amylase activity is found in various bodily fluids including semen, tears, milk, perspiration, and vaginal secretions. Most amylase present in normal serum consists of HPA and HSA. The amylases are small molecules and can pass through the glomeruli of the kidney (Chapter 17).

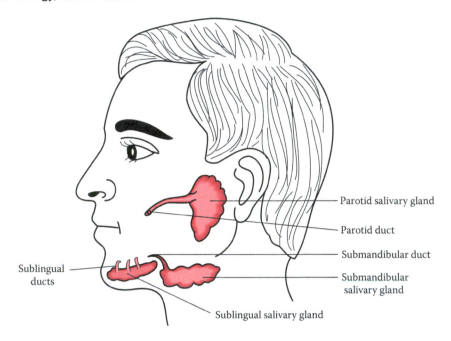

Figure 15.1   Human salivary glands. (© Richard C. Li.)

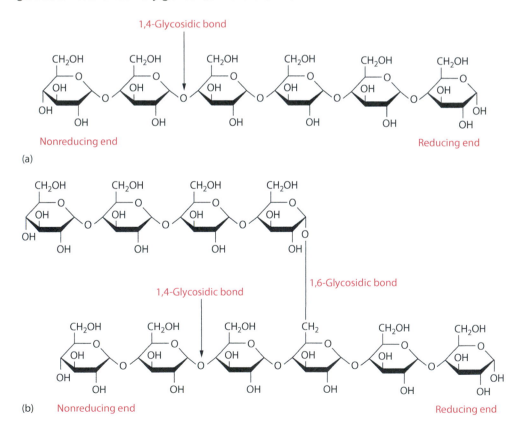

Figure 15.2 Chemical structures of polysaccharides found in starch: (a) amylose and (b) amylopectin.

Thus, the amylase present in urine is derived from plasma. Amylase can be inactivated under boiling temperatures and strong acidic and alkaline conditions. Based on various studies, its stability varies from a few weeks to several months.

15.2 Analytical Techniques for Identification of Saliva

The use of alternate light sources facilitates the search for and the visual examination of saliva stains. The identification of saliva is largely based on detecting the presence of amylase in a sample. Two types of amylase assays can be utilized. The first type measures the enzymatic activity of total amylase. This type of assay cannot distinguish HSA from other amylases including HPA and nonhuman amylases, such as those from plants, animals, and microorganisms. The second type of assay, which includes direct detection of HSA proteins and RNA-based assays, is more confirmatory than enzymatic assays.

15.2.1 Presumptive Assays
15.2.1.1 Visual Examination

The lighting techniques used to search for semen stains can be utilized in searching for saliva stains. For example, a 470 nm excitation wavelength can be used with orange goggles to allow visualization of fluorescence. However, the fluorescence of a saliva stain is usually less intense than that of a seminal stain (Figure 15.3). Microscopic examination with proper histological staining can also be performed to identify the buccal epithelial cells, indicating the presence of a saliva stain (Figure 15.4).

15.2.1.2 Determination of Amylase Activity
15.2.1.2.1 Starch–Iodine Assay

Iodine (I_2) is used to test for the presence of starch. The amylose in starch reacts strongly with iodine to form a dark blue complex, while amylopectin develops a reddish-purple color. In the presence of amylases, starch is broken down to mono- or disaccharides. Consequently, such colors do not develop when iodine is added.

Figure 15.3   Examining saliva stains using an ALS technique. (© Richard C. Li.)

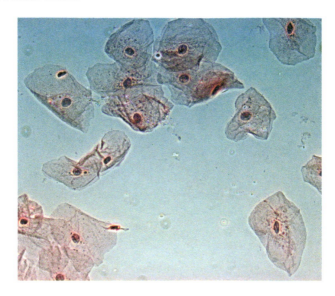

Figure 15.4   Buccal epithelial cells. (© Richard C. Li.)

A common configuration of the method is the radial diffusion assay (Figure 15.5). An agar gel containing starch is prepared. A sample well is created by punching a hole in the gel and an extract of the questioned sample is placed into the well. If amylase is present in the sample, it diffuses from the sample well and hydrolyzes the starch in the gel. The gel is then stained using an iodine solution. A clear area in the gel indicates amylase activity, and the size of the clear area is proportional to the amount of amylase in the sample. A linear standard curve

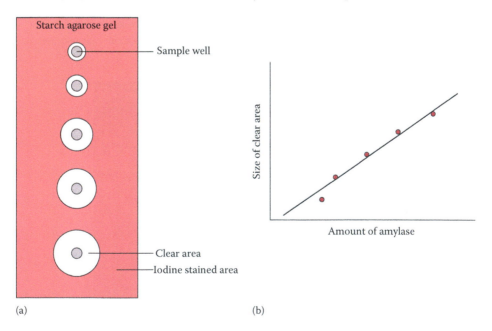

Figure 15.5   Radial diffusion assay for the identification of amylase. (a) Known amounts of amylase standards are applied to the well and allowed to diffuse. (b) The size of the clear area arising from amylase activity is plotted. The standard curve can be used to determine the amount of amylase in a questioned sample. (© Richard C. Li.)

(in log scale) can be prepared using a series of standard amylases with known concentrations. The amount of amylase can be quantified by comparing the results with the standard curve. However, this assay is not specific to HSA and can produce false-positive results. Moreover, the quantities of amylase can be used for determining the amount of sample needed for Y-STR analysis (Chapter 19), which now can be determined by using Y-chromosome-specific quantitative PCR assays (Chapter 6).

15.2.1.2.2 Colorimetric Assays

Dye-labeled amylase substrates such as dye-conjugated amylose or amylopectin are utilized. These substrates are not soluble in water. In the presence of amylase, the dye-containing moieties are cleaved and are soluble in water to produce a color. The degree of coloration, which can be measured colorimetrically by spectrophotometric methods, is proportional to the amount of amylase in the sample. Most of these assays are not HSA specific, although their specificity can be tested by using inhibitors that preferentially inhibit HSA, such as α-amylase inhibitors derived from the seeds of the wheat plant, *Triticum aestivum*. These amylase activity assays are considered presumptive, which means they are not conclusive for the presence of saliva in a sample.

While many substrates are available, Phadebas reagent (Pharmacia) is usually used in forensic laboratories. Produced in a tablet form, it is used to detect α-amylase in specimens for clinical diagnostic purposes. A small portion (approximately 3 mm²) of a sample is cut and placed in a tube and incubated for 5 min at 37°C. One Phadebas tablet is added to each tube and mixed. Samples are then incubated for 15 min at 37°C, and the reaction is stopped at an alkaline pH by the addition of sodium hydroxide. The amylase substrate is an insoluble blue dye conjugated to starch. Amylase hydrolyzes the substrate to generate a blue color that can be measured at 620 nm using a spectrophotometer. The optical density of the supernatant is read and can be converted to amylase units by comparing to a standard curve.

The amylase assay can also be used for amylase mapping as a method of searching for possible saliva stains (Figures 15.6 and 15.7). These assays are based on the principle that amylase is water soluble and can be transferred from evidence to filter paper and then analyzed via colorimetric assay. This procedure is also referred to as a *press test*. The sensitivity of the method is similar to that of the test tube method.

The substrate can be prepared by evenly spraying the Phadebas reagent on a sheet of filter paper and allowing it to air-dry. The dried substrate-containing paper can be used immediately or stored until needed. To perform amylase mapping, a piece of paper is placed over the entire area to be tested (the item to be tested must be fairly flat to ensure good contact with the paper). The paper is dampened slightly by spraying with distilled water. An outline may be drawn on the paper to aid in locating stains. A piece of plastic wrap is placed on top to prevent the paper from drying during the assay, and a weight is applied to ensure close contact of substrate and evidence. The test is observed every minute for the first 10 min, and every 5 min thereafter up to 40 min, when a positive reaction should appear as a light blue area.

The SALIgAE® kit (Abacus Diagnostics), another commercially available colorimetric assay, has been validated for saliva identification (Figure 15.8). Its manufacturer also produces the SALIgAE spray kit, which can be used for amylase mapping.

15.2.2 Confirmatory Assays
15.2.2.1 Identification of Human Salivary α-Amylase
15.2.2.1.1 Immunochromatographic Assays

Commercially produced immunochromatographic kits include the RSID®-Saliva kit (Independent Forensics). A labeled monoclonal anti-HSA antibody is contained in a sample well. A second monoclonal anti-HSA antibody is immobilized onto a test zone of a membrane, and an antiglobulin that recognizes the antibody is immobilized onto a control zone (Figures 15.9 and 15.10).

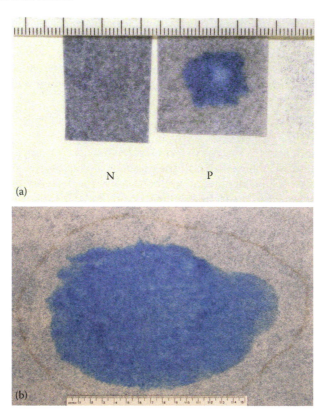

Figure 15.6   Amylase colorimetric assay using Phadebas reagent. (a) A spot test for saliva and (b) amylase mapping result showing a saliva-stained area. N, a negative result; P, a positive result. (© Richard C. Li.)

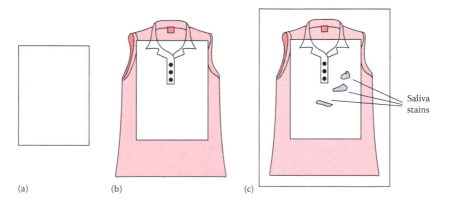

Figure 15.7   Amylase mapping for saliva stains. (a) Amylase substrate is sprayed on a sheet of filter paper. (b) Substrate-containing paper is placed over the area to be tested. The orientation of the filter paper is marked to aid in locating the stain. (c) The filter paper is dampened by spraying it with water, and plastic wrap is placed on top to prevent the paper from drying. Blue color indicates a positive reaction. (© Richard C. Li.)

Figure 15.8 Saliva identification using SALIgAE reagent. P, a positive result; N, a negative result; S, a positive control. (© Richard C. Li.)

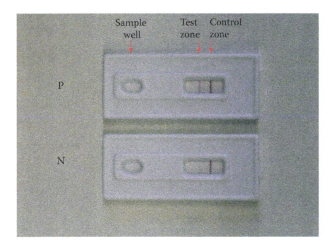

Figure 15.9 Identification of saliva using immunochromatographic assays. P, a positive result; N, a negative result. (© Richard C. Li.)

A sample can be prepared by cutting out a small portion of a stain or a swab. Each sample is extracted for 1–2 h in 200–300 μL of an extraction buffer. Approximately 10% of the extract is removed and mixed with a running buffer. The assay is carried out by loading an extracted sample into the sample well where the antigen in the sample binds to the labeled anti-HSA antibody in the well to form a labeled antibody–antigen complex. The complex then diffuses across the membrane to the test zone, where the solid-phase anti-HSA antibody binds with the labeled complex to form a labeled antibody–antigen–antibody sandwich.

The presence of antigen in the sample results in a pink line at the test zone. In the control zone, unbound labeled anti-HSA antibody binds to the solid-phase antiglobulin. The labeled antibody–antiglobulin complex at the control zone also results in a pink line. The test is

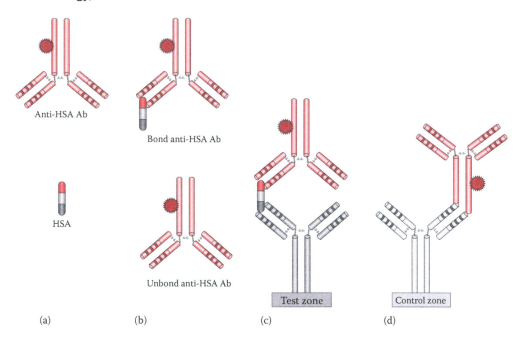

Figure 15.10   Immunochromatographic assay for the identification of HSA in saliva. (a) Sample containing amylases is loaded in a sample well. (b) Antigen binds to a labeled anti-HSA Ab to form a labeled Ab–HSA complex. (c) At the test zone, the labeled complex binds to an immobilized antihuman HSA Ab to form a labeled Ab–HSA–Ab sandwich. (d) At the control zone, the labeled anti-HSA Ab binds to an immobilized antiglobulin and is captured at the control zone. Ab, an antibody; HSA, human salivary α-amylase. (© Richard C. Li.)

considered valid only if the line in the control zone is observed. The presence of HSA results in a pink line at both the test zone and the control zone, while the absence of HSA results in a pink line in the control zone only. A result can be read after 10 min.

The sensitivity of the RSID®-Saliva kit can be as low as 1 μL of saliva. Additionally, the assay is responsive to samples extracted from saliva stains on both smooth and porous surfaces. In terms of species specificity, the kit has no cross-reactivity with various animal species, including monkeys (tamarin and callimico). As for bodily fluid specificity, it has also been shown that it is not responsive to human blood, semen, or urine. The high-dose hook effect, which creates an artifact that may cause false-negative results as described previously (Chapter 11), is not observed when up to 50 μL of saliva is tested. This method is rapid, specific, and sensitive and can be used in both laboratory and field analysis.

15.2.2.1.2 Enzyme-Linked Immunosorbent Assay (ELISA)

This method can be used to detect and to quantify a sample with the use of an anti-HSA antibody. The most common configuration in forensic serology is the antibody–antigen–antibody sandwich (Figure 15.11). ELISA utilizes reporting enzymes to produce colorimetric or fluorometric signals. The intensity of the signal can be detected spectrophotometrically and is proportional to the amount of bound antigen. The amount of HSA can be quantified by comparison with a standard of known concentration. This method is specific and highly sensitive in detecting HSA, but it is time-consuming. Chapter 11 discusses the ELISA principle in further detail.

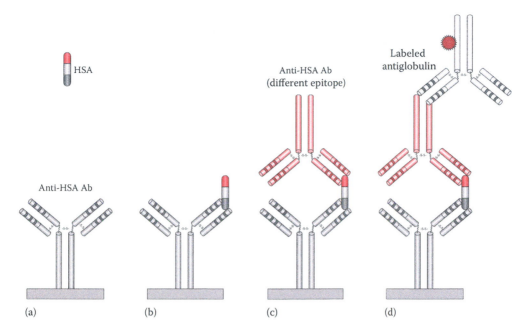

Figure 15.11 ELISA for identification of HSA in saliva. (a) Sample containing antigen is applied to polystyrene tubes where anti-HSA Ab is immobilized. (b) Antigen binds to immobilized Ab to form HSA–Ab complex. (c) Second anti-HSA Ab, specific for a different epitope of HSA, is added to form Ab–HSA–Ab sandwich. (d) Labeled antiglobulin then binds to the sandwich. The bound antiglobulin can be detected by various reporting schemes. Ab, an antibody; HSA, human salivary α-amylase. (© Richard C. Li.)

15.2.2.2 RNA-Based Assays

RNA-based assays (Chapter 11) have been developed recently for the identification of saliva. They are based on the expression of certain genes in certain cell or tissue types. Thus, the techniques used in the identification of saliva are based on the detection of specific types of mRNA expressed exclusively in certain cells in the oral cavity. These assays utilize reverse transcriptase polymerase chain reaction (RT-PCR; see Chapter 7) methods to detect gene expression levels of mRNAs for saliva identification. Table 15.1 summarizes the tissue-specific genes utilized for saliva identification. Compared to conventional assays used for saliva identification, RNA-based assays present higher specificity and are amenable to automation. However, one limitation is that the RNA is unstable because of degradation by endogenous ribonucleases.

Table 15.1	Application of RT-PCR Assay for Saliva Identification		
Gene Symbol	**Gene Product**	**Description**	**Further Reading**
HTN3	Histatin 3	Histidine-rich protein involved in nonimmune host defense in oral cavity	Sabatini et al. (1993)
STATH	Statherin	Inhibitor of precipitation of calcium phosphate salts in oral cavity	Sabatini et al. (1990)

Source: Adapted from Juusola, J. and Ballantyne, J., *Forensic Sci Int*, 152, 1–12, 2005.

Bibliography

Akutsu, T., et al., Applicability of ELISA detection of statherin for forensic identification of saliva. *Int J Legal Med*, 2010, **124**(5): 493–498.

Ali, M.M., et al., PCR applications in identification of saliva samples exposed to different conditions (streptococci detection based). *Pak J Biol Sci*, 2013, **16**(12): 575–579.

Aps, J.K. and L.C. Martens, Review: The physiology of saliva and transfer of drugs into saliva. *Forensic Sci Int*, 2005, **150**(2–3): 119–131.

Auvdel, M.J., Amylase levels in semen and saliva stains. *J Forensic Sci*, 1986, **31**(2): 426–431.

Auvdel, M.J., Comparison of laser and high-intensity quartz arc tubes in the detection of body secretions. *J Forensic Sci*, 1988, **33**(4): 929–945.

Barni, F., et al., Alpha-amylase kinetic test in bodily single and mixed stains. *J Forensic Sci*, 2006, **51**(6): 1389–1396.

Casey, D.G. and J. Price, The sensitivity and specificity of the RSID-saliva kit for the detection of human salivary amylase in the Forensic Science Laboratory, Dublin, Ireland. *Forensic Sci Int*, 2010, **194**(1–3): 67–71.

Courts, C. and B. Madea, Specific micro-RNA signatures for the detection of saliva and blood in forensic body-fluid identification. *J Forensic Sci*, 2011, **56**(6): 1464–1470.

Culliford, B.J., Precipitin reactions in forensic problems. A new method for precipitin reactions on forensic blood, semen and saliva stains. *Nature*, 1964, **201**: 1092–1093.

DeForest, P., R. Gaensslen, and H.C. Lee, *Forensic Science: An Introduction to Criminalistics*, 1983. New York: McGraw-Hill Book Company.

De Leo, D., et al., A sensitive and simple assay of saliva on stamps. *Z Rechtsmed*, 1985, **95**(1): 27–33.

Eckersall, P.D., et al., The production and evaluation of an antiserum for the detection of human saliva. *J Forensic Sci Soc*, 1981, **21**(4): 293–300.

Gaensslen, R.E., *Sourcebook in Forensic Serology, Immunology, and Biochemistry*, 1983. Washington, DC: US Government Printing Office.

Greenfield, A. and M. Sloan, Identification of biological fluids and stains, in S. James and J. Nordby (eds), *Forensic Science: An Introduction to Scientific and Investigative Techniques*, 2nd edn., 2005. CRC Press.

Haas, C., et al., RNA/DNA co-analysis from human saliva and semen stains—Results of a third collaborative EDNAP exercise. *Forensic Sci Int Genet*, 2013, **7**(2): 230–239.

Hedman, J., et al., Evaluation of amylase testing as a tool for saliva screening of crime scene trace swabs. *Forensic Sci Int Genet*, 2011, **5**(3): 194–198.

Hochmeister, M.N., et al., High levels of alpha-amylase in seminal fluid may represent a simple artifact in the collection process. *J Forensic Sci*, 1997, **42**(3): 535–536.

Hsu, L., et al., Amplification of oral streptococcal DNA from human incisors and bite marks. *Curr Microbiol*, 2012, **65**(2): 207–211.

Jones, E., The identification of semen and other body fluids, in R. Saferstein (ed.), *Forensic Science Handbook*, 2005. New Jersey: Pearson Prentice Hall.

Jones, E.L. Jr. and J.A. Leon, Lugol's test reexamined again: Buccal cells. *J Forensic Sci*, 2004, **49**(1): 64–67.

Juusola, J. and J. Ballantyne, Messenger RNA profiling: A prototype method to supplant conventional methods for body fluid identification. *Forensic Sci Int*, 2003, **135**(2): 85–96.

Juusola, J. and J. Ballantyne, Multiplex mRNA profiling for the identification of body fluids. *Forensic Sci Int*, 2005, **152**(1): 1–12.

Keating, S.M. and D.F. Higgs, The detection of amylase on swabs from sexual assault cases. *J Forensic Sci Soc*, 1994, **34**(2): 89–93.

Kennedy, D.M., et al., Microbial analysis of bite marks by sequence comparison of streptococcal DNA. *PLoS One*, 2012, **7**(12): e51757.

Lee, H.C., Identification and grouping of bloodstains, in R. Saferstein (ed.), *Forensic Science Handbook*, 1982. Englewood Cliffs, NJ: Prentice Hall.

Martin, N.C., N.J. Clayson, and D.G. Scrimger, The sensitivity and specificity of red-starch paper for the detection of saliva. *Sci Justice*, 2006, **46**(2): 97–105.

Miller, D.W. and J.C. Hodges, Validation of Abacus SALIgAE® test for the forensic identification of saliva, 2005. Available from: http://www.dnalabsinternational.com/SalivaValidation.pdf. Retrieved April 21, 2007.

Myers, J.R. and W.K. Adkins, Comparison of modern techniques for saliva screening. *J Forensic Sci*, 2008, **53**(4): 862–867.

Nakanishi, H., et al., A novel method for the identification of saliva by detecting oral streptococci using PCR. *Forensic Sci Int*, 2009, **183**(1–3): 20–23.

Nakanishi, H., et al., A simple identification method of saliva by detecting Streptococcus salivarius using loop-mediated isothermal amplification. *J Forensic Sci*, 2011, **56**(Suppl 1): S158–S161.

Nussbaumer, C., E. Gharehbaghi-Schnell, and I. Korschineck, Messenger RNA profiling: A novel method for body fluid identification by real-time PCR. *Forensic Sci Int*, 2006, **157**(2–3): 181–186.

Old, J.B., et al., Developmental validation of RSID-saliva: A lateral flow immunochromatographic strip test for the forensic detection of saliva. *J Forensic Sci*, 2009, **54**(4): 866–873.

Omelia, E.J., M.L. Uchimoto, and G. Williams, Quantitative PCR analysis of blood- and saliva-specific microRNA markers following solid-phase DNA extraction. *Anal Biochem*, 2013, **435**(2): 120–122.

Pandeshwar, P. and R. Das, Role of oral fluids in DNA investigations. *J Forensic Leg Med*, 2014, **22C**: 45–50.

Pang, B.C. and B.K. Cheung, Applicability of two commercially available kits for forensic identification of saliva stains. *J Forensic Sci*, 2008, **53**(5): 1117–1122.

Power, D.A., et al., PCR-based detection of salivary bacteria as a marker of expirated blood. *Sci Justice*, 2010, **50**(2): 59–63.

Pretty, I.A., The barriers to achieving an evidence base for bitemark analysis. *Forensic Sci Int*, 2006, **159**(Suppl 1): S110–S120.

Quarino, L., et al., Differentiation of α-amylase from various sources: An approach using selective inhibitors. *J Forensic Sci Soc*, 1993, **33**(2): 87–94.

Quarino, L., et al., An ELISA method for the identification of salivary amylase. *J Forensic Sci*, 2005, **50**(4): 873–876.

Ricci, U., I. Carboni, and F. Torricelli, False-positive results with amylase testing of citrus fruits. *J Forensic Sci*, 2014, **59**(5):1410–1412.

Rushton, C., et al., The distribution and significance of amylase-containing stains on clothing. *J Forensic Sci Soc*, 1979, **19**(1): 53–58.

Sabatini, L.M., Y.Z. He, and E.A. Azen, Structure and sequence determination of the gene encoding human salivary statherin. *Gene*, 1990, **89**(2): 245–251.

Sabatini, L.M., T. Ota, and E.A. Azen, Nucleotide sequence analysis of the human salivary protein genes HIS1 and HIS2, and evolution of the STATH/HIS gene family. *Mol Biol Evol*, 1993, **10**(3): 497–511.

Sakurada, K., et al., Evaluation of mRNA-based approach for identification of saliva and semen. *Leg Med (Tokyo)*, 2009, **11**(3): 125–128.

Searcy, R.L., et al., The interaction of human serum protein fractions with the starch–iodine complex. *Clin Chim Acta*, 1965, **12**(6): 631–638.

Shaler, R.C., Modern forensic biology, in R. Saferstein (ed.), *Forensic Science Handbook*, 2nd edn., 2002. Upper Saddle River, NJ: Pearson Education.

Soukos, N.S., et al., A rapid method to detect dried saliva stains swabbed from human skin using fluorescence spectroscopy. *Forensic Sci Int*, 2000, **114**(3): 133–138.

Sweet, D., et al., An improved method to recover saliva from human skin: The double swab technique. *J Forensic Sci*, 1997, **42**(2): 320–322.

Uchimoto, M.L., et al., Considering the effect of stem-loop reverse transcription and real-time PCR analysis of blood and saliva specific microRNA markers upon mixed body fluid stains. *Forensic Sci Int Genet*, 2013, **7**(4): 418–421.

Vandenberg, N. and R.A. van Oorschot, The use of Polilight in the detection of seminal fluid, saliva, and bloodstains and comparison with conventional chemical-based screening tests. *J Forensic Sci*, 2006, **51**(2): 361–370.

Vennemann, M., et al., Sensitivity and specificity of presumptive tests for blood, saliva and semen. *Forensic Sci Med Pathol*, 2014, **10**(1): 69–75.

Wawryk, J. and M. Odell, Fluorescent identification of biological and other stains on skin by the use of alternative light sources. *J Clin Forensic Med*, 2005, **12**(6): 296–301.

Whitehead, P.H. and A.E. Kipps, A test paper for detecting saliva stains. *J Forensic Sci Soc*, 1975, **15**(1): 39–42.

Whitehead, P.H. and A.E. Kipps, The significance of amylase in forensic investigations of body fluids. *Forensic Sci*, 1975, **6**(3): 137–144.

Willott, G.M., An improved test for the detection of salivary amylase in stains. *J Forensic Sci Soc*, 1974, **14**(4): 341–344.

Willott, G.M. and M. Griffiths, A new method for locating saliva stains: Spotty paper for spotting spit. *Forensic Sci Int*, 1980, **15**(1): 79–83.

Zimmermann, B.G., N.J. Park, and D.T. Wong, Genomic targets in saliva. *Ann N Y Acad Sci*, 2007, **1098**: 184–191.

Zubakov, D., et al., Stable RNA markers for identification of blood and saliva stains revealed from whole genome expression analysis of time-wise degraded samples. *Int J Legal Med*, 2008, **122**(2): 135–142.

Zubakov, D., et al., New markers for old stains: Stable mRNA markers for blood and saliva identification from up to 16-year-old stains. *Int J Legal Med*, 2009, **123**(1): 71–74.

Identification of Vaginal Secretions and Menstrual Blood

The identification of vaginal secretions and menstrual blood is important for the investigation of sexual assault cases. Such identification can help to corroborate allegations of sexual assault. For example, in a sexual assault investigation, a stain was observed after examining the suspect's clothing. Subsequently, forensic DNA analysis revealed that the DNA of the stain originated from the victim, establishing a link between the victim and the suspect. However, the defendant may assert that the victim's DNA originated from a sweat stain as a result of casual contact and deny any criminal act. If vaginal secretions were found in the stain, the evidence has probative value to corroborate an allegation of a sexual act. Additionally, postcoital drainage stains on clothing or bedding are often recovered at crime scenes. These stains usually consist of a mixture of semen and vaginal secretions. In such cases, the presence of vaginal secretions in these stains indicates the occurrence of sexual intercourse. Sometimes, vaginal secretions can be transferred onto a perpetrator during a sexual assault. For example, the presence of vaginal secretions on a suspect's genital area can indicate the occurrence of sexual intercourse. Furthermore, the presence of vaginal secretions on an object can corroborate an allegation of vaginal rape with a foreign object.

If a sexual assault victim is in menses when an assault occurred, blood evidence, such as the victim's bloodstains located on the suspect's clothing, may be recovered at the scene. The defense may argue that the bloodstains resulted from an injury and deny that any sexual act occurred. In this case, the identification of menstrual blood would corroborate an alleged rape. Therefore, it is necessary to distinguish between peripheral and menstrual blood in investigations of sexual assault where blood evidence is found at the scene.

16.1 Identification of Vaginal Stratified Squamous Epithelial Cells

A normal human vagina is covered by the *squamous mucosa*, which is composed of stratified squamous epithelial tissue (Figure 16.1a). Lying under the squamous mucosa is the *submucosa*, which contains an abundance of connective tissue and capillaries. Below the submucosa is the *muscularis*, which is made up of smooth muscle.

The squamous mucosa consists of multiple layers of cells (Figure 16.1b). At the *basal layer* of the squamous mucosa, basal cells are anchored to the basement membrane that separates the squamous mucosa from the submucosa. The basal cells are small in size with relatively large nuclei and are highly proliferative. As the cells migrate up from the basal layer to the *parabasal layer*, the cells undergo differentiation. At the *intermediate layer*, the cells are flattened and their nuclei are compressed. As the epithelial cells reach the apical layer, the *superficial layer*, the cells are fully differentiated with small and dense nuclei.

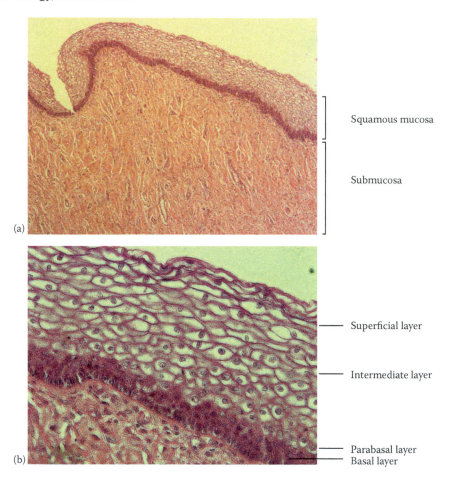

Figure 16.1   Histological structure of human vaginal tissues. (a) The squamous mucosa and the submucosa. (b) Multiple layers of cells in the squamous mucosa. (© Richard C. Li.)

Vaginal epithelial cells do not accumulate keratin, which is different from skin cells (Chapter 4). The intermediate and superficial layer cells contain abundant glycogen in their cytoplasm. The presence of glycogen in these cells is an indication of normal development and the differentiation of the vaginal epithelial cells. The apical surface of the vaginal squamous mucosa is usually covered by mucus that is secreted from glands that are located deep in the epithelium. The cells of the apical layers are eventually sloughed and are continuously replaced by the cells of deeper layers. When the lining of the vagina is swabbed during the collection of evidence, glycogenated cells with small numbers of parabasal cells are usually recovered (Figure 16.2). In addition to the vagina, glycogenated squamous epithelial cells are found in the linings of the oral cavity, pharynx, esophagus, anus, and the apex of the urethra. The differentiation of the vaginal epithelial cells requires estrogen. However, in premenarche and postmenopausal women, estrogen levels are very low; thus, the vaginal epithelial cells only differentiate to the parabasal cells. Primarily, parabasal cells are found in specimens from these individuals.

16.1.1 Lugol's Iodine Staining and Periodic Acid–Schiff Method

Lugol's iodine solution, named after the French physician Jean Lugol, is originally used as an antiseptic that is applied to skin or tissue to prevent infection. In forensic applications, it is utilized for the identification of glycogenated vaginal epithelial cells. The technique is based on the

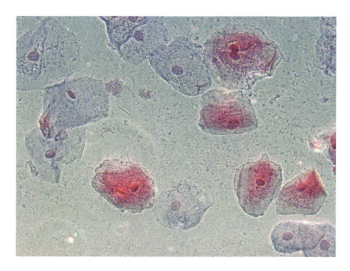

Figure 16.2   Vaginal stratified squamous epithelial cells. (© Richard C. Li.)

principle that iodine reacts with intracellular glycogen to exhibit a color. Glycogen is the principal storage form of glucose in animal and human cells and is found in the granules in the cytoplasm of the cells of many tissues. In addition to squamous epithelial cells, glycogen is also found in hepatocytes, which have the highest glycogen content, as well as muscle cells. Glycogen is a polysaccharide composed of D-glucose units. Similar to the amylopectin in plant cells (Chapter 15), glycogen is a branched polysaccharide consisting of linear and branched chains (Figure 16.3). The D-glucose residues of the linear chain are connected by α1→4 glycosidic bonds, while the branching points are connected by α1→6 glycosidic bonds. However, glycogen is more branched than

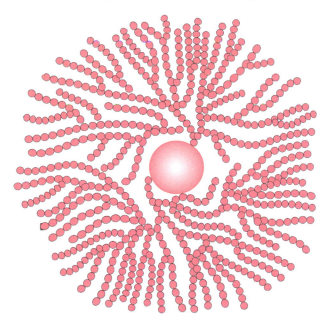

Figure 16.3   Structure of glycogen. Glycogen is a highly branched polysaccharide composed of glucose. Glycogen has a similar structure to amylopectin in starch (see Chapter 15). However, glycogen is more branched than starch. Additionally, glycogen contains a protein known as glycogenin at the center of its structure. (© Richard C. Li.)

amylopectin. Branch points occur approximately every 10 glucose residues in glycogen while they occur every 25 glucose residues in amylopectin. Lugol's stock solution is an aqueous solution of 5% iodine (I_2) and 10% potassium iodide (KI). Potassium iodide allows the iodine to be soluble in water through the formation of the triiodide ion. For staining the vaginal epithelial cells, 5% of the stock solution is usually used as the working solution. Iodine atoms fit into the helices of glycogen to form a dark brown glycogen–iodine complex (Figures 16.4 and 16.5). Vaginal epithelial cells can also be stained using the periodic acid–Schiff method. The cytoplasm of the vaginal epithelial cells is stained magenta, and the nucleus is stained purple.

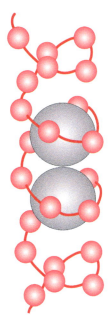

Figure 16.4   Diagram of a glycogen–iodine complex. In the complex, the glycogen chain forms a helix structure with six monosaccharide residues (red) per turn. Iodine molecules (gray) fit in the helix to form the glycogen–iodine complex, exhibiting a color. (© Richard C. Li.)

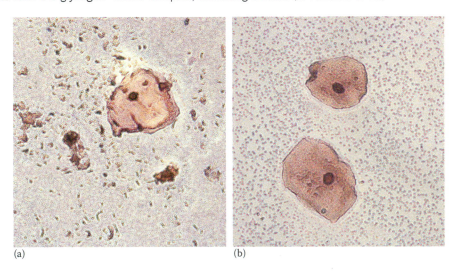

(a)   (b)

Figure 16.5   Epithelial cells stained with Lugol's iodine solution. (a) Vaginal and (b) buccal epithelial cells. (© Richard C. Li.)

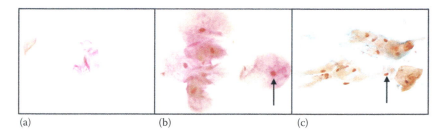

Figure 16.6   Staining of epithelial cells with Dane's method. The results obtained using methanol fixation followed by Dane's staining on skin (a), buccal (b), and vaginal epithelial (c) cells are shown. The arrows indicate red nuclei in buccal and orange nuclei in vaginal cells. (From French, C.E., et al., *Forensic Sci Int*, 178, 1–6, 2008. With permission.)

16.1.2 Dane's Staining Method

Evidence containing skin, buccal, and vaginal epithelial cells is often recovered in forensic investigations, particularly in sexual assault cases. Thus, it is necessary to distinguish between these cells. Skin, buccal, and vaginal epithelial cells belong to stratified squamous epithelium. Differentiated skin epithelial cells are keratinized and are classified as keratinizing squamous epithelial cells, while buccal and vaginal epithelial cells are nonkeratinizing squamous epithelial cells. Additionally, the skin epithelial cells lose nuclei and other cellular organelles during differentiation. In contrast, buccal and vaginal cells contain nuclei. Based on their morphology, it is possible to distinguish skin epithelial cells from buccal and vaginal epithelial cells. However, buccal and vaginal cells are morphologically indistinguishable from each other. Although Lugol's iodine solution and periodic acid–Schiff can stain vaginal epithelial cells, these stains cannot distinguish vaginal from other glycogenated epithelial cells in the oral mucosa and the urinary tract. Recently, Dane's staining method has been developed to distinguish all three types of cells. Skin cells are stained red and orange; buccal cells are stained predominantly orange-pink with red nuclei; and vaginal cells are stained bright orange with orange nuclei (Figure 16.6).

16.2 Identification of Vaginal Acid Phosphatase

Acid phosphatases are a group of enzymes that are capable of hydrolyzing a variety of small organic phosphomonoesters under acidic conditions. To date, at least five different acid phosphatase isoenzymes have been identified in human tissues: erythroid acid phosphatase (encoded by the *ACP1* gene), lysosomal acid phosphatase (encoded by the *ACP2* gene), prostate acid phosphatase (encoded by the *ACPP* gene, also known as *ACP3*), macrophage acid phosphatase (encoded by the *ACP5* gene), and testicular acid phosphatase (encoded by the *ACPT* gene). Human prostatic acid phosphatase is found in large quantities in seminal fluid and is used as a biomarker for semen identification (Chapter 14). The prostatic acid phosphatase is a homodimer containing two identical subunits with a molecular weight of 50 kDa. Small amounts of acid phosphatase can be detected in vaginal fluid, which is produced in normal cervical epithelial cells. However, the molecular characteristics of vaginal acid phosphatase are still not known. Historically, vaginal acid phosphatase has been used as a biomarker for the identification of vaginal secretions using acid phosphatase catalytic assays (Chapter 14). In sexual assault investigations, it is important to distinguish vaginal acid phosphatase from prostate acid phosphatase originating from semen exposure. These two enzymes have identical molecular weights, enzymatic specificities, and responses to the same inhibitors. Nevertheless, vaginal and prostate acid phosphatases can be distinguished using agarose electrophoresis. Based on their electrophoretic mobility, bands of vaginal and prostate acid phosphatases can be separated. The prostate acid phosphatase

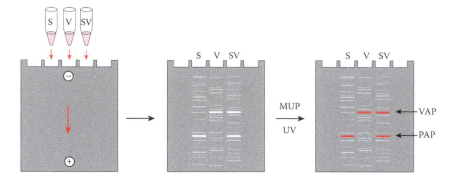

Figure 16.7 Distinguishing prostate and vaginal acid phosphatases using polyacrylamide gel electrophoresis. Bands of acid phosphatases are detected using 4-methylumbelliferyl phosphate (MUP). MUP is a substrate of acid phosphatases, resulting in a product that fluoresces in UV illumination (see Chapter 14). The prostate acid phosphatase has higher electrophoretic mobility toward an anode (a positively charged electrode) than vaginal acid phosphatase. S, semen; V, vaginal secretions; SV, mixture of semen and vaginal secretions; PAP, prostate acid phosphatase; VAP, vaginal acid phosphatase; MUP, 4-methylumbelliferyl phosphate; UV, ultraviolet illumination. (© Richard C. Li.)

has higher electrophoretic mobility toward an anode (a positively charged electrode) than the vaginal acid phosphatase (Figure 16.7). Thus, the presence of vaginal acid phosphatase can be determined.

16.3 Identification of Vaginal Bacteria

Lactobacillus can be found in the respiratory, the gastrointestinal, and the urogenital tract of healthy humans and animals. *Lactobacillus* taxa are the predominant bacteria in the vagina of women of reproductive age, and they play an important role in protecting the host against invasive pathogenic organisms. *Lactobacillus* consists of rod-shaped, nonmotile, and non-spore-forming gram-positive bacteria (Figure 16.8). Since these bacteria survive on carbohydrates, *Lactobacillus* bacteria produce lactic acid. As a result, a low pH environment is established in

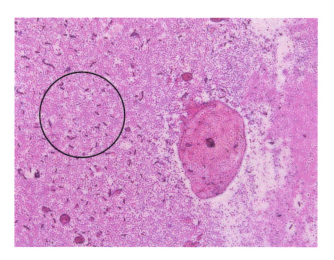

Figure 16.8 Vaginal *Lactobacillus* bacteria (circled). (© Richard C. Li.)

the lumen of the vagina, which restricts the growth of pathogenic organisms. Such vaginal lactic acid–producing bacteria were thought to be *Lactobacillus acidophilus*. In the 1980s, it was determined that *L. acidophilus* was not a single species, but actually a group of related species known as the *L. acidophilus* complex. The species of the complex can now be distinguished based on their DNA sequences. The most frequently occurring *Lactobacillus* species found in the vagina are *L. iners*, *L. crispatus*, *L. gasseri*, and *L. jensenii*. Among them, *L. iners* is the most common species of *Lactobacillus* in women. *Lactobacillus* can be identified based on the sequences of DNA markers such as the 16S rRNA gene and the intergenic spacer region between the 16S rRNA and 23S rRNA genes (Chapter 11). In forensic applications, *L. iners*, *L. crispatus*, *L. gasseri*, and *L. jensenii* can be detected in vaginal secretions. Thus, the identification of *Lactobacillus* taxa can potentially be utilized for the forensic identification of vaginal fluid. However, the presence of *Lactobacillus* taxa is not specific enough to vaginal fluid. Some studies show that *Lactobacillus* taxa are present in semen while others show that *Lactobacillus* can be found in female urine. The openings of the female urogenital system are in close proximity; thus, it is possible to have partially overlapping microbiota between the urine and vaginal secretions. Thus, the identification of multiple bacterial species, at least the four *Lactobacillus* species previously mentioned, should be carried out to distinguish these samples.

16.4 Outlook for Confirmatory Assays of Vaginal Secretions

The identification of the vaginal stratified squamous epithelial cells provides important probative evidence in forensic investigations. However, the existing methods described earlier can sometimes give false-negative or false-positive results. In some situations, these assays also cross-react with other types of bodily fluids. Thus, none of these assays is confirmatory. A useful identification method for vaginal secretions should be able to distinguish vaginal secretions from other bodily fluids and should be easy to perform. For example, nondestructive confirmatory identification methods such as fluorescence spectroscopy and Raman spectroscopy (Chapter 11) can potentially be useful for the identification of vaginal secretions. Recently, the analysis of tissue-specific gene expression has been utilized for the identification of vaginal secretions. Using the reverse transcription polymerase chain reaction (RT-PCR) technique (Chapter 7), the mRNAs of the tissue-specific genes of vaginal epithelial cells can be detected. For example, two commonly used markers for vaginal secretion identification are *MUC4* and *HBD1*. *MUC4* encodes a mucin protein that is a major component of vaginal mucus, and *HBD1*, the human β defensin 1, encodes a vaginal antimicrobial peptide. Both *MUC4* and *HBD1* are expressed in vaginal epithelial cells and are considered reliable markers of vaginal fluid. Additional mRNA and miRNA markers are described in Tables 16.1 and 16.2.

16.5 Menstruation

Menstruation is the periodic discharge of blood and the elimination of the degenerated lining of the endometrium from the uterus of nonpregnant women. From menarche to menopause, women may menstruate up to 400 times during their reproductive age. The uterus plays an important role in preparing the uterine endometrium for the possible implantation of a developing embryo. The linings of the uterus are composed of the myometrium and the endometrium. The *myometrium* consists of the muscle fibers of the uterus. The *endometrium* consists of the simple columnar epithelium and the stroma (Figure 16.9). The simple columnar epithelium is formed by single-layered elongated cells located at the apical surface of the endometrium. The stroma consists of connective tissues as well as spiral arteries. Spiral arteries are small arteries that ascend through the endometrium and form a coil-like structure, which supplies blood to the endometrium.

Table 16.1 Representative Markers of mRNA-Based Assays for Vaginal Secretions and Menstrual Blood Identification

Biological Fluid	Gene Symbol	Description
Vaginal secretions	CYP2B7P1	Cytochrome P450, family 2, subfamily B, polypeptide 7, pseudogene 1
	DKK4	Dickkopf homolog 4
	FUT6	Fucosyltransferase 6
	HBD1	β Defensin 1
	IL19	Interleukin 19
	MUC4	Mucin 4
	MYOZ1	Myozenin 1
	SFTA2	Surfactant associated 2
Menstrual blood	MMP7	Matrix metalloproteinase 7
	MMP11	Matrix metalloproteinase 11

Source: Adapted from Bauer, M. and Patzelt, D., *J Forensic Sci*, 47, 1278–1282, 2002; Hanson, E.K. and Ballantyne, J., *Sci Justice*, 53, 14–22, 2013; Juusola, J. and Ballantyne, J., *Forensic Sci Int*, 152, 1–12, 2005; Nussbaumer, C., Gharehbaghi-Schnell, E., and Korschineck, I., *Forensic Sci Int*, 157, 181–186, 2006.

Table 16.2 miRNA Markers for Vaginal Secretions and Menstrual Blood Identification

Bodily Fluid	miRNA Marker	Sequence
Vaginal secretions	miR124a	UAAGGCACGCGGUGAAUGCC
	miR372	AAAGUGCUGCGACAUUUGAGCGU
	miR617	AGACUUCCCAUUUGAAGGUGGC
	miR891a	UGCAACGAACCUGAGCCACUGA
Menstrual blood	miR214	UGCCUGUCUACACUUGCUGUGC
	miR412	ACUUCACCUGGUCCACUAGCCGU
	miR451	AAACCGUUACCAUUACUGAGUU

Source: Adapted from Hanson, E.K., Lubenow, H., and Ballantyne, J., *Anal Biochem*, 387, 303–314, 2009; Wang, Z., et al., *Forensic Sci Int Genet*, 7, 116–123, 2013.

16.5.1 Uterine Cycle

The endometrium can be divided into two zones: the functionalis and the basalis (Figure 16.10). The *functionalis* is the luminal part of the endometrium. It is the zone of cyclic changes in the endometrium and is shed during menstruation. The *basalis* is the basal part of the endometrium and is not shed during menstruation. This zone produces cells to regenerate the functionalis during the next cycle.

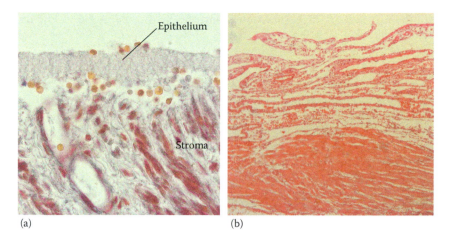

Figure 16.9 Uterus endometrium. (a) The proliferative phase and (b) the menstrual phase. (© Richard C. Li.)

During the uterine cycle, repetitive physiological changes occur in the functionalis (Figure 16.10). The cycle is divided into three phases: the menstrual, proliferative, and secretory phases. The first day of menstrual bleeding is considered the onset of the menstrual phase. During the menstrual phase, the functionalis degenerates and is sloughed off from the uterine wall and bleeding occurs, known as menses. During the proliferative phase, the functionalis begins regeneration in which the spiral arteries are proliferated. During the secretory phase, the spiral arteries are further developed and coiled. In the absence of pregnancy, a decrease in the progesterone level leads to the constriction of the spiral arteries. As a result, the functionalis becomes ischemic (insufficient blood flow), leading to hypoxia (low levels of oxygen in tissue). In addition, the activation of an enzymatic degradation process causes the destruction and

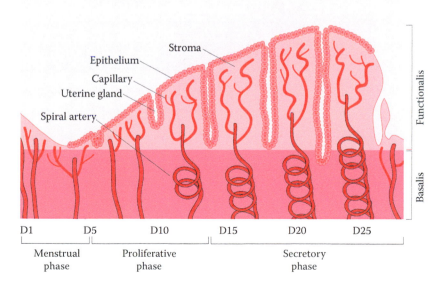

Figure 16.10 Changes in the functionalis of the uterine mucosa during a uterine cycle. D, day. (© Richard C. Li.)

shedding of the functionalis during menses. Menstrual fluid contains blood, the functionalis layer tissue, and mucus.

16.5.2 Uterine Endometrial Hemostasis

The cessation of menstrual bleeding is achieved by endometrial *hemostasis* that is initiated when injury occurs due to the shedding of the endometrium (Figure 16.11). Hemostasis begins with platelet activation and aggregation to form platelet plugs at the site of injury. Additionally, the blood coagulation cascade is activated to produce *thrombin*. Thrombin, a serine protease, converts soluble fibrinogen into fibrin. Fibrin, a protein involved in blood clotting, aggregates with the platelet plugs and leads to the cessation of bleeding by forming blood clots, known as *thrombi*. Under normal physiological conditions, uterine endometrial hemostasis is a balanced process between blood coagulation and clot dissolution to control blood loss and to prevent clot accumulation within the uterus. As a result, the balance of these two processes allows the removal of tissue fragments from the uterus cavity in order to reduce the risks of infection. Blood clots are prevented from accumulating during menstruation by forming low amounts of platelet plugs and synthesizing coagulation factor inhibitors that inhibit blood coagulation. Additionally, *fibrinolysis* is activated, during which thrombus is broken down by a protease known as *plasmin*. Plasmin cleaves fibrin, generating soluble degradation products. As a result, fibrinolysis can inhibit blood clot formation.

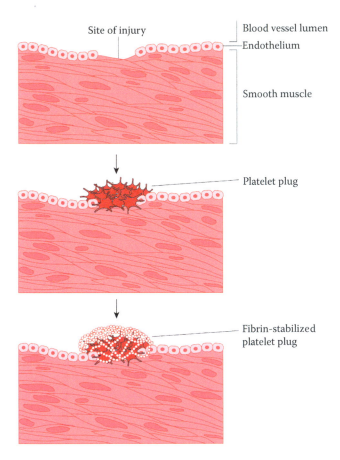

Figure 16.11 Diagram of hemostasis. Hemostasis occurs at the site of injury of a blood vessel. A platelet plug is formed as a result of the aggregation of platelets. The plug is further stabilized by the formation of a fibrin clot over the platelet plug. (© Richard C. Li.)

16.6 D-dimer Assay

During fibrinolysis, cross-linked fibrin is cleaved by plasmin, producing a degradation product known as *D-dimer* (Figure 16.12). Assays for D-dimer fragments have been utilized in the forensic identification of menstrual blood. A number of different formats of D-dimer assays can be used. In an enzyme-linked immunosorbent assay (ELISA; Chapter 11), antibodies bind to the D-dimer antigens on the solid phase. The D-dimer–antibody complex is subsequently analyzed using an antibody-based detection system. This method is highly sensitive, but is time-consuming. Latex agglutination assays are based on the interaction of antibodies and D-dimers that are located on carriers to form aggregates during the agglutination process (Chapter 11). However, the magnitude of the agglutination response is manually read and conclusions are based on subjective judgments. Immunochromatographic assays (Chapter 11) utilize monoclonal antibodies specific to D-dimers, which have been developed recently (Figures 16.13 and 16.14). This immunochromatographic assay is very specific, sensitive, and rapid, and can be completed within minutes. The immunochromatographic devices are portable, and thus can potentially be used at crime scenes. The D-dimer assays can positively identify menstrual blood samples. Although peripheral blood contains low levels of D-dimer, these assays do not show positive reactions with peripheral blood. Thus, menstrual blood can be distinguished from peripheral blood using D-dimer assays. However, postmortem blood also contains these D-dimers, which are detected by these assays. Although the detection of postmortem blood would complicate the interpretation of results, postmortem blood is not often encountered in sexual assault cases.

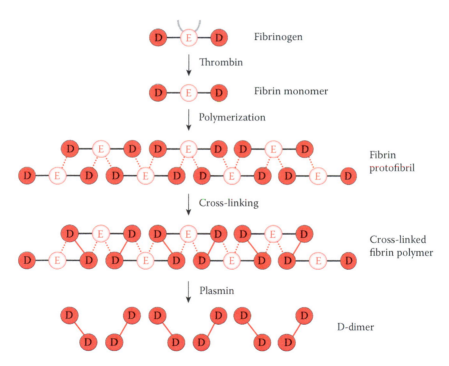

Figure 16.12 The formation and the degradation of a fibrin polymer. The activation of coagulation ultimately generates thrombin, which catalyzes the conversion of fibrinogen to fibrin by cleaving the fibrinopeptides (gray). A fibrin monomer contains an E domain and two D domains. The fibrin monomers are held together by noncovalent bonds (dotted red) between the D domains and E domain to form a fibrin polymer. Fibrin polymers are then covalently linked (solid red) to form a cross-linked fibrin polymer, which plays a role in forming clots. During fibrinolysis, plasmin cleaves the cross-linked fibrin at multiple sites giving rise to fibrin degradation products including D-dimer. (© Richard C. Li.)

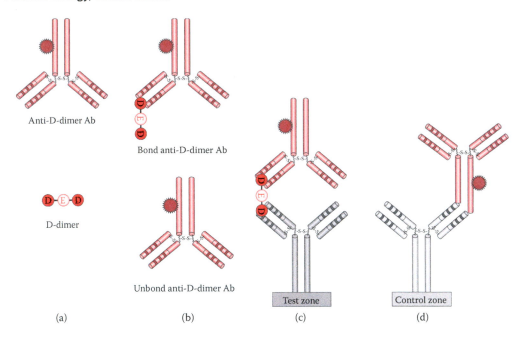

Anti-D-dimer Ab

Bond anti-D-dimer Ab

D-dimer

Unbond anti-D-dimer Ab

Test zone

Control zone

(a)   (b)   (c)   (d)

Figure 16.13 Immunochromatographic assays for the identification of D-dimer in menstrual blood. (a) In a sample well, D-dimer antigen in a menstrual blood sample is mixed with labeled anti-D-dimer Ab. (b) The D-dimer binds to the labeled anti-D-dimer Ab to form a labeled Ab–antigen complex. (c) At the test zone, the labeled complex binds to an immobilized anti-D-dimer Ab to form a labeled Ab–antigen–Ab sandwich. (d) At the control zone, the labeled anti-D-dimer Ab binds to an immobilized antiglobulin and is captured at the control zone. Ab, antibody. (© Richard C. Li.)

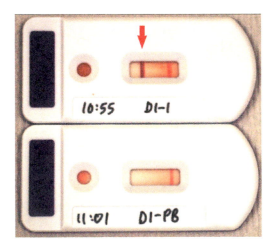

Figure 16.14 Detecting D-dimer using immunochromatographic assays. The results obtained using the Clearview Simplify D-dimer devices (Alere, Cheshire, UK) on menstrual (top) and peripheral blood (bottom) are shown. A positive result (arrow) indicates the presence of D-dimer. (From Baker, D.J., Grimes, E.A., and Hopwood, A.J., *Forensic Sci Int*, 212, 210–214, 2011. With permission.)

16.7 Lactate Dehydrogenase Assay

Lactate dehydrogenase (LDH) is an enzyme that plays an important role in glycolysis. In humans, LDH catalyzes the reversible reduction of pyruvate into lactate when the amount of oxygen is limited. LDH is a tetrameric enzyme consisting of three different types of subunits. The A subunit, also known as the M subunit, is encoded by the *LDHA* gene and is primarily expressed in skeletal muscle. The B subunit, also known as the H subunit, is encoded by the *LDHB* gene and is primarily expressed in cardiac muscle. The C subunit, encoded by the *LDHC* gene, is expressed restrictively in the testes. LDHs are found in various human tissues. Five isoenzymes can be found in blood (Figure 16.15). LDH1 consists of four identical B subunits; LDH2 consists of one A and three B subunits; LDH3 consists of two A and two B subunits; LDH4 consists of three A and one B subunits; and LDH5 consists of four identical A subunits. The five isoenzymes can be separated using electrophoresis (Figure 16.16) and detected using a colorimetric assay (Figures 16.17 and 16.18). According to their electrophoretic mobility, five bands can be identified. LDH1 has the highest electrophoretic mobility (toward an anode that is a positively charged electrode) and LDH5 has the lowest electrophoretic mobility. In peripheral blood, LDH1, 2, and 3 are the predominant forms of the isoenzymes and LDH4 and 5 are the minor forms of the isoenzymes. In contrast, LDH4 and 5 are consistently the predominant isoenzymes in menstrual blood, while the amounts of LDH1, LDH2, and LDH3 vary. Thus, menstrual blood can be distinguished from peripheral blood. LDH was historically used as a marker for the forensic identification of menstrual blood.

Figure 16.15   Isozymes of lactate dehydrogenase. Human lactate dehydrogenases are composed of four subunits. Five types of isozymes with their subunits are shown. A, LDH A subunit; B, LDH B subunit.

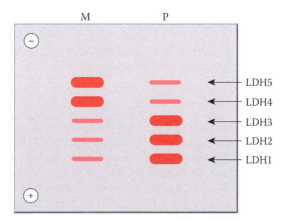

Figure 16.16   Electrophoretic separation of lactate dehydrogenase (LDH) isozymes. Extracted blood samples are loaded onto a cellulose acetate membrane. Electrophoresis is then carried out. Separated bands of LDH isozymes are detected using a colorimetric assay (see Figure 16.15). M, menstrual blood; P, peripheral blood. (© Richard C. Li.)

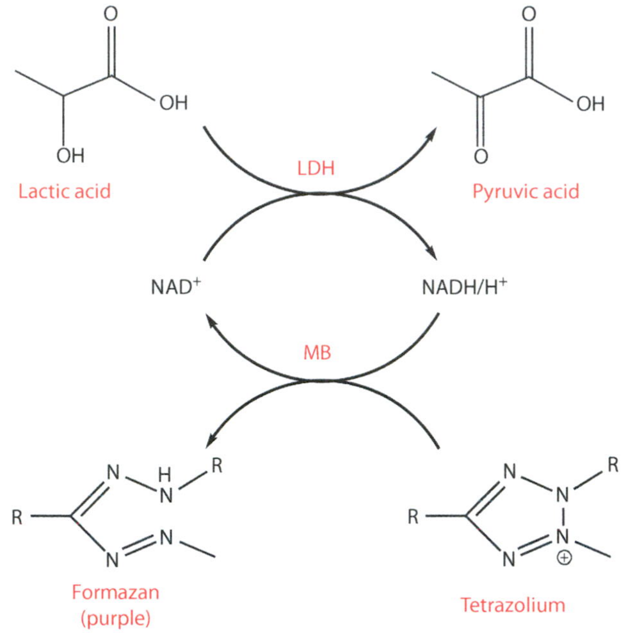

Figure 16.17 Detecting LDH activity using a colorimetric assay forming a colored dye. At the start of a cascade reaction, the LDH catalyzes the conversion of lactic acid to pyruvic acid. Nicotinamide adenine dinucleotide (NAD$^+$) is then reduced to NADH/H$^+$. Subsequently, Meldola's Blue (MB; 8-dimethylamino-2-benzophenoxazine), an electron carrier, transfers H/H$^+$ from NADH/H$^+$ to a tetrazolium salt. The pale yellow tetrazolium salt, nitroblue tetrazolium (NBT), is reduced to form a purple formazan dye.

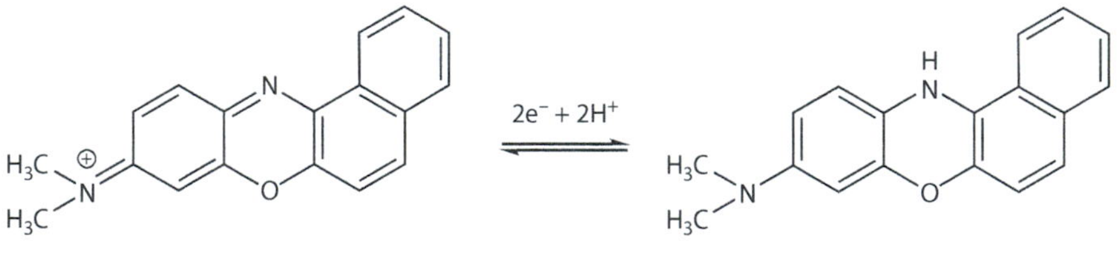

Figure 16.18 The reduction and oxidation of Meldola's Blue. Meldola's Blue is an electron carrier, which transfers hydrogen atoms (coupled to electron transfer).

16.8 RNA-Based Assays

Matrix metalloproteinase (MMP) genes are considered tissue-specific markers for human endometrium tissues. MMPs are zinc-dependent endopeptidases that degrade extracellular matrix components. Additionally, they cleave other proteins such as cytokines, chemokines, and growth factors. The *extracellular matrix* (ECM) is the extracellular space of tissue that is filled by macromolecules such as collagens, laminins, fibronectins, and proteoglycans. The ECM can be divided into two categories: the interstitial matrix and the basement membrane. The interstitial matrices are located in the intercellular spaces. The basement membranes are thin layers of macromolecule fibers that usually lie under the epithelium and the endothelium. Both the interstitial matrices and the basement membranes provide structural support to the cells. It is proposed that MMPs play an important role in the degradation of ECM, leading to the destruction of endometrium tissues during the uterine cycle. To date, a total of 23 MMPs, divided into five subgroups, have been found in humans. The most commonly used markers for the forensic identification of menstrual blood are *MMP7* and *MMP11* (Table 16.1).

Many MMPs are expressed in human endometrium throughout the uterine cycle. The patterns of the MMP gene expressions are correlated with their functions in endometrium tissue breakdown during menstruation. *MMP7* is predominantly expressed in epithelial cells, while *MMP11* is expressed in the stromal cells of the endometrium. Both *MMP7* and *MMP11* mRNA expressions are elevated at the menstrual phase and remain at high levels during the proliferative phase. It is also known that MMPs' mRNA may be elevated in postpartum, wound healing, and metastatic cancer conditions, which may potentially lead to a false-positive identification of menstrual blood.

In menstrual blood samples, among all of the MMP genes tested, *MMP11* is the most sensitive and specific marker for distinguishing menstrual blood from peripheral blood. Using the RT-PCR technique (Chapter 7), *MMP11* mRNA can be detected in menstrual blood from the first to the eighth day of menstruation but it is absent in peripheral blood and vaginal secretions. Therefore, *MMP11* can be used as a marker for the identification of menstrual blood. Likewise, *MMP7* is also a useful marker for menstrual blood identification. However, the *MMP7* mRNA level in menstrual blood is less than that of *MMP11*. Additional markers such as miRNA markers for menstrual blood identification are included in Table 16.2.

Bibliography

Ablett, P.J., The identification of the precise conditions for seminal acid phosphatase (SAP) and vaginal acid phosphatase (VAP) separation by isoelectric focusing patterns. *J Forensic Sci Soc*, 1983, **23**(3): 255–256.

Adams, E.G. and Wraxall, B.G., Phosphatases in body fluids: The differentiation of semen and vaginal secretion. *Forensic Sci*, 1974, **3**: 57–62.

Akutsu, T., et al., Evaluation of latex agglutination tests for fibrin-fibrinogen degradation products in the forensic identification of menstrual blood. *Leg Med (Tokyo)*, 2012, **14**(1): 51–54.

Akutsu, T., et al., Detection of bacterial 16S ribosomal RNA genes for forensic identification of vaginal fluid. *Leg Med (Tokyo)*, 2012, **14**(3): 160–162.

Albrecht, K., et al., Immunohistochemical distribution of cyclic nucleotide phosphodiesterase (PDE) isoenzymes in the human vagina: A potential forensic value? *J Forensic Leg Med*, 2007, **14**(5): 270–274.

Alsawaf, K. and A.T. Tu, Isotachophoretic analysis of bloodstains: Differentiation of human, menstrual, bovine, and ovine bloods. *J Forensic Sci*, 1985, **30**(3): 922–930.

Baker, D.J., E.A. Grimes, and A.J. Hopwood, D-dimer assays for the identification of menstrual blood. *Forensic Sci Int*, 2011, **212**(1–3): 210–214.

Bauer, M. and D. Patzelt, Evaluation of mRNA markers for the identification of menstrual blood. *J Forensic Sci*, 2002, **47**(6): 1278–1282.

Bauer, M. and D. Patzelt, Identification of menstrual blood by real time RT-PCR: Technical improvements and the practical value of negative test results. *Forensic Sci Int*, 2008, **174**(1): 55–59.

Benschop, C.C., et al., Vaginal microbial flora analysis by next generation sequencing and microarrays; can microbes indicate vaginal origin in a forensic context? *Int J Legal Med*, 2012, **126**(2): 303–310.

Divall, G.B. and M. Ismail, Lactate dehydrogenase isozymes in vaginal swab extracts: A problem for the identification of menstrual blood. *Forensic Sci Int*, 1983, **21**(2): 139–147.

Ferri, G., et al., Successful identification of two years old menstrual bloodstain by using MMP-11 shorter amplicons. *J Forensic Sci*, 2004, **49**(6): 1387.

Fleming, R.I. and S. Harbison, The use of bacteria for the identification of vaginal secretions. *Forensic Sci Int Genet*, 2010, **4**(5): 311–315.

French, C.E., et al., A novel histological technique for distinguishing between epithelial cells in forensic casework. *Forensic Sci Int*, 2008, **178**(1): 1–6.

Fujita, Y., I. Tokunaga, and S.I. Kubo, Forensic identification of a vaginal fluid and saliva mixture through DNA analysis. *Acta Criminologiae et Medicinae Legalis Japonica*, 2003, **69**(2): 48–52.

Giampaoli, S., et al., Molecular identification of vaginal fluid by microbial signature. *Forensic Sci Int Genet*, 2012, **6**(5): 559–564.

Gray, D., N. Frascione, and B. Daniel, Development of an immunoassay for the differentiation of menstrual blood from peripheral blood. *Forensic Sci Int*, 2012, **220**(1–3): 12–18.

Hanson, E.K. and J. Ballantyne, Highly specific mRNA biomarkers for the identification of vaginal secretions in sexual assault investigations. *Sci Justice*, 2013, **53**(1): 14–22.

Hanson, E.K., Lubenow, H., and Ballantyne, J., Identification of forensically relevant body fluids using a panel of differentially expressed microRNAs. *Anal Biochem*, 2009, **387**: 303–314.

Hausmann, R. and B. Schellmann, Forensic value of the Lugol's staining method: Further studies on glycogenated epithelium in the male urinary tract. *Int J Legal Med*, 1994, **107**(3): 147–151.

Jakubowska, J., et al., mRNA profiling for vaginal fluid and menstrual blood identification. *Forensic Sci Int Genet*, 2013, **7**(2): 272–278.

Jones, E., The identification of semen and other body fluids, in R. Saferstein (ed.), *Forensic Science Handbook*, pp. 329–399, 2005. Upper Saddle River, NJ: Pearson Prentice Hall.

Juusola, J. and J. Ballantyne, Multiplex mRNA profiling for the identification of body fluids. *Forensic Sci Int*, 2005, **152**(1): 1–12.

Kaur, G. and V.K. Sharma, The determination of esterase D in menstrual blood stains. *Adli Tip Dergisi*, 1988, **4**(1–2): 3–9.

Metropolitan Police Forensic Science Laboratory, *Biology Methods Manual*, p. 5, 1978. London: Metropolitan Police Forensic Science Laboratory.

Miyaishi, S., et al., Identification of menstrual blood by the simultaneous determination of FDP-D-dimer and myoglobin contents. *Nihon Hoigaku Zasshi*, 1996, **50**(6): 400–403.

Nussbaumer, C., E. Gharehbaghi-Schnell, and I. Korschineck, Messenger RNA profiling: A novel method for body fluid identification by real-time PCR. *Forensic Sci Int*, 2006, **157**(2–3): 181–186.

Omelia, E.J., M.L. Uchimoto, and G. Williams, Quantitative PCR analysis of blood- and saliva-specific microRNA markers following solid-phase DNA extraction. *Anal Biochem*, 2013, **435**(2): 120–122.

Peabody, A.J., R.M. Burgess, and R.E. Stockdale, Re-examination of the Lugol's iodine test, pp. 1–18, 1981. Aldermaston: Home Office Central Research Establishment.

Peri, S. and P.K. Chattopadhyay, Differentiation of menstrual blood from other blood stains. *J Forensic Med Toxicol*, 2000, **17**(1): 6–11.

Peri, S. and P.K. Chattopadhyay, Studies on menstrual bloodstains: For species origin and blood groups. *Int J Med Toxicol Leg Med*, 2000, **2**(2): 1–8.

Plac-Bobula, E. and B. Turowska, Identification of menstrual bloodstains by determination of lactate dehydrogenase isoenzymes. *Arch Med Sadowej Kryminol*, 1979, **29**(4): 263–267.

Randall, B., Glycogenated squamous epithelial cells as a marker of foreign body penetration in sexual assault. *J Forensic Sci*, 1988, **33**(2): 511–514.

Sakurada, K., et al., Expression of statherin mRNA and protein in nasal and vaginal secretions. *Leg Med (Tokyo)*, 2011, **13**(6): 309–313.

Sikirzhytskaya, A., V. Sikirzhytski, and I.K. Lednev, Raman spectroscopic signature of vaginal fluid and its potential application in forensic body fluid identification. *Forensic Sci Int*, 2012, **216**(1–3): 44–48.

Simons, J.L. and S.K. Vintiner, Efficacy of several candidate protein biomarkers in the differentiation of vaginal from buccal epithelial cells. *J Forensic Sci*, 2012, **57**(6): 1585–1590.

Stombaugh Jr, P.M. and J.J. Kearney, Factors affecting the use of lactate dehydrogenase as a means of bloodstain differentiation. *J Forensic Sci*, 1978, **23**(1): 94–105.

Takasaka, T., et al., Trials of the detection of semen and vaginal fluid RNA using the genome profiling method. *Leg Med (Tokyo)*, 2011, **13**(5): 265–267.

Thompson, J. and K. Belschner, Differentiation of various menstrual bloodstains using LDH. *Calif Depart Justice Tie-Line*, 1997, **21**: 34–39.

Wang, Y.X., et al., A new method of identifying the peripheral blood and the menstrual blood. *Fa Yi Xue Za Zhi*, 2012, **28**(5): 359–361.

Wang, Z., et al., Screening and confirmation of microRNA markers for forensic body fluid identification. *Forensic Sci Int Genet*, 2013, **7**: 116–123.

Whitehead, P.H. and G.B. Divall, Assay of "soluble fibrinogen" in bloodstain extracts as an aid to identification of menstrual blood in forensic science. Preliminary findings. *Clin Chem*, 1973, **19**(7): 762–765.

Yang, H., et al., Proteomic analysis of menstrual blood. *Mol Cell Proteomics*, 2012, **11**(10): 1024–1035.

Yao, Y.N., H.L. Lu, and S. Chen, Preparation of anti-human MMP-11 polyclonal antibody for forensic application. *Chin J Forensic Med*, 2007, **22**(5): 305–307, 310.

Yao, Y.N., et al., Detection of MMP-11 from menstrual blood using immunohistochemistry. *Fa Yi Xue Za Zhi*, 2008, **24**(1): 32–33.

Zhang, Y.Q., H.L. Lu, and Y.N. Yao, Detection of matrix metalloproteinase-11 in menstrual blood by enhanced chemiluminescence method. *Fa Yi Xue Za Zhi*, 2012, **28**(2): 109–111.

Zheng, J., H.L. Lu, and S. Chen, Menstrual blood identification and matrix metalloproteinase-11. *Fa Yi Xue Za Zhi*, 2006, **22**(1): S1–S3.

Identification of Urine, Sweat, Fecal Matter, and Vomitus

17.1 Identification of Urine

The identification of the presence of urine is useful for the forensic investigation of an alleged sexual assault and harassment involving urination. More importantly, the identification of urine stains can aid in the investigation of homicides involving either ligature or manual strangulation. In these incidents, strangulation victims often involuntarily excrete urine prior to death. The locations of urine stains at the crime scene provide useful information to determine the site where the violence has occurred.

17.1.1 Urine Formation

In humans, the urinary excretory system eliminates soluble toxic wastes that are cellular metabolic by-products. The urinary system consists of the kidney, the ureter, the urinary bladder, and the urethra (Figure 17.1). The formation of urine takes place in the kidneys and, in particular, in the nephrons. The *nephron* is the basic structural and functional unit of the kidney (Figures 17.2 and 17.3). Each nephron is composed of a glomerulus, a Bowman's capsule, and a renal tubule. The glomerulus is formed by a network of capillaries and is surrounded by a Bowman's capsule. *Filtration* is the first step in urine formation. As blood flows through the glomeruli, much of its fluid, except cells and large molecules, is filtered through the capillaries into the Bowman's capsule. The glomerular filtrate, the preliminary form of urine, consists of water, salts (largely sodium and potassium ions), glucose, and waste products such as urea. The filtrate is then passed through the renal tubule where *reabsorption* occurs. During the reabsorption process, water, glucose, nutrients, and ions such as sodium are reabsorbed back into the blood. The last process of urine formation is *secretion*. Secretion occurs at the distal and the collecting tubules of the nephron where ions (such as hydrogen and potassium ions), ammonia, and certain metabolites are secreted from the blood into the lumen of the renal tubule to be eliminated in the urine. The urine is drained from the kidneys through the ureters and is stored in the bladder before it is finally excreted through the urethra.

Urine is an aqueous solution consisting largely of water. Urea is the most abundant waste product in urine, resulting from the elimination of ammonia that is produced from the metabolic process of amino acids. The average urea concentration in human urine is approximately 9 g/L. Other major components are creatinine, uric acid, and ions such as phosphate, sulfate,

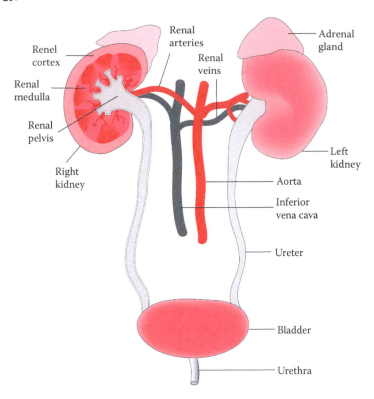

Figure 17.1   Human urinary system. A longitudinal view of the right kidney is shown. (© Richard C. Li.)

chloride, sodium, and potassium. Healthy individuals also excrete a small amount of protein. Urine has a characteristic yellow color. Subsequent to excretion from the body, urine acquires an odor that results from the ammonia that is released from the breakdown of urea by bacteria.

17.1.2 Presumptive Assays

In forensic identification, urine stains can be located by visual examination based on the characteristic yellow color of urine and the detection of the distinctive odor of urine stains. Under alternative light sources, urine stains emit a fluorescent light that facilitates the locating of urine stains in clothing and bedding. Chemical analysis can be carried out to detect the major inorganic anions in urine such as phosphate and sulfate as well as the major organic compounds in urine such as urea, creatinine, and uric acid. These assays are summarized in Table 17.1. However, these assays are not specific to urine. Other bodily fluids, such as sweat, also contain these chemical components.

17.1.2.1 The Identification of Urea

The *para-dimethylaminocinnamaldehyde* (DMAC) assay is simple and rapid and is the most commonly used presumptive assay for the forensic identification of urine stains (Figure 17.4). The DMAC assay can be performed using two different methods: the colorimetric and the fluorometric methods (Figure 17.5).

In the colorimetric method, a portion of a stain (~1 cm²) is cut and is extracted with 1 mL of distilled water. The extraction is transferred onto a piece of filter paper and is allowed to dry. One drop of 0.1% DMAC solution is then added to the filter paper. DMAC reacts specifically with urea, if present, producing a pink-colored (or magenta-colored) product. DMAC does not react with creatinine, ammonia, or uric acid. The appearance of a pink color within 30 min after applying the DMAC reagent is considered a positive reaction. No color change within 30 min is

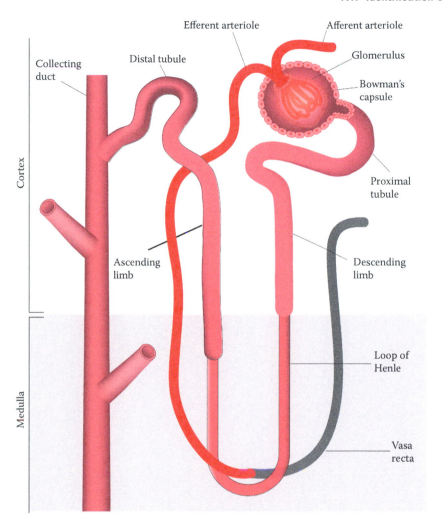

Figure 17.2　Diagram of the human nephron. (© Richard C. Li.)

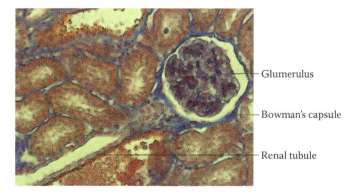

Figure 17.3　Section view of the human nephron. (© Richard C. Li.)

Table 17.1 Assays for Identification of Urine and Sweat

Biological Fluid	Assays	Identification	Assay	Method	Further Reading
Urine	Presumptive assays	Locating stains	Visual examination	Alternate light source	Wawryk and Odell (2005)
		Urea	DMAC assay	Colorimetric method	Ong et al. (2012)
				Fluorometric method	Farrugia et al. (2012)
			Microscopic crystal assay	Microscopic examination of urea crystals and urea nitrate crystals converted from urea	Nickolls (1956)
				Microscopic examination of crystals using xanthydrol	Biles and Ziobro (1998)
			Urease assay	The enzyme catalyzes the breakdown of urea and releases ammonia and carbon dioxide. The ammonia is detected using an acid-base indicator, bromthymol blue, which exhibits a blue color	Bedrosian et al. (1984)
				The ammonia is detected using Mn and Ag nitrates, which exhibit a black color	Gaensslen (1983)
			Chromatography	Gas chromatograph–mass spectrometer, thin-layer chromatography	Lum (1991)
		Creatinine	Colorimetric assay	Jaffe's test, using picric acid to convert creatinine to a red creatinine picrate compound	Greenwald (1930)
				Using potassium ferricyanide to produce a blue nitroprusside compound	O'Leary et al. (1992)
			Chromatography	Gas chromatograph–mass spectrometer	Lum (1991)

		Uric acid	Uricase assay	Measuring the disappearance of the absorption of uric acid at 293 nm after treatment with uricase that catalyzes the oxidation of uric acid to 5-hydroxyisourate	Kageyama (1971)
	Confirmatory assays	Tamm–Horsfall glycoprotein	Immunological assays	RIA, crossover electrophoresis, Ouchterlony, ELISA, and immunochromatographic methods	Tsutsumi et al. (1988)
		17-ketosteroids	Mass spectrometer	Liquid chromatograph–mass spectrometer	Nakazono et al. (2002)
Sweat	Presumptive assays	Lactate, lactic acid, urea, and single amino acids	Various	SEM coupled with EDX and Raman microspectroscopy	Sikirzhytski et al. (2011)
	Confirmatory assays	Dermcidin	Immunological assay	ELISA	Sagawa et al. (2003)
			RNA-based assay	RT-PCR	Sakurada et al. (2010)

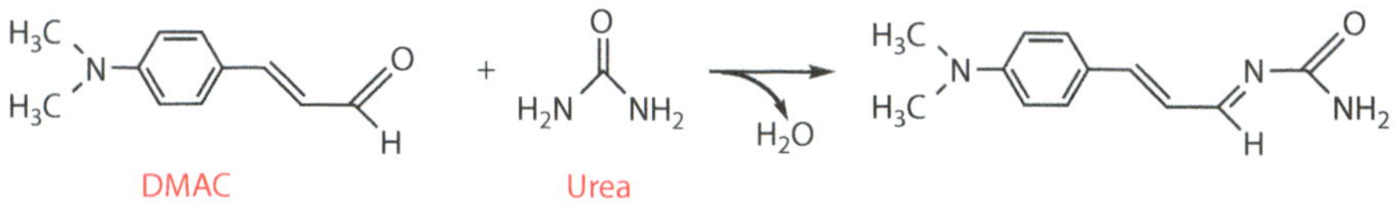

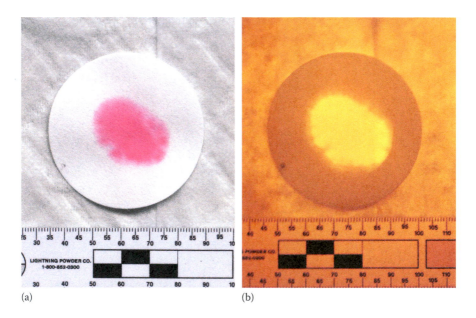

Figure 17.4   Chemical reaction of DMAC assay.

(a)   (b)

Figure 17.5   Positive reaction of DMAC assay of a urine stain. (a) Colorimetric and (b) fluorometric method. (© Richard C. Li.)

considered a negative reaction. However, this method is not specific to urine, as other bodily fluids such as saliva, semen, sweat, and vaginal secretions can also give positive reactions. A diluted DMAC solution, from 0.1% to 0.05%, can maintain appropriate sensitivity to urine stains and minimize false-positive reactions caused by the detection of low levels of urea that are present in other bodily fluids.

The fluorometric method is useful for locating urine stains on large pieces of evidence such as clothing and bedding (Figure 17.6). Additionally, this method can detect patterns of urine stains (Figure 17.7), which can be useful in crime scene reconstructions. In the fluorometric method, the evidence to be examined, such as a garment, is covered by a sheet of filter paper that has been preabsorbed with the DMAC solution. The evidence and the filter are then wrapped together in a sheet of aluminum foil and are left overnight in a press, ensuring that the evidence is in close contact with the filter paper. Alternatively, the layers can be heated for 30 s using an iron. Using a light source at 473–548 nm, the DMAC-treated urine stain fluoresces. The fluoresced stain is best observed with a 549 nm filter. However, colored fabrics interfere with the assay since dyes and pigments can inhibit the fluorescence.

In addition to the DMAC test, the identification of urea that is present in urine stains can be carried out by urease assays. Ureases catalyze the breakdown of urea, thus releasing ammonia and carbon dioxide (Figure 17.8). The ammonia is detected using an acid-base indicator, bromthymol blue, which exhibits a blue color. Alternatively, the ammonia can be detected by manganese and silver nitrates, which exhibit a black color. Additional assays such as microscopic crystal and chromatographic assays are summarized in Table 17.1.

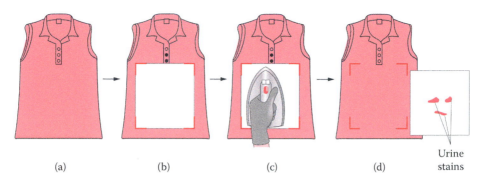

Figure 17.6   Fluorometric DMAC assay of urine stains. Evidence item (a) is closely covered by a piece of moistened filter paper containing DMAC. The orientation of the paper is marked (b). The layers are pressed to allow the transfer of a small amount of the stain onto the paper and are heated using an iron (c). The paper is then lifted. The DMAC-treated urine stain fluoresces under a light source at 473–548 nm (d). (© Richard C. Li.)

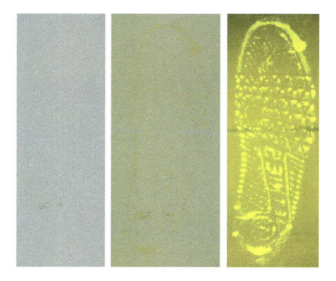

Figure 17.7   Fluorometric DMAC enhancement of a footwear impression in urine on white cotton fabric. From left to right: urine impression, DMAC-treated urine stains under white light, and DMAC-treated urine stains using an excitation light source. (From Farrugia, K.J., et al., *Forensic Sci Int*, 214, 67–81, 2012. With permission.)

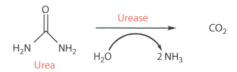

Figure 17.8   Biochemical reaction of urease.

17.1.2.2 Identification of Creatinine

Creatinine is produced during normal muscle cell metabolism (Figure 17.9). During this metabolism, phosphocreatine, an energy-storing molecule in muscle cells, breaks down to form creatine. Creatine is then metabolized to creatinine, which is released from the muscle cells into the blood. Serum creatinine is largely filtrated by the renal glomeruli. A small amount of creatinine is secreted by the renal distal tubules. The amount of creatinine excreted in urine is proportional to the muscle mass of an individual. However, creatinine is not only present in urine. It is also present in other bodily fluids such as blood and semen.

The creatinine present in urine can be detected using the Jaffe color test (Figure 17.10). In this test, picric acid is used to convert creatinine, under alkaline conditions, to form creatinine picrate, which is a bright red product. Additional tests for urine creatinine are summarized in Table 17.1. Recently, a Uritrace device (Abacus Diagnostics) has been made commercially available for the detection of creatinine.

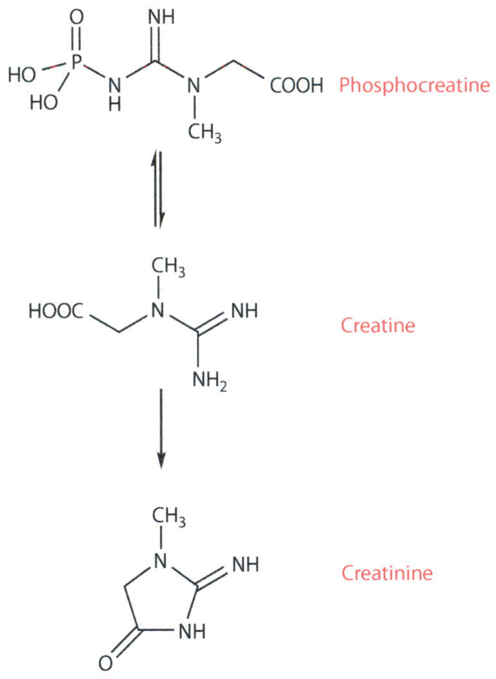

Figure 17.9 Schematic view of the formation of creatinine.

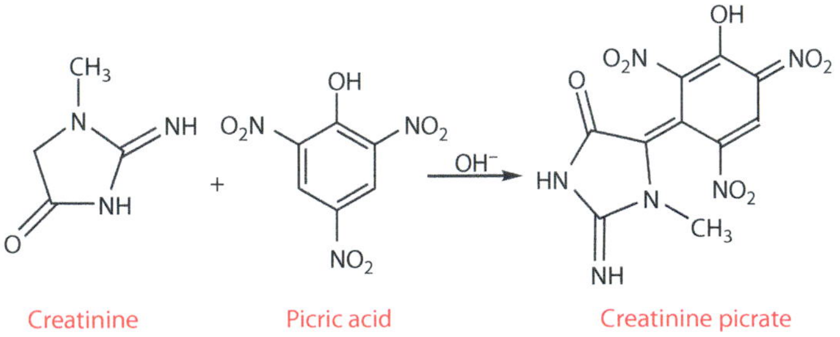

Figure 17.10 Chemical reaction of Jaffe's assay.

17.1.3 Confirmative Assays

17.1.3.1 Identification of Tamm–Horsfall Protein

Tamm–Horsfall protein (THP), also known as *uromodulin*, is the most abundant protein in urine, and accounts for 40% of the urine proteins. THP is exclusively synthesized in the epithelial cells of Henle's loop. THP is secreted from the apical plasma membrane of the epithelial cells into the lumen. Under physiological conditions, an adult excretes 20–100 mg of THP into urine daily. The biological function of THP is not fully understood. It is speculated that it prevents the body from contracting urinary tract infections and from forming renal stones.

THP is a urine-specific biomarker for forensic urine identification. It can be detected using an enzyme-linked immunosorbent assay (ELISA). More recently, it can be detected using an immunochromatographic assay, RSID-Urine, which utilizes a polyclonal rabbit antibody that is specific to THP (Figures 17.11 and 17.12). This test is rapid and simple and thus can be used as a screening test in laboratories and as a field test at crime scenes to identify urine. The detection limit of RSID-Urine for THP is 0.5 μL of urine. Although the sensitivity of RSID-Urine is lower than that of ELISA detection of THP, it is sufficient for forensic applications. These methods are specific to THP that is present only in urine while no detection is found in other bodily fluids such as plasma, saliva, semen, vaginal fluid, or sweat.

17.1.3.2 Identification of 17-Ketosteroids

In humans, urine contains derivatives of 17-ketosteroids such as androsterone, dehydroepiandrosterone (DHEA), and etiocholanolone. *Androsterone* is a steroid hormone with a weak potency of testosterone. *DHEA* is a metabolic intermediate in the biosynthesis of the gonadal steroids. DHEA also has a potential function as a steroid hormone. *Etiocholanolone* is a metabolite of testosterone. These compounds are present in urine as conjugates, in which 17-ketosteroids

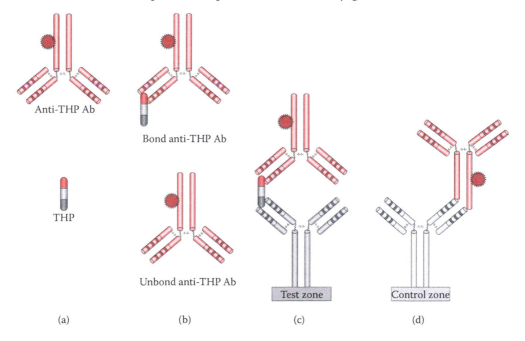

Anti-THP Ab

Bond anti-THP Ab

THP

Unbond anti-THP Ab

Test zone

Control zone

(a) (b) (c) (d)

Figure 17.11 Immunochromatographic assays for the identification of THP in urine. (a) In a sample well, THP in a urine sample is mixed with labeled anti-THP Ab. (b) The THP binds to the labeled anti-THP Ab to form a labeled Ab–THP complex. (c) At the test zone, the labeled complex binds to an immobilized anti-THP Ab to form a labeled Ab–THP–Ab sandwich. (d) At the control zone, the labeled anti-THP Ab binds to an immobilized antiglobulin and is captured at the control zone. Ab, antibody; THP, Tamm–Horsfall protein antigen. (© Richard C. Li.)

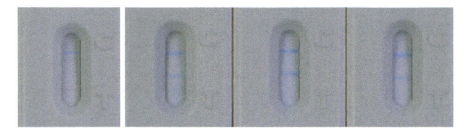

Figure 17.12   Results obtained using immunochromatographic assays (RSID™-Urine; Independent Forensics, Lombard, IL) for the identification of THP in urine. From left to right: negative control, undiluted, 1:10 diluted, and 1:100 diluted urine samples. A visible blue line at the test (T) zone and control (C) zone indicates a positive result. The blue line at the control zone only indicates a negative result. (From Akutsu, T., Watanabe, K., and Sakurada, K., *J Forensic Sci*, 57, 1570–1573, 2012. With permission.)

are modified in the liver through glucuronidation and sulfation. In glucuronidation, glucuronic acids are added to 17-ketosteroids while in sulfation, sulfates are transferred to 17-ketosteroids. As a result, these conjugates, containing charged moieties, are more water soluble than the non-conjugated 17-ketosteroids. Thus, they are excreted into urine to be eliminated from the body. The five major components of the 17-ketosteroid conjugates present in human urine are andros-terone glucuronide, androsterone sulfate, DHEA sulfate, etiocholanolone glucuronide, and etio-cholanolone sulfate (Figure 17.13). Thus, the analysis of 17-ketosteroid conjugates in urine stains is useful for the identification of human urine stains. These five 17-ketosteroid conjugates can be identified using liquid chromatography–mass spectrometry (LC-MS). However, some of these 17-ketosteroid conjugates are also detected in serum. Therefore, the presence of all five conju-gated 17-ketosteroids in a sample is required to identify a urine stain. Additionally, the profiles of the 17-ketosteroid conjugates are human specific and are distinguishable between humans and animals.

17.2 Identification of Sweat

Sweat is the least common bodily fluid analyzed in forensic laboratories compared with others that have been mentioned in previous chapters. However, sweat identification is still useful for forensic investigations. For example, forensic DNA analysis allows the generation of DNA pro-files from trace biological evidence such as fingerprints. Identifying sweat can be important for the analysis of these types of evidence.

17.2.1 Biology of Perspiration

Humans have two types of secretory sweat glands: the eccrine and the apocrine sweat glands (Figure 17.14). *Eccrine* sweat glands are distributed almost all over the body and are controlled by the sympathetic nervous system. Eccrine sweat glands play a role in regulating body tem-perature. When the body temperature rises, eccrine sweat glands secrete a watery sweat to the skin's surface where heat is carried away through the evaporation of the sweat to main-tain normal body temperature. In humans, *apocrine* sweat glands, which are associated with hair follicles, are usually restricted to the underarm and genital areas and are controlled by emotional stress. Apocrine sweat glands are inactive until puberty. This type of sweat gland secretes an oily sweat that is odorous after being processed by skin bacteria. The majority of sweat evidence that is analyzed in forensic laboratories is sweat stains secreted from eccrine glands. Sweat contains water, minerals, lactate, and urea. Its biochemical composition varies

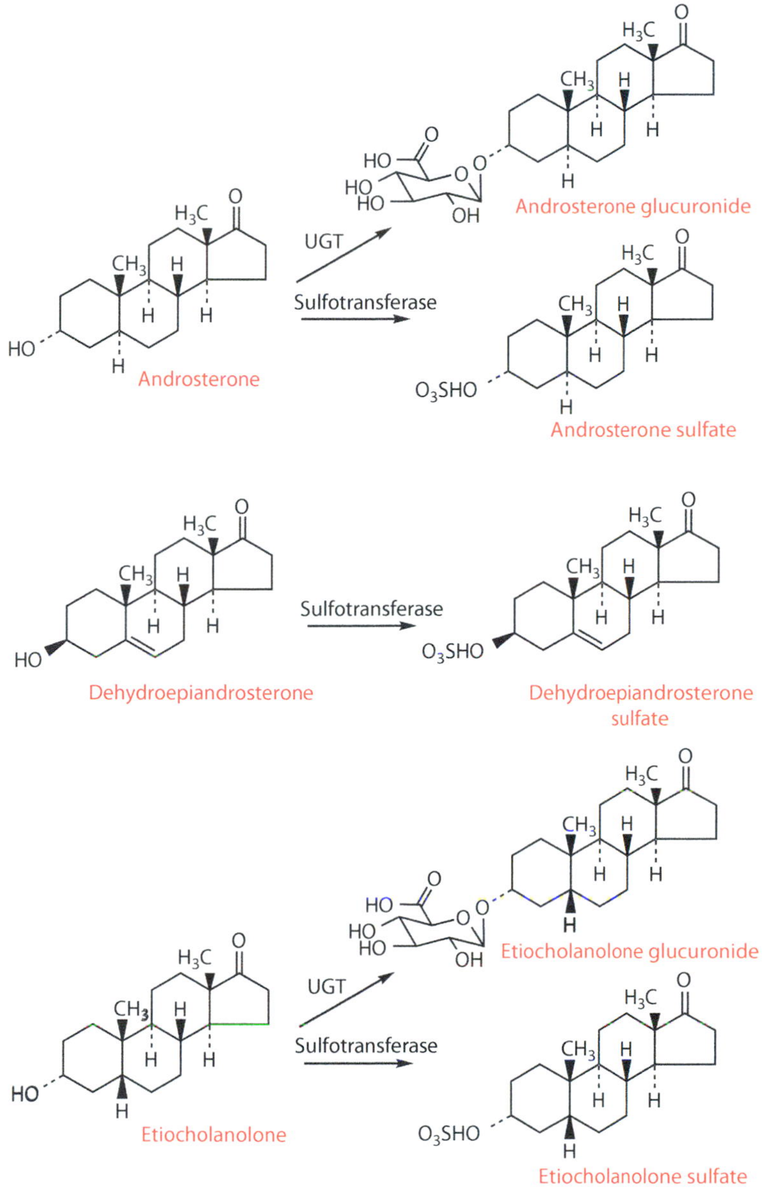

Figure 17.13 Structural formula of 17-ketosteroids and 17-ketosteroid conjugates present in human urine. UGT, UDP glucuronosyltransferase. (© Richard C. Li.)

among individuals and their physical activities. Sweat contains low levels of constituents that are also present in other bodily fluids such as urine. Thus, it has been considered a difficult bodily fluid to identify.

17.2.2 Sweat Identification Assays

Sweat evidence has been analyzed using presumptive assays (Table 17.1) such as elemental analysis using scanning electron microscopes coupled with energy dispersive x-ray spectroscopy in the detection of lactic acid. Since sweat contains many inorganic and organic compounds that are also present in other bodily fluids, these assays are not specific to sweat. Raman

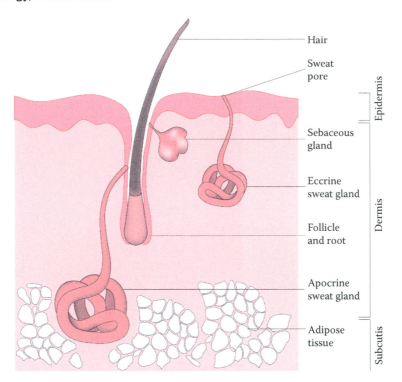

Figure 17.14   Diagram of the human eccrine and apocrine sweat glands. (© Richard C. Li.)

microspectroscopy is potentially useful for the identification of sweat for forensic purposes, which is based largely on the profiles of lactate, lactic acid, urea, and single amino acids in urine.

Recently, *dermcidin* has been identified as a potential biomarker of human sweat. Dermcidin belongs to a class of human antimicrobial peptides of the innate immune defense system and plays an important role in protecting epithelial barriers from infections. Dermcidin, specifically expressed in eccrine sweat glands (Figure 17.15), is secreted into the sweat and is transferred to the epidermal surface. In forensic applications, dermcidin can potentially be utilized as a biomarker for the confirmatory assay of sweat identification. The detection of dermcidin in sweat stains can be performed using ELISA assays utilizing antibodies specific to human dermcidin. This method is highly sensitive and is able to detect dermcidin in sweat samples that are diluted 10,000-fold. Dermcidin is encoded by the *DCD* gene. Its mRNA can be detected using reverse transcription polymerase chain reaction (RT-PCR) assays (Chapter 7) that can detect *DCD* mRNA in 10 μL of sweat sample. Dermcidin assays are also specific to sweat as dermcidin is not detected in other bodily fluids such as semen, saliva, and urine.

17.3 Identification of Fecal Matter

The examination of feces has been used in criminal investigations for over a century. Specifically in 1948, a burglary case was reported using fecal analysis to link the shoes of a suspect to a crime scene. One of the aspects of fecal analysis is to determine a common origin of the reference sample and the fecal evidence, thus potentially linking a suspect to a crime scene. Today, the individual characteristics of a fecal sample can be effectively determined using forensic DNA analysis of sloughed intestinal epithelial cells that are present in fecal matter. The identification of fecal matter is valuable in providing important information for a criminal investigation. For example, the presence of fecal matter may corroborate a sexual

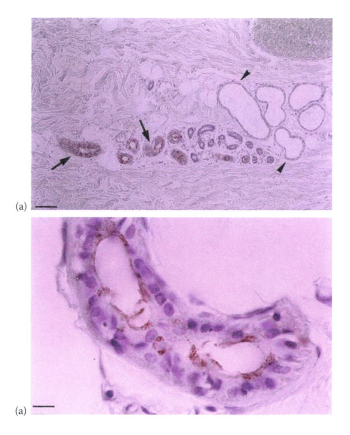

(a)

(a)

Figure 17.15 Localization of the dermcidin peptide in the sweat glands. Skin tissue sections are treated with a dermcidin antibody and stained. (a) Both the eccrine and apocrine sweat glands are shown. Only cells of the eccrine gland in the skin express dermcidin (brown stained). Arrows: eccrine sweat glands. Arrowheads: apocrine sweat glands. Scale bar: 100 μm. (b) Close-up view of the eccrine gland. The presence of the dermcidin peptide is observed (brown). Scale bar: 10 μm. (From Sagawa, K., et al., *Int J Legal Med*, 117, 90–95, 2003. With permission.)

assault involving sodomy, assault with fecal matter, vandalism, and burglary during which the perpetrator defecated at the scene.

17.3.1 Fecal Formation

Feces are a type of waste matter that is the direct result of food that has been processed by the digestive system (Figure 17.16). Human feces contain undigested foodstuffs, sloughed intestinal epithelial cells, intestinal bacteria, bile pigments, electrolytes, and water. Feces are formed in the intestines during the last phase of digestion. Feces first enter the colon in liquid form. Most of the nutrients are absorbed on the surface area of the small intestine. In the large intestine, water, sodium, and chloride are absorbed on the surface of the lumen. The remaining luminal contents are converted into feces. Food stays for approximately 2–6 h in the stomach. It takes an additional 3–5 h to travel through the small intestine and 12–24 h to travel through the large intestine.

17.3.2 Fecal Matter Identification Assays

A fecal analysis includes macroscopic and microscopic examination, chemical tests, and fecal bacterial identification.

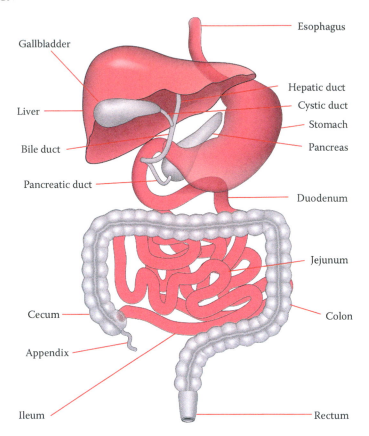

Figure 17.16 Diagram of the human digestive system. The esophagus, stomach, liver, gallbladder, pancreas, small intestine (duodenum, jejunum, and ileum), and large intestine (cecum, colon, appendix, and rectum) are shown. (© Richard C. Li.)

17.3.2.1 Macroscopic and Microscopic Examination

The color and odor of human feces are useful characteristics for fecal identification. The normal brown color of feces primarily results from the presence of urobilinoids, which are heme catabolic by-products. The characteristic odor of feces is caused by the metabolic by-products of the intestinal bacterial flora. Indole, skatole, and hydrogen sulfide are the compounds that are responsible for the odor of feces.

When fecal stains are analyzed, the microscopic examination of the feces can be performed on an aliquot of fecal suspension. The presence of characteristic undigested foodstuffs can indicate human feces. Fecal matter can be transferred from samples of clothing by scraping with a sterile stainless steel spatula. The fecal matter is then hydrated in 6% formalin solution for 24–48 h prior to microscopic examination. Undigested foodstuffs such as vegetable fragments and meat fibers are often present in human feces. Vegetable fragments (Figure 17.17) are often undigested vegetable dermal tissues that cover and protect the plant and fragments of vascular tissues that play roles in transporting water and nutrients throughout the plant. The types of vegetables can be identified by comparing the observed fragments with known plants. Meat fibers (Figure 17.18) are undigested animal skeletal muscle fibers. These fibers have characteristic striations, usually rectangular in shape, that are used for comparison with known animal tissues.

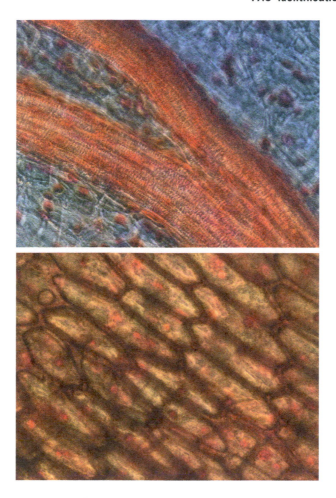

Figure 17.17 Scallion vascular tissue (top) and dermal tissue (bottom). (© Richard C. Li.)

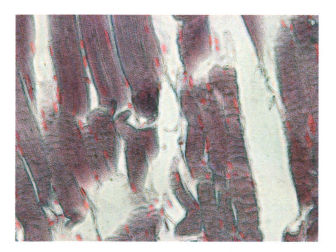

Figure 17.18 Cattle meat fibers. (© Richard C. Li.)

17.3.2.2 Urobilinoids Tests

The forensic analysis of fecal matter often involves the identification of fecal stains on swabs and clothing from which small amounts of sample may be available. Chemical tests can be useful for the analysis of fecal stains. By far the most common chemical test performed on feces is the detection of urobilinoids. *Urobilinoids*, including urobilin and stercobilin, are generated from the degradation of heme and are excreted into feces.

The average lifetime of erythrocytes is approximately 3–4 months. Erythrocytes are continuously undergoing hemolysis in which erythrocytes are naturally broken down and are usually processed in the reticuloendothelial system of the spleen. Hemoglobin is released daily during the hemolysis process and is degraded into heme, globin, and iron. Other sources of heme are derived from the degradation of erythrocyte precursors in bone marrow and other heme-containing proteins such as myoglobin and cytochromes. In the peripheral tissues, heme undergoes catabolism to form *bilirubin* (Figures 17.19 and 17.20). Bilirubin is further converted to *urobilinogen* in the intestine. A portion of urobilinogen is reduced to *stercobilinogen*. In the large intestine, the spontaneous oxidation of urobilinogen and stercobilinogen results in the formation of

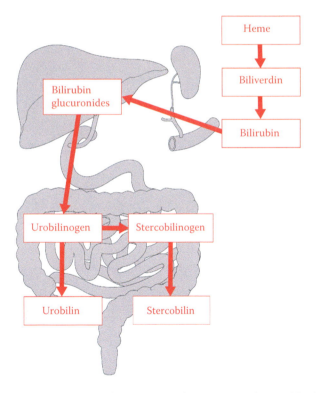

Figure 17.19 The formation of urobilinoids. Aged erythrocytes are disposed in the spleen, releasing hemoglobin that is broken down to heme. The heme is converted to biliverdin and is subsequently reduced to bilirubin. The bilirubin is then released into the bloodstream where it is bound to albumin, which cannot be filtrated at the glomeruli. The bilirubin is transported through the bloodstream to the liver where it is conjugated with glucuronic acid, forming water-soluble bilirubin monoglucuronide and diglucuronide. The bilirubin glucuronides are excreted into the bile and are subsequently excreted into the small intestine. In the intestines, the glucuronic acid of the conjugated bilirubin is removed. The unconjugated bilirubin is metabolized by intestinal bacteria, forming urobilinogen. A portion of the urobilinogen is further metabolized to stercobilinogen. The urobilinogen and the stercobilinogen are oxidized by intestinal bacteria, forming urobilin and stercobilin, respectively, which are excreted into the feces. (© Richard C. Li.)

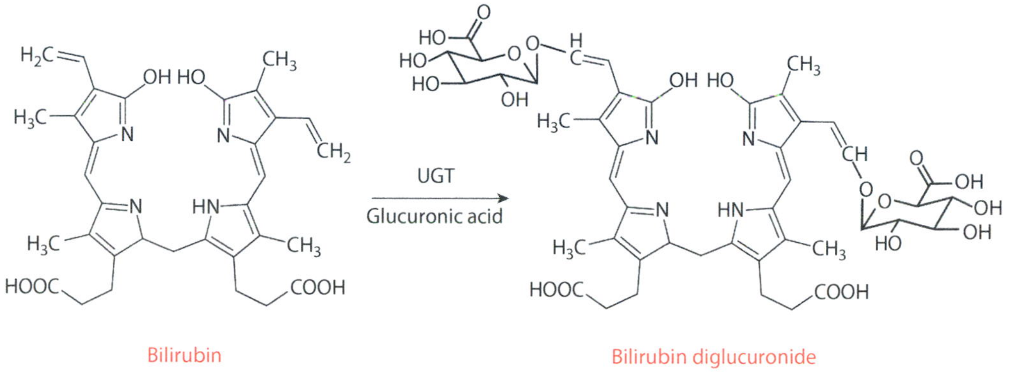

Figure 17.20 The formation of bilirubin glucuronides in the liver. UGT, UDP glucuronosyltransferases.

urobilin and stercobilin (Figure 17.21), respectively. These compounds are brown colored and are responsible for the characteristic color of feces.

The urobilinoids can be detected using the Schlesinger and Edelman tests. In the *Schlesinger* test, a sample is mixed with saturated zinc acetate in ethanol solution to form aurobilinoid–zinc chelation complex that emits a characteristic green fluorescence under ultraviolet light. The *Edelman* test is a variation of the Schlesinger test. A sample is treated with a mercuric salt solution to yield a pink-colored compound. Further treatment with a zinc salt produces fluorescence. However, less fluorescence is observed in the Edelman test than in the Schlesinger test. Inconclusive and inconsistent results are often obtained using these tests where fecal material sometimes gives no visible fluorescence. Additionally, the intensity of the fluorescence observed varies between samples. The reliability and selectivity of the tests can be increased using a spectrometric measurement of the fluorescence detection of fecal urobilinoids based on the principle of the Schlesinger test. A dry sample is treated with 1 mL of zinc acetate solution (1% zinc acetate methoxyethanol solution and 0.2% Tris). The suspension is then sonicated for 5 min, heated at 100°C for 10 min, cooled, and centrifuged. The presence of urobilinoids can be detected using excitation and emission maxima at 507 and 514 nm, respectively.

The disadvantages of the Schlesinger and the Edelman tests are their low species specificity as both tests cannot distinguish between human and other mammalian fecal materials. Moreover, under normal circumstances, up to 5% of urobilinogen is transported to the kidney and oxidized to urobilin in urine. Under some pathological conditions such as hepatic function disorders, the level of urobilin in the urine can be very high. Both the Schlesinger and Edelman tests respond to the urobilin of urine stains as well.

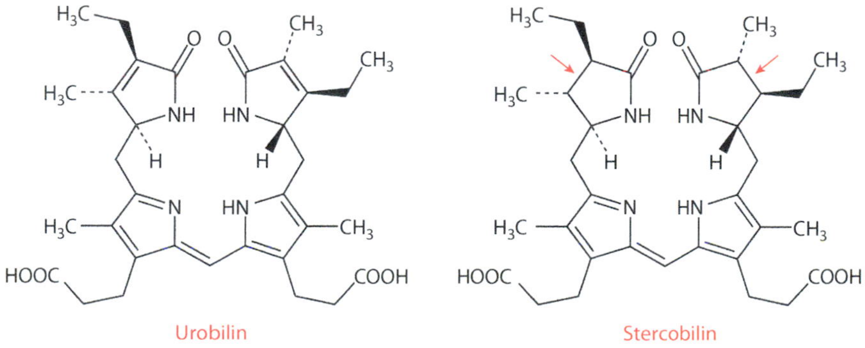

Figure 17.21 Urobilinoids: urobilin and stercobilin. Arrow, site of reduction.

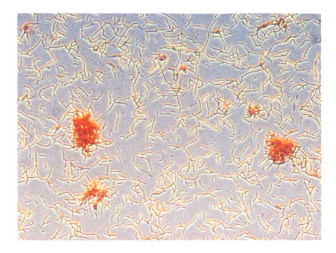

Figure 17.22 Intestinal bacteria. (© Richard C. Li.)

17.3.2.3 Fecal Bacterial Identification

The human intestinal microbiota contains more than 4000 bacterial species (Figure 17.22). Traditional methods for detecting fecal materials utilize the cultivation of fecal indicator bacteria such as *Escherichia coli* or *Enterococci* spp. However, these fecal indicator bacteria constitute only a small portion of the fecal microbiota, and thus may not be adequate for forensic identification purposes. *Bacteroides*, however, accounts for approximately 30% of fecal microbiota that are the predominant bacteria in human feces. *Bacteroides* can potentially be used for the forensic identification of feces. *Bacteroides* is a genus of rod-shaped, anaerobic gram-negative bacteria. These bacteria play a role in digesting complex carbohydrates and other substances that cannot be digested by human enzymes. The identification of *bacteroides* can be carried out by detecting specific DNA sequences of the *rpoB* gene, which encodes the β subunit of bacterial RNA polymerase (Chapter 11). The presence of the species-specific DNA sequence can be detected by RT-PCR utilizing primers that are specific to the target species but not to other species. Thus, only a targeted species can be amplified if it is present. Two fecal predominant bacteroides species, *B. uniformis* and *B. vulgatus*, can be detected in feces. *B. uniformis* is not detectable in blood, saliva, semen, urine, vaginal fluids, or on skin surfaces. Therefore, *B. uniformis* is considered as a specific indicator bacterium for forensic fecal identification. Sometimes, *B. vulgatus* can also be detected in vaginal fluid samples. Therefore, precaution should be taken in interpreting the results obtained using a *B. vulgatus* assay. Additionally, this method alone cannot discriminate between human and animal feces. Furthermore, fecal microbial populations can be affected by the host's diet. Individuals who consume saturated fats and proteins, which are abundant in Western diets, have predominantly *Bacteroides* species in their feces. However, individuals who consume a low-fat and carbohydrate-rich diet have predominantly *Prevotella* species, also a genus of gram-negative bacteria, in their feces.

17.4 Identification of Vomitus

17.4.1 Biology of Gastric Fluid

Gastric fluid can be found in stains derived from stomach wounds. Most often, gastric fluid is from vomitus found at a crime scene or as dried stains on clothing. Vomiting is the forceful expulsion of the contents of the stomach through the mouth (Figure 17.23). It is usually preceded by salivation, sweating, and the sensation of nausea. Vomiting usually begins with a deep

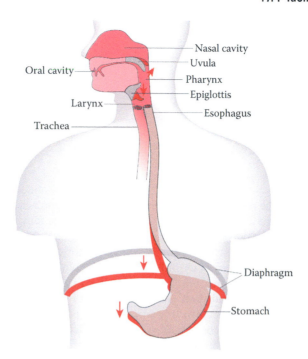

Figure 17.23   Act of vomiting. The contraction of the diaphragm and the stomach during the phase of vomiting is shown (in red). Additionally, the positions of the epiglottis and the uvula during vomiting are shown (in red). As a result, the stomach contents are expelled through the mouth. (© Richard C. Li.)

inhalation and the closure of the glottis. Subsequent contractions of the diaphragm and the abdominal muscles compress the stomach. The gastric contents are then forced upward through the relaxed sphincters and the esophagus, and are expelled through the mouth. Since the glottis is closed, vomitus usually does not enter the respiratory tract. Although the uvula is usually raised to close the nasal cavity, vomitus sometimes enters the nose. Vomiting can be caused by a wide variety of pathological conditions. Vomiting may be a specific response to acute intoxication in homicidal poisoning cases or it may be caused by trauma in a violent assault. Thus, the forensic identification of the gastric fluid can corroborate a criminal act.

The stomach stores ingested food until it can be emptied into the small intestines. When food enters the stomach, hydrochloric acid is secreted in large quantities, which facilitates the initial degradation of proteins. The stomach also secretes mucus that lubricates the gastric surface to protect the epithelium from acidic environments. Hormones such as gastrin, which are found in the gastric fluid, regulate acid secretion and gastric movement. A number of enzymes are secreted into the gastric fluid, including lipase, which plays a role in lipid hydrolysis, and gelatinase, which can hydrolyze gelatin. The stomach also secretes pepsinogens, which are enzyme precursors, into the gastric fluid. In the stomach, pepsinogens are activated by hydrochloric acid into pepsin, which is largely responsible for the digestion of proteins.

17.4.2 Vomitus Identification Assays

Vomitus is highly acidic and tends to be malodorous. The color of vomitus may be of forensic interest. Fresh blood in the vomit is usually bright red and suggests bleeding due to injuries, while dark red blood clots suggest bleeding in the stomach due to pathological conditions such as an ulcer. A microscopic examination (Section 17.3.2.1) can be performed to identify recently ingested food particles that are present in a sample. The presence of gastric fluid in vomitus can also be identified by the detection of pepsins secreted from the stomach. This identification test

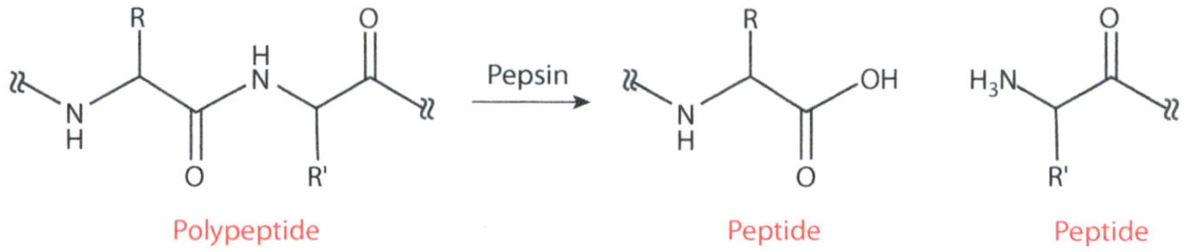

Figure 17.24 The cleavage of polypeptide by pepsin. Pepsin cleaves polypeptide at sites of aromatic amino acids. R and R′: tryptophan, phenylalanine, or tyrosine.

is based on the proteolytic activity of pepsins. Pepsins are endopeptidases that cleave primarily on peptide bonds in the middle of the protein. Aromatic amino acids such as tryptophan, phenylalanine, and tyrosine are the preferred targeted amino acids for the cleavage reaction by pepsins (Figure 17.24). In the pepsin-proteolytic assay, fibrin blue is used as a substrate for pepsin. Fibrin blue is an insoluble protein–dye complex that is colorless. In the presence of vomitus, pepsin cleaves fibrin blue and releases a chromophore that is soluble in water and exhibits a blue color. In the assay, a fibrin blue–containing agarose gel is utilized. The sample from vomitus is loaded onto the gel plate. After incubation, a blue ring around the sample can be observed as a result of the enzymatic reactivity of pepsin. The amounts of pepsin in a sample can be quantified. The results on the gel plate can be photographed and the dried gel plate can also be preserved as evidence. This method can determine the pepsin content of fresh and aged forensic samples. Bodily fluids other than vomitus do not show positive reactions with the use of this method. However, this method cannot distinguish the vomitus of humans from that of other vertebrates.

Bibliography

Akutsu, T., K. Watanabe, and K. Sakurada, Specificity, sensitivity, and operability of RSID-urine for forensic identification of urine: Comparison with ELISA for Tamm–Horsfall protein. *J Forensic Sci*, 2012, **57**(6): 1570–1573.

Akutsu, T., et al., Evaluation of Tamm–Horsfall protein and uroplakin III for forensic identification of urine. *J Forensic Sci*, 2010, **55**(3): 742–746.

Bedrosian, J.L., M.D. Stolorow, and M.A. Tahir, Development of a radial gel diffusion technique for the identification of urea in urine stains. *J Forensic Sci*, 1984, **29**(2): 601–606.

Biles, P.V. and G.C. Ziobro, Identification of the source of reagent variability in the xanthydrol/urea method. *J Assoc Official Anal Chem Int*, 1998, **81**(6): 1155–1161.

Bock, J.H., M.A. Lane, and D.O. Norris, *Identifying Plant Food Cells in Gastric Contents for Use in Forensic Investigations: A Laboratory Manual*, 1988. Washington, DC: US Department of Justice, National Institute of Justice.

Brunzel, N.A., *Fundamentals of Urine and Body Fluid Analysis*, 2nd edn., 2004. Philadelphia, PA: W.B. Saunder.

De Araújo, W.R., M.O. Salles, and T.R.L.C. Paixão, Development of an enzymeless electroanalytical method for the indirect detection of creatinine in urine samples. *Sensor Actuat B-Chem*, 2012, **173**: 847–851.

Farrugia, K.J., et al., Chemical enhancement of footwear impressions in urine on fabric. *Forensic Sci Int*, 2012, **214**(1–3): 67–81.

Farrugia, K.J., et al., A comparison of enhancement techniques for footwear impressions on dark and patterned fabrics. *J Forensic Sci*, 2013, **58**(6): 1472–1485.

Fujishiro, M., et al., Identification of human urinary stains by the uric acid/creatinine quotient and HPLC chromatogram. *J Showa Med Assoc*, 2008, **68**(3): 175–181.

Gaensslen, R.E., Identification of urine, in *Sourcebook in Forensic Serology, Immunology, and Biochemistry*, pp. 191–195, 1983. Washington, DC: US Government Printing Office.

Greenwald, I., The chemistry of Jaffe's reaction for creatinine: VI. A compound of picric acid with two molecules of creatinine. Its combinations with acid and alkali. *J Biol Chem*, 1930, **86**: 333–343.

Iida, R., T. Yasuda, and K. Kishi, Identification of novel fibronectin fragments detected specifically in juvenile urine. *Fed Eur Biochem Soc J*, 2007, **274**(15): 3939–3947.

Johnson, D.J., L.R. Martin, and K.A. Roberts, STR-typing of human DNA from human fecal matter using the QIAGEN QIAamp® stool mini kit. *J Forensic Sci*, 2005, **50**(4): 802–808.

Jones, E., The identification of semen and other body fluids, in R. Saferstein (ed.), *Forensic Science Handbook*, pp. 329–399, 2005. Upper Saddle River, NJ: Prentice Hall.

Juusola, J. and J. Ballantyne, Multiplex mRNA profiling for the identification of body fluids. *Forensic Sci Int*, 2005, **152**(1): 1–12.

Kageyama, N., A direct colorimetric determination of uric acid in serum and urine with uricase-catalase system. *Clin Chim Acta*, 1971, **31**(2): 421–426.

Knight, B.H., The significance of the postmortem discovery of gastric contents in the air passages. *Forensic Sci*, 1975, **6**(3): 229–234.

Lai, Y., C.H. Wang, and J. Qin, A study on STR typing of human sweat latent fingerprints developed by various methods. *Chin J Forensic Med*, 2007, **22**(6): 366–368.

Lee, H.C., et al., Enzyme assays for the identification of gastric fluid. *J Forensic Sci*, 1985, **30**(1): 97–102.

Lloyd, J.B. and N.T. Weston, A spectrometric study of the fluorescence detection of fecal urobilinoids. *J Forensic Sci*, 1982, **27**(2): 352–365.

Lum, P., Seven month old substituted urine sample. *Calif Depart Justice Tie-Line*, 1991, **16**: 79–82.

Metropolitan Police Forensic Science Laboratory, *Biology Methods and Manual*, p. 5, 1978. London: Metropolitan Police Forensic Science Laboratory.

Nakanishi, H., et al., Identification of feces by detection of bacteroides genes. *Forensic Sci Int Genet*, 2013, **7**(1): 176–179.

Nakazono, T., et al., Identification of human urine stains by HPLC analysis of 17-ketosteroid conjugates. *J Forensic Sci*, 2002, **47**(3): 568–572.

Nakazono, T., et al., Dual examinations for identification of urine as being of human origin and for DNA-typing from small stains of human urine. *J Forensic Sci*, 2008, **53**(2): 359–363.

Nickolls, L.C., Urine, in *The Scientific Investigation of Crime*, pp. 209–210, 1956. London: Butterworth.

Norris, D.O. and J.H. Bock, Use of fecal material to associate a suspect with a crime scene: Report of two cases. *J Forensic Sci*, 2000, **45**(1): 184–187.

Nussbaumer, C., E. Gharehbaghi-Schnell, and I. Korschineck, Messenger RNA profiling: A novel method for body fluid identification by real-time PCR. *Forensic Sci Int*, 2006, **157**(2–3): 181–186.

O'Leary, N., A. Pembroke, and P.F. Duggan, A simplified procedure for eliminating the negative interference of bilirubin in the Jaffe reaction for creatinine. *Clin Chem*, 1992, **38**(9): 1749–1751.

Ong, S.Y., et al., Forensic identification of urine using the DMAC test: A method validation study. *Sci Justice*, 2012, **52**(2): 90–95.

Poon, H.H.L., Identification of human urine by immunological techniques. *Can Soc Forensic Sci J*, 1984, **17**: 81–89.

Reddy, K. and E.J. Lowenstein, Forensics in dermatology: Part I. *J Am Acad Dermatol*, 2011, **64**(5): 801–808.

Rhodes, E.F. and J.I. Thornton, DNAC test for urine stains. *J Police Sci Admin*, 1976, **4**: 88–89.

Roy, R., Analysis of human fecal material for autosomal and Y chromosome STRs. *J Forensic Sci*, 2003, **48**(5): 1035–1040.

Sagawa, K., et al., Production and characterization of a monoclonal antibody for sweat-specific protein and its application for sweat identification. *Int J Legal Med*, 2003, **117**(2): 90–95.

Sakurada, K., et al., Detection of dermcidin for sweat identification by real-time RT-PCR and ELISA. *Forensic Sci Int*, 2010, **194**(1–3): 80–84.

Sato, I., Detection of α S1-casein in vomit from bottle-fed babies by enzyme-linked immunosorbent assay. *Int J Legal Med*, 1992, **105**(3): 127–131.

Sato, K., et al., Identification of human urinary stains by the quotient uric acid/urea nitrogen. *Forensic Sci Int*, 1990, **45**(1–2): 27–38.

Sikirzhytski, V., A. Sikirzhytskaya, and I.K. Lednev, Multidimensional Raman spectroscopic signatures as a tool for forensic identification of body fluid traces: A review. *Appl Spectrosc*, 2011, **65**(11): 1223–1232.

Sikirzhytski, V., A. Sikirzhytskaya, and I.K. Lednev, Multidimensional Raman spectroscopic signature of sweat and its potential application to forensic body fluid identification. *Anal Chim Acta*, 2012, **718**: 78–83.

Srch, M., Examination of vomit [in Czech]. *Ceskoslovenska Patologie*, 1975, **11**(3): 46–47.

Srch, M., Examination of vomits. *Vysetrovánizvratků*, 1975, **20**(3): 46–48.

Taylor, M.C. and J.S. Hunt, Forensic identification of human urine by radioimmunoassay for Tamm–Horsfall urinary glycoprotein. *J Forensic Sci Soc*, 1983, **23**(1): 67–72.

Tsutsumi, H., et al., Identification of human urinary stains by enzyme-linked immunosorbent assay for human uromucoid. *J Forensic Sci*, 1988, **33**(1): 237–243.

Virkler, K. and I.K. Lednev, Analysis of body fluids for forensic purposes: From laboratory testing to non-destructive rapid confirmatory identification at a crime scene. *Forensic Sci Int*, 2009, **188**(1–3): 1–17.

Wawryk, J. and M. Odell, Fluorescent identification of biological and other stains on skin by the use of alternative light sources. *J Clin Forensic Med*, 2005, **12**(6): 296–301.

Yamada, S., et al., Vomit identification by a pepsin assay using a fibrin blue-agarose gel plate. *Forensic Sci Int*, 1992, **52**(2): 215–221.

Yan, P., et al., DNA-based species identification for faecal samples: An application on the mammalian survey in Mountain Huangshan scenic spot. *Afr J Biotechnol*, 2011, **10**(57): 12134–12141.

Zhang, S.H., et al., Genotyping of urinary samples stored with EDTA for forensic applications. *Genet Mol Res*, 2012, **11**(3): 3007–3012.

SECTION IV

Individualization of Biological Evidence

Blood Group Typing and Protein Profiling

18.1 Blood Group Typing

18.1.1 Blood Groups

For the purposes of this text, blood groups are defined as antigen polymorphisms present on erythrocyte surfaces. Human erythrocyte surface membranes contain a variety of blood group antigens. Transfusion reactions occur when an incompatible type of blood is transfused into an individual, which can lead to severe symptoms or even death. Karl Landsteiner discovered the first blood group, known as the ABO system, in the early 1900s, while studying transfusion and transplantation. The discovery made blood transfusions feasible, and Landsteiner was awarded the Nobel Prize in 1930.

The International Society of Blood Transfusion currently recognizes 29 blood group systems, which include hundreds of antigen polymorphisms (Table 18.1). From the 1950s to the 1970s, the structures and biosynthesis pathways of many blood group antigens were determined. The genes for most of these blood group systems have been identified as well. The isolation of blood group genes has made it possible to understand the molecular mechanisms of the antigenic characteristics of the blood group systems.

The ABO system of antigens in human erythrocytes is the most commonly used blood group system for forensic applications. Forensic laboratories also use others, including the Rh, MNS, Kell, Duffy, and Kidd systems.

18.1.2 ABO Blood Group System

In the ABO blood group system, two types of antigens, designated A and B, give rise to four blood types:

- Type A individuals have the A antigen.
- Type B individuals have the B antigen.
- Type AB individuals have both A and B antigens.
- Type O individuals have neither A nor B antigens.

The antigens may be found in other bodily fluids as well as blood, such as amniotic fluid, saliva, and semen as well as many organs including the kidney, pancreas, liver, and lungs.

Table 18.1	Blood Group Systems				
Number	Name	Symbol	Number of Antigens	Gene Names	Chromosomal Location
001	ABO	ABO	4	*ABO*	9q34.2
002	MNS	MNS	43	*GYPA, GYPB, GYPE*	4q31.21
003	P	P1	1		22q11.2-qter
004	Rh	RH	49	*RHD, RHCE*	1p36.11
005	Lutheran	LU	20	*LU*	19q13.32
006	Kell	KEL	25	*KEL*	7q34
007	Lewis	LE	6	*FUT3*	19p13.3
008	Duffy	FY	6	*FY*	1q23.2
009	Kidd	JK	3	*SLC41A1*	18q12.3
010	Diego	DI	21	*SLC4A1*	17q21.31
011	Yt	YT	2	*ACHE*	7q22.1
012	Xg	XG	2	*XG, MIC2*	Xp22.33, Yp11.3
013	Scianna	SC	5	*ERMAP*	1p34.2
014	Dombrock	DO	5	*DO*	12p12.3
015	Colton	CO	3	*AQP1*	7p14.3
016	Landsteiner-Wiener	LW	3	*ICAM4*	19p13.2
017	Chido/Rodgers	CH/RG	9	*C4A, C4B*	6p21.3
018	H	H	1	*FUT1*	19q13.33
019	Kx	XK	1	*XK*	Xp21.1
020	Gerbich	GE	8	*GYPC*	2q14.3
021	Cromer	CROM	12	*DAF*	1q32.2
022	Knops	KN	8	*CR1*	1q32.2
023	Indian	IN	2	*CD44*	11p13
024	Ok	OK	1	*BSG*	19p13.3
025	Raph	RAPH	1	*CD151*	11p15.5
026	John Milton Hagen	JMH	1	*SEMA7A*	15q24.1
027	I	I	1	*GCNT2*	6p24.2
028	Globoside	GLOB	1	*B3GALT3*	3q26.1
029	Gill	GIL	1	*AQP3*	9p13.3

Source: Adapted from Daniels, G.L., et al., *Vox Sang*, 87, 304–316, 2004.

18.1.2.1 Biosynthesis of Antigens

All individuals generate the *O antigen*, also known as the *H antigen*. The O antigen is synthesized by *fucosyltransferase*, a fucose transferase encoded by the *FUT* genes, which adds a fucose on the end of a glycolipid (in erythrocytes) or glycoprotein (in tissues). An additional monosaccharide (Figure 18.1) is then transferred to the O antigen by a transferase encoded by the *ABO* locus. The specificity of this enzyme determines the ABO blood type (Figure 18.2):

- The *A* allele produces the A-transferase, which transfers N-acetylgalactosamine to the O antigen and thus synthesizes the A antigen.

- The *B* allele produces the B-transferase, which transfers galactose to the O antigen and thus synthesizes the B antigen.

- The *O* allele has a mutation (small deletion), which eliminates transferase activity, and no modification of the O antigen occurs.

As a result, the A and B antigens differ in their terminal sugar molecules. Subgroups of blood types A and B have been described. The most important are the A_1 and A_2 antigens. Both A_1 and A_2 (and A_1B and A_2B) cells react with anti-A antibodies. However, A_1 cells react more strongly than A_2 cells. The apparent difference between A_1 and A_2 is that each A_1 cell contains more copies of the A antigen than A_2 cells.

18.1.2.2 Molecular Basis of the ABO System

A- and B-transferases are encoded by a single gene, *ABO*, on chromosome 9. The *ABO* gene (approximately 20 kb) is organized into seven exons. Most of its coding regions are located in exons 6 and 7 of the *ABO* locus, including the domain responsible for catalytic activity (Figure 18.3). The gene products of the *A* and *B* alleles differ by four amino acid substitutions (Table 18.2). In particular, amino acid residues at positions 266 and 268 are more important in determining the enzymatic property of a transferase.

The A^1 allele and A^2 allele differ in a single nucleotide deletion upstream from the translation stop codon. The resulting reading-frame shift in the A^2 allele abolishes the stop codon, yielding a product with an extra 21–amino acid residue at the C-terminus.

Subgroups of blood types O have also been reported. The sequence of the O^1 allele has a deletion of a single nucleotide at exon 6. This nucleotide deletion leads to a reading-frame shift generating a truncated protein, which lacks the catalytic domain. While the O^{1y} allele also has a single nucleotide deletion, it differs from O^1 by nine nucleotides within the coding sequence. O^1 and O^{1y} have identical phenotypes. There is also an O^2 allele, which is inactivated by a substitution

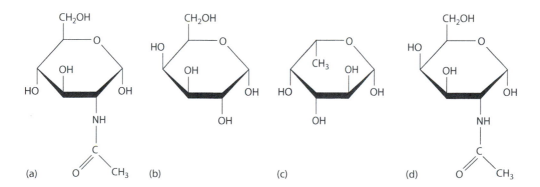

Figure 18.1 Chemical structures of (a) *N*-acetylglucosamine, (b) galactose, (c) fucose, and (d) *N*-acetylgalactosamine.

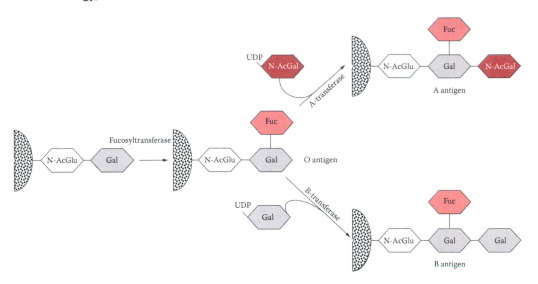

Figure 18.2 Biosynthesis of ABO antigens. O-antigen biosynthesis is catalyzed by fucosyltransferase. A-antigen biosynthesis is catalyzed by the A-transferase that transfers the N-acetylgalactosamine from the donor and uridine diphosphate (UDP)-N-acetylgalactosamine to the O antigen. B-antigen biosynthesis is catalyzed by the B-transferase that transfers the galactose from UDP-galactose to the O antigen. N-AcGlu, *N*-acetytglucosamine; Gal, galactose; Fuc, fucose; N-AcGal, *N*-acetylgalactosamine. (© Richard C. Li.)

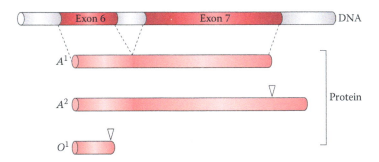

Figure 18.3 Structure of ABO gene and variants. Exons 6 and 7 are shown. The deletion mutation in A^2 and O^1 variants is indicated by an inverted triangle and leads to their A_2 and O phenotypes, respectively. (© Richard C. Li.)

Table 18.2 Amino Acid Substitutions at Four Positions in Human ABO Variants A^1, *B*, and O^2				
	Amino Acid Position			
ABO Variant	**176**	**235**	**266**	**268**
A^1	Arginine	Glycine	Leucine	Glycine
B	Glycine	Serine	Methionine	Alanine
O^2	Glycine	Glycine	Leucine	Arginine

mutation at glycine (position 268) by arginine. Additionally, a few dozen other rare O alleles, which yield inactive proteins, have also been documented.

18.1.2.3 Secretors

In addition to erythrocytes, individuals whose A, B, and O antigens can be found in other types of bodily fluids are referred to as *secretors*. Eighty percent of Caucasians are secretors. As described earlier, the O antigen is the substrate for the A- and B-transferase because the A- and B-transferase can only utilize a fucosylated substrate. The O antigen is synthesized by fucosylation of the terminal galactosyl residue catalyzed by the fucosyltransferase, which is encoded by *FUT* genes.

Chromosome 19 contains two homologous genes: *FUT1* and *FUT2*. *FUT1* is expressed in tissues of mesodermal origin (embryonic tissues that serve as precursors of hemopoietic tissues, muscle, the skeleton, and internal organs) and is responsible for the synthesis of the O antigen in erythrocytes. *FUT2* is expressed in tissues of endodermal origin (embryonic tissues that are precursors of the gut and other internal organs); it is responsible for the synthesis of the O antigen in secretions.

About 20% of Caucasian individuals (called *nonsecretors*) are homozygous for a nonsense mutation in *FUT2* at amino acid position 143, resulting in a truncated protein. Bodily fluids such as the semen of type A or B nonsecretors (who carry homozygous *FUT2* mutations) contain no A or B antigens despite containing active A- or B-transferases (Figure 18.4). This can be a problem in investigating sexual assault cases when the blood type of the seminal evidence needs to be determined. However, nonsecretors have O antigens on erythrocytes synthesized by *FUT1* and thus have A or B antigens in blood. Individuals carrying very rare homozygous *FUT1* mutations produce erythrocyte O-deficient phenotypes in which the erythrocytes express no O antigens and thus express neither A nor B antigens, regardless of *ABO* genotype. Individuals

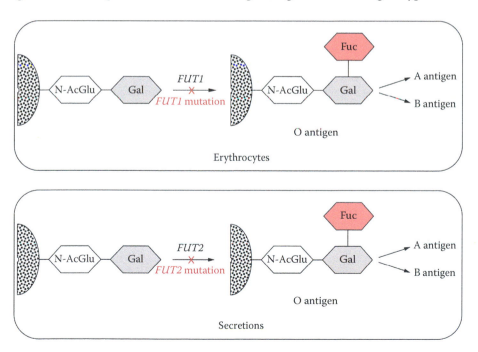

Figure 18.4 Tissue-specific O-antigen biosynthesis by *FUT1* and *FUT2* gene products. Tissue-specific O-antigen biosynthesis in erythrocytes is catalyzed by the *FUT1* gene product; in secretions, it is catalyzed by the *FUT2* gene product. The mutations abolishing the biosynthesis of O antigens are indicated. The *FUT2* mutation produces a nonsecretor phenotype. (© Richard C. Li.)

who carry both *FUT1* and *FUT2* mutations have no O antigens (nor A nor B antigens) in their erythrocytes and other bodily fluids and are known as Bombay (O_h) phenotypes.

18.1.2.4 Inheritance of A and B Antigens

A and *B* alleles are dominant. For *AO* and *BO* heterozygotes, the corresponding transferase synthesizes the A or B antigen. *A* and *B* alleles are codominant in *AB* heterozygotes because both transferase activities are expressed. The *OO* homozygote produces neither transferase activity and therefore lacks both antigens. The inheritance of *A* and *B* alleles obeys Mendelian principles (Chapter 25). For example, an individual with type B blood may have inherited a *B* allele from each parent or a *B* allele from one parent and an *O* allele from the other; thus, an individual whose phenotype is B may have the *BB* (homozygous) or *BO* (heterozygous) genotype. Conversely, if the blood types of the parents are known, the possible genotypes of their children can be determined. When both parents are type B (heterozygous), they may produce children with the genotype *BB* (B antigens from both parents), *BO* (B antigen from one parent, O from the other heterozygous parent), or *OO* (O antigens from parents who are both heterozygous). Thus, blood group typing can be used for paternity testing.

18.1.3 Forensic Applications of Blood Group Typing

The application and usefulness of blood typing in forensic identification are based on the ability to group individuals into four different types using the ABO blood system, allowing individuals to be identified. For example, if one crime scene blood sample is type B and a suspect has type A, the crime scene sample must have a different origin. However, if both the sample and the suspect are type A, the sample may have come from the same origin or from a different origin that happened to be type A.

Unfortunately, the probability that any two randomly chosen individuals have an identical blood type is very high. Approximately 42% of Caucasians have type A blood. The frequency of other blood types within the ABO system is shown in Figure 18.5. Multiple blood group systems were utilized to decrease the probability of a coincident match.

The A and B antigens are very stable and can be identified in dried blood even after many years. They can also be found in semen and other bodily fluids of secretors. Thus, in sexual assault cases, for example, the ABO type of a semen sample can be examined to identify a perpetrator.

18.1.4 Blood Group Typing Techniques

The most common assays used in forensic serology involve agglutination and include the Lattes crust and absorption–elution assays.

18.1.4.1 Lattes Crust Assay

In the early 1900s, Karl Landsteiner used his blood and blood obtained from his laboratory coworkers to test the effects of serum on erythrocytes. He discovered that naturally occurring

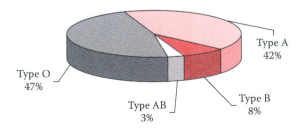

Figure 18.5 Frequency distributions of ABO types observed in American Caucasians. Different human populations may exhibit different frequencies of the four blood types. (© Richard C. Li.)

antibodies in serum caused agglutination of certain erythrocytes, and the agglutination patterns observed were designated A, B, and O. Each pattern indicated the presence or absence of a particular antigen on erythrocytes.

Shortly after birth, newborn infants develop antibodies against antigens that are not present in their own bodies. For example, type A individuals develop anti-B antibodies, type B individuals develop anti-A antibodies, type O individuals develop both types of antibodies, and type AB individuals do not develop anti-A or anti-B antibodies. When the plasma of a type A individual is mixed with type B cells, the anti-B antibodies from the type A individual cause the type B cells to agglutinate. This result forms the basis for blood group typing.

The Lattes crust assay relies on the principles of Landsteiner's experiments. It is an agglutination-based assay that utilizes the A, B, and O indicator cells to test the agglutination reaction with its corresponding naturally occurring serum antibodies in a questioned sample. The procedure for the Lattes crust assay is described in Box 18.1 and illustrated in Figure 18.6. Typical results are summarized in Table 18.3 and illustrated in Figure 18.7. Type A blood contains naturally occurring anti-B antibodies that agglutinate only with B cells. Likewise, type B blood agglutinates only with A cells, type O blood agglutinates with both A and B cells, and type AB blood does not agglutinate with any cells.

The Lattes crust assay is simple and rapid. However, one limitation is that the assay is not very sensitive and requires a large quantity of blood. Recall that successful agglutination reactions usually require intact cells. The agglutination assay of forensic samples is, therefore, difficult to carry out because blood cells lyse when they are dry. Therefore, this method is not reliable for testing old stains.

BOX 18.1   LATTES CRUST ASSAY PROCEDURE

1. Place small quantities of blood crust from a specimen on a microscopic slide and place a cover slide over the crusts. Prepare slides for A, B, and O cells separately.
2. Prepare cell suspensions with saline (0.85% NaCl in phosphate buffer, pH 7.4) for the A, B, and O cells separately.
3. Apply a few drops of the A-cell suspension and allow the cells to diffuse under the cover slip. Repeat this step for B cells and O cells.
4. Incubate the slides in a moisture chamber at room temperature for 2 h.
5. Examine results under a microscope.

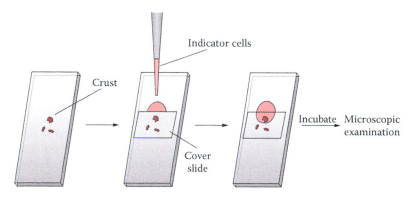

Figure 18.6   Lattes crust assay. (© Richard C. Li.)

Table 18.3 Representative Results of Lattes Crust Assay		
Blood Type	Serum Antibody	Agglutination Reaction Observed
A	Anti-B	B cells
B	Anti-A	A cells
O	Anti-A, anti-B	A cells, B cells
AB	None	None

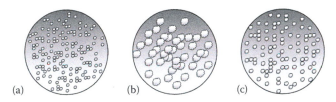

(a) (b) (c)

Figure 18.7 Diagram of Lattes crust assay results. (a) Indicator cells added before incubation. (b) Strong agglutination: large clumps are observed after incubation. (c) Negative agglutination: a cloudy background may be observed after incubation.

18.1.4.2 Absorption–Elution Assay

The absorption–elution assay is highly sensitive and can be used for testing dried bloodstains. This method indirectly detects the presence of antigens. The antigens are immobilized in a solid phase (Figure 18.8). At low temperatures, the antigens bind to their corresponding antibodies: anti-A antibodies, anti-B antibodies, or anti-O lectins. (The anti-O lectin is isolated from plants and reacts strongly with the O antigen present in type O blood, but has some cross-reaction with the A antigen). The excess unbound antibodies are removed by washing, and the bound antibodies are then eluted at higher temperatures (recall that antigen–antibody binding can be affected by temperature; Chapter 13). The eluted antibodies can then be identified by an agglutination assay using A, B, and O indicator cells.

Typical results of an absorption–elution assay are summarized in Table 18.4. The bloodstains containing the A antigen can bind to anti-A antibodies. The eluted anti-A antibody can form agglutination with A cells. Likewise, for type B blood, the eluted anti-B antibody can form agglutination with B cells; for type AB blood, the eluted antibodies can form agglutination with both A and B cells; and with type O blood, the eluted anti-O lectins can form agglutination with O cells.

18.2 Forensic Protein Profiling

Because of the limitations of blood group systems, inherited protein polymorphic markers have been utilized to decrease the chances of matches between two unrelated individuals. The amino acid sequences of many proteins vary in the human population. An estimated 20%–30% of the proteins in humans are polymorphic. Some of the variations in amino acid sequences affect the function of proteins, but many of them exert little or no effect on protein function. Thus, individuals can be divided into groups based on the types of protein polymorphisms. A combination of the blood group systems and protein polymorphic markers can be used for

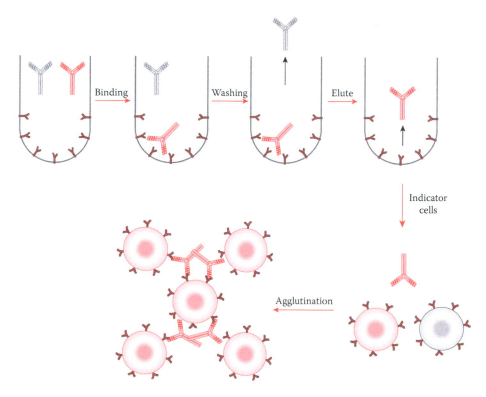

Figure 18.8 Absorption–elution assay. An antigen sample is immobilized on a solid-phase matrix. Antibodies (two different types are shown) are added. The antibody that is specific to the antigen binds. Unbound antibody is washed away. The bound antibody is then eluted. The eluted antibody is tested with indicator cells (two different types are shown). The indicator cell that is specific to the antibody eluted shows a positive agglutination reaction. (© Richard C. Li.)

Table 18.4 Representative Results of Absorption–Elution Assay		
Blood Type (Stain)	**Antibody Bound and Eluted**	**Agglutination Reaction Observed**
A	Anti-A	A cells
B	Anti-B	B cells
O	Anti-O	O cells
AB	Anti-A, anti-B	A cells, B cells

criminal investigations and paternity testing. The probability that results for two unrelated persons would match is decreased to one in several hundred through use of the blood-typing and protein-profiling techniques.

18.2.1 Methods

Identification of protein polymorphisms is performed through electrophoretic separation based on the molecular weights (Mr) and charges of the protein variants.

18.2.1.1 Matrices Supporting Protein Electrophoresis

Electrophoresis of proteins is generally carried out in a support material, also called the matrix, to separate various macromolecules. The matrix also reduces the effects of diffusion and convection on the macromolecules. Historically, protein profiling for forensic application utilizes two types of matrices: papers such as cellulose acetate; and gels composed of starch, agar, agarose, or polyacrylamide. The first polymorphic protein marker, phosphoglucomutase, was characterized by starch-gel electrophoresis. However, agarose and polyacrylamide became more commonly used in electrophoresis due to good reproducibility and reliability (Table 18.5).

18.2.1.2 Separation by Molecular Weight

An electrophoretic method is frequently utilized to resolve various proteins based on their molecular weights. Native electrophoresis, also known as nondenaturing electrophoresis, can be used to isolate proteins for studying the functions of proteins. Biological activity of the protein can be retained for further analysis. However, some proteins are not well separated in electrophoresis in their native form. Thus, it may be necessary to denature the proteins in order for them to be resolved better during separation. This process is called *denaturing protein electrophoresis*. The following additives can be used:

18.2.1.2.1 Reducing Agents

It is common to include reducing agents such as mercaptoethanol (ME), dithiothreitol (DTT), or sodium mercaptoethane sulfonate (MESNA) to denature proteins. Reducing agents cleave the disulfide bonds of proteins. As a result, protein shape becomes unfolded and linear. These agents can be used during sample preparation and can also be added to the electrophoresis buffer.

18.2.1.2.2 Detergents

Detergents disrupt noncovalent interactions within the structures of native proteins. The procedure is generally performed with sodium dodecylsulfate (SDS), a strong anion detergent that binds to most proteins in amounts proportional to the molecular weight of the protein (approximately one molecule of SDS for two amino acids). The bound SDS contributes a large net negative charge on the protein, which masks any surface charges of the native protein. As a result, the charge-to-mass ratio of the protein becomes a constant. As with reducing agents, the various native conformations of proteins change to a more uniformly linear shape when SDS is bound.

Table 18.5 Properties of Matrices Supporting Protein Electrophoresis

Supporting Matrix	Pore Size	EEO	Reproducibility	Strength	Preparation	Toxicity
Cellulose acetate	Large	High	Poor	Good	Simple	Nontoxic
Starch	Large	High	Poor	Fragile	Simple	Nontoxic
Agar	Large	High	Poor	Fragile	Simple	Nontoxic
Agarose	Large	Low	Good	Fragile	Simple	Nontoxic
Polyacrylamide	Small	Very Low	Good	High; tolerates high electronic field	Complex	Toxic

Note: Electroendosmosis (EEO) occurs when fixed charges of the supporting matrix cause liquid flow toward the electrodes. A matrix with high EEO may affect the mobilities and separation performances of proteins during electrophoresis.

Electrophoretic mobility in the presence of SDS, therefore, becomes based on Mr rather than both Mr and the charge. Smaller proteins move through the pores of the gel matrix more rapidly than larger proteins. As a result, the larger the size of the protein, the smaller its electrophoretic mobility.

SDS gel electrophoresis can also be used to determine the Mr of an unidentified protein based on its electrophoretic mobility on the gel. Standard marker proteins of known molecular weight are run on the same gel and allow the estimation of the Mr of an unknown protein. A linear plot of log Mr values of marker proteins versus relative migration during electrophoresis allows the molecular weight of the unknown protein to be determined from the graph.

18.2.1.3 Separation by Isoelectric Point

The *isoelectric focusing (IEF)* technique can be used to separate proteins according to their *iso-electric points (pI)*. The pI is the pH value at which the net electric charge of an amino acid is zero. All proteins are composed of amino acids, and each has its own characteristic pI at which its net electric charge is zero and does not migrate in an electric field.

In IEF electrophoresis, a pH gradient is created in a gel between the electrodes, and a protein sample is placed in a well on the gel. With an applied electric field, proteins enter the gel and migrate until they reach a pH equivalent to their pI values, at which they lose mobility (Figure 18.9). IEF, based on molecular charge, is capable of producing sharper bands than denaturing protein electrophoresis and thus has a higher resolving power. The technique can detect very low quantities of proteins in samples. A pH gradient in the gel is established by utilizing materials such as carrier ampholytes or immobilines that are dispersed in the gel. Carrier ampholytes are synthetic amphoteric compounds that contain multiple weak ionizable moieties acting as either acids or bases. To establish a pH gradient, a mixture of ampholytes with slightly different pIs is directly added to an IEF gel. The pH gradient is generated by applying an electric field on the ampholyte-containing gel. Under the electric field, the negatively charged ampholytes migrate toward the anode, and the positively charged ampholytes migrate to the cathode. As a result, a gradual pH gradient is created between the anodal end of the gel (acidic) and the cathodal side of the gel (basic). Immobilines are a series of modified acrylamide monomers that can be acidic or basic. The pH gradient of an IEF gel can also be established using a gradient-forming device that changes the proportion of the immobilines added to the gel matrix mixture as it is loaded into the gel-casting apparatus.

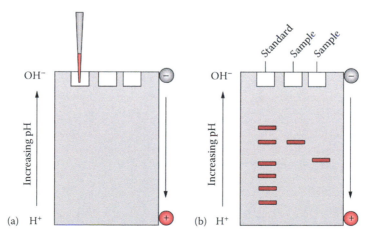

Figure 18.9 Isoelectric focusing. (a) A pH gradient is established by allowing low-molecular-weight organic acids and bases to distribute themselves in an electric field across a gel. A sample containing a protein mixture is loaded into a sample well. (b) During the electrophoresis, each protein migrates until it matches its pI. Proteins with different pIs are separated. (© Richard C. Li.)

18.2.2 Erythrocyte Protein Polymorphisms

18.2.2.1 Erythrocyte Isoenzymes

The human erythrocyte contains a number of *isoenzymes*, which are multiple forms of an enzyme that catalyze the same reaction but differ in their amino acid sequences. Individuals can be divided into groups on the basis of the different isoenzymes present in their erythrocytes. The isoenzyme type is also inherited according to Mendelian principles.

The polymorphism of erythrocyte phosphoglucomutase (PGM) was first described in the 1960s and was later successfully applied to the testing of bloodstains. PGM, an important metabolic enzyme, catalyzes the reversible conversion of glucose-1-phosphate and glucose-6-phosphate. The PGM found in erythrocytes is encoded at the *PGM1* locus at chromosome 1. The PGM encoded by *PGM1* can also be found in semen and thus can be utilized for the testing of semen samples in sexual assault cases. The protein polymorphisms of the PGM have two alleles, which result in three different phenotypes, depending on the combination of the two alleles. The success in the forensic application of PGM led to the similar use of many other erythrocyte isoenzyme polymorphisms. The most commonly used erythrocyte isoenzyme systems are listed in Table 18.6.

18.2.2.2 Hemoglobin

Recall that the use of hemoglobin (Hb) in screening and confirmatory blood tests was discussed in Chapter 12. Adult human Hb consists of two α chains and two β chains. Each polypeptide chain contains a heme group involved in oxygen binding. A very small portion of blood possesses a form of the human adult Hb consisting of two α chains and two δ chains.

More than 200 Hb variants have been identified and can be useful as markers for forensic applications. In particular, two types of human Hb variants are important in forensic testing: fetal Hb and sickle-cell Hb (Hb S). Hb S is the factor responsible for sickle-cell disease (Figure 18.10). Hb variants can be resolved using electrophoresis (Figure 18.11).

18.2.2.2.1 Fetal Hemoglobin

Humans have three forms of Hb during their development: embryonic, fetal, and adult Hb. In adults, the Hb tetramer consists of two identical α and two identical β chains. Embryonic erythrocytes contain Hb tetramers that are different from the adult form. Each embryonic Hb consists of two identical α-like chains and two identical β-like chains. The embryonic Hb is gradually replaced during pregnancy (approximately 3 months after conception) by fetal Hb, which comprises approximately 70% of the Hb in fetal blood. The fetal Hb has two identical α chains and two identical γ chains.

TABLE 18.6 Common Isoenzymes Used for Forensic Protein Profiling		
Erythrocyte Isoenzyme	**Protein Symbol**	**Number of Alleles**
Phosphoglucomutase	PGM	2[a]
Erythrocyte acid phosphatase	ACP/EAP	3
Esterase D	ESD	2
Adenylate kinase	AK	2
Glyoxalase I	GLO	2
Adenosine deaminase	ADA	2

[a] Ten alleles can be observed using IEF electrophoresis.

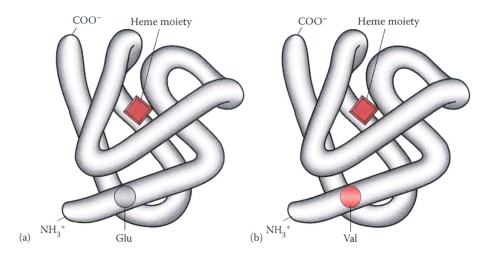

Figure 18.10   Normal and sickle-cell hemoglobin β chains. (a) Normal hemoglobin β chain contains a glutamic acid residue (Glu) at position 6 of the N-terminal of the protein. (b) At position 6 of the sickle-cell hemoglobin β chain, the glutamic acid residue is replaced by a valine (Val). (© Richard C. Li.)

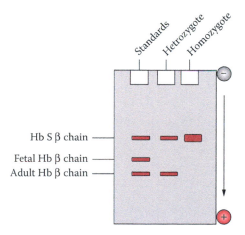

Figure 18.11   Hemoglobins resolved by isoelectric focusing electrophoresis. (© Richard C. Li.)

The embryonic and fetal Hbs have higher affinities to oxygen, required to provide the embryo and fetus with sufficient amounts of oxygen taken from maternal blood. Fetal Hb is replaced by adult Hb approximately 6 months after birth. These Hbs are encoded by their corresponding genes located at the globin gene clusters. The detection of fetal Hb in a blood-stain via electrophoresis can provide important evidence in cases of infanticide and concealed birth.

18.2.2.2.2 Hemoglobin S

Hb S polymorphism has forensic importance in identifying individuals. The Hb S polymorphism is observed in high frequencies among those of African heritage and some Hispanic populations. Such a protein polymorphic marker can provide investigational leads for the indication of the ethnic origin of a perpetrator. Hb S transports oxygen much less efficiently than normal Hb. Individuals who are homozygous for Hb S usually die early after suffering from sickle-cell anemia and related complications. However, a heterozygous individual (an individual with a copy of the wild-type Hb allele from one parent and a copy of the Hb S allele from the other) can survive. This condition is known as the sickle-cell trait.

In the 1950s, Vernon Ingram of Cambridge University discovered the molecular mechanism of the Hb S defect. His work revealed that the Hb S bears a mutation, which changes the glutamic acid in wild type to a valine at the sixth amino acid from the N-terminal end of the β chain. This substitution of amino acids causes a major change in the structure of the β chain, which in turn results in sickle-cell anemia.

18.2.3 Serum Protein Polymorphisms

The serum portion of blood consists of a large number of proteins. The work on serum proteins for forensic purposes started in the 1950s, when variations in serum proteins were found useful for distinguishing individuals. Over the years, a number of serum proteins were characterized and applied for forensic testing. Haptoglobin (Hp) was the most widely used of the polymorphic serum proteins in forensic biology (Figure 18.12). Haptoglobin is a protein that binds and transports Hb from the bloodstream to the liver for the recycling of the iron contained in the Hb.

Immunoglobulin (Chapter 10) accounts for approximately 15% of serum protein and has been found to be highly variable. Two immunoglobulin proteins are utilized for forensic application. The γ chain protein (G_m) is the heavy chain of immunoglobulin G and the κ chain protein (K_m) is one of two types of the light chain of all immunoglobulins. Table 18.7 lists common serum group systems. All exhibit genetic variations and can be detected in bloodstains. The variants of these proteins can be determined by electrophoresis or serological methods.

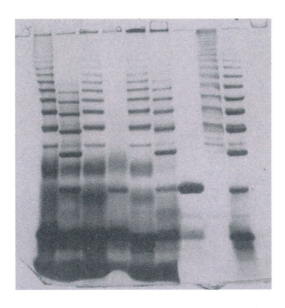

Figure 18.12 Polyacrylamide gel electrophoresis of haptoglobin proteins. From left to right: Hp2, Hp2-1, Hp2, Hp1, Hp2, Hp2-1, Hp1, Hp2, Hp2-1. An anode is at the bottom. (From James, S. and Nordby, J.J., *Forensic Science: An Introduction to Scientific and Investigative Techniques*, CRC Press, Boca Raton, FL, 2005. With permission.)

TABLE 18.7 Serum Proteins Used for Forensic Protein Profiling

Serum Protein	Protein Symbol	Gene Symbol	Chromosomal Location	Number of Amino Acids	Number of Alleles
Haptoglobin	Hp	*HP*	16q22.1	328/387	2
Group-specific component	Gc	*GC*	4q12–13	458	2
Transferrin	Tf	*TF*	3q22.1	679	3
Protease inhibitor (α1-antitrypsin)	Pi	*PI*	14q32.1	394	Many

Source: Adapted from Yuasa, I. and Umetsu, K., *Leg Med*, 7, 251–254, 2005.

Bibliography

Blood Groups

Achermann, F.J., et al., Soluble type A substance in fresh-frozen plasma as a function of ABO and secretor genotypes and Lewis phenotype. *Transfus Apher Sci*, 2005, **32**(3): 255–262.

Aki, K., A. Izumi, and E. Hosoi, The evaluation of histo-blood group ABO typing by flow cytometric and PCR-amplification of specific alleles analyses and their application in clinical laboratories. *J Med Invest*, 2012, **59**(1–2): 143–151.

Allen, F.H. Jr. and R.E. Rosenfield, Review of Rh serology. Eight new antigens in nine years. *Haematologia* (Budap), 1972, **6**(1): 113–120.

Anstee, D.J. and M.J. Tanner, Separation of ABH, I, Ss antigenic activity from the MN-active sialoglycoprotein of the human erythrocyte membrane. *Vox Sang*, 1975, **29**(5): 378–389.

Asari, M., et al., A new method for human ABO genotyping using a universal reporter primer system. *J Forensic Sci*, 2010, **55**(6): 1576–1581.

Bargagna, M. and M. Pereira, A study of absorption–elution as a method of identification of Rhesus antigens in dried bloodstains. *J Forensic Sci Soc*, 1967, **7**(3): 123–130.

Bargagna, M., M. Sabelli, and C. Giacomelli, The detection of Rh antigens (D,C,c,E,e) on bloodstains by a micro-elution technique using low ionic strength solution (LISS) and papain-treated red cells. *Forensic Sci Int*, 1982, **19**(2): 197–203.

Bassler, G., Determination of the Lewis blood group substances in stains of forensically relevant body fluids. *Forensic Sci Int*, 1986, **30**(1): 29–35.

Boorman, K.E., B.E. Dodd, and P.J. Lincoln, *Blood Group Serology*, 5th edn., vol. 4, p. 495, 1977. London: Churchill Livingstone.

Brauner, P., DNA typing and blood transfusion. *J Forensic Sci*, 1996, **41**(5): 895–897.

Brewer, C.A., P.L. Cropp, and L.E. Sharman, A low ionic strength, hemagglutinating, autoanalyzer for rhesus typing of dried bloodstains. *J Forensic Sci*, 1976, **21**(4): 811–815.

Busuttil, A., et al., Assessment of Lewis blood group antigens and secretor status in autopsy specimens. *Forensic Sci Int*, 1993, **61**(2–3): 133–140.

Camp, F.R. Jr., Forensic serology in the United States. I. Blood grouping and blood transfusion—Historical aspects. *Am J Forensic Med Pathol*, 1980, **1**(1): 47–55.

Cartron, J.P., et al., "Weak A" phenotypes. Relationship between red cell agglutinability and antigen site density. *Immunology*, 1974, **27**(4): 723–727.

Cayzer, I. and P.H. Whitehead, The use of sensitized latex in the identification of human bloodstains. *J Forensic Sci Soc*, 1973, **13**(3): 179–181.

Chatterji, P.K., A simplified mixed agglutination technique for ABO grouping of dried bloodstains using cellulose acetate sheets. *J Forensic Sci Soc*, 1978, **17**(2–3): 143–144.

Chester, M.A. and M.L. Olsson, The ABO blood group gene: A locus of considerable genetic diversity. *Transfus Med Rev*, 2001, **15**(3): 177–200.

Daniels, G., *Human Blood Groups*, 2nd edn., p. 576, 2002. Oxford: Blackwell Science.

Daniels, G.L., et al., Blood group terminology 2004: From the International Society of Blood Transfusion committee on terminology for red cell surface antigens. *Vox Sang*, 2004, **87**(4): 304–316.

DeForest, P., R. Gaensslen, and H.C. Lee, Blood, in *Forensic Science: An Introduction to Criminalistics*, pp. 230–263, 1983. New York: McGraw-Hill.

De Soyza, K. and D.G. Garland, Studies and observations on Lewis grouping of body fluids and stains. *Forensic Sci Int*, 1988, **38**(1–2): 129–137.

Denomme, G.A., The structure and function of the molecules that carry human red blood cell and platelet antigens. *Transfus Med Rev*, 2004, **18**(3): 203–231.

Erskine, A.G., *The Principles and Practices of Blood Grouping*, 1973. St. Louis, MO: Mosby.

Fayrouz, I.N., N. Farida, and A.H. Irshad, Relation between fingerprints and different blood groups. *J Forensic Leg Med*, 2012, **19**(1): 18–21.

Ferri, G. and S. Pelotti, Multiplex ABO genotyping by minisequencing. *Methods Mol Biol*, 2009, **496**: 51–58.

Fiori, A. and P. Benciolini, The ABO grouping of stains from body fluids. *Z Rechtsmed*, 1972, **70**(4): 214–222.

Fiori, A., M. Marigo, and P. Benciolini, Modified absorption–elution method Siracusa for ABO and MN grouping of blood-stains. *J Forensic Sci*, 1963, **8**(3): 419–445 contd.

Gaensslen, R.E., et al., Evaluation of antisera for bloodstain grouping. II. Ss, Kell, Duffy, Kidd, and G_m/K_m. *J Forensic Sci*, 1985, **30**(3): 655–676.

Gardas, A. and J. Koscielak, Megaloglycolipids—Unusually complex glycosphingolipids of human erythrocyte membrane with A, B, H and I blood group specificity. *Fed Eur Biochem Soc Lett*, 1974, **42**(1): 101–104.

Grunbaum, B.W., Some new approaches to the individualization of fresh and dried bloodstains. *J Forensic Sci*, 1976, **21**(3): 488–509.

Haak, W., J. Burger, and K.W. Alt, ABO genotyping by PCR-RFLP and cloning and sequencing. *Anthropol Anz*, 2004, **62**(4): 397–410.

Hakomori, S. and A. Kabata, Blood group antigens, in *The Antigens*, 1974. New York: Academic Press.

Hamaguchi, H. and H. Cleve, Solubilization of human erythrocyte membrane glycoproteins and separation of the MN glycoprotein from a glycoprotein with I, S, and A activity. *Biochim Biophys Acta*, 1972, **278**(2): 271–280.

Harrington, J.J., R. Martin, and L. Kobilinsky, Detection of hemagglutinins in dried saliva stains and their potential use in blood typing. *J Forensic Sci*, 1988, **33**(3): 628–637.

Harrington, J.J., et al., Chemically sensitized erythrocytes for hemagglutination reactions. *J Forensic Sci*, 1990, **35**(5): 1115–1124.

Henry, S., et al., Molecular basis for erythrocyte Le(a+ b+) and salivary ABH partial-secretor phenotypes: Expression of a *FUT2* secretor allele with an A→T mutation at nucleotide 385 correlates with reduced alpha (1,2) fucosyltransferase activity. *Glycoconj J*, 1996, **13**(6): 985–993.

Hughes-Jones, N.C. and B. Gardner, The Kell system studied with radioactively-labelled anti-K. *Vox Sang*, 1971, **21**(2): 154–158.

Hughes-Jones, N.C., B. Gardner, and P.J. Lincoln, Observations of the number of available c,D, and E antigen sites on red cells. *Vox Sang*, 1971, **21**(3): 210–216.

Huh, J.Y., et al., A rapid long PCR-direct sequencing analysis for ABO genotyping. *Ann Clin Lab Sci*, 2011, **41**(4): 340–345.

Jaff, M.S., Higher frequency of secretor phenotype in O blood group—Its benefits in prevention and/or treatment of some diseases. *Int J Nanomed*, 2010, **5**: 901–905.

Jiang, X., et al., An integrated system of ABO typing and multiplex STR testing for forensic DNA analysis. *Forensic Sci Int Genet*, 2012, **6**(6): 785–797.

Kabat, E.A., *The Blood Group Substances*, p. 330, 1956. New York: Academic Press.

Kaneko, M., et al., Molecular characterization of a human monoclonal antibody to B antigen in ABO blood type. *Immunol Lett*, 2003, **86**(1): 45–51.

Karpoor, C., et al., Study of secretors and non-secretors in normal healthy population—Its forensic implication in human identification. *Indian J Forensic Med Tox*, 2010, **4**(1): 11–13.

Kelly, R.J., et al., Molecular basis for H blood group deficiency in Bombay (Oh) and para-Bombay individuals. *Proc Natl Acad Sci U S A*, 1994, **91**(13): 5843–5847.

Kelly, R.J., et al., Sequence and expression of a candidate for the human secretor blood group alpha (1,2) fucosyltransferase gene (*FUT2*). Homozygosity for an enzyme-inactivating nonsense mutation commonly correlates with the non-secretor phenotype. *J Biol Chem*, 1995, **270**(9): 4640–4649.

Kimura, H. and S. Matsuzawa, Lewis blood group determination in bloodstains by planimetric measurement of eluted monoclonal antibodies. *J Forensic Sci*, 1991, **36**(4): 999–1009.

Kimura, A., et al., ABO blood grouping of semen from mixed body fluids with monoclonal antibody to tissue-specific epitopes on seminal ABO blood group substance. *Int J Legal Med*, 1991, **104**(5): 255–258.

Kimura, A., et al., Blood group A glycosphingolipid accumulation in the hair of patients with alpha-*N*-acetylgalactosaminidase deficiency. *Life Sci*, 2005, **76**(16): 1817–1824.

Kind, S.S., Absorption–elution grouping of dried blood smears. *Nature*, 1960, **185**: 397–398.

Kind, S.S., Absorption–elution grouping of dried blood-stains on fabrics. *Nature*, 1960, **187**: 789–790.

Kind, S.S. and R.M. Cleevely, The fluorescent antibody technique. Its application to the detection of blood group antigens in stains. *J Forensic Med*, 1970, **17**(3): 121–129.

Kobilinsky, L. and J.J. Harrington, Detection and use of salivary hemagglutinins for forensic blood grouping. *J Forensic Sci*, 1988, **33**(2): 396–403.

Koda, Y., M. Soejima, and H. Kimura, The polymorphisms of fucosyltransferases. *Leg Med (Tokyo)*, 2001, **3**(1): 2–14.

Koda, Y., et al., Missense mutation of *FUT1* and deletion of *FUT2* are responsible for Indian Bombay phenotype of ABO blood group system. *Biochem Biophys Res Commun*, 1997, **238**(1): 21–25.

Korchagina, E.Y., et al., Design of the blood group AB glycotope. *Glycoconj J*, 2005, **22**(3): 127–133.

Ladd, C., et al., A PCR-based strategy for ABO genotype determination. *J Forensic Sci*, 1996, **41**(1): 134–137.

Lappas, N.T. and M.E. Fredenburg, The identification of human bloodstains by means of a micro-thin-layer immunoassay procedure. *J Forensic Sci*, 1981, **26**(3): 564–569.

Lattes, L., *The Individuality of Blood*, p. 413, 1932. London: Oxford University Press.

Lee, H.C., Identification and grouping of bloodstains, in R. Saferstein (ed.), *Forensic Science Handbook*, 1982. Englewood Cliffs, NJ: Prentice Hall.

Lee, H.C., et al., Genetic markers in human bone: II. Studies on ABO (and IGH) grouping. *J Forensic Sci*, 1991, **36**(3): 639–655.

Lee, H.Y., et al., Rapid direct PCR for ABO blood typing. *J Forensic Sci*, 2011, **56**(Suppl 1): S179–S182.

Lee, J.C., et al., ABO genotyping by mutagenically separated polymerase chain reaction. *Forensic Sci Int*, 1996, **82**(3): 227–232.

Lee, J.C., et al., ABO genotyping by single strand conformation polymorphism—Using CE. *Electrophoresis*, 2009, **30**(14): 2544–2548.

Lee, J.C., et al., ABO genotyping by capillary electrophoresis. *Methods Mol Biol*, 2013, **919**: 113–120.

Lee, S.H., et al., Rapid ABO genotyping using whole blood without DNA purification. *Korean J Lab Med*, 2009, **29**(3): 231–237.

Lincoln, P.J. and B.E. Dodd, The application of a micro-elution technique using anti-human globulin for the detection of the S, s, K, Fya, Fyb and Jka antigens in stains. *Med Sci Law*, 1975, **15**(2): 94–101.

Liu, Y.H., et al., Distribution of H type 1 and of H type 2 antigens of ABO blood group in different cells of human submandibular gland. *J Histochem Cytochem*, 1998, **46**(1): 69–76.

Maeda, K., et al., ABO genotyping by TaqMan assay and allele frequencies in a Japanese population. *Leg Med (Tokyo)*, 2013, **15**(2): 57–60.

Martin, P.D., A manual method for the detection of Rh antigens in dried bloodstains. *J Forensic Sci Soc*, 1978, **17**(2–3): 139–142.

McDowall, M.J., P.J. Lincoln, and B.E. Dodd, Increased sensitivity of tests for the detection of blood group antigens in stains using a low ionic strength medium. *Med Sci Law*, 1978, **18**(1): 16–23.

McDowall, M.J., P.J. Lincoln, and B.E. Dodd, Observations on the use of an autoanalyser and a manual technique for the detection of the red cell antigens C, D, E, c, K and S in bloodstains. *Forensic Sci*, 1978, **11**(2): 155–164.

Miller, C.H., et al., Measurement of von Willebrand factor activity: Relative effects of ABO blood type and race. *J Thromb Haemost*, 2003, **1**(10): 2191–2197.

Mollicone, R., A. Cailleau, and R. Oriol, Molecular genetics of H, Se, Lewis and other fucosyltransferase genes. *Transfus Clin Biol*, 1995, **2**(4): 235–242.

Moureau, P., Determination of blood groups in bloodstains, in F. Lundquist (ed.), *Methods of Forensic Science*, p. 137, 1963. New York: Interscience.

Muro, T., et al., Determination of ABO genotypes by real-time PCR using allele-specific primers. *Leg Med (Tokyo)*, 2012, **14**(1): 47–50.

Nakanishi, H., et al., Preparation of latex reagents combined with IgM and its F(ab')2 fragment from commercial ABO blood grouping reagent. *Colloid Surf B Biointerf*, 2007, **54**(1): 114–117.

Nickolls, L.C. and M. Pereira, A study of modern method of grouping dried bloodstains. *Med Sci Law*, 1962, **2**: 172.

Ohmori, T., et al., Monoclonal antibodies against blood group A secretors and nonsecretors saliva. *Hybrid Hybridomics*, 2003, **22**(3): 183–186.

Okiura, T., et al., A-elute alleles of the ABO blood group system in Japanese. *Leg Med (Tokyo)*, 2003, **5**(Suppl 1): S207–S209.

Oriol, R., ABO, Hh, Lewis, and secretion: Serology, genetics, and tissue distribution, in J.-P. Cartron and P. Rouger (eds), *Blood Cell Chemistry*, pp. 36–73, 1995. New York: Plenum Press.

Oriol, R., J.J. Candelier, and R. Mollicone, Molecular genetics of H. *Vox Sang*, 2000, **78**(Suppl 2): 105–108.

Outteridge, R.A., Absorption–elution grouping of bloodstains: Modification and development. *Nature*, 1962, **194**: 385.

Outteridge, R.A., Absorption–elution method of grouping blood-stains. *Nature*, 1962, **195**: 818–819.

Outteridge, R.A., Recent advances in the grouping of dried blood and secretion stains. A.S. Curry (ed.), *Methods of Forensic Science*, 1965. New York: Interscience.

Painter, T.J., W.M. Watkins, and W.T. Morgan, Serologically active fucose-containing oligosaccharides isolated from human blood-group A and B substances. *Nature*, 1965, **206**(984): 594–597.

Prokop, O. and G. Uhlenbruck, *Human Blood and Serum Groups*, 2nd edn., p. 891, 1969. New York: Wiley Interscience.

Reid, M.E. and C. Lomas-Francis, *The Blood Group Antigen Facts Book*, 2nd edn., 2004. London: Academic Press.

Reid, M.E. and N. Mohandas, Red blood cell blood group antigens: Structure and function. *Semin Hematol*, 2004, **41**(2): 93–117.

Rosenfield, R.E., F.H. Allen Jr., and P. Rubinstein, Genetic model for the Rh blood-group system. *Proc Natl Acad Sci U S A*, 1973, **70**(5): 1303–1307.

Ruan, L., H. Zhao, and Q. Li, Multicolor real-time PCR genotyping of ABO system using displacing probes. *J Forensic Sci*, 2010, **55**(1): 19–24.

Sasaki, M. and H. Shiono, ABO genotyping of suspects from sperm DNA isolated from postcoital samples in sex crimes. *J Forensic Sci*, 1996, **41**(2): 275–278.

Schleyer, F., Investigation of biological stains with regard to species origin, in F. Lundquist (ed.), *Methods of Forensic Science*, p. 291, 1962. New York: Interscience.

Shaler, R.C., A.M. Hagins, and C.E. Mortimer, MN determination in bloodstains–selective destruction of cross-reacting activity. *J Forensic Sci*, 1978, **23**(3): 570–576.

Shintani-Ishida, K., et al., A new method for ABO genotyping to avoid discrepancy between genetic and serological determinations. *Int J Legal Med*, 2008, **122**(1): 7–9.

Smeets, B., H. van de Voorde, and P. Hooft, ABO bloodgrouping on tooth material. *Forensic Sci Int*, 1991, **50**(2): 277–284.

Snyder, L.H., *Blood Groups*, p. 35, 1973. Minneapolis: Burgess.

Springer, G.F. and P.R. Desai, Human blood-group MN and precursor specificities: Structural and biological aspects. *Carbohydr Res*, 1975, **40**(1): 183–192.

Storry, J.R. and M.L. Olsson, Genetic basis of blood group diversity. *Br J Haematol*, 2004, **126**(6): 759–771.

Styles, W.M., B.E. Dodd, and R.R. Coombs, Identification of human bloodstains by means of the mixed antiglobulin reaction on separate cloth fibrils. *Med Sci Law*, 1963, **20**: 257–267.

Tsuji, A., et al., A familial case of ABO phenotype-genotype discrepancy. *Fukuoka Igaku Zasshi*, 2013, **104**(2): 40–45.

Watanabe, K., et al., A novel method for ABO genotyping using a DNA chip. *J Forensic Sci*, 2011, **56**(Suppl 1): S183–S187.

Watkins, W.M., Blood-group substances. *Science*, 1966, **152**(3719): 172–181.

Watkins, W.M., Commemoration of the centenary of the discovery of the ABO blood group system. *Transfu Med*, 2001, **11**: 239–351.

Wynbrandt, F. and W.J. Chisum, Determination of the ABO blood group in hair. *J Forensic Sci Soc*, 1971, **11**(3): 201–204.

Xingzhi, X., et al., ABO blood grouping on dental tissue. *J Forensic Sci*, 1993, **38**(4): 956–960.

Yamada, M., et al., Determination of ABO genotypes with DNA extracted from formalin-fixed, paraffin-embedded tissues. *Int J Legal Med*, 1994, **106**(6): 285–287.

Yamamoto, F., et al., Molecular genetic basis of the histo-blood group ABO system. *Nature*, 1990, **345**(6272): 229–233.

Yu, Q., et al., Congenital tetragametic blood chimerism explains a case of questionable paternity. *J Forensic Sci*, 2011, **56**(5): 1346–1348.

Protein Profiling

Allen, R.C., R.A. Harley, and R.C. Talamo, A new method for determination of alpha-1-antitrypsin phenotypes using isoelectric focusing on polyacrylamide gel slabs. *Am J Clin Pathol*, 1974, **62**(6): 732–739.

Bark, J.E., M.J. Harris, and M. Firth, Typing of the common phosphoglucomutase variants using isoelectric focusing—A new interpretation of the phosphoglucomutase system. *J Forensic Sci Soc*, 1976, **16**(2): 115–120.

Blake, N.M. and K. Omoto, Phosphoglucomutase types in the Asian-Pacific area: A critical review including new phenotypes. *Ann Hum Genet*, 1975, **38**(3): 251–273.

Budowle, B., A method for subtyping group-specific component in bloodstains. *Forensic Sci Int*, 1987, **33**(3): 187–196.

Budowle, B., A method to increase the volume of sample applied to isoelectric focusing gels. *Forensic Sci Int*, 1984, **24**(4): 273–277.

Budowle, B. and E. Scott, Transferrin subtyping of human bloodstains. *Forensic Sci Int*, 1985, **28**(3–4): 269–275.

Budowle, B. and P. Eberhardt, Ultrathin-layer polyacrylamide gel isoelectric focusing for the identification of hemoglobin variants. *Hemoglobin*, 1986, **10**(2): 161–172.

Budowle, B., S. Sundaram, and R.E. Wenk, Population data on the forensic genetic markers: Phosphoglucomutase-1, esterase D, erythrocyte acid phosphatase and glyoxylase I. *Forensic Sci Int*, 1985, **28**(2): 77–81.

Burdett, P.E. and P.H. Whitehead, The separation of the phenotypes of phosphoglucomutase, erythrocyte acid phosphatase, and some haemoglobin variants by isoelectric focusing. *Anal Biochem*, 1977, **77**(2): 419–428.

Burdett, P.E., Isoelectric focusing in agarose: Phosphoglucomutase (*PGM* locus 1) typing. *J Forensic Sci*, 1981, **26**(2): 405–409.

Carracedo, A. and L. Concheiro, The typing of alpha-1-antitrypsin in human bloodstains by isoelectric focusing. *Forensic Sci Int*, 1982, **19**(2): 181–184.

Carracedo, A., et al., A silver staining method for the detection of polymorphic proteins in minute bloodstains after isoelectric focusing. *Forensic Sci Int*, 1983, **23**(2–3): 241–248.

Constans, J. and M. Viau, Group-specific component: Evidence for two subtypes of the Gc1 gene. *Science*, 1977, **198**(4321): 1070–1071.

Constans, J., et al., Analysis of the Gc polymorphism in human populations by isoelectrofocusing on polyacrylamide gels. Demonstration of subtypes of the Gc allele and of additional Gc variants. *Hum Genet*, 1978, **41**(1): 53–60.

Cox, D.W., S. Smyth, and G. Billingsley, Three new rare variants of alpha-1-antitrypsin. *Hum Genet*, 1982, **61**(2): 123–126.

Divall, G.B., Studies on the use of isoelectric focusing as a method of phenotyping erythrocyte acid phosphatase. *Forensic Sci Int*, 1981, **18**(1): 67–78.

Divall, G.B., The esterase D polymorphism as revealed by isoelectric focusing in ultra-thin polyacrylamide gels. *Forensic Sci Int*, 1984, **26**(4): 255–267.

Divall, G.B. and M. Ismail, Studies and observations on the use of isoelectric focusing in ultra-thin polyacrylamide gels as a method of typing human red cell phosphoglucomutase. *Forensic Sci Int*, 1983, **22**(2–3): 253–263.

Dorrill, M. and P.H. Whitehead, The species identification of very old human blood-stains. *Forensic Sci Int*, 1979, **13**(2): 111–116.

Dykes, D.D. and H.F. Polesky, Review of isoelectric focusing for Gc, PGM1, Tf, and Pi subtypes: Population distributions. *Crit Rev Clin Lab Sci*, 1984, **20**(2): 115–151.

Frants, R.R. and A.W. Eriksson, Alpha-1-antitrypsin: Common subtypes of Pi M. *Hum Hered*, 1976, **26**(6): 435–440.

Gorg, A., W. Postel, and R. Westermeier, Ultrathin-layer isoelectric focusing in polyacrylamide gels on cellophane. *Anal Biochem*, 1978, **89**(1): 60–70.

Grunbaum, B.W. and P.L. Zajac, Rapid phenotyping of the group specific component by immunofixation on cellulose acetate. *J Forensic Sci*, 1977, **22**(3): 586–589.

Grunbaum, B.W. and P.L. Zajac, Phenotyping of erythrocyte acid phosphatase in fresh blood and in blood-stains on cellulose acetate. *J Forensic Sci*, 1978, **23**(1): 84–88.

Hoste, B., Group-specific component (Gc) and transferrin (Tf) subtypes ascertained by isoelectric focusing. A simple nonimmunological staining procedure for Gc. *Hum Genet*, 1979, **50**(1): 75–79.

Itoh, Y. and S. Matsuzawa, Detection of human hemoglobin A (HbA) and human hemoglobin F (HbF) in biological stains by microtiter latex agglutination-inhibition test. *Forensic Sci Int*, 1990, **47**(1): 79–89.

James, S. and Nordby, J.J., *Forensic Science: An Introduction to Scientific and Investigative Techniques*, 2005. Boca Raton, FL: CRC Press.

Jones, D.A., Blood samples: Probability of discrimination. *J Forensic Sci Soc*, 1972, **12**(2): 355–359.

Khalap, S. and G.B. Divall, Gm(5) grouping of dried bloodstains. *Med Sci Law*, 1979, **19**(2): 86–88.

Kido, A., et al., A stability study on Gc subtyping in bloodstains: Comparison by two different techniques. *Forensic Sci Int*, 1984, **26**(1): 39–43.

Kimura, H., et al., The typing of group-specific component (Gc protein) in human blood stains. *Forensic Sci Int*, 1983, **22**(1): 49–55.

Kipps, A.E., G_m and K_m typing in forensic science—A methods monograph. *J Forensic Sci Soc*, 1979, **19**(1): 27–47.

Kipps, A.E., V.E. Quarmby, and P.H. Whitehead, The detection of mixtures of blood and other body secretions in stains. *J Forensic Sci Soc*, 1978, **18**(3–4): 189–191.

Kueppers, F. and B. Harpel, Group-specific component (Gc) "subtypes" of Gc1 by isoelectric focusing in US blacks and whites. *Hum Hered*, 1979, **29**(4): 242–249.

Kuhnl, P. and W. Spielmann, A third common allele in the transferrin system, TfC3, detected by isoelectric focusing. *Hum Genet*, 1979, **50**(2): 193–198.

Kuhnl, P. and W. Spielmann, Transferrin: Evidence for two common subtypes of the TfC allele. *Hum Genet*, 1978, **43**(1): 91–95.

Kuhnl, P., U. Schmidtmann, and W. Spielmann, Evidence for two additional common alleles at the *PGM1* locus (phosphoglucomutase—E.C.: 2.7.5.1). A comparison by three different techniques. *Hum Genet*, 1977, **35**(2): 219–223.

Lamm, L.U., Family studies of red cell acid phosphatase types. Report of a family with the D variant. *Hum Hered*, 1970, **20**(3): 329–335.

Lincoln, P.J. and B.E. Dodd, An evaluation of factors affecting the elution of antibodies from bloodstains. *J Forensic Sci Soc*, 1973, **13**(1): 37–45.

Markert, C.L., The molecular basis for isozymes. *Ann N Y Acad Sci*, 1968, **151**(1): 14–40.

Miscicka, D., T. Dobosz, and S. Raszeja, Determination of phenotypes of phosphoglucomutase (PGM1) in bloodstains by cellulose acetate electrophoresis. *Z Rechtsmed*, 1977, **79**(4): 297–300.

Murch, R.S. and B. Budowle, Applications of isoelectric focusing in forensic serology. *J Forensic Sci*, 1986, **31**(3): 869–880.

Neilson, D.M., et al., Simultaneous electrophoresis of peptidase A, phosphoglucomutase, and adenylate kinase. *J Forensic Sci*, 1976, **21**(3): 510–513.

Olaisen, B., et al., The ESD polymorphism: Further studies of the *ESD2* and *ESD5* allele products. *Hum Genet*, 1981, **57**(4): 351–353.

Randall, T., W.A. Harland, and J.W. Thorpe, A method of phenotyping erythrocyte acid phosphatase by isoelectric focusing. *Med Sci Law*, 1980, **20**(1): 43–47.

Rees, B., T.J. Rothwell, and J. Bonnar, Correlation of phosphoglucomutase isoenzymes in blood and semen. *Lancet*, 1974, **1**(7861): 783.

Sensabaugh, G.F., Biochemical markers for individuality, in R. Saferstein (ed.), *Forensic Science Handbook*, 1982. Englewood Cliffs, NJ: Prentice Hall.

Shaler, R.C., Interpretation of Gm testing results: Two case histories. *J Forensic Sci*, 1982, **27**(1): 231–235.

Sonneborn, H.H., Comments on the determination of isoenzyme polymorphism (ADA, AK, 6-PGD, PGM) by cellulose acetate electrophoresis. *Humangenetik*, 1972, **17**(1): 49–55.

Sorensen, S.A., Agarose gel electrophoresis of the human red cell acid phosphatase. *Vox Sang*, 1974, **27**(6): 556–563.

Sorensen, S.A., Report and characterization of a new variant, EB, of human red cell acid phosphatase. *Am J Hum Genet*, 1975, **27**(1): 100–109.

Stedman, R., Human population frequencies in twelve blood grouping systems. *J Forensic Sci Soc*, 1972, **12**(2): 379–413.

Stolorow, M.D. and B.G. Wraxall, An efficient method to eliminate streaking in the electrophoretic analysis of haptoglobin in bloodstains. *J Forensic Sci*, 1979, **24**(4): 856–863.

Sutton, J.G. and R. Burgess, Genetic evidence for four common alleles at the phosphoglucomutase-1 locus (*PGM1*) detectable by isoelectric focusing. *Vox Sang*, 1978, **34**(2): 97–103.

Teige, B., et al., Forensic aspects of haptoglobin: Electrophoretic patterns of haptoglobin allotype products and an evaluation of typing procedure. *Electrophoresis*, 1988, **9**(8): 384–392.

Thogmartin, J.R., et al., Sickle cell trait-associated deaths: A case series with a review of the literature. *J Forensic Sci*, 2011, **56**(5): 1352–1360.

Turowska, B. and M. Bogusz, The rare silent gene Po of human red cell acid phosphatase in a second family in Poland. *Forensic Sci*, 1978, **11**(3): 175–176.

Twibell, J. and P.H. Whitehead, Enzyme typing of human hair roots. *J Forensic Sci*, 1978, **23**(2): 356–360.

Whitehead, P.H., L.A. King, and D.J. Werrett, New information from bloodstains. *Naturwissenschaften*, 1979, **66**(9): 446–451.

Whitehead, P.H., et al., The examination of bloodstains by Laurell electrophoresis (antigen-antibody crossed electrophoresis). *J Forensic Sci Soc*, 1970, **10**(2): 83–90.

Wraxall, B.G. and E.G. Emes, Erythrocyte acid phosphatase in bloodstains. *J Forensic Sci Soc*, 1976, **16**(2): 127–132.

Wrede, B., E. Koops, and B. Brinkmann, Determination of three enzyme polymorphisms in one electrophoretic step on horizontal polyacrylamide gels. *Humangenetik*, 1971, **13**(3): 250–252.

Yuasa, I. and K. Umetsu, Molecular aspects of biochemical markers. *Leg Med*, 2005, **7**: 251–254.

Variable Number Tandem Repeat Profiling

Tandem repeats are abundant in the human genome. *Minisatellites* were first defined as a class of tandem repeats in the 1980s. Some of these repeats share a GC-rich core sequence. Subsequently, tandem repeats with higher AT contents of core sequence have also been characterized. The minisatellites are also called *variable number tandem repeats* (*VNTRs*), as shown in Figure 19.1. The repeat unit length of a VNTR can range from several to hundreds of base pairs (bp). The tandem repeat arrays can be kilobases (kb, corresponding to 10^3 bp) long, and the numbers of tandem repeat units in some VNTR loci are highly variable, leading to variable lengths of DNA fragments. A genotype is defined by a particular number of tandem repeat units at a given locus.

Table 19.1 lists the common VNTR loci used for forensic testing. To achieve high discriminating power, the VNTR loci should not be linked, which means that they should be inherited independently of each other. For example, loci located on different chromosomes or far apart on the same chromosome can be used (Section 21.2). Many VNTR loci used for forensic applications are highly polymorphic, and as many as hundreds of different genotypes per locus can be observed among the population. The discriminating power of VNTR loci used for forensic testing can be measured by *population match probability* (P_m; Chapter 25). The lower the P_m, the less likely a match will occur between two randomly chosen individuals. A P_m of up to 10^{-12} can be achieved by testing several VNTR loci.

19.1 Restriction Fragment Length Polymorphism

VNTR profiling utilizes RFLP—the first historical method used in forensic DNA testing (Figure 19.2). It utilizes *restriction endonucleases* that recognize and cleave specific sites along the DNA sequence. Cleavage of a DNA sample with a particular restriction endonuclease results in a reproducible set of restriction fragments of various lengths. Appropriate restriction endonucleases should be selected so that the genomic DNA is cleaved at sites that flank the VNTR core repeat region. The resulting fragments are then separated according to their sizes by gel electrophoresis through a standard agarose gel (Chapter 9).

The DNA is then processed using the *Southern transfer and hybridization* technique. The DNA is denatured and transferred from the gel to a supporting matrix such as a nylon or nitrocellulose membrane. The DNA immobilized on the membrane is then hybridized with a labeled *probe*. Only bands of DNA that have complementary sequences to the probe are recognized by detection systems such as autoradiography (Section 19.1.4). Using the RFLP technique, the

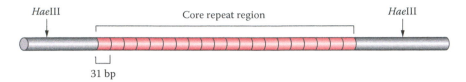

Figure 19.1   VNTR locus D2S44 (2q21.3–2q22). Each repeat unit consists of 31 bp. *Hae*III represents the *Hae*III restriction site.

Table 19.1	Common VNTR Loci			
Locus	**Chromosome Location**	**Repeat Unit Length (bp)**	***Hae*III Fragment Size (kb)**	**Probe**
D1S7	1	9	0.5–12	MS1
D2S44	2	31	0.7–8.5	yNH24
D4S139	4	31	2–12	pH30
D10S28	10	33	0.4–10	pTBQ7
D14S13	14	15	0.7–12	pCMM101
D16S85	16	17	0.2–5	3′HVR
D17S26	17	18	0.7–11	pEFD52
D17S79	17	38	0.5–3	V1

Source: Adapted from Budowle, B., et al., *DNA Typing Protocols: Molecular Biology and Forensic Analysis*, Eaton Publishing, Natick, MA, 2000; Office of Justice Programs, Future of forensic DNA testing: Predictions of the Research and Development Working Group. National Institute of Justice, US Department of Justice, 2000.

length variations among restriction sites can be detected. Most forensic applications focus on the length variations of VNTR regions located between two restriction sites.

In summary, the RFLP method includes several steps: (1) genomic DNA preparation, (2) restriction endonuclease digestion of the genomic DNA into fragments, (3) agarose gel electrophoretic separation of the DNA fragments according to size, (4) transfer of DNA fragments using Southern transfer, (5) hybridization with locus-specific probes, and (6) detection of locus-specific bands by autoradiography or chemiluminescence.

19.1.1 Restriction Endonuclease Digestion

Restriction endonucleases are enzymes that cleave the phosphodiester bond of DNA at or near specific recognition nucleotide sequences known as restriction sites. A restriction site usually is a short motif that is 4–8 bp in length. It often has a specific palindromic recognition sequence, that is, a segment of double-stranded DNA in which the nucleotide's sequence is identical with an inverted sequence in the complementary strand. Thus, double-stranded DNA is required to be cleaved by most restriction endonucleases. As a result, both sticky ends and blunt ends of restriction fragments can be generated after the cleavage (Figure 19.3).

To date, hundreds of restriction endonucleases have been described. They are traditionally classified into three types on the basis of subunit composition and enzymatic properties. Type II restriction endonucleases are most commonly used in molecular biology applications. Type II restriction endonucleases, requiring magnesium as a cofactor, usually cleave DNA at defined positions within their recognition sequences. The Enzyme Commission (EC) number of type II

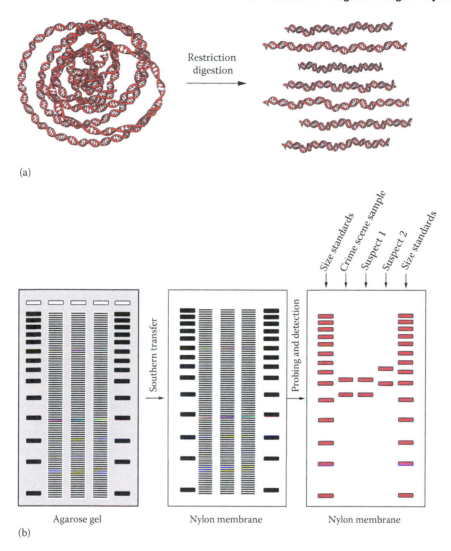

Figure 19.2   RFLP. (a) Restriction digestion generates restriction fragments with various lengths of genomic DNA. (b) Restriction fragments are separated by gel electrophoresis. DNA is transferred to a solid phase and probed. The signal is detected and the DNA fragment of interest can be observed. Band patterns of heterozygous loci of individuals are shown. (© Richard C. Li.)

restriction endonucleases is EC 3.1.21.4. The EC number is a numerical classification system of nomenclature based on the chemical reaction that is catalyzed by the enzyme. In contrast, type I (EC 3.1.21.3) and type III (EC 3.1.21.5) restriction endonucleases cleave at sites remote from their recognition site, which are not utilized for RFLP applications.

Restriction endonucleases are isolated from various bacteria. Each enzyme is named using a nomenclature system after the bacterium from which it was isolated. For example, in the restriction enzyme *Hae*III, H is from the genus name *Haemophilus*, ae is from the species name *aegypticus*, and III stands for the third endonuclease isolated from the *Haemophilus aegyptius* bacteria.

The restriction endonucleases play a role in protecting the bacteria from phage (bacterial virus) infections by using their endonucleases to destroy foreign DNA molecules. Bacterial DNA is usually methylated. In prokaryotes, the *Dam* methylase transfers a methyl group to the N6 position of the adenine in the sequence GmATC and the *Dcm* methylase transfers a methyl

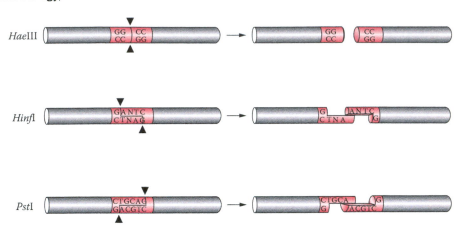

Figure 19.3 Restriction sites for *Hae*III, *Hinf*I, and *Pst*I. *Hae*III digestion produces a blunt end DNA fragment. *Hinf*I and *Pst*I digestions produce sticky ends. *N* represents any nucleotide. (© Richard C. Li.)

group to the C5 position of cytosine in the sequences CmCAGG and CmCTGG. The activities of restriction endonucleases can be influenced by DNA methylation. Many restriction endonucleases cannot cleave methylated DNA. Therefore, bacterial DNA is distinguishable from foreign DNA by the bacteria's restriction endonucleases. This phenomenon protects bacterial DNA from digestion by their own endonucleases. Note that because DNA methylation also occurs in the human genome (Chapter 11), it is important to choose the restriction endonucleases that are not affected by the methylation of human genomic DNA for RFLP analysis.

The type II restriction endonucleases were used in RFLP analysis for forensic DNA testing (Figure 19.3). In order to perform this analysis, the preferred restriction endonucleases for RFLP were those that cleave at the flanking regions of VNTR repeat units but not within the core repeat sequences of the VNTR. Most forensic laboratories used a single restriction endonuclease for a panel of VNTR loci because the DNA in evidence samples was often insufficient for performing multiple tests with different restriction endonucleases. For instance, *Hinf*I-based RFLP was commonly used in European forensic laboratories, and *Hae*III-based RFLP was used in North American and some European forensic laboratories. Other restriction endonuclease-based RFLPs such as *Pst*I were also used.

The use of these common restriction endonucleases allows the comparison of data of various laboratories. Several VNTR loci are suitable for RFLP analysis with these restriction endonucleases. For example, *Hae*III presents several advantages for forensic RFLP analysis. It recognizes a four-base sequence, 5′-GGCC-3′, and cleaves the DNA between the internal G and C residues of the recognition site (GG/CC). *Hinf*I recognizes a five-base restriction site, and *Pst*I recognizes a six-base restriction site. Hypothetically, four-base restriction sites are likely to occur more often than five- and six-base restriction sites in the human genome. Thus, *Hae*III restriction sites occur more frequently than *Hinf*I and *Pst*I sites. As a result, *Hae*III-cleaved DNA fragments are smaller than those of *Hinf*I and *Pst*I. The *Hae*III-generated VNTR allele sizes are easier to separate using conventional agarose gels, also called analytic gel electrophoresis. After electrophoresis, a smear of various sizes of DNA fragments can be observed. The analytic gel is then processed for Southern transfer. Moreover, the enzymatic activity of *Hae*III is not affected by the methylation of human genomic DNA. Its enzymatic activities also appear unaffected when a reaction proceeds under nonoptimal conditions. Additionally, low star activity is observed (Section 19.1.5.2.2).

19.1.2 Southern Transfer

Also known as *Southern blotting*, this technique was named after Sir Edwin Southern, who developed it in the United Kingdom in the mid-1970s. The method can be used to transfer

DNA from an agarose gel to a solid matrix so that it can be detected with a hybridization probe (Section 19.1.3). This method is still used today in many research laboratories. Prior to the transfer of DNA, the DNA in the gel must be denatured, under alkaline conditions such as treatment with sodium hydroxide, into single-stranded DNA. The single-stranded DNA in the gel is then transferred by capillary action to a solid matrix such as a piece of nylon membrane. The single-stranded DNA fragments transferred can be immobilized on a nylon membrane by an ultraviolet cross-linking process (Figure 19.4).

19.1.3 Hybridization with Probes

A hybridization probe of RFLP is a small segment of labeled DNA that is usually several hundred to a thousand bases in length containing the VNTR sequence. It is utilized to detect the presence of much longer target DNA sequences, in this case the VNTR sequences that are complementary to the nucleotide sequences of the probe. The probe is first denatured by heating or by exposure to alkaline conditions into single-stranded DNA. The hybridization process allows complementary pairing between the probe and the target sequence. Two types of probe techniques were developed for VNTR analysis: the multilocus probe and single-locus probe techniques.

19.1.3.1 Multilocus Probe Technique

The *multilocus probe* (*MLP*) technique can detect multiple VNTR loci simultaneously (Figure 19.5). Some VNTRs in the human genome share a short GC-rich core sequence of 10–15 bp. The MLP consists of this core sequence and hybridizes to multiple VNTRs that share these core sequences. As a result, the utilization of MLP produces a complex bar-code-like band pattern from alleles of multiple VNTR loci (Figure 19.6).

The MLP technique was pioneered by Sir Alec Jeffreys in 1984 at the University of Leicester in the United Kingdom and was called *DNA fingerprinting*. Because of its excellent discriminating power, the method was used for parentage testing in immigration disputes with great success.

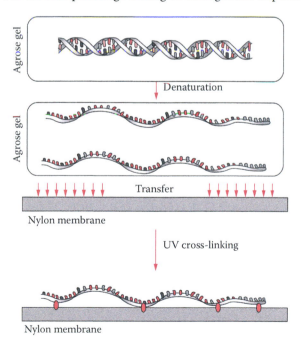

Figure 19.4 Southern blotting. DNA in agarose gel is denatured into single-stranded DNA and transferred to a solid-phase membrane where the single-stranded DNA is immobilized by ultraviolet cross-linking. (© Richard C. Li.)

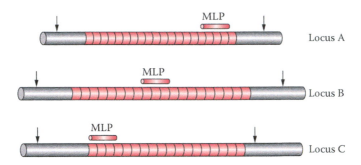

Figure 19.5   VNTR analysis using the MLP method. The technique can detect multiple VNTR loci simultaneously. Restriction sites are indicated by arrows. (© Richard C. Li.)

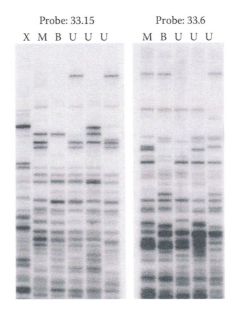

Figure 19.6   First application of DNA fingerprinting. The MLP method was used to analyze samples for an immigration case. M, mother; U, three undisputed children; B, male child in dispute; X, an unrelated individual. All bands in B can be traced back to M or U. (From Jeffreys, A.J., *Nat Med*, 11, 1035, 2005. With permission.)

However, one of the disadvantages of the MLP approach is that the interpretation of a mixed DNA sample from more than one individual is nearly impossible due to its complex DNA finger-printing patterns. Therefore, MLP analysis was not widely utilized in forensic DNA laboratories.

19.1.3.2 Single-Locus Probe Technique

To resolve the disadvantages encountered in the MLP technique, probes that recognize the genomic DNA at the flanking regions of specific VNTR loci can be used. The probe only hybrid-izes to a single VNTR locus and the technique is called the *single-locus probe* (*SLP*), as depicted in Figure 19.7. SLP generates a simple pattern called a *DNA profile*, consisting of one band for a homozygote and two bands for a heterozygote per locus. In order to improve the discriminating power of the test, SLP analyses of different VNTR loci can be performed by using different probes sequentially with a single locus at one time. SLP can analyze mixed DNA samples from two or more contributors. The sizes of fragments can be estimated and converted into a numerical form suitable for databasing. Therefore, DNA profiles can be compared among different laboratories.

Figure 19.7   VNTR analysis using the SLP method. The technique can detect a single VNTR locus. Restriction sites are indicated by arrows. (© Richard C. Li.)

This technique led to the solving of a double murder case in Leicestershire in the 1980s. The case was the first to apply DNA evidence to a criminal investigation. DNA profiling identified the true perpetrator and also excluded an innocent suspect (Figure 19.8). In 1983 and 1986, two girls were raped and murdered. Crime scene evidence suggested that the two cases were committed by the same perpetrator. A young local man, Richard Buckland, was the suspect. However, the DNA evidence revealed that the semen samples from both crimes did not originate from Buckland. To solve the crimes, an investigation was carried out in which 5000 local men were asked to volunteer DNA samples for testing. Several months after the investigation, a witness tipped off police that a man named Pitchfork had paid someone for giving a blood sample as Pitchfork's. In 1987, Pitchfork was arrested. It was discovered that Pitchfork's DNA profile matched that of the crime scene evidence. He was sentenced to life imprisonment. This case demonstrated the great potential of DNA profiling in forensic investigations. Consequently, SLP became a common method in most forensic laboratories in the late 1980s–1990s.

19.1.4 Detection

To detect VNTR loci, a labeled SLP probe is hybridized to the target sequence of DNA, which has been immobilized on a solid matrix such as a piece of nylon membrane (see Southern

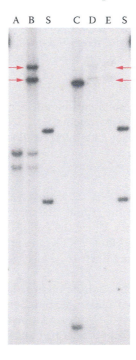

Figure 19.8   First application of DNA profiling in a criminal investigation using the SLP method. A, hair roots from the first victim; B, a mixture of semen and vaginal fluid from the first victim; C, blood from second victim; D, a vaginal swab from the second victim; E, a semen stain on clothing from the second victim; S, blood from the suspect. Alleles (arrows) are matched with the profiles of the two cases but not with the suspect profile. (From Jeffreys, A.J., *Nat Med*, 11, 1035, 2005. With permission.)

transfer, Section 19.1.2). Any unbound probes are washed away so that they do not interfere with the signal. Two types of detection systems are used for VNTR analysis. Radioisotope labeling, such as with a ^{32}P-labeled probe, can be used. The hybridized probe can be detected by exposing the membrane to a sheet of x-ray film to generate an autoradiograph. Alternatively, an enzyme-conjugated probe can also be used. Alkaline phosphatase (Section 9.2.2) is an example of an enzyme used in this type of probe. Its enzymatic activity can be detected with chemiluminescent substrates. A chemiluminescent signal can be detected by exposure to x-ray film as well.

In RFLP analysis, several loci are commonly analyzed sequentially using the same membrane. This approach has the advantage of not consuming additional DNA samples, which are often limited in forensic cases. When probing for multiple loci, multiple probes for each locus are sequentially hybridized and removed one at a time. Once the analysis of the first probe is completed, the probe is removed by a procedure called probe stripping, which is carried out under conditions such as high-temperature washing to denature the DNA strands (Chapter 17). The probe for the next locus to be analyzed is then hybridized to the same membrane, and the process is repeated for each probe to be tested.

Typically, a size standard is utilized on each gel. Band sizes can thus be estimated by comparison to these standards. However, the VNTR alleles that differ by only one or two repeat units are usually not distinguishable. For this reason, genotypes can be determined by bins but not discrete alleles. A *bin* is a range of DNA fragments that differ by only a few repeat units. A sample with a known VNTR genotype is also utilized on each gel as a positive control where historically a genomic DNA sample from cell line K562 (a human erythroleukemic cell line) was used for the positive control. DNA samples to be compared can be loaded side by side on the same gel. As a result, the patterns of VNTR fragments can be compared from sample to sample. The following possible conclusions can be made. If the VNTR fragments are at corresponding positions (profiles match), they are considered to be a match (inclusion). Chapter 25 evaluates and discusses the strengths of the results. If the profiles are different, the two DNA samples are considered to have come from different origins (exclusion).

19.1.5 Factors Affecting RFLP Results

The accuracy of VNTR profiling results can be affected by certain factors such as sample conditions, genetic mutations, and experimental artifacts appearing during the procedure. Consequently, these factors can impact data interpretation and are explained in the following sections.

19.1.5.1 DNA Degradation

RFLP analysis requires the genomic DNA to be intact. DNA degradation results in damage such as creating nicks and breaks in the strand. The more severe the degradation, the smaller the average size of the DNA fragments. When the average size of DNA fragments becomes too small, the allele may not be detected. Many VNTR tandem arrays can span several kilobases in length. In theory, large alleles are more likely to be affected by degradation than smaller alleles at a different locus.

A two-banded heterozygous profile can be observed as a one-banded homozygous profile if the larger band is not detected due to degradation. This artifact could lead to a false determination of exclusion. However, DNA degradation can be detected prior to conducting RFLP by the use of agarose gel electrophoresis, also known as a *yield gel*, used for evaluating the yield and integrity of the isolated genomic DNA. High-molecular-weight genomic DNA bands are usually observed for a typical genomic DNA sample. In contrast, a smear of low-molecular-weight DNA bands can be observed if DNA is degraded. The sizes of the DNA can be estimated by comparison to a size standard run on the same gel. Additionally, the yield of DNA can be estimated by comparing the intensity of the size standards with a known quantity of DNA.

19.1.5.2 Restriction Digestion–Related Artifacts
19.1.5.2.1 Partial Restriction Digestion

Complete restriction digestion should be achieved for RFLP analysis. If partial digestion occurs, the partially cleaved DNA strands are longer than the cleaved fragments (Figure 19.9). Thus, partial digestion results in a mixture of fragments with correct sizes and slightly larger fragments. Under these conditions, a larger uncleaved band, usually lower in intensity than the true bands, can be observed. The multibanded pattern due to the partial digestion can be observed at multiple loci analyzed in the same nylon membrane.

Detection of more than two bands in an RFLP profile may lead to a false interpretation and be incorrectly concluded to be a mixture. However, partial digestion can be detected after restriction digestion. DNA cleavage by restriction endonuclease digestion can be examined using agarose gel electrophoresis. A small portion of a sample can be analyzed. After separation using electrophoresis, a smear of various sizes of cleaved DNA fragments can usually be observed if restriction digestion is completed. Conversely, high-molecular-weight genomic DNA can still be observed from partially digested DNA samples. Additionally, comparisons can be made between the sample and uncleaved and completely cleaved standard samples of DNA. This quick assay is also called a *test gel*, used to determine if the DNA was cleaved to completion.

If partial digestion occurs, procedures such as additional purification of the DNA sample can be carried out. Additionally, optimal amounts of DNA, restriction enzymes, buffer, and proper incubation conditions should be used in achieving complete digestion.

19.1.5.2.2 Star Activity

Star activity refers to a deviation of the specificity of a cleavage site of a restriction endonuclease under certain conditions, such as a high concentration ratio of enzyme to DNA, the use of non-optimal buffers for restriction digestion, prolonged digestion time, the substitution of Mg^{2+} with other divalent cations, and the presence of organic solvents such as ethanol and concentrated glycerol. For instance, *Hae*III cleaves at the GGCC DNA sequence. When star activity occurs, the enzymatic specificity is reduced and cleaves at a sequence slightly different from GGCC. If the start site is presented at an internal location of a VNTR locus, the enzyme would cleave the GGCC sequences and additionally cleave at the internal star site (Figure 19.10). This would result in an additional band smaller than the true alleles, although the intensity of this band is usually not the same as other bands. Thus, a multiband pattern is observed. However, the star

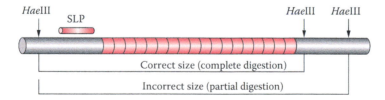

Figure 19.9 Effects of partial restriction digestion on the RFLP profile. Only the restriction fragments detectable by the probe are shown. *Hae*III restriction sites are indicated by arrows. (© Richard C. Li.)

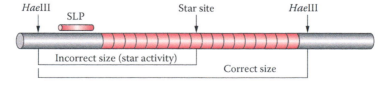

Figure 19.10 Star effects on RFLP profiles. Only the restriction fragments detectable by the probe are shown. *Hae*III restriction sites and star sites are indicated by arrows. (© Richard C. Li.)

site presented externally to a VNTR locus cannot be detected by the probe and does not affect the profiling results. Star activity can usually be avoided by carrying out restriction digestion reactions under conditions recommended by their manufacturers.

19.1.5.2.3 Point Mutations

A *point mutation* is caused by the substitution, deletion, or insertion of a single nucleotide. Point mutations at a restriction site within flanking regions may abolish the site, and the result is a band slightly larger than the true allele (Figure 19.11). The point mutation may also be present internally in a VNTR sequence. If such a point mutation creates a restriction enzyme site, the enzyme will cleave at the regular site and at the mutation site and yield two smaller bands. If the created restriction site is located internal to the probe binding region, both bands will be detected for that allele. These rare mutations obey Mendelian inheritance.

19.1.5.3 Electrophoresis and Blotting Artifacts

19.1.5.3.1 Partial Stripping

If more than one VNTR locus is analyzed sequentially using the same membrane, a probe must be removed by the stripping process before the application of the next probe. Any probe remaining on the membrane due to partial stripping may generate additional bands when the next probe is analyzed. However, bands due to partial stripping are usually faint and have the same electrophoretic mobility as the previous autoradiograph.

19.1.5.3.2 Separation Resolution Limits and Band Shifting

Agarose gel electrophoresis cannot resolve restriction fragments that differ by one or a few repeat units, especially for high-molecular-weight fragments. These bands may not be

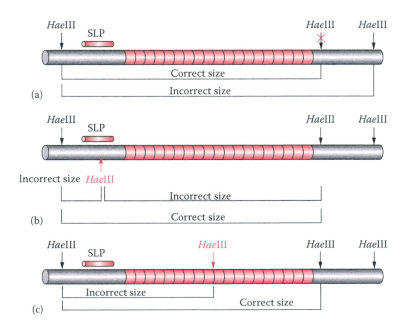

Figure 19.11 Effects of point mutations on the RFLP profile. Only the restriction fragments detectable by the probe are shown. *Hae*III restriction sites are indicated by arrows. (a) Point mutation (in red) abolishes the *Hae*III restriction site. (b) A point mutation (in red) creates an internal *Hae*III restriction site residing within the probe-binding region. (c) A point mutation (in red) creates an internal *Hae*III restriction site residing outside the probe-binding region. (© Richard C. Li.)

separated and will appear as a single band. This may lead to a false interpretation as a homozygous profile. Additionally, minor variations in the electrophoretic mobility of DNA fragments, known as band shifting, can cause two samples from the same individual to appear different.

19.1.5.3.3 Bands Running off Gel

The commonly used VNTR loci generate bands from hundreds of base pairs to 20 kb in length. The small bands have higher electrophoretic mobility and may run off the front edge of a gel during electrophoresis and fail to be detected. This phenomenon may also lead to a false interpretation as a homozygous profile. To prevent these DNA fragments from running off the gel into the buffer tank, a longer gel can be used. Alternatively, the electrophoresis can be stopped before the dye front, the furthest extent that dyes migrate, reaches the front edge of the gel.

19.2 Amplified Fragment Length Polymorphism

The RFLP analysis of VNTR profiling does not perform well for degraded or limited quantities of DNA from crime scene samples. For these reasons, an improved VNTR method was developed. Some VNTR loci have relatively short alleles (<1 kb). These loci are suitable for PCR amplification. This technique is called *amplified fragment length polymorphism (AFLP)*. One locus, D1S80, was used by forensic DNA laboratories for AFLP analysis. Fragments in the range of 14–42 repeat units (16 bp per repeat) were amplified using the AFLP method (Figure 19.12). The amplified DNA fragments were commonly separated according to size using polyacrylamide gel electrophoresis and detected using a silver stain (Figure 19.13).

D1S80 loci are detected as discrete alleles and thus can be compared directly to an allelic ladder (a collection of common alleles used as a standard) on the same gel. This technique represented an improvement over the RFLP system. RFLP allele sizing cannot be performed with precision and the resolution limits of agarose gel electrophoresis are much lower than those of the polyacrylamide gels.

The AFLP technique requires less DNA than the RFLP method and performs better for degraded samples. The AFLP method at the D1S80 locus can be analyzed in a multiplex fashion with an amelogenin locus (Chapter 21). The amelogenin gene is used for forensic sex-typing applications. Typing the amelogenin gene enables the determination of the sex of the contributor of a biological sample.

Due to the wide variation in allele sizes at the D1S80 locus, preferential amplification may be observed. Under certain conditions, the larger alleles may not be as consistently amplified as the small alleles, which may cause lower signal intensity of the larger allele. Additionally, only one locus is analyzed in this system. Furthermore, the D1S80 locus contains two alleles that are very common in some populations. Thus, the discriminating power is reduced compared to RFLP. D1S80 was gradually replaced by multiplex STR systems in the late 1990s.

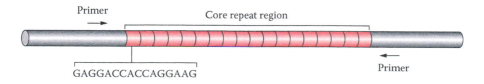

Figure 19.12 VNTR locus D1S80 (chromosome 1p). Each repeat unit is 16 bp long. PCR primers are indicated to amplify the core repeat region.

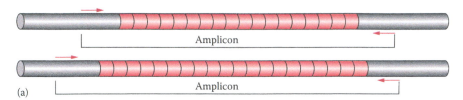

(a)

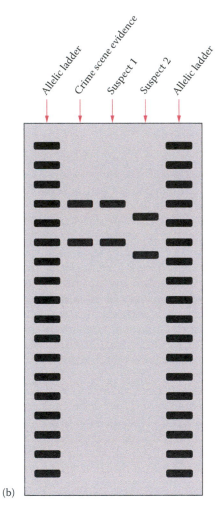

(b)

Figure 19.13 AFLP analysis. (a) Heterozygous D1S80 loci amplified by PCR. PCR primers are indicated as *arrows*. (b) Silver-stained polyacrylamide gel showing D1S80 amplicons along with allelic ladders. (© Richard C. Li.)

Bibliography

RFLP

Balazs, I., et al., Human population genetic studies of five hypervariable DNA loci. *Am J Hum Genet*, 1989, **44**(2): 182–190.

Benzinger, E.A., et al., Time course and inhibitors of *Hae*III digestion in the forensic laboratory. *Appl Theor Electrophor*, 1995, **4**(4): 179–188.

Benzinger, E.A., et al., Products of partial digestion with *Hae*III. Part 1. Characterization, casework experience and confirmation of the theory of three-, four- and five-banded RFLP pattern origins using partial digestion. *J Forensic Sci*, 1997, **42**(5): 850–863.

Benzinger, E.A., et al., An illustrated guide to RFLP troubleshooting. *J Forensic Sci*, 1998, **43**(3): 665–679.

Bragg, T., et al., Isolation and mapping of a polymorphic DNA sequence (cTBQ7) on chromosome 10 [D10S28]. *Nucleic Acids Res*, 1988, **16**(23): 11395.

Budowle, B., et al., *Hae*III—A suitable restriction endonuclease for restriction fragment length polymorphism analysis of biological evidence samples. *J Forensic Sci*, 1990, **35**(3): 530–536.

Budowle, B., et al., Fixed-bin analysis for statistical evaluation of continuous distributions of allelic data from VNTR loci, for use in forensic comparisons. *Am J Hum Genet*, 1991, **48**(5): 841–855.

Budowle, B., et al., Simple protocols for typing forensic biological evidence: Chemiluminescent detection for human DNA quantitation and restriction fragment length polymorphism (RFLP) analyses and manual typing of polymerase chain reaction (PCR) amplified polymorphisms. *Electrophoresis*, 1995, **16**(9): 1559–1567.

Budowle, B., et al., The assessment of frequency estimates of *Hae*III-generated VNTR profiles in various reference databases. *J Forensic Sci*, 1994, **39**(2): 319–352.

Budowle, B., et al., *DNA Typing Protocols: Molecular Biology and Forensic Analysis*, 2000. Natick, MA: Eaton Publishing.

Duewer, D.L., K.L. Richie, and D.J. Reeder, RFLP band size standards: NIST standard reference material 2390. *J Forensic Sci*, 2000, **45**(5): 1093–1105.

Duewer, D.L., et al., Interlaboratory comparison of autoradiographic DNA profiling measurements. 4. Protocol effects. *Anal Chem*, 1997, **69**(10): 1882–1892.

Elder, J.K. and E.M. Southern, Measurement of DNA length by gel electrophoresis II: Comparison of methods for relating mobility to fragment length. *Anal Biochem*, 1983, **128**(1): 227–231.

Eriksen, B. and O. Svensmark, Comparison of two molecular weight markers used in DNA-profiling. *Int J Legal Med*, 1992, **105**(3): 145–148.

Gary, K.T., D.L. Duewer, and D.J. Reeder, Graphical tools for RFLP measurement quality assurance: Laboratory performance charts. *J Forensic Sci*, 1999, **44**(5): 978–982.

Gill, P. and D.J. Werrett, Exclusion of a man charged with murder by DNA fingerprinting. *Forensic Sci Int*, 1987, **35**(2–3): 145–148.

Giusti, A.M. and B. Budowle, Effect of storage conditions on restriction fragment length polymorphism (RFLP) analysis of deoxyribonucleic acid (DNA) bound to positively charged nylon membranes. *J Forensic Sci*, 1992, **37**(2): 597–603.

Hartmann, J.M., et al., The effect of sampling error and measurement error and its correlation on the estimation of multi-locus fixed-bin VNTR RFLP genotype probabilities. *J Forensic Sci*, 1997, **42**(2): 241–245.

Hau, P. and N. Watson, Sequencing and four-state minisatellite variant repeat mapping of the D1S7 locus (MS1) by fluorescence detection. *Electrophoresis*, 2000, **21**(8): 1478–1483.

Jeffreys, A.J., Genetic fingerprinting. *Nat Med*, 2005, **11**: 1035.

Jeffreys, A.J., V. Wilson, and S.L. Thein, Hypervariable "minisatellite" regions in human DNA. *Biotechnology*, 1985, **24**: 467–472.

Jeffreys, A.J., V. Wilson, and S.L. Thein, Individual-specific "fingerprints" of human DNA. *Nature*, 1985, **316**(6023): 76–79.

Jeffreys, A.J., M. Turner, and P. Debenham, The efficiency of multilocus DNA fingerprint probes for individualization and establishment of family relationships, determined from extensive casework. *Am J Hum Genet*, 1991, **48**(5): 824–840.

Kanter, E., et al., Analysis of restriction fragment length polymorphisms in deoxyribonucleic acid (DNA) recovered from dried bloodstains. *J Forensic Sci*, 1986, **31**(2): 403–408.

Kasai, K., Y. Nakamura, and R. White, Amplification of a variable number of tandem repeats (VNTR) locus (pMCT118) by the polymerase chain reaction (PCR) and its application to forensic science. *J Forensic Sci*, 1990, **35**(5): 1196–1200.

Laber, T.L., et al., Evaluation of four deoxyribonucleic acid (DNA) extraction protocols for DNA yield and variation in restriction fragment length polymorphism (RFLP) sizes under varying gel conditions. *J Forensic Sci*, 1992, **37**(2): 404–424.

Laber, T.L., et al., Validation studies on the forensic analysis of restriction fragment length polymorphism (RFLP) on LE agarose gels without ethidium bromide: Effects of contaminants, sunlight, and the electrophoresis of varying quantities of deoxyribonucleic acid (DNA). *J Forensic Sci*, 1994, **39**(3): 707–730.

Laber, T.L., et al., The evaluation and implementation of match criteria for forensic analysis of DNA. *J Forensic Sci*, 1995, **40**(6): 1058–1064.

Lander, E.S., DNA fingerprinting on trial. *Nature*, 1989, **339**(6225): 501–505.

Lewis, M.E., et al., Restriction fragment length polymorphism DNA analysis by the FBI laboratory protocol using a simple, convenient hardware system. *J Forensic Sci*, 1990, **35**(5): 1186–1190.

McNally, L., et al., The effects of environment and substrata on deoxyribonucleic acid (DNA): The use of casework samples from New York City. *J Forensic Sci*, 1989, **34**(5): 1070–1077.

Milner, E.C., et al., Isolation and mapping of a polymorphic DNA sequence pH30 on chromosome 4 [HGM provisional no. D4S139]. *Nucleic Acids Res*, 1989, **17**(10): 4002.

Nakamura, Y., et al., Isolation and mapping of a polymorphic DNA sequence pYNH24 on chromosome 2 (D2S44). *Nucleic Acids Res*, 1987, **15**(23): 10073.

Office of Justice Programs, Future of forensic DNA testing: Predictions of the Research and Development Working Group, 2000. National Institute of Justice, US Department of Justice.

Rankin, D.R., et al., Restriction fragment length polymorphism (RFLP) analysis on DNA from human compact bone. *J Forensic Sci*, 1996, **41**(1): 40–46.

Reed, K.C. and D.A. Mann, Rapid transfer of DNA from agarose gels to nylon membranes. *Nucleic Acids Res*, 1985, **13**(20): 7207–7221.

Richie, K.L., et al., Long PCR for VNTR analysis. *J Forensic Sci*, 1999, **44**(6): 1176–1185.

Rossi, U., International recommendations (as of March, 1988) on the application of methods involving DNA-polymorphisms in forensic haematology. *Haematologica*, 1989, **74**(2): 219–221.

Sharma, B.R., et al., A comparative study of genetic variation at five VNTR loci in three ethnic groups of Houston, Texas. *J Forensic Sci*, 1995, **40**(6): 933–942.

Stolorow, A.M., et al., Interlaboratory comparison of autoradiographic DNA profiling measurements. 3. Repeatability and reproducibility of restriction fragment length polymorphism band sizing, particularly bands of molecular size >10K base pairs. *Anal Chem*, 1996, **68**(11): 1941–1947.

Tahir, M.A., et al., Restriction fragment length polymorphism (RFLP) typing of DNA extracted from nasal secretions. *J Forensic Sci*, 1995, **40**(3): 459–463.

Tahir, M.A., et al., Deoxyribonucleic acid profiling by restriction fragment length polymorphism analysis—A compilation of validation studies. *Sci Justice*, 1996, **36**(3): 173–182.

Tully, G., K.M. Sullivan, and P. Gill, Analysis of 6 VNTR loci by "multiplex" PCR and automated fluorescent detection. *Hum Genet*, 1993, **92**(6): 554–562.

Wyman, A.R. and R. White, A highly polymorphic locus in human DNA. *Proc Natl Acad Sci U S A*, 1980, **77**(11): 6754–6758.

AFLP

Baechtel, F.S., K.W. Presley, and J.B. Smerick, D1S80 typing of DNA from simulated forensic specimens. *J Forensic Sci*, 1995, **40**(4): 536–545.

Baechtel, F.S., et al., Multigenerational amplification of a reference ladder for alleles at locus D1S80. *J Forensic Sci*, 1993, **38**(5): 1176–1182.

Budowle, B., et al., Analysis of the VNTR locus D1S80 by the PCR followed by high-resolution PAGE. *Am J Hum Genet*, 1991, **48**(1): 137–144.

Cosso, S. and R. Reynolds, Validation of the AmpliFLP D1S80 PCR amplification kit for forensic casework analysis according to TWGDAM guidelines. *J Forensic Sci*, 1995, **40**(3): 424–434.

Gross, A.M., G. Carmody, and R.A. Guerrieri, Validation studies for the genetic typing of the D1S80 locus for implementation into forensic casework. *J Forensic Sci*, 1997, **42**(6): 1140–1146.

Kasai, K., Y. Nakamura, and R. White, Amplification of a variable number of tandem repeats (VNTR) locus (pMCT118) by the polymerase chain reaction (PCR) and its application to forensic science. *J Forensic Sci*, 1990, **35**(5): 1196–1200.

Kloosterman, A.D., B. Budowle, and P. Daselaar, PCR-amplification and detection of the human D1S80 VNTR locus. Amplification conditions, population genetics and application in forensic analysis. *Int J Legal Med*, 1993, **105**(5): 257–264.

Koseler, A., A. Atalay, and E.O. Atalay, Allele frequency of VNTR locus D1S80 observed in Denizli Province of Turkey. *Biochem Genet*, 2009, **47**(7–8): 540–546.

Latorra, D. and M.S. Schanfield, Analysis of human specificity in AFLP systems APOB, PAH, and D1S80. *Forensic Sci Int*, 1996, **83**(1): 15–25.

Nakamura, Y., et al., Isolation and mapping of a polymorphic DNA sequence (pMCT118) on chromosome 1p [D1S80]. *Nucleic Acids Res*, 1988, **16**(19): 9364.

Sajantila, A., et al., Amplification of reproducible allele markers for amplified fragment length polymorphism analysis. *Biotechniques*, 1992, **12**(1): 16, 18, 20–22.

Skinker, D.M., et al., DNA typing of azoospermic semen at the D1S80 locus. *J Forensic Sci*, 1997, **42**(4): 718–720.

Skowasch, K., P. Wiegand, and B. Brinkmann, pMCT 118 (D1S80): A new allelic ladder and an improved electrophoretic separation lead to the demonstration of 28 alleles. *Int J Legal Med*, 1992, **105**(3): 165–168.

Walsh, S.J. and C. Eckhoff, Australian Aboriginal population genetics at the D1S80 VNTR locus. *Ann Hum Biol*, 2007, **34**(5): 557–565.

Watanabe, G., et al., Nucleotide substitution in the 5′ flanking region of D1S80 locus. *Forensic Sci Int*, 1997, **89**(1–2): 75–80.

20

Autosomal Short Tandem Repeat Profiling

A *short tandem repeat* (STR) is a region of genomic DNA containing an array of short repeating sequences. STRs are also called *microsatellites* or *simple sequence repeats*. A STR repeat unit can be several base pairs (bp) in length. Arrays range from several to approximately a hundred *repeat units*, which are the component of repetition. The number of STR repeat units varies among individuals. The most commonly used STR loci are 100–500 bp in length, which are shorter than the smallest variable number tandem repeats (VNTRs) (approximately 1000 bp). Thus, STR loci have many advantages compared to VNTR loci:

- STR loci can be amplified by PCR.

- STR profiling can be carried out for degraded DNA samples.

- Preferential amplification is reduced at STR loci.

- The resolution of electrophoretic separation of STR fragments is superior.

- STR loci are suitable for multiplex amplification.

Additionally, STR profiling, as with VNTR profiling, is suitable for the interpretation of mixed DNA profiles from multiple individuals. Thus, STR loci are better candidates for forensic DNA testing than VNTR loci. This chapter will discuss autosomal STR profiling. Male-specific Y chromosomal STR will be discussed in Chapter 21.

20.1 Characteristics of STR Loci

More than 10^5 STRs exist in the human genome. Many STRs have been characterized and used in various types of studies such as genetic mapping and linkage analysis. Some STRs have been characterized specifically for forensic DNA profiling.

20.1.1 Core Repeat and Flanking Regions

The *core repeat region* of each STR locus contains tandemly repeated sequences. The designation of genotypes for human identification is based on the number of tandem repeat units at a STR locus, which varies among human individuals (Figure 20.1). The *flanking regions* surrounding the core repeat region are also needed for STR analysis. PCR primers complementary to these flanking regions are used, allowing the core repeat regions to be amplified.

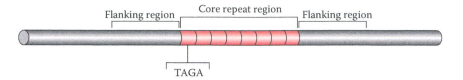

Figure 20.1   Core repeat and flanking regions of CSF1PO STR locus. It consists of eight repeating units of tetrameric nucleotides (TAGA); thus, it is designated as allele 8. (© Richard C. Li.)

20.1.2 Repeat Unit Length

Repeat unit length is the number of nucleotides in a single repeat unit. Dimeric, trimeric, tetrameric, pentameric, and hexameric repeat units appear in the human genome. For example, dimeric and trimeric repeats are very abundant, but they are not usually used for forensic applications. High frequencies of stutter peaks (Section 20.4.2.1) that interfere with genotype interpretation are observed when dimeric and trimeric repeats are amplified. On the contrary, only a few thousand pentameric repeats and a few hundred hexameric repeats exist in the human genome. The pentameric and hexameric repeats are very polymorphic. Only a few pentameric and hexameric repeats are used for forensic applications because they are less abundant in the human genome. The human genome has at least 10^4 tetrameric repeats representing approximately 9% of the total STRs. The STRs with tetrameric repeats are very polymorphic. When they are amplified by PCR, this category of STRs exhibits fewer frequencies of stutter peaks than STRs with dimeric and trimeric repeats. Therefore, the most commonly used STR loci for forensic DNA profiling are the STRs with tetrameric repeats.

20.1.3 Repeat Unit Sequences

STR loci compatible for forensic use can be divided into several classes based on their repeat unit sequences. Figure 20.2 shows representative examples of core repeat sequences. *Simple repeats* consist of tandem repeats with identical repeat unit sequences (Figure 20.2a). Allele designation is based on the number of repeat units in the core repeat region. For example, a D5S818 allele consisting of ten repeating units of the tetrameric nucleotide sequence AGAT is designated as allele 10. *Compound repeats* consist of more than one type of simple repeat (Figure 20.2b). *Complex repeats* contain several clusters of different tandem repeats with intervening sequences (Figure 20.2c).

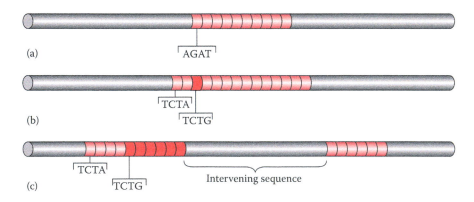

Figure 20.2   Examples of core repeat sequences. (a) A simple repeat in which D5S818 [AGAT]$_{10}$ is designated as allele 10, consisting of 10 repeating units of the tetrameric nucleotides, AGAT. (b) Compound repeats. Allele 14 of D8S1179 consists of two types of repeating units: [TCTA]$_2$, [TCTG]$_1$, and [TCTA]$_{11}$. (c) Complex repeats. Allele 24 of D21S11 contains several clusters of different tandem repeats, [TCTA]$_4$, [TCTG]$_6$, and [TCTA]$_6$, with a 43 bp intervening sequence: [TCTA]$_3$ TA [TCTA]$_3$ TCA [TCTA]$_2$ TCCATA. (© Richard C. Li.)

Nonconsensus alleles with partial repeat units also appear in the population. These nonconsensus alleles, also known as *microvariants*, differ from common alleles by one or more nucleotides. They are designated by the number of consensus repeats, followed by a decimal point and the number of nucleotides of the partial repeat, for example, the TH01 allele 9.3 is 1 nucleotide shorter than allele 10.

Another type of nonconsensus allele can result from a limitation of STR analysis. These alleles have the same number of tandem repeats as common alleles but contain different sequences. These microvariants cannot be distinguished by STR profiling because their length is identical to the lengths of common alleles.

20.2 STR Loci Commonly Used for Forensic DNA Profiling

In the early 1990s, STR loci were initially utilized for genetic studies and were later applied to forensic DNA profiling. The first STR multiplex system, known as the quadruplex, was developed by Forensic Science Services in the United Kingdom. It consisted of four STR loci (F13A1, FES, TH01, and VWA) with a *population match probability* (P_m) of 10^{-4} (Figure 20.3). P_m measures the discriminating power of an STR locus used for forensic DNA analysis. The lower the P_m (i.e., the higher the discriminating power), the less likely a match will occur between two randomly chosen profiles from different individuals (Chapter 25). In 1995, the first national DNA database was established in the United Kingdom. It contained six STR loci, also known as the *second-generation multiplex* (SGM), consisting of D8S1179, D18S51, D21S11, FGA, TH01, and VWA, with a P_m value of 10^{-7}. The SGM system also included the amelogenin locus (Section 21.3.1) for determining the sexes of DNA contributors. Subsequently, four additional loci were added to SGM with a P_m as low as 10^{-13} (SGM Plus).

To allow for international data exchange, the European DNA Profiling Group (EDNAP) recommended the use of TH01 and VWA loci for all participating European laboratories in 1996. In 1998, the *European Standard Set* (ESS) of loci was established and included TH01, VWA, FGA, and D21S11 for forensic use in Europe. Thereafter, D3S1358, D8S1179, and D18S51 loci were added to the ESS. Other loci are used as well, such as SE33, which is used in Germany's database. In 1998, the US Federal Bureau of Investigation established the *Combined DNA Index System* (CODIS). It contains 13 core STR loci plus the amelogenin sex-typing locus with a P_m

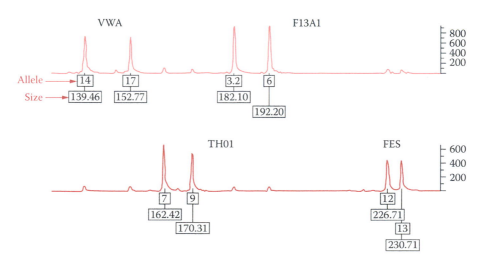

Figure 20.3   DNA profile obtained using the first STR multiplex system: the quadruplex. F13A1, FES, TH01, and VWA loci are shown. (© Richard C. Li.)

Table 20.1	Common STR Loci				
Locus Symbol	Repeat Motif	Repeat Category	Cytogenetic Map Location on Chromosome	Distance from pter (Mb)	Structural Gene
CSFIPO	TAGA	Simple	5q33.1	149.4	Intron 6 of c-fms protooncogene
FGA	[CTTT] [CCTT]	Compound	4q31.3	155.5	Intron 3 of fibrinogen α chain gene
THO1	AATG	Simple	11p15.5	2.2	Intron 1 of tyrosine hydroxylase gene
TPOX	TGAA	Simple	2p25.3	1.5	Intron 10 of thyroid peroxidase gene
SE33/ ACTBP2	AAAG	Complex	6q14	89	5′ Flanking sequence of β-actin-related pseudogene 2 gene
VWA	[TCTG] [TCTA]	Compound	12p13.31	6.1	Intron 40 of von Willebrand factor gene
D1S1656	[TAGA] [TAGC]	Compound	1q42.2	230.9	Anonymous
D2S441	[TCTA] [TCAA]	Compound	2p14	68.2	Anonymous
D2S1338	[TGCC] [TTCC]	Compound	2q35	218.9	Anonymous
D3S1358	[TCTG] [TCTA]	Compound	3p21.31	45.6	Anonymous
D5S818	AGAT	Simple	5q23.2	123.1	Anonymous
D7S820	GATA	Simple	7q21.11	83.8	Anonymous
D8S1179	[TCTA] [TCTG]	Compound	8q24.13	125.9	Anonymous
D10S1248	GGAA	Simple	10q26	—[a]	Anonymous
D12S391	[AGAT] [AGAC]	Compound	12p13.2	12.5	Anonymous
D13S317	TATC	Simple	13q31.1	82.7	Anonymous
D16S539	GATA	Simple	16q24.1	86.4	Anonymous
D18S51	AGAA	Simple	18q21.33	60.9	Anonymous
D19S433	[AAGG] [TAGG]	Compound	19q12	30.4	Anonymous
D21S11	[TCTA] [TCTG]	Complex	21q21.1	20.6	Anonymous
D22S1045	ATT	Simple	22q12.3	37.5	Anonymous

Source: Ensembl Homo sapiens version 75.37 (GRCh37). Mb, megabase.
[a] Locus is not mapped to the assembly in the current Ensembl database.

of 10^{-15}. As the database grows rapidly, the chances of finding incidental matches among DNA profiles are increasing. Recently, additional loci have been added to CODIS and ESS core loci to reduce the likelihood of adventitious matches, as well as to facilitate international data sharing among law enforcement agencies and to improve the discrimination power of forensic STR analysis.

Commonly used STR loci characterized for forensic DNA profiling are summarized in Tables 20.1 and 20.2. To achieve low P_m in forensic STR profiling, desired STR loci should possess certain characteristics as described below. First, the alleles of STR loci selected should be highly variable among individuals. Second, if more than one locus is selected, the loci should not be linked to each other or inherited together (Section 21.2). The STR loci utilized are usually located at different chromosomes to ensure that they are not linked. However, loci that are far

TABLE 20.2	Core STR Loci					
Locus	**SGM**	**SGM Plus**	**ESS**	**ESS-Extended**	**CODIS**	**CODIS-Extended**
Amel	☑	☑	☑	☑	☑	☑
CSF1PO					☑	☑
D1S1656				☑		☑
D2S441				☑		☑
D2S1338		☑				☑
D3S1358		☑	☑	☑	☑	☑
D5S818					☑	☑
D7S820					☑	☑
D8S1179	☑	☑	☑	☑	☑	☑
D10S1248				☑		☑
D12S391				☑		☑
D13S317					☑	☑
D16S539		☑			☑	☑
D18S51	☑	☑	☑	☑	☑	☑
D19S433		☑				☑
D21S11	☑	☑	☑	☑	☑	☑
D22S1045				☑		2
DYS391						☑
FGA	☑	☑	☑	☑	☑	☑
TH01	☑	☑	☑	☑	☑	☑
TPOX					☑	1
VWA	☑	☑	☑	☑	☑	☑
SE33						3

Source: Hares, D.R., *Forensic Sci Int Genet*, 6, e135, 2012; Gill, P., et al., *Forensic Sci Int*, 156, 242–244, 2006.

☑, minimum required loci; 1–3, recommended loci in ranked order of preference.

enough apart on the same chromosome can still be used (Figures 20.4 and 20.5), since they are not linked. Additionally, STR loci with fewer amplification artifacts such as stutter products are desired. Stutters can complicate the interpretation of profiles derived from a mixed DNA sample from more than one contributor. Moreover, STR loci with short amplicon (amplified product) lengths are preferred for multiplex STR analysis and the testing of degraded DNA samples.

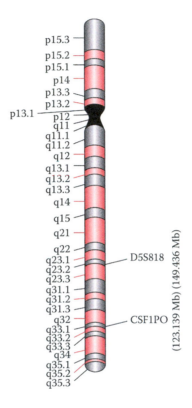

Figure 20.4 Cytogenetic map showing the locations of STR markers on chromosome 5. CSF1PO and D5S818 are separated by 26 Mb (megabases). (© Richard C. Li.)

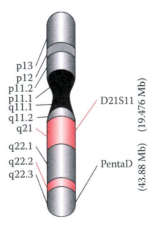

Figure 20.5 Cytogenetic map showing the locations of STR markers on chromosome 21. D21S11 and PentaD are separated by 24 Mb (megabases). (© Richard C. Li.)

20.3 Forensic STR Analysis

STR loci are amplified using fluorescent dye-labeled primers. A multiplex STR system utilizes multiple fluorescent dyes to label each amplicon. The amplicons are separated via electrophoresis. The different fluorescent dye colors are resolved by the detector, and the signals corresponding to each DNA fragment are identified using specialized computer software. The data collection process generates an *electropherogram* that shows a profile of peaks corresponding to each DNA fragment. The positions of these peaks represent the electrophoretic mobility of the DNA fragments. A small fragment, migrating faster than a large one, peaks earlier in the electropherogram than a longer fragment. The DNA fragments are sized by comparison to an internal size standard (Figure 20.6; Chapter 8). Figure 20.7 summarizes the work flow of a forensic STR analysis.

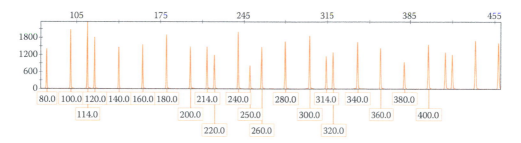

Figure 20.6 Electropherogram of GeneScan™ 500 size standard (Applied Biosystems). RFU represents relative fluorescence unit. (© Richard C. Li.)

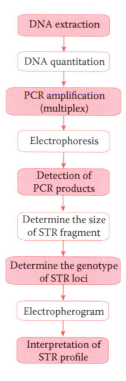

Figure 20.7 Capillary electrophoresis separation of amplified STR products. Fluorescent dye-labeled amplification products are separated and subsequently detected. Various fluorescent dye colors are resolved by the detector. The peaks corresponding to each DNA fragment are identified. (© Richard C. Li.)

The area or amplitude of the peak, expressed as relative fluorescence units (RFU) (Section 9.3.1.3), reflects the fluorescent signal intensity. The RFU value of peak height is proportional to the amount of DNA amplicons being analyzed. When the RFU value is very low, it is difficult to distinguish a signal from background noise. The manufacturer of commonly used instruments for forensic DNA analysis recommends 150 RFU as the threshold of detection. Peaks below 150 RFU should be interpreted with caution. Some forensic laboratories use lower thresholds, as low as 50 RFU, based on their own validation studies. In contrast, when the RFU value is too high, it saturates the sensitivity of an instrument as well as causing artifacts such as pull-up signals (Section 20.4.3.1). The maximum RFU allowed is usually 6000 RFU.

20.3.1 Determining the Genotypes of STR Fragments

As noted earlier, electropherograms are usually plotted as fluorescent signal intensity (in RFU) versus the sizes of the DNA fragments. The data in an electropherogram can then be converted into a genotype. The genotype for a specific STR locus is defined as the number of repeat units of the allele. STR genotype data generated from different laboratories can be compared easily and are suitable for databasing.

The genotype is determined by using an allelic ladder, which is important to achieve accurate genotype profiling. An allelic ladder is a collection of synthetic fragments corresponding to common alleles observed in the human population for a given set of STR loci (Figures 20.8 and 20.9). The ladders are compared to data obtained from an electropherogram of a questioned sample to determine the genotype. Thus, each allele in a ladder must be resolved properly in order to determine correct STR alleles for a sample. The sizes of DNA fragments of a sample are correlated to sizes of fragments for each allele in an allelic ladder in order to determine the allele designation (genotype) of a questioned sample (Figures 20.10 and 20.11). If a rare allele fails to match alleles within an allelic ladder, it is considered an *off-ladder allele*. If an off-ladder allele is present, the sample should be reanalyzed. The presence of a rare allele can be confirmed by repeating the electrophoresis process based on the characteristic electrophoretic mobility of the rare allele.

20.3.2 Interpretation of STR Profiling Results

General guidelines for the interpretation and the reporting of STR profile results were set by the Scientific Working Group on DNA Analysis Methods (SWGDAM) and the DNA Commission of the International Society of Forensic Genetics (ISFG). Typically, conclusions are categorized as inclusion, exclusion, or inconclusive result.

20.3.2.1 Inclusion (Match)

Peaks of compared STR loci, such as those between the profiles of suspect and crime scene evidence or victim and crime scene evidence, show identical genotypes. The strength of this conclusion can be evaluated via statistical analysis and is usually cited in the case report (Chapter 25).

20.3.2.2 Exclusion

The genotypes of two or more samples differ, and the profile of the sample is determined to be an exclusion, meaning that the profiles originated from different sources.

20.3.2.3 Inconclusive Result

The data do not support a conclusion of inclusion or exclusion. In other words, insufficient information is available to reach a conclusion.

20.4 Factors Affecting Genotyping Results

A number of genetics-, amplification-, and electrophoresis-related factors may affect the accuracy of genotypic profiles.

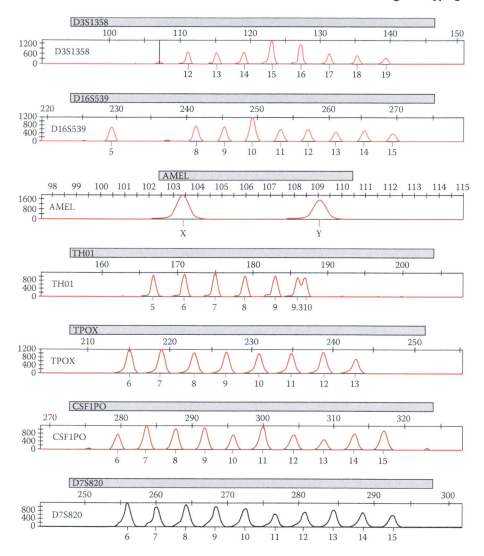

Figure 20.8 Electropherogram of the allelic ladder of the AmpFISTR® COfiler® PCR Amplification Kit (Applied Biosystems). AMEL, CSF1PO, D3S1358, D7S820, D16S539, TH01, and TPOX loci are shown. (© Richard C. Li.)

20.4.1 Mutations

STR loci with low mutation frequencies are desired, in particular, for human identification in mass disasters and for missing person and paternity cases. However, STR mutations do occur, which can affect the profiling results.

20.4.1.1 Mutations at STR Core Repeat Regions

Mutations, usually resulting in a gain or a loss of a single repeat unit, are observed at STR loci. If a mutation occurs in the germ cells (cells that form gametes), the mutant allele will be transmitted to and be present in all cell types of the progeny. This type of inheritable mutation in germ cell lineage is called a *germ-line mutation*. The frequency of germ-line mutation can be measured by the *mutation rate*, expressed as the number of mutations per generation (germ-line transmission). The average mutation rate of commonly used STR loci is about 10^{-4} mutations per germ-line transmission. However, the mutation rate may vary among different STR loci.

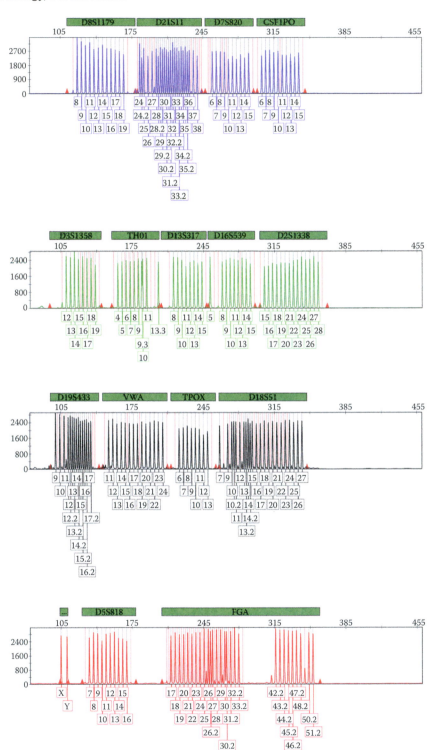

Figure 20.9 Electropherogram of the allelic ladder of the AmpFlSTR® Identifiler® Plus Kit (Applied Biosystems). AMEL, CSF1PO, D2S1338, D3S1358, D5S818, D7S820, D8S1179, D13S317, D16S539, D18S51, D19S433, D21S11, FGA, TH01, TPOX, and VWA loci are shown. (© Richard C. Li.)

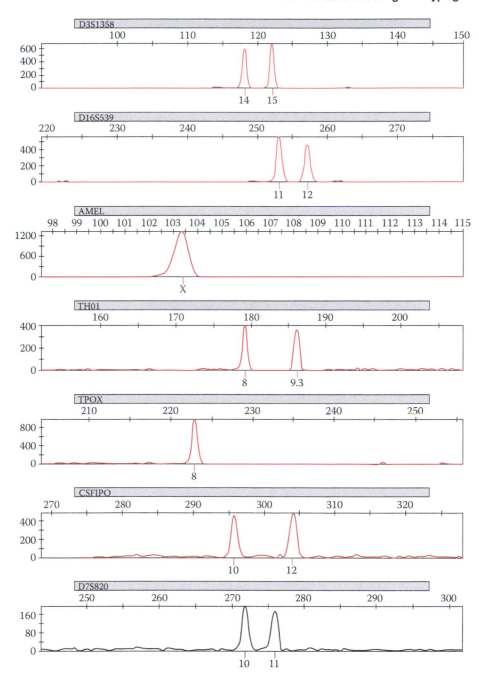

Figure 20.10 Individual DNA profile (Cofiler). The genotype of the DNA profile is shown. AMEL: X, X. CSF1PO: 10, 12. D3S1358: 14, 15. D16S539: 11, 12. D7S820:10, 11. TH01:8, 9.3. TPOX: 8, 8. (© Richard C. Li.)

In contrast, *somatic mutations* involve the mutation of only somatic cells. The germ cells are not affected, and thus a mutant allele is not transmitted to the progeny. A somatic mutation occurring at the core repeat region of an STR locus can be detected and compared to the wild-type allele. The ratio of the signal intensities of the wild-type and mutant alleles varies, depending on the number

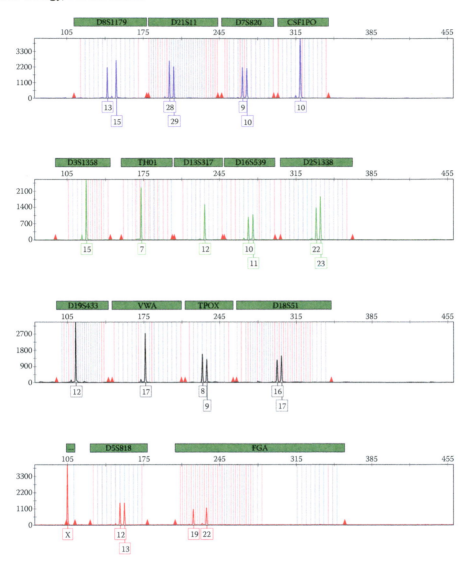

Figure 20.11   Individual DNA profile (Identifiler). The genotype of the DNA profile is shown. AMEL: X, X. CSF1PO: 10, 10. D2S1338: 22, 23. D3S1358: 15, 15. D5S818: 12, 13. D7S820: 9, 10. D8S1179: 13, 15. D13S317: 12, 12. D16S539: 10, 11. D18S51: 16, 17. D19S433: 12, 12. D21S11: 28, 29. FGA: 9, 22. TH01: 7, 7. TPOX: 8, 9. VWA: 17, 17. (© Richard C. Li.)

of mutation-carrying cells in the tissue. Somatic mutations are usually tissue specific. STR profiles from different tissues of the same individual can be compared if a somatic mutation is suspected.

20.4.1.2 Chromosomal and Gene Duplications

Duplicating one of the homologous chromosomes results in a total of three copies of a particular chromosome. This condition, called *trisomy*, is rare and often associated with genetic diseases such as Down's syndrome (chromosome 21 duplication). Duplications have also been observed in chromosomes 13, 18, and X. Other anomalies include duplication of a portion of a chromosome and a single or group of genes instead of an entire chromosome.

A duplication bearing a mutation within the STR core repeat region can affect the number of tandem repeat units. If the duplicated locus is mutated, a *triallelic* or three-peaks pattern can be detected at a single locus, but not at loci located at other chromosomes in a multiplex STR profile (Figure 20.12). The three alleles usually have equal signal intensity (peak amplitude or peak area). Triallelic patterns at STR loci commonly used for forensic DNA analysis have been documented. Many occur at the TPOX, FGA, and CSF1PO loci. If the duplicate locus is not mutated, only two alleles will be observed in a heterozygote. However, the ratio of the peak amplitude of the alleles will be 1:2 (one copy vs. two copies including the duplicate) at that particular locus. However, STR profiles at the loci located at other chromosomes are not affected.

20.4.1.3 Point Mutations

Point mutations involve the changing of a nucleotide sequence through nucleotide substitution, insertion, or deletion. Insertion or deletion mutations affect the lengths of the core repeat regions and the amplified flanking regions of STR loci and thus affect STR profiles. Nucleotide substitution mutations (except those residing within the primer-binding regions) do not affect the length of DNA and thus do not affect STR profiles.

However, mutations occurring at the primer-binding sequences of the flanking regions of STR loci may affect genotype results. If a mutation at a primer-binding sequence prevents the primer from annealing to the template, this leads to a complete failure of the amplification of the allele. This phenomenon is known as a *null allele* or silent allele (Figure 20.13). To overcome the consequences of a null allele, an alternative primer annealing to a flanking region away from the mutated sequence can be used. Additionally, a primer with the sequence that is complementary to the known mutation can also be used. If the mutation does not completely prevent the primer from annealing but reduces the efficiency of the amplification, the resulting signal intensity of the allele is usually decreased. This problem may be solved by modifying the condition of amplification for the mutant allele.

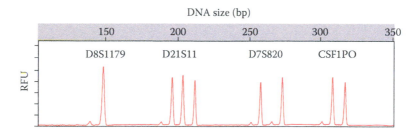

Figure 20.12 Triallele. In this example, the triallele is observed only at D21S11 and not at other STR loci. (© Richard C. Li.)

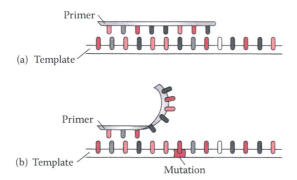

Figure 20.13 Null allele. An allele present in the sample failed to be amplified by one of the primer sets as a result of a rare mutation at the primer-binding sequence of the flanking region: (a) wild type and (b) mutation. (© Richard C. Li.)

20.4.2 Amplification Artifacts

20.4.2.1 Stuttering

A stutter is a minor allele peak, also known as a stutter peak, whose repeat units are shorter or longer than the parental allele peak (Figure 20.14). Less stuttering is observed with pentameric and hexameric repeat unit loci compared to shorter repeat unit loci. The loci that contain complex repeat sequences usually exhibit reduced stuttering. At a given STR locus, large alleles appear to yield more stutter than smaller alleles.

Commonly observed stutters are one repeat unit shorter than the parental allele. It is believed that stuttering is due to the slippage of polymerase, which may have occurred during amplification reactions (Figure 20.15). Stutters with repeat units longer than the parental allele peak can also be observed, but are very rare. The *stutter ratio* is defined as the area of the stutter peak divided by the area of the parental peak. The stutter ratio is usually less than 0.15. A ratio of greater than 0.15 should be interpreted with caution due to the potential presence of DNA from more than one contributor.

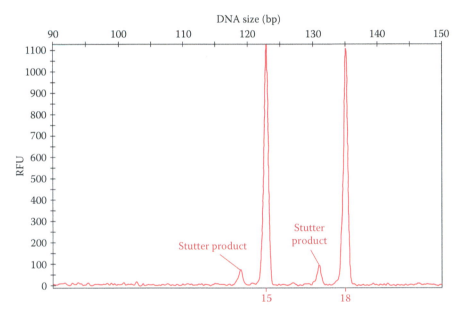

Figure 20.14 Stutter products. (© Richard C. Li.)

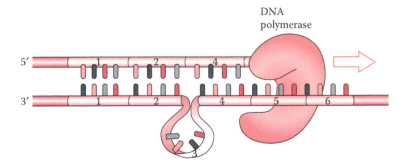

Figure 20.15 Proposed mechanism for stutter products. During the DNA synthesis step of PCR amplification, a DNA polymerase slips, and a region of the primer–template complex becomes unpaired, causing the template strand to form a loop. The consequence of this one-repeat loop is a shortened PCR product smaller than the template by a single repeat unit. (© Richard C. Li.)

20.4.2.2 Nontemplate Adenylation

During PCR amplification, DNA polymerase often adds an extra nucleotide, usually an adenosine, to the 3′-end of an amplicon. Such a phenomenon is referred to as a *nontemplate addition* resulting in an amplicon that is one base pair longer than the parental allele (designated the +A peak), as shown in Figure 20.16. As a result, an amplicon has both the –A amplicon, which corresponds to the size of the parental allele, and the +A amplicon, which represents the amplicon with the nontemplate addition. To simplify the analysis, the most commonly used multiplex STR kits utilize amplification conditions that favor the adenylation of amplicons. Thus, most amplicons in a sample contain an additional adenosine at the 3′ end (+A peak). However, partial nontemplate addition can occur when too much DNA template is utilized in PCR amplification. As a result, a mixture of –A and +A peaks is usually observed.

20.4.2.3 Heterozygote Imbalance

Heterozygote imbalance occurs when one of the alleles has greater peak area or amplitude than the other allele within the same locus in which the two alleles of a heterozygote are compared (Figure 20.17). It is believed that heterozygote imbalance may arise if the DNA sample contains unequal copies of DNA template of the two alleles for the heterozygote, or the two alleles of a heterozygote are unequally amplified, a condition known as *preferential amplification*. Preferential amplification usually refers to an event where a smaller allele is amplified more efficiently than larger ones. As a result, the presence of heterozygote imbalance interferes with the interpretation of samples with a DNA mixture derived from more than one contributor.

20.4.2.4 Allelic Dropout

Allelic dropout occurs when an allele, usually one of the heterozygote alleles, fails to be detected. To date, our understanding of what causes the dropout is very limited. The occurrence of allelic dropout can be the result of an extreme situation of preferential amplification or heterozygote imbalance. Additionally, certain mutations leading to amplification failure (Section 20.4.1.3) can cause allelic dropout.

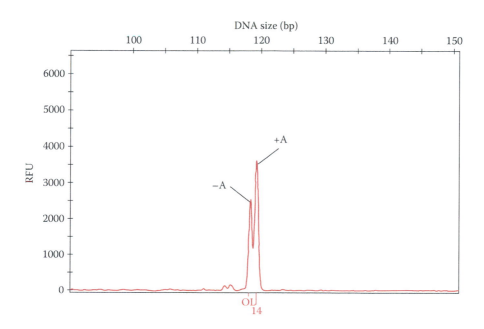

Figure 20.16   Nontemplate adenylation. OL represents the off-ladder allele. (© Richard C. Li.)

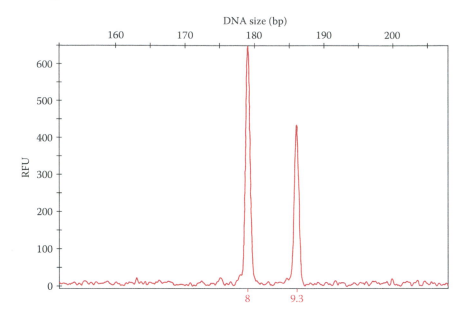

Figure 20.17　Heterozygote imbalance. The signal intensity of one allele is greater than that of the other allele within the same locus. (© Richard C. Li.)

20.4.3 Electrophoretic Artifacts

20.4.3.1 Pull-Up Peaks

A *pull-up peak* occurs when a minor peak of one color on an electropherogram is pulled up from a major allelic peak in another color (Figure 20.18) when the colors have overlapping spectra. For example, a green peak may pull up a yellow peak, or a blue peak may pull up a green peak. A pull-up peak may contribute to the inaccuracy of a profile if the position of a pull-up peak corresponds to the position of an allele. A pull-up peak often occurs when a sample is overloaded or a matrix file (a spectral calibration) is not updated. Thus, loading a proper amount of sample or installing an appropriate matrix file can prevent the occurrence of pull-up peaks.

20.4.3.2 Spikes

Spikes are sharp peaks, with similar signal intensities, that are present in all color panels of an electropherogram (Figure 20.19). Spikes are caused by air bubbles and urea crystals in the capillary of an electrophoretic platform. Voltage spikes can also contribute to spike peaks. The spikes are electrophoretic artifacts and are not reproducible. Thus, electrophoresis can be repeated to verify that the spikes occurred in a previous electrophoresis result.

20.5 Genotyping of Challenging Forensic Samples

20.5.1 Degraded DNA

Environmental exposure, such as high humidity and temperature, of biological evidence can lead to DNA degradation such as the breaking of DNA molecules into small fragments. The more severe the degradation, the more intensive the fragmentation. In forensic DNA analysis, the size range of STR amplicons is usually 100–500 bp in length. When a sample experiences some degradation, large alleles are less likely to be amplified than small alleles (Figures 20.20 and 20.21). As a result, the dropout of large alleles often occurs, leading to a partial DNA profile or even a failure in obtaining a DNA profile. To address this issue, the PCR primers can be

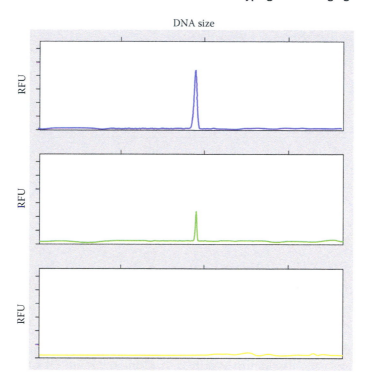

Figure 20.18 Pull-up peaks. The peaks with overlapping spectra observed in the top and middle panels are not observed in the bottom panel. (© Richard C. Li.)

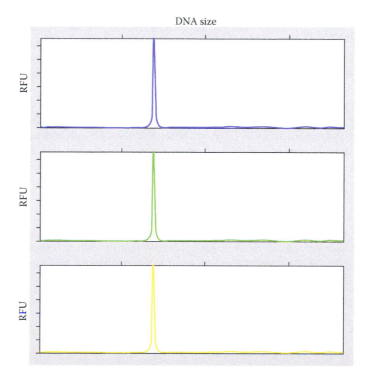

Figure 20.19 Spike peaks can be observed in various intensities. (© Richard C. Li.)

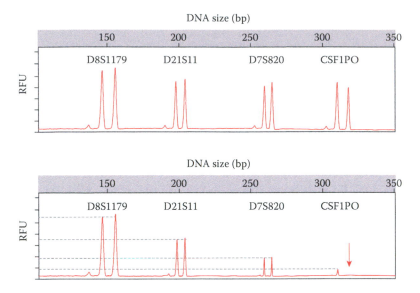

Figure 20.20 Diagram of STR profiles of samples: without DNA degradation (top) and with degradation (bottom). Arrow: allelic dropout. (© Richard C. Li.)

redesigned to anneal more proximally to the STR core repeat region than standard STR primers, yielding small amplicons also known as miniSTRs. Using the miniSTR strategy, more alleles can be detected in degraded DNA samples than using the standard STR primers.

20.5.2 Low Copy Number DNA Testing

Low copy number (LCN) DNA analysis involves the testing of very small amounts of DNA (<100 pg) in a sample. LCN DNA analysis is often needed for samples derived from evidence such as fingerprints and tools and weapons handled by perpetrators. STR analysis of extremely low levels of human DNA can be achieved by increasing the number of PCR cycles (e.g., increasing from 28 to 34 cycles) to improve the yield of amplicons, thus improving the sensitivity of the analysis.

However, this approach also increases the appearance of artifacts that can make interpretation difficult. For instance, the occurrence of allele dropout, heterozygote imbalance, and stuttering is frequently observed in LCN DNA analysis. Additionally, allele drop-in can arise from contamination. The phenomenon of allele drop-in is usually not reproducible. In an LCN DNA analysis, genotypes can be determined if identical alleles can be detected from two independent amplification reactions.

20.5.3 Mixtures

Samples of DNA from two or more contributors are commonly encountered in forensic cases such as sexual assaults in which the evidence recovered from a victim is mixed with a suspect's bodily fluids (Figure 20.22). The interpretation of DNA profiles of mixed stains is known as *mixture interpretation*, which is described below:

1. *To determine the presence of a mixture*: First, determine whether the source of the DNA in the sample came from one or more individuals by examining the number of alleles at multiple loci. The characteristics listed below usually indicate a mixture:

   a. Severe heterozygote imbalance.

   b. Stutter ratio above 0.15.

   c. Presence of three or more alleles per locus at multiple loci.

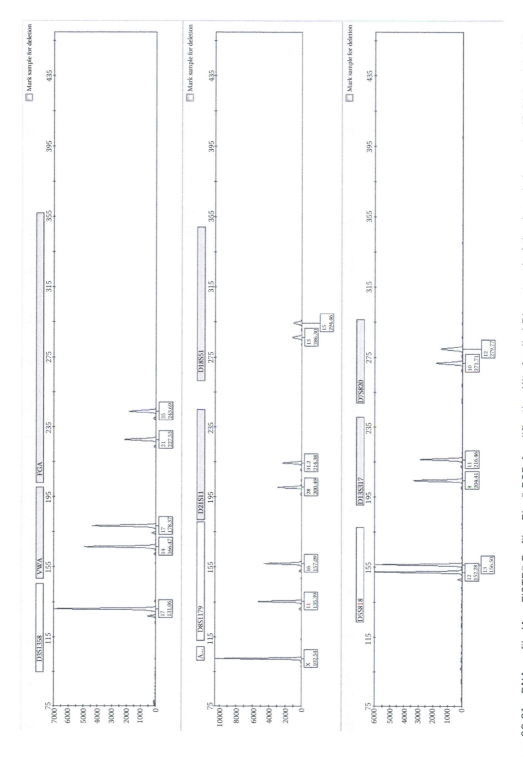

Figure 20.21 DNA profile (AmpFISTR® Profiler Plus® PCR Amplification Kit; Applied Biosystems) of the degraded sample. (© Richard C. Li.)

Additionally, caution should be taken to distinguish the presence of a mixture and various artifacts such as stutters and nontemplate adenylation.

2. *To determine the genotypes of all alleles and to identify the number of contributors:* Note that the maximum number of alleles at any given locus is two per individual. In the case of homozygous or allele overlap, the number of alleles observed can be less than two per individual.

3. *To estimate the ratios of the contributions:* Determine the relative ratios of the contributions to the mixture made by each individual by comparing the peak areas or

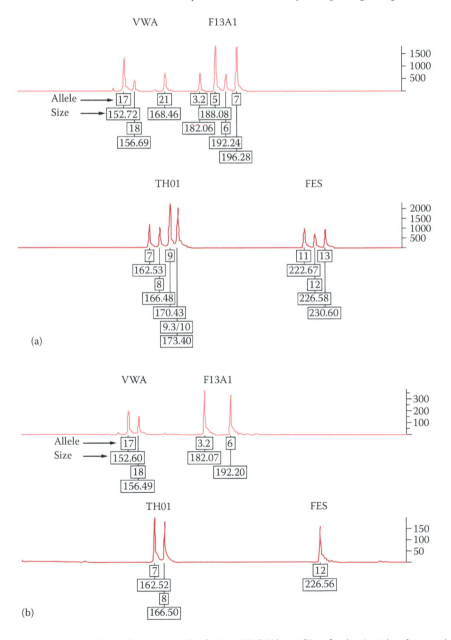

Figure 20.22 DNA profiles of mixed bodily fluids. (a) DNA profile of mixed stains from evidence, (b) DNA profile of the victim. (© Richard C. Li.)

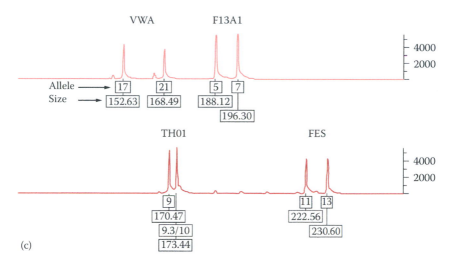

Figure 20.22 (Continued) (c) DNA profile of the suspect.

amplitudes. Amelogenin, a sex-typing marker, is useful in determining the genders of DNA contributors.

4. *To consider all possible genotype combinations*: This may be done by pair-wise comparisons to determine the allele combinations that belong to the minor contributor and those that belong to the major contributor.

5. *To compare reference samples*: The final step is to compare the genotype profiles with the genotypes of reference samples from a suspect and victim. If the DNA profile of the suspect's reference sample matches a major or minor component of the mixture, the suspect cannot be excluded as a contributor.

Detailed guidance for basic steps in mixture interpretation is provided in a number of published guidelines. For example, the Interpretation Guidelines for Autosomal STR Typing by Forensic DNA Testing Laboratories was recently established by the Scientific Working Group on DNA Analysis Methods (SWGDAM), which provides direction on DNA profiling and mixture interpretation. Additionally, guidance for the statistical evaluations of the mixture analysis results is also provided.

Bibliography
STR Profiling

Andersen, J., et al., Report on the third EDNAP collaborative STR exercise. *Forensic Sci Int*, 1996, **78**(2): 83–93.

Anker, R., T. Steinbrueck, and H. Donis-Keller, Tetranucleotide repeat polymorphism at the human thyroid peroxidase (hTPO) locus. *Hum Mol Genet*, 1992, **1**(2): 137.

Asamura, H., et al., MiniSTR multiplex systems based on non-CODIS loci for analysis of degraded DNA samples. *Forensic Sci Int*, 2007, **173**(1): 7–15.

Asamura, H., et al., MiniY-STR quadruplex systems with short amplicon lengths for analysis of degraded DNA samples. *Forensic Sci Int Genet*, 2007, **1**(1): 56–61.

Aznar, J.M., et al., I-DNASE21 system: Development and SWGDAM validation of a new STR 21-plex reaction. *Forensic Sci Int Genet*, 2014, **8**(1): 10–19.

Balding, D.J., Evaluation of mixed-source, low-template DNA profiles in forensic science. *Proc Natl Acad Sci U S A*, 2013, **110**(30): 12241–12246.

Ballantyne, J., E.K. Hanson, and M.W. Perlin, DNA mixture genotyping by probabilistic computer interpretation of binomially-sampled laser captured cell populations: Combining quantitative data for greater identification information. *Sci Justice*, 2013, **53**(2): 103–114.

Bar, W., et al., DNA recommendations. Further report of the DNA commission of the ISFH regarding the use of short tandem repeat systems. International Society for Forensic Haemogenetics. *Int J Legal Med*, 1997, **110**(4): 175–176.

Barbaro, A., et al., Casework application of a stand-alone pentaplex assay of extended-ESS STRs. *Leg Med (Tokyo)*, 2013, **15**(4): 217–221.

Barber, M.D. and B.H. Parkin, Sequence analysis and allelic designation of the two short tandem repeat loci D18S51 and D8S1179. *Int J Legal Med*, 1996, **109**(2): 62–65.

Barber, M.D., B.J. McKeown, and B.H. Parkin, Structural variation in the alleles of a short tandem repeat system at the human alpha fibrinogen locus. *Int J Legal Med*, 1996, **108**(4): 180–185.

Becker, D., et al., New alleles and mutational events at 14 STR loci from different German populations. *Forensic Sci Int Genet*, 2007, **1**(3–4): 232–237.

Benschop, C., H. Haned, and T. Sijen, Consensus and pool profiles to assist in the analysis and interpretation of complex low template DNA mixtures. *Int J Legal Med*, 2013, **127**(1): 11–23.

Benschop, C.C., et al., Assessment of mock cases involving complex low template DNA mixtures: A descriptive study. *Forensic Sci Int Genet*, 2012, **6**(6): 697–707.

Betz, T., et al., "Paterniplex", a highly discriminative decaplex STR multiplex tailored for investigating special problems in paternity testing. *Electrophoresis*, 2007, **28**(21): 3868–3874.

Bill, M., et al., PENDULUM—A guideline-based approach to the interpretation of STR mixtures. *Forensic Sci Int*, 2005, **148**(2–3): 181–189.

Bornman, D.M., et al., Short-read, high-throughput sequencing technology for STR genotyping. *Biotechniques*, 2012, 10.2144/000113857: 1–6.

Brinkmann, B., E. Meyer, and A. Junge, Complex mutational events at the HumD21S11 locus. *Hum Genet*, 1996, **98**(1): 60–64.

Brookes, C., et al., Characterising stutter in forensic STR multiplexes. *Forensic Sci Int Genet*, 2012, **6**(1): 58–63.

Budimlija, Z.M. and T.A. Caragine, Interpretation guidelines for multilocus STR forensic profiles from low template DNA samples. *Methods Mol Biol*, 2012, **830**: 199–211.

Budowle, B., et al., Validation studies of the CTT STR multiplex system. *J Forensic Sci*, 1997, **42**(4): 701–707.

Budowle, B., et al., *DNA Typing Protocols: Molecular Biology and Forensic Analysis*, 2000. Natick, MA: Eaton Publishing.

Budowle, B., et al., STR primer concordance study. *Forensic Sci Int*, 2001, **124**(1): 47–54.

Buse, E.L., et al., Performance evaluation of two multiplexes used in fluorescent short tandem repeat DNA analysis. *J Forensic Sci*, 2003, **48**(2): 348–357.

Butler, J.M., Constructing STR multiplex assays. *Methods Mol Biol*, 2005, **297**: 53–66.

Butler, J.M., Genetics and genomics of core short tandem repeat loci used in human identity testing. *J Forensic Sci*, 2006, **51**(2): 253–265.

Butler, J.M., Y. Shen, and B.R. McCord, The development of reduced size STR amplicons as tools for analysis of degraded DNA. *J Forensic Sci*, 2003, **48**(5): 1054–1064.

Caragine, T., et al., Validation of testing and interpretation protocols for low template DNA samples using AmpFlSTR Identifiler. *Croat Med J*, 2009, **50**(3): 250–267.

Castella, V., J. Gervaix, and D. Hall, DIP-STR: Highly sensitive markers for the analysis of unbalanced genomic mixtures. *Hum Mutat*, 2013, **34**(4): 644–654.

Chakraborty, R., et al., The utility of short tandem repeat loci beyond human identification: Implications for development of new DNA typing systems. *Electrophoresis*, 1999, **20**(8): 1682–1696.

Chen, P.Y., et al., Substitution mutation induced migration anomaly of a D10S2325 allele on capillary electrophoresis. *Int J Legal Med*, 2013, **127**(2): 363–368.

Clayton, T.M., et al., A genetic basis for anomalous band patterns encountered during DNA STR profiling. *J Forensic Sci*, 2004, **49**(6): 1207–1214.

Coble, M.D. and J.M. Butler, Characterization of new miniSTR loci to aid analysis of degraded DNA. *J Forensic Sci*, 2005, **50**(1): 43–53.

Coble, M.D., Capillary electrophoresis of miniSTR markers to genotype highly degraded DNA samples. *Methods Mol Biol*, 2012, **830**: 31–42.

Collins, J.R., et al., An exhaustive DNA micro-satellite map of the human genome using high performance computing. *Genomics*, 2003, **82**(1): 10–19.

Collins, P.J., et al., Developmental validation of a single-tube amplification of the 13 CODIS STR loci, D2S1338, D19S433, and amelogenin: The AmpFlSTR Identifiler PCR amplification kit. *J Forensic Sci*, 2004, **49**(6): 1265–1277.

Cotton, E.A., et al., Validation of the AMPFlSTR SGM plus system for use in forensic casework. *Forensic Sci Int*, 2000, **112**(2–3): 151–161.

Cowell, R.G., Validation of an STR peak area model. *Forensic Sci Int Genet*, 2009, **3**(3): 193–199.

Cowell, R.G., S.L. Lauritzen, and J. Mortera, Identification and separation of DNA mixtures using peak area information. *Forensic Sci Int*, 2007, **166**(1): 28–34.

Cowen, S., et al., An investigation of the robustness of the consensus method of interpreting low-template DNA profiles. *Forensic Sci Int Genet*, 2011, **5**(5): 400–406.

Crouse, C.A. and J. Schumm, Investigation of species specificity using nine PCR-based human STR systems. *J Forensic Sci*, 1995, **40**(6): 952–956.

Cupples, C.M., et al., STR profiles from DNA samples with "undetected" or low quantifiler results. *J Forensic Sci*, 2009, **54**(1): 103–107.

Dauber, E.M., et al., Germline mutations of STR-alleles include multi-step mutations as defined by sequencing of repeat and flanking regions. *Forensic Sci Int Genet*, 2012, **6**(3): 381–386.

Davis, C., et al., Variants observed for STR locus SE33: A concordance study. *Forensic Sci Int Genet*, 2012, **6**(4): 494–497.

Debernardi, A., et al., One year variability of peak heights, heterozygous balance and inter-locus balance for the DNA positive control of AmpFlSTR© Identifiler© STR kit. *Forensic Sci Int Genet*, 2011, **5**(1): 43–49.

Dib, C., et al., A comprehensive genetic map of the human genome based on 5,264 microsatellites. *Nature*, 1996, **380**(6570): 152–154.

Divne, A.M., H. Edlund, and M. Allen, Forensic analysis of autosomal STR markers using pyrosequencing. *Forensic Sci Int Genet*, 2010, **4**(2): 122–129.

Drabek, J., et al., Concordance study between Miniplex assays and a commercial STR typing kit. *J Forensic Sci*, 2004, **49**(4): 859–860.

Duewer, D.L., et al., NIST mixed stain study 3: Signal intensity balance in commercial short tandem repeat multiplexes. *Anal Chem*, 2004, **76**(23): 6928–6934.

Edwards, A., et al., DNA typing and genetic mapping with trimeric and tetrameric tandem repeats. *Am J Hum Genet*, 1991, **49**(4): 746–756.

Edwards, A., et al., Genetic variation at five trimeric and tetrameric tandem repeat loci in four human population groups. *Genomics*, 1992, **12**(2): 241–253.

Ellegren, H., Microsatellites: Simple sequences with complex evolution. *Nat Rev Genet*, 2004, **5**(6): 435–445.

Ensenberger, M.G., et al., Developmental validation of the PowerPlex 16 HS system: An improved 16-locus fluorescent STR multiplex. *Forensic Sci Int Genet*, 2010, **4**(4): 257–264.

Ensenberger, M.G., et al., Developmental validation of the PowerPlex® 21 system. *Forensic Sci Int Genet*, 2014, **9**: 169–178.

Flores, S., et al., Internal validation of the GlobalFiler express PCR amplification kit for the direct amplification of reference DNA samples on a high-throughput automated workflow. *Forensic Sci Int Genet*, 2014, **10C**: 33–39.

Fordyce, S.L., et al., High-throughput sequencing of core STR loci for forensic genetic investigations using the Roche Genome Sequencer FLX platform. *Biotechniques*, 2011, **51**(2): 127–133.

Fregeau, C.J. and R.M. Fourney, DNA typing with fluorescently tagged short tandem repeats: A sensitive and accurate approach to human identification. *Biotechniques*, 1993, **15**(1): 100–119.

Fregeau, C.J., K.L. Bowen, and R.M. Fourney, Validation of highly polymorphic fluorescent multiplex short tandem repeat systems using two generations of DNA sequencers. *J Forensic Sci*, 1999, **44**(1): 133–166.

Fregeau, C.J., et al., AmpFlSTR Profiler Plus short tandem repeat DNA analysis of casework samples, mixture samples, and nonhuman DNA samples amplified under reduced PCR volume conditions (25 microL). *J Forensic Sci*, 2003, **48**(5): 1014–1034.

Ge, J., A. Eisenberg, and B. Budowle, Developing criteria and data to determine best options for expanding the core CODIS loci. *Investig Genet*, 2012, **3**: 1.

Gibb, A.J., et al., Characterisation of forward stutter in the AmpFlSTR SGM Plus PCR. *Sci Justice*, 2009, **49**(1): 24–31.

Gil, A., et al., Linkage between HPRTB STR alleles and Lesch-Nyhan syndrome inside a family: Implications in forensic casework. *Forensic Sci Int Genet*, 2013, **7**(1): e5–e6.

Gilder, J.R., et al., Run-specific limits of detection and quantitation for STR-based DNA testing. *J Forensic Sci*, 2007, **52**(1): 97–101.

Gilder, J.R., et al., Magnitude-dependent variation in peak height balance at heterozygous STR loci. *Int J Legal Med*, 2011, **125**(1): 87–94.

Gill, P. and J. Buckleton, Commentary on: Budowle B, Onorato AJ, Callaghan TF, Della Manna A, Gross AM, Guerrieri RA, Luttman JC, McClure DL. Mixture interpretation: Defining the relevant features for guidelines for the assessment of mixed DNA profiles in forensic casework. J Forensic Sci 2009;54(4):810–821. *J Forensic Sci*, 2010, **55**(1): 265–268; author reply 269–272.

Gill, P., A. Kirkham, and J. Curran, LoComatioN: A software tool for the analysis of low copy number DNA profiles. *Forensic Sci Int*, 2007, **166**(2–3): 128–138.

Gill, P., et al., The evolution of DNA databases—Recommendations for new European STR loci. *Forensic Sci Int*, 2006, **156**(2–3): 242–244.

Gill, P., et al., Interpretation of complex DNA profiles using empirical models and a method to measure their robustness. *Forensic Sci Int Genet*, 2008, **2**(2): 91–103.

Gill, P., et al., DNA commission of the International Society of Forensic Genetics: Recommendations on the evaluation of STR typing results that may include drop-out and/or drop-in using probabilistic methods. *Forensic Sci Int Genet*, 2012, **6**(6): 679–688.

Gill, P., Role of short tandem repeat DNA in forensic casework in the UK—Past, present, and future perspectives. *Biotechniques*, 2002, **32**(2): 366–368, 370, 372, passim.

Goodwin, W. and C. Peel, Theoretical value of the recommended expanded European standard set of STR loci for the identification of human remains. *Med Sci Law*, 2012, **52**(3): 162–168.

Graydon, M., F. Cholette, and L.K. Ng, Inferring ethnicity using 15 autosomal STR loci—Comparisons among populations of similar and distinctly different physical traits. *Forensic Sci Int Genet*, 2009, **3**(4): 251–254.

Green, R.L., et al., Developmental validation of the AmpFlSTR® NGM SElect PCR amplification kit: A next-generation STR multiplex with the SE33 locus. *Forensic Sci Int Genet*, 2013, **7**(1): 41–51.

Greenspoon, S.A., et al., Validation of the PowerPlex 1.1 loci for use in human identification. *J Forensic Sci*, 2000, **45**(3): 677–683.

Greenspoon, S.A., et al., Validation and implementation of the PowerPlex 16 BIO System STR multiplex for forensic casework. *J Forensic Sci*, 2004, **49**(1): 71–80.

Griffiths, R.A., et al., New reference allelic ladders to improve allelic designation in a multiplex STR system. *Int J Legal Med*, 1998, **111**(5): 267–272.

Guo, F., et al., Development of a 24-locus multiplex system to incorporate the core loci in the combined DNA index system (CODIS) and the European standard set (ESS). *Forensic Sci Int Genet*, 2014, **8**(1): 44–54.

Hammond, H.A., et al., Evaluation of 13 short tandem repeat loci for use in personal identification applications. *Am J Hum Genet*, 1994, **55**(1): 175–189.

Haned, H., et al., Estimating drop-out probabilities in forensic DNA samples: A simulation approach to evaluate different models. *Forensic Sci Int Genet*, 2011, **5**(5): 525–531.

Harder, M., et al., STR-typing of ancient skeletal remains: Which multiplex-PCR kit is the best? *Croat Med J*, 2012, **53**(5): 416–422.

Hares, D.R., Addendum to expanding the CODIS core loci in the United States. *Forensic Sci Int Genet*, 2012, **6**(5): e135.

Hares, D.R., Expanding the CODIS core loci in the United States. *Forensic Sci Int Genet*, 2012, **6**(1): e52–e54.

Heinrich, M., et al., Allelic drop-out in the STR system ACTBP2 (SE33) as a result of mutations in the primer binding region. *Int J Legal Med*, 2004, **118**(6): 361–363.

Hering, S., J. Edelmann, and J. Dressler, Sequence variations in the primer binding regions of the highly polymorphic STR system SE33. *Int J Legal Med*, 2002, **116**(6): 365–367.

Hering, S., et al., Complex variability of intron 40 of the von Willebrand factor (vWF) gene. *Int J Legal Med*, 2008, **122**(1): 67–71.

Hill, C.R., Capillary electrophoresis and 5-channel LIF detection of a 26plex autosomal STR assay for human identification. *Methods Mol Biol*, 2012, **830**: 17–29.

Hill, C.R., et al., Concordance study between the AmpFlSTR MiniFiler PCR amplification kit and conventional STR typing kits. *J Forensic Sci*, 2007, **52**(4): 870–873.

Hill, C.R., et al., Characterization of 26 miniSTR loci for improved analysis of degraded DNA samples. *J Forensic Sci*, 2008, **53**(1): 73–80.

Holland, M.M. and W. Parson, GeneMarker® HID: A reliable software tool for the analysis of forensic STR data. *J Forensic Sci*, 2011, **56**(1): 29–35.

Holt, C.L., et al., TWGDAM validation of AmpFlSTR PCR amplification kits for forensic DNA casework. *J Forensic Sci*, 2002, **47**(1): 66–96.

Huang, Y.M., et al., Assessment of application value of 19 autosomal short tandem repeat loci of GoldenEye 20A kit in forensic paternity testing. *Int J Legal Med*, 2013, **127**(3): 587–590.

Irwin, J.A., et al., Application of low copy number STR typing to the identification of aged, degraded skeletal remains. *J Forensic Sci*, 2007, **52**(6): 1322–1327.

Jamieson, A., et al., Two-, three-, and four-person mixtures in forensic casework: Difficulties and questions. *Croat Med J*, 2011, **52**(5): 653–654; author reply 654–656.

Jiang, X., et al., Development of a 20-locus fluorescent multiplex system as a valuable tool for national DNA database. *Forensic Sci Int Genet*, 2013, **7**(2): 279–289.

Jobling, M.A., In the name of the father: Surnames and genetics. *Trends Genet*, 2001, **17**(6): 353–357.

Junge, A., et al., Validation of the multiplex kit genRESMPX-2 for forensic casework analysis. *Int J Legal Med*, 2003, **117**(6): 317–325.

Kadash, K., et al., Validation study of the TrueAllele automated data review system. *J Forensic Sci*, 2004, **49**(4): 660–667.

Katsanis, S.H. and J.K. Wagner, Characterization of the standard and recommended CODIS markers. *J Forensic Sci*, 2013, **58**(Suppl 1): S169–S172.

Kimpton, C., A. Walton, and P. Gill, A further tetranucleotide repeat polymorphism in the vWF gene. *Hum Mol Genet*, 1992, **1**(4): 287.

Kimpton, C.P., et al., Automated DNA profiling employing multiplex amplification of short tandem repeat loci. *PCR Methods Appl*, 1993, **3**(1): 13–22.

Kimpton, C., et al., Evaluation of an automated DNA profiling system employing multiplex amplification of four tetrameric STR loci. *Int J Legal Med*, 1994, **106**(6): 302–311.

Kline, M.C., et al., STR sequence analysis for characterizing normal, variant, and null alleles. *Forensic Sci Int Genet*, 2011, **5**(4): 329–332.

Konarzewska, M., et al., Population data and sequence analysis of a "new" microsatellite locus HumHUU (D16S3433). *Forensic Sci Int Genet*, 2010, **4**(5): e143–e144.

Krenke, B.E., et al., Validation of a 16-locus fluorescent multiplex system. *J Forensic Sci*, 2002, **47**(4): 773–785.

Laberke, P.J., et al., Method to predict the chance of developing a male profile out of mixtures of male and female DNA. *Int J Legal Med*, 2012, **126**(1): 157–160.

Laird, R., P.M. Schneider, and S. Gaudieri, Forensic STRs as potential disease markers: A study of VWA and von Willebrand's disease. *Forensic Sci Int Genet*, 2007, **1**(3–4): 253–261.

Lareu, M.V., et al., A highly variable STR at the D12S391 locus. *Int J Legal Med*, 1996, **109**(3): 134–138.

Laurin, N., A. DeMoors, and C. Fregeau, Performance of Identifiler Direct and PowerPlex 16 HS on the applied biosystems 3730 DNA analyzer for processing biological samples archived on FTA cards. *Forensic Sci Int Genet*, 2012, **6**(5): 621–629.

Leclair, B., et al., Systematic analysis of stutter percentages and allele peak height and peak area ratios at heterozygous STR loci for forensic casework and database samples. *J Forensic Sci*, 2004, **49**(5): 968–980.

Lederer, T., et al., Characterization of two unusual allele variants at the STR locus ACTBP2 (SE33). *Forensic Sci Med Pathol*, 2008, **4**(3): 164–166.

Leibelt, C., et al., Identification of a D8S1179 primer binding site mutation and the validation of a primer designed to recover null alleles. *Forensic Sci Int*, 2003, **133**(3): 220–227.

Levedakou, E.N., et al., Characterization and validation studies of powerPlex 2.1, a nine-locus short tandem repeat (STR) multiplex system and penta D monoplex. *J Forensic Sci*, 2002, **47**(4): 757–772.

Li, H., et al., Three tetranucleotide polymorphisms for loci: D3S1352; D3S1358; D3S1359. *Hum Mol Genet*, 1993, **2**(8): 1327.

Lohmueller, K.E., Graydon et al. provide no new evidence that forensic STR loci are functional. *Forensic Sci Int Genet*, 2010, **4**(4): 273–274.

Luce, C., et al., Validation of the AmpFlSTR MiniFiler PCR amplification kit for use in forensic casework. *J Forensic Sci*, 2009, **54**(5): 1046–1054.

Lucy, D., et al., The probability of achieving full allelic representation for LCN-STR profiling of haploid cells. *Sci Justice*, 2007, **47**(4): 168–171.

Lygo, J.E., et al., The validation of short tandem repeat (STR) loci for use in forensic casework. *Int J Legal Med*, 1994, **107**(2): 77–89.

Manabe, S., et al., Mixture interpretation: Experimental and simulated reevaluation of qualitative analysis. *Leg Med (Tokyo)*, 2013, **15**(2): 66–71.

Mardini, A.C., et al., Mutation rate estimates for 13 STR loci in a large population from Rio Grande do Sul, Southern Brazil. *Int J Legal Med*, 2013, **127**(1): 45–47.

Margolis-Nunno, H., et al., A new allele of the short tandem repeat (STR) locus, CSF1PO. *J Forensic Sci*, 2001, **46**(6): 1480–1483.

Martinez-Gonzalez, L.J., et al., Intentional mixed buccal cell reference sample in a paternity case. *J Forensic Sci*, 2007, **52**(2): 397–399.

Masibay, A., T.J. Mozer, and C. Sprecher, Promega corporation reveals primer sequences in its testing kits. *J Forensic Sci*, 2000, **45**(6): 1360–1362.

Mills, K.A., D. Even, and J.C. Murray, Tetranucleotide repeat polymorphism at the human alpha fibrinogen locus (FGA). *Hum Mol Genet*, 1992, **1**(9): 779.

Miozzo, M.C., et al., Characterization of the variant allele 9.2 of Penta D locus. *J Forensic Sci*, 2007, **52**(5): 1073–1076.

Mitchell, A.A., et al., Validation of a DNA mixture statistics tool incorporating allelic drop-out and drop-in. *Forensic Sci Int Genet*, 2012, **6**(6): 749–761.

Moller, A., E. Meyer, and B. Brinkmann, Different types of structural variation in STRs: HumFES/FPS, HumVWA and HumD21S11. *Int J Legal Med*, 1994, **106**(6): 319–323.

Moretti, T.R., et al., Validation of short tandem repeats (STRs) for forensic usage: Performance testing of fluorescent multiplex STR systems and analysis of authentic and simulated forensic samples. *J Forensic Sci*, 2001, **46**(3): 647–660.

Mulero, J.J., et al., Development and validation of the AmpFlSTR MiniFiler PCR amplification kit: A MiniSTR multiplex for the analysis of degraded and/or PCR inhibited DNA. *J Forensic Sci*, 2008, **53**(4): 838–852.

Muller, K., et al., Q8—A short amplicon multiplex including the German DNA database systems. *Forensic Sci Int Genet*, 2007, **1**(2): 205–207.

Muller, K., et al., Validation of the short amplicon multiplex q8 including the German DNA database systems. *J Forensic Sci*, 2009, **54**(4): 862–865.

Muller, K., et al., Casework testing of the multiplex kits AmpFlSTR SEfiler Plus PCR amplification kit (AB), PowerPlex S5 system (Promega) and AmpFlSTR MiniFiler PCR amplification kit (AB). *Forensic Sci Int Genet*, 2010, **4**(3): 200–205.

Narkuti, V., et al., Microsatellite mutation in the maternally/paternally transmitted D18S51 locus: Two cases of allele mismatch in the child. *Clin Chim Acta*, 2007, **381**(2): 171–175.

Narkuti, V., et al., Mother-child double incompatibility at vWA and D5S818 loci in paternity testing. *Clin Chem Lab Med*, 2007, **45**(10): 1288–1291.

Narkuti, V., et al., Single and double incompatibility at vWA and D8S1179/D21S11 loci between mother and child: Implications in kinship analysis. *Clin Chim Acta*, 2008, **395**(1–2): 162–165.

Nelson, M.S., et al., Detection of a primer-binding site polymorphism for the STR locus D16S539 using the Powerplex 1.1 system and validation of a degenerate primer to correct for the polymorphism. *J Forensic Sci*, 2002, **47**(2): 345–349.

Odriozola, A., et al., Development and validation of I-DNA1: A 15-Loci multiplex system for identity testing. *Int J Legal Med*, 2011, **125**(5): 685–694.

Odriozola, A., et al., Development and validation for identity testing of I-DNADuo, a combination of I-DNA1 and a new multiplex system, I-DNA2. *Int J Legal Med*, 2012, **126**(1): 167–172.

Odriozola, A., et al., Recent advances and considerations for the future in forensic analysis of degraded DNA by autosomic miniSTR multiplex genotyping. *Recent Pat DNA Gene Seq*, 2011, **5**(2): 110–116.

Oh, C.S., et al., Autosomal short tandem repeat analysis of ancient DNA by coupled use of mini- and conventional STR kits. *J Forensic Sci*, 2012, **57**(3): 820–825.

Oostdik, K., et al., Developmental validation of the PowerPlex® 18D system, a rapid STR multiplex for analysis of reference samples. *Forensic Sci Int Genet*, 2013, **7**(1): 129–135.

Opel, K.L., et al., The application of miniplex primer sets in the analysis of degraded DNA from human skeletal remains. *J Forensic Sci*, 2006, **51**(2): 351–356.

Opel, K.L., et al., Developmental validation of reduced-size STR miniplex primer sets. *J Forensic Sci*, 2007, **52**(6): 1263–1271.

Park, H., et al., Detection of very large off-ladder alleles at the PentaE locus in a 15 locus autosomal STR database of 199 Korean individuals. *Forensic Sci Int Genet*, 2012, **6**(6): e189–e191.

Parsons, T.J., et al., Application of novel "mini-amplicon" STR multiplexes to high volume casework on degraded skeletal remains. *Forensic Sci Int Genet*, 2007, **1**(2): 175–179.

Pascali, V.L. and S. Merigioli, Joint Bayesian analysis of forensic mixtures. *Forensic Sci Int Genet*, 2012, **6**(6): 735–748.

Perez, J., et al., Estimating the number of contributors to two-, three-, and four-person mixtures containing DNA in high template and low template amounts. *Croat Med J*, 2011, **52**(3): 314–326.

Perlin, M.W., et al., Validating TrueAllele® DNA mixture interpretation. *J Forensic Sci*, 2011, **56**(6): 1430–1447.

Perlin, M.W., J.L. Belrose, and B.W. Duceman, New York State TrueAllele® casework validation study. *J Forensic Sci*, 2013, **58**(6): 1458–1466.

Perlin, M.W., et al., TrueAllele casework on Virginia DNA mixture evidence: Computer and manual interpretation in 72 reported criminal cases. *PLoS One*, 2014, **9**(3): e92837.

Petricevic, S., et al., Validation and development of interpretation guidelines for low copy number (LCN) DNA profiling in New Zealand using the AmpFlSTR SGM Plus multiplex. *Forensic Sci Int Genet*, 2010, **4**(5): 305–310.

Pfeifer, C.M., et al., Comparison of different interpretation strategies for low template DNA mixtures. *Forensic Sci Int Genet*, 2012, **6**(6): 716–722.

Phillips, C., et al., Development of a novel forensic STR multiplex for ancestry analysis and extended identity testing. *Electrophoresis*, 2013, **34**(8): 1151–1162.

Picanco, J.B., et al., Tri-allelic pattern at the TPOX locus: A familial study. *Gene*, 2014, **535**(2): 353–358.

Pitterl, F., et al., Increasing the discrimination power of forensic STR testing by employing high-performance mass spectrometry, as illustrated in indigenous South African and Central Asian populations. *Int J Legal Med*, 2010, **124**(6): 551–558.

Pitterl, F., et al., The next generation of DNA profiling—STR typing by multiplexed PCR—Ion-pair RP LC-ESI time-of-flight MS. *Electrophoresis*, 2008, **29**(23): 4739–4750.

Planz, J.V., et al., Automated analysis of sequence polymorphism in STR alleles by PCR and direct electrospray ionization mass spectrometry. *Forensic Sci Int Genet*, 2012, **6**(5): 594–606.

Poetsch, M., et al., Powerplex® ES versus Powerplex® S5—Casework testing of the new screening kit. *Forensic Sci Int Genet*, 2011, **5**(1): 57–63.

Poltl, R., C. Luckenbach, and H. Ritter, The short tandem repeat locus D3S1359. *Forensic Sci Int*, 1998, **95**(2): 163–168.

Poltl, R., et al., Typing of the short tandem repeat D8S347 locus with different fluorescence markers. *Electrophoresis*, 1997, **18**(15): 2871–2873.

Polymeropoulos, M.H., et al., Tetranucleotide repeat polymorphism at the human tyrosine hydroxylase gene (TH). *Nucleic Acids Res*, 1991, **19**(13): 3753.

Puch-Solis, R., et al., Evaluating forensic DNA profiles using peak heights, allowing for multiple donors, allelic dropout and stutters. *Forensic Sci Int Genet*, 2013, **7**(5): 555–563.

Puers, C., et al., Identification of repeat sequence heterogeneity at the polymorphic short tandem repeat locus HUMTH01[AATG]n and reassignment of alleles in population analysis by using a locus-specific allelic ladder. *Am J Hum Genet*, 1993, **53**(4): 953–958.

Reichenpfader, B., R. Zehner, and M. Klintschar, Characterization of a highly variable short tandem repeat polymorphism at the D2S1242 locus. *Electrophoresis*, 1999, **20**(3): 514–517.

Repnikova, E.A., et al., Characterization of copy number variation in genomic regions containing STR loci using array comparative genomic hybridization. *Forensic Sci Int Genet*, 2013, **7**(5): 475–481.

Ricci, U., et al., A single mutation in the FGA locus responsible for false homozygosities and discrepancies between commercial kits in an unusual paternity test case. *J Forensic Sci*, 2007, **52**(2): 393–396.

Rockenbauer, E., et al., Characterization of mutations and sequence variants in the D21S11 locus by next generation sequencing. *Forensic Sci Int Genet*, 2014, **8**(1): 68–72.

Ruitberg, C.M., D.J. Reeder, and J.M. Butler, STRBase: A short tandem repeat DNA database for the human identity testing community. *Nucleic Acids Res*, 2001, **29**(1): 320–322.

Schilz, F., S. Hummel, and B. Herrmann, Design of a multiplex PCR for genotyping 16 short tandem repeats in degraded DNA samples. *Anthropol Anz*, 2004, **62**(4): 369–378.

Schneider, H.R., et al., ACTBP2-nomenclature recommendations of GEDNAP. *Int J Legal Med*, 1998, **111**(2): 97–100.

Schumm, J.W., et al., A 27-locus STR assay to meet all United States and European law enforcement agency standards. *J Forensic Sci*, 2013, **58**(6): 1584–1592.

Seidl, C., et al., Sequence analysis and population data of short tandem repeat polymorphisms at loci D8S639 and D11S488. *Int J Legal Med*, 1999, **112**(6): 355–359.

Sharma, V. and M. Litt, Tetranucleotide repeat polymorphism at the D21S11 locus. *Hum Mol Genet*, 1992, **1**(1): 67.

Slooten, K. and F. Ricciardi, Estimation of mutation probabilities for autosomal STR markers. *Forensic Sci Int Genet*, 2013, **7**(3): 337–344.

Smith, R.N., Accurate size comparison of short tandem repeat alleles amplified by PCR. *Biotechniques*, 1995, **18**(1): 122–128.

Sparkes, R., et al., The validation of a 7-locus multiplex STR test for use in forensic casework. (I). Mixtures, ageing, degradation and species studies. *Int J Legal Med*, 1996, **109**(4): 186–194.

Spathis, R. and J.K. Lum, An updated validation of Promega's PowerPlex 16 system: High throughput data-basing under reduced PCR volume conditions on applied biosystem's 96 capillary 3730xl DNA analyzer. *J Forensic Sci*, 2008, **53**(6): 1353–1357.

Sun, H.Y., et al., A paternity case with mutations at three CODIS core STR loci. *Forensic Sci Int Genet*, 2012, **6**(1): e61–e62.

Takayama, T., et al., Identification of a rare mutation in a TH01 primer binding site. *Leg Med (Tokyo)*, 2007, **9**(6): 289–292.

Tomsey, C.S., et al., Comparison of PowerPlex 16, PowerPlex 1.1/2.1, and ABI AmpFiSTR Profiler Plus/COfiler for forensic use. *Croat Med J*, 2001, **42**(3): 239–243.

Tucker, V.C., A.J. Kirkham, and A.J. Hopwood, Forensic validation of the PowerPlex® ESI 16 STR multiplex and comparison of performance with AmpFlSTR® SGM Plus®. *Int J Legal Med*, 2012, **126**(3): 345–356.

Tucker, V.C., et al., Developmental validation of the PowerPlex® ESI 16 and PowerPlex® ESI 17 systems: STR multiplexes for the new European standard. *Forensic Sci Int Genet*, 2011, **5**(5): 436–448.

Tucker, V.C., et al., Developmental validation of the PowerPlex® ESX 16 and PowerPlex® ESX 17 systems. *Forensic Sci Int Genet*, 2012, **6**(1): 124–131.

Tvedebrink, T., et al., Identifying contributors of DNA mixtures by means of quantitative information of STR typing. *J Comput Biol*, 2012, **19**(7): 887–902.

Tvedebrink, T., et al., Performance of two 17 locus forensic identification STR kits-applied biosystems's AmpFlSTR® NGMSElect™ and Promega's PowerPlex® ESI17 kits. *Forensic Sci Int Genet*, 2012, **6**(5): 523–531.

Tvedebrink, T., et al., Statistical model for degraded DNA samples and adjusted probabilities for allelic dropout. *Forensic Sci Int Genet*, 2012, **6**(1): 97–101.

Urquhart, A., C.P. Kimpton, and P. Gill, Sequence variability of the tetranucleotide repeat of the human beta-actin related pseudogene H-beta-Ac-psi-2 (ACTBP2) locus. *Hum Genet*, 1993, **92**(6): 637–638.

Urquhart, A., et al., Variation in short tandem repeat sequences—A survey of twelve microsatellite loci for use as forensic identification markers. *Int J Legal Med*, 1994, **107**(1): 13–20.

Vallone, P.M., C.R. Hill, and J.M. Butler, Demonstration of rapid multiplex PCR amplification involving 16 genetic loci. *Forensic Sci Int Genet*, 2008, **3**(1): 42–45.

Van Neste, C., et al., Forensic STR analysis using massive parallel sequencing. *Forensic Sci Int Genet*, 2012, **6**(6): 810–818.

Vanderheyden, N., et al., Identification and sequence analysis of discordant phenotypes between AmpFlSTR SGM Plus™ and PowerPlex® 16. *Int J Legal Med*, 2007, **121**(4): 297–301.

Voskoboinik, L. and A. Darvasi, Forensic identification of an individual in complex DNA mixtures. *Forensic Sci Int Genet*, 2011, **5**(5): 428–435.

Walsh, P.S., H.A. Erlich, and R. Higuchi, Preferential PCR amplification of alleles: Mechanisms and solutions. *PCR Methods Appl*, 1992, **1**(4): 241–250.

Walsh, P.S., N.J. Fildes, and R. Reynolds, Sequence analysis and characterization of stutter products at the tetranucleotide repeat locus vWA. *Nucleic Acids Res*, 1996, **24**(14): 2807–2812.

Wang, D.Y., et al., Development and validation of the AmpFlSTR® Identifiler® direct PCR amplification kit: A multiplex assay for the direct amplification of single-source samples. *J Forensic Sci*, 2011, **56**(4): 835–845.

Wang, D.Y., et al., Developmental validation of the AmpFlSTR® Identifiler® Plus PCR amplification kit: An established multiplex assay with improved performance. *J Forensic Sci*, 2012, **57**(2): 453–465.

Warshauer, D.H., et al., STRait Razor: A length-based forensic STR allele-calling tool for use with second generation sequencing data. *Forensic Sci Int Genet*, 2013, **7**(4): 409–417.

Weber, J.L. and P.E. May, Abundant class of human DNA polymorphisms which can be typed using the polymerase chain reaction. *Am J Hum Genet*, 1989, **44**(3): 388–396.

Welch, L.A., et al., European Network of Forensic Science Institutes (ENFSI): Evaluation of new commercial STR multiplexes that include the European standard set (ESS) of markers. *Forensic Sci Int Genet*, 2012, **6**(6): 819–826.

Westen, A.A., et al., Assessment of the stochastic threshold, back- and forward stutter filters and low template techniques for NGM. *Forensic Sci Int Genet*, 2012, **6**(6): 708–715.

Westen, A.A., et al., Combining results of forensic STR kits: HDplex validation including allelic association and linkage testing with NGM and Identifiler loci. *Int J Legal Med*, 2012, **126**(5): 781–789.

Wetton, J.H., et al., Analysis and interpretation of mixed profiles generated by 34 cycle SGM Plus® amplification. *Forensic Sci Int Genet*, 2011, **5**(5): 376–380.

Wiegand, P., et al., D18S535, D1S1656 and D10S2325: Three efficient short tandem repeats for forensic genetics. *Int J Legal Med*, 1999, **112**(6): 360–363.

Wiegand, P., et al., Forensic validation of the STR systems SE 33 and TC 11. *Int J Legal Med*, 1993, **105**(6): 315–320.

Yeung, S.H., et al., Rapid and high-throughput forensic short tandem repeat typing using a 96-lane microfabricated capillary array electrophoresis microdevice. *J Forensic Sci*, 2006, **51**(4): 740–747.

Yoshida, K., et al., Efficacy of extended kinship analyses utilizing commercial STR kit in establishing personal identification. *Leg Med (Tokyo)*, 2011, **13**(1): 12–15.

Yuasa, I., et al., A hypervariable STR polymorphism in the CFI gene: Mutation rate and no linkage disequilibrium with FGA. *Leg Med (Tokyo)*, 2013, **15**(3): 161–163.

Zhang, S., et al., A new multiplex assay of 17 autosomal STRs and amelogenin for forensic application. *PLoS One*, 2013, **8**(2): e57471.

Zhou, Y., et al., DNA profiling in blood, buccal swabs and hair follicles of patients after allogeneic peripheral blood stem cells transplantation. *Leg Med (Tokyo)*, 2011, **13**(1): 47–51.

Zupanic Pajnic, I., et al., Highly efficient nuclear DNA typing of the World War II skeletal remains using three new autosomal short tandem repeat amplification kits with the extended European standard set of loci. *Croat Med J*, 2012, **53**(1): 17–23.

Genetic Population Data

Abrahams, Z., et al., Allele frequencies of six non-CODIS miniSTR loci (D1S1627, D3S4529, D5S2500, D6S1017, D8S1115 and D9S2157) in three South African populations. *Forensic Sci Int Genet*, 2011, **5**(4): 354–355.

Abu Halima, M.S., L.P. Bernal, and F.A. Sharif, Genetic variation of 15 autosomal short tandem repeat (STR) loci in the Palestinian population of Gaza Strip. *Leg Med (Tokyo)*, 2009, **11**(4): 203–204.

Aguiar, V.R., et al., Updated Brazilian STR allele frequency data using over 100,000 individuals: An analysis of CSF1PO, D3S1358, D5S818, D7S820, D8S1179, D13S317, D16S539, D18S51, D21S11, FGA, Penta D, Penta E, TH01, TPOX and vWA loci. *Forensic Sci Int Genet*, 2012, **6**(4): 504–509.

Akhteruzzaman, S., et al., Forensic evaluation of 11 non-standard STR loci in Bangladeshi population. *Leg Med (Tokyo)*, 2013, **15**(2): 106–108.

Al-Enizi, M., et al., Population genetic analyses of 15 STR loci from seven forensically-relevant populations residing in the state of Kuwait. *Forensic Sci Int Genet*, 2013, **7**(4): e106–e107.

Alenizi, M., et al., STR data for the AmpFlSTR Identifiler loci in Kuwaiti population. *Leg Med (Tokyo)*, 2008, **10**(6): 321–325.

Alenizi, M.A., et al., Concordance between the AmpFlSTR MiniFiler and AmpFlSTR Identifiler PCR amplification kits in the Kuwaiti population. *J Forensic Sci*, 2009, **54**(2): 350–352.

Ali, I., et al., Allele frequencies of 15 STR loci using AmpF/STR Identifiler kit in the Maldivian population. *Forensic Sci Int Genet*, 2012, **6**(5): e136.

Andreassen, R., S. Jakobsen, and B. Mevaag, Norwegian population data for the 10 autosomal STR loci in the AmpFlSTR® SGM Plus™ system. *Forensic Sci Int*, 2007, **170**(1): 59–61.

Andreassen, R., et al., Icelandic population data for the STR loci in the AmpFlSTR SGM Plus system and the PowerPlex Y-system. *Forensic Sci Int Genet*, 2010, **4**(4): e101–e103.

Andreini, E., et al., Allele frequencies for nine STR loci (D3S1358, vWA, FGA, D8S1179, D21S11, D18S51, D5S818, D13S317, D7S820) in the Italian population. *Forensic Sci Int*, 2007, **168**(1): e13–e16.

Aranda, X.G., et al., Genetic composition of six miniSTR in a Brazilian Mulatto sample population. *J Forensic Leg Med*, 2011, **18**(4): 184–186.

Asamura, H., M. Ota, and H. Fukushima, Population data on 10 non-CODIS STR loci in Japanese population using a newly developed multiplex PCR system. *J Forensic Leg Med*, 2008, **15**(8): 519–523.

Bagdonavicius, A., et al., Western Australian sub-population data for the thirteen AmpFiSTR Profiler Plus and COfiler STR loci. *J Forensic Sci*, 2002, **47**(5): 1149–1153.

Barni, F., et al., Allele frequencies of 15 autosomal STR loci in the Iraq population with comparisons to other populations from the middle-eastern region. *Forensic Sci Int*, 2007, **167**(1): 87–92.

Bindu, G.H., R. Trivedi, and V.K. Kashyap, Allele frequency distribution based on 17 STR markers in three major Dravidian linguistic populations of Andhra Pradesh, India. *Forensic Sci Int*, 2007, **170**(1): 76–85.

Borosky, A., L. Catelli, and C. Vullo, Analysis of 17 STR loci in different provinces of Argentina. *Forensic Sci Int Genet*, 2009, **3**(3): e93–e95.

Branham, A., R. Wenk, and F. Chiafari, Allele frequencies of fifteen STR loci in U.S. immigrants from Haiti compared with African Americans and Afro-Caribbeans. *Forensic Sci Int Genet*, 2012, **6**(1): e3–e4.

Bright, J.A., J.S. Buckleton, and C.E. McGovern, Allele frequencies for the four major sub-populations of New Zealand for the 15 Identifiler loci. *Forensic Sci Int Genet*, 2010, **4**(2): e65–e66.

Brisighelli, F., et al., Allele frequencies of fifteen STRs in a representative sample of the Italian population. *Forensic Sci Int Genet*, 2009, **3**(2): e29–e30.

Budowle, B., D.A. Defenbaugh, and K.M. Keys, Genetic variation at nine short tandem repeat loci in Chamorros and Filipinos from Guam. *Leg Med (Tokyo)*, 2000, **2**(1): 26–30.

Budowle, B., et al., Population genetic analyses of the NGM STR loci. *Int J Legal Med*, 2011, **125**(1): 101–109.

Cai, J.F., et al., Population genetics of two STR loci D2S1322 and D2S1356 in a Chinese Han population in Hohhot. *J Forensic Sci*, 2007, **52**(4): 1001.

Camacho, M.V., C. Benito, and A.M. Figueiras, Allelic frequencies of the 15 STR loci included in the AmpFlSTR Identifiler PCR amplification kit in an autochthonous sample from Spain. *Forensic Sci Int*, 2007, **173**(2–3): 241–245.

Cassar, M., C. Farrugia, and C. Vidal, Allele frequencies of 14 STR loci in the population of Malta. *Leg Med (Tokyo)*, 2008, **10**(3): 153–156.

Castro, E.A., et al., Genetic polymorphism and forensic parameters of nine short tandem repeat loci in Ngobe and Embera Amerindians of Panama. *Hum Biol*, 2007, **79**(5): 563–577.

Chang, Y.F., et al., Genetic polymorphism of 17 STR loci in Chinese population from Hunan province in Central South China. *Forensic Sci Int Genet*, 2012, **6**(5): e151–e153.

Chen, J.G., et al., Population genetic data of 15 autosomal STR loci in Uygur ethnic group of China. *Forensic Sci Int Genet*, 2012, **6**(6): e178–e179.

Chouery, E., et al., Population genetic data for 17 STR markers from Lebanon. *Leg Med (Tokyo)*, 2010, **12**(6): 324–326.

Chula, F.G., et al., 15 STR loci frequencies with mutation rates in the population from Rio Grande do Sul, Southern Brazil. *Forensic Sci Int Genet*, 2009, **3**(2): e35–e38.

Chung, U., et al., Population data of nine miniSTR loci in Koreans. *Forensic Sci Int*, 2007, **168**(2–3): e51–e53.

Clark, D., et al., STR data for the AmpFlSTR SGM Plus loci from two South Asian populations. *Leg Med (Tokyo)*, 2009, **11**(2): 97–100.

Cortellini, V., N. Cerri, and A. Verzeletti, Genetic variation at 5 new autosomal short tandem repeat markers (D10S1248, D22S1045, D2S441, D1S1656, D12S391) in a population-based sample from Maghreb region. *Croat Med J*, 2011, **52**(3): 368–371.

Cortellini, V., N. Cerri, and A. Verzeletti, Population data on 5 non-CODIS STR loci (D10S1248, D22S1045, D2S441, D1S1656, D12S391) in a population sample from Brescia county (Northern Italy). *Forensic Sci Int Genet*, 2011, **5**(3): e97–e98.

Coudray, C., et al., Allele frequencies of 15 short tandem repeats (STRs) in three Egyptian populations of different ethnic groups. *Forensic Sci Int*, 2007, **169**(2–3): 260–265.

Coudray, C., et al., Allele frequencies of 15 tetrameric short tandem repeats (STRs) in Andalusians from Huelva (Spain). *Forensic Sci Int*, 2007, **168**(2–3): e21–e24.

Coudray, C., et al., Population genetic data of 15 tetrameric short tandem repeats (STRs) in Berbers from Morocco. *Forensic Sci Int*, 2007, **167**(1): 81–86.

Curic, G., et al., Genetic parameters of five new European standard set STR loci (D10S1248, D22S1045, D2S441, D1S1656, D12S391) in the population of Eastern Croatia. *Croat Med J*, 2012, **53**(5): 409–415.

da Costa Francez, P.A., et al., Allelic frequencies and statistical data obtained from 12 codis STR loci in an admixed population of the Brazilian Amazon. *Genet Mol Biol*, 2011, **34**(1): 35–39.

de Assis Poiares, L., et al., Allele frequencies of 15 STRs in a representative sample of the Brazilian population. *Forensic Sci Int Genet*, 2010, **4**(2): e61–e63.

Demeter, S.J., et al., Genetic variation at 15 polymorphic, autosomal, short tandem repeat loci of two Hungarian populations in Transylvania, Romania. *Croat Med J*, 2010, **51**(6): 515–523.

Deng, Q., et al., Genetic relationships among four minorities in Guangxi revealed by analysis of 15 STRs. *J Genet Genomics*, 2007, **34**(12): 1072–1079.

Deng, Y., et al., Genetic polymorphisms of 15 STR loci of Chinese Dongxiang and Salar ethnic minority living in Qinghai Province of China. *Leg Med (Tokyo)*, 2007, **9**(1): 38–42.

Deng, Y.J., et al., Genetic polymorphism analysis of 15 STR loci in Chinese Hui ethnic group residing in Qinghai province of China. *Mol Biol Rep*, 2011, **38**(4): 2315–2322.

Deng, Y.J., et al., Polymorphic analysis of 21 new STR loci in Chinese Uigur group. *Forensic Sci Int Genet*, 2013, **7**(3): e97–e98.

Di Cristofaro, J., et al., Genetic data of 15 STR loci in five populations from Afghanistan. *Forensic Sci Int Genet*, 2012, **6**(1): e44–e45.

Dognaux, S., et al., Allele frequencies for the new European standard set (ESS) loci and D1S1677 in the Belgian population. *Forensic Sci Int Genet*, 2012, **6**(2): e75–e77.

dos Santos, L.L., et al., Allele distribution of six STR/miniSTR loci (CD4, FABP2, D12S391, D14S1434, D22S1045 and D10S1248) for forensic purposes in Southeastern Brazil. *Ann Hum Biol*, 2011, **38**(1): 110–113.

Du, B., et al., Allele frequencies for two STR loci D6S1274 and D17S1299 in Chinese and Thai populations. *J Forensic Sci*, 2007, **52**(1): 225.

Dubey, B., et al., Forensic STR profile of two endogamous populations of Madhya Pradesh, India. *Leg Med (Tokyo)*, 2009, **11**(1): 41–44.

Eaaswarkhanth, M., S. Roy, and I. Haque, Allele frequency distribution for 15 autosomal STR loci in two Muslim populations of Tamilnadu, India. *Leg Med (Tokyo)*, 2007, **9**(6): 332–335.

Eaaswarkhanth, M., T.S. Vasulu, and I. Haque, Genetic affinity between diverse ethnoreligious communities of Tamil Nadu, India: A microsatellite study. *Hum Biol*, 2008, **80**(6): 601–609.

Eaaswarkhanth, M., et al., Diverse genetic origin of Indian Muslims: Evidence from autosomal STR loci. *J Hum Genet*, 2009, **54**(6): 340–348.

Eaaswarkhanth, M., et al., Microsatellite diversity delineates genetic relationships of Shia and Sunni Muslim populations of Uttar Pradesh, India. *Hum Biol*, 2009, **81**(4): 427–445.

Eckhoff, C., S.J. Walsh, and J.S. Buckleton, Population data from sub-populations of the Northern Territory of Australia for 15 autosomal short tandem repeat (STR) loci. *Forensic Sci Int*, 2007, **171**(2–3): 237–249.

El Andari, A., et al., Population genetic data for 23 STR markers from Lebanon. *Forensic Sci Int Genet*, 2013, **7**(4): e108–e113.

El Ossmani, H., et al., Assessment of phylogenetic structure of Berber-speaking population of Azrou using 15 STRs of Identifiler kit. *Leg Med (Tokyo)*, 2010, **12**(1): 52–56.

Elmrghni, S., et al., Genetic data provided by 15 autosomal STR loci in the Libyan population living in Benghazi. *Forensic Sci Int Genet*, 2012, **6**(3): e93–e94.

Erkol, Z., et al., STR data for the AmpFlSTR identifier loci from an old settlement in Northwestern Turkey. *Forensic Sci Int*, 2007, **173**(2–3): 238–240.

Ferdous, A., et al., Forensic evaluation of STR data for the PowerPlex 16 system loci in a Bangladeshi population. *Leg Med (Tokyo)*, 2009, **11**(4): 198–199.

Forward, B.W., et al., AmpPFlSTR Identifiler STR allele frequencies in Tanzania, Africa. *J Forensic Sci*, 2008, **53**(1): 245.

Fridman, C., et al., Brazilian population profile of 15 STR markers. *Forensic Sci Int Genet*, 2008, **2**(2): e1–e4.

Fujihara, J., et al., Allele frequencies for nine STR loci in Ovambo population using AmpFlSTR Profiler Kit. *Forensic Sci Int*, 2007, **169**(1): e7–e9.

Gazi, N.N., et al., Genetic polymorphisms of 15 autosomal STR loci in three isolated tribal populations of Bangladesh. *Forensic Sci Int Genet*, 2010, **4**(4): 265–266.

Goetz, R., et al., Population data from the New South Wales Aboriginal Australian sub-population for the Profiler Plus autosomal short tandem repeat (STR) loci. *Forensic Sci Int*, 2008, **175**(2–3): 235–237.

Gomes, A.V., et al., 13 STR loci frequencies in the population from Paraiba, Northeast Brazil. *Forensic Sci Int*, 2007, **173**(2–3): 231–234.

Gomes, V., et al., Population data defined by 15 autosomal STR loci in Karamoja population (Uganda) using AmpFlSTR Identifiler kit. *Forensic Sci Int Genet*, 2009, **3**(2): e55–e58.

Gomez, M.A., et al., Allele frequencies of 15 STR loci in the population of the city of Quito, Ecuador. *J Forensic Sci*, 2008, **53**(2): 510–511.

Gorostiza, A., et al., Allele frequencies of the 15 AmpFlSTR Identifiler loci in the population of Metztitlan (Estado de Hidalgo), Mexico. *Forensic Sci Int*, 2007, **166**(2–3): 230–232.

Grubwieser, P., et al., Evaluation of an extended set of 15 candidate STR loci for paternity and kinship analysis in an Austrian population sample. *Int J Legal Med*, 2007, **121**(2): 85–89.

Gutierrez, C.C., et al., Population genetic data for 15 STR loci (PowerPlex 16 kit) in Nicaragua. *Forensic Sci Int Genet*, 2011, **5**(5): 563–564.

Hatzer-Grubwieser, P., et al., Allele frequencies and concordance study of 16 STR loci—Including the new European standard set (ESS) loci—In an Austrian population sample. *Forensic Sci Int Genet*, 2012, **6**(1): e50–e51.

Havas, D., et al., Population genetics of 15 AmpFlSTR Identifiler loci in Macedonians and Macedonian Romani (Gypsy). *Forensic Sci Int*, 2007, **173**(2–3): 220–224.

He, J. and F. Guo, Allele frequencies for fifteen autosomal STR loci in a Xibe population from Liaoning Province, Northeast China. *Forensic Sci Int Genet*, 2013, 7(3): e80–e81.

He, J. and F. Guo, Genetic variation of fifteen autosomal STR loci in a Manchu population from Jilin Province, Northeast China. *Forensic Sci Int Genet*, 2013, 7(2): e45–e46.

Herrera-Paz, E.F., et al., Allele frequencies distributions for 13 autosomal STR loci in 3 Black Carib (Garifuna) populations of the Honduran Caribbean coasts. *Forensic Sci Int Genet*, 2008, **3**(1): e5–e10.

Hill, C.R., et al., U.S. population data for 29 autosomal STR loci. *Forensic Sci Int Genet*, 2013, **7**(3): e82–e83.

Hong, S.B., et al., Korean population genetic data and concordance for the PowerPlex® ESX 17, AmpFlSTR Identifiler®, and PowerPlex® 16 systems. *Forensic Sci Int Genet*, 2013, **7**(3): e47–e51.

Hou, G., et al., Genetic distribution on 15 STR loci from a Han population of Shenyang region in northeast China. *Forensic Sci Int Genet*, 2013, **7**(3): e86–e87.

Hou, Y.P., et al., D20S161 data for three ethnic populations and forensic validation. *Int J Legal Med*, 1999, **112**(6): 400–402.

Huang, Q., et al., Genetic polymorphism of 15 STR loci in Chinese Han population from Shanghai municipality in East China. *Forensic Sci Int Genet*, 2013, **7**(2): e31–e34.

Huang, S., et al., Genetic variation analysis of 15 autosomal STR loci of AmpFlSTR Sinofiler PCR amplification kit in Henan (central China) Han population. *Leg Med (Tokyo)*, 2010, **12**(3): 160–161.

Hwa, H.L., et al., Fifteen non-CODIS autosomal short tandem repeat loci multiplex data from nine population groups living in Taiwan. *Int J Legal Med*, 2012, **126**(4): 671–675.

Hwa, H.L., et al., Fourteen non-CODIS autosomal short tandem repeat loci multiplex data from Taiwanese. *Int J Legal Med*, 2011, **125**(2): 219–226.

Illeperuma, R.J., et al., Genetic profile of 11 autosomal STR loci among the four major ethnic groups in Sri Lanka. *Forensic Sci Int Genet*, 2009, **3**(3): e105–e106.

Illeperuma, R.J., et al., Genetic variation at 11 autosomal STR loci in the aboriginal people, the Veddahs of Sri Lanka. *Forensic Sci Int Genet*, 2010, **4**(2): 142.

Jacewicz, R., et al., Population database on 15 autosomal STR loci in 1000 unrelated individuals from the Lodz region of Poland. *Forensic Sci Int Genet*, 2008, **2**(1): e1–e3.

Jedrzejczyk, M., et al., Genetic population studies on 15 NGM STR loci in central Poland population. *Forensic Sci Int Genet*, 2012, **6**(4): e119–e120.

Jeran, N., et al., Genetic diversity of 15 STR loci in a population of Montenegro. *Coll Antropol*, 2007, **31**(3): 847–852.

Jin, H.J., et al., Forensic and population genetic analyses of eighteen non-CODIS miniSTR loci in the Korean population. *J Forensic Leg Med*, 2013, **20**(8): 1093–1097.

Juarez-Cedillo, T., et al., Genetic admixture and diversity estimations in the Mexican Mestizo population from Mexico City using 15 STR polymorphic markers. *Forensic Sci Int Genet*, 2008, **2**(3): e37–e39.

Kalpana, D., et al., Pentaplex typing of new European standard set (ESS) STR loci in Indian population. *Forensic Sci Int Genet*, 2012, **6**(3): e86–e89.

Kang, L., et al., Genetic polymorphisms of 15 STR loci in two Tibetan populations from Tibet Changdu and Naqu, China. *Forensic Sci Int*, 2007, **169**(2–3): 239–243.

Kang, L.L., et al., Allele frequencies of 15 STR loci of Tibetan lived in Tibet Lassa. *Forensic Sci Int*, 2007, **168**(2–3): 236–240.

Kee, B.P., et al., Genetic data for 15 STR loci in a Kadazan-Dusun population from East Malaysia. *Genet Mol Res*, 2011, **10**(2): 739–743.

Khodjet-el-Khil, H., et al., Allele frequencies for 15 autosomal STR markers in the Libyan population. *Ann Hum Biol*, 2012, **39**(1): 80–83.

Kido, A., et al., Population data on the AmpFlSTR Identifiler loci in Africans and Europeans from South Africa. *Forensic Sci Int*, 2007, **168**(2–3): 232–235.

Korzebor, A., et al., Statistical analysis of six STR loci located in MHC region in Iranian population for pre-implantation genetic diagnosis. *Int J Immunogenet*, 2007, **34**(6): 441–443.

Kraaijenbrink, T., et al., Allele frequency distribution for 21 autosomal STR loci in Bhutan. *Forensic Sci Int*, 2007, **170**(1): 68–72.

Kraaijenbrink, T., et al., Allele frequency distribution for 21 autosomal STR loci in Nepal. *Forensic Sci Int*, 2007, **168**(2–3): 227–231.

Krithika, S., et al., Allele frequency distribution at 15 autosomal STR loci in Panggi, Komkar and Padam sub tribes of Adi, a Tibeto-Burman speaking population of Arunachal Pradesh, India. *Leg Med (Tokyo)*, 2007, **9**(4): 210–217.

Lee, M.H., et al., Genetic variation of three autosomal STR loci D21S1435, D21S1411, and D21S1412 in Korean population. *Mol Biol Rep*, 2010, **37**(1): 99–104.

Li, B., et al., Population genetic analyses of the STR loci of the AmpFlSTR NGM SElect kit for Han population in Fujian Province, China. *Int J Legal Med*, 2013, **127**(2): 345–346.

Li, C., et al., Genetic polymorphism of 17 STR loci for forensic use in Chinese population from Shanghai in East China. *Forensic Sci Int Genet*, 2009, **3**(4): e117–e118.

Li, S., et al., Allele frequencies of nine non-CODIS STR loci in Chinese Uyghur ethnic minority group. *Forensic Sci Int Genet*, 2012, **6**(1): e11–e12.

Liao, G., et al., Population genetic study of 15 STR loci in a Chinese population. *J Forensic Sci*, 2008, **53**(1): 252–253.

Liu, C., et al., Genetic polymorphisms of 9 non-CODIS short tandem repeat loci in two ethnic minority populations in Southern China. *Forensic Sci Int Genet*, 2013, **7**(4): e114–e115.

Liu, J., et al., Allele frequencies of 19 autosomal STR loci in Manchu population of China with phylogenetic structure among worldwide populations. *Gene*, 2013, **529**(2): 282–287.

Lu, D.J., Q.L. Liu, and H. Zhao, Genetic data of nine non-CODIS STRs in Chinese Han population from Guangdong Province, Southern China. *Int J Legal Med*, 2011, **125**(1): 133–137.

Manamperi, A., et al., STR polymorphisms in Sri Lanka: Evaluation of forensic utility in identification of individuals and parentage testing. *Ceylon Med J*, 2009, **54**(3): 85–89.

Manta, F.S., et al., Terena Amerindian group autosomal STR data: Comparison studies with other Brazilian populations. *Mol Biol Rep*, 2012, **39**(4): 4455–4459.

Marian, C., et al., STR data for the 15 AmpFlSTR identifiler loci in the Western Romanian population. *Forensic Sci Int*, 2007, **170**(1): 73–75.

Martin, P., et al., Allele frequencies of six miniSTR loci (D10S1248, D14S1434, D22S1045, D4S2364, D2S441 and D1S1677) in a Spanish population. *Forensic Sci Int*, 2007, **169**(2–3): 252–254.

Martinez, B., J.J. Builes, and L. Caraballo, Genetic data analysis of nine STRs in two Caribbean Colombian populations: Cesar and Guajira. *J Forensic Sci*, 2008, **53**(1): 254–255.

Martinez-Cortes, G., et al., Origin and genetic differentiation of three native Mexican groups (Purepechas, Triquis and Mayas): Contribution of CODIS-STRs to the history of human populations of Mesoamerica. *Ann Hum Biol*, 2010, **37**(6): 801–819.

Martins, J.A., et al., Genetic data of 15 autosomal STR loci: An analysis of the Araraquara population colonization (Sao Paulo, Brazil). *Mol Biol Rep*, 2011, **38**(8): 5397–5403.

Maruyama, S., et al., Population data on 15 STR loci using AmpFlSTR Identifiler kit in a Malay population living in and around Kuala Lumpur, Malaysia. *Leg Med (Tokyo)*, 2008, **10**(3): 160–162.

Mastana, S.S., et al., Genetic variation of 13 STR loci in the four endogamous tribal populations of Eastern India. *Forensic Sci Int*, 2007, **169**(2–3): 266–273.

Melo, M.M., et al., Genetic study of 15 STRs loci of Identifiler system in Angola population. *Forensic Sci Int Genet*, 2010, **4**(5): e153–e157.

Montelius, K., A.O. Karlsson, and G. Holmlund, STR data for the AmpFlSTR Identifiler loci from Swedish population in comparison to European, as well as with non-European population. *Forensic Sci Int Genet*, 2008, **2**(3): e49–e52.

Monterrosa, J.C., et al., Population genetic data for 16 STR loci (PowerPlex ESX-17 kit) in El Salvador. *Forensic Sci Int Genet*, 2012, **6**(5): e134.

Mornhinweg, E., et al., D3S1358: Sequence analysis and gene frequency in a German population. *Forensic Sci Int*, 1998, **95**(2): 173–178.

Munoz, A., et al., Allele frequencies of 15 STRs in the Calchaqui Valleys population (North-Western Argentina). *Forensic Sci Int Genet*, 2012, **6**(1): e58–e60.

Nascimento, E., E. Cerqueira, and L. Gusmao, Population database defined by 13 autosomal STR loci in a representative sample from Bahia, Northeast Brazil. *Forensic Sci Int Genet*, 2011, **5**(1): e38–e40.

Nasibov, E., et al., Allele frequencies of 15 STR loci in Azerbaijan population. *Forensic Sci Int Genet*, 2013, **7**(3): e99–e100.

Nie, S., et al., Genetic data of 15 STR loci in Chinese Yunnan Han population. *Forensic Sci Int Genet*, 2008, **3**(1): e1–e3.

Noor, S., S. Roy, and I. Haque, An autosomal STR database of Muslims: The largest minority community, Uttar Pradesh, India. *Forensic Sci Int Genet*, 2011, **5**(4): e117–e118.

Noor, S., et al., Allele frequency distribution for 15 autosomal STR loci in Afridi Pathan population of Uttar Pradesh, India. *Leg Med (Tokyo)*, 2009, **11**(6): 308–311.

Novokmet, N., et al., Forensic efficiency parameters for the 15 STR loci in the population of the island of Cres (Croatia). *Coll Antropol*, 2009, **33**(4): 1319–1322.

Ocampos, M., et al., 15 STR loci frequencies in the population from Santa Catarina, Southern Brazil. *Forensic Sci Int Genet*, 2009, **3**(4): e129–e131.

Omran, G.A., G.N. Rutty, and M.A. Jobling, Genetic variation of 15 autosomal STR loci in Upper (Southern) Egyptians. *Forensic Sci Int Genet*, 2009, **3**(2): e39–e44.

Ossmani, H.E., et al., Allele frequencies of 15 autosomal STR loci in the Southern Morocco population with phylogenetic structure among worldwide populations. *Leg Med (Tokyo)*, 2009, **11**(3): 155–158.

Ota, M., et al., Allele frequencies for 15 STR loci in Tibetan populations from Nepal. *Forensic Sci Int*, 2007, **169**(2–3): 234–238.

Park, J.H., et al., Genetic variation of 23 autosomal STR loci in Korean population. *Forensic Sci Int Genet*, 2013, **7**(3): e76–e77.

Parys-Proszek, A., et al., Genetic variation of 15 autosomal STR loci in a population sample from Poland. *Leg Med (Tokyo)*, 2010, **12**(5): 246–248.

Pepinski, W., et al., Polymorphism of 11 non-CODIS STRs in a population sample of Lithuanian minority residing in northeastern Poland. *Forensic Sci Int Genet*, 2011, **5**(1): e37.

Pepinski, W., et al., Polymorphism of 11 non-CODIS STRs in a population sample of religious minority of Old Believers residing in Northeastern Poland. *Adv Med Sci*, 2010, **55**(2): 328–332.

Petric, G., et al., Genetic variation at 15 autosomal STR loci in the Hungarian population of Vojvodina Province, Republic of Serbia. *Forensic Sci Int Genet*, 2012, **6**(6): e163–e165.

Phillips, C., et al., Analysis of global variability in 15 established and 5 new European standard set (ESS) STRs using the CEPH human genome diversity panel. *Forensic Sci Int Genet*, 2011, **5**(3): 155–169.

Piatek, J., et al., Population genetics of 15 autosomal STR loci in the population of Pomorze Zachodnie (NW Poland). *Forensic Sci Int Genet*, 2008, **2**(3): e41–e43.

Pinto, L.M., et al., Molecular characterization and population genetics of non-CODIS microsatellites used for forensic applications in Brazilian populations. *Forensic Sci Int Genet*, 2014, **9**: e16–e17.

Poetsch, M. and N. von Wurmb-Schwark, Allele frequencies for the 16 short tandem repeats of the Powerplex ESX17 kit in a population from Turkey. *Int J Legal Med*, 2013, **127**(3): 591–592.

Poetsch, M., et al., First experiences using the new Powerplex® ESX17 and ESI17 kits in casework analysis and allele frequencies for two different regions in Germany. *Int J Legal Med*, 2011, **125**(5): 733–739.

Poiares Lde, A., et al., 15 STR loci frequencies in the population from Parana, Southern Brazil. *Forensic Sci Int Genet*, 2009, **4**(1): e23–e24.

Pontes, M.L. and M.F. Pinheiro, Population data of the AmpFISTR® NGM STR loci in a North of Portugal sample. *Forensic Sci Int Genet*, 2012, **6**(5): e127–e128.

Porras, L., et al., Genetic polymorphism of 15 STR loci in central Western Colombia. *Forensic Sci Int Genet*, 2008, **2**(1): e7–e8.

Projic, P., et al., Allele frequencies for 15 short tandem repeat loci in representative sample of Croatian population. *Croat Med J*, 2007, **48**(4): 473–477.

Raczek, E., Population data on the 11 STR loci in the Upper Silesia (Poland). *Forensic Sci Int*, 2007, **168**(1): 68–72.

Rak, S.A., et al., Population genetic data on 15 STR loci in the Hungarian population. *Forensic Sci Int Genet*, 2011, **5**(5): 543–544.

Rakha, A., et al., Genetic analysis of Kashmiri Muslim population living in Pakistan. *Leg Med (Tokyo)*, 2008, **10**(4): 216–219.

Rakha, A., et al., Population genetic data on 15 autosomal STRs in a Pakistani population sample. *Leg Med (Tokyo)*, 2009, **11**(6): 305–307.

Rangel-Villalobos, H., et al., Evaluation of forensic and anthropological potential of D9S1120 in Mestizos and Amerindian populations from Mexico. *Croat Med J*, 2012, **53**(5): 423–431.

Rangel-Villalobos, H., et al., Forensic parameters for 15 STRs in eight Amerindian populations from the north and west of Mexico. *Forensic Sci Int Genet*, 2013, **7**(3): e62–e65.

Rebala, K., et al., Belarusian population genetic database for 15 autosomal STR loci. *Forensic Sci Int*, 2007, **173**(2–3): 235–237.

Riccardi, L.N., et al., Genetic profiling of Bolivian population using 15 STR markers of forensic importance. *Leg Med (Tokyo)*, 2009, **11**(3): 149–151.

Rocchi, A., et al., Italian data of 23 STR loci amplified in a single multiplex reaction. *Forensic Sci Int Genet*, 2012, **6**(6): e157–e158.

Rodenbusch, R., et al., Allele frequencies of the five new generation forensic STR (D1S1656, D2S441, D10S1248, D12S391 and D22S1045) in the population from Rio Grande do Sul, Southern Brazil. *Forensic Sci Int Genet*, 2012, **6**(1): e55–e57.

Rodrigues, E.L., et al., Genetic data on 15 STR autosomal loci for a sample population of the Northern Region of the State of Rio de Janeiro, Brazil. *Forensic Sci Int Genet*, 2009, **4**(1): e25–e26.

Rodrigues, E.M., J. Palha Tde, and S.E. dos Santos, Allele frequencies data and statistic parameters for 13 STR loci in a population of the Brazilian Amazon Region. *Forensic Sci Int*, 2007, **168**(2–3): 244–247.

Rodriguez, A., et al., Population genetic data for 18 STR loci in Costa Rica. *Forensic Sci Int*, 2007, **168**(1): 85–88.

Roy, S., et al., Autosomal STR variations in three endogamous populations of West Bengal, India. *Leg Med (Tokyo)*, 2008, **10**(6): 326–332.

Rubi-Castellanos, R., et al., Genetic data of 15 autosomal STRs (Identifiler kit) of three Mexican Mestizo population samples from the States of Jalisco (West), Puebla (Center), and Yucatan (Southeast). *Forensic Sci Int Genet*, 2009, **3**(3): e71–e76.

Sanchez-Diz, P., et al., 16 STR data of a Greek population. *Forensic Sci Int Genet*, 2008, **2**(4): e71–e72.

Sanchez-Diz, P., et al., Population data on 15 autosomal STRs in a sample from Colombia. *Forensic Sci Int Genet*, 2009, **3**(3): e81–e82.

Santos, C., et al., Genetic structure of the Azores Islands: A study using 15 autosomal short tandem repeat loci. *Coll Antropol*, 2009, **33**(4): 991–999.

Schlebusch, C.M., H. Soodyall, and M. Jakobsson, Genetic variation of 15 autosomal STR loci in various populations from southern Africa. *Forensic Sci Int Genet*, 2012, **6**(1): e20–e21.

Shi, M., et al., Genetic polymorphism of 14 non-CODIS STR loci for forensic use in Southeast China population. *Forensic Sci Int*, 2008, **174**(1): 77–80.

Shotivaranon, J., et al., DNA database of populations from different parts in the Kingdom of Thailand. *Forensic Sci Int Genet*, 2009, **4**(1): e37–e38.

Silva, B.C., et al., Genetic diversity and statistical parameters of 15 autosomal STR loci in the Pomeranian subpopulation of Espirito Santo State, Brazil. *Mol Biol Rep*, 2011, **38**(5): 3013–3016.

Silva, M.B., et al., Allele frequencies of fifteen STR loci in a population from Central Brazil. *Forensic Sci Int Genet*, 2010, **4**(5): e151–e152.

Simkova, H., et al., Allele frequency data for 17 short tandem repeats in a Czech population sample. *Forensic Sci Int Genet*, 2009, **4**(1): e15–e17.

Smith, B.G., et al., Population data for 15 STR loci (Identifiler kit) in a Filipino population. *Leg Med (Tokyo)*, 2009, **11**(3): 159–161.

Soltyszewski, I., et al., Analysis of forensically used autosomal short tandem repeat markers in Polish and neighboring populations. *Forensic Sci Int Genet*, 2008, **2**(3): 205–211.

Sotak, M., et al., Genetic variation analysis of 15 autosomal STR loci in Eastern Slovak Caucasian and Romany (Gypsy) population. *Forensic Sci Int Genet*, 2008, **3**(1): e21–e25.

Sotak, M., et al., Population database of 17 autosomal STR loci from the four predominant Eastern Slovakia regions. *Forensic Sci Int Genet*, 2011, **5**(3): 262–263.

Stanciu, F., O.R. Popescu, and I.M. Stoian, Allele frequencies of 15 STR loci in Moldavia region (NE Romania). *Forensic Sci Int Genet*, 2009, **4**(1): e39–e40.

Stanciu, F., O.R. Popescu, and I.M. Stoian, STR data for the AmpFlSTR Identifiler from Dobruja region (SE Romania). *Forensic Sci Int Genet*, 2009, **3**(2): 146–147.

Stanciu, F., I.M. Stoian, and O.R. Popescu, Comprehensive STR data for the AmpFlSTR Identifiler from Transylvania (NW Romania). *Leg Med (Tokyo)*, 2009, **11**(1): 48–50.

Stanciu, F., I.M. Stoian, and O.R. Popescu, Population data for 15 short tandem repeat loci from Wallachia Region, South Romania. *Croat Med J*, 2009, **50**(3): 321–325.

Stanciu, F., et al., Genetic parameters and allele frequencies of five new European standard set STR loci (D10S1248, D22S1045, D2S441, D1S1656, D12S391) in the population of Romania. *Croat Med J*, 2013, **54**(3): 232–237.

Steinlechner, M., et al., Population genetics of ten STR loci (AmpFlSTR SGM plus) in Austria. *Int J Legal Med*, 2001, **114**(4–5): 288–290.

Stepanov, V.A., et al., Genetic variability of 15 autosomal STR loci in Russian populations. *Leg Med (Tokyo)*, 2010, **12**(5): 256–258.

Suadi, Z., et al., STR data for the AmpFlSTR Identifiler loci from the three main ethnic indigenous population groups (Iban, Bidayuh, and Melanau) in Sarawak, Malaysia. *J Forensic Sci*, 2007, **52**(1): 231–234.

Talledo, M., et al., Comparative allele distribution at 16 STR loci between the Andean and coastal population from Peru. *Forensic Sci Int Genet*, 2010, **4**(4): e109–e117.

Tamaki, K., et al., Evaluation of tetranucleotide repeat locus D7S809 (wg1g9) in the Japanese population. *Forensic Sci Int*, 1996, **81**(2–3): 133–140.

Tang, J., et al., Genetic analyzing of 15 STR loci in a Han population of Jinan (Northern China). *Leg Med (Tokyo)*, 2009, **11**(3): 144–146.

Taylor, D.A., J.M. Henry, and S.J. Walsh, South Australian Aboriginal sub-population data for the nine AmpFlSTR Profiler Plus short tandem repeat (STR) loci. *Forensic Sci Int Genet*, 2008, **2**(2): e27–e30.

Teng, Y., et al., Genetic variation of new 21 autosomal short tandem repeat loci in a Chinese Salar ethnic group. *Mol Biol Rep*, 2012, **39**(2): 1465–1470.

Thangaraj, K., et al., Autosomal STR data on the enigmatic Andaman Islanders. *Forensic Sci Int*, 2007, **169**(2–3): 247–251.

Tillmar, A.O., G. Backstrom, and K. Montelius, Genetic variation of 15 autosomal STR loci in a Somali population. *Forensic Sci Int Genet*, 2009, **4**(1): e19–e20.

Tong da, Y., et al., Polymorphism analysis and evaluation of nine non-CODIS STR loci in the Han population of Southern China. *Ann Hum Biol*, 2010, **37**(6): 820–826.

Tong, D., et al., Polymorphism analysis of 15 STR loci in a large sample of the Han population in Southern China. *Forensic Sci Int Genet*, 2009, **4**(1): e27–e29.

Tong, D., et al., Polymorphism analysis and evaluation of 19 STR loci in the Han population of Southern China. *Ann Hum Biol*, 2013, **40**(2): 191–196.

Toscanini, U., et al., Testing for genetic structure in different urban Argentinian populations. *Forensic Sci Int*, 2007, **165**(1): 35–40.

Turrina, S., et al., Evaluation of genetic parameters of 22 autosomal STR loci (PowerPlex® Fusion System) in a population sample from Northern Italy. *Int J Legal Med*, 2014, **128**(2): 281–283.

Uchihi, R., et al., Population data of six STR loci using hexaplex PCR amplification and typing system, "Midi-6" in four regional populations in Japan. *Forensic Sci Int*, 2007, **169**(2–3): 255–259.

Untoro, E., et al., Allele frequency of CODIS 13 in Indonesian population. *Leg Med (Tokyo)*, 2009, **11**(Suppl 1): S203–S205.

Vargas-Diaz, R., et al., A genetic study of the Identifiler(™)System 15 STR loci in the general population of Nicaragua, Central America. *Leg Med (Tokyo)*, 2011, **13**(4): 213–214.

Vergara, I.A., et al., Autosomal STR allele frequencies for the CODIS system from a large random population sample in Chile. *Forensic Sci Int Genet*, 2012, **6**(3): e83–e85.

Veselinovic, I.S., et al., Serbian population data at the CODIS core STR loci. *J Forensic Sci*, 2007, **52**(6): 1426–1427.

Vieira, T.C., et al., Allelic frequencies and statistical data obtained from 15 STR loci in a population of the Goias State. *Genet Mol Res*, 2013, **12**(1): 23–27.

Vullo, C., et al., Frequency data for 12 mini STR loci in Argentina. *Forensic Sci Int Genet*, 2010, **4**(3): e79–e81.

Walsh, S.J. and J.S. Buckleton, Autosomal microsatellite allele frequencies for 15 regionally defined Aboriginal Australian population datasets. *Forensic Sci Int*, 2007, **168**(2–3): e29–e42.

Walsh, S.J. and J.S. Buckleton, Autosomal microsatellite allele frequencies for a nationwide dataset from the Australian Caucasian sub-population. *Forensic Sci Int*, 2007, **168**(2–3): e47–e50.

Wang, H.D., et al., Allelic frequency distributions of 21 non-combined DNA index system STR loci in a Russian ethnic minority group from Inner Mongolia, China. *J Zhejiang Univ Sci B*, 2013, **14**(6): 533–540.

Wang, J., et al., Allele frequencies of nine non-CODIS STR loci in Western Chinese Han population. *Forensic Sci Int Genet*, 2013, **7**(3): e88–e89.

Wang, R., et al., Genetic distribution on 15 STR loci from a population of Southern Liaoning in Northeast of China. *Forensic Sci Int Genet*, 2008, **2**(2): e25–e26.

Wei, Y., et al., Genetic diversity of five STR loci in Southeastern China. *J Forensic Sci*, 2007, **52**(4): 999–1000.

Wu, Y.M., et al., Genetic polymorphisms of 15 STR loci in Chinese Han population living in Xi'an city of Shaanxi Province. *Forensic Sci Int Genet*, 2008, **2**(2): e15–e18.

Xing, J., et al., Genetic polymorphism of 15 STR loci in a Manchu population in Northeast China. *Forensic Sci Int Genet*, 2011, **5**(3): e93–e95.

Xu, L.N., S.P. Hu, and G.Y. Feng, STR polymorphisms of the Henan population and investigation of the Central Plains Han origin of Chaoshanese. *Biochem Genet*, 2009, **47**(7–8): 569–581.

Yan, J., et al., Genetic analysis of 15 STR loci on Chinese Tibetan in Qinghai Province. *Forensic Sci Int*, 2007, **169**(1): e3–e6.

Yan, J., et al., Genetic polymorphism of two novel STR loci AY639920 and AY639923 in a Chinese population. *J Forensic Sci*, 2007, **52**(3): 754.

Yong, R.Y., et al., Allele frequencies of six miniSTR loci of three ethnic populations in Singapore. *Forensic Sci Int*, 2007, **166**(2–3): 240–243.

Yoo, S.Y., et al., A large population genetic study of 15 autosomal short tandem repeat loci for establishment of Korean DNA profile database. *Mol Cells*, 2011, **32**(1): 15–19.

Yuan, L., et al., Population data of 21 non-CODIS STR loci in Han population of Northern China. *Int J Legal Med*, 2012, **126**(4): 659–664.

Zeng, Z., et al., Genetic data of six STR loci of Han population in Henan Province (Central China). *J Forensic Sci*, 2008, **53**(2): 515–516.

Zeng, Z.S., et al., Population genetic data of 15 STR loci in Han population of Henan province (central China). *Leg Med (Tokyo)*, 2007, **9**(1): 30–32.

Zha, L., et al., Genetic polymorphism of 21 non-CODIS STR loci in the Chinese Mongolian ethnic minority. *Forensic Sci Int Genet*, 2014, **9**: e32–e33.

Zhang, H., et al., Analysis of 15 STR loci in Chinese population from Sichuan in West China. *Forensic Sci Int*, 2007, **171**(2–3): 222–225.

Zhang, H.B., et al., Nine polymorphic STR loci in the HLA region in the Shaanxi Han population of China. *Genet Mol Res*, 2012, **11**(3): 2534–2538.

Zhang, S., et al., Genetic polymorphisms in 12 autosomal STRs in a Shanghai Han population from China. *Electrophoresis*, 2013, **34**(4): 613–617.

Zheng, X., et al., Genetic data of 9 STR loci from Henan Province (central China). *Forensic Sci Int*, 2007, **169**(2–3): 244–246.

Zhivotovsky, L.A., et al., A comprehensive population survey on the distribution of STR frequencies in Belarus. *Forensic Sci Int*, 2007, **172**(2–3): 156–160.

Zhivotovsky, L.A., et al., A reference database on STR allele frequencies in the Belarus population developed from paternity cases. *Forensic Sci Int Genet*, 2009, **3**(3): e107–e109.

Zhivotovsky, L.A., et al., An STR database on the Volga-Ural population. *Forensic Sci Int Genet*, 2009, **3**(4): e133–e136.

Zhivotovsky, L.A., et al., Developing STR databases on structured populations: The native South Siberian population versus the Russian population. *Forensic Sci Int Genet*, 2009, **3**(4): e111–e116.

Zhou, A., et al., Population genetic data of the NGM SElect STR loci in Chinese Han population from Zhejiang region, China. *Int J Legal Med*, 2013, **127**(2): 377–378.

Zhu, B., et al., Population genetic analysis of 15 STR loci of Chinese Tu ethnic minority group. *Forensic Sci Int*, 2008, **174**(2–3): 255–258.

Zhu, B.F., et al., Genetic diversities of 21 non-CODIS autosomal STRs of a Chinese Tibetan ethnic minority group in Lhasa. *Int J Legal Med*, 2011, **125**(4): 581–585.

Zhu, B.F., et al., Population data of 15 STR loci of Chinese Yi ethnic minority group. *Leg Med (Tokyo)*, 2008, **10**(4): 220–224.

Zhu, B.F., et al., Population genetic analysis of 15 autosomal STR loci in the Russian population of Northeastern Inner-Mongolia, China. *Mol Biol Rep*, 2010, **37**(8): 3889–3895.

Zhu, J., et al., Genetic polymorphisms of 15 STR in Chinese Salar ethnic minority group. *Forensic Sci Int*, 2007, **173**(2–3): 210–213.

Zhu, Y., et al., Genetic analysis of 15 STR loci in the population of Zhejiang Province (Southeast China). *Forensic Sci Int Genet*, 2009, **3**(4): e139–e140.

Sex Chromosome Haplotyping and Gender Identification

21.1 Y Chromosome Haplotyping

21.1.1 Human Y Chromosome Genome

The Y chromosome is inherited from the father and is passed on to all male offspring; this is known as *patrilineage* (Figure 21.1). Thus, the Y chromosome is unique to males. The chromosome encodes dozens of genes required for male-specific functions, including sex determination and spermatogenesis. The human Y chromosome genome contains approximately 59 million base pairs (bp) and likely contains 50–60 genes. The Y chromosome can be divided into two regions: the pseudoautosomal region and the male-specific Y region (Figure 21.2).

21.1.1.1 Pseudoautosomal Region

The *pseudoautosomal regions (PARs)* are homologous nucleotide sequences that are present on the X and Y chromosomes. There are two PARs on each X and Y chromosome: PAR1 and PAR2. In the Y chromosome, PAR1 is located on the terminal region of the short arm. The PAR1 comprises 2.6 Mb. Twenty-four genes have been identified within the PAR1. The PAR2 of the Y chromosome is located at the tip of the long arm. The PAR2 comprises 320 kb, with only four genes identified so far.

The PARs play an important role in proper segregation of the X and Y chromosomes during meiosis. During meiosis in males, the PARs allow the Y chromosome to pair with the X chromosome. Crossing over and recombination (Chapter 25) within the PARs of the X and Y chromosomes can occur. As a result, males can inherit an allele originally present on the PARs of the X chromosome, and females can inherit an allele originally present on the PARs of the Y chromosome. In particular, PAR1 plays a major role in X–Y chromosome pairing. The deletion of PAR1 results in failure of pairing and leads to male infertility. PAR2 is much shorter than PAR1, and thus shows a lower frequency of pairing than PAR1. The deletion of PAR2 exhibits less severe phenotypes than that of PAR1.

21.1.1.2 Male-Specific Y Region

The remainder of the Y chromosome is known as the *male-specific Y (MSY)* region. It was previously called the *nonrecombining Y (NRY)* region. The bulk of this region does not participate in homologous recombination. However, certain sections involve intrachromosomal gene conversion, which is the nonreciprocal transfer of genomic DNA between a pair of repeated genes on a single chromosome; in this case, the Y chromosome. About 40 megabases (Mb) within the

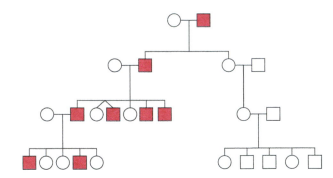

Figure 21.1   Human family pedigree showing inheritance of the Y chromosome. Females and males are denoted by circles and squares, respectively. Red symbols indicate individuals who inherited the same Y chromosome.

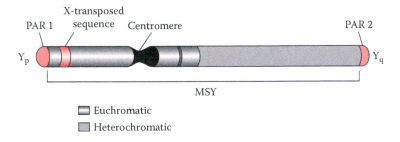

Figure 21.2   Human Y chromosome structure. PAR, the pseudoautosomal region; MSY, male-specific Y region; Y_p, the short arm of the Y chromosome; Y_q, the long arm of the Y chromosome. (© Richard C. Li.)

MSY region are heterochromatic (highly repetitive sequences) including the centromeric region and the bulk of the distal long arm. The euchromatic region, which comprises transcriptional active genes, is approximately 23 Mb. Certain sections of the euchromatic region share some homology with the X chromosome. For instance, X-transposed sequences of the Y chromosome are 99% identical to sequences within Xq21 (a band in the long arm of the X chromosome). The X-transposed sequences are the sequences, a total of approximately 3.4 Mb in length, that were transposed from the X chromosome to the Y chromosome several million years ago. Additionally, dozens of genes located in the euchromatic region share 60%–96% homology with their X chromosome counterparts. The regions sharing homology with the X chromosome should be avoided when selecting Y chromosome–specific markers for forensic DNA profiling.

21.1.1.3 Polymorphic Sequences

The Y chromosome contains an abundance of polymorphic markers. DYF155S1, also known as MSY1, is the first characterized variable number of tandem repeats (VNTR) or minisatellite at the human Y chromosome. It consists of an array of AT-rich repeats at 25 bp per unit repeat. The DYF155S1 locus is highly polymorphic. The unit repeat sequence varies through base substitution. At least five different variant types of unit repeat sequences exist. Additionally, the numbers of these units also vary, ranging from 50 to 115 repeats. Thus, the length of alleles range from approximately 1200 to 2800 bp. DYF155S1 is of considerable interest as a potential marker in forensic testing. However, the analysis method is labor intensive, making it difficult to implement in forensic casework, such as the investigation of sex crimes. Moreover, many single nucleotide polymorphisms (SNPs) and mobile elements exist at the Y chromosome. However, the discrimination power of SNPs and mobile elements is considerably less than that of Y chromosome short tandem repeats (Y-STRs). To date, Y-STRs are usually used for Y chromosome DNA testing due to the high-throughput analysis and good discrimination power (Section 21.1.2).

21.1.2 Y-STR

Y chromosome loci are very important for forensic DNA profiling, and this chapter will discuss such applications. For instance, the Y-STRs used in forensic DNA testing are male specific (for humans and certain higher primates) and are thus useful in the investigation of sexual assault cases involving male suspects. The evidence gathered in such cases usually consists of a mixture of high levels of female DNA and low levels of male DNA. The Y chromosome–specific loci can be examined without interference from large amounts of female DNA; differential extraction of sperm and nonsperm cells may not be needed. Furthermore, the Y-STR system is useful for determining the numbers of unrelated male perpetrators in sexual assault cases. The Y-STR loci used for forensic applications are located in the nonrecombining section of the Y chromosome so that patrilineage can be established. The technique can be used for paternity testing and the identification of missing persons. Finally, data interpretation can be simplified by the use of a single allele per Y-STR locus profile at most loci. Reference databases are available for estimating Y-STR haplotype frequencies among various human populations for statistical analysis of profiling results.

The major disadvantage of Y-STR loci is that their discriminating power is lowerthan that of autosomal loci. Because Y chromosome loci are linked, the product rule for statistical calculations for profile probability does not apply. Chapter 25 discusses the statistical evaluation of the strength of the matches. Also, the current Y-STR profiling cannot distinguish individuals with the same patrilineage.

More than 400 STR loci have been identified in the Y chromosome genome. The precise locations of these loci have been sequentially mapped using human genome sequencing data. Most Y-STR loci, approximately 60% of the 400 identified, are located on the long arm of the chromosome; about 22% are located on the short arm and a few are found in the centromeric region (Figure 21.3). Y-STRs in the telomeric region have yet to be identified. Only about 5% of Y-STRs are located within 5′ untranslated or intron regions of protein-coding genes. The

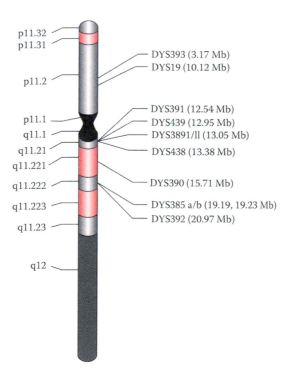

Figure 21.3 Human cytogenetic map of the Y chromosome. The Y-STRs and positions are shown (Mb=megabase). Cytogenetic patterns with alternating dark and light bands are shown. (© Richard C. Li.)

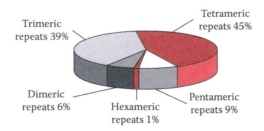

Figure 21.4 Human Y-STRs with different repeat unit length. About 400 Y-STRs have been identified and categorized according to repeat unit length. (Adapted from Hanson, E.K. and Ballantyne, J., *Leg Med (Tokyo)*, 8, 110, 2006.)

repeat unit lengths of identified Y-STRs have been analyzed. Among the 400 Y-STRs, 6% are dimeric repeats, 39% are trimeric, 45% are tetrameric, 9% are pentameric, and 1% are hexameric (Figure 21.4).

Fewer than half of the Y-STRs have been characterized. Some loci are polymorphic and are useful for forensic applications and developing new Y-STR multiplex systems. Since homologous recombination does not occur on the majority of the Y chromosome, alleles of Y-STR loci are linked (inherited together); they are referred to as *haplotypes*. As a result, the discrimination power of Y-STRs is much lower than that of autosomal STRs. The most commonly used Y-STR loci for forensic testing are described below.

21.1.2.1 Core Y-STR Loci

In 1997, the *European minimal haplotype* locus set, also known as the minimal haplotype loci, was recommended by the International Y-STR User Group for forensic applications (Table 21.1; Figure 21.5). This haplotype set includes a core set of nine Y-STR loci: DYS19, DYS385a and b, DYS389I, DYS389II, DYS390, DYS391, DYS392, and DYS393. In 2003, two additional loci were recommended by the Scientific Working Group for DNA Analysis Methods (SWGDAM) for forensic DNA analysis. The SWGDAM loci include the European minimal haplotype loci plus two additional loci: DYS438 and DYS439 (Table 21.1).

The application Y-STR for forensic casework has been expanded with additional Y-STR markers. The use of Y-STR loci has been facilitated by commercially available PCR amplification

Table 21.1	Common Y-STR Core Loci		
Locus	**EMH**	**US Haplotype Loci**	**Repeat Motif**
DYS19	☑	☑	TAGA
DYS385a/b	☑	☑	GAAA
DYS389 I	☑	☑	TCTA
DYS389 II	☑	☑	[TCTG] [TCTA]
DYS390	☑	☑	[TCTG] [TCTA]
DYS391	☑	☑	TCTA
DYS392	☑	☑	TAT
DYS393	☑	☑	AGAT
DYS438		☑	TTTTC
DYS439		☑	GATA

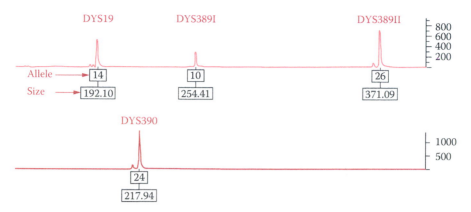

Figure 21.5   Electropherogram of Y-STR profile using an early generation of multiplex of four loci. The genotype of the DNA profile is shown (DYS19: 14. DYS389I: 10. DYS389II: 26. DYS390: 24). (© Richard C. Li.)

kits in multiplex Y-STR systems. To improve discriminating power, multiplex systems including new Y-STR loci are desired. Many new Y-STR loci are being characterized for developing new multiplex systems. Commercially available kits with more Y-STR loci are now available and have been validated for forensic use (Table 21.2; Figures 21.6 and 21.7). Some of the additional Y-STR loci are highly discriminating, allowing for further distinction between unrelated male individuals. Additionally, Y-STR loci, such as DYS570 and DYS576, with high mutation rates (see rapidly mutating Y-STR loci; Section 21.1.2.3) are included, which are useful for the discrimination of related individuals.

21.1.2.2 Multilocal Y-STR Loci

DYS385 and DYS389 are *multilocal Y-STR loci* (*MLL*). The MLL designation refers to the presence of a particular STR at more than one site on the Y chromosome due to duplication. To date, about 50 such MLL Y-STRs have been identified. Further MLL subdivisions are designated bilocal, trilocal, and so on. DYS385 and DYS389 are bilocal.

The DYS385 locus has two inverted duplicated clusters and is separated by a 4×10^4 bp interstitial region (Figure 21.8). It can be amplified by a single set of primers. One allele is observed if the duplicates are the same length. If the duplicated clusters have different lengths, they can generate two different alleles when amplified. The smaller allele is designated "a" and the larger allele is designated "b." Moreover, the DYS389 locus has two duplicated clusters with the same orientation (Figure 21.9). In a single set of PCR primers, there are two binding sites for the same forward primer at each 5′ flanking sequence of the core repeat region of DYS389. These binding sites between DYS389I and DYS389II are about 120 bp apart. Therefore, two amplicons are produced. DYS389I is designated for the smaller allele, and DYS389II is designated for the larger allele.

21.1.2.3 Rapidly Mutating Y-STR

Current Y-STR loci implemented in forensic casework have adequate resolution of males from different patrilineages. However, most Y-STR sets have limited abilities to differentiate related males who belong to the same patrilineage. Thus, current forensic Y-STR profiling is not able to exclude patrilineal relatives of the suspect. Recently, rapidly mutating Y-STR (RM Y-STR) loci have been identified and characterized. The mutation rates of RM Y-STR are above 10^{-2}, which is considerably higher than the average mutation rates of Y-STRs. The average mutation rate for the majority of the Y-STRs characterized, including the core Y-STR loci currently used in forensic casework, is approximately 10^{-4}–10^{-3} per generation. Due to the high mutation rates, RM Y-STRs can improve patrilineage resolution. For use in forensic casework, the RM Y-STR loci have potential abilities to differentiate both paternally related and unrelated males.

Table 21.2   Y-STR Loci Covered by Representative Commercial Kits

	Loci	Yfiler[a]	PowerPlexY23[b]
Core loci	DYS19	☑	☑
	DYS385a/b	☑	☑
	DYS389 I	☑	☑
	DYS389 II	☑	☑
	DYS390	☑	☑
	DYS391	☑	☑
	DYS392	☑	☑
	DYS393	☑	☑
	DYS438	☑	☑
	DYS439	☑	☑
Additional loci	DYS437	☑	☑
	DYS448	☑	☑
	DYS456	☑	☑
	DYS458	☑	☑
	DYS481		☑
	DYS533		☑
	DYS549		☑
	DYS570		☑
	DYS576		☑
	DYS635	☑	☑
	DYS643		☑
	Y-GATA-H4	☑	☑

[a] AmpFlSTR® Yfiler® PCR Amplification Kit (Applied Biosystems).
[b] PowerPlex® Y23 System (Promega).

21.2 X Chromosome Haplotyping

X-chromosomal STR (X-STR) profiling is a useful tool in kinship testing in forensic investigations. For example, males under usual circumstances have only one X chromosome; thus males are hemizygous for the X-STR loci on the X chromosome. Homologous recombination between the X and Y chromosomes is restricted to the homologous PARs (Figure 21.2). The paternal X chromosome is inherited by daughters as haplotypes. Thus, father–daughter kinship is easier to determine using X-STRs than autosomal STRs. In females, there are two copies of the X chromosomes. Homologous recombination can occur between two X chromosomes in the mother–child transmission. Although it is possible to determine mother–child kinship using X-STR analysis, the level of certainty is less than that of father–daughter kinship analysis. Nevertheless, sharing of rare X-STR haplotypes can strengthen an indication of kinship. The most useful

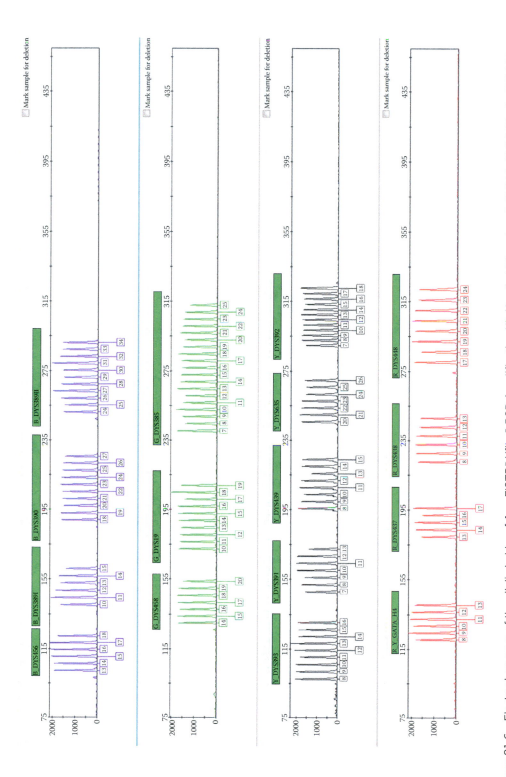

Figure 21.6   Electropherogram of the allelic ladder of AmpFlSTR® Yfiler® PCR Amplification Kit (Applied Biosystems). DYS19, DYS385 a/b, DYS389 I, DYS389 II, DYS390, DYS391, DYS392, DYS393, DYS437, DYS438, DYS439, DYS448, DYS456, DYS458, DYS635, and Y-GATA-H4 loci are shown. (© Richard C. Li.)

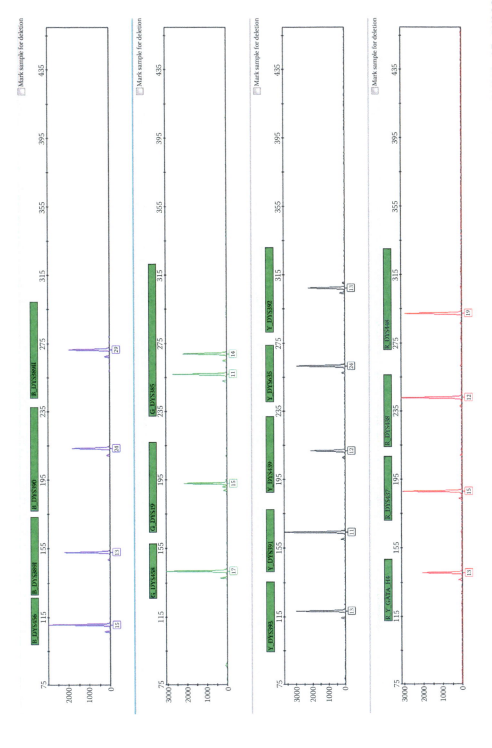

Figure 21.7 Individual Y-STR profile (AmpFISTR® Yfiler® PCR Amplification Kit). The genotype of the DNA profile is shown. DYS19: 15. DYS385 a/b: 11, 14. DYS389 I: 13. DYS389 II: 29. DYS390: 24. DYS391: 11. DYS392: 13. DYS393: 13. DYS437: 15. DYS438: 12. DYS439: 12. DYS448: 19. DYS456: 15. DYS458: 17. DYS635: 24. Y-GATA-H4: 13. (© Richard C. Li.)

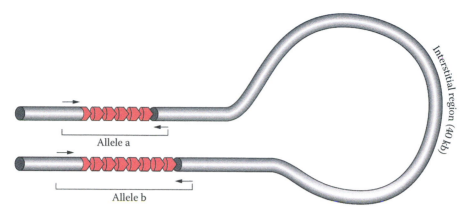

Figure 21.8   MLL Y-STR locus DYS385. At the DYS385 locus, note the two inverted duplicated regions of the Y chromosome with an interstitial region of 40 kilobases (kb). These inverted regions can be amplified with a single pair of primers (indicated by arrows) in one PCR reaction. The allele designations are described in the text. (© Richard C. Li.)

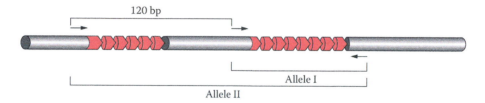

Figure 21.9   MLL Y-STR locus DYS389. At the DYS389 locus, note the two duplicated regions of the Y chromosome with the same orientation. These duplicated regions can be amplified with a single pair of primers (indicated by arrows) in one PCR reaction. The allele designations are described in the text. (© Richard C. Li.)

application of X-STR profiling, however, is in situations where autosomal STR profiling fails to determine kinship with high levels of certainty, and where Y-chromosomal or mitochondrial DNA profiling results are inconclusive.

Many X-STR loci have been identified spanning the entire human X chromosome. Increasing the number of X-STR loci to be analyzed in the inherited region would increase the degree of certainty in determining kinship. However, it is necessary to take into consideration the linkage between X-STR loci on the X chromosome. Based on Gregor Mendel's first law, the different alleles of two loci segregate independently. This is definitely true for loci localized on different chromosomes. The segregation can also be observed for loci located on the same chromosome, including the X chromosome, due to homologous recombination and meiotic crossing-over. Generally speaking, the farther apart the two loci are on the chromosome in physical distance, the more likely those loci are to segregate independently. When two loci are very close to each other, the alleles of the two loci are inherited together as a haplotype. Therefore, including additional closely linked STR loci does not necessarily improve the discrimination power in kinship analysis.

Homologous recombination events at different regions of the genome may occur with different probabilities. Thus, the tightness of linkage between two X-STR loci cannot be accurately measured based on the physical distance. In order to determine how closely two different loci are linked, the *recombination fraction* (or recombination frequency) can be used, which is the percentage of recombinants resulting from chromosomal crossover between two loci during meiosis among all the offspring. The recombination fraction reflects how often the loci are inherited together or how often they are separated. The recombination fraction ranges from 0% to 50%. The minimum recombination fraction is 0, indicating perfect linkage (no recombinants). The maximum is 50%, meaning complete independence of the two loci (they are actually located on different chromosomes or on the same chromosome with great distance between them).

Multiplex X-STR assays have also been developed. Commonly used X-STR loci can be grouped into linkage groups. Each linkage group is presumably transmitted independently from one another. These loosely linked loci can be treated as if they were located on different chromosomes. The distribution of alleles at these X-STR loci should be in accordance with the Hardy–Weinberg principle (Chapter 25). Within each linkage group, the occurrence of recombination is negligible. Thus, loci within groups are generally considered as closely linked and are thus treated as a haplotype.

21.3 Sex Typing for Gender Identification

Sex typing of a biological sample is useful in forensic investigation. For example, the application of sex typing facilitates the identification of the victim and the offender's DNA evidence in sexual assault cases and the remains of victims in mass disasters or missing persons investigations. One commonly used sex-typing marker is the amelogenin (*AMEL*) locus. The use of multiplex PCR systems with an additional amelogenin marker, a non-STR marker, leads to potential gender determination.

21.3.1 Amelogenin Locus

This region encodes extracellular matrix proteins involved in tooth enamel formation (Table 21.3). Mutations in the *AMEL* gene can lead to an enamel defect known as amelogenesis imperfecta. Amelogenesis imperfecta is a disorder that causes abnormal formation of tooth enamel in both primary and permanent teeth. The formed tooth enamel is soft and thin; therefore it is easily damaged. The *AMEL* locus has two homologous genes: *AMELX* (Xp22.2) is located on the human X chromosome (Figure 21.10) and *AMELY* (Yp11.2) is located on the human Y chromosome (Figure 21.11). Although the genes constitute a homologous pair, they differ in size and sequence. The coding sequence of the *AMELX* gene has a size of 2872 bp and the *AMELY* gene has a size of 3272 bp in length.

The most commonly used sex-typing method at the *AMEL* locus is the detection of a 6 bp deletion at intron 3 of *AMELX* (Figures 21.12 and 21.13). This deletion is not present in *AMELY*. Sex typing can be performed using primers specifically amplifying the region of the *AMEL* locus. Primer sets were developed to amplify both alleles in a single PCR. The two most commonly

Table 21.3 Sex-Typing Makers and Homologous Genes on the Human X and Y Chromosomes

Sex-Typing Markers					
X Chromosome			Y Chromosome		
Gene Symbol	Chromosomal Location	Distance from Xpter (Mb)	Gene Symbol	Chromosomal Location	Distance from Ypter (Mb)
AMELX	Xp22.2	11.3	*AMELY*	Yp11.2	6.7
DXYS156	Xq21.31	88.9	DXYS156	Yp11.2	3.2
SOX3	Xq27.1	139.6	*SRY*	Yp11.31	2.7
STS	Xp22.31	7.1–7.3	STSP1	Yq11.221	17.7
TSPYL2	Xp11.2	53.1	*TSPY1*	Yp11.2	9.2–9.3

Source: Ensembl Homo sapiens version 75.37 (GRCh37). Mb, megabase.

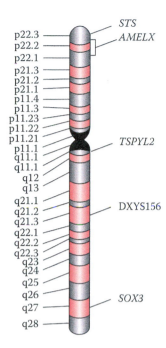

Figure 21.10   Cytogenetic map of the human X chromosome and sex-typing loci shown with physical positions (Mb = megabases). Cytogenetic patterns with alternating dark and light bands are shown. (© Richard C. Li.)

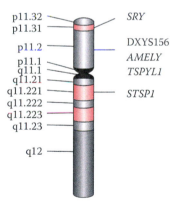

Figure 21.11   Cytogenetic map of the human Y chromosome and sex-typing loci shown with physical positions (Mb = megabases). Cytogenetic patterns with alternating dark and light bands are shown. (© Richard C. Li.)

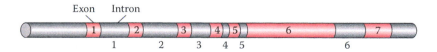

Figure 21.12   Structure of human *AMELY* gene. Exons 1 through 7 and introns 1 through 6 are numbered. (© Richard C. Li.)

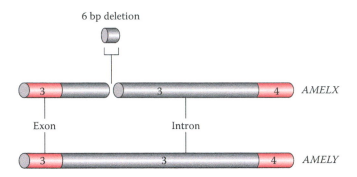

Figure 21.13   Sex typing using *AMEL* markers. A 6-bp deletion in intron 3 is present in *AMELX* but not in *AMELY* and can be resolved using electrophoresis as described in the text. (© Richard C. Li.)

used amelogenin primer sets generate amplicons of 106 and 112 bp or 212 and 218 bp for *AMELX* and *AMELY* loci, respectively. The amplicons generated from *AMELX* and *AMELY* are separated by electrophoresis. The observation of the *AMELX* fragment alone indicates a female, whereas the observation of both *AMELX* and *AMELY* fragments indicates a male (Figure 20.8). Nevertheless, DNA from primates and some rudiments can be amplified as well, but the amplicon sizes vary. The *AMEL* locus has been coamplified with other markers to provide a combined sex and identity test. Such combined tests have been used in historical D1S80 AFLP profiling (Chapter 19) and various current STR multiplex analyses.

Cases of *AMELY* null mutations have been reported where only the *AMELX* fragment is detected in *AMELY* null males. Many of them are phenotypically normal but present the *AMELX* sex types of females. Various interstitial deletions at the Y chromosome short arm and point mutations within the primer-binding sites have been identified as the possible cause of some *AMELY* null sex typing. The frequency of *AMELY* null males is rare, but is higher in Sri Lanka and India. However, the mutations may lead to incorrect identification of sex if the mutations occurred in the DNA evidence of criminal investigations such as sexual assaults or in the identification of human remains in mass disasters. In the case of a female individual, an *AMELX* mutation, which is rare, usually affects one copy of *AMELX*; however, the wild-type copy of the *AMELX* is detected. Thus, the *AMELX* mutation is less problematic in females than the *AMELY* mutation in sex typing in males.

21.3.2 Other Loci

To solve the incorrect identification problem with null *AMEL* mutations, additional candidate genetic markers are used in combination with *AMEL* analysis (Table 21.3). One useful marker is the sex-determining region Y (*SRY*) gene located on the Yp11.31 of the Y chromosome. *SRY* encodes a transcription factor that plays a role in the regulation of sex determination toward male development. The *SRY* protein contains a DNA-binding domain known as the HMG box. *SOX3*, an *SRY*-related HMG box-containing gene, has been identified at Xq27.1 of the X chromosome, which shares sequence homology with *SRY*. *SRY*-specific amplification is achieved using *SRY*-specific primers.

Another marker is the *TSPY* locus located on Yp11.2 of the Y chromosome. The *TSPY* locus encodes the testis-specific protein Y-encoded gene that is only expressed in the testis and may play a role in spermatogenesis. The *TSPY* structural gene is approximately 20 kb in length, consisting of 6 exons and 5 introns. The *TSPY* gene has multiple copies. A single Y chromosome contains several dozen copies of *TSPY* genes that are organized as a tandem array. The *TSPY* gene cluster shows copy number variation (CNV) between individuals in a population ranging from 23 to 64 copies. *DYS14* is a marker utilized for the characterization of the *TSPY* locus. The

DYS14 locus includes the partial sequence of the first exon and intron of the *TSPY* gene and shares approximately 98% sequence homology with other members of the *TSPY* gene cluster.

A *TSPY*-like (*TSPYL*) gene has been identified on the short arm of the X chromosome and is designated as *TSPYL2*. *TSPYL2* is a single-copy gene per X chromosome and is 6.3 kb in length, consisting of 7 exons and 6 introns. *TSPYL2* shares homology with *TSPY*. Additionally, several *TSPYL* genes have been identified on the autosomes. Primers are designed to amplify *TSPY* only and not amplify *TSPYL2* and other *TSPYL* gene sequences. The sensitivity of detecting *TSPY*, due to multiple copies, can be considerably higher than any single copy genetic maker such as *SRY*.

Furthermore, the DXYS156 locus, located at the pseudoautosomal region of both X and Y chromosomes, is another candidate marker used for sex typing. DXYS156 is a polymorphic pentanucleotide STR. The DXYS156 Y alleles have an additional adenine insertion in the repeat units of STR. Thus, the DXYS156 Y alleles can be distinguished from the DXYS156 X alleles because of the insertion. Additionally, the DXYS156 alleles are ethnically distributed. The DXYS156 profiling may potentially indicate the ethnic origin of an individual as an investigative lead.

Moreover, the steroid sulfatase (*STS*) gene encodes an enzyme that catalyzes the conversion of sulfated steroid precursors to biologically active steroids such as estrogens and androgens. The *STS* gene is located on the distal part of the short arm of the X chromosome within Xp22.31. The *STS* gene is 146 kb in length, consisting of 10 exons and 9 introns. The size of exons is between 120 and 160 bp long. However, the sizes of introns vary greatly, ranging from 102 bp to 35 kb. The entire coding sequence of the *STS* gene is 1542 bp in length. Deletions and point mutations of the *STS* gene have been associated with X-linked ichthyosis, a skin disease that affects males. There is an *STS* pseudogene, 100 kb in length, on the long arm of the Y chromosome known as *STSP1*. Although *STSP1* shares some sequence homologies with the *STS* gene, the pseudogene does not encode a functional gene. Primer sets are designed to amplify the target sequences within the first intron (35 kb in length) of the *STS* gene and its homologous *STSP1* pseudogene sequence. Although the homology of these target sequences of the *STS* gene and *STSP1* pseudogene is approximately 80%, amplicons with two different sizes are produced to identify *STS* and *STSP1* alleles for sex typing.

Bibliography

Y Chromosome Haplotyping

Alves, C., et al., Evaluating the informative power of Y-STRs: A comparative study using European and new African haplotype data. *Forensic Sci Int*, 2003, **134**(2–3): 126–133.

Ambers, A., et al., Autosomal and Y-STR analysis of degraded DNA from the 120-year-old skeletal remains of Ezekiel Harper. *Forensic Sci Int Genet*, 2014, **9**: 33–41.

Andersen, M.M., et al., Estimating stutter rates for Y-STR alleles. *Forensic Sci Int Genet Suppl*, 2011, **3**(1): e192–e193.

Andersen, M.M., et al., Estimating Y-STR allelic drop-out rates and adjusting for interlocus balances. *Forensic Sci Int Genet*, 2013, 7(3): 327–336.

Asamura, H., et al., Evaluation of miniY-STR multiplex PCR systems for extended 16 Y-STR loci. *Int J Legal Med*, 2008, **122**(1): 43–49.

Ayadi, I., et al., Combining autosomal and Y-chromosomal short tandem repeat data in paternity testing with male child: Methods and application. *J Forensic Sci*, 2007, **52**(5): 1068–1072.

Ayub, Q., et al., Identification and characterisation of novel human Y-chromosomal microsatellites from sequence database information. *Nucleic Acids Res*, 2000, **28**(2): e8.

Balaresque, P., et al., Genomic complexity of the Y-STR DYS19: Inversions, deletions and founder lineages carrying duplications. *Int J Legal Med*, 2009, **123**(1): 15–23.

Ballantyne, K.N., et al., A new future of forensic Y-chromosome analysis: Rapidly mutating Y-STRs for differentiating male relatives and paternal lineages. *Forensic Sci Int Genet*, 2012, **6**(2): 208–218.

Ballantyne, K.N., et al., Mutability of Y-chromosomal microsatellites: Rates, characteristics, molecular bases, and forensic implications. *Am J Hum Genet*, 2010, **87**(3): 341–353.

Ballard, D.J., et al., A study of mutation rates and the characterisation of intermediate, null and duplicated alleles for 13 Y chromosome STRs. *Forensic Sci Int*, 2005, **155**(1): 65–70.

Beleza, S., et al., Extending STR markers in Y chromosome haplotypes. *Int J Legal Med*, 2003, **117**(1): 27–33.

Bosch, E., et al., High resolution Y chromosome typing: 19 STRs amplified in three multiplex reactions. *Forensic Sci Int*, 2002, **125**(1): 42–51.

Bouakaze, C., et al., First successful assay of Y-SNP typing by SNaPshot minisequencing on ancient DNA. *Int J Legal Med*, 2007, **121**(6): 493–499.

Brion, M., et al., Hierarchical analysis of 30 Y-chromosome SNPs in European populations. *Int J Legal Med*, 2005, **119**(1): 10–15.

Burgarella, C. and M. Navascues, Mutation rate estimates for 110 Y-chromosome STRs combining population and father-son pair data. *Eur J Hum Genet*, 2011, **19**(1): 70–75.

Butler, J.M., et al., A novel multiplex for simultaneous amplification of 20 Y chromosome STR markers. *Forensic Sci Int*, 2002, **129**(1): 10–24.

Caglia, A., et al., Increased forensic efficiency of a STR-based Y-specific haplotype by addition of the highly polymorphic DYS385 locus. *Int J Legal Med*, 1998, **111**(3): 142–146.

Capelli, C., et al., Phylogenetic evidence for multiple independent duplication events at the DYS19 locus. *Forensic Sci Int Genet*, 2007, **1**(3–4): 287–290.

Carboni, I. and U. Ricci, Unexpected patterns in Y-STR analyses and implications for profile identification. *Forensic Sci Int Genet Suppl*, 2009, **2**(1): 55–56.

Carracedo, A., et al., Results of a collaborative study of the EDNAP group regarding the reproducibility and robustness of the Y-chromosome STRs DYS19, DYS389 I and II, DYS390 and DYS393 in a PCR pentaplex format. *Forensic Sci Int*, 2001, **119**(1): 28–41.

Cerri, N., et al., Mixed stains from sexual assault cases: Autosomal or Y-chromosome short tandem repeats? *Croat Med J*, 2003, **44**(3): 289–292.

Corach, D., et al., Routine Y-STR typing in forensic casework. *Forensic Sci Int*, 2001, **118**(2–3): 131–135.

D'Amato, M.E., V.B. Bajic, and S. Davison, Design and validation of a highly discriminatory 10-locus Y-chromosome STR multiplex system. *Forensic Sci Int Genet*, 2011, **5**(2): 122–125.

Davis, C., et al., Prototype PowerPlex® Y23 system: A concordance study. *Forensic Sci Int Genet*, 2013, **7**(1): 204–208.

de Knijff, P., et al., Chromosome Y microsatellites: Population genetic and evolutionary aspects. *Int J Legal Med*, 1997, **110**(3): 134–149.

Decker, A.E., et al., The impact of additional Y-STR loci on resolving common haplotypes and closely related individuals. *Forensic Sci Int Genet*, 2007, **1**(2): 215–217.

Decker, A.E., et al., Analysis of mutations in father-son pairs with 17 Y-STR loci. *Forensic Sci Int Genet*, 2008, **2**(3): e31–e35.

Dekairelle, A.F. and B. Hoste, Application of a Y-STR-pentaplex PCR (DYS19, DYS389I and II, DYS390 and DYS393) to sexual assault cases. *Forensic Sci Int*, 2001, **118**(2–3): 122–125.

Delfin, F.C., et al., Y-STR analysis for detection and objective confirmation of child sexual abuse. *Int J Legal Med*, 2005, **119**(3): 158–163.

Deng, Z.H., et al., Application of 17 Y-chromosome specific STR loci in paternity testing. *Zhongguo Shi Yan Xue Ye Xue Za Zhi*, 2008, **16**(3): 699–703.

Djelloul, S. and V. Sarafian, Validation of a 17-locus Y-STR multiplex system. *Forensic Sci Int Genet Suppl*, 2008, **1**(1): 198–199.

Dupuy, B.M., et al., Y-chromosomal microsatellite mutation rates: Differences in mutation rate between and within loci. *Hum Mutat*, 2004, **23**(2): 117–124.

Edlund, H. and M. Allen, Y chromosomal STR analysis using pyrosequencing technology. *Forensic Sci Int Genet*, 2009, **3**(2): 119–124.

Ge, J., B. Budowle, and R. Chakraborty, Interpreting Y chromosome STR haplotype mixture. *Leg Med*, 2010, **12**(3): 137–143.

Ge, J., et al., Mutation rates at Y chromosome short tandem repeats in Texas populations. *Forensic Sci Int Genet*, 2009, **3**(3): 179–184.

Ge, J., et al., US forensic Y-chromosome short tandem repeats database. *Leg Med (Tokyo)*, 2010, **12**(6): 289–295.

Geppert, M., J. Edelmann, and R. Lessig, The Y-chromosomal STRs DYS481, DYS570, DYS576 and DYS643. *Leg Med*, 2009, **11**(Suppl 1): S109–S110.

Geppert, M., et al., Hierarchical Y-SNP assay to study the hidden diversity and phylogenetic relationship of native populations in South America. *Forensic Sci Int Genet*, 2011, **5**(2): 100–104.

Gill, P., et al., DNA commission of the International Society of Forensic Genetics: Recommendations on forensic analysis using Y-chromosome STRs. *Int J Legal Med*, 2001, **114**(6): 305–309.

Goedbloed, M., et al., Comprehensive mutation analysis of 17 Y-chromosomal short tandem repeat poly-morphisms included in the AmpFSTR® Yfiler® PCR amplification kit. *Int J Legal Med*, 2009, **123**(6): 471–482.

Gonzalez-Neira, A., et al., Sequence structure of 12 novel Y chromosome microsatellites and PCR amplifica-tion strategies. *Forensic Sci Int*, 2001, **122**(1): 19–26.

Gross, A.M., et al., Internal validation of the AmpFlSTR Yfiler amplification kit for use in forensic casework. *J Forensic Sci*, 2008, **53**(1): 125–134.

Gusmao, L., C. Alves, and A. Amorim, Molecular characteristics of four human Y-specific microsatellites (DYS434, DYD437, DYS438, DYS439) for population and forensic studies. *Ann Hum Genet*, 2001, **65**(Pt 3): 285–291.

Gusmao, L., et al., Robustness of the Y STRs DYS19, DYS389 I and II, DYS390 and DYS393: Optimization of a PCR pentaplex. *Forensic Sci Int*, 1999, **106**(3): 163–172.

Gusmao, L., et al., Alternative primers for DYS391 typing: Advantages of their application to forensic genet-ics. *Forensic Sci Int*, 2000, **112**(1): 49–57.

Hall, A. and J. Ballantyne, Novel Y-STR typing strategies reveal the genetic profile of the semen donor in extended interval post-coital cervicovaginal samples. *Forensic Sci Int*, 2003, **136**(1–3): 58–72.

Hall, T.A., et al., Human Y-STR profiling using fully automated electrospray ionization time of flight mass spectrometry. *Forensic Sci Int Genet Suppl*, 2011, **3**(1): e451–e452.

Hammer, M.F., et al., Hierarchical patterns of global human Y-chromosome diversity. *Mol Biol Evol*, 2001, **18**(7): 1189–1203.

Hammer, M.F., et al., Population structure of Y chromosome SNP haplogroups in the United States and forensic implications for constructing Y chromosome STR databases. *Forensic Sci Int*, 2006, **164**(1): 45–55.

Hanson, E., et al., Performance evaluation and optimization of multiplex PCRs for the highly discriminating OSU 10-locus set Y-STRs. *J Forensic Sci*, 2012, **57**(1): 52–59.

Hanson, E.K. and J. Ballantyne, Comprehensive annotated STR physical map of the human Y chromosome: Forensic implications. *Leg Med (Tokyo)*, 2006, **8**(2): 110–120.

Hanson, E.K. and J. Ballantyne, An ultra-high discrimination Y chromosome short tandem repeat multiplex DNA typing system. *PLoS One*, 2007, **2**(8): e688.

Hartzell, B., K. Graham, and B. McCord, Response of short tandem repeat systems to temperature and sizing methods. *Forensic Sci Int*, 2003, **133**(3): 228–234.

He, M., et al., Geographical affinities of the HapMap samples. *PLoS One*, 2009, **4**(3): e4684.

Hedman, M., et al., Dissecting the Finnish male uniformity: The value of additional Y-STR loci. *Forensic Sci Int Genet*, 2011, **5**(3): 199–201.

Henke, J., et al., Application of Y-chromosomal STR haplotypes to forensic genetics. *Croat Med J*, 2001, **42**(3): 292–297.

Heyer, E., et al., Estimating Y chromosome specific microsatellite mutation frequencies using deep rooting pedigrees. *Hum Mol Genet*, 1997, **6**(5): 799–803.

Holtkemper, U., et al., Mutation rates at two human Y-chromosomal microsatellite loci using small pool PCR techniques. *Hum Mol Genet*, 2001, **10**(6): 629–633.

Honda, K., The Ashikaga case of Japan—Y-STR testing used as the exculpatory evidence to free a convicted felon after 17.5 years in prison. *Forensic Sci Int Genet*, 2013, **7**(1): e1–e2.

Huang, D., et al., Y-haplotype screening of local patrilineages followed by autosomal STR typing can detect likely perpetrators in some populations. *J Forensic Sci*, 2011, **56**(5): 1340–1342.

Huang, Y.M., et al., Assessment of application value of 19 autosomal short tandem repeat loci of GoldenEye™ 20A kit in forensic paternity testing. *Int J Legal Med*, 2013, **127**(3): 587–590.

Hurles, M.E. and M.A. Jobling, A singular chromosome. *Nat Genet*, 2003, **34**(3): 246–247.

Jobling, M.A., Y-chromosomal SNP haplotype diversity in forensic analysis. *Forensic Sci Int*, 2001, **118**(2–3): 158–162.

Jobling, M.A., A. Pandya, and C. Tyler-Smith, The Y chromosome in forensic analysis and paternity testing. *Int J Legal Med*, 1997, **110**(3): 118–124.

Jobling, M.A. and C. Tyler-Smith, Fathers and sons: The Y chromosome and human evolution. *Trends Genet*, 1995, **11**(11): 449–456.

Jobling, M.A. and C. Tyler-Smith, New uses for new haplotypes the human Y chromosome, disease and selection. *Trends Genet*, 2000, **16**(8): 356–362.

Jobling, M.A. and C. Tyler-Smith, The human Y chromosome: An evolutionary marker comes of age. *Nat Rev Genet*, 2003, **4**(8): 598–612.

Jobling, M.A., et al., Y-chromosome-specific microsatellite mutation rates re-examined using a minisatellite, MSY1. *Hum Mol Genet*, 1999, **8**(11): 2117–2120.

Kayser, M. and A. Sajantila, Mutations at Y-STR loci: Implications for paternity testing and forensic analysis. *Forensic Sci Int*, 2001, **118**(2–3): 116–121.

Kayser, M., et al., Evaluation of Y-chromosomal STRs: A multicenter study. *Int J Legal Med*, 1997, **110**(3): 125–133, 141–149.

Kayser, M., et al., A comprehensive survey of human Y-chromosomal microsatellites. *Am J Hum Genet*, 2004, **74**(6): 1183–1197.

Kayser, M., et al., Characteristics and frequency of germ-line mutations at microsatellite loci from the human Y chromosome, as revealed by direct observation in father/son pairs. *Am J Hum Genet*, 2000, **66**(5): 1580–1588.

Krawczak, M., Forensic evaluation of Y-STR haplotype matches: A comment. *Forensic Sci Int*, 2001, **118**(2–3): 114–115.

Krenke, B.E., et al., Validation of a male-specific, 12-locus fluorescent short tandem repeat (STR) multiplex. *Forensic Sci Int*, 2005, **148**(1): 1–14.

Lagoa, A.M., T. Magalhães, and M.F. Pinheiro, Autosomic STR, Y-STR and miniSTR markers evaluation for genetic analysis of fingerprints. *Forensic Sci Int Genet Suppl*, 2008, **1**(1): 435–436.

Lahn, B.T., N.M. Pearson, and K. Jegalian, The human Y chromosome, in the light of evolution. *Nat Rev Genet*, 2001, **2**(3): 207–216.

Laouina, A., et al., Mutation rate at 17 Y-STR loci in "father/son" pairs from Moroccan population. *Leg Med (Tokyo)*, 2013, **15**(5): 269–271.

Lareu, M., et al., The use of the LightCycler for the detection of Y chromosome SNPs. *Forensic Sci Int*, 2001, **118**(2–3): 163–168.

Lim, S.K., et al., Variation of 52 new Y-STR loci in the Y chromosome consortium worldwide panel of 76 diverse individuals. *Int J Legal Med*, 2007, **121**(2): 124–127.

Ma, Y., et al., A Y-chromosomal haplotype with two short tandem repeat mutations. *J Forensic Sci*, 2012, **57**(6): 1630–1633.

Malsom, S., et al., The prevalence of mixed DNA profiles in fingernail samples taken from couples who cohabit using autosomal and Y-STRs. *Forensic Sci Int Genet*, 2009, **3**(2): 57–62.

Marjanović, D., et al., Identification of skeletal remains of communist armed forces victims during and after World War II: Combined Y-chromosome short tandem repeat (STR) and miniSTR approach. *Croat Med J*, 2009, **50**(3): 296–304.

Maybruck, J.L., et al., A comparative analysis of two different sets of Y-chromosome short tandem repeats (Y-STRs) on a common population panel. *Forensic Sci Int Genet*, 2009, **4**(1): 11–20.

Mayntz-Press, K.A. and J. Ballantyne, Performance characteristics of commercial Y-STR multiplex systems. *J Forensic Sci*, 2007, **52**(5): 1025–1034.

Mayntz-Press, K.A., et al., Y-STR profiling in extended interval (≥3 days) postcoital cervicovaginal samples. *J Forensic Sci*, 2008, **53**(2): 342–348.

Mulero, J.J., C.W. Chang, and L.K. Hennessy, Characterization of the N + 3 stutter product in the trinucleotide repeat locus DYS392. *J Forensic Sci*, 2006, **51**(5): 1069–1073.

Naitoh, S., et al., Assignment of Y-chromosomal SNPs found in Japanese population to Y-chromosomal haplogroup tree. *J Hum Genet*, 2013, **58**(4): 195–201.

Nebel, A., et al., Haplogroup-specific deviation from the stepwise mutation model at the microsatellite loci DYS388 and DYS392. *Eur J Hum Genet*, 2001, **9**(1): 22–26.

Niederstatter, H., et al., Separate analysis of DYS385a and b versus conventional DYS385 typing: Is there forensic relevance? *Int J Legal Med*, 2005, **119**(1): 1–9.

Olofsson, J., et al., Evaluation of Y-STR analyses of sperm cell negative vaginal samples. *Forensic Sci Int Genet Suppl*, 2011, **3**(1): e141–e142.

Onofri, V., L. Buscemi, and A. Tagliabracci, Evaluating Y-chromosome STRs mutation rates: A collaborative study of the Ge.F.I.-ISFG Italian group. *Forensic Sci Int Genet Suppl*, 2009, **2**(1): 419–420.

Oz, C., et al., A Y-chromosome STR marker should be added to commercial multiplex STR kits. *J Forensic Sci*, 2008, **53**(4): 858–861.

Paracchini, S., et al., Hierarchical high-throughput SNP genotyping of the human Y chromosome using MALDI-TOF mass spectrometry. *Nucleic Acids Res*, 2002, **30**(6): e27.

Park, M.J., et al., Characterization of deletions in the DYS385 flanking region and null alleles associated with AZFc microdeletions in Koreans. *J Forensic Sci*, 2008, **53**(2): 331–334.

Park, M.J., et al., Y-SNP miniplexes for East Asian Y-chromosomal haplogroup determination in degraded DNA. *Forensic Sci Int Genet*, 2013, **7**(1): 75–81.

Park, M.J., et al., Y-STR analysis of degraded DNA using reduced-size amplicons. *Int J Legal Med*, 2007, **121**(2): 152–157.

Parson, W., et al., When autosomal short tandem repeats fail: Optimized primer and reaction design for Y-chromosome short tandem repeat analysis in forensic casework. *Croat Med J*, 2001, **42**(3): 285–287.

Parson, W., et al., Improved specificity of Y-STR typing in DNA mixture samples. *Int J Legal Med*, 2003, **117**(2): 109–114.

Parson, W., et al., Y-STR analysis on DNA mixture samples—Results of a collaborative project of the ENFSI DNA working group. *Forensic Sci Int Genet*, 2008, **2**(3): 238–242.

Pascali, V.L., M. Dobosz, and B. Brinkmann, Coordinating Y-chromosomal STR research for the courts. *Int J Legal Med*, 1999, **112**(1): 1.

Prinz, M. and M. Sansone, Y chromosome-specific short tandem repeats in forensic casework. *Croat Med J*, 2001, **42**(3): 288–291.

Prinz, M., et al., Multiplexing of Y chromosome specific STRs and performance for mixed samples. *Forensic Sci Int*, 1997, **85**(3): 209–218.

Prinz, M., et al., Validation and casework application of a Y chromosome specific STR multiplex. *Forensic Sci Int*, 2001, **120**(3): 177–188.

Raina, A., et al., Misinterpretation of results in medico-legal cases due to microdeletion in the Y-chromosome. *Mol Cell Probes*, 2010, **24**(6): 418–420.

Ravid-Amir, O. and S. Rosset, Maximum likelihood estimation of locus-specific mutation rates in Y-chromosome short tandem repeats. *Bioinformatics*, 2010, **26**(18): i440–i445.

Redd, A.J., S.L. Clifford, and M. Stoneking, Multiplex DNA typing of short-tandem-repeat loci on the Y chromosome. *Biol Chem*, 1997, **378**(8): 923–927.

Redd, A.J., et al., Forensic value of 14 novel STRs on the human Y chromosome. *Forensic Sci Int*, 2002, **130**(2–3): 97–111.

Rodig, H., et al., Evaluation of haplotype discrimination capacity of 35 Y-chromosomal short tandem repeat loci. *Forensic Sci Int*, 2008, **174**(2–3): 182–188.

Roewer, L., Y chromosome STR typing in crime casework. *Forensic Sci Med Pathol*, 2009, **5**(2): 77–84.

Roewer, L. and M. Geppert, Interpretation guidelines of a standard Y-chromosome STR 17-plex PCR-CE assay for crime casework. *Methods Mol Biol*, 2012, **830**: 43–56.

Roewer, L., et al., A new method for the evaluation of matches in non-recombining genomes: Application to Y-chromosomal short tandem repeat (STR) haplotypes in European males. *Forensic Sci Int*, 2000, **114**(1): 31–43.

Roewer, L., et al., Online reference database of European Y-chromosomal short tandem repeat (STR) haplotypes. *Forensic Sci Int*, 2001, **118**(2–3): 106–113.

Rolf, B., et al., Paternity testing using Y-STR haplotypes: Assigning a probability for paternity in cases of mutations. *Int J Legal Med*, 2001, **115**(1): 12–15.

Rozen, S., et al., Abundant gene conversion between arms of palindromes in human and ape Y chromosomes. *Nature*, 2003, **423**(6942): 873–876.

Sanchez, J.J., et al., Multiplex PCR and minisequencing of SNPs—A model with 35 Y chromosome SNPs. *Forensic Sci Int*, 2003, **137**(1): 74–84.

Sanchez-Diz, P., et al., Results of the GEP-ISFG collaborative study on two Y-STRs tetraplexes: GEPY I (DYS461, GATA C4, DYS437 and DYS438) and GEPY II (DYS460, GATA A10, GATA H4 and DYS439). *Forensic Sci Int*, 2003, **135**(2): 158–162.

Santos, F.R., N.O. Bianchi, and S.D. Pena, Worldwide distribution of human Y-chromosome haplotypes. *Genome Res*, 1996, **6**(7): 601–611.

Santos, F.R., S.D. Pena, and J.T. Epplen, Genetic and population study of a Y-linked tetranucleotide repeat DNA polymorphism with a simple non-isotopic technique. *Hum Genet*, 1993, **90**(6): 655–656.

Schneider, P.M., et al., Results of collaborative study regarding the standardization of the Y-linked STR system DYS385 by the European DNA Profiling (EDNAP) group. *Forensic Sci Int*, 1999, **102**(2–3): 159–165.

Schoske, R., et al., High-throughput Y-STR typing of U.S. populations with 27 regions of the Y chromosome using two multiplex PCR assays. *Forensic Sci Int*, 2004, **139**(2–3): 107–121.

Schoske, R., et al., Multiplex PCR design strategy used for the simultaneous amplification of 10 Y chromosome short tandem repeat (STR) loci. *Anal Bioanal Chem*, 2003, **375**(3): 333–343.

Seo, Y., et al., A method for genotyping Y chromosome-linked DYS385a and DYS385b loci. *Leg Med (Tokyo)*, 2003, **5**(4): 228–232.

Sibille, I., et al., Y-STR DNA amplification as biological evidence in sexually assaulted female victims with no cytological detection of spermatozoa. *Forensic Sci Int*, 2002, **125**(2–3): 212–216.

Sinha, S.K., et al., Development and validation of a multiplexed Y-chromosome STR genotyping system, Y-PLEX 6, for forensic casework. *J Forensic Sci*, 2003, **48**(1): 93–103.

Sinha, S.K., et al., Development and validation of the Y-PLEX 5, a Y-chromosome STR genotyping system, for forensic casework. *J Forensic Sci*, 2003, **48**(5): 985–1000.

Skaletsky, H., et al., The male-specific region of the human Y chromosome is a mosaic of discrete sequence classes. *Nature*, 2003, **423**(6942): 825–837.

Sturk, K.A., et al., Evaluation of modified Yfiler™ amplification strategy for compromised samples. *Croat Med J*, 2009, **50**(3): 228–238.

Szibor, R., M. Kayser, and L. Roewer, Identification of the human Y-chromosomal microsatellite locus DYS19 from degraded DNA. *Am J Forensic Med Pathol*, 2000, **21**(3): 252–254.

Thompson, J.M., et al., Developmental validation of the PowerPlex® Y23 System: A single multiplex Y-STR analysis system for casework and database samples. *Forensic Sci Int Genet*, 2013, **7**(2): 240–250.

Tilford, C.A., et al., A physical map of the human Y chromosome. *Nature*, 2001, **409**(6822): 943–945.

Tsuji, A., et al., Personal identification using Y-chromosomal short tandem repeats from bodily fluids mixed with semen. *Am J Forensic Med Pathol*, 2001, **22**(3): 288–291.

Tsukada, K., et al., DNA typing using AmpFlSTR Y-filer for very long-term stored specimens. *Leg Med (Tokyo)*, 2009, **11**(Suppl 1): S466–S467.

Tsutsumi, H., et al., A case of personal identification due to detection of rare DNA types from seminal stain. *J Oral Sci*, 2009, **51**(4): 645–650.

Underhill, P.A., et al., Y chromosome sequence variation and the history of human populations. *Nat Genet*, 2000, **26**(3): 358–361.

Underhill, P.A., et al., The phylogeography of Y chromosome binary haplotypes and the origins of modern human populations. *Ann Hum Genet*, 2001, **65**(Pt 1): 43–62.

Vallone, P.M. and J.M. Butler, Y-SNP typing of U.S. African American and Caucasian samples using allele-specific hybridization and primer extension. *J Forensic Sci*, 2004, **49**(4): 723–732.

Valverde, L., et al., Improving the analysis of Y-SNP haplogroups by a single highly informative 16 SNP multiplex PCR-minisequencing assay. *Electrophoresis*, 2013, **34**(4): 605–612.

van Oven, M., A. Ralf, and M. Kayser, An efficient multiplex genotyping approach for detecting the major worldwide human Y-chromosome haplogroups. *Int J Legal Med*, 2011, **125**(6): 879–885.

Vermeulen, M., et al., Improving global and regional resolution of male lineage differentiation by simple single-copy Y-chromosomal short tandem repeat polymorphisms. *Forensic Sci Int Genet*, 2009, **3**(4): 205–213.

Wei, W., et al., Exploring of new Y-chromosome SNP loci using Pyrosequencing and the SNaPshot methods. *Int J Legal Med*, 2012, **126**(6): 825–833.

White, P.S., et al., New, male-specific microsatellite markers from the human Y chromosome. *Genomics*, 1999, **57**(3): 433–437.

Willuweit, S., L. Roewer, and International Forensic Y Chromosome User Group, Y chromosome haplotype reference database (YHRD): Update. *Forensic Sci Int Genet*, 2007, **1**(2): 83–87.

Zhivotovsky, L.A., et al., The effective mutation rate at Y chromosome short tandem repeats, with application to human population-divergence time. *Am J Hum Genet*, 2004, **74**(1): 50–61.

Zuccarelli, G., et al., Rapid screening for Native American mitochondrial and Y-chromosome haplogroups detection in routine DNA analysis. *Forensic Sci Int Genet*, 2011, **5**(2): 105–108.

Genetic Population Data of Y Chromosome Haplotyping

Aboukhalid, R., et al., Haplotype frequencies for 17 Y-STR loci (AmpFlSTRY-filer) in a Moroccan population sample. *Forensic Sci Int Genet*, 2010, **4**(3): e73–e74.

Achakzai, N.M., et al., Y-chromosomal STR analysis in the Pashtun population of Southern Afghanistan. *Forensic Sci Int Genet*, 2012, **6**(4): e103–e105.

Alakoc, Y.D., et al., Y-chromosome and autosomal STR diversity in four proximate settlements in Central Anatolia. *Forensic Sci Int Genet*, 2010, **4**(5): e135–e137.

Alam, S., et al., Haplotype diversity of 17 Y-chromosomal STR loci in the Bangladeshi population. *Forensic Sci Int Genet*, 2010, **4**(2): e59–e60.

Ali, N., et al., Announcement of population data: Genetic data for 17 Y-STR AmpFlSTR® Yfiler™ markers from an immigrant Pakistani population in the UK (British Pakistanis). *Forensic Sci Int Genet*, 2013, **7**(2): e40–e42.

Alshamali, F., et al., Local population structure in Arabian Peninsula revealed by Y-STR diversity. *Hum Hered*, 2009, **68**(1): 45–54.

Ambrosio, B., et al., Y-STR genetic diversity in autochthonous Andalusians from Huelva and Granada provinces (Spain). *Forensic Sci Int Genet*, 2012, **6**(2): e66–e71.

Andersen, M.M., P.S. Eriksen, and N. Morling, The discrete Laplace exponential family and estimation of Y-STR haplotype frequencies. *J Theor Biol*, 2013, **329**: 39–51.

Andersen, M.M., et al., Estimating trace-suspect match probabilities for singleton Y-STR haplotypes using coalescent theory. *Forensic Sci Int Genet*, 2013, **7**(2): 264–271.

Baeza, C., et al., Population data for 15 Y-chromosome STRs in a population sample from Quito (Ecuador). *Forensic Sci Int*, 2007, **173**(2–3): 214–219.

Bai, R., et al., Haplotype diversity of 17 Y-STR loci in a Chinese Han population sample from Shanxi Province, Northern China. *Forensic Sci Int Genet*, 2013, **7**(1): 214–216.

Balamurugan, K., et al., Y chromosome STR allelic and haplotype diversity in five ethnic Tamil populations from Tamil Nadu, India. *Leg Med (Tokyo)*, 2010, **12**(5): 265–269.

Bembea, M., et al., Y-chromosome STR haplotype diversity in three ethnically isolated population from North-Western Romania. *Forensic Sci Int Genet*, 2011, **5**(3): e99–e100.

Bento, A.M., et al., Distribution of Y-chromosomal haplotypes in the Central Portuguese population using 17-STRs. *Forensic Sci Int Genet*, 2009, **4**(1): e35–e36.

Bing, L., et al., Population genetics for 17 Y-STR loci(AmpFlSTR® Y-filer™) in Luzhou Han ethnic group. *Forensic Sci Int Genet*, 2013, **7**(2): e23–e26.

Borjas, L., et al., Usefulness of 12 Y-STRs for forensic genetics evaluation in two populations from Venezuela. *Leg Med (Tokyo)*, 2008, **10**(2): 107–112.

Brisighelli, F., et al., Patterns of Y-STR variation in Italy. *Forensic Sci Int Genet*, 2012, **6**(6): 834–839.

Budowle, B., et al., Texas population substructure and its impact on estimating the rarity of Y STR haplotypes from DNA evidence. *J Forensic Sci*, 2009, **54**(5): 1016–1021.

Budowle, B., et al., The effects of Asian population substructure on Y STR forensic analyses. *Leg Med (Tokyo)*, 2009, **11**(2): 64–69.

Builes, J.J., et al., Y chromosome STR haplotypes in the Caribbean city of Cartagena (Colombia). *Forensic Sci Int*, 2007, **167**(1): 62–69.

Caine, L.M., M.M. de Pancorbo, and F. Pinheiro, Y-chromosomal STR haplotype diversity in males from Santa Catarina, Brazil. *J Forensic Leg Med*, 2010, **17**(2): 92–95.

Carvalho, M., et al., Paternal and maternal lineages in Guinea-Bissau population. *Forensic Sci Int Genet*, 2011, **5**(2): 114–116.

Chang, Y.M., et al., Haplotype diversity of 16 Y-chromosomal STRs in three main ethnic populations (Malays, Chinese and Indians) in Malaysia. *Forensic Sci Int*, 2007, **167**(1): 70–76.

Cloete, K., et al., Analysis of seventeen Y-chromosome STR loci in the Cape Muslim population of South Africa. *Leg Med (Tokyo)*, 2010, **12**(1): 42–45.

Coble, M.D., C.R. Hill, and J.M. Butler, Haplotype data for 23 Y-chromosome markers in four U.S. population groups. *Forensic Sci Int Genet*, 2013, **7**(3): e66–e68.

Cortellini, V., et al., Y-chromosome polymorphisms and ethnic group—A combined STR and SNP approach in a population sample from northern Italy. *Croat Med J*, 2013, **54**(3): 279–285.

D'Amato, M.E., M. Benjeddou, and S. Davison, Evaluation of 21 Y-STRs for population and forensic studies. *Forensic Sci Int Genet Suppl*, 2009, **2**(1): 446–447.

Davis, C., et al., Y-STR loci diversity in native Alaskan populations. *Int J Legal Med*, 2011, **125**(4): 559–563.

Diaz-Lacava, A., et al., Geostatistical inference of main Y-STR-haplotype groups in Europe. *Forensic Sci Int Genet*, 2011, 5(2): 91–94.

Diaz-Lacava, A., et al., Spatial assessment of Argentinean genetic admixture with geographical information systems. *Forensic Sci Int Genet*, 2011, 5(4): 297–302.

Ehler, E., R. Marvan, and D. Vanek, Evaluation of 14 Y-chromosomal short tandem repeat haplotype with focus on DYS449, DYS456, and DYS458: Czech population sample. *Croat Med J*, 2010, **51**(1): 54–60.

Elmrghni, S., et al., Population genetic data for 17 Y STR markers from Benghazi (East Libya). *Forensic Sci Int Genet*, 2012, **6**(2): 224–227.

Ferri, G., et al., Y-STR variation in Albanian populations: Implications on the match probabilities and the genetic legacy of the minority claiming an Egyptian descent. *Int J Legal Med*, 2010, **124**(5): 363–370.

Frank, W.E., H.C. Ralph, and M.A. Tahir, Y chromosome STR haplotypes and allele frequencies in a southern Indian male population. *J Forensic Sci*, 2008, **53**(1): 248–251.

Gayden, T., et al., Y-chromosomal microsatellite diversity in three culturally defined regions of historical Tibet. *Forensic Sci Int Genet*, 2012, **6**(4): 437–446.

Gois, C.C., et al., Genetic population data of 12 STR loci of the PowerPlex Y system in the state of Sao Paulo population (Southeast of Brazil). *Forensic Sci Int*, 2008, **174**(1): 81–86.

Gonzalez-Andrade, F., et al., Y-chromosome STR haplotypes in three different population groups from Ecuador (South America). *J Forensic Sci*, 2008, **53**(2): 512–514.

Gonzalez-Andrade, F., et al., Y-STR variation among ethnic groups from Ecuador: Mestizos, Kichwas, Afro-Ecuadorians and Waoranis. *Forensic Sci Int Genet*, 2009, **3**(3): e83–e91.

Grskovic, B., et al., Population genetic analysis of haplotypes based on 17 short tandem repeat loci on Y chromosome in population sample from eastern Croatia. *Croat Med J*, 2010, **51**(3): 202–208.

Grskovic, B., et al., Genetic polymorphisms of 17 short tandem repeat loci on Y chromosome in central Croatian population. *Forensic Sci Med Pathol*, 2011, **7**(2): 155–161.

Grskovic, B., et al., Population data for 17 short tandem repeat loci on Y chromosome in northern Croatia. *Mol Biol Rep*, 2011, **38**(3): 2203–2209.

Guo, H., et al., Genetic polymorphisms for 17 Y-chromosomal STRs haplotypes in Chinese Hui population. *Leg Med (Tokyo)*, 2008, **10**(3): 163–169.

Haliti, N., et al., Evaluation of population variation at 17 autosomal STR and 16 Y-STR haplotype loci in Croatians. *Forensic Sci Int Genet*, 2009, **3**(4): e137–e138.

Hallenberg, C., et al., Y-chromosome STR haplotypes in males from Greenland. *Forensic Sci Int Genet*, 2009, **3**(4): e145–e146.

Hara, M., et al., Genetic data for 16 Y-chromosomal STR loci in Japanese. *Leg Med (Tokyo)*, 2007, **9**(3): 161–170.

Hashiyada, M., et al., Population genetics of 17 Y-chromosomal STR loci in Japanese. *Forensic Sci Int Genet*, 2008, **2**(4): e69–e70.

He, J. and F. Guo, Population genetics of 17 Y-STR loci in Chinese Manchu population from Liaoning Province, northeast China. *Forensic Sci Int Genet*, 2013, **7**(3): e84–e85.

Hedman, M., et al., Dissecting the Finnish male uniformity: The value of additional Y-STR loci. *Forensic Sci Int Genet*, 2011, **5**(3): 199–201.

Huang, T.Y., et al., Polymorphism of 17 Y-STR loci in Taiwan population. *Forensic Sci Int*, 2008, **174**(2–3): 249–254.

Hwa, H.L., et al., Seventeen Y-chromosomal short tandem repeat haplotypes in seven groups of population living in Taiwan. *Int J Legal Med*, 2010, **124**(4): 295–300.

Hwang, J.H., et al., Haplotypes for 12 Y-chromosomal STR loci in a Korean population (the central region). *Forensic Sci Int*, 2007, **168**(1): 73–84.

Illeperuma, R.J., et al., Haplotype data for 12 Y-chromosome STR loci of Sri Lankans. *Forensic Sci Int Genet*, 2010, **4**(4): e119–e120.

Jakovski, Z., et al., Genetic data for 17 Y-chromosomal STR loci in Macedonians in the Republic of Macedonia. *Forensic Sci Int Genet*, 2011, **5**(4): e108–e111.

Julieta Avila, S., I. Briceno, and A. Gomez, Genetic population analysis of 17 Y-chromosomal STRs in three states (Valle del Cauca, Cauca and Narino) from Southwestern Colombia. *J Forensic Leg Med*, 2009, **16**(4): 204–211.

Kang, L.L., K. Liu, and Y. Ma, Y chromosome STR haplotypes of Tibetan Living Tibet Lassa. *Forensic Sci Int*, 2007, **172**(1): 79–83.

Katsaloulis, P., et al., Genetic population study of 11 Y chromosome STR loci in Greece. *Forensic Sci Int Genet*, 2013, **7**(3): e56–e58.

Kayser, M., et al., Relating two deep-rooted pedigrees from central Germany by high-resolution Y-STR haplotyping. *Forensic Sci Int Genet*, 2007, **1**(2): 125–128.

Kim, S.H., et al., Genetic polymorphisms of 16 Y chromosomal STR loci in Korean population. *Forensic Sci Int Genet*, 2008, **2**(2): e9–e10.

Kim, S.H., et al., Population genetics and mutational events at 6 Y-STRs in Korean population. *Forensic Sci Int Genet*, 2009, **3**(2): e53–e54.

Kim, S.H., et al., Y chromosome homogeneity in the Korean population. *Int J Legal Med*, 2010, **124**(6): 653–657.

Kim, S.H., et al., Forensic genetic data of 6 Y-STR loci: An expanded Korean population database. *Forensic Sci Int Genet*, 2012, **6**(1): e35–e36.

Kovacevic, L., et al., Haplotype data for 23 Y-chromosome markers in a reference sample from Bosnia and Herzegovina. *Croat Med J*, 2013, **54**(3): 286–290.

Kovatsi, L., J.L. Saunier, and J.A. Irwin, Population genetics of Y-chromosome STRs in a population of Northern Greeks. *Forensic Sci Int Genet*, 2009, **4**(1): e21–e22.

Kumagai, R., et al., Haplotype analysis of 17 Y-STR loci in a Japanese population. *Forensic Sci Int*, 2007, **172**(1): 72–78.

Lacau, H., et al., Y-STR profiling in two Afghanistan populations. *Leg Med (Tokyo)*, 2011, **13**(2): 103–108.

Laouina, A., et al., Allele frequencies and population data for 17 Y-STR loci (The AmpFlSTR® Y-filer™) in Casablanca resident population. *Forensic Sci Int Genet*, 2011, **5**(1): e1–e3.

Larmuseau, M.H., et al., Micro-geographic distribution of Y-chromosomal variation in the central-western European region Brabant. *Forensic Sci Int Genet*, 2011, **5**(2): 95–99.

Leat, N., et al., Properties of novel and widely studied Y-STR loci in three South African populations. *Forensic Sci Int*, 2007, **168**(2–3): 154–161.

Lecerf, M., et al., Allele frequencies and haplotypes of eight Y-short tandem repeats in Bantu population living in Central Africa. *Forensic Sci Int*, 2007, **171**(2–3): 212–215.

Lee, H.Y., et al., Haplotypes and mutation analysis of 22 Y-chromosomal STRs in Korean father-son pairs. *Int J Legal Med*, 2007, **121**(2): 128–135.

Lee, J., et al., Y chromosomal STRs haplotypes in two populations from Bolivia. *Leg Med (Tokyo)*, 2007, **9**(1): 43–47.

Lim, S.K., et al., Variation of 52 new Y-STR loci in the Y chromosome consortium worldwide panel of 76 diverse individuals. *Int J Legal Med*, 2007, **121**(2): 124–127.

Ljubkovic, J., et al., Y-chromosomal short tandem repeat haplotypes in southern Croatian male population defined by 17 loci. *Croat Med J*, 2008, **49**(2): 201–206.

Lowery, R.K., et al., Sub-population structure evident in forensic Y-STR profiles from Armenian geographical groups. *Leg Med (Tokyo)*, 2013, **15**(2): 85–90.

Maybruck, J.L., et al., A comparative analysis of two different sets of Y-chromosome short tandem repeats (Y-STRs) on a common population panel. *Forensic Sci Int Genet*, 2009, **4**(1): 11–20.

Melo, M.M., et al., Y-STR haplotypes in three ethnic linguistic groups of Angola population. *Forensic Sci Int Genet*, 2011, **5**(3): e83–e88.

Mendes-Junior, C.T., et al., Y-chromosome STR haplotypes in a sample from Sao Paulo State, southeastern Brazil. *J Forensic Sci*, 2007, **52**(2): 495–497.

Mertens, G., et al., Twelve-locus Y-STR haplotypes in the Flemish population. *J Forensic Sci*, 2007, **52**(3): 755–757.

Mielnik-Sikorska, M., et al., Genetic data from Y chromosome STR and SNP loci in Ukrainian population. *Forensic Sci Int Genet*, 2013, **7**(1): 200–203.

Mizuno, N., et al., 16 Y chromosomal STR haplotypes in Japanese. *Forensic Sci Int*, 2008, **174**(1): 71–76.

Mizuno, N., et al., A forensic method for the simultaneous analysis of biallelic markers identifying Y chromosome haplogroups inferred as having originated in Asia and the Japanese archipelago. *Forensic Sci Int Genet*, 2010, **4**(2): 73–79.

Mršić, G., et al., Croatian national reference Y-STR haplotype database. *Mol Biol Rep*, 2012, **39**(7): 7727–7741.

Muro, T., et al., Simultaneous determination of seven informative Y chromosome SNPs to differentiate East Asian, European, and African populations. *Leg Med (Tokyo)*, 2011, **13**(3): 134–141.

Nagy, M., et al., Searching for the origin of Romanies: Slovakian Romani, Jats of Haryana and Jat Sikhs Y-STR data in comparison with different Romani populations. *Forensic Sci Int*, 2007, **169**(1): 19–26.

Nonaka, I., K. Minaguchi, and N. Takezaki, Y-chromosomal binary haplogroups in the Japanese population and their relationship to 16 Y-STR polymorphisms. *Ann Hum Genet*, 2007, **71**(Pt 4): 480–495.

Nunez, C., et al., Y chromosome haplogroup diversity in a Mestizo population of Nicaragua. *Forensic Sci Int Genet*, 2012, **6**(6): e192–e195.

Onofri, V., et al., Y-chromosome genetic structure in sub-Apennine populations of Central Italy by SNP and STR analysis. *Int J Legal Med*, 2007, **121**(3): 234–237.

Padilla-Gutierrez, J.R., et al., Population data and mutation rate of nine Y-STRs in a mestizo Mexican population from Guadalajara, Jalisco, Mexico. *Leg Med (Tokyo)*, 2008, **10**(6): 319–320.

Palet, L., et al., Y-STR genetic diversity in Moroccans from the Figuig oasis. *Forensic Sci Int Genet*, 2010, **4**(5): e139–e141.

Palha Tde, J., E.M. Rodrigues, and S.E. Dos Santos, Y-chromosomal STR haplotypes in a population from the Amazon region, Brazil. *Forensic Sci Int*, 2007, **166**(2–3): 233–239.

Palha, T., et al., Fourteen short tandem repeat loci Y chromosome haplotypes: Genetic analysis in populations from northern Brazil. *Forensic Sci Int Genet*, 2012, **6**(3): 413–418.

Palha, T.J., E.M. Rodrigues, and S.E. dos Santos, Y-STR haplotypes of Native American populations from the Brazilian Amazon region. *Forensic Sci Int Genet*, 2010, **4**(5): e121–e123.

Palo, J.U., et al., High degree of Y-chromosomal divergence within Finland—Forensic aspects. *Forensic Sci Int Genet*, 2007, **1**(2): 120–124.

Palo, J.U., et al., The effect of number of loci on geographical structuring and forensic applicability of Y-STR data in Finland. *Int J Legal Med*, 2008, **122**(6): 449–456.

Park, S.W., et al., Development of a Y-STR 12-plex PCR system and haplotype analysis in a Korean population. *J Genet*, 2009, **88**(3): 353–358.

Parkin, E.J., et al., Diversity of 26-locus Y-STR haplotypes in a Nepalese population sample: Isolation and drift in the Himalayas. *Forensic Sci Int*, 2007, **166**(2–3): 176–181.

Parvathy, S.N., A. Geetha, and C. Jagannath, Haplotype analysis of the polymorphic 17 YSTR markers in Kerala nontribal populations. *Mol Biol Rep*, 2012, **39**(6): 7049–7059.

Petrejcikova, E., et al., Allele frequencies and population data for 11 Y-chromosome STRs in samples from Eastern Slovakia. *Forensic Sci Int Genet*, 2011, **5**(3): e53–e62.

Piglionica, M., et al., Population data for 17 Y-chromosome STRs in a sample from Apulia (Southern Italy). *Forensic Sci Int Genet*, 2013, **7**(1): e3–e4.

Ploski, R., et al., Homogeneity and distinctiveness of Polish paternal lineages revealed by Y chromosome microsatellite haplotype analysis. *Hum Genet*, 2002, **110**(6): 592–600.

Pokupcic, K., et al., Y-STR genetic diversity of Croatian (Bayash) Roma. *Forensic Sci Int Genet*, 2008, **2**(2): e11–e13.

Pontes, M.L., et al., Allele frequencies and population data for 17 Y-STR loci (AmpFlSTR® Y-filer™) in a Northern Portuguese population sample. *Forensic Sci Int*, 2007, **170**(1): 62–67.

Rafiee, M.R., et al., Analysis of Y-chromosomal short tandem repeat (STR) polymorphism in an Iranian Sadat population. *Genetika*, 2009, **45**(8): 1105–1109.

Rapone, C., et al., Y chromosome haplotypes in central-south Italy: Implication for reference database. *Forensic Sci Int*, 2007, **172**(1): 67–71.

Rebala, K., et al., Forensic analysis of polymorphism and regional stratification of Y-chromosomal microsatellites in Belarus. *Forensic Sci Int Genet*, 2011, **5**(1): e17–e20.

Rebala, K., et al., Y-STR variation among Slavs: Evidence for the Slavic homeland in the middle Dnieper basin. *J Hum Genet*, 2007, **52**(5): 406–414.

Roewer, L., et al., Y-chromosomal STR haplotypes in Kalmyk population samples. *Forensic Sci Int*, 2007, **173**(2–3): 204–209.

Roewer, L., et al., Analysis of Y chromosome STR haplotypes in the European part of Russia reveals high diversities but non-significant genetic distances between populations. *Int J Legal Med*, 2008, **122**(3): 219–223.

Roewer, L., et al., A Y-STR database of Iranian and Azerbaijanian minority populations. *Forensic Sci Int Genet*, 2009, **4**(1): e53–e55.

Romero, R.E., et al., A Colombian Caribbean population study of 16 Y-chromosome STR loci. *Forensic Sci Int Genet*, 2008, **2**(2): e5–e8.

Salvador, J.M., K.A. Tabbada, and M.C. De Ungria, Population data of 10 Y-chromosomal STR loci in Cebu province, Central Visayas (Philippines). *J Forensic Sci*, 2008, **53**(1): 256–258.

Sanchez, C., et al., Haplotype frequencies of 16 Y-chromosome STR loci in the Barcelona metropolitan area population using Y-Filer kit. *Forensic Sci Int*, 2007, **172**(2–3): 211–217.

Schwengber, S.P., et al., Population data of 17 Y-STR loci from Rio Grande do Sul state (South Brazil). *Forensic Sci Int Genet*, 2009, **4**(1): e31–e33.

Seong, K.M., et al., Population genetic polymorphisms of 17 Y-chromosomal STR loci in South Koreans. *Forensic Sci Int Genet*, 2011, **5**(5): e122–e123.

Serin, A., et al., Haplotype frequencies of 17 Y-chromosomal short tandem repeat loci from the Cukurova region of Turkey. *Croat Med J*, 2011, **52**(6): 703–708.

Shi, M., et al., Southwest China Han population data for nine Y-STR loci by multiplex polymerase chain reaction. *J Forensic Sci*, 2007, **52**(1): 228–230.

Shi, M.S., et al., Haplotypes of 20 Y-chromosomal STRs in a population sample from southeast China (Chaoshan area). *Int J Legal Med*, 2007, **121**(6): 455–462.

Shi, M., et al., Population genetics for Y-chromosomal STRs haplotypes of Chinese Tujia ethnic group. *Forensic Sci Int Genet*, 2008, **2**(4): e65–e68.

Shi, M., et al., Haplotype diversity of 22 Y-chromosomal STRs in a southeast China population sample (Chaoshan area). *Forensic Sci Int Genet*, 2009, **3**(2): e45–e47.

Shi, M., et al., Population genetics for Y-chromosomal STRs haplotypes of Chinese Xibe ethnic group. *Forensic Sci Int Genet*, 2011, **5**(5): e119–e121.

Sims, L.M., D. Garvey, and J. Ballantyne, Sub-populations within the major European and African derived haplogroups R1b3 and E3a are differentiated by previously phylogenetically undefined Y-SNPs. *Hum Mutat*, 2007, **28**(1): 97.

Siriboonpiputtana, T., et al., Y-chromosomal STR haplotypes in central Thai population. *Forensic Sci Int Genet*, 2010, **4**(3): e71–e72.

Soltyszewski, I., et al., Y-chromosomal haplotypes for the AmpFlSTR Yfiler PCR amplification kit in a population sample from Central Poland. *Forensic Sci Int*, 2007, **168**(1): 61–67.

Stanciu, F., et al., Population data for Y-chromosome haplotypes defined by 17 STRs in South-East Romania. *Leg Med (Tokyo)*, 2010, **12**(5): 259–264.

Stevanovic, M., et al., Human Y-specific STR haplotypes in population of Serbia and Montenegro. *Forensic Sci Int*, 2007, **171**(2–3): 216–221.

Taylor, D., et al., An investigation of admixture in an Australian Aboriginal Y-chromosome STR database. *Forensic Sci Int Genet*, 2012, **6**(5): 532–538.

Taylor, D.A. and J.M. Henry, Haplotype data for 16 Y-chromosome STR loci in Aboriginal and Caucasian populations in South Australia. *Forensic Sci Int Genet*, 2012, **6**(6): e187–e188.

Thangaraj, K., et al., Y-chromosomal STR haplotypes in two endogamous tribal populations of Karnataka, India. *J Forensic Sci*, 2007, **52**(3): 751–753.

Theves, C., et al., Population genetics of 17 Y-chromosomal STR loci in Yakutia. *Forensic Sci Int Genet*, 2010, **4**(5): e129–e130.

Tian-Xiao, Z., Y. Li, and L. Sheng-Bin, Y-STR haplotypes and the genetic structure from eight Chinese ethnic populations. *Leg Med (Tokyo)*, 2009, **11**(Suppl 1): S198–S200.

Triki-Fendri, S., et al., Population genetics of 17 Y-STR markers in West Libya (Tripoli region). *Forensic Sci Int Genet*, 2013, **7**(3): e59–e61.

Trynova, E.G., et al., Presentation of 17 Y-chromosomal STRs in the population of the Sverdlovsk region. *Forensic Sci Int Genet*, 2011, **5**(3): e101–e104.

Valverde, L., et al., 17 Y-STR haplotype data for a population sample of residents in the Basque Country. *Forensic Sci Int Genet*, 2012, **6**(4): e109–e111.

Valverde, L., et al., Y-STR variation in the Basque diaspora in the Western USA: Evolutionary and forensic perspectives. *Int J Legal Med*, 2012, **126**(2): 293–298.

Veselinovic, I.S., et al., Allele frequencies and population data for 17 Y-chromosome STR loci in a Serbian population sample from Vojvodina province. *Forensic Sci Int*, 2008, **176**(2–3): e23–e28.

Volgyi, A., et al., Hungarian population data for 11 Y-STR and 49 Y-SNP markers. *Forensic Sci Int Genet*, 2009, **3**(2): e27–e28.

Weng, W., et al., Mutation rates at 16 Y-chromosome STRs in the South China Han population. *Int J Legal Med*, 2013, **127**(2): 369–372.

Willuweit, S., et al., Y-STR frequency surveying method: A critical reappraisal. *Forensic Sci Int Genet*, 2011, **5**(2): 84–90.

Wolanska-Nowak, P., et al., A population data for 17 Y-chromosome STR loci in South Poland population sample—Some DYS458.2 variants uncovered and sequenced. *Forensic Sci Int Genet*, 2009, **4**(1): e43–e44.

Wolfgramm Ede, V., et al., Genetic analysis of 15 autosomal and 12 Y-STR loci in the Espirito Santo State population, Brazil. *Forensic Sci Int Genet*, 2011, **5**(3): e41–e43.

Wozniak, M., et al., Continuity of Y chromosome haplotypes in the population of Southern Poland before and after the Second World War. *Forensic Sci Int Genet*, 2007, **1**(2): 134–140.

Wu, F.C., et al., Y-chromosomal STRs haplotypes in the Taiwanese Paiwan population. *Int J Legal Med*, 2011, **125**(1): 39–43.

Wu, W., et al., Population genetics of 17 Y-STR loci in a large Chinese Han population from Zhejiang Province, Eastern China. *Forensic Sci Int Genet*, 2011, **5**(1): e11–e13.

Yadav, B., A. Raina, and T.D. Dogra, Genetic polymorphisms for 17 Y-chromosomal STR haplotypes in Jammu and Kashmir Saraswat Brahmin population. *Leg Med (Tokyo)*, 2010, **12**(5): 249–255.

Yadav, B., A. Raina, and T.D. Dogra, Haplotype diversity of 17 Y-chromosomal STRs in Saraswat Brahmin community of North India. *Forensic Sci Int Genet*, 2011, **5**(3): e63–e70.

Yanmei, Y., et al., Genetic polymorphism of 11 Y-chromosomal STR loci in Yunnan Han Chinese. *Forensic Sci Int Genet*, 2010, **4**(2): e67–e69.

Yoshida, Y. and S. Kubo, Y-SNP and Y-STR analysis in a Japanese population. *Leg Med (Tokyo)*, 2008, **10**(5): 243–252.

Zahra, N., et al., The analysis of UAE populations using AmpFlSTR® Y Filer™: Identification of novel and null alleles. *Forensic Sci Int Genet Suppl*, 2008, **1**(1): 255–256.

Zastera, J., et al., Assembly of a large Y-STR haplotype database for the Czech population and investigation of its substructure. *Forensic Sci Int Genet*, 2010, **4**(3): e75–e78.

Zhang, D., et al., Haplotypes of six miniY-STR loci in the Han population from Sichuan province and the Zhuang population in Guangxi Zhuang autonomous region. *Forensic Sci Int Genet*, 2009, **3**(2): e49–e51.

Zhang, G.Q., et al., Structure and polymorphism of 16 novel Y-STRs in Chinese Han population. *Genet Mol Res*, 2012, **11**(4): 4487–4500.

Zhang, H., et al., Haplotype of 12 Y-STR loci of the PowerPlex Y-system in Sichuan Han ethnic group in west China. *Forensic Sci Int*, 2008, **175**(2–3): 244–249.

Zhang, Y. and J. Lee, Genetic polymorphism of Y-chromosomal STR loci in South Korean population. *Forensic Sci Int*, 2007, **173**(2–3): 225–230.

Zhang, Y., et al., Population genetics for Y-chromosomal STRs haplotypes of Chinese Korean ethnic group in northeastern China. *Forensic Sci Int*, 2007, **173**(2–3): 197–203.

Zhang, Y., et al., Allele frequencies of 12 Y-chromosomal STRs in Chinese Tuvans in the Altay region. *Forensic Sci Int Genet*, 2013, **7**(1): e7–e8.

Zhang, Y.J., et al., Allele frequencies of DYS19, DYS385, DYS388, DYS389 (I and II), DYS390, DYS391, DYS392, and DYS393 STR loci in a Korean population. *J Forensic Sci*, 2007, **52**(6): 1424–1425.

Zhang, Y.L., et al., Genetic diversity of Y-chromosome microsatellites in the Fujian Han and the Sichuan Han populations of China. *Anthropol Anz*, 2007, **65**(1): 1–14.

Zhu, B., et al., Genetic analysis of 17 Y-chromosomal STRs haplotypes of Chinese Tibetan ethnic group residing in Qinghai province of China. *Forensic Sci Int*, 2008, **175**(2–3): 238–243.

Zhu, B., et al., Population genetic polymorphisms for 17 Y-chromosomal STRs haplotypes of Chinese Salar ethnic minority group. *Leg Med (Tokyo)*, 2007, **9**(4): 203–209.

X Chromosome Haplotyping

Castaneda, M., et al., Haplotypic blocks of X-linked STRs for forensic cases: Study of recombination and mutation rates. *J Forensic Sci*, 2012, **57**(1): 192–195.

Chen, D.P., et al., Use of X-linked short tandem repeats loci to confirm mutations in parentage caseworks. *Clin Chim Acta*, 2009, **408**(1–2): 29–33.

Diegoli, T.M. and M.D. Coble, Development and characterization of two mini-X chromosomal short tandem repeat multiplexes. *Forensic Sci Int Genet*, 2011, **5**(5): 415–421.

Edelmann, J., et al., Characterisation of the STR markers DXS10146, DXS10134 and DXS10147 located within a 79.1 kb region at Xq28. *Forensic Sci Int Genet*, 2008, **2**(1): 41–46.

Edelmann, J., et al., Validation of six closely linked STRs located in the chromosome X centromere region. *Int J Legal Med*, 2010, **124**(1): 83–87.

Freitas, N.S., et al., X-linked insertion/deletion polymorphisms: Forensic applications of a 33-markers panel. *Int J Legal Med*, 2010, **124**(6): 589–593.

Gusmao, L., et al., Capillary electrophoresis of an X-chromosome STR decaplex for kinship deficiency cases. *Methods Mol Biol*, 2012, **830**: 57–71.

Hering, S., et al., X chromosomal recombination—A family study analysing 39 STR markers in German three-generation pedigrees. *Int J Legal Med*, 2010, **124**(5): 483–491.

Li, C., et al., Development of 11 X-STR loci typing system and genetic analysis in Tibetan and Northern Han populations from China. *Int J Legal Med*, 2011, **125**(5): 753–756.

Li, L., et al., Analysis of 14 highly informative SNP markers on X chromosome by TaqMan SNP genotyping assay. *Forensic Sci Int Genet*, 2010, **4**(5): e145–e148.

Liu, Q.L., D.J. Lu, and H. Zhao, The establishment of a nine X-chromosome short tandem repeat loci multiplex PCR and detection of the polymorphism of these loci. *Chin J Med Genet*, 2009, **26**(6): 664–669.

Liu, Q.L., et al., Development and population study of the 12 X-STR loci multiplexes PCR systems. *Int J Legal Med*, 2012, **126**(4): 665–670.

Liu, Q.L., et al., Development of multiplex PCR system with 15 X-STR loci and genetic analysis in three nationality populations from China. *Electrophoresis*, 2012, **33**(8): 1299–1305.

Nothnagel, M., et al., Collaborative genetic mapping of 12 forensic short tandem repeat (STR) loci on the human X chromosome. *Forensic Sci Int Genet*, 2012, **6**(6): 778–784.

Penna, L.S., et al., Development of two multiplex PCR systems for the analysis of 14 X-chromosomal STR loci in a southern Brazilian population sample. *Int J Legal Med*, 2012, **126**(2): 327–330.

Pepinski, W., et al., X-chromosomal polymorphism data for the ethnic minority of Polish Tatars and the religious minority of Old Believers residing in northeastern Poland. *Forensic Sci Int Genet*, 2007, **1**(2): 212–214.

Pereira, R., et al., A method for the analysis of 32 X chromosome insertion deletion polymorphisms in a single PCR. *Int J Legal Med*, 2012, **126**(1): 97–105.

Turrina, S., et al., Development and forensic validation of a new multiplex PCR assay with 12 X-chromosomal short tandem repeats. *Forensic Sci Int Genet*, 2007, **1**(2): 201–204.

Zarrabeitia, M.T., et al., Analysis of 10 X-linked tetranucleotide markers in mixed and isolated populations. *Forensic Sci Int Genet*, 2009, **3**(2): 63–66.

Genetic Population Data of X Chromosome Haplotyping

Aler, M., et al., Genetic data of 10 X-STRs in a Spanish population sample. *Forensic Sci Int*, 2007, **173**(2–3): 193–196.

Barbaro, A., et al., Distribution of 8 X-chromosomal STR loci in an Italian population sample (Calabria). *Forensic Sci Int Genet*, 2012, **6**(6): e174–e175.

Becker, D., et al., Population genetic evaluation of eight X-chromosomal short tandem repeat loci using Mentype Argus X-8 PCR amplification kit. *Forensic Sci Int Genet*, 2008, **2**(1): 69–74.

Caine, L.M., et al., Genetic data of a Brazilian population sample (Santa Catarina) using an X-STR decaplex. *J Forensic Leg Med*, 2010, **17**(5): 272–274.

Fracasso, T., et al., An X-STR meiosis study in Kurds and Germans: Allele frequencies and mutation rates. *Int J Legal Med*, 2008, **122**(4): 353–356.

Gao, S., et al., Allele frequencies for 10 X-STR loci in Nu population of Yunnan, China. *Leg Med (Tokyo)*, 2007, **9**(5): 284–286.

Gomes, I., et al., Analysis of 10 X-STRs in three African populations. *Forensic Sci Int Genet*, 2007, **1**(2): 208–211.

Gomes, I., et al., The Karimojong from Uganda: Genetic characterization using an X-STR decaplex system. *Forensic Sci Int Genet*, 2009, **3**(4): e127–e128.

Grskovic, B., et al., Analysis of 8 X-chromosomal markers in the population of central Croatia. *Croat Med J*, 2013, **54**(3): 238–247.

Hedman, M., J.U. Palo, and A. Sajantila, X-STR diversity patterns in the Finnish and the Somali population. *Forensic Sci Int Genet*, 2009, **3**(3): 173–178.

Horvath, G., et al., A genetic study of 12 X-STR loci in the Hungarian population. *Forensic Sci Int Genet*, 2012, **6**(1): e46–e47.

Hou, Q.F., B. Yu, and S.B. Li, Genetic polymorphisms of nine X-STR loci in four population groups from Inner Mongolia, China. *Genom Proteom Bioinformat*, 2007, **5**(1): 59–65.

Hwa, H.L., et al., Genetic analysis of eight population groups living in Taiwan using a 13 X-chromosomal STR loci multiplex system. *Int J Legal Med*, 2011, **125**(1): 33–37.

Hwa, H.L., et al., Thirteen X-chromosomal short tandem repeat loci multiplex data from Taiwanese. *Int J Legal Med*, 2009, **123**(3): 263–269.

Illescas, M.J., et al., Population genetic data for 10 X-STR loci in autochthonous Basques from Navarre (Spain). *Forensic Sci Int Genet*, 2012, **6**(5): e146–e148.

Inturri, S., et al., Linkage and linkage disequilibrium analysis of X-STRs in Italian families. *Forensic Sci Int Genet*, 2011, **5**(2): 152–154.

Laouina, A., et al., Genetic data of X-chromosomal STRs in a Moroccan population sample (Casablanca) using an X-STR decaplex. *Forensic Sci Int Genet*, 2013, **7**(3): e90–e92.

Li, C., et al., Genetic analysis of the 11 X-STR loci in Uigur population from China. *Forensic Sci Int Genet*, 2012, **6**(5): e139–e140.

Li, H., et al., A multiplex PCR for 4 X chromosome STR markers and population data from Beijing Han ethnic group. *Leg Med (Tokyo)*, 2009, **11**(5): 248–250.

Lim, E.J., et al., Genetic polymorphism and haplotype analysis of 4 tightly linked X-STR duos in Koreans. *Croat Med J*, 2009, **50**(3): 305–312.

Liu, Q.L., et al., Development of the nine X-STR loci typing system and genetic analysis in three nationality populations from China. *Int J Legal Med*, 2011, **125**(1): 51–58.

Luczak, S., et al., Diversity of 15 human X chromosome microsatellite loci in Polish population. *Forensic Sci Int Genet*, 2011, **5**(3): e71–e77.

Luo, H.B., et al., Characteristics of eight X-STR loci for forensic purposes in the Chinese population. *Int J Legal Med*, 2011, **125**(1): 127–131.

Martins, J.A., et al., X-chromosome genetic variation in Sao Paulo State (Brazil) population. *Ann Hum Biol*, 2010, **37**(4): 598–603.

Mukerjee, S., et al., Genetic variation of 10 X chromosomal STR loci in Indian population. *Int J Legal Med*, 2010, **124**(4): 327–330.

Mukerjee, S., et al., Population genetic data for 11 X-STR loci in eleven populations of India. *Leg Med (Tokyo)*, 2012, **14**(3): 163–165.

Nadeem, A., et al., Development of pentaplex PCR and genetic analysis of X chromosomal STRs in Punjabi population of Pakistan. *Mol Biol Rep*, 2009, **36**(7): 1671–1675.

Nakamura, Y. and K. Minaguchi, Sixteen X-chromosomal STRs in two octaplex PCRs in Japanese population and development of 15-locus multiplex PCR system. *Int J Legal Med*, 2010, **124**(5): 405–414.

Nakamura, Y., et al., Multiplex PCR for 18 X-chromosomal STRs in Japanese population. *Leg Med (Tokyo)*, 2013, **15**(3): 164–170.

Nishi, T., et al., Application of a novel multiplex polymerase chain reaction system for 12 X-chromosomal short tandem repeats to a Japanese population study. *Leg Med (Tokyo)*, 2013, **15**(1): 43–46.

Pereira, V., et al., Study of 25 X-chromosome SNPs in the Portuguese. *Forensic Sci Int Genet*, 2011, **5**(4): 336–338.

Poetsch, M., et al., Allele frequencies of 11 X-chromosomal loci in a population sample from Ghana. *Int J Legal Med*, 2009, **123**(1): 81–83.

Poetsch, M., et al., Allele frequencies of 11 X-chromosomal loci of two population samples from Africa. *Int J Legal Med*, 2011, **125**(2): 307–314.

Ribeiro Rodrigues, E.M., et al., A multiplex PCR for 11 X chromosome STR markers and population data from a Brazilian Amazon region. *Forensic Sci Int Genet*, 2008, **2**(2): 154–158.

Ribeiro-Rodrigues, E.M., et al., Extensive survey of 12 X-STRs reveals genetic heterogeneity among Brazilian populations. *Int J Legal Med*, 2011, **125**(3): 445–452.

Samejima, M., Y. Nakamura, and K. Minaguchi, Population genetic study of six closely linked groups of X-STRs in a Japanese population. *Int J Legal Med*, 2011, **125**(6): 895–900.

Samejima, M., et al., Genetic study of 12 X-STRs in Malay population living in and around Kuala Lumpur using Investigator Argus X-12 kit. *Int J Legal Med*, 2012, **126**(4): 677–683.

Sim, J.E., et al., Population genetic study of four closely-linked X-STR trios in Koreans. *Mol Biol Rep*, 2010, **37**(1): 333–337.

Sun, R., et al., Genetic polymorphisms of 10 X-STR among four ethnic populations in northwest of China. *Mol Biol Rep*, 2012, **39**(4): 4077–4081.

Tariq, M.A., et al., Allele frequency distribution of 13 X-chromosomal STR loci in Pakistani population. *Int J Legal Med*, 2008, **122**(6): 525–528.

Tie, J., S. Uchigasaki, and S. Oshida, Genetic polymorphisms of eight X-chromosomal STR loci in the population of Japanese. *Forensic Sci Int Genet*, 2010, **4**(4): e105–e108.

Tillmar, A.O., et al., Analysis of linkage and linkage disequilibrium for eight X-STR markers. *Forensic Sci Int Genet*, 2008, **3**(1): 37–41.

Tomas, C., et al., X-chromosome SNP analyses in 11 human Mediterranean populations show a high overall genetic homogeneity except in north-west Africans (Moroccans). *BMC Evol Biol*, 2008, **8**: 75.

Uchigasaki, S., J. Tie, and D. Takahashi, Genetic analysis of twelve X-chromosomal STRs in Japanese and Chinese populations. *Mol Biol Rep*, 2013, **40**(4): 3193–3196.

Wu, W., et al., Allele frequencies of seven X-linked STR loci in Chinese Han population from Zhejiang Province. *Forensic Sci Int Genet*, 2009, **4**(1): e41–e42.

Xu, J., et al., Population data of 12 X-chromosomal STR loci in Chinese Han samples from Hebei Province. *Forensic Sci Int Genet*, 2013, **7**(2): e43–e44.

Zalan, A., et al., Hungarian population data of eight X-linked markers in four linkage groups. *Forensic Sci Int*, 2008, **175**(1): 73–78.

Zeng, X., A. Rakha, and S. Li, Genetic polymorphisms of 10 X-chromosome STR loci in Chinese Daur ethnic minority group. *Leg Med (Tokyo)*, 2009, **11**(3): 152–154.

Zeng, X.P., et al., Genetic polymorphisms of twelve X-chromosomal STR loci in Chinese Han population from Guangdong Province. *Forensic Sci Int Genet*, 2011, **5**(4): e114–e116.

Zhang, S., et al., Genetic polymorphisms of 12 X-STR for forensic purposes in Shanghai Han population from China. *Mol Biol Rep*, 2012, **39**(5): 5705–5707.

Zhang, S.H., et al., Genetic polymorphism of eight X-linked STRs of Mentype® Argus X-8 kit in Chinese population from Shanghai. *Forensic Sci Int Genet*, 2011, **5**(1): e21–e24.

Sex Typing for Gender Identification

Blagodatskikh, E.G., et al., Sex determination in biological specimens using the DYS14 marker. *Mol Biol*, 2010, **44**(4): 568–570.

Buel, E., G. Wang, and M. Schwartz, PCR amplification of animal DNA with human X-Y amelogenin primers used in gender determination. *J Forensic Sci*, 1995, **40**(4): 641–644.

Cadenas, A.M., et al., Male amelogenin dropouts: Phylogenetic context, origins and implications. *Forensic Sci Int*, 2007, **166**(2–3): 155–163.

Chang, Y.M., L.A. Burgoyne, and K. Both, Higher failures of amelogenin sex test in an Indian population group. *J Forensic Sci*, 2003, **48**(6): 1309–1313.

Chang, Y.M., et al., A distinct Y-STR haplotype for amelogenin negative males characterized by a large Y(p)11.2 (DYS458-MSY1-AMEL-Y) deletion. *Forensic Sci Int*, 2007, **166**(2–3): 115–120.

Davis, C., et al., A case of amelogenin Y-null: A simple primer binding site mutation or unusual genetic anomaly? *Leg Med (Tokyo)*, 2012, **14**(6): 320–323.

Esteve Codina, A., H. Niederstatter, and W. Parson, "GenderPlex" a PCR multiplex for reliable gender determination of degraded human DNA samples and complex gender constellations. *Int J Legal Med*, 2009, **123**(6): 459–464.

Fincham, A.G., et al., Human developing enamel proteins exhibit a sex-linked dimorphism. *Calcif Tissue Int*, 1991, **48**(4): 288–290.

Frances, F., et al., Amelogenin test: From forensics to quality control in clinical and biochemical genomics. *Clin Chim Acta*, 2007, **386**(1–2): 53–56.

Gibbon, V., et al., Novel methods of molecular sex identification from skeletal tissue using the amelogenin gene. *Forensic Sci Int Genet*, 2009, **3**(2): 74–79.

Graham, E.A.M., Sex determination. *Forensic Sci Med Pathol*, 2006, **2**(4): 283–286.

Haas-Rochholz, H. and G. Weiler, Additional primer sets for an amelogenin gene PCR-based DNA-sex test. *Int J Legal Med*, 1997, **110**(6): 312–315.

Jacot, T.A., et al., TSPY4 is a novel sperm-specific biomarker of semen exposure in human cervicovaginal fluids; potential use in HIV prevention and contraception studies. *Contraception*, 2013, **88**(3): 387–395.

Kashyap, V.K., et al., Deletions in the Y-derived amelogenin gene fragment in the Indian population. *BMC Med Genet*, 2006, 7: 37.

Kastelic, V., B. Budowle, and K. Drobnic, Validation of SRY marker for forensic casework analysis. *J Forensic Sci*, 2009, **54**(3): 551–555.

Kim, K.Y., et al., A real-time PCR-based amelogenin Y allele dropout assessment model in gender typing of degraded DNA samples. *Int J Legal Med*, 2013, **127**(1): 55–61.

Kumagai, R., et al., DNA analysis of family members with deletion in Yp11.2 region containing amelogenin locus. *Leg Med (Tokyo)*, 2008, **10**(1): 39–42.

Lattanzi, W., et al., A large interstitial deletion encompassing the amelogenin gene on the short arm of the Y chromosome. *Hum Genet*, 2005, **116**(5): 395–401.

Lau, E.C., H.C. Slavkin, and M.L. Snead, Analysis of human enamel genes: Insights into genetic disorders of enamel. *Cleft Palate J*, 1990, **27**(2): 121–130.

Lau, E.C., et al., Human and mouse amelogenin gene loci are on the sex chromosomes. *Genomics*, 1989, **4**(2): 162–168.

Luptáková, L., et al., Sex determination of early medieval individuals through nested PCR using a new primer set in the SRY gene. *Forensic Sci Int*, 2011, **207**(1–3): 1–5.

Ma, Y., et al., Y chromosome interstitial deletion induced Y-STR allele dropout in AMELY-negative individuals. *Int J Legal Med*, 2012, **126**(5): 713–724.

Maciejewska, A. and R. Pawlowski, A rare mutation in the primer binding region of the amelogenin X homologue gene. *Forensic Sci Int Genet*, 2009, **3**(4): 265–267.

Mannucci, A., et al., Forensic application of a rapid and quantitative DNA sex test by amplification of the X-Y homologous gene amelogenin. *Int J Legal Med*, 1994, **106**(4): 190–193.

Mitchell, R.J., et al., An investigation of sequence deletions of amelogenin (AMELY), a Y-chromosome locus commonly used for gender determination. *Ann Hum Biol*, 2006, **33**(2): 227–240.

Morikawaa, T., Y. Yamamoto, and S. Miyaish, A new method for sex determination based on detection of SRY, STS and amelogenin gene regions with simultaneous amplification of their homologous sequences by a multiplex PCR. *Acta Medica Okayama*, 2011, **65**(2): 113–122.

Mukerjee, S., et al., Differential pattern of genetic variability at the DXYS156 locus on homologous regions of X and Y chromosomes in Indian population and its forensic implications. *Int J Legal Med*, 2013, **127**(1): 1–6.

Nakahori, Y., O. Takenaka, and Y. Nakagome, A human X-Y homologous region encodes "amelogenin". *Genomics*, 1991, **9**(2): 264–269.

Ou, X., et al., Null alleles of the X and Y chromosomal amelogenin gene in a Chinese population. *Int J Legal Med*, 2012, **126**(4): 513–518.

Oz, C., et al., A Y-chromosome STR marker should be added to commercial multiplex STR kits. *J Forensic Sci*, 2008, **53**(4): 858–861.

Parton, J., et al., Sex identification of ancient DNA samples using a microfluidic device. *J Archaeol Sci*, 2013, **40**(1): 705–711.

Quincey, D., et al., Difficulties of sex determination from forensic bone degraded DNA: A comparison of three methods. *Sci Justice*, 2013, **53**(3): 253–260.

Roffey, P.E., C.I. Eckhoff, and J.L. Kuhl, A rare mutation in the amelogenin gene and its potential investigative ramifications. *J Forensic Sci*, 2000, **45**(5): 1016–1019.

Sire, J.Y., S. Delgado, and M. Girondot, The amelogenin story: Origin and evolution. *Eur J Oral Sci*, 2006, **114**(Suppl 1): 64–77; discussion 93–95, 379–380.

Snead, M.L., et al., Of mice and men: Anatomy of the amelogenin gene. *Connect Tissue Res*, 1989, **22**(1–4): 101–109.

Steinlechner, M., et al., Rare failures in the amelogenin sex test. *Int J Legal Med*, 2002, **116**(2): 117–120.

Takayama, T., et al., Determination of deleted regions from Yp11.2 of an amelogenin negative male. *Leg Med (Tokyo)*, 2009, **11**(Suppl 1): S578–S580.

Thangaraj, K., A.G. Reddy, and L. Singh, Is the amelogenin gene reliable for gender identification in forensic casework and prenatal diagnosis? *Int J Legal Med*, 2002, **116**(2): 121–123.

Tomas, C., et al., Typing of 48 autosomal SNPs and amelogenin with GenPlex SNP genotyping system in forensic genetics. *Forensic Sci Int Genet*, 2008, **3**(1): 1–6.

Tozzo, P., et al., Deletion of amelogenin Y-locus in forensics: Literature revision and description of a novel method for sex confirmation. *J Forensic Leg Med*, 2013, **20**(5): 387–391.

Turrina, S., G. Filippini, and D. De Leo, Evaluation of deleted region from Yp11.2 of two amelogenin negative related males. *Forensic Sci Int Genet Suppl*, 2009, **2**(1): 240–241.

Umeno, M., et al., A rapid and simple system of detecting deletions on the Y chromosome related with male infertility using multiplex PCR. *J Med Invest*, 2006, **53**(1–2): 147–152.

von Wurmb-Schwark, N., H. Bosinski, and S. Ritz-Timme, What do the X and Y chromosomes tell us about sex and gender in forensic case analysis? *J Forensic Leg Med*, 2007, **14**(1): 27–30.

22

Single Nucleotide Polymorphism Profiling

22.1 Basic Characteristics of SNPs

Human genomes contain sequence polymorphisms, which are the variations in nucleotide sequences among individuals. One type of sequence polymorphism is called *single-nucleotide polymorphism* (*SNP*). An SNP constitutes a single-base-pair change originating from a spontaneous mutation that can be a base substitution, insertion, or deletion at a single site. An estimated 10 million SNPs may exist in the human genome, and among them, over a million SNPs have been identified. Thus, SNPs are the most frequent form of DNA polymorphism observed in humans. Most SNPs appear in noncoding regions and some SNPs are found in coding regions of genes (Figure 22.1).

The vast majority of SNPs are biallelic, although very rare triallelic and tetraallelic SNPs also exist. If an SNP originating from a spontaneous mutation occurs in the germ line, it can be inherited by offspring and can spread in the human population. As a result, both the wild type (the typical form of an allele occurring in nature) and a mutant allele are produced. This type of SNP is referred to as a biallelic SNP. Subsequently, a third mutation, a very rare event, may occur at the same nucleotide site. This SNP has a wild type and two different mutant alleles, which results in a triallelic SNP. It is also possible that a triallelic SNP is created by introducing two of the mutations simultaneously at a single site, which is also a very rare event.

Generally speaking, there are a number of advantages to utilizing SNP loci as markers for forensic applications. First, SNPs are abundant within the human genome; therefore, a sufficient number of SNP loci that are suitable for human identification can be selected. Second, in SNP analysis, amplified fragments (amplicons) are usually 50–100 base pairs in length and are smaller than that in short tandem repeat (STR) analysis (Chapter 20). Therefore, SNP profiling can be more useful than STR for analyzing degraded DNA samples in which genomic DNA is fragmented (Section 20.5.1). Third, SNP loci have lower mutation rates than STRs and are, therefore, useful for specialized human identification such as paternity testing. Lastly, it is possible to achieve high-throughput SNP analysis by utilizing multiplex SNP assays, which are amendable for automation.

The application of SNPs for forensic DNA analysis also has some disadvantages, in that SNP loci are not as polymorphic as STR loci. It is estimated that the analysis of 50–60 SNP loci is needed to achieve a similar level of population match probability (P_m—the lower the P_m, the less likely a match occurs between two randomly chosen individuals; see Chapter 25) using 13

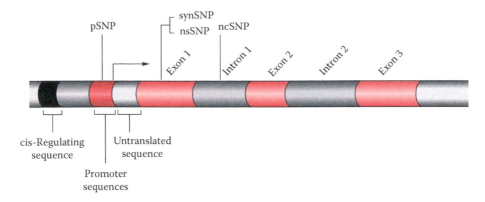

Figure 22.1 SNPs fall into several classes. Most reside in the noncoding regions of DNA and are designated as noncoding SNPs (ncSNPs). A subset of ncSNPs can also be found in introns. SNPs residing in exons are further divided into two types: The synonymous type (synSNP) is an exonic SNP that does not change the amino acid composition of the encoded polypeptide. Conversely, a nonsynonymous type (nsSNP) changes the encoded amino acid. SNPs in the promoter regions of the genome are known as promoter SNPs (pSNPs). Arrow: transcription start site. (© Richard C. Li.)

STR loci in CODIS. Moreover, most SNPs are biallelic; therefore, it is difficult to resolve mixed DNA profiles when a shared common allele exists in a mixture of more than one individual. Furthermore, most DNA databases contain STR profiles instead of SNP profiles, which means that it is not possible to carry out a database search for a matching SNP profile in cases where no suspect has yet been identified.

22.2 Forensic Applications of SNP Profiling

22.2.1 *HLA-DQA1* Locus

The first use of SNP-based profiling for forensic application involved sequence polymorphisms at the *HLA-DQA1* locus (formerly called the DQα locus). The *HLA-DQA1* gene is a member of the human leukocyte antigen (HLA) family, which contains a large number of genes involved in the immune response in humans. The *HLA-DQA1* locus is located within human HLA gene clusters on chromosome 6. The region tested for forensic use is highly polymorphic and is located at the second exon of the gene (Table 22.1).

22.2.1.1 DQα AmpliType and Polymarker Assays

The SNP profiling of the *HLA-DQA1* locus was carried out using the DQα AmpliType® kit. It was the first commercial kit, developed in the late 1980s by Cetus Corporation in Emeryville, CA. The *HLA-DQA1* panel can distinguish the following: alleles 1 (subtyped as 1.1, 1.2, or 1.3), 2, 3, and 4 (subtyped as 4.1 and 4.2/4.3, in which the 4.2 and 4.3 alleles are combined and cannot be distinguished). Therefore, a total of 28 possible genotypes from combinations of these alleles can be distinguished. Although the P_m of this SNP profiling is high (approximately 5×10^{-2}), it is useful as a preliminary test to quickly exclude suspects.

In addition to the *HLA-DQA1* locus, five additional loci—*LDLR, GYPA, HBGG,* D7S8, and *GC*—were utilized for forensic application in 1993 (Table 22.1). These loci were included in the second generation of the kit known as the AmpliType® PM PCR amplification and typing kit (also known as Polymarker), manufactured by Perkin-Elmer (Norwalk, CT). It consisted of one panel for the testing of *HLA-DQAl* and another panel for an additional five loci (Figure 22.2). Among these five additional loci, *LDLR, GYPA,* and D7S8 each have two alleles (designated

Table 22.1   Chromosomal Locations of SNP Loci Used in AmpliType® PM PCR Amplification and Typing Kit

Locus	Gene Product	Chromosome Location	Number of Alleles	Further Reading
HLA-DQA1	HLA-DQA1	6p21.3	7	Gyllensten and Erlich (1988)
LDLR	Low-density lipoprotein receptor	19p 13.1-13.3	2	Yamamoto et al. (1984)
GYPA	Glycophorin A	4q28-31	2	Siebert and Fukuda (1987)
HBGG	Hemoglobin G gammaglobin	11 p 15.5	3	Slightom et al. (1980)
D7S8	Anonymous	7q22-31.1	2	Horn et al. (1990)
GC	Group-specific component	4q11-13	3	Yang et al. (1985)

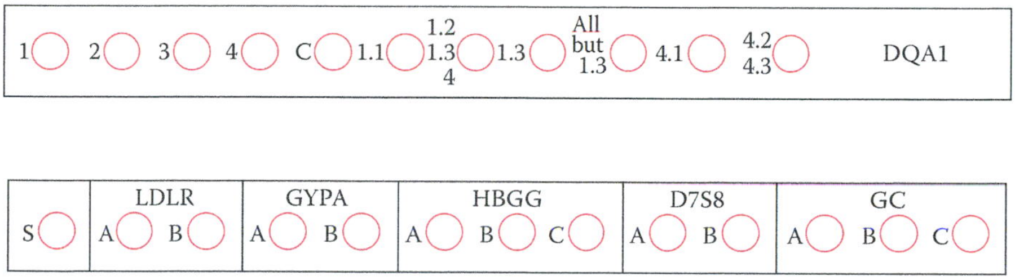

Figure 22.2   Panels of immobilized probes in the Polymarker kit. Top: *HLA-DQA1*. Bottom: additional five loci (*LDLR, GYPA, HBGG*, D7S8, and *GC*). C and S represent threshold control dots. (© Richard C. Li.)

A and B), while *HBGG* and *GC* each have three alleles (A, B, and C). As a result, the P_m of the Polymarker panels decreased to 10^{-4}.

The DNA profiles of the Polymarker system have been accepted in US courts. The Polymarker system has a number of advantages, particularly compared to variable number tandem repeat analysis (VNTR; see Chapter 19). First, the Polymarker system is a PCR-based method and is capable of analyzing a small amount of DNA sample (approximately 2 ng per analysis). Therefore, it is more sensitive than the method used in VNTR analysis. Second, the Polymarker system is an SNP assay with short amplicon sizes and, therefore, can analyze degraded DNA samples, which is not possible in VNTR analysis. Third, the Polymarker system is more rapid and less laborious than VNTR analysis. Lastly, the amplicon sizes of alleles at a given locus have identical lengths and, therefore, this assay does not exhibit preferential amplification as in amplified fragment length polymorphism (Chapter 19). However, the Polymarker system has its limitations compared to VNTR analysis. For example, compared to the VNTR analysis, the P_m of the Polymarker system is high, resulting in poor discrimination power in the comparison of two DNA profiles. Therefore, the Polymarker system is merely useful for excluding a suspect. Moreover, the SNP loci utilized in the Polymarker system are less polymorphic than VNTR loci. The limited number of alleles per SNP locus makes identifying the components of mixtures more difficult than with VNTR analysis in situations when contributors of a mixture sample share common alleles. For these reasons, the Polymarker system was replaced by STR profiling in the late 1990s.

22.2.1.2 Allele-Specific Oligonucleotide Hybridization

Both DQα AmpliType® and Polymarker kits utilize the *allele-specific oligonucleotide (ASO) hybridization* technique. This technique analyzes single-nucleotide variations, such as SNPs, at a given locus. It is based on the principle that ASO probes, usually 14–17 bases in length, hybridize to their complementary DNA sequences to distinguish known polymorphic alleles. ASO probes for multiple alleles at several loci can be arranged on the same panel to establish the presence or absence of specific alleles in PCR-amplified fragments of a DNA sample (Figure 22.3). Thus, the genotypes can be determined.

In the DQα AmpliType® and Polymarker kits, the oligonucleotides representing different alleles are immobilized to a solid matrix consisting of nylon membrane strips. Each immobilized probe at a particular site on the membrane is utilized to detect corresponding SNPs. Since the probe rather than the target DNA (as with regular blot format; see Chapter 9) is immobilized to the solid phase, the configuration used here is known as a *reverse blot* format.

The regions of the DNA in question are amplified by PCR. One of each pair of the primers is conjugated with biotin at the 5′ end (Figure 22.4). Thus, the amplicons are biotinylated for purposes of detection. This kit established multiplexing of a six-locus system, allowing simultaneous amplification of the *HLA-DQA1* locus along with *LDLR, GYPA, HBGG*, D7S8, and *GC* in

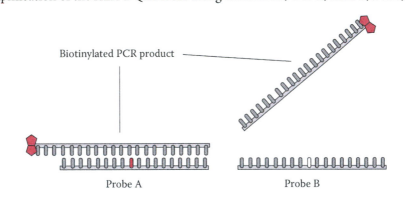

Figure 22.3 Hybridization with ASO. Two probes are incubated with the target DNA containing the SNP. Only the perfectly matched probe–target DNA can hybridize under optimal conditions.

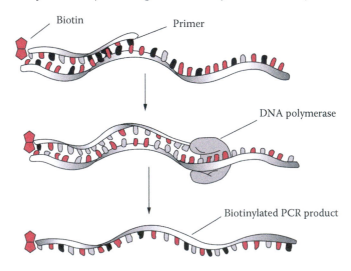

Figure 22.4 Amplification of DNA using biotinylated primers. The biotin is conjugated at the 5′ end of the primer. The amplified products are biotinylated. (© Richard C. Li.)

a single reaction for each sample. Following the denaturation of the PCR product to separate the two complementary DNA strands of the amplicons, the biotinylated strands are hybridized to the immobilized probes. Hybridization and washing conditions are established to ensure proper hybridization of the ASO probe and its target sequence. Unbound amplicons are washed away.

The presence of a PCR product bound to a specific probe can be detected by a colorimetric detection system (Figure 22.5). Since the amplicons are biotinylated, a horseradish peroxidase–conjugated streptavidin complex is utilized to bind to the biotinylated amplicons. The horseradish peroxidase then catalyzes the oxidation reaction of the colorless substrate, tetramethyl benzidine (TMB), into a blue precipitate at a designated location on the membrane strip, allowing the genotype of the sample to be determined (Figures 22.6 and 22.7).

The kit uses a threshold control (C dot in the *HLA-DQA1* panel and S dot in the panel of the other five loci) to distinguish between signal and background noise and to determine whether a

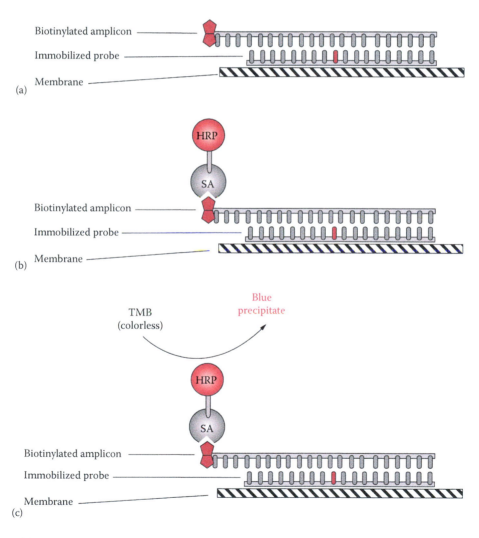

Figure 22.5 Reverse blot assay. (a) The probe is immobilized onto a solid-phase membrane and hybridized with a biotinylated PCR product having the target sequence. (b) The detection of hybridization is carried out by streptavidin (SA) and horseradish peroxidase (HRP) conjugate. (c) Colorimetric reaction is catalyzed by HRP using TMB as a substrate. (© Richard C. Li.)

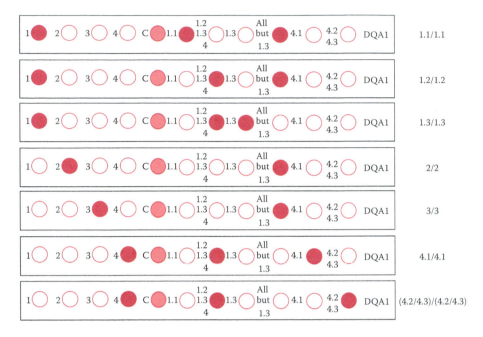

Figure 22.6 DNA typing of homozygous alleles at the *HLA-DQA1* locus using the reverse blot assay (Polymarker kit). The genotype of each sample is noted on the right. (© Richard C. Li.)

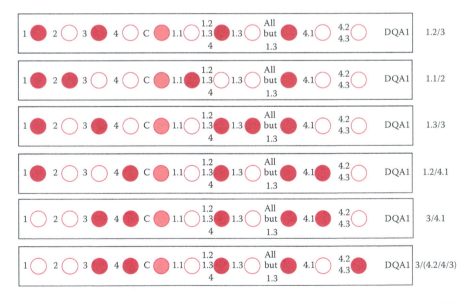

Figure 22.7 DNA typing of heterozygous alleles at the *HLA-DQA1* locus using the reverse blot assay (Polymarker kit). The representative genotypes of the samples are noted on the right. (© Richard C. Li.)

sufficient amount of DNA has been amplified to detect all the alleles present in a sample. If the signal intensities of allele dots are greater than or equal to the threshold control, the alleles are considered true. If the allele dots are less intense than the threshold control, they are considered inconclusive for the determination of full genotypes because an allele may not have been detected due to a low level of DNA template.

22.2.2 Current and Potential Applications of SNP Analysis

22.2.2.1 Application of SNP Analysis for Forensic Identification

The applications of SNP analysis for forensic human identification are summarized in Table 22.2. Autosomal SNP panels can be used for the most common types of forensic testing, including analysis of degraded DNA samples. The SNP panels for mitochondrial DNA (mtDNA) profiling are useful for identifying human remains by comparing the DNA profiles in question to those of potential relatives. This approach may serve as an alternative method to the DNA sequencing method of mtDNA analysis, which is time-consuming and laborious. SNP loci on the Y chromosome are also potentially useful markers for paternity testing because of low mutation rates. SNP loci such as ancestry informative markers (AIM) can be used to determine the ethnic origins of questioned samples to generate leads for investigations.

Table 22.2   Examples of Forensic Applications of SNP Profiling

Testing	Candidate SNP Loci	Application	Further Reading
Identity	Autosomal SNPs	Human identification	Kidd et al. (2006); Sanchez et al. (2006)
		Human identification via degraded DNA	Budowle (2004)
	mtDNA SNPs	Human identification	Grignani et al. (2006)
	Y chromosome SNPs	Paternity testing	Hammer et al. (2006)
Biogeographical origin	Ancestry informative markers (AIMs)	Ethnic group identification	Frudakis et al. (2003); Shriver and Kittles (2004)
Physical characteristics	*MC1R* (melanocortin 1 receptor gene)	Hair color identification (investigative lead)	Grimes et al. (2001)
	P (gene has role in pigmentation)	Eye color identification (investigative lead)	Rebbeck et al. (2002)
Pathology	*KCNH2* (cardiac potassium channel gene)	Determining cause of sudden death from cardiac arrhythmia long QT syndrome	Lunetta et al. (2002)
	SCN5A (encodes cardiac sodium channel gene)	Determining cause of sudden death from cardiac arrhythmia long QT syndrome	Burke et al. (2005)
Toxicology	*CYP2C19, CYP2D6, CYP3A4, CYP2E1* (drug metabolizing enzyme genes)	Investigation of drug overdose (including death)	Kupiec et al. (2006)

22.2.2.2 Potential Applications of SNPs for Phenotyping

One potential application of SNP analysis is in determining phenotypic information, also known as *phenotyping* (Table 22.2). The relevant SNP loci are usually *nonsynonymous SNPs* (*nsSNPs*); they reside in the exon and change the encoded amino acid, which leads to an altered phenotype. Phenotyping of a questioned sample can reveal physical characteristics of an individual, such as hair and eye color, to provide leads for investigations. A number of SNPs residing within the melanocortin 1 receptor gene (*MC1R*) are associated with red hair, fair skin, and freckles, while SNPs residing within the *P* gene, which play a role in pigmentation, are associated with eye color variations.

Phenotyping can also be employed in the area of forensic pathology. Cardiac arrhythmia long QT syndrome (LQTS) can cause sudden death. A number of LQTS-associated SNPs—for example, SNPs in *KCNH2* and *SCN5A* genes—have been shown to correlate to such deaths. Thus, these SNPs are potentially useful for investigating the causes of death. Finally, phenotyping also has applications in forensic toxicology. A number of SNPs in genes, such as *CYP2D6*, that are responsible for metabolizing drugs can serve as potential markers for postmortem investigations of drug-overdose cases.

22.3 SNP Techniques

Over the years, various techniques of SNP analysis have been developed and can be divided into several groups based on the mechanisms used: allele-specific hybridization, primer extension, oligonucleotide ligation, and invasive cleavage. In *allele-specific hybridization*, allele discrimination is based on an optimal condition allowing only the perfectly matched probe–target hybridization to form. *Primer extension* methods are based on the ability of DNA polymerase to incorporate specific deoxynucleotides (dNTPs) complementary to the sequence of the template DNA. *Allele-specific oligonucleotide ligation* is based on the condition that only the allelic probe perfectly matched to the target is ligated. In the *invasive cleavage* method, allelic discrimination is based on DNA sequence-specific cleavage by endonucleases. A number of detection methods can be utilized in SNP analysis, such as the measurements of fluorescence, luminescence, and molecular mass. Most assays are carried out in solutions or on solid matrices such as glass slides, chips, or beads. Table 22.3 summarizes the representative assays for SNP typing.

For decades, Sanger sequencing, using chain-termination chemistry, has been the standard method for DNA sequencing (Chapter 23). In recent years, *next-generation sequencing* (NGS), a rapidly developing technology, has had a profound impact on biology. Compared to Sanger sequencing, NGS is advantageous in that it can achieve substantially higher throughput, and at lower cost, than the Sanger method. For example, a human genome can now be sequenced in several days using NGS technologies. Although NGS is not yet widely utilized for forensic applications, it has great potential for forensic DNA analysis, particularly in SNP analysis.

22.3.1 Next-Generation Sequencing Technologies

NGS technologies have two categories of application: *de novo* sequencing and resequencing. In *de novo* genome sequencing, uncharacterized genomes or characterized genomes with substantial structural variations are sequenced. Sequence reads are assembled without any reference sequence. In *resequencing* applications, characterized genomes are sequenced. Sequence reads are assembled against an existing reference sequence to identify sequence polymorphisms. Thus, resequencing can potentially be used for forensic applications to detect polymorphisms associated with human identification. *Target resequencing* is a useful method of resequencing that can be utilized for forensic applications. Prior to sequencing, the genomic regions of interest from a DNA sample are selectively isolated through a method known as *enrichment*. Several target-enrichment strategies have been developed. PCR is the most widely used enrichment method.

Table 22.3 Techniques for SNP Assay

Basis of Technique	Representative Assay	Detection	Format	Notes	Further Reading
Allele-specific hybridization	Reverse Blot	Colorimetry	Membrane based	Useful for screen assay	Saiki et al. (1989)
	LightCycler® (Roche)	Fluorescence	Solution based	Applicable for automation; limited multiplexing capability	Lareu et al. (2001)
	TaqMan® (Applied Biosystems)	Fluorescence	Solution based	Applicable for automation; limited multiplexing capability	De la Vega et al. (2005)
	Molecular beacons	Fluorescence	Solution based	Applicable for automation; limited multiplexing capability	Tyagi et al. (1998)
	GeneChip® (Affymetrix)	Fluorescence	Chip based	Designed for genotyping large number of SNPs; exceeds forensic needs	Mei et al. (2000)
Primer extension	SNaPshot™ (Applied Biosystems)	Fluorescence	Solution based	Applicable for capillary electrophoresis and multiplexing	Sanchez et al. (2006)
	PinPoint (Applied Biosystems)	Mass spectrometry	Solution based	Multiplex capability lower than SNaPshot™	Haff and Smirnov (1997)
	Arrayed primer extension (APEX)	Fluorescence	Chip based	Low reproducibility	Pastinen et al. (1997)
	Pyrosequencing	Fluorescence	Solution based	May be applicable for analysis of mixed profiles; not applicable for automation; limited multiplexing capability	Ronaghi et al. (1998)
Allele-specific oligonucleotide ligation	SNPstream® UHT (Orchid-Gene Screen)	Fluorescence	Chip based	Developed for mass disaster cases	Bell et al. (2002)
	SNPlex™ (Applied Biosystems)	Fluorescence	Solution based	Applicable for multiplexing; requires higher amount of target DNA template than other PCR-based methods	De la Vega et al. (2005)
Invasive cleavage	Invader® (Third Wave Technology)	Fluorescence	Solution based	PCR not required; large amount of target DNA required	Lyamichev et al. (1999)

Source: Adapted from Sobrino, B., M. Brion, and A. Carracedo, SNPs in forensic genetics: a review on SNP typing methodologies. Forensic Sci Int, 2005. 154(2-3): p. 181–94.; Budowle, B., SNP typing strategies. Forensic Sci Int, 2004. 146 Suppl: p. S139–42.

The PCR-based approach is highly effective in targeting genomic regions that are small in size. Nevertheless, a typical NGS involves several major steps, including sample, library, and template DNA preparation; template amplification; sequencing and detection; and base calling; as well as data analysis (Figure 22.8).

22.3.2 DNA Samples, Sequencing Library, and Template Preparation

For most sequencing applications, micrograms of purified DNA are needed. This requirement is still a challenge for forensic casework applications where nanograms of DNA are often obtained. Human autosomal genome sequencing usually requires converting a DNA sample into a sequencing library. Two types of sequencing libraries are usually used: mate-pair libraries and fragment libraries. A mate-pair library is often used in *de novo* sequencing applications, while the fragment libraries can be used for resequencing and forensic applications. To construct a fragment library, a DNA sample is fragmented using mechanical methods such as sonication, nebulization, or shearing; endonuclease digestion; or a transposon-based method. Subsequently, sequencing adapters containing primer-binding sites for universal PCR primers are ligated to both ends of the DNA fragments. In resequencing applications, multiplex sequencing of pooled samples is often carried out, which can improve the efficiency and reduce the costs of sequencing. For multiplex sequencing, an index tag containing a bar code can be ligated to each DNA fragment to allow it to be identified after sequencing. Indexing can minimize the risk of sample mix-ups and contaminations during the sample preparation.

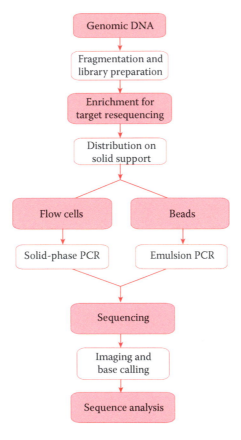

Figure 22.8 NGS work flow for potential forensic applications. (© Richard C. Li.)

Most imaging systems of sequencing platforms are not equipped to detect single-molecule fluorescence. Thus, DNA templates are amplified using either emulsion or solid-phase PCR. In the *emulsion PCR*, single-stranded DNA templates are bound to beads. DNA-bound beads are suspended in a water-in-oil emulsion, where each bead is placed in a single aqueous droplet. DNA fragments are amplified using PCR. As a result, each bead is coated with thousands of copies of the same template sequence. Subsequently, beads carrying amplified DNA are deposited on a glass slide or into individual fiber-optic wells for sequencing. In *solid-phase PCR*, both forward and reverse primers are attached to a slide (Figure 22.9). DNA templates are hybridized to immobilized PCR primers. The extended primer bends and hybridizes to a second immobilized primer, forming a bridge. During the "bridge amplification," the single-stranded template is amplified to form clusters. Solid-phase amplification can produce millions of spatially separated template clusters with free ends, allowing universal sequencing primers to hybridize for sequencing.

22.3.3 NGS Chemistry

The amplified template fragments are then sequenced. At present, there are two major categories of NGS chemistry that are routinely used: *sequencing by synthesis* and *sequencing by ligation*. The *pyrosequencing* technology is one example of NGS using the sequencing-by-synthesis chemistry. During the pyrosequencing, each nucleotide substrate is introduced one at a time. Only the correct nucleotide corresponding to the template is incorporated, and a pyrophosphate is released (Figure 22.10a). Pyrophosphate is then converted to ATP, catalyzed by sulfurylase (Figure 22.10b). ATP is then utilized by luciferase to convert luciferin to oxyluciferin, and the reaction emits light that is detected by a charge-coupled device camera (Figure 22.10c). This technology allows the generation of sequences with long read lengths. *Cyclic reversible termination* is another example of NGS using sequencing-by-synthesis chemistry. In this method, chain terminators (fluorescent-labeled dideoxynucleotides) are used to extend a primer sequence complementary to the template DNA. Each of four nucleotides is labeled with a different fluorophore. After incorporation and fluorescence detection, the terminating and fluorescent moieties are cleaved and removed (Figure 22.11). As a result, the next sequencing cycle is carried out. This process is repeated until the sequence is completed. This is the most commonly used method throughout the field of NGS.

In the sequencing-by-ligation chemistry, probes (eight nucleotides in length) are utilized for ligation reactions. Each probe contains five specific nucleotides that are complementary to the template and three nucleotides that are universal. Probes containing all possible combinations of the first five nucleotides are utilized. Each probe is fluorescently labeled according to the first two bases of the probe so that it can be identified. During the first round of sequencing reaction, only the probe that is complementary to the template sequence can hybridize. The probe is then ligated to the 3′ end of the primer. After detection, three universal nucleotides, including fluorophore, are cleaved. The ligation process is repeated several times using a new set of probes each time. The newly made complementary DNA strand is then stripped off. In the next round of sequencing, new primers that are one nucleotide shorter than the previous primer are utilized, and the ligation reaction is repeated. In this round of sequencing, different nucleotide positions are read. This process is repeated for several rounds with different primers, until all nucleotides in the template have been sequenced twice. As a result, this sequencing method has a low error rate.

22.3.4 NGS Coverage

In NGS, sequencing reads are usually not distributed evenly over the genomic regions of interest. As a result, some nucleotides will be covered by fewer or more reads than the average. Therefore, multiple reads for each nucleotide are necessary to obtain a reliable sequence. The average number of times that each nucleotide in the genomic regions of interest is sequenced is known as the

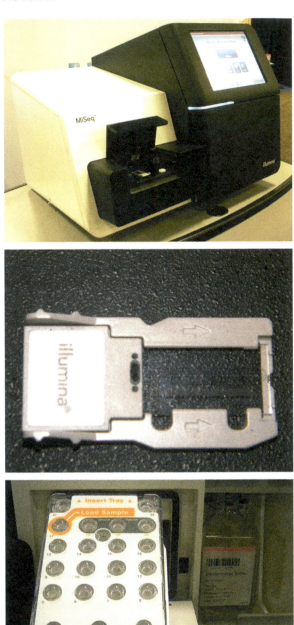

Figure 22.9   A photo of an NGS platform. A MiSeq (Illumna) desktop sequencer (top), a flow-cell device (middle) where DNA templates are distributed for PCR, and a reagent cartridge (bottom) are shown. (© Richard C. Li.)

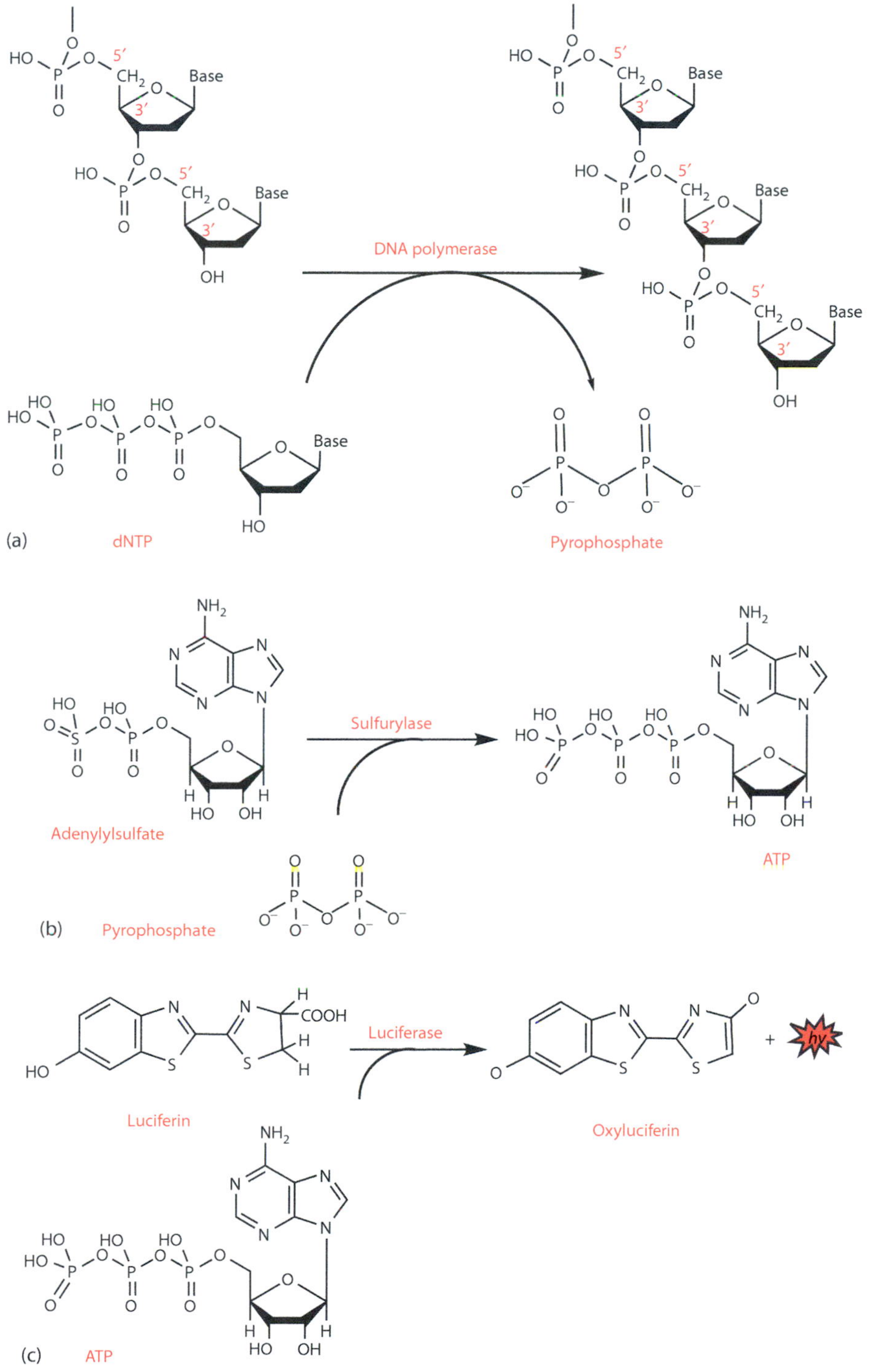

Figure 22.10 Pyrosequencing chemistry. (a) During the pyrosequencing, a nucleotide substrate is incorporated, releasing a pyrophosphate. (b) Pyrophosphate is then converted to ATP. (c) ATP is utilized by luciferase, converting luciferin to oxyluciferin, which emits light.

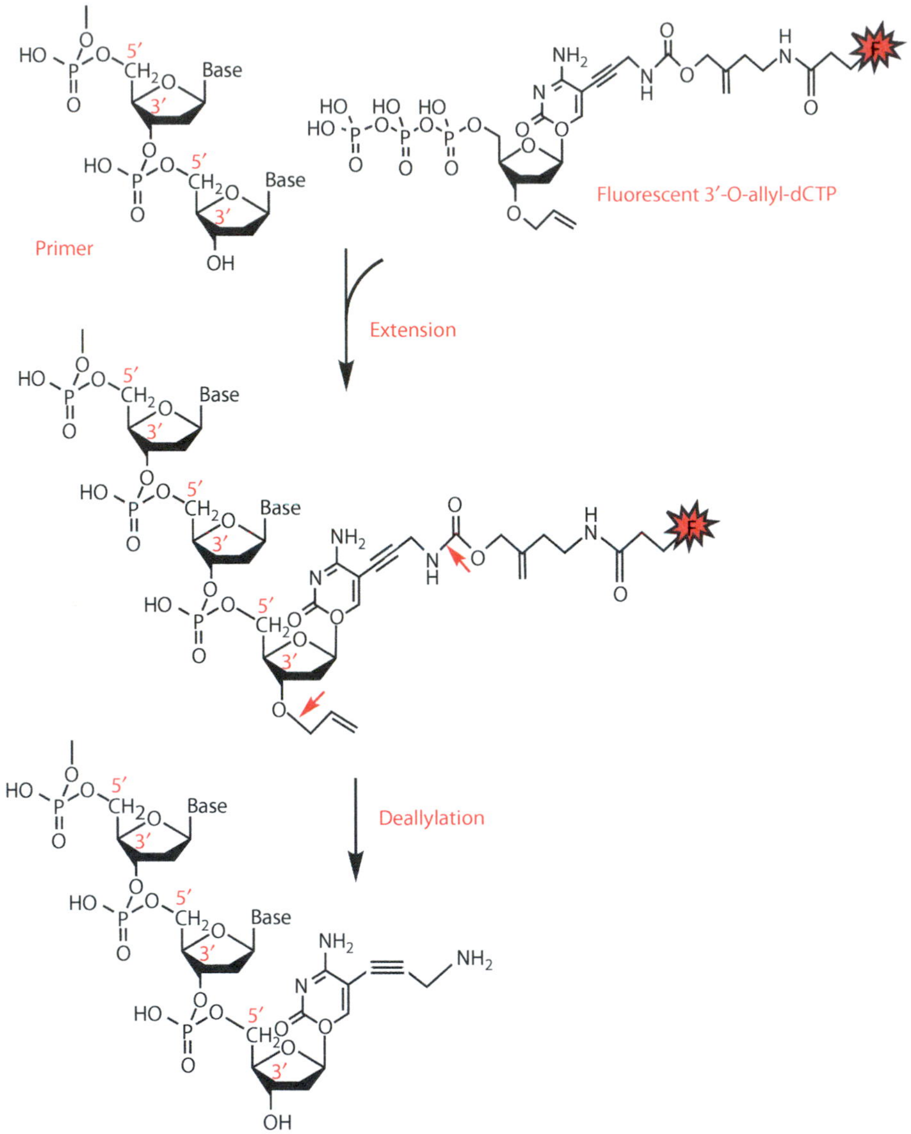

Figure 22.11 Cyclic reversible termination. A single sequencing cycle is shown. The incorporation of a C is illustrated as an example.

coverage or the *sequencing depth*. The level of coverage depends on the types of applications. For example, for human genome SNP analysis, the level of coverage is from 10 to 30. The coverage can be calculated using the following equation:

$$C = \frac{LN}{G}$$

where:
  C = coverage
  G = haploid genome length
  L = read length
  N = number of reads

Currently, NGS has not been widely used in forensic applications. One disadvantage of NGS is that the error rate of NGS is higher than that of Sanger sequencing. Although the error rate of NGS is not yet acceptable for forensic casework analysis, it can potentially be overcome by increasing the sequencing coverage.

Bibliography

Ackerman, M.J., D.J. Tester, and D.J. Driscoll, Molecular autopsy of sudden unexplained death in the young. *Am J Forensic Med Pathol*, 2001, **22**(2): 105–111.

Ali, M.E., et al., HLA-A, -B and -DRB1 allele frequencies in the Bangladeshi population. *Tissue Antigens*, 2008, **72**(2): 115–119.

Alkhayat, A., F. Alshamali, and B. Budowle, Population data on the PCR-based loci LDLR GYPA, HBGG, D7S8, Gc, HLA-DQA1, and D1S80 from Arabs from Dubai. *Forensic Sci Int*, 1996, **81**(1): 29–34.

Allah, R., L. Yang, and S.B. Li, SNPs and forensic DNA typing. *Fa Yi Xue Za Zhi*, 2007, **23**(5): 373–379.

Allwood, J.S. and S. Harbison, SNP model development for the prediction of eye colour in New Zealand. *Forensic Sci Int Genet*, 2013, **7**(4): 444–452.

Altshuler, D., et al., An SNP map of the human genome generated by reduced representation shotgun sequencing. *Nature*, 2000, **407**(6803): 513–516.

Amigo, J., et al., The SNPforID browser: An online tool for query and display of frequency data from the SNPforID project. *Int J Legal Med*, 2008, **122**(5): 435–440.

Arnestad, M., et al., Prevalence of long-QT syndrome gene variants in sudden infant death syndrome. *Circulation*, 2007, **115**(3): 361–367.

Babol-Pokora, K. and J. Berent, SNP-minisequencing as an excellent tool for analysing degraded DNA recovered from archival tissues. *Acta Biochim Pol*, 2008, **55**(4): 815–819.

Bandelt, H.J. and A. Salas, Current next generation sequencing technology may not meet forensic standards. *Forensic Sci Int Genet*, 2012, **6**(1): 143–145.

Barbaro, A., et al., Genetic variability of the SNPforID 52-plex identification SNP panel in Italian population samples. *Forensic Sci Int Genet*, 2012, **6**(6): e185–e186.

Bell, B., et al., Distribution of types for six PCR-based loci; LDLR, GYPA, HBGG, D7S8, GC and HLA-DQA1 in central Pyrenees and Teruel (Spain). *J Forensic Sci*, 1997, **42**(3): 510–513.

Bell, P.A., et al., SNP stream UHT: Ultra-high throughput SNP genotyping for pharmacogenomics and drug discovery. *Bio Techniques*, 2002, **74**(Suppl 70-2): 76–77.

Birngruber, C., F. Ramsthaler, and M.A. Verhoff, The color(s) of human hair—Forensic hair analysis with SpectraCube. *Forensic Sci Int*, 2009, **185**(1–3): e19–e23.

Blake, E., et al., Polymerase chain reaction (PCR) amplification and human leukocyte antigen (HLA)-DQ alpha oligonucleotide typing on biological evidence samples: Casework experience. *J Forensic Sci*, 1992, **37**(3): 700–726.

Blanco-Verea, A., et al., Forensic validation and implementation of Y-chromosome SNP multiplexes. *Forensic Sci Int Genet Suppl*, 2008, **1**(1): 181–183.

Borsting, C. and N. Morling, Mutations and/or close relatives? Six case work examples where 49 autosomal SNPs were used as supplementary markers. *Forensic Sci Int Genet*, 2011, **5**(3): 236–241.

Borsting, C. and N. Morling, Reinvestigations of six unusual paternity cases by typing of autosomal single-nucleotide polymorphisms. *Transfusion*, 2012, **52**(2): 425–430.

Borsting, C., C. Tomas, and N. Morling, SNP typing of the reference materials SRM 2391b 1-10, K562, XY1, XX74, and 007 with the SNPforID multiplex. *Forensic Sci Int Genet*, 2011, **5**(3): e81–e82.

Borsting, C., C. Tomas, and N. Morling, Typing of 49 autosomal SNPs by single base extension and capillary electrophoresis for forensic genetic testing. *Methods Mol Biol*, 2012, **830**: 87–107.

Borsting, C., E. Rockenbauer, and N. Morling, Validation of a single nucleotide polymorphism (SNP) typing assay with 49 SNPs for forensic genetic testing in a laboratory accredited according to the ISO 17025 standard. *Forensic Sci Int Genet*, 2009, **4**(1): 34–42.

Borsting, C., H.S. Mogensen, and N. Morling, Forensic genetic SNP typing of low-template DNA and highly degraded DNA from crime case samples. *Forensic Sci Int Genet*, 2013, **7**(3): 345–352.

Borsting, C., et al., Performance of the SNPforID 52 SNP-plex assay in paternity testing. *Forensic Sci Int Genet*, 2008, **2**(4): 292–300.

Bouakaze, C., et al., Pigment phenotype and biogeographical ancestry from ancient skeletal remains: Inferences from multiplexed autosomal SNP analysis. *Int J Legal Med*, 2009, **123**(4): 315–325.

Branicki, W., U. Brudnik, and A. Wojas-Pelc, Interactions between HERC2, OCA2 and MC1R may influence human pigmentation phenotype. *Ann Hum Genet*, 2009, **73**(2): 160–170.

Branicki, W., et al., Determination of phenotype associated SNPs in the MC1R gene. *J Forensic Sci*, 2007, **52**(2): 349–354.

Branicki, W., et al., Model-based prediction of human hair color using DNA variants. *Hum Genet*, 2011, **129**(4): 443–454.

Brenner, C.H. and B.S. Weir, Issues and strategies in the DNA identification of World Trade Center victims. *Theor Popul Biol*, 2003, **63**(3): 173–178.

Brito, R.M., et al., South Portuguese population data on the loci HLA-DQA1, LDLR, GYPA, HBGG, D7S8 and Gc. *J Forensic Sci*, 1998, **43**(5): 1031–1036.

Brown, R.J., et al., Distribution of the HLA-DQA1 and polymarker alleles in the Basque population of Spain. *Forensic Sci Int*, 2000, **108**(2): 145–151.

Budowle, B. and A. van Daal, Forensically relevant SNP classes. *Biotechniques*, 2008, **44**(5): 603–608, 610.

Budowle, B., F.S. Baechtel, and R. Fejeran, Polymarker, HLA-DQA1, and D1S80 allele frequency data in Chamorro and Filipino populations from Guam. *J Forensic Sci*, 1998, **43**(6): 1195–1198.

Budowle, B., SNP typing strategies. *Forensic Sci Int*, 2004, **146**(Suppl): S139–S142.

Budowle, B., et al., PCR-based analysis: Allele-specific oligonucleotide assays, in *DNA Typing Protocols: Molecular Biology and Forensic Analysis*, pp. 95–118, 2000. Eaton: Natick, MA.

Budowle, B., et al., Validation and population studies of the loci LDLR, GYPA, HBGG, D7S8, and Gc (PM loci), and HLA-DQ alpha using a multiplex amplification and typing procedure. *J Forensic Sci*, 1995, **40**(1): 45–54.

Bunten, H., et al., OPRM1 and CYP2B6 gene variants as risk factors in methadone-related deaths. *Clin Pharmacol Ther*, 2010, **88**(3): 383–389.

Burke, A., et al., Role of SCN5A Y1102 polymorphism in sudden cardiac death in blacks. *Circulation*, 2005, **112**(6): 798–802.

Caratti, S., et al., Subtyping of Y-chromosomal haplogroup E-M78 (E1b1b1a) by SNP assay and its forensic application. *Int J Legal Med*, 2009, **123**(4): 357–360.

Chen, X., K.J. Livak, and P.Y. Kwok, A homogeneous, ligase-mediated DNA diagnostic test. *Genome Res*, 1998, **8**(5): 549–556.

Comey, C.T. and B. Budowle, Validation studies on the analysis of the HLA DQ alpha locus using the polymerase chain reaction. *J Forensic Sci*, 1991, **36**(6): 1633–1648.

Comey, C.T., et al., PCR amplification and typing of the HLA DQ alpha gene in forensic samples. *J Forensic Sci*, 1993, **38**(2): 239–249.

Consortium, I.H., A haplotype map of the human genome. *Nature*, 2005, **437**(7063): 1299–1320.

Cowland, J.B., H.O. Madsen, and N. Morling, HLA-DQA1 typing in Danes by two polymerase chain reaction (PCR) based methods. *Forensic Sci Int*, 1995, **73**(1): 1–13.

De la Vega, F.M., et al., Assessment of two flexible and compatible SNP genotyping platforms: TaqMan SNP genotyping assays and the SNPlex senotyping system. *Mutat Res*, 2005, **573**(1–2): 111–135.

de Paula Careta, F. and G.G. Paneto, Recent patents on high-throughput single nucleotide polymorphism (SNP) genotyping methods. *Recent Pat DNA Gene Seq*, 2012, **6**(2): 122–126.

Dimo-Simonin, N. and C. Brandt-Casadevall, Evaluation and usefulness of reverse dot blot DNA-PolyMarker typing in forensic case work. *Forensic Sci Int*, 1996, **81**(1): 61–72.

Dixon, L.A., et al., Validation of a 21-locus autosomal SNP multiplex for forensic identification purposes. *Forensic Sci Int*, 2005, **154**(1): 62–77.

Doktycz, M.J., et al., Analysis of polymerase chain reaction-amplified DNA products by mass spectrometry using matrix-assisted laser desorption and electrospray: Current status. *Anal Biochem*, 1995, **230**(2): 205–214.

Drobnic, K., et al., Typing of 49 autosomal SNPs by SNaPshot in the Slovenian population. *Forensic Sci Int Genet*, 2010, **4**(5): e125–e127.

Duffy, D.L., et al., Interactive effects of MC1R and OCA2 on melanoma risk phenotypes. *Hum Mol Genet*, 2004, **13**(4): 447–461.

Edelmann, J., et al., Long QT syndrome mutation detection by SNaPshot technique. *Int J Legal Med*, 2012, **126**(6): 969–973.

Eiberg, H. and J. Mohr, Assignment of genes coding for brown eye colour (BEY2) and brown hair colour (HCL3) on chromosome 15q. *Eur J Hum Genet*, 1996, **4**(4): 237–241.

Erlich, H.A., E.L. Sheldon, and G. Horn, HLA typing using DNA probes. *Biotechnology*, 1986, **4**: 7.

Erlich, H.A., HlA-DQ alpha typing of forensic specimens. *Forensic Sci Int*, 1992, **53**(2): 227–228.

Erlich, H.A., et al., Reliability of the HLA-DQ alpha PCR-based oligonucleotide typing system. *J Forensic Sci*, 1990, **35**(5): 1017–1019.

Fan, J.B., et al., Parallel genotyping of human SNPs using generic high-density oligonucleotide tag arrays. *Genome Res*, 2000, **10**(6): 853–860.

Fei, Z., T. Ono, and L.M. Smith, MALDI-TOF mass spectrometric typing of single nucleotide polymorphisms with mass-tagged ddNTPs. *Nucleic Acids Res*, 1998, **26**(11): 2827–2828.

Fildes, N. and R. Reynolds, Consistency and reproducibility of AmpliType PM results between seven laboratories: Field trial results. *J Forensic Sci*, 1995, **40**(2): 279–286.

Flanagan, N., et al., Pleiotropic effects of the melanocortin 1 receptor (MC1R) gene on human pigmentation. *Hum Mol Genet*, 2000, **9**(17): 2531–2537.

Fondevila, M., et al., Forensic performance of two insertion-deletion marker assays. *Int J Legal Med*, 2012, **126**(5): 725–737.

Fondevila, M., et al., Revision of the SNPforID 34-plex forensic ancestry test: Assay enhancements, standard reference sample genotypes and extended population studies. *Forensic Sci Int Genet*, 2013, 7(1): 63–74.

Freire-Aradas, A., et al., A new SNP assay for identification of highly degraded human DNA. *Forensic Sci Int Genet*, 2012, **6**(3): 341–349.

Friis, S.L., et al., Typing of 30 insertion/deletions in Danes using the first commercial indel kit—Mentype® DIPplex. *Forensic Sci Int Genet*, 2012, **6**(2): e72–e74.

Frudakis, T., et al., Sequences associated with human iris pigmentation. *Genetics*, 2003, **165**(4): 2071–2083.

Gene, M., et al., Population study of Aymara Amerindians for the PCR-DNA polymorphisms HUMTH01, HUMVWA31A, D3S1358, D8S1179, D18S51, D19S253, YNZ22 and HLA-DQ alpha. *Int J Legal Med*, 2000, **113**(2): 126–128.

Geppert, M. and L. Roewer, SNaPshot® minisequencing analysis of multiple ancestry-informative Y-SNPs using capillary electrophoresis. *Methods Mol Biol*, 2012, **830**: 127–140.

Giardina, E., et al., Frequency assessment of SNPs for forensic identification in different populations. *Forensic Sci Int Genet*, 2007, **1**(3–4): e1–e3.

Gill, P., et al., An assessment of whether SNPs will replace STRs in national DNA databases—Joint considerations of the DNA working group of the European Network of Forensic Science Institutes (ENFSI) and the Scientific Working Group on DNA Analysis Methods (SWGDAM). *Sci Justice*, 2004, **44**(1): 51–53.

Gino, S., et al., LDLR, GYPA, HBGG, D7S8 and GC allele and genotype frequencies in the northwest Italian population. *J Forensic Sci*, 1999, **44**(1): 171–174.

Giroti, R. and V.K. Kashyap, Detection of the source of mislabeled biopsy tissue paraffin block and histopathological section on glass slide. *Diagn Mol Pathol*, 1998, **7**(6): 331–334.

Giroti, R.I., R. Biswas, and K. Mukherjee, Restriction fragment length polymorphism and polymerase chain reaction: HLA-DQA1 and polymarker analysis of blood samples from transfusion recipients. *Am J Clin Pathol*, 2002, **118**(3): 382–387.

Gra, O., et al., Microarray-based detection of CYP1A1, CYP2C9, CYP2C19, CYP2D6, GSTT1, GSTM1, MTHFR, MTRR, NQO1, NAT2, HLA-DQA1, and AB0 allele frequencies in native Russians. *Genet Test Mol Biomarkers*, 2010, **14**(3): 329–342.

Graf, J., R. Hodgson, and A. van Daal, Single nucleotide polymorphisms in the MATP gene are associated with normal human pigmentation variation. *Hum Mutat*, 2005, **25**(3): 278–284.

Graham, E.A., DNA reviews: Predicting phenotype. *Forensic Sci Med Pathol*, 2008, **4**(3): 196–199.

Grignani, P., et al., Subtyping mtDNA haplogroup H by SNaPshot minisequencing and its application in forensic individual identification. *Int J Legal Med*, 2006, **120**(3): 151–156.

Grimes, E.A., et al., Sequence polymorphism in the human melanocortin 1 receptor gene as an indicator of the red hair phenotype. *Forensic Sci Int*, 2001, **122**(2–3): 124–129.

Gross, A.M. and R.A. Guerrieri, HLA DQA1 and Polymarker validations for forensic casework: Standard specimens, reproducibility, and mixed specimens. *J Forensic Sci*, 1996, **41**(6): 1022–1026.

Grossman, P.D., et al., High-density multiplex detection of nucleic acid sequences: Oligonucleotide ligation assay and sequence-coded separation. *Nucleic Acids Res*, 1994, **22**(21): 4527–4534.

Gyllensten, U.B. and H.A. Erlich, Generation of single-stranded DNA by the polymerase chain reaction and its application to direct sequencing of the HLA-DQA locus. *Proc Natl Acad Sci U S A*, 1988, **85**(20): 7652–7656.

Haff, L.A. and I.P. Smirnov, Multiplex genotyping of PCR products with MassTag-labeled primers. *Nucleic Acids Res*, 1997, **25**(18): 3749–3750.

Hall, J.G., et al., Sensitive detection of DNA polymorphisms by the serial invasive signal amplification reaction. *Proc Natl Acad Sci U S A*, 2000, **97**(15): 8272–8277.

Hallenberg, C. and N. Morling, A report of the 1997, 1998 and 1999 Paternity Testing Workshops of the English Speaking Working Group of the International Society for Forensic Genetics. *Forensic Sci Int*, 2001, **116**(1): 23–33.

Hammer, M.F., et al., Population structure of Y chromosome SNP haplogroups in the United States and forensic implications for constructing Y chromosome STR databases. *Forensic Sci Int*, 2006, **164**(1): 45–55.

Hardenbol, P., et al., Multiplexed genotyping with sequence-tagged molecular inversion probes. *Nat Biotechnol*, 2003, **21**(6): 673–678.

Hart, K.L., et al., Improved eye- and skin-color prediction based on 8 SNPs. *Croat Med J*, 2013, **54**(3): 248–256.

Hashiyada, M., et al., Development of a spreadsheet for SNPs typing using Microsoft EXCEL. *Leg Med*, 2009, **11**(Suppl 1): S453–S454.

Heinrich, M., et al., Reduced-volume and low-volume typing of Y-chromosomal SNPs to obtain Finnish Y-chromosomal compound haplotypes. *Int J Legal Med*, 2009, **123**(5): 413–418.

Helmuth, R., et al., HLA-DQ alpha allele and genotype frequencies in various human populations, determined by using enzymatic amplification and oligonucleotide probes. *Am J Hum Genet*, 1990, **47**(3): 515–523.

Herrin, G., N. Fildes, and R. Reynolds, Evaluation of the AmpliType PM DNA test system on forensic case samples. *J Forensic Sci*, 1994, **39**: 9.

Hill, C.R., J.M. Butler, and P.M. Vallone, A 26plex autosomal STR assay to aid human identity testing*. *J Forensic Sci*, 2009, **54**(5): 1008–1015.

Hirschhorn, J.N., et al., SBE-TAGS: An array-based method for efficient single-nucleotide polymorphism genotyping. *Proc Natl Acad Sci U S A*, 2000, **97**(22): 12164–12169.

Hochmeister, M.N., et al., A method for the purification and recovery of genomic DNA from an HLA DQA1 amplification product and its subsequent amplification and typing with the AmpliType PM PCR amplification and typing kit. *J Forensic Sci*, 1995, **40**(4): 649–733.

Hochmeister, M.N., et al., Swiss population data on the loci HLA-DQ alpha, LDLR, GYPA, HBGG, D7S8, Gc and D1S80. *Forensic Sci Int*, 1994, **67**(3): 175–184.

Horn, G.T., et al., Characterization and rapid diagnostic analysis of DNA polymorphisms closely linked to the cystic fibrosis locus. *Clin Chem*, 1990, **36**(9): 1614–1619.

Hromadnikova, I., et al., Analysis of paternal alleles in nucleated red blood cells enriched from maternal blood. *Folia Biol*, 2001, **47**(1): 36–39.

Hsu, T.M., et al., Genotyping single-nucleotide polymorphisms by the invader assay with dual-color fluorescence polarization detection. *Clin Chem*, 2001, **47**(8): 1373–1377.

Huang, N.E. and B. Budowle, Chinese population data on the PCR-based loci HLA-DQ alpha, low-density-lipoprotein receptor, glycophorin A, hemoglobin gamma G, D7S8, and group-specific component. *Hum Hered*, 1995, **45**(1): 34–40.

Huckenbeck, W., et al., German population data on the loci low-density-lipoprotein receptor, glycophorin A, hemoglobin gamma G, D7S8 and group-specific component. *Anthropol Anz*, 1995, **53**(3): 193–198.

Inagaki, S., et al., A new 39-plex analysis method for SNPs including 15 blood group loci. *Forensic Sci Int*, 2004, **144**(1): 45–57.

Jankowski, L.B., et al., New Jersey Caucasian, African American, and Hispanic population data on the PCR-based loci HLA-DQA1, LDLR, GYPA, HBGG, D7S8, and Gc. *J Forensic Sci*, 1998, **43**(5): 1037–1040.

Johansen, P., et al., Evaluation of the iPLEX® sample ID Plus Panel designed for the Sequenom MassARRAY® system. A SNP typing assay developed for human identification and sample tracking based on the SNPforID panel. *Forensic Sci Int Genet*, 2013, **7**(5): 482–487.

Jung, J.M., et al., Extraction strategy for obtaining DNA from bloodstains for PCR amplification and typing of the HLA-DQ alpha gene. *Int J Legal Med*, 1991, **104**(3): 145–148.

Kastelic, V. and K. Drobnic, A single-nucleotide polymorphism (SNP) multiplex system: The association of five SNPs with human eye and hair color in the Slovenian population and comparison using a Bayesian network and logistic regression model. *Croat Med J*, 2012, **53**(5): 401–408.

Kastelic, V., et al., Prediction of eye color in the Slovenian population using the IrisPlex SNPs. *Croat Med J*, 2013, **54**(4): 381–386.

Keating, B., et al., First all-in-one diagnostic tool for DNA intelligence: Genome-wide inference of biogeographic ancestry, appearance, relatedness, and sex with the Identitas v1 Forensic Chip. *Int J Legal Med*, 2013, **127**(3): 559–572.

Kersbergen, P., et al., Developing a set of ancestry-sensitive DNA markers reflecting continental origins of humans. *BMC Genet*, 2009, **10**: 69.

Khodjet-el-Khil, H., et al., Genetic analysis of the SNPforID 34-plex ancestry informative SNP panel in Tunisian and Libyan populations. *Forensic Sci Int Genet*, 2011, **5**(3): e45–e47.

Kidd, K.K., et al., Developing a SNP panel for forensic identification of individuals. *Forensic Sci Int*, 2006, **164**(1): 20–32.

Kiesler, K.M. and P.M. Vallone, Allele frequencies for 40 autosomal SNP loci typed for US population samples using electrospray ionization mass spectrometry. *Croat Med J*, 2013, **54**(3): 225–231.

Kim, J.J., et al., Development of SNP-based human identification system. *Int J Legal Med*, 2010, **124**(2): 125–131.

Kimura, R., et al., Gene flow and natural selection in oceanic human populations inferred from genome-wide SNP typing. *Mol Biol Evol*, 2008, **25**(8): 1750–1761.

Kingback, M., et al., Influence of CYP2D6 genotype on the disposition of the enantiomers of venlafaxine and its major metabolites in postmortem femoral blood. *Forensic Sci Int*, 2012, **214**(1–3): 124–134.

Kitano, T., et al., Allele frequencies of a SNP and a 27-bp deletion that are the determinant of earwax type in the ABCC11 gene. *Leg Med*, 2008, **10**(2): 113–114.

Kling, D., et al., DNA microarray as a tool in establishing genetic relatedness—Current status and future prospects. *Forensic Sci Int Genet*, 2012, **6**(3): 322–329.

Kostrikis, L.G., et al., Spectral genotyping of human alleles. *Science*, 1998, **279**(5354): 1228–1229.

Krawczak, M., Informativity assessment for biallelic single nucleotide polymorphisms. *Electrophoresis*, 1999, **20**(8): 1676–1681.

Krjutskov, K., et al., Development of a single tube 640-plex genotyping method for detection of nucleic acid variations on microarrays. *Nucleic Acids Res*, 2008, **36**(12): e75.

Krjutskov, K., et al., Evaluation of the 124-plex SNP typing microarray for forensic testing. *Forensic Sci Int Genet*, 2009, **4**(1): 43–48.

Kubo, S., et al., Personal identification from skeletal remain by D1S80, HLA DQA1, TH01 and polymarker analysis. *J Med Invest*, 2002, **49**(1–2): 83–86.

Kupiec, T.C., V. Raj, and N. Vu, Pharmacogenomics for the forensic toxicologist. *J Anal Toxicol*, 2006, **30**(2): 65–72.

Kuppuswamy, M.N., et al., Single nucleotide primer extension to detect genetic diseases: Experimental application to hemophilia B (factor IX) and cystic fibrosis genes. *Proc Natl Acad Sci U S A*, 1991, **88**(4): 1143–1147.

Lai, E., Application of SNP technologies in medicine: Lessons learned and future challenges. *Genome Res*, 2001, **11**(6): 927–929.

Lamason, R.L., et al., SLC24A5, a putative cation exchanger, affects pigmentation in zebrafish and humans. *Science*, 2005, **310**(5755): 1782–1786.

Landegren, U., et al., A ligase-mediated gene detection technique. *Science*, 1988, **241**(4869): 1077–1080.

Lareu, M., et al., The use of the LightCycler for the detection of Y chromosome SNPs. *Forensic Sci Int*, 2001, **118**(2–3): 163–168.

Larsen, M.K., et al., Molecular autopsy in young sudden cardiac death victims with suspected cardiomyopathy. *Forensic Sci Int*, 2012, **219**(1–3): 33–38.

Larsen, M.K., et al., Postmortem genetic testing of the ryanodine receptor 2 (RYR2) gene in a cohort of sudden unexplained death cases. *Int J Legal Med*, 2013, **127**(1): 139–144.

LaRue, B.L., et al., A validation study of the Qiagen investigator DIPplex® kit; an INDEL-based assay for human identification. *Int J Legal Med*, 2012, **126**(4): 533–540.

Levo, A., et al., Post-mortem SNP analysis of CYP2D6 gene reveals correlation between genotype and opioid drug (tramadol) metabolite ratios in blood. *Forensic Sci Int*, 2003, **135**(1): 9–15.

Li, C., et al., Genetic polymorphism of 29 highly informative InDel markers for forensic use in the Chinese Han population. *Forensic Sci Int Genet*, 2011, **5**(1): e27–e30.

Li, C., et al., Selection of 29 highly informative InDel markers for human identification and paternity analysis in Chinese Han population by the SNPlex genotyping system. *Mol Biol Rep*, 2012, **39**(3): 3143–3152.

Lou, C., et al., A SNaPshot assay for genotyping 44 individual identification single nucleotide polymorphisms. *Electrophoresis*, 2011, **32**(3–4): 368–378.

Lunetta, P., et al., Death in bathtub revisited with molecular genetics: A victim with suicidal traits and a LQTS gene mutation. *Forensic Sci Int*, 2002, **130**(2–3): 122–124.

Lyamichev, V., et al., Polymorphism identification and quantitative detection of genomic DNA by invasive cleavage of oligonucleotide probes. *Nat Biotechnol*, 1999, **17**(3): 292–296.

Manta, F., et al., Indel markers: Genetic diversity of 38 polymorphisms in Brazilian populations and application in a paternity investigation with post mortem material. *Forensic Sci Int Genet*, 2012, **6**(5): 658–661.

Martin, P., et al., Population genetic data of 30 autosomal indels in central Spain and the Basque Country populations. *Forensic Sci Int Genet*, 2013, **7**(2): e27–e30.

Martinez-Cadenas, C., et al., Gender is a major factor explaining discrepancies in eye colour prediction based on HERC2/OCA2 genotype and the IrisPlex model. *Forensic Sci Int Genet*, 2013, **7**(4): 453–460.

Medintz, I., et al., HLA-DQA1 and polymarker allele frequencies in two New York City Jewish populations. *J Forensic Sci*, 1997, **42**(5): 919–922.

Mei, R., et al., Genome-wide detection of allelic imbalance using human SNPs and high-density DNA arrays. *Genome Res*, 2000, **10**(8): 1126–1137.

Mein, C.A., et al., Evaluation of single nucleotide polymorphism typing with invader on PCR amplicons and its automation. *Genome Res*, 2000, **10**(3): 330–343.

Mengel-From, J., et al., Human eye colour and HERC2, OCA2 and MATP. *Forensic Sci Int Genet*, 2010, **4**(5): 323–328.

Mikkelsen, M., et al., Frequencies of 33 coding region mitochondrial SNPs in a Danish and a Turkish population. *Forensic Sci Int Genet*, 2011, **5**(5): 559–560.

Muddiman, D.C., et al., Length and base composition of PCR-amplified nucleic acids using mass measurements from electrospray ionization mass spectrometry. *Anal Chem*, 1997, **69**(8): 1543–1549.

Musgrave-Brown, E., et al., Forensic validation of the SNPforID 52-plex assay. *Forensic Sci Int Genet*, 2007, **1**(2): 186–190.

Myakishev, M.V., et al., High-throughput SNP genotyping by allele-specific PCR with universal energy-transfer-labeled primers. *Genome Res*, 2001, **11**(1): 163–169.

Nakahara, H., et al., Automated SNPs typing system based on the Invader assay. *Leg Med*, 2009, **11**(Suppl 1): S111–S114.

Nakajima, T., et al., Evaluation of 7 DNA markers (D1S80, HLA-DQ alpha, LDLR, GYPA, HBGG, D7S8 and GC) in a Japanese population. *Int J Legal Med*, 1996, **109**(1): 47–48.

Neitzke-Montinelli, V., et al., Polymorphisms upstream of the melanocortin-1 receptor coding region are associated with human pigmentation variation in a Brazilian population. *Am J Hum Biol*, 2012, **24**(6): 853–855.

Neuvonen, A.M., et al., Discrimination power of investigator DIPplex loci in Finnish and Somali populations. *Forensic Sci Int Genet*, 2012, **6**(4): e99–e102.

Nicklas, J.A. and E. Buel, A real-time multiplex SNP melting assay to discriminate individuals. *J Forensic Sci*, 2008, **53**(6): 1316–1324.

Odriozola, A., et al., SNPSTR rs59186128_D7S820 polymorphism distribution in European Caucasoid, Hispanic, and Afro-American populations. *Int J Legal Med*, 2009, **123**(6): 527–533.

Osawa, M., et al., SNP association and sequence analysis of the NOS1AP gene in SIDS. *Leg Med*, 2009, **11**(Suppl 1): S307–S308.

Pai, C.Y., et al., Flow chart HLA-DQA1 genotyping and its application to a forensic case. *J Forensic Sci*, 1995, **40**(2): 228–235.

Pakstis, A.J., et al., Candidate SNPs for a universal individual identification panel. *Hum Genet*, 2007, **121**(3–4): 305–317.

Parson, W., et al., Evaluation of next generation mtGenome sequencing using the Ion Torrent Personal Genome Machine (PGM). *Forensic Sci Int Genet*, 2013, **7**(5): 543–549.

Pastinen, T., et al., Minisequencing: A specific tool for DNA analysis and diagnostics on oligonucleotide arrays. *Genome Res*, 1997, **7**(6): 606–614.

Pastinen, T., et al., A system for specific, high-throughput genotyping by allele-specific primer extension on microarrays. *Genome Res*, 2000, **10**(7): 1031–1042.

Pereira, R. and L. Gusmao, Capillary electrophoresis of 38 noncoding biallelic mini-Indels for degraded samples and as complementary tool in paternity testing. *Methods Mol Biol*, 2012, **830**: 141–157.

Pereira, R., et al., A new multiplex for human identification using insertion/deletion polymorphisms. *Electrophoresis*, 2009, **30**(21): 3682–3690.

Phillips, C., SNP databases. *Methods Mol Biol*, 2009, **578**: 43–71.

Phillips, C., M. Fondevila, and M.V. Lareau, A 34-plex autosomal SNP single base extension assay for ancestry investigations. *Methods Mol Biol*, 2012, **830**: 109–126.

Phillips, C., et al., Evaluation of the Genplex SNP typing system and a 49plex forensic marker panel. *Forensic Sci Int Genet*, 2007, **1**(2): 180–185.

Phillips, C., et al., Inferring ancestral origin using a single multiplex assay of ancestry-informative marker SNPs. *Forensic Sci Int Genet*, 2007, **1**(3–4): 273–280.

Phillips, C., et al., Resolving relationship tests that show ambiguous STR results using autosomal SNPs as supplementary markers. *Forensic Sci Int Genet*, 2008, **2**(3): 198–204.

Phillips, C., et al., Eurasiaplex: A forensic SNP assay for differentiating European and South Asian ancestries. *Forensic Sci Int Genet*, 2013, **7**(3): 359–366.

Pilgrim, J.L., et al., Characterization of single nucleotide polymorphisms of cytochrome p450 in an Australian deceased sample. *Curr Drug Metab*, 2012, **13**(5): 679–692.

Pimenta, J.R. and S.D. Pena, Efficient human paternity testing with a panel of 40 short insertion-deletion polymorphisms. *Genet Mol Res*, 2010, **9**(1): 601–607.

Pinto, N., et al., Assessing paternities with inconclusive STR results: The suitability of bi-allelic markers. *Forensic Sci Int Genet*, 2013, **7**(1): 16–21.

Podini, D. and P.M. Vallone, SNP genotyping using multiplex single base primer extension assays. *Methods Mol Biol*, 2009, **578**: 379–391.

Poetsch, M., et al., Prediction of people's origin from degraded DNA—Presentation of SNP assays and calculation of probability. *Int J Legal Med*, 2013, **127**(2): 347–357.

Pomeroy, R., et al., A low-cost, high-throughput, automated single nucleotide polymorphism assay for forensic human DNA applications. *Anal Biochem*, 2009, **395**(1): 61–67.

Porras, L., et al., Genetic variability of the SNPforID 52-plex identification-SNP panel in Central West Colombia. *Forensic Sci Int Genet*, 2009, **4**(1): e9–e10.

Pulker, H., et al., Finding genes that underlie physical traits of forensic interest using genetic tools. *Forensic Sci Int Genet*, 2007, **1**(2): 100–104.

Rebbeck, T.R., et al., P gene as an inherited biomarker of human eye color. *Cancer Epidemiol Biomarkers Prev*, 2002, **11**(8): 782–784.

Rees, J.L., Genetics of hair and skin color. *Annu Rev Genet*, 2003, **37**: 67–90.

Reich, D.E., et al., Human genome sequence variation and the influence of gene history, mutation and recombination. *Nat Genet*, 2002, **32**(1): 135–142.

Riccardi, L.N., et al., Development of a tetraplex PCR assay for CYP2D6 genotyping in degraded DNA samples. *J Forensic Sci*, 2013, **59**(3): 690–695.

Rockenbauer, E., et al., Characterization of mutations and sequence variants in the D21S11 locus by next generation sequencing. *Forensic Sci Int Genet*, 2014, **8**(1): 68–72.

Rodriguez-Calvo, M.S., et al., Population data on the loci LDLR, GYPA, HBGG, D7S8, and GC in three southwest European populations. *J Forensic Sci*, 1996, **41**(2): 291–296.

Rolf, B., N. Bulander, and P. Wiegand, Insertion-/deletion polymorphisms close to the repeat region of STR loci can cause discordant genotypes with different STR kits. *Forensic Sci Int Genet*, 2011, **5**(4): 339–341.

Romanini, C., et al., Typing short amplicon binary polymorphisms: Supplementary SNP and Indel genetic information in the analysis of highly degraded skeletal remains. *Forensic Sci Int Genet*, 2012, **6**(4): 469–476.

Romero, R.L., et al., The applicability of formalin-fixed and formalin fixed paraffin embedded tissues in forensic DNA analysis. *J Forensic Sci*, 1997, **42**(4): 708–714.

Ronaghi, M., M. Uhlen, and P. Nyren, A sequencing method based on real-time pyrophosphate. *Science*, 1998, **281**(5375): 363, 365.

Ronaghi, M., et al., Real-time DNA sequencing using detection of pyrophosphate release. *Anal Biochem*, 1996, **242**(1): 84–89.

Ross, P., et al., High level multiplex genotyping by MALDI-TOF mass spectrometry. *Nat Biotechnol*, 1998, **16**(13): 1347–1351.

Ruiz, Y., et al., Analysis of the SNPforID 52-plex markers in four Native American populations from Venezuela. *Forensic Sci Int Genet*, 2012, **6**(5): e142–e145.

Ruiz, Y., et al., Further development of forensic eye color predictive tests. *Forensic Sci Int Genet*, 2013, **7**(1): 28–40.

Sachidanandam, R., et al., A map of human genome sequence variation containing 1.42 million single nucleotide polymorphisms. *Nature*, 2001, **409**(6822): 928–933.

Saiki, R.K., et al., Analysis of enzymatically amplified beta-globin and HLA-DQ alpha DNA with allele-specific oligonucleotide probes. *Nature*, 1986, **324**(6093): 163–166.

Saiki, R.K., et al., Genetic analysis of amplified DNA with immobilized sequence-specific oligonucleotide probes. *Proc Natl Acad Sci U S A*, 1989, **86**(16): 6230–6234.

Sanchez, J.J., et al., A multiplex assay with 52 single nucleotide polymorphisms for human identification. *Electrophoresis*, 2006, **27**(9): 1713–1724.

Santos, C., et al., A study of East Timor variability using the SNPforID 52-plex SNP panel. *Forensic Sci Int Genet*, 2011, **5**(1): e25–e26.

Sato, Y., et al., HLA typing of aortic tissues from unidentified bodies using hot start polymerase chain reaction-sequence specific primers. *Leg Med (Tokyo)*, 2003, **5**(Suppl 1): S191–S193.

Sauer, S., et al., A novel procedure for efficient genotyping of single nucleotide polymorphisms. *Nucleic Acids Res*, 2000, **28**(5): E13.

Schmid, D., B. Bayer, and K. Anslinger, Comparison of telogen hair analyses: GenRES MPX-2SP kit versus genRES MPX-SP1 and genRES MPX-SP2 kits. *Forensic Sci Int Genet*, 2008, **3**(1): 22–26.

Schneider, P.M. and C. Rittner, Experience with the PCR-based HLA-DQ alpha DNA typing system in routine forensic casework. *Int J Legal Med*, 1993, **105**(5): 295–299.

Scholl, S., et al., Navajo, Pueblo, and Sioux population data on the loci HLA-DQA1, LDLR, GYPA, HBGG, D7S8, Gc, and D1S80. *J Forensic Sci*, 1996, **41**(1): 47–51.

Senge, T., et al., STRs, mini STRs and SNPs—A comparative study for typing degraded DNA. *Leg Med*, 2011, **13**(2): 68–74.

Sharafi Farzad, M., et al., Analysis of 49 autosomal SNPs in three ethnic groups from Iran: Persians, Lurs and Kurds. *Forensic Sci Int Genet*, 2013, **7**(4): 471–473.

Shen, M., Y. Shi, and P. Xiang, CYP3A4 and CYP2C19 genetic polymorphisms and zolpidem metabolism in the Chinese Han population: A pilot study. *Forensic Sci Int*, 2013, **227**(1–3): 77–81.

Shi, M., et al., 6 Y-SNP typing of China and Korean samples using primer extension and DHPLC. *J Forensic Sci*, 2007, **52**(1): 235–236.

Shi, Y., et al., Analysis of 50 SNPs in CYP2D6, CYP2C19, CYP2C9, CYP3A4 and CYP1A2 by MALDI-TOF mass spectrometry in Chinese Han population. *Forensic Sci Int*, 2011, **207**(1–3): 183–187.

Shriver, M.D. and R.A. Kittles, Genetic ancestry and the search for personalized genetic histories. *Nat Rev Genet*, 2004, **5**(8): 611–618.

Siebert, P.D. and M. Fukuda, Molecular cloning of a human glycophorin B cDNA: Nucleotide sequence and genomic relationship to glycophorin A. *Proc Natl Acad Sci U S A*, 1987, **84**(19): 6735–6739.

Slightom, J.L., A.E. Blechl, and O. Smithies, Human fetal G gamma- and A gamma-globin genes: Complete nucleotide sequences suggest that DNA can be exchanged between these duplicated genes. *Cell*, 1980, **21**(3): 627–638.

Soares-Vieira, J.A., et al., Brazilian population data on the polymerase chain reaction-based loci LDLR, GYPA, HBGG, D7S8, and Gc. *Am J Forensic Med Pathol*, 2003, **24**(3): 283–287.

Sobrino, B., M. Brion, and A. Carracedo, SNPs in forensic genetics: A review on SNP typing methodologies. *Forensic Sci Int*, 2005, **154**(2–3): 181–194.

Sokolov, B.P., Primer extension technique for the detection of single nucleotide in genomic DNA. *Nucleic Acids Res*, 1990, **18**(12): 3671.

Spichenok, O., et al., Prediction of eye and skin color in diverse populations using seven SNPs. *Forensic Sci Int Genet*, 2011, **5**(5): 472–478.

Spinella, A., et al., Italian population allele and genotype frequencies for the AmpliType PM and the HLA-DQ-alpha loci. *J Forensic Sci*, 1997, **42**(3): 514–518.

Syvanen, A.C., et al., A primer-guided nucleotide incorporation assay in the genotyping of apolipoprotein E. *Genomics*, 1990, **8**(4): 684–692.

Tahir, M.A., et al., DNA typing of samples for polymarker, DQA1, and nine STR loci from a human body exhumed after 27 years. *J Forensic Sci*, 2000, **45**(4): 902–907.

Takada, Y., et al., Application of SNPs in forensic casework: Identification of pathological and autoptical specimens due to sample mix-up. *Leg Med (Tokyo)*, 2009, **11**(Suppl 1): S196–S197.

Taylor, M.S., A. Challed-Spong, and E.A. Johnson, Co-amplification of the amelogenin and HLA DQ alpha genes: Optimization and validation. *J Forensic Sci*, 1997, **42**(1): 130–136.

Templeton, J.E., et al., DNA capture and next-generation sequencing can recover whole mitochondrial genomes from highly degraded samples for human identification. *Investig Genet*, 2013, **4**(1): 26.

Tomas, C., C. Borsting, and N. Morling, A 48-plex autosomal SNP GenPlex assay for human individualization and relationship testing. *Methods Mol Biol*, 2012, **830**: 73–85.

Tomas, C., et al., Typing of 48 autosomal SNPs and amelogenin with GenPlex SNP genotyping system in forensic genetics. *Forensic Sci Int Genet*, 2008, **3**(1): 1–6.

Tomas, C., et al., Autosomal SNP typing of forensic samples with the GenPlex HID system: Results of a collaborative study. *Forensic Sci Int Genet*, 2011, **5**(5): 369–375.

Tomas, C., et al., Analysis of 49 autosomal SNPs in an Iraqi population. *Forensic Sci Int Genet*, 2013, **7**(1): 198–199.

Tsongalis, G.J. and M.M. Berman, Application of forensic identity testing in a clinical setting. Specimen identification. *Diagn Mol Pathol*, 1997, **6**(2): 111–114.

Tyagi, S. and F.R. Kramer, Molecular beacons: Probes that fluoresce upon hybridization. *Nat Biotechnol*, 1996, **14**(3): 303–308.

Tyagi, S., D.P. Bratu, and F.R. Kramer, Multicolor molecular beacons for allele discrimination. *Nat Biotechnol*, 1998, **16**(1): 49–53.

Valenzuela, R.K., et al., Predicting phenotype from genotype: Normal pigmentation. *J Forensic Sci*, 2010, **55**(2): 315–322.

Van Geystelen, A., et al., *In silico* detection of phylogenetic informative Y-chromosomal single nucleotide polymorphisms from whole genome sequencing data. *Electrophoresis*, 2014 (in press) 10.1002/elps.201300459.

Van Neste, C., et al., Forensic STR analysis using massive parallel sequencing. *Forensic Sci Int Genet*, 2012, **6**(6): 810–818.

Van Neste, C., et al., My-Forensic-Loci-queries (MyFLq) framework for analysis of forensic STR data generated by massive parallel sequencing. *Forensic Sci Int Genet*, 2014, **9**: 1–8.

Vu, N.T., A.K. Chaturvedi, and D.V. Canfield, Genotyping for DQA1 and PM loci in urine using PCR-based amplification: Effects of sample volume, storage temperature, preservatives, and aging on DNA extraction and typing. *Forensic Sci Int*, 1999, **102**(1): 23–34.

Vural, B., et al., Turkish population data on the HLA-DO alpha, LDLR, GYPA, HBGG, D7s8, and GC loci. *Int J Legal Med*, 1998, **111**(1): 43–45.

Walsh, S., et al., Developmental validation of the IrisPlex system: Determination of blue and brown iris colour for forensic intelligence. *Forensic Sci Int Genet*, 2011, **5**(5): 464–471.

Walsh, S., et al., IrisPlex: A sensitive DNA tool for accurate prediction of blue and brown eye colour in the absence of ancestry information. *Forensic Sci Int Genet*, 2011, **5**(3): 170–180.

Walsh, S., et al., DNA-based eye colour prediction across Europe with the IrisPlex system. *Forensic Sci Int Genet*, 2012, **6**(3): 330–340.

Walsh, S., et al., The HIrisPlex system for simultaneous prediction of hair and eye colour from DNA. *Forensic Sci Int Genet*, 2013, **7**(1): 98–115.

Watanabe, Y., et al., Japanese population DNA typing data for the loci LDLR, GYPA, HBGG, D7S8, and GC. *J Forensic Sci*, 1997, **42**(5): 911–913.

Weber-Lehmann, J., et al., Finding the needle in the haystack: Differentiating "identical" twins in paternity testing and forensics by ultra-deep next generation sequencing. *Forensic Sci Int Genet*, 2014, **9**: 42–46.

Wei, Y.L., et al., Forensic identification using a multiplex assay of 47 SNPs. *J Forensic Sci*, 2012, **57**(6): 1448–1456.

Westen, A.A., et al., Tri-allelic SNP markers enable analysis of mixed and degraded DNA samples. *Forensic Sci Int Genet*, 2009, **3**(4): 233–241.

Wilson, R.B., et al., Guidelines for internal validation of the HLA-DQ alpha DNA typing system. *Forensic Sci Int*, 1994, **66**(1): 9–22.

Woo, K.M. and B. Budowle, Korean population data on the PCR-based loci LDLR, GYPA, HBGG, D7S8, Gc, HLA-DQA1, and D1S80. *J Forensic Sci*, 1995, **40**(4): 645–648.

Yamamoto, T., et al., The human LDL receptor: A cysteine-rich protein with multiple Alu sequences in its mRNA. *Cell*, 1984, **39**(1): 27–38.

Yang, F., et al., Human group-specific component (Gc) is a member of the albumin family. *Proc Natl Acad Sci U S A*, 1985, **82**(23): 7994–7998.

Yang, H.C., et al., Integrative analysis of single nucleotide polymorphisms and gene expression efficiently distinguishes samples from closely related ethnic populations. *BMC Genomics*, 2012, **13**: 346.

Zaumsegel, D., M.A. Rothschild, and P.M. Schneider, A 21 marker insertion deletion polymorphism panel to study biogeographic ancestry. *Forensic Sci Int Genet*, 2013, **7**(2): 305–312.

Zeng, Z., et al., Genome-wide screen for individual identification SNPs (IISNPs) and the confirmation of six of them in Han Chinese with pyrosequencing technology. *J Forensic Sci*, 2010, **55**(4): 901–907.

Zeng, Z., et al., Evaluation of 96 SNPs in 14 populations for worldwide individual identification. *J Forensic Sci*, 2012, **57**(4): 1031–1035.

Zha, L., et al., Exploring of tri-allelic SNPs using pyrosequencing and the SNaPshot methods for forensic application. *Electrophoresis*, 2012, **33**(5): 841–848.

Zhao, S.M., S.H. Zhang, and C.T. Li, InDel-typer30: A multiplex PCR system for DNA identification among five Chinese populations. *Fa Yi Xue Za Zhi*, 2010, **26**(5): 343–348+356.

Zidkova, A., et al., Application of the new insertion-deletion polymorphism kit for forensic identification and parentage testing on the Czech population. *Int J Legal Med*, 2013, **127**(1): 7–10.

23

Mitochondrial DNA Profiling

Forensic mitochondrial DNA (mtDNA) analysis is an important tool for human identification and is especially useful for identifying victims such as missing persons and individuals in mass fatality cases. Because mtDNA is maternally inherited, the mtDNA profiles of these individuals can be compared to those of their maternal relatives, and thus these individuals can be identified. Additionally, cells contain a much higher copy number of the mtDNA genome than that of the nuclear genome. Therefore, mtDNA testing is frequently used to analyze evidence samples, such as hair shafts, that contain low amounts of nuclear DNA. Furthermore, buried bones and decomposed tissues, in which nuclear DNA may be degraded, can be tested with mtDNA analysis.

23.1 Human Mitochondrial Genome

Mitochondria are cellular organelles that serve as the energy-generating components of cells (Figure 23.1). Each cell contains hundreds of mitochondria, which have their own extrachromosomal genomes separated from the nuclear genomes. Although each human mitochondrion contains multiple copies of the mtDNA genome, the exact copy number varies for each cell. However, it is estimated that hundreds of copies of the mtDNA genome exist in most cells.

23.1.1 Genetic Contents of Mitochondrial Organelle Genomes

The human mitochondrial genome was first sequenced by Fred Sanger's laboratory at Cambridge University and was published in 1981; it is known as the *Cambridge reference sequence* (CRS). The sequence was largely derived from a placental sample from an individual of European descent and also partially from HeLa cells (a cell line derived from cervical cancer cells), as well as from bovine cells. It was later discovered, by resequencing the original mtDNA sample, that the CRS contains substitution errors at 10 nucleotide positions. The *revised Cambridge reference sequence* (rCRS) was published in 1999 and presented corrections to these substitution errors. Additionally, CRS contains a cytosine dimer at nucleotide positions 3106 and 3107, which is in fact a single cytosine nucleotide. This error is not corrected in the rCRS in order to retain the original nucleotide numbering system of CRS and thus to avoid inconsistency with the previous literature. The human mitochondrial DNA genome is circular with no beginning and end, which can make sequence comparison a potential problem. Therefore, the rCRS, with its nucleotide numbering system, is used as a reference when aligning with other mitochondrial sequences for comparison purposes.

Organelle genomes are usually much smaller than their nuclear counterparts. The human mtDNA genome consists of 16,569 base pairs (bp) containing 37 genes (Figure 23.2). Thirteen of

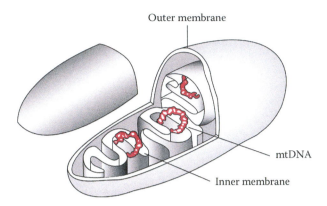

Figure 23.1   Mitochondrion. mtDNA, mitochondrial DNA. (© Richard C. Li.)

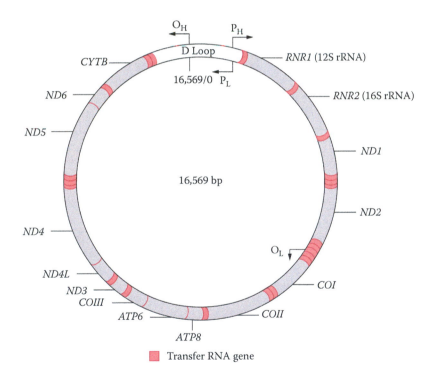

Figure 23.2   Human circular mitochondrial genome. The transcription direction for the H (heavy) and L (light) strands are indicated by arrows (P_H, P_L, respectively). The origins of replication are labeled O_H for the heavy strand and O_L for the light strand. The mitochondrial DNA genome encodes genes. ND, NADH coenzyme Q oxidoreductase complex; CO, cytochrome c oxidase complex; CYTB, cytochrome b; ATP, ATP synthase; rRNA, ribosomal RNA. Transfer RNA genes are shown as indicated. (© Richard C. Li.)

these genes code for proteins involved in the respiratory complex, a main energy-generating component in mitochondria. The other 24 specify noncoding RNA molecules required for the expression of the mitochondrial genome. The genes in the human mitochondrial genome are much more closely packed than in the nuclear genome and contain no introns. A *control region*, also known as a *displacement loop* (*D loop*), contains the origin of replication for one of the mtDNA strands but does not code for any gene products (Figure 23.2). An asymmetric distribution of

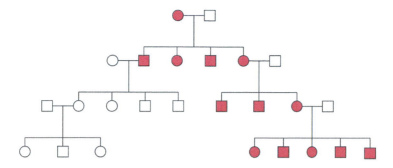

Figure 23.3 Pedigree of a human family showing inheritance of mtDNA. Females and males are denoted by circles and squares, respectively. Red symbols indicate individuals who inherited the same mtDNA.

nucleotides gives rise to light (L) and heavy (H) strands. The H strand contains a greater number of guanine nucleotides and a higher molecular weight in comparison to the L strand.

23.1.2 Maternal Inheritance of mtDNA

Maternal inheritance is typically observed for the mtDNA genome (Figure 23.3), which is inherited differently from nuclear genes. The inheritance of the mtDNA genome does not obey the rules of Mendelian inheritance and is thus called non-Mendelian inheritance.

The mitochondria of a spermatozoon are located in the midpiece (Chapter 14). At conception, only the head portion of a spermatozoon (containing a nucleus but no mitochondria) enters the egg. The fertilized egg contains the maternal mitochondria, which are transmitted to the progeny. Occasionally, the paternal mitochondria can enter the cell. However, paternal mitochondria in the spermatozoon that enter the egg are usually destroyed by the egg cell after fertilization (Figure 23.4). Therefore, the coinheritance of maternal and paternal mtDNA in a single individual is extremely rare in humans. The mtDNA sequence is identical for relatives within the same maternal lineage, a property that is useful when identifying individuals by comparing their mtDNA with that of maternal relatives.

Homologous DNA recombination (Chapter 25) has not been observed in the mtDNA genome. Thus, an mtDNA profile, also referred to as the *mitotype*, is considered a haplotype treated as a single locus. The mitochondrial genome has a higher mutation rate (up to 10 times higher) than its nuclear counterpart. The presence of mutations can be problematic in victim identification when comparing the mtDNA profiles of a victim with the relatives of the victim.

23.2 mtDNA Polymorphic Regions

23.2.1 Hypervariable Regions

The most polymorphic region of mtDNA is located within the D-loop (Figure 23.5). The three *hypervariable regions* in the D-loop are designated hypervariable region I (HV1: 16,024–16,365; 342 bp), hypervariable region II (HV2: 73–340; 268 bp), and hypervariable region III (HV3: 438–574; 137 bp). The most common polymorphic regions of the human mtDNA genome analyzed for forensic purposes are HV1 and HV2.

23.2.2 Heteroplasmy

Heteroplasmy occurs when an individual carries more than one mtDNA haplotype. Heteroplasmy may be observed with one kind of tissue and may be absent in other kinds of tissues; for example,

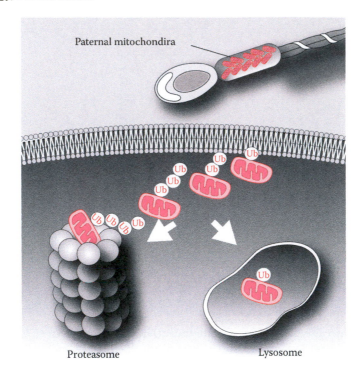

Figure 23.4 A model of uniparental mtDNA inheritance in humans. Paternal mitochondria occasionally enter the egg cell. Paternal mitochondria inside fertilized eggs are tagged by ubiquitin protein (Ub). It is proposed that the ubiquitination of sperm mitochondria leads to the degradation of paternal mitochondria in fertilized eggs. Tagged Ub can be recognized by proteasomes and lysosomes, which are cellular degradation machineries. A polyubiquitin chain with at least four ubiquitin units is needed to be recognized by the proteasomes. Ubiquitin with less than four ubiquitin units can be processed by the lysosomes. (© Richard C. Li.)

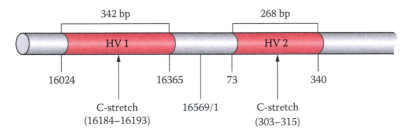

Figure 23.5 Hypervariable regions of the D-loop in mtDNA (with nucleotide positions). (© Richard C. Li.)

mtDNA heteroplasmy is commonly observed in hair samples (Chapter 4). Additionally, an individual may exhibit one mitotype in one tissue and a different mitotype in another. Thus, it is necessary to obtain and process additional samples to confirm the heteroplasmy when it is observed in a questioned sample but not in a known sample or vice versa. The two types of heteroplasmies are sequence and length heteroplasmy.

Sequence heteroplasmy is defined as the presence of two different nucleotides at a single position shown as overlapping peaks in a sequence electropherogram (Figure 23.6). Heteroplasmy usually occurs at one position, but on rare occasions it can be observed at more than one position. Hot spots for heteroplasmy have been documented at both HV1 and HV2 regions. Heteroplasmy may complicate the interpretation of mtDNA results, but its presence can also

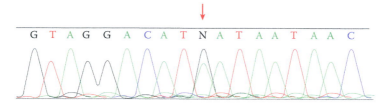

Figure 23.6 Electropherogram showing mtDNA sequence heteroplasmy at position 234R (A/G) as indicated by an arrow. N, unresolved sequence. (© Richard C. Li.)

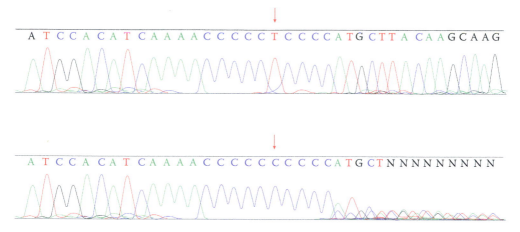

Figure 23.7 Length heteroplasmy. Electropherogram showing mtDNA length heteroplasmy at the C stretch of the HV1 region where position 16,189 is a T (top) and a C (bottom) as indicated by an arrow. N, unresolved sequence. (© Richard C. Li.)

improve the strength of a match. Both HV1 and HV2 of the human mtDNA D-loop region contain homopolymeric cytosine sequences known as *C stretches*. The HVI region contains a C stretch between nucleotide positions 16,184 and 16,193, interrupted by a thymine at position 16,189. If a base transition from T to C occurs at position 16,189 (a variant present in approximately 20% of the population), it results in an uninterrupted C stretch. A similar C stretch resides between positions 303 and 315 of the HV2 region.

Length heteroplasmies are often observed at the uninterrupted C stretches in sequencing, in which sequencing products with various lengths of polymeric cytosine residues are present. As a result, sequences downstream from the C stretch cannot be resolved (Figure 23.7). It is not clear whether the length heteroplasmy is due to replication slippage at the C stretches or results from the presence of a mixture of length variants in the cells. If length heteroplasmy occurs, alternative sequencing primers that anneal at the downstream of C stretches can be used to obtain the downstream sequences of the C stretches.

23.3 Forensic mtDNA Testing
23.3.1 General Considerations

mtDNA analysis is often used on samples derived from skeletal or decomposed remains. The surface of the sample should be cleaned to remove any adhering debris or contaminants. Bones and teeth are pulverized to facilitate extraction of the mtDNA (Chapter 5). Duplicate extractions (e.g., two sections of a single hair) are recommended if sufficient sample material is available.

mtDNA is extracted using a similar method to nuclear DNA (nuclear DNA is coextracted with mtDNA). The amount of mtDNA can therefore be estimated from the quantity of nuclear DNA obtained. Alternatively, mtDNA-specific quantization methods using real-time PCR (Chapter 6) can also be used to directly obtain measurements of mtDNA extracted.

For mtDNA sequencing, the analysis of both strands of the mtDNA in a given region must be performed to ensure accuracy. Due to the high sensitivity of mtDNA analysis, it is essential to minimize risks of contamination during the procedure. Contamination must be strictly monitored using proper controls such as extraction reagent blanks (Section 7.5.3) and amplification negative controls (samples containing all reagents except DNA template).

Finally, a positive control must also be used to monitor the success of the analysis. It should be introduced at the amplification step and remain through the sequencing process. A positive control consists of a DNA template of known sequence, such as DNA purified from the HL60 cell line.

23.3.2 mtDNA Screen Assay

One example of the assay for screening mtDNA variations is the allele-specific oligonucleotide (ASO) assay. It allows initial screening of mtDNA sequence polymorphisms and has the potential to reduce the number of samples required for mtDNA sequencing. This method is also useful for excluding or eliminating suspects from a case. However, HV1 and HV2 sequencing should be performed to obtain complete sequence information for the targeted HV regions to confirm a match.

The commercial Linear Array™ mtDNA HV1/HV2 region sequence typing kit (Roche Applied Sciences, Indianapolis) utilizes reverse ASO configuration with a panel of immobilized ASO probes that detect common polymorphic sites (Figures 23.8 and 23.9). The mtDNA is amplified at both HV1 (444 bp amplicon) and HV2 (415 bp amplicon) regions and the forward primers are biotin labeled at the 5′ ends of the oligonucleotides. Thus, the amplified PCR product (amplicon) is biotinylated. A horseradish peroxidase–conjugated streptavidin complex then binds to the biotinylated amplicon (see Chapter 9). Finally, colorimetric detection is carried out with tetramethylbenzidine (TMB) as the substrate to produce a colored precipitate at the designated location. The typing kit detects sequence variations in 19 positions within the HV1 and HV2 regions.

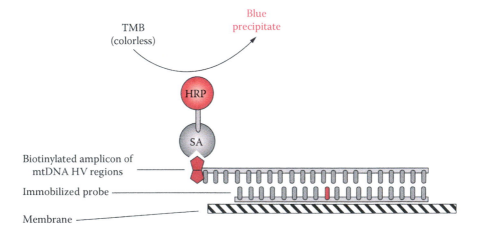

Figure 23.8 Reverse blot assay employed in mtDNA screen. A probe is immobilized onto a solid-phase membrane, and then hybridized with a biotinylated amplicon of the mtDNA HV sequences. Hybridization is detected by a streptavidin (SA) and horseradish peroxidase (HRP) conjugate. A colorimetric reaction is catalyzed by HRP using tetramethylbenzidine (TMB) as a substrate. (© Richard C. Li.)

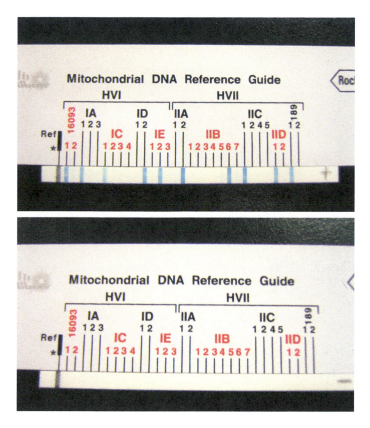

Figure 23.9   Linear array mtDNA assay results (top) and negative control (bottom). (© Richard C. Li.)

23.3.3 mtDNA Sequencing

To sequence a specific region of mtDNA, a combination of PCR amplification and DNA sequencing techniques is utilized to reduce the time and labor needed to obtain DNA sequences from genomic DNA templates. mtDNA sequencing usually consists of (1) PCR amplification, (2) DNA sequencing reactions, (3) separation using electrophoresis, and (4) data collection and sequence analysis (Figure 23.10). Chapter 25 describes the evaluation of the strength of the results via statistical analysis.

23.3.3.1 PCR Amplification

The extracted DNA samples must be amplified to yield sufficient quantities of template for sequencing reactions. PCR amplification of all or a part of the D-loop region can be carried out with various primer sets. If a sample contains intact mtDNA, the HV1 and HV2 regions can be amplified as two amplicons, each of about 350–400 bp in length. If an mtDNA sample is fragmented due to degradation, the hypervariable regions can be amplified as smaller amplicons. PCR amplification of mtDNA is usually done in 34–38 cycles. Protocols for highly degraded DNA specimens sometimes require 42 cycles. The use of higher PCR cycle numbers can improve the yield of the amplicons.

Following mtDNA amplification, a purification step is necessary to remove excess primers and deoxynucleotide triphosphates (dNTPs). This step can be performed using filtration devices such as a Microcon® to remove small molecules from the sample or using nuclease digestion with shrimp alkaline phosphatase or exonuclease I to degrade remaining primers and dNTPs. The concentration of the amplicons is important for an optimal sequencing reaction in the next phase of mtDNA sequencing. The quality and quantity of the mtDNA amplicon must be evaluated to confirm the presence or absence of amplicons and their concentrations. This can be done

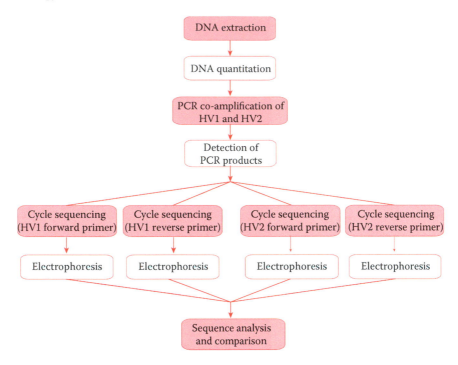

Figure 23.10   Sanger sequencing work flow for the mtDNA HV regions. (© Richard C. Li.)

using an agarose yield gel to visualize the amplicons of the sample or via capillary electrophoresis using a modular microfluidic chip device (Figure 8.12), a more informative method, for quantifying amplicons.

23.3.3.2 DNA Sequencing Reactions
The best-known DNA sequencing techniques are the *chain-termination* method and the *chemical degradation* method developed by Sanger and Gilbert (who shared the Nobel Prize for their work), respectively, in 1977. Over the years, the chain-termination method has become more common because it is suitable for automation and does not require the toxic chemicals necessary for the chemical degradation method.

23.3.3.2.1 Chain-Termination or Sanger Method
Using the chain-termination method, an oligonucleotide primer that can anneal to a single-stranded DNA template is utilized. A sequencing reaction contains DNA polymerase and the four dNTPs in order to carry out extension. The reaction also contains small quantities of dideoxynucleotide triphosphates (ddNTPs, Figure 23.11). Thus, a sequencing reaction involves a combination of extension and termination of the chain (Figure 23.12). If a ddNTP molecule is incorporated into a growing DNA chain, the absence of a 3′ OH group in the ddNTP molecule prevents the formation of a phosphodiester bond and thus disrupts the extension of the oligonucleotide chain.

The ratio of ddNTPs to dNTPs has been optimized to result in a collection of DNA fragments varying in length by one nucleotide from the primer length to the full length of the sequencing reaction product. As a result, the products of the sequencing reaction consist of a pool of various lengths of oligonucleotide chains terminated by ddNTPs. By using the four different ddNTPs, populations of DNA fragments are generated that terminate at positions occupied by every A, C, G, or T in the template strand.

The sequencing product of chain termination can be labeled with the *dye-terminator* system, in which the terminator is labeled, or with the *dye-primer* system, in which the primer is labeled.

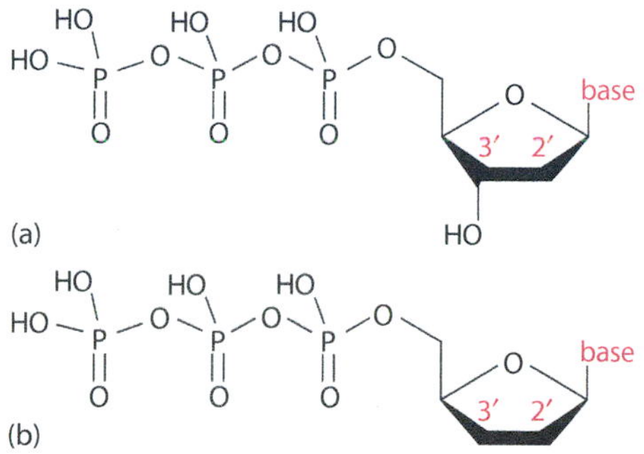

Figure 23.11 Chemical structures of dNTP and ddNTP: (a) dNTP; (b) ddNTP. Both hydroxyl groups attached to the 2′ and 3′ carbons of ribose are replaced by hydrogens.

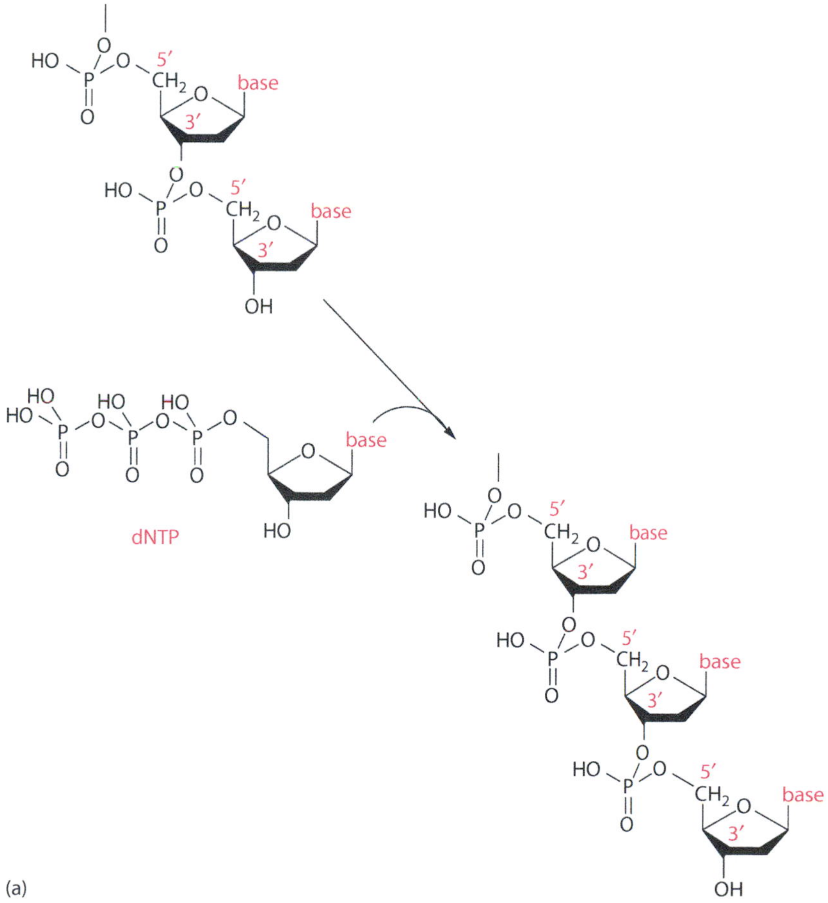

Figure 23.12 Competition between extension and termination. (a) Growing chain in extension.

(b)

Figure 23.12 (Continued) (b) DNA strand with labeled ddNTP incorporated into a growing DNA chain that interrupts the incorporation of new nucleotides. The ddNTP usually is labeled with a fluorescent dye used for detection.

The dye-terminator system is commonly used for mtDNA sequencing in forensic laboratories. With the dye-terminator system, the ddNTPs are labeled with four different fluorescent dyes, each with a distinct spectrum. Thus, the sequencing with all four ddNTPs can be carried out in a single reaction. The labeled products of sequencing reactions are then resolved during electrophoresis and the sequencing data can be collected (Section 23.3.3.).

23.3.3.2.2 Cycle Sequencing

The chain-termination reaction is carried out using a *cycle sequencing* technique commonly used in forensic laboratories for mtDNA sequencing. Cycle sequencing, developed in the late 1980s, utilizes thermal cycling to generate a single-stranded template for chain-termination sequencing reactions. The application of thermal cycling in a sequencing reaction greatly increases the signal intensity and thus the sensitivity of the sequencing.

The sequencing reactions are carried out with multiple rounds of thermal cycling. Each cycle consists of three steps: denaturation of the double-stranded DNA template, annealing of a sequencing primer to its target sequence, and the extension of the annealed primer by DNA polymerase. Cycle sequencing utilizes only a single primer per reaction. During the extension of cycle sequencing, the extension of the strand is terminated with the incorporation of a ddNTP (Figure 23.13). The resulting partially double-stranded hybrid, consisting of the full-length template strand and its complementary chain-terminated product, is denatured during the first step of the next cycle, thereby liberating the template strand for another round of annealing, extension, and termination.

23.3.3.3 Electrophoresis, Sequence Analysis, and Mitotype Designations

The cycle sequencing products can be separated using electrophoresis in a 4% polyacrylamide denatured gel or a POP-6 polymer (Applied Biosystems) as the matrix for capillary electrophoresis (Figure 23.14). Following data collection, sequence data analysis can be performed with the Sequencher™ software (Gene Codes Corporation, Ann Arbor). Figure 23.15 shows sequence data.

Sequencing of a region of the mtDNA genome should be performed twice. Additionally, both strands of a region of the mtDNA genome should be sequenced to reduce ambiguities in sequence determination. The sequences of evidence samples and reference samples such as that of the victim or suspect can be compared. The nomenclature used in reporting should be compatible with International Union of Pure and Applied Chemistry (IUPAC) codes.

23.3.3.3.1 Reporting Format

The rCRS is used as a reference standard to facilitate the designation of mitotypes. For reporting purposes, sequence differences relative to the rCRS are listed in data format. When a difference between an individual's sequence and that of the rCRS sequence is observed, only the position (designated by a number) and the nucleotide differing from the reference standard are recorded. In this format, nucleotides identical to the rCRS are not listed. For example, at position 228 (HV2), the rCRS has a G. If a mitotype carries an A at position 228, the individual's mtDNA sequence is described as 228A. If an unresolved sequence ambiguity is observed at a position,

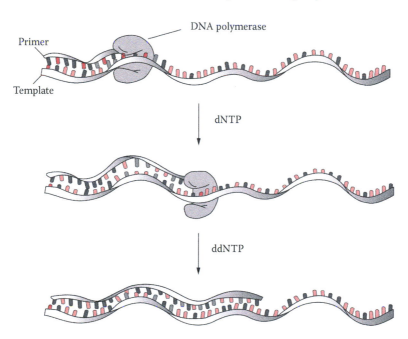

Figure 23.13 Cycle sequencing reaction. The reaction requires DNA polymerase, a template, and a primer. During DNA synthesis, the dNTP is incorporated by a new phosphodiester bond with the primer. The incorporation of ddNTP blocks further DNA synthesis of the growing chain. (© Richard C. Li.)

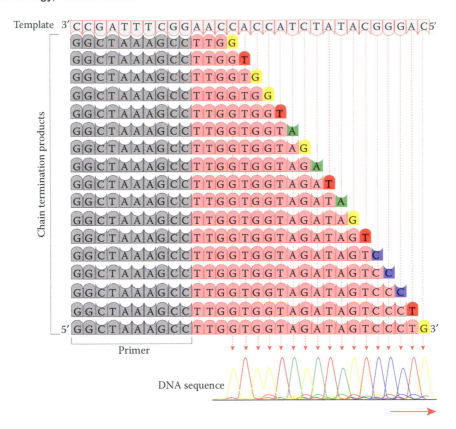

Figure 23.14 Diagram of the Sanger sequencing products. The DNA chain lengths are determined by the addition of ddNTP at different positions. The fluorescent dye-labeled cycle-sequencing products are separated using capillary electrophoresis. The fluorescent dyes are resolved by a detector, and the peaks corresponding to each DNA fragment are identified. (© Richard C. Li.)

the base number for the position is listed followed by an N. For example, 228N means an unresolved sequence ambiguity was observed at position 228.

23.3.3.3.2 Insertions

The insertion site is described by noting the position (at 5′ to the insertion) followed by a decimal point and a number. The number indicates the order of the insertions (e.g., 1 indicates the first insertion, 2 indicates the second, etc.). The base calling following the number indicates the inserted nucleotide (e.g., 524.1A, 524.2C).

23.3.3.3.3 Deletions

The deletion site designation is followed by the letter *d*. For example, a deletion at position 16,296 is recorded as 16,296d.

23.3.3.3.4 Heteroplasmic Sites

The IUPAC codes for base calling can be applied to heteroplasmic sites. For example, an A/G heteroplasmy can be designated as R, and a C/T heteroplasmy can be designated as Y.

23.3.4 Interpretation of mtDNA Profiling Results

Interpretation guidelines are used when comparing sequencing results between evidence and reference samples. General guidelines were set forth by the Scientific Working Group on DNA

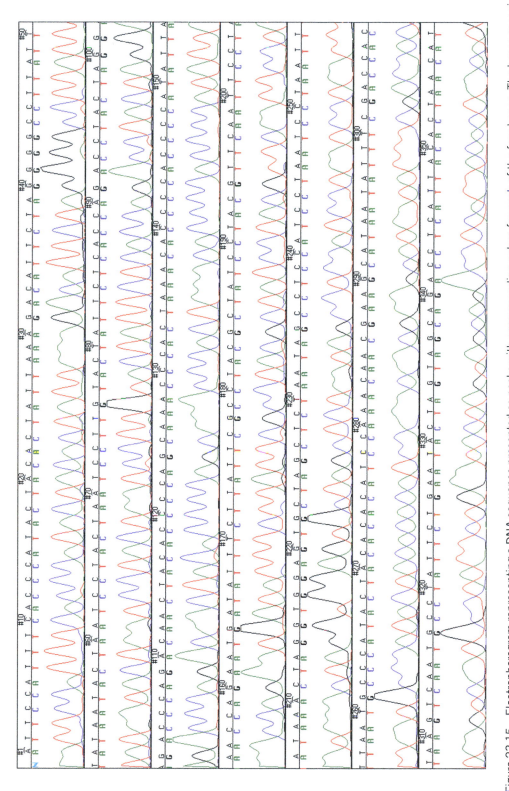

Figure 23.15 Electropherogram representing a DNA sequence presented as peaks with corresponding colors for each of the four bases. The bases associated with each peak are noted. The sequence is read from left to right (5′ to 3′). (© Richard C. Li.)

Analysis Methods (SWGDAM) and the DNA Commission of the International Society of Forensic Genetics (ISFG). In reporting mtDNA profiling results, the most common categories of conclusions are the following: cannot exclude, exclusion, and inconclusive result.

23.3.4.1 Exclusion

If the sequences of questioned and known samples are different, then the samples can be excluded as originating from the same source. It should be taken into account that higher mutation rates are found with the mtDNA genome than are found with the nuclear genome. The SWGDAM's guidelines define that the conclusion of exclusion can be made if there are two or more nucleotide differences between the questioned and known samples. Additionally, mutations seem to be more common in certain tissues. For this reason, the sources of the tissues analyzed should be taken into consideration.

23.3.4.2 Cannot Exclude

If the sequences are the same, the reference sample and evidence cannot be excluded as arising from the same source. When an mtDNA profile cannot be excluded, it is desirable to evaluate the weight of the evidence. In cases where the same heteroplasmy is observed in both questioned and known samples, its presence increases the strength of the evidence. However, if heteroplasmy is observed in a questioned sample but not in a known sample or vice versa, a maternal lineage still cannot be excluded.

23.3.4.3 Inconclusive Result

If the questioned and known samples differ by a single nucleotide, and no heteroplasmy is present, the results are considered to be inconclusive.

Bibliography
Mitochondrial DNA Profiling

Adachi, N., K. Umetsu, and H. Shojo, Forensic strategy to ensure the quality of sequencing data of mitochondrial DNA in highly degraded samples. *Leg Med (Tokyo)*, 2014, **16**(1): 52–55.

Anderson, S., et al., Sequence and organization of the human mitochondrial genome. *Nature*, 1981, **290**(5806): 457–465.

Andreasson, H., et al., Forensic mitochondrial coding region analysis for increased discrimination using pyrosequencing technology. *Forensic Sci Int Genet*, 2007, **1**(1): 35–43.

Andrews, R.M., et al., Reanalysis and revision of the Cambridge reference sequence for human mitochondrial DNA. *Nat Genet*, 1999, **23**(2): 147.

Anjum, G.M., et al., Pyrosequencing-based strategy for a successful SNP detection in two hypervariable regions: HV-I/HV-II of the human mitochondrial displacement loop. *Electrophoresis*, 2010, **31**(2): 309–314.

Arnestad, M., et al., A mitochondrial DNA polymorphism associated with cardiac arrhythmia investigated in sudden infant death syndrome. *Acta Paediatr*, 2007, **96**(2): 206–210.

Asari, M., et al., Differences in tissue distribution of HV2 length heteroplasmy in mitochondrial DNA between mothers and children. *Forensic Sci Int*, 2008, **175**(2–3): 155–159.

Bandelt, H.J. and W. Parson, Consistent treatment of length variants in the human mtDNA control region: A reappraisal. *Int J Legal Med*, 2008, **122**(1): 11–21.

Bandelt, H.J., M. van Oven, and A. Salas, Haplogrouping mitochondrial DNA sequences in legal medicine/forensic genetics. *Int J Legal Med*, 2012, **126**(6): 901–916.

Bendall, K.E. and B.C. Sykes, Length heteroplasmy in the first hypervariable segment of the human mtDNA control region. *Am J Hum Genet*, 1995, **57**(2): 248–256.

Bendall, K.E., V.A. Macaulay, and B.C. Sykes, Variable levels of a heteroplasmic point mutation in individual hair roots. *Am J Hum Genet*, 1997, **61**(6): 1303–1308.

Bendall, K.E., et al., Heteroplasmic point mutations in the human mtDNA control region. *Am J Hum Genet*, 1996, **59**(6): 1276–1287.

Berger, C. and W. Parson, Mini-midi-mito: Adapting the amplification and sequencing strategy of mtDNA to the degradation state of crime scene samples. *Forensic Sci Int Genet*, 2009, **3**(3): 149–153.

Berger, C., et al., Evaluating sequence-derived mtDNA length heteroplasmy by amplicon size analysis. *Forensic Sci Int Genet*, 2011, **5**(2): 142–145.

Bintz, B.J., G.B. Dixon, and M.R. Wilson, Simultaneous detection of human mitochondrial DNA and nuclear-inserted mitochondrial-origin sequences (NumtS) using forensic mtDNA amplification strategies and pyrosequencing technology. *J Forensic Sci*, 2014, **59**(4): 1064–1073.

Brandon, M.C., et al., MITOMASTER: A bioinformatics tool for the analysis of mitochondrial DNA sequences. *Hum Mutat*, 2009, **30**(1): 1–6.

Brandstatter, A., et al., Dissection of mitochondrial superhaplogroup H using coding region SNPs. *Electrophoresis*, 2006, **27**(13): 2541–2550.

Buckleton, J.S., J.M. Curran, and P. Gill, Towards understanding the effect of uncertainty in the number of contributors to DNA stains. *Forensic Sci Int Genet*, 2007, **1**(1): 20–28.

Budowle, B., et al., Mitochondrial DNA regions HVI and HVII population data. *Forensic Sci Int*, 1999, **103**(1): 23–35.

Budowle, B., et al., Forensics and mitochondrial DNA: Applications, debates, and foundations. *Annu Rev Genom Hum Genet*, 2003, **4**: 119–141.

Budowle, B., et al., Addressing the use of phylogenetics for identification of sequences in error in the SWGDAM mitochondrial DNA database. *J Forensic Sci*, 2004, **49**(6): 1256–1261.

Budowle, B., et al., Mixture interpretation: Defining the relevant features for guidelines for the assessment of mixed DNA profiles in forensic casework. *J Forensic Sci*, 2009, **54**(4): 810–821.

Budowle, B., et al., Automated alignment and nomenclature for consistent treatment of polymorphisms in the human mitochondrial DNA control region. *J Forensic Sci*, 2010, **55**(5): 1190–1195.

Butler, J.M. and B.C. Levin, Forensic applications of mitochondrial DNA. *Trends Biotechnol*, 1998, **16**(4): 158–162.

Calloway, C.D., et al., The frequency of heteroplasmy in the HVII region of mtDNA differs across tissue types and increases with age. *Am J Hum Genet*, 2000, **66**(4): 1384–1397.

Carracedo, A., et al., DNA commission of the international society for forensic genetics: Guidelines for mitochondrial DNA typing. *Forensic Sci Int*, 2000, **110**(2): 79–85.

Cassandrini, D., et al., A new method for analysis of mitochondrial DNA point mutations and assess levels of heteroplasmy. *Biochem Biophys Res Commun*, 2006, **342**(2): 387–393.

Cerezo, M., et al., Applications of MALDI-TOF MS to large-scale human mtDNA population-based studies. *Electrophoresis*, 2009, **30**(21): 3665–3673.

Chemale, G., et al., Development and validation of a D-loop mtDNA SNP assay for the screening of specimens in forensic casework. *Forensic Sci Int Genet*, 2013, **7**(3): 353–358.

Clayton, D.A., Structure and function of the mitochondrial genome. *J Inherit Metab Dis*, 1992, **15**(4): 439–447.

Comas, D., S. Paabo, and J. Bertranpetit, Heteroplasmy in the control region of human mitochondrial DNA. *Genome Res*, 1995, **5**(1): 89–90.

Crespillo, M., et al., Results of the 2003–2004 GEP-ISFG collaborative study on mitochondrial DNA: Focus on the mtDNA profile of a mixed semen-saliva stain. *Forensic Sci Int*, 2006, **160**(2–3): 157–167.

Curran, J.M., A MCMC method for resolving two person mixtures. *Sci Justice*, 2008, **48**(4): 168–177.

Curtis, P.C., et al., Optimization of primer-specific filter metrics for the assessment of mitochondrial DNA sequence data. *Mitochondrial DNA*, 2010, **21**(6): 191–197.

D'Eustachio, P., High levels of mitochondrial DNA heteroplasmy in human hairs by Budowle et al. *Forensic Sci Int*, 2002, **130**(1): 63–67; author reply 68–70.

Danielson, P.B., et al., Resolving mtDNA mixtures by denaturing high-performance liquid chromatography and linkage phase determination. *Forensic Sci Int Genet*, 2007, **1**(2): 148–153.

Danielson, P.B., et al., Separating human DNA mixtures using denaturing high-performance liquid chromatography. *Expert Rev Mol Diagn*, 2005, **5**(1): 53–63.

Desmyter, S. and B. Hoste, Influence of the electrophoresis-resequencing method on the forensic mtDNA profiling quality. *Forensic Sci Int Genet*, 2007, **1**(2): 199–200.

Eduardoff, M., et al., Mass spectrometric base composition profiling: Implications for forensic mtDNA databasing. *Forensic Sci Int Genet*, 2013, **7**(6): 587–592.

Eichmann, C. and W. Parson, "Mitominis": Multiplex PCR analysis of reduced size amplicons for compound sequence analysis of the entire mtDNA control region in highly degraded samples. *Int J Legal Med*, 2008, **122**(5): 385–388.

Fan, L. and Y.G. Yao, An update to MitoTool: Using a new scoring system for faster mtDNA haplogroup determination. *Mitochondrion*, 2013, **13**(4): 360–363.

Fattorini, P., et al., Estimating the integrity of aged DNA samples by CE. *Electrophoresis*, 2009, **30**(22): 3986–3995.

Fendt, L., et al., Sequencing strategy for the whole mitochondrial genome resulting in high quality sequences. *BMC Genom*, 2009, **10**: 139.

Fernández, C. and A. Alonso, Microchip capillary electrophoresis protocol to evaluate quality and quantity of mtDNA amplified fragments for DNA sequencing in forensic genetics. *Methods Mol Biol*, 2012, **830**: 367–379.

Forster, L., et al., Evaluating length heteroplasmy in the human mitochondrial DNA control region. *Int J Legal Med*, 2010, **124**(2): 133–142.

Gabriel, M.N., et al., Improved mtDNA sequence analysis of forensic remains using a "mini-primer set" amplification strategy. *J Forensic Sci*, 2001, **46**(2): 247–253.

Gidlof, O., et al., Complete discrimination of six individuals based on high-resolution melting of hypervariable regions I and II of the mitochondrial genome. *Biotechniques*, 2009, **47**(2): 671–672, 674, 676, passim.

Gill, P., et al., National recommendations of the technical UK DNA working group on mixture interpretation for the NDNAD and for court going purposes. *Forensic Sci Int Genet*, 2008, **2**(1): 76–82.

Grignani, P., et al., Subtyping mtDNA haplogroup H by SNaPshot minisequencing and its application in forensic individual identification. *Int J Legal Med*, 2006, **120**(3): 151–156.

Grignani, P., et al., A mini-primer set covering the mtDNA hypervariable regions for the genetic typing of old skeletal remains. *Forensic Sci Int Genet Suppl*, 2009, **2**(1): 265–266.

Grignani, P., et al., Multiplex mtDNA coding region SNP assays for molecular dissection of haplogroups U/K and J/T. *Forensic Sci Int Genet*, 2009, **4**(1): 21–25.

Grzybowski, T. and U. Rogalla, Mitochondria in anthropology and forensic medicine. *Adv Exp Med Biol*, 2012, **942**: 441–453.

Hall, T.A., et al., Base composition profiling of human mitochondrial DNA using polymerase chain reaction and direct automated electrospray ionization mass spectrometry. *Anal Chem*, 2009, **81**(18): 7515–7526.

Hartmann, A., et al., Validation of microarray-based resequencing of 93 worldwide mitochondrial genomes. *Hum Mutat*, 2009, **30**(1): 115–122.

Hauswirth, W.W. and D.A. Clayton, Length heterogeneity of a conserved displacement-loop sequence in human mitochondrial DNA. *Nucleic Acids Res*, 1985, **13**(22): 8093–8104.

He, Y., et al., Heteroplasmic mitochondrial DNA mutations in normal and tumour cells. *Nature*, 2010, **464**(7288): 610–614.

Holland, M.M., M.R. McQuillan, and K.A. O'Hanlon, Second generation sequencing allows for mtDNA mixture deconvolution and high resolution detection of heteroplasmy. *Croat Med J*, 2011, **52**(3): 299–313.

Holland, M.M., Molecular analysis of the human mitochondrial DNA control region for forensic identity testing. *Curr Protoc Hum Genet*, 2012, (Suppl 74) 10.1002/0471142905.hg1407s74.

Howard, R., et al., Comparative analysis of human mitochondrial DNA from World War I bone samples by DNA sequencing and ESI-TOF mass spectrometry. *Forensic Sci Int Genet*, 2013, **7**(1): 1–9.

Hwan, Y.L., et al., A modified mini-primer set for analyzing mitochondrial DNA control region sequences from highly degraded forensic samples. *BioTechniques*, 2008, **44**(4): 555–558.

Irwin, J.A., et al., mtGenome reference population databases and the future of forensic mtDNA analysis. *Forensic Sci Int Genet*, 2011, **5**(3): 222–225.

Jarman, P.G., S.L. Fentress, and D.E. Katz, Mitochondrial DNA validation in a state laboratory. *J Forensic Sci*, 2009, **54**(1): 95–102.

Just, R.S., et al., Complete mitochondrial genome sequences for 265 African American and U.S. "Hispanic" individuals. *Forensic Sci Int Genet*, 2008, **2**(3): e45–e48.

Just, R.S., et al., The use of mitochondrial DNA single nucleotide polymorphisms to assist in the resolution of three challenging forensic cases. *J Forensic Sci*, 2009, **54**(4): 887–891.

Just, R.S., et al., Development of forensic-quality full mtGenome haplotypes: Success rates with low template specimens. *Forensic Sci Int Genet*, 2014, **10**: 73–79.

Kim, N.Y., et al., Modified midi- and mini-multiplex PCR systems for mitochondrial DNA control region sequence analysis in degraded samples. *J Forensic Sci*, 2013, **58**(3): 738–743.

Kohnemann, S. and H. Pfeiffer, Application of mtDNA SNP analysis in forensic casework. *Forensic Sci Int Genet*, 2011, **5**(3): 216–221.

Kohnemann, S., C. Hohoff, and H. Pfeiffer, An economical mtDNA SNP assay detecting different mitochondrial haplogroups in identical HVR 1 samples of Caucasian ancestry. *Mitochondrion*, 2009, **9**(5): 370–375.

Kohnemann, S., et al., A rapid mtDNA assay of 22 SNPs in one multiplex reaction increases the power of forensic testing in European Caucasians. *Int J Legal Med*, 2008, **122**(6): 517–523.

Kristinsson, R., S.E. Lewis, and P.B. Danielson, Comparative analysis of the HV1 and HV2 regions of human mitochondrial DNA by denaturing high-performance liquid chromatography. *J Forensic Sci*, 2009, **54**(1): 28–36.

Kurelac, I., et al., Searching for a needle in the haystack: Comparing six methods to evaluate heteroplasmy in difficult sequence context. *Biotechnol Adv*, 2012, **30**(1): 363–371.

Lacan, M., et al., Detection of the A189G mtDNA heteroplasmic mutation in relation to age in modern and ancient bones. *Int J Legal Med*, 2009, **123**(2): 161–167.

Lacan, M., et al., Detection of age-related duplications in mtDNA from human muscles and bones. *Int J Legal Med*, 2011, **125**(2): 293–300.

Lee, H.Y., et al., A modified mini-primer set for analyzing mitochondrial DNA control region sequences from highly degraded forensic samples. *Biotechniques*, 2008, **44**(4): 555–556, 558.

Lee, H.Y., et al., DNA typing for the identification of old skeletal remains from Korean War victims. *J Forensic Sci*, 2010, **55**(6): 1422–1429.

Lee, H.Y., et al., A one step multiplex PCR assay for rapid screening of East Asian mtDNA haplogroups on forensic samples. *Leg Med (Tokyo)*, 2013, **15**(1): 50–54.

Linch, C.A., D.A. Whiting, and M.M. Holland, Human hair histogenesis for the mitochondrial DNA forensic scientist. *J Forensic Sci*, 2001, **46**(4): 844–853.

Loreille, O.M. and J.A. Irwin, Capillary electrophoresis of human mtDNA control region sequences from highly degraded samples using short mtDNA amplicons. *Methods Mol Biol*, 2012, **830**: 283–299.

Mabuchi, T., et al., Typing the 1.1 kb control region of human mitochondrial DNA in Japanese individuals. *J Forensic Sci*, 2007, **52**(2): 355–363.

Malyarchuk, B.A., Improving the reconstructed sapiens reference sequence of mitochondrial DNA. *Forensic Sci Int Genet*, 2013, **7**(3): e74–e75.

Maricic, T., M. Whitten, and S. Paabo, Multiplexed DNA sequence capture of mitochondrial genomes using PCR products. *PLoS One*, 2010, **5**(11): e14004.

Marquez, M.C., Interpretation guidelines of mtDNA control region sequence electropherograms in forensic genetics. *Methods Mol Biol*, 2012, **830**: 301–319.

Melton, T., et al., Forensic mitochondrial DNA analysis of 691 casework hairs. *J Forensic Sci*, 2005, **50**(1): 73–80.

Miller, K.W. and B. Budowle, A compendium of human mitochondrial DNA control region: Development of an international standard forensic database. *Croat Med J*, 2001, **42**(3): 315–327.

Montesino, M. and L. Prieto, Capillary electrophoresis of big-dye terminator sequencing reactions for human mtDNA control region haplotyping in the identification of human remains. *Methods Mol Biol*, 2012, **830**: 267–281.

Montesino, M., et al., Analysis of body fluid mixtures by mtDNA sequencing: An inter-laboratory study of the GEP-ISFG working group. *Forensic Sci Int*, 2007, **168**(1): 42–56.

Moore, C.A., J. Gudikote, and G.C. Van Tuyle, Mitochondrial DNA rearrangements, including partial duplications, occur in young and old rat tissues. *Mutat Res*, 1998, **421**(2): 205–217.

Morling, N., et al., Interpretation of DNA mixtures—European consensus on principles. *Forensic Sci Int Genet*, 2007, **1**(3–4): 291–292.

Mosquera-Miguel, A., et al., Testing the performance of mtSNP minisequencing in forensic samples. *Forensic Sci Int Genet*, 2009, **3**(4): 261–264.

Nakahara, H., et al., Heteroplasmies detected in an amplified mitochondrial DNA control region from a small amount of template. *J Forensic Sci*, 2008, **53**(2): 306–311.

Nara, A., et al., Sequence analysis for HV I region of mitochondrial DNA using WGA (whole genome amplification) method. *Leg Med (Tokyo)*, 2009, **11**(Suppl 1): S115–S118.

Naue, J., et al., Factors affecting the detection and quantification of mitochondrial point heteroplasmy using Sanger sequencing and SNaPshot minisequencing. *Int J Legal Med*, 2011, **125**(3): 427–436.

Nelson, K. and T. Melton, Forensic mitochondrial DNA analysis of 116 casework skeletal samples. *J Forensic Sci*, 2007, **52**(3): 557–561.

Nelson, T.M., et al., Development of a multiplex single base extension assay for mitochondrial DNA haplogroup typing. *Croat Med J*, 2007, **48**(4): 460–472.

Nicklas, J.A., T. Noreault-Conti, and E. Buel, Development of a real-time method to detect DNA degradation in forensic samples. *J Forensic Sci*, 2012, **57**(2): 466–471.

Nilsson, M., et al., Evaluation of mitochondrial DNA coding region assays for increased discrimination in forensic analysis. *Forensic Sci Int Genet*, 2008, **2**(1): 1–8.

Oberacher, H., H. Niederstatter, and W. Parson, Liquid chromatography-electrospray ionization mass spectrometry for simultaneous detection of mtDNA length and nucleotide polymorphisms. *Int J Legal Med*, 2007, **121**(1): 57–67.

Oberacher, H., et al., Increased forensic efficiency of DNA fingerprints through simultaneous resolution of length and nucleotide variability by high-performance mass spectrometry. *Hum Mutat*, 2008, **29**(3): 427–432.

Pakendorf, B. and M. Stoneking, Mitochondrial DNA and human evolution. *Annu Rev Genom Hum Genet*, 2005, **6**: 165–183.

Paneto, G.G., et al., Heteroplasmy in hair: Study of mitochondrial DNA third hypervariable region in hair and blood samples. *J Forensic Sci*, 2010, **55**(3): 715–718.

Parson, W. and A. Dur, EMPOP—A forensic mtDNA database. *Forensic Sci Int Genet*, 2007, **1**(2): 88–92.

Parsons, T.J. and M.D. Coble, Increasing the forensic discrimination of mitochondrial DNA testing through analysis of the entire mitochondrial DNA genome. *Croat Med J*, 2001, **42**(3): 304–309.

Polanskey, D., et al., Comparison of mitotyper rules and phylogenetic-based mtDNA nomenclature systems. *J Forensic Sci*, 2010, **55**(5): 1184–1189.

Prieto, L., et al., 2006 GEP-ISFG collaborative exercise on mtDNA: Reflections about interpretation, artefacts, and DNA mixtures. *Forensic Sci Int Genet*, 2008, **2**(2): 126–133.

Prieto, L., et al., GHEP-ISFG proficiency test 2011: Paper challenge on evaluation of mitochondrial DNA results. *Forensic Sci Int Genet*, 2013, **7**(1): 10–15.

Quintans, B., et al., Typing of mitochondrial DNA coding region SNPs of forensic and anthropological interest using SNaPshot minisequencing. *Forensic Sci Int*, 2004, **140**(2–3): 251–257.

Roberts, K.A. and C. Calloway, Characterization of mitochondrial DNA sequence heteroplasmy in blood tissue and hair as a function of hair morphology. *J Forensic Sci*, 2011, **56**(1): 46–60.

Rock, A., et al., SAM: String-based sequence search algorithm for mitochondrial DNA database queries. *Forensic Sci Int Genet*, 2011, **5**(2): 126–132.

Salas, A., et al., Phylogeographic investigations: The role of trees in forensic genetics. *Forensic Sci Int*, 2007, **168**(1): 1–13.

Salas, A., et al., A cautionary note on switching mitochondrial DNA reference sequences in forensic genetics. *Forensic Sci Int Genet*, 2012, **6**(6): e182–e184.

Schneider, P.M., et al., The German stain commission: Recommendations for the interpretation of mixed stains. *Int J Legal Med*, 2009, **123**(1): 1–5.

Seo, S.B., et al., Alterations of length heteroplasmy in mitochondrial DNA under various amplification conditions. *J Forensic Sci*, 2010, **55**(3): 719–722.

Serizawa, Y., et al., Detection of mitochondrial DNA polymorphisms from human hair shafts and formalin fixed tissue using whole genome amplification. *Int Med J*, 2008, **15**(3): 163–167.

Sosa, C., et al., A preliminary study on the incidence of heteroplasmy in mitochondrial DNA from vitreous humour. *Leg Med (Tokyo)*, 2009, **11**(Suppl 1): S460–S462.

Stringer, P., et al., Interpretation of DNA mixtures—Australian and New Zealand consensus on principles. *Forensic Sci Int Genet*, 2009, **3**(2): 144–145.

Sutovsky, P., et al., Ubiquitin tag for sperm mitochondria. *Nature*, 1999, **402**: 371–372.

Tobe, S.S. and A.M. Linacre, A technique for the quantification of human and non-human mammalian mitochondrial DNA copy number in forensic and other mixtures. *Forensic Sci Int Genet*, 2008, **2**(4): 249–256.

Tully, G., et al., Considerations by the European DNA profiling (EDNAP) group on the working practices, nomenclature and interpretation of mitochondrial DNA profiles. *Forensic Sci Int*, 2001, **124**(1): 83–91.

Tully, G., et al., Results of a collaborative study of the EDNAP group regarding mitochondrial DNA heteroplasmy and segregation in hair shafts. *Forensic Sci Int*, 2004, **140**(1): 1–11.

Tully, L.A. and B.C. Levin, Human mitochondrial genetics. *Biotechnol Genet Eng Rev*, 2000, **17**: 147–177.

Turchi, C., et al., Italian mitochondrial DNA database: Results of a collaborative exercise and proficiency testing. *Int J Legal Med*, 2008, **122**(3): 199–204.

Umetsu, K. and I. Yuasa, Recent progress in mitochondrial DNA analysis. *Leg Med (Tokyo)*, 2005, **7**(4): 259–262.

Vallone, P.M., Capillary electrophoresis of an 11-plex mtDNA coding region SNP single base extension assay for discrimination of the most common Caucasian HV1/HV2 mitotype. *Methods Mol Biol*, 2012, **830**: 159–167.

Vallone, P.M., J.P. Jakupciak, and M.D. Coble, Forensic application of the Affymetrix human mitochondrial resequencing array. *Forensic Sci Int Genet*, 2007, **1**(2): 196–198.

van Oven, M. and M. Kayser, Updated comprehensive phylogenetic tree of global human mitochondrial DNA variation. *Hum Mutat*, 2009, **30**(2): E386–E394.

Warshauer, D.H., et al., Validation of the PLEX-ID™ mass spectrometry mitochondrial DNA assay. *Int J Legal Med*, 2013, **127**(2): 277–286.

Wilson, M.R., et al., Extraction, PCR amplification and sequencing of mitochondrial DNA from human hair shafts. *Biotechniques*, 1995, **18**(4): 662–669.

Wilson, M.R., et al., Recommendations for consistent treatment of length variants in the human mitochondrial DNA control region. *Forensic Sci Int*, 2002, **129**(1): 35–42.

Wilson, M.R., et al., Validation of mitochondrial DNA sequencing for forensic casework analysis. *Int J Legal Med*, 1995, **108**(2): 68–74.

Wu, J., et al., Multiplex mutagenically separated polymerase chain reaction assay for rapid detection of human mitochondrial DNA variations in coding area. *Croat Med J*, 2008, **49**(1): 32–38.

Xiu-Cheng Fan, A., et al., A rapid and accurate approach to identify single nucleotide polymorphisms of mitochondrial DNA using MALDI-TOF mass spectrometry. *Clin Chem Lab Med*, 2008, **46**(3): 299–305.

Genetic Population Data

Afonso Costa, H., et al., Mitochondrial DNA sequence analysis of a native Bolivian population. *J Forensic Leg Med*, 2010, **17**(5): 247–253.

Alshamali, F., et al., Mitochondrial DNA control region variation in Dubai, United Arab Emirates. *Forensic Sci Int Genet*, 2008, **2**(1): e9–e10.

Alvarez, J.C., et al., Characterization of human control region sequences for Spanish individuals in a forensic mtDNA data set. *Leg Med (Tokyo)*, 2007, **9**(6): 293–304.

Alvarez-Iglesias, V., et al., Coding region mitochondrial DNA SNPs: Targeting East Asian and native American haplogroups. *Forensic Sci Int Genet*, 2007, **1**(1): 44–55.

Al-Zahery, N., et al., Characterization of mitochondrial DNA control region lineages in Iraq. *Int J Legal Med*, 2013, **127**(2): 373–375.

Asari, M., et al., Utility of haplogroup determination for forensic mtDNA analysis in the Japanese population. *Leg Med (Tokyo)*, 2007, **9**(5): 237–240.

Ballantyne, K.N., et al., MtDNA SNP multiplexes for efficient inference of matrilineal genetic ancestry within Oceania. *Forensic Sci Int Genet*, 2012, **6**(4): 425–436.

Barbosa, A.B., et al., Mitochondrial DNA control region polymorphism in the population of Alagoas state, north-eastern Brazil. *J Forensic Sci*, 2008, **53**(1): 142–146.

Bobillo, M.C., et al., Amerindian mitochondrial DNA haplogroups predominate in the population of Argentina: Towards a first nationwide forensic mitochondrial DNA sequence database. *Int J Legal Med*, 2010, **124**(4): 263–268.

Bodner, M., et al., Southeast Asian diversity: First insights into the complex mtDNA structure of Laos. *BMC Evol Biol*, 2011, **11**: 49.

Brandstatter, A., et al., Generating population data for the EMPOP database—An overview of the mtDNA sequencing and data evaluation processes considering 273 Austrian control region sequences as example. *Forensic Sci Int*, 2007, **166**(2–3): 164–175.

Brandstatter, A., et al., Mitochondrial DNA control region variation in Ashkenazi Jews from Hungary. *Forensic Sci Int Genet*, 2008, **2**(1): e4–e6.

Cardoso, S., et al., Mitochondrial DNA control region variation in an autochthonous Basque population sample from the Basque Country. *Forensic Sci Int Genet*, 2012, **6**(4): e106–e108.

Cardoso, S., et al., Mitochondrial DNA control region data reveal high prevalence of nNative American lineages in Jujuy province, NW Argentina. *Forensic Sci Int Genet*, 2013, **7**(3): e52–e55.

Cardoso, S., et al., Variability of the entire mitochondrial DNA control region in a human isolate from the Pas Valley (northern Spain). *J Forensic Sci*, 2010, **55**(5): 1196–1201.

Castro de Guerra, D., et al., Sequence variation of mitochondrial DNA control region in North Central Venezuela. *Forensic Sci Int Genet*, 2012, **6**(5): e131–e133.

Chen, F., et al., Analysis of mitochondrial DNA polymorphisms in Guangdong Han Chinese. *Forensic Sci Int Genet*, 2008, **2**(2): 150–153.

Chen, F., et al., Genetic polymorphism of mitochondrial DNA HVS-I and HVS-II of Chinese Tu ethnic minority group. *J Genet Genom*, 2008, **35**(4): 225–232.

Chen, F., et al., Sequence-length variation of mtDNA HVS-I C-stretch in Chinese ethnic groups. *J Zhejiang Univ Sci B*, 2009, **10**(10): 711–720.

Diegoli, T.M., et al., Mitochondrial control region sequences from an African American population sample. *Forensic Sci Int Genet*, 2009, **4**(1): e45–e52.

Egyed, B., et al., Mitochondrial control region sequence variations in the Hungarian population: Analysis of population samples from Hungary and from Transylvania (Romania). *Forensic Sci Int Genet*, 2007, **1**(2): 158–162.

Fendt, L., et al., Mitochondrial DNA control region data from indigenous Angolan Khoe-San lineages. *Forensic Sci Int Genet*, 2012, **6**(5): 662–663.

Fendt, L., et al., MtDNA diversity of Ghana: A forensic and phylogeographic view. *Forensic Sci Int Genet*, 2012, **6**(2): 244–249.

Grzybowski, T., et al., Complex interactions of the eastern and western Slavic populations with other European groups as revealed by mitochondrial DNA analysis. *Forensic Sci Int Genet*, 2007, **1**(2): 141–147.

Hedman, M., et al., Finnish mitochondrial DNA HVS-I and HVS-II population data. *Forensic Sci Int*, 2007, **172**(2–3): 171–178.

Huang, D., et al., Typing of 24 mtDNA SNPs in a Chinese population using SNaPshot minisequencing. *J Huazhong Univ Sci Technol Med Sci*, 2010, **30**(3): 291–298.

Irwin, J., et al., Hungarian mtDNA population databases from Budapest and the Baranya county Roma. *Int J Legal Med*, 2007, **121**(5): 377–383.

Irwin, J., et al., Mitochondrial control region sequences from northern Greece and Greek Cypriots. *Int J Legal Med*, 2008, **122**(1): 87–89.

Irwin, J.A., et al., Development and expansion of high-quality control region databases to improve forensic mtDNA evidence interpretation. *Forensic Sci Int Genet*, 2007, **1**(2): 154–157.

Irwin, J.A., et al., Mitochondrial DNA control region variation in a population sample from Hong Kong, China. *Forensic Sci Int Genet*, 2009, **3**(4): e119–e125.

Irwin, J.A., et al., The mtDNA composition of Uzbekistan: A microcosm of Central Asian patterns. *Int J Legal Med*, 2010, **124**(3): 195–204.

Kato, H., et al., Molecular analysis of mitochondrial hypervariable region 1 in 394 Japanese individuals. *Leg Med (Tokyo)*, 2009, **11**(Suppl 1): S443–S445.

Lander, N., et al., Haplotype diversity in human mitochondrial DNA hypervariable regions I-III in the city of Caracas (Venezuela). *Forensic Sci Int Genet*, 2008, **2**(4): e61–e64.

Lehocky, I., et al., A database of mitochondrial DNA hypervariable regions I and II sequences of individuals from Slovakia. *Forensic Sci Int Genet*, 2008, **2**(4): e53–e59.

Liu, C., et al., Mitochondrial DNA polymorphisms in Gelao ethnic group residing in Southwest China. *Forensic Sci Int Genet*, 2011, **5**(1): e4–e10.

Maruyama, S., et al., MtDNA control region sequence polymorphisms and phylogenetic analysis of Malay population living in or around Kuala Lumpur in Malaysia. *Int J Legal Med*, 2010, **124**(2): 165–170.

Maruyama, S., et al., Analysis of human mitochondrial DNA polymorphisms in the Japanese population. *Biochem Genet*, 2013, **51**(1–2): 33–70.

Mikkelsen, M., et al., Mitochondrial DNA HV1 and HV2 variation in Danes. *Forensic Sci Int Genet*, 2010, **4**(4): e87–e88.

Mikkelsen, M., et al., Forensic and phylogeographic characterisation of mtDNA lineages from Somalia. *Int J Legal Med*, 2012, **126**(4): 573–579.

Nohira, C., S. Maruyama, and K. Minaguchi, Phylogenetic classification of Japanese mtDNA assisted by complete mitochondrial DNA sequences. *Int J Legal Med*, 2010, **124**(1): 7–12.

Nur Haslindawaty, A.R., et al., Sequence polymorphisms of mtDNA HV1, HV2, and HV3 regions in the Malay population of Peninsular Malaysia. *Int J Legal Med*, 2010, **124**(5): 415–426.

Palencia, L., et al., Mitochondrial DNA diversity in a population from Santa Catarina (Brazil): Predominance of the European input. *Int J Legal Med*, 2010, **124**(4): 331–336.

Parson, W., et al., Identification of West Eurasian mitochondrial haplogroups by mtDNA SNP screening: Results of the 2006–2007 EDNAP collaborative exercise. *Forensic Sci Int Genet*, 2008, **2**(1): 61–68.

Powell, G.T., et al., The population history of the Xibe in northern China: A comparison of autosomal, mtDNA and Y-chromosomal analyses of migration and gene flow. *Forensic Sci Int Genet*, 2007, **1**(2): 115–119.

Rakha, A., et al., Forensic and genetic characterization of mtDNA from Pathans of Pakistan. *Int J Legal Med*, 2011, **125**(6): 841–848.

Ribeiro-dos-Santos, A.K., et al., Nucleotide variability of HV-I in Afro-descendents populations of the Brazilian Amazon region. *Forensic Sci Int*, 2007, **167**(1): 77–80.

Saunier, J.L., et al., Mitochondrial control region sequences from a U.S. "Hispanic" population sample. *Forensic Sci Int Genet*, 2008, **2**(2): e19–e23.

Saunier, J.L., et al., Mitochondrial control region sequences from an Egyptian population sample. *Forensic Sci Int Genet*, 2009, **3**(3): e97–e103.

Scheible, M., et al., Mitochondrial DNA control region variation in a Kuwaiti population sample. *Forensic Sci Int Genet*, 2011, **5**(4): e112–e113.

Szibor, R., et al., Mitochondrial D-loop $(CA)_n$ repeat length heteroplasmy: Frequency in a German population sample and inheritance studies in two pedigrees. *Int J Legal Med*, 2007, **121**(3): 207–213.

Tetzlaff, S., et al., Mitochondrial DNA population data of HVS-I and HVS-II sequences from a northeast German sample. *Forensic Sci Int*, 2007, **172**(2–3): 218–224.

Tillmar, A.O., et al., Homogeneity in mitochondrial DNA control region sequences in Swedish subpopulations. *Int J Legal Med*, 2010, **124**(2): 91–98.

Turchi, C., et al., Polymorphisms of mtDNA control region in Tunisian and Moroccan populations: An enrichment of forensic mtDNA databases with Northern Africa data. *Forensic Sci Int Genet*, 2009, **3**(3): 166–172.

Wilson, J.L., et al., Forensic analysis of mtDNA haplotypes from two rural communities in Haiti reflects their population history. *J Forensic Sci*, 2012, **57**(6): 1457–1466.

Wong, H.Y., et al., Sequence polymorphism of the mitochondrial DNA hypervariable regions I and II in 205 Singapore Malays. *Leg Med (Tokyo)*, 2007, **9**(1): 33–37.

Yan, J., et al., Allele frequencies of mitochondrial D-loop $(CA)_n$ repeat polymorphism in six Chinese ethnic groups. *Leg Med (Tokyo)*, 2007, **9**(6): 330–331.

Yang, Y., et al., A new strategy for the discrimination of mitochondrial DNA haplogroups in Han population. *J Forensic Sci*, 2011, **56**(3): 586–590.

Zgonjanin, D., et al., Sequence polymorphism of the mitochondrial DNA control region in the population of Vojvodina Province, Serbia. *Leg Med (Tokyo)*, 2010, **12**(2): 104–107.

Zimmermann, B., et al., Mitochondrial DNA control region population data from Macedonia. *Forensic Sci Int Genet*, 2007, **1**(3–4): e4–e9.

Zimmermann, B., et al., Forensic and phylogeographic characterization of mtDNA lineages from northern Thailand (Chiang Mai). *Int J Legal Med*, 2009, **123**(6): 495–501.

SECTION V

Forensic Issues

Forensic DNA Databases
Tools for Crime Investigations

24.1 Brief History of Forensic DNA Databases

Forensic DNA databases are networks for exchanging information among law enforcement agencies to assist in solving crimes. For example, forensic DNA databases allow forensic laboratories to search DNA profiles against the databases to identify criminals. In 1995, the United Kingdom established the world's first national DNA database, NDNAD, in England and Wales. Scotland and Northern Ireland have their own databases but also submit their profiles to NDNAD. NDNAD demonstrated initial success in solving crimes. Three years later, the United States introduced its national *Combined DNA Index System* or CODIS. By the end of 1998, other countries (such as Austria, Germany, the Netherlands, New Zealand, and Slovenia) had also introduced national DNA databases. Table 24.1 describes some of these national DNA databases. This chapter will focus on CODIS.

24.2 Infrastructure of CODIS

In the United States, a pilot project of DNA databasing was initiated by the Federal Bureau of Investigation (FBI) and 14 participating state and local laboratories. Subsequently in 1994, the Congressional DNA Identification Act authorized the FBI to establish a national DNA database including "DNA identification profiles of persons convicted of crimes, and analyses of DNA samples recovered from crime scenes and from unidentified human remains." By 1997, 13 STR loci were selected and in 1998 were implemented as the core loci for the national database, known today as CODIS. All 50 states, the District of Columbia, the federal laboratories, the US Army Criminal Investigation Laboratory, and Puerto Rico contribute to CODIS.

CODIS has three hierarchical levels: the Local DNA Index System (LDIS), the State DNA Index System (SDIS), and the National DNA Index System (NDIS) (Figure 24.1). The LDIS is maintained at crime laboratories operated by police departments, sheriff's offices, and local agencies. All forensic DNA profiles originating at the local level are stored in the LDIS and are transmitted to the SDIS and NDIS. Each state maintains a SDIS, which is typically operated by a designated state laboratory. An SDIS also stores the DNA profiles generated from state laboratories. The quality assurance standards for a qualified laboratory were set up by the DNA Advisory Board in 1998. The periodic revision of the standards is now carried out by the

Table 24.1	Characteristics of National DNA Databases			
Country	**Year Established**	**Suspect Entry Criteria**	**Convicted Offender Entry Criteria**	**Removal Criteria**
United Kingdom	1995	Any recordable offense that leads to imprisonment	Entered as suspect	Never removed
New Zealand	1996	No suspects entered	Relevant offense (≥7 years in prison)	Never removed unless conviction quashed
Austria	1997	Any recordable offense that leads to imprisonment	Entered as suspect	Only after acquittal
The Netherlands	1997	No suspects entered except when suspect's DNA is tested for case	Offense leading to >4 years in prison	20–30 years after conviction
Germany	1998	Offense leading to >1 year in prison	After court decision	After acquittal or 5–10 years after conviction if prognosis good
Slovenia	1998	Any recordable offense that leads to imprisonment	Entered as suspect	Depends on severity of crime
United States	1998	No suspects entered; under revision	Depends on state law	Depends on state law
Finland	1999	Offense leading to >1 year in prison	Entered as suspect	Only after acquittal
Sweden	2000	No suspects entered	Offense leading to >2 years in prison	10 years after release from prison
Switzerland	2000	Any recordable offense that leads to imprisonment	Entered as suspect	After acquittal or 5–30 years after conviction
France	2001	No suspects entered	Sexual assaults and serious crimes	40 years after conviction

Source: Adapted from Jobling, M.A. and Gill, P., *Nat Rev Genet*, 5, 739–751, 2004.

Scientific Working Group of DNA Analysis and Methods (SWGDAM). The SDIS enables local laboratories within that state to compare DNA profiles. SDIS also serves as the communication path through which the LDIS and NDIS are able to exchange messages. Communication is mediated using a secured network with encryption. Only authorized personnel, approved by the FBI, have access to CODIS network terminals and servers. NDIS is the highest level of the CODIS infrastructure, which contains DNA profiles contributed by participating local and state

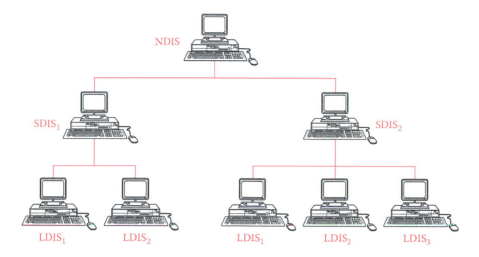

Figure 24.1 CODIS infrastructure. All DNA profiles originate at LDIS, and then enter SDIS and NDIS. SDIS allows laboratories within states to exchange DNA profiles. NDIS is the highest level of the infrastructure. It allows the participating laboratories to exchange DNA profiles on a national level.

laboratories. Additionally, NDIS stores DNA profiles generated by federal laboratories, such as those generated by the FBI and the US Army Criminal Investigation Laboratory. NDIS enables qualified state laboratories to compare DNA profiles. Searches of DNA profiles can be conducted at the national level. All DNA profiles submitted to NDIS are automatically searched weekly against the DNA profiles from other states. NDIS is administered by the FBI, which provides software, training, and support for all participating laboratories. Many law enforcement laboratories from other countries utilize the CODIS software for their own databases. However, laboratories in foreign countries do not have any access to the CODIS system. Nevertheless, a search through the CODIS database may be requested either from the FBI or International Criminal Police Organization (Interpol). Conversely, a search of an international DNA database can be arranged through Interpol.

24.3 Indexes of CODIS

The DNA profiles entered in CODIS are organized into categories known as indexes (Figure 24.2). The Convicted Offender Index contains DNA profiles of individuals convicted of crimes. The Arrestee Index contains DNA profiles of arrested individuals. It varies in SDIS databases based on each state's law permitting the collection of DNA samples from arrestees. The Forensic Index contains DNA profiles, also known as forensic profiles, derived from crime scene evidence, potentially originating from perpetrators but not including suspects.

Additionally, the FBI also established the National Missing Person DNA Database (NMPDD) Program for the identification of missing and unidentified persons at the national level. The DNA profiles entered into the NMPDD are categorized into three indexes that can be searched against each other. The Missing Person Index and the Unidentified Human Remains Index contain DNA profiles from missing persons and unidentified human remains, respectively. The Biological Relatives of the Missing Person Index contains DNA profiles voluntarily contributed from relatives. This index may also store patrilineal or matrilineal DNA profiles from the relative such as a biological father, mother, or child of the missing person to assist investigations. Additionally, a pedigree chart (a diagram showing the relationship between the missing person and relatives) can be created.

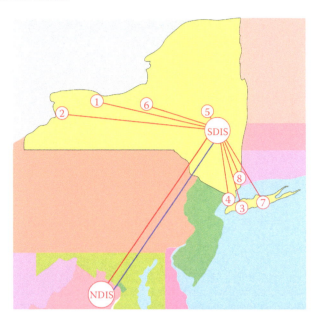

Figure 24.2 Example of an SDIS maintaining the Forensic Index (red) and Convicted Offender Index (blue). In this example, the New York State Police Forensic Investigation Center serves as the SDIS laboratory for New York State. LDIS sites: 1. Erie County Department of Central Police Services Forensic Laboratory, 2. Monroe County Public Safety Laboratory, 3. Nassau County Office of the Medical Examiner Division of Forensic Services, 4. New York City Office of the Chief Medical Examiner Department of Forensic Biology, 5. New York State Police Forensic Investigation Center, 6. Onondaga County Center for Forensic Sciences, 7. Suffolk County Crime Laboratory, and 8. Westchester County Department of Laboratories & Research Division of Forensic Sciences. (© Richard C. Li.)

In 2010, the FBI established a Rapid DNA Program Office for the purpose of developing rapid DNA technology. Rapid DNA technology is a fully automated process of performing STR analysis within 1–2 h to generate a CODIS core STR profile from a reference sample such as a buccal swab (Figure 24.3). The Rapid DNA Index System (RDIS) is the proposed index of NDIS. It shall be an integrated system capable of applying rapid DNA technology and carrying out database searches from police custody or booking units by trained police officers. The entire process, including obtaining any match from a database search, shall take less than 2 h.

The forensic DNA analysis of a reference sample, taken from an individual, takes 2–3 days if processed immediately in a forensic laboratory. The entire process includes the extraction of DNA from a sample, DNA quantification, PCR amplification, electrophoresis, and data collection and analysis. After being transported to a laboratory for processing, samples are usually stored and batched in laboratories prior to analysis. A typical turnaround time for analysis of such a sample is 1–3 months. While the sample is processed, the suspect is often released from custody. Perpetrators released on police bail may commit another crime. Thus, it is desirable to develop new technology that is capable of completing the forensic DNA analysis and database search while the suspect remains in custody. This technology would also facilitate rapid exclusion of a suspect, thus redirecting the investigation.

Rapid DNA instruments are portable, compact instruments designed to be deployed into field testing. These instruments are fully automated for processing reference samples in order to generate a DNA profile in less than 2 h. Several versions of the instruments have recently been developed by manufacturers and some of them are currently commercially available. In addition to rapid DNA instruments, rapid DNA profiling may be achieved through alternative processes. First, it can be achieved through rapid services that offer a quick turnaround time (<2 h). Second,

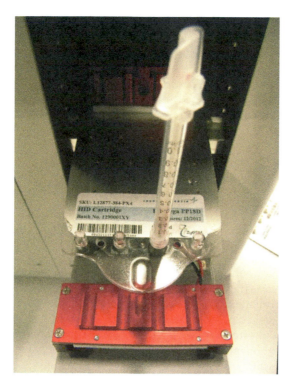

Figure 24.3   A rapid DNA instrument. A compact instrument that is designed to process swab samples and to produce DNA profiles within 2 h in the field. (© Richard C. Li.)

rapid DNA techniques can be applied using standard laboratory equipment but implementing specialized protocols to generate CODIS-compatible profiles in less than 2 h.

The primary goal of the rapid DNA initiative is to produce a CODIS-compatible DNA profile of a sample taken from an arrestee in police custody and to search DNA databases during the booking process in less than 2 h. This technology could be used, for example, to quickly identify other unsolved crimes or eliminate suspects, which can aid in investigative decision making. It can also potentially be useful for other investigations, such as identifying human remains in mass disasters; identifying detainees in counterterrorism applications; and, for immigration and border agents, confirming and verifying individual identifications or family relationships.

24.4 Database Entries

Currently, over 190 public law enforcement laboratories participate in CODIS across the United States. As of April 2013, CODIS contains the DNA profiles of more than 10 million convicted felons, over 1 million arrestees, and half a million crime scene samples. The proportion of the population represented in the database is approximately 3%. The NDIS is one of the largest DNA databases containing DNA profiles in terms of absolute numbers. Each CODIS entry consists of the DNA profile of the sample and the specimen identification number, as well as the information of the laboratory submitting the DNA profile and the laboratory analyst that performed the DNA analysis. The entry does not include case information or the personal information of the offenders or arrestees. Access to CODIS profiles is restricted to criminal justice agencies for law enforcement identification purposes.

The SDIS database retains samples collected from convicted offenders after DNA profiles are obtained. State policies vary in the retention of samples and DNA profiles in situations of dismissed charges, acquittal, or no charges for arrestees. Some states, such as Virginia, destroy the DNA sample and expunge the DNA profile from the database. Other states, such as California, require a petition to destroy the sample and expunge the DNA profile from the database. There are two major reasons for sample retention. First, the sample is needed for confirmatory processes for the purpose of quality assurance and control. In confirmatory processes, the DNA analysis of the original sample is repeated and compared against the prior analysis. Second, the retention of samples allows for possible retesting if new technology becomes available. It also allows retesting of a sample for purposes of updating with expanded loci. Considerable debate has surrounded the retention of samples. It can be argued from the opposite perspective that the database could reveal private genetic information that could then be misused. The objection is based on concern for the protection of privacy rights. If a sample were made available to an unauthorized person, confidential information could be disclosed.

24.5 Database Expansion

Currently, all 50 states have authorized the collection of samples from convicted felons for DNA databasing. Over the years, the demand for the utilization of databases has increased sharply (Figure 24.4). More jurisdictions are incorporating more felonies into the lists of crimes that require DNA profiles, and some jurisdictions plan to include all felonies in such databases. Additionally, the database system is projected to include the profiles of minor criminals, because statistics show that most offenders found guilty of serious crimes were previously convicted for minor crimes. Broadening the size of the database and including samples from more types of crimes could lead to the assumption that the number of crimes solved would also increase. Although state laws vary, each state has its own statute governing the entry criteria of the database samples. More and more states are authorizing the collection of additional types of DNA samples, including individuals convicted of misdemeanor crimes and adult felony arrestees who have not yet been convicted for the offense. It is known that offenders tend to commit multiple

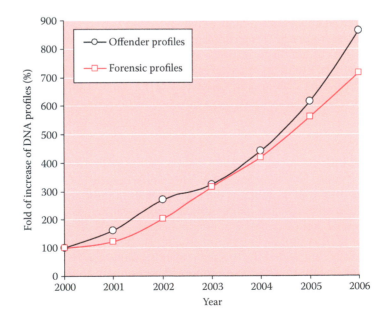

Figure 24.4 Growth of the CODIS database. In 2000, 460,365 offender profiles and 22,484 forensic profiles were entered.

crimes. Thus, including DNA profiles of arrestees in the database is beneficial in catching serial offenders and potentially preventing future crimes. Virginia is the first state to collect DNA from violent-felony arrestees, among other crimes. Over the years, more than half of states in the United States have amended their law to include all felony arrestees or a subset of arrestees for felonies that involve violence or sexual assault.

In 2009, Alonzo King was arrested in Wicomico County, Maryland, on violent assault charges. Maryland's DNA Collection Act authorizes law enforcement agencies to collect DNA samples from an individual who is arrested for violent crimes. During the booking process, under state law, King's buccal swabs were collected for forensic DNA identification. When King's DNA profile was entered into Maryland's SDIS, it matched to a crime scene DNA profile from an unsolved rape case in 2003. Based on the DNA evidence, King was indicted for first-degree rape by a grand jury. The defendant filed a motion to suppress the DNA evidence. King argued that using his DNA to investigate his connection to the 2003 rape was an unreasonable search and seizure under the Fourth Amendment, since the police had no reason to suspect his involvement in the 2003 rape. During the trial at Wicomico County Circuit Court, King's motion was denied. He was convicted of rape and sentenced to life in prison. The Court of Appeals of Maryland, the state's highest court, upheld that the DNA Collection Act was constitutional. The police had probable cause to arrest King on the assault charge, and thus collecting King's DNA sample was a reasonable search. However, the court overturned the low court and reversed King's conviction. It stated that the DNA Collection Act was inappropriately applied to King in this case. Thus, investigating King without probable cause for the unrelated rape was unconstitutional. To determine whether the Fourth Amendment permits states to collect and analyze the DNA of arrestees, a divided Supreme Court ruled 5–4 and reversed the Court of Appeals. The Supreme Court's judgment ruled that Maryland has the right to collect DNA evidence from individuals arrested for serious crimes. The ruling reflects the Supreme Court's view on the balance between the interest of criminal justice systems in solving violent crimes and an individual's interest in the Fourth Amendment, which protects them from unreasonable searches.

Some jurisdictions, such as the United Kingdom, allow DNA samples to be taken from individuals suspected of committing recordable offenses that may lead to prison sentences. In contrast, DNA profiles from suspects are not eligible for entry into CODIS. In some jurisdictions, it has been suggested that databases include many more offenders and suspects, as well as the general public. One advantage of including an entire population in a database is the ability to identify missing, kidnapped, and abducted individuals in addition to victims of major accidents and mass fatalities. Nevertheless, debates concerning the need to balance the benefits and dangers of developing a broader database will inevitably continue.

24.6 DNA Profiles

Currently, the CODIS software supports the storage and search of DNA profiles of short tandem repeat (STR), Y chromosome STR (Y-STR), and mitochondrial DNA (mtDNA). Y-STR and mtDNA profiles may only be searched within NMPDD-related indexes. The CODIS software no longer supports searches of DNA profiles generated by restriction fragment length polymorphism (RFLP) analysis. The 13 core CODIS STR loci are CSF1PO, FGA, THO1, TPOX, VWA, D3S1358, D5S818, D7S820, D8S1179, D13S317, D16S539, D18S51, and D21S11. DNA profiles are entered into one of the indexes, such as convicted offender, arrestee, forensic, unidentified human remains, missing person, or a relative of a missing person. There is a minimum loci requirement for each DNA profile entering CODIS: 13 core CODIS loci are required for a DNA profile entering the Convicted Offender Index and the Arrestee Index; 13 core CODIS loci and amelogenin, a sex-typing marker, are required for a DNA profile entering the Relatives of Missing Person Index; at least 10 CODIS loci are required (all 13 core loci must be tested) for the Forensic Index; and at least 8 loci and amelogenin are required (all 13 core loci must be tested) for the Missing Person and Unidentified Human Remains Index. The DNA profiles must be generated by an accredited and audited laboratory in accordance with the FBI's Quality Assurance Standards using approved commercially available kits. As the database is growing rapidly, the chances of finding incidental matches among DNA profiles is increasing. To increase discrimination power, it has recently been proposed that CODIS core loci be expanded to include 20 or more markers (Chapter 20). Additionally, the expanded core loci share additional loci compatible with international standards for forensic DNA analysis, which facilitates the exchange of information with international law enforcement agencies.

24.7 Routine Database Searches for Forensic Investigations

The ultimate goal of the database utilization is to provide investigative leads for law enforcement in solving crimes, particularly in cases where no suspect has yet been identified. Currently, DNA profiles uploaded to NDIS are automatically searched once a week (Figure 24.5). A hit is a match made from the information provided by comparing a target DNA profile against the DNA profiles contained in the database. There are two types of CODIS hits: an *offender hit* provides the identity of a potential suspect of a crime, while a *forensic hit* reveals the linkage between two or more crime scenes. Once a hit is identified, the match is verified by the laboratories that originally processed the evidence. A verified CODIS hit can be utilized as probable cause to allow law enforcement to obtain a court order to collect a DNA sample from a suspect. Collecting DNA with a warrant ensures admissibility in court. *Investigation-aided* cases are those assisted by CODIS hits, including case-to-case matches as well as case-to-offender matches. The number of investigations aided is a useful measure of the successful application of the database (Figure 24.6). As of April 2013, over 207,800 hits had been made with CODIS assistance in more than 199,200 investigations.

24.7.1 Case-to-Offender Searches

Matches of profiles from the Forensic Index and the Offender Index reveal the identities of perpetrators of crimes. For example, in 1998, a Florida man, Leon Dundas, became a suspect in a rape case but refused to provide his DNA reference sample for testing. Dundas was killed a year later in an illegal drug deal. Thus, a postmortem sample of Dundas was obtained and his DNA profile was compared with the Forensic Index of CODIS. It was discovered that Dundas's profile

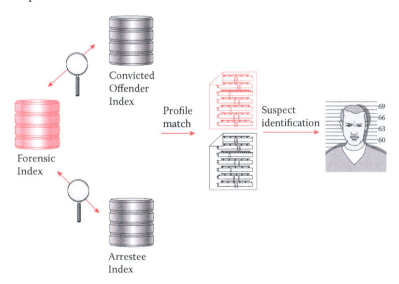

Figure 24.5   Example of a weekly routine CODIS search. (© Richard C. Li.)

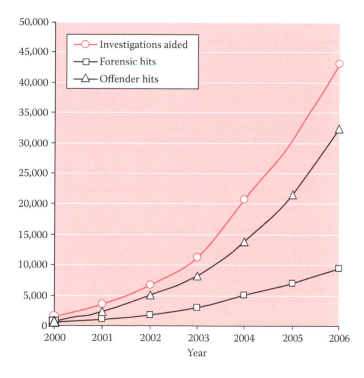

Figure 24.6   Number of hits and investigations aided by the CODIS database.

matched the DNA sample found in the rape case. Additionally, his DNA profile was linked to several other unsolved rapes in Jacksonville and Washington, DC. Such cases can now be solved by utilizing DNA databases.

24.7.2 Case-to-Case Searches

A target DNA profile from a crime scene is also searched against the profiles stored in the Forensic Index of the database. Matches of profiles among the target profile and the profile in the Forensic Index can link separate crime scenes and aid in identifying serial offenders. This helps law enforcement agencies in multiple jurisdictions to coordinate their investigations and share leads. For example, in 1996, two young girls were abducted from bus stops in St. Louis. Both girls were raped and DNA samples were collected. Both DNA profiles pointed to the same perpetrator. In 1999, the St. Louis police decided to reanalyze the samples using new STR technology through the CODIS database. The database found a match to a different rape case, to which the perpetrator, Dominic Moore, had already confessed, thus identifying him as the perpetrator of the 1996 rapes.

24.7.3 Search Stringency and Partial Matches

Database searches are carried out using the CODIS software with three stringency levels, which allow the search of complex forensic profiles against offender profiles (Figure 24.7). A high-stringency match requires an exact match in which all alleles are matched at each locus between the target and candidate profiles. A moderate-stringency match, as defined by the FBI, is a candidate match "between two single source profiles having at each locus all of the alleles of one sample represented in the other sample." In a moderate-stringency match, allelic drop-outs in a target or candidate profile are allowed, possibly resulting in a partial match at some loci. As a result, high-stringency matches are automatically included in a moderate-stringency

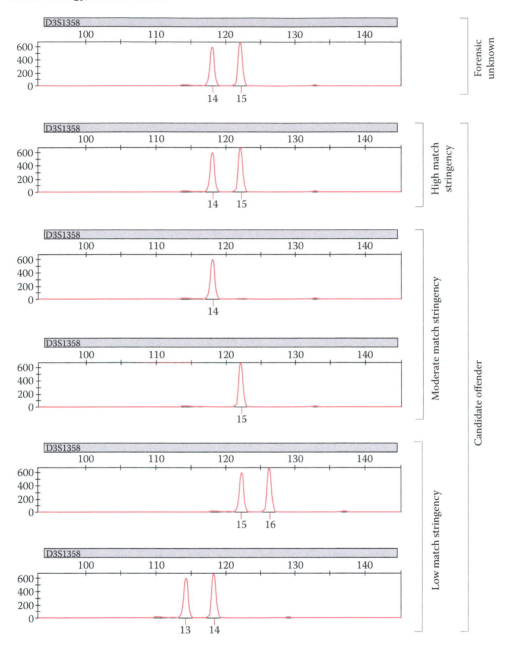

Figure 24.7   Diagram of match stringencies. A heterozygote (allele 14 and 15) at locus D3S1358 is utilized as the genotype for the forensic unknown profile. (© Richard C. Li.)

search. In a low-stringency search, both mismatches and allelic dropouts are allowed. Presently, NDIS searches are only carried out at moderate stringency. During forensic DNA analysis, DNA degradation may prevent full DNA profiles from being processed, producing only partial profiles. Additionally, mixture profiles derived from forensic samples containing DNA contributed by more than one individual may be encountered. Furthermore, due to mutations, null alleles may occur in the profiles produced with some primer sets but not other primers. Therefore, searching at moderate stringency allows the detection of matches under the situations described above.

Such moderate-stringency searches between the Forensic Index and Offender Index may occasionally generate partial match profiles. Since a partial match is not an exact match of the two profiles, further investigation is needed. For instance, additional Y-STR and mtDNA analysis is needed to eliminate unrelated individuals.

24.8 Familial Searches

Familial search, initiated in the United Kingdom, is a new method of applying databases in criminal investigations. It is known that, in the United States, nearly half of prison inmates have close relatives who have also been incarcerated. Familial search is based on the assumption that close relatives share more alleles of DNA profiles than unrelated individuals. Thus, databases may be utilized to identify perpetrators by finding a close relative, if the close relative has been convicted of a crime and is listed in the database. Familial search is an intentional search of a target crime scene profile against an offender database to obtain a list of candidate profiles that are similar to the target profile. This list may include the profile of a close relative of the perpetrator, who is the source of crime scene evidence. These matches most frequently involve siblings, parents, or children. The investigative leads produced by familial searches allow law enforcement to conduct further investigations to identify the perpetrator.

The first familial search leading to a successful prosecution was conducted in Surrey, England. In 2003, a truck driver was killed after a brick was thrown through his windshield from a bridge. The perpetrator's DNA profile was obtained from the brick. A search of the UK's national DNA database revealed no match for the perpetrator. Next, a familial search of the database was conducted. The system identified a close relative that led police to identify the perpetrator, Craig Harman, who was then convicted of manslaughter.

24.8.1 Legal and Ethical Issues of Familial Search

The use of forensic databases involves a balance of individual civil rights and the interests of the criminal justice systems. Many concerns have been raised, including the potential for these searches to violate the privacy of unrelated people whose genetic profiles happen to resemble those of individuals included in the databases. In the United States, the Fourth Amendment protects against unreasonable searches and seizures. The permissibility of familial searching under the Fourth Amendment is yet to be addressed by courts. The collection of biological material for the initial creation of a profile for law enforcement purposes is subject to the Fourth Amendment implications. In terms of familial searches, some legal experts argue that the biological materials are not collected directly from the individuals for a familial search, and that these individuals may thus be protected under the Fourth Amendment. Others argue, however, that the Fourth Amendment protects the initial creation of the profile, including the sample collection, forensic DNA analysis, and the databasing. They argue that the Fourth Amendment may not protect subsequent investigations of DNA profiles during familial searches. In addition, the CODIS database consists of a high percentage of profiles from individuals of racial minority groups, including African Americans. Familial searches can disproportionately focus on a specific racial group.

Familial searches are rare in the United States. Two jurisdictions, Maryland and the District of Columbia, have laws prohibiting the use of familial searches. Familial searches are not conducted at the NDIS. While familial searches are now being performed in several jurisdictions in the United States, policies on familial searches vary among jurisdictions. The major issues relate to criteria for privacy, information release, search approval, and the types of crimes eligible for familial searches. Currently, California, Colorado, Texas, and Virginia have state legislation permitting familial searches. Familial searches are initiated when a specific suspect is not known, and the cause for the search needs to be justified. Familial searches are usually conducted for crimes that pose a substantial threat to public safety, typically those cases involving the most serious offenses. Additionally, familial searches are only conducted after a routine search of a DNA profile has yielded no match in the database.

In 2003, for example, a crime scene DNA sample of a closed rape and murder case, committed decades previously, was reanalyzed using forensic DNA techniques. The crime scene DNA profile was compared to the profile of a man, Darryl Hunt, who was then imprisoned for the crime. Surprisingly, the crime scene DNA profile did not match either that of Hunt or those of convicted felons in the database. However, the database search revealed a close relative of the true perpetrator and thus led law enforcement to identify Willard Brown as the perpetrator. Brown was sentenced to life imprisonment. Additionally, Hunt was exonerated after 18 years in prison.

24.8.2 Familial Search Strategies

The familial search is usually carried out using specially designed software. Although the CODIS software is not designed for familial search, it can be used for familial search through a low-stringency search, which may result in a list of candidate profiles including close relatives of individuals, such as parent–offspring or full-sibling relatives. In a large offender DNA database, similar DNA profiles from unrelated people are often observed due to shared alleles. Therefore, the candidate profile list may also include unrelated individuals whose DNA profiles are similar to the target profile. As a result, a familial search can provide a list of potential candidates consisting of hundreds of profiles, which would be too labor intensive to pursue through further investigations. Several methods can be used to determine a cutoff analytical threshold in order to limit a pool of candidate profiles and exclude unrelated individuals from familial searches (Figure 24.8).

24.8.2.1 Identity-by-State and Kinship Index Method

The *identity-by-state* (IBS) method compares the number of shared alleles and loci between a target forensic profile and the offender profiles in a database but does not take into account allele frequencies. The analytical threshold for a familial search is determined by a preset number of shared alleles or loci in order to prompt further investigation. For example, some states use 15 shared alleles as the analytical threshold to be considered as a candidate, while other states require at least one shared allele for each locus. Additionally, this method can prioritize a pool of candidates based on the highest to the lowest number of shared alleles for investigation.

The *Kinship Index* (KI) method is a likelihood ratio–based method that evaluates the familial match by comparing the probability that two DNA profiles are from related individuals to the probability that they are unrelated. The KI method analyzes the allele frequency data, including all CODIS core loci, to calculate the Combined Kinship Index. The KI may vary based on the allele frequency data across the population. Thus, the accuracy of the KI method relies on the relevance of the population data set analyzed. False inclusions or exclusions may occur if nonrelevant population data are utilized. The KI method also allows the generation of a ranked candidate list according to the probability that the individuals are related. Generally speaking, the accuracy of the KI method is higher than that of the IBS method. However, using both IBS and KI methods is better than using a single method alone for familial search.

24.8.2.2 Focusing on Rare Alleles

During a familial search, target DNA profiles may contain rare alleles with low allelic frequencies. The chance of sharing rare alleles for two close relatives is higher than for two unrelated individuals. If a target profile carries a rare allele, it can be used to narrow the pool of candidates.

For example, Jeffrey Gafoor, then 23, had been living in a suburban neighborhood in Cardiff, Wales. He had a reputation for being a loner. Besides working in his family-owned shop, he spent most of his time at home. On February 13, 1988, Gafoor entered 7 James Street in the Butetown area of Cardiff. The first-floor unit was temporarily occupied by Lynette White, a 20-year-old

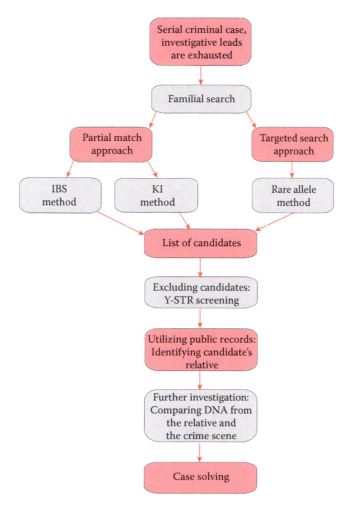

Figure 24.8   Example of a familial search workflow. (© Richard C. Li.)

woman who was working as a prostitute in the area. Gafoor went there for the purpose of receiving sexual services. He left the scene in the early hours of Valentine's Day. On Valentine's night, White's body was discovered. She suffered from more than 50 stab, cut, and slash wounds as well as defensive wounds on her hands. The body was dragged to the corner of the room adjacent to the bed, the only piece of furniture in the room.

Gafoor left blood and semen evidence at the scene. He bled during the murder and deposited bloodstains at the scene. Additionally, he left seminal stains at the scene; the semen had no measurable level of sperm and most likely came from an individual who had a medical condition known as azoospermia (Chapter 14). After the murder, Gafoor lived at the same place as he did beforehand and kept the same job for many years.

Two years later, three local men, known as the "Cardiff Three," were convicted of White's murder and sentenced to life in prison. In 1992, however, their convictions were reversed by the court of appeal. Subsequently, the crime scene evidence was tested using new techniques in forensic DNA analysis that were not available in 1988. A DNA profile was obtained from the crime scene. A familial search revealed a similar profile from a 14-year-old boy in the UK National DNA Database. In particular, there was a rare allele match between the crime scene and the candidate DNA profiles. This search led to the further investigation of Gafoor, who was the uncle of the boy. Gafoor was charged with murder.

24.8.2.3 Excluding Candidates through Y-STR Screening

The vast majority of DNA profiles in CODIS come from males. Patrilineal markers, such as Y-STR loci, are successfully utilized to screen the candidates to verify the relationship between two individuals. This analysis helps to identify first-order relatives as well as paternal half-siblings. STR loci beyond the CODIS core loci and maternal lineage makers, such as mitochondrial DNA typing systems, can be used to narrow the pool of candidates and eliminate coincidental partial matches. One or more of these methods can be incorporated to evaluate highly ranked candidates.

For example, Lonnie Franklin lived in a southern Los Angeles neighborhood. Throughout the 1980s, Franklin worked for the city as a maintenance worker and a sanitation truck operator. His neighbors recalled that he often brought prostitutes to a camper parked in the backyard of his house. He also took photos of nude women, which he kept in his garage. Nevertheless, to many of his neighbors, Franklin seemed to be a friendly person who often chatted with them.

Since 1985, the bodies of many women, most of which were prostitutes, had been dumped in an alley running along Western Avenue in a southern Los Angeles neighborhood. The victims had been shot and some had been strangled after sexual contact. DNA and ballistics analysis revealed that at least 10 of the murders had been committed by the same perpetrator. The perpetrator was known to the general public as the "Grim Sleeper" because he had taken a 14-year break during the period in which these 10 murders had taken place. The crime scene DNA profile did not match any profile in the database.

In 2008, the first familial search of the "Grim Sleeper" case was conducted by the California Department of Justice. California state law allows familial searches for high-profile cases if all other leads have been exhausted. However, all candidate DNA profiles were excluded through Y-STR screening. As a result, no candidate was identified as a possible relative of the "Grim Sleeper." In 2010, a second search targeted partial matches that shared at least 15 alleles. It produced a list of 100 candidate profiles ranked by the likelihood that they were related. Y-STR was used as a screen tool to identify possible patrilineal relatives. This time, the search generated a match to a new DNA profile of a felon, which had recently been entered into California's SDIS, in 2009. The search suggested that the candidate was a relative of the source of the DNA from the crime scenes. This search led to the identification of Franklin as a suspect in the "Grim Sleeper" murders. Police detectives conducted surveillance and collected a discarded slice of pizza and a cup used by Franklin at a local restaurant for DNA identification. The DNA profile was identical to the crime scene DNA profiles. Franklin was arrested and was charged with ten counts of murder and one of attempted murder.

Bibliography

Alonso, A., et al., Challenges of DNA profiling in mass disaster investigations. *Croat Med J*, 2005, **46**(4): 540–548.

Aoki, Y., et al., Comparison of the likelihood ratio and identity-by-state scoring methods for analyzing sib-pair test cases: A study using computer simulation. *Tohoku J Exp Med*, 2001, **194**(4): 241–250.

Asplen, C. and S.A. Lane, International perspectives on forensic DNA databases. *Forensic Sci Int*, 2004, **146**(Suppl): S119–S121.

Baker, L.E. and E.J. Baker, Reuniting families: An online database to aid in the identification of undocumented immigrant remains. *J Forensic Sci*, 2008, **53**(1): 50–53.

Balding, D.J., The DNA database search controversy. *Biometrics*, 2002, **58**(1): 241–244.

Balding, D.J., et al., Decision-making in familial database searching: KI alone or not alone? *Forensic Sci Int Genet*, 2013, **7**(1): 52–54.

Barash, M., et al., A search for obligatory paternal alleles in a DNA database to find an alleged rapist in a fatherless paternity case. *J Forensic Sci*, 2012, **57**(4): 1098–1101.

Barber, M.D. and B.H. Parkin, Sequence analysis and allelic designation of the two short tandem repeat loci D18S51 and D8S1179. *Int J Legal Med*, 1996, **109**(2): 62–65.

Benschop, C.C., et al., Low template STR typing: Effect of replicate number and consensus method on genotyping reliability and DNA database search results. *Forensic Sci Int Genet*, 2011, **5**(4): 316–328.

Bereano, P.L., DNA identification systems: Social policy and civil liberties concerns. *J Int Bioethique*, 1990, **1**(3): 146–155.

Bieber, F.R. and D. Lazer, Guilt by association: Should the law be able to use one person's DNA to carry out surveillance on their family? Not without a public debate. *New Sci*, 2004, **184**(2470): 20.

Bieber, F.R., Turning base hits into earned runs: Improving the effectiveness of forensic DNA data bank programs. *J Law Med Ethics*, 2006, **34**(2): 222–233.

Bingham, R., A database of the innocent? *Splice Life*, 2001, **7**(2–3): 8–9.

Birmingham, K., UK leads on science and technology bioethics. *Nat Med*, 1998, **4**(6): 651–652.

Bradford, L., et al., Disaster victim investigation recommendations from two simulated mass disaster scenarios utilized for user acceptance testing CODIS 6.0. *Forensic Sci Int Genet*, 2011, **5**(4): 291–296.

Buckleton, J., J.A. Bright, and S.J. Walsh, Database crime to crime match rate calculation. *Forensic Sci Int Genet*, 2009, **3**(3): 200–201.

Budowle, B., et al., CODIS and PCR-based short tandem repeat loci: Law enforcement tools, in *Second European Symposium on Human Identification*, 1998. Innsbruck, Austria: Promega Corporation.

Burk, D.L., DNA identification testing: Assessing the threat to privacy. *Univ Toledo Law Rev*, 1992, **24**(1): 87–102.

Burk, D.L. and J.A. Hess, Genetic privacy: Constitutional considerations in forensic DNA testing. *Geoge Mason Univ Civ Rights Law J*, 1994, **5**(1): 1–53.

Cash, H.D., J.W. Hoyle, and A.J. Sutton, Development under extreme conditions: Forensic bioinformatics in the wake of the World Trade Center disaster. *Pac Symp Biocomput*, 2003, **8**: 638–653.

Chow-White, P.A. and T. Duster, Do health and forensic DNA databases increase racial disparities? *PLoS Med*, 2011, **8**(10): e1001100.

Chung, Y.K. and W.K. Fung, Identifying contributors of two-person DNA mixtures by familial database search. *Int J Legal Med*, 2013, **127**(1): 25–33.

Chung, Y.K., Y.Q. Hu, and W.K. Fung, Evaluation of DNA mixtures from database search. *Biometrics*, 2010, **66**(1): 233–238.

Corte-Real, F., Forensic DNA databases. *Forensic Sci Int*, 2004, **146**(Suppl): S143–S144.

Curran, J.M. and J.S. Buckleton, Effectiveness of familial searches. *Sci Justice*, 2008, **48**(4): 164–167.

Dawid, A.P., Comment on Stockmarr's "Likelihood ratios for evaluating DNA evidence when the suspect is found through a database search". *Biometrics*, 2001, **57**(3): 976–980.

Duster, T., Explaining differential trust of DNA forensic technology: Grounded assessment or inexplicable paranoia? *J Law Med Ethics*, 2006, **34**(2): 293–300.

Evett, I.W., J. Scranage, and R. Pinchin, Efficient retrieval from DNA databases: Based on the second European DNA Profiling Group collaborative experiment. *Forensic Sci Int*, 1992, **53**(1): 45–50.

Evison, M., DNA database could end problem of identity fraud. *Nature*, 2002, **420**(6914): 359.

Gabriel, M., C. Boland, and C. Holt, Beyond the cold hit: Measuring the impact of the national DNA data bank on public safety at the city and county level. *J Law Med Ethics*, 2010, **38**(2): 396–411.

Gamero, J.J., et al., A study of Spanish attitudes regarding the custody and use of forensic DNA databases. *Forensic Sci Int Genet*, 2008, **2**(2): 138–149.

Ge, J., et al., Comparisons of familial DNA database searching strategies. *J Forensic Sci*, 2011, **56**(6): 1448–1456.

Gershaw, C.J., et al., Forensic utilization of familial searches in DNA databases. *Forensic Sci Int Genet*, 2011, **5**(1): 16–20.

Gill, P., et al., Databases, quality control and interpretation of DNA profiling in the hHome Office fForensic sScience sService. *Electrophoresis*, 1991, **12**(2–3): 204–209.

Gill, P., et al., A comparison of adjustment methods to test the robustness of an STR DNA database comprised of 24 European populations. *Forensic Sci Int*, 2003, **131**(2–3): 184–196.

Gill, P., et al., The evolution of DNA databases—Recommendations for new European STR loci. *Forensic Sci Int*, 2006, **156**(2–3): 242–244.

Gosline, A., Will DNA profiling fuel prejudice? *New Sci*, 2005, **186**(2494): 12–13.

Guillen Vazquez, M., C. Pestoni, and A. Carracedo, DNA databases for criminal investigation purposes: Technical aspects and ethical–legal problems. *Law Hum Genome Rev*, 1998, (8): 137–158.

Guillen, M., et al., Ethical-legal problems of DNA databases in criminal investigation. *J Med Ethics*, 2000, **26**(4): 266–271.

Haimes, E., Social and ethical issues in the use of familial searching in forensic investigations: Insights from family and kinship studies. *J Law Med Ethics*, 2006, **34**(2): 263–276.

Harbison, S.A., J.F. Hamilton, and S.J. Walsh, The New Zealand DNA databank: Its development and significance as a crime solving tool. *Sci Justice*, 2001, **41**(1): 33–37.

Harding, H.W. and R. Swanson, DNA database size. *J Forensic Sci*, 1998, **43**(1): 248–249.

Hedman, J., et al., A fast analysis system for forensic DNA reference samples. *Forensic Sci Int Genet*, 2008, **2**(3): 184–189.

Heninger, M. and R. Hanzlick, The Fulton County Medical Examiner's experience with the Federal Bureau of iInvestigation National Missing Person DNA Database Program, 2004–2007. *Am J Forensic Med Pathol*, 2011, **32**(1): 78–82.

Hepple, B., Forensic databases: Implications of the cases of S and Marper. *Med Sci Law*, 2009, **49**(2): 77–87.

Herkenham, M.D., Retention of offender DNA samples necessary to ensure and monitor quality of forensic DNA efforts: Appropriate safeguards exist to protect the DNA samples from misuse. *J Law Med Ethics*, 2006, **34**(2): 380–384.

Hibbert, M., DNA databanks: Law enforcement's greatest surveillance tool? *Wake Forest Law Rev*, 1999, **34**(3): 767–825.

Hicks, T., et al., Use of DNA profiles for investigation using a simulated national DNA database: Part I. Partial SGM Plus profiles. *Forensic Sci Int Genet*, 2010, **4**(4): 232–238.

Hicks, T., et al., Use of DNA profiles for investigation using a simulated national DNA database: Part II. Statistical and ethical considerations on familial searching. *Forensic Sci Int Genet*, 2010, **4**(5): 316–322.

Hoyle, R., Forensics. The FBI's national DNA database. *Nat Biotechnol*, 1998, **16**(11): 987.

Huang, D., et al., Y-haplotype screening of local patrilineages followed by autosomal STR typing can detect likely perpetrators in some populations. *J Forensic Sci*, 2011, **56**(5): 1340–1342.

Jobling, M.A. and Gill, P., Encoded evidence: DNA in forensic analysis. *Nat Rev Genet*, 2004, **5**: 739–751.

Karjala, D.S., A legal research agenda for the human genome initiative. *Jurimetrics*, 1992, **32**(2): 121–222.

Kim, J., et al., Policy implications for familial searching. *Investig Genet*, 2011, **2**: 22.

Kimel, C.W., DNA profiles, computer searches, and the Fourth Amendment. *Duke Law J*, 2013, **62**(4): 933–973.

Kirby, R., The many dangers of relying on a DNA database. *Nature*, 2002, **419**(6904): 247.

Lakra, R. and F. Dimitriadis, Forensic DNA data banks: Considerations for the health sector. *Health Law Can*, 2002, **22**(3): 57–70.

Lehrman, S., Prisoner's DNA database ruled unlawful. *Nature*, 1998, **394**(6696): 818.

Lehrman, S., Partial to crime: Families become suspects as rules on DNA matches relax. *Sci Am*, 2006, **295**(6): 28–29.

Levitt, M., Forensic databases: Benefits and ethical and social costs. *Br Med Bull*, 2007, **83**: 235–248.

Linacre, A., The UK National DNA Database. *Lancet*, 2003, **361**(9372): 1841–1842.

Love, S.H., Allowing new technology to erode constitutional protections: A Fourth Amendment challenge to non-consensual DNA testing of prisoners. *Villanova Law Rev*, 1993, **38**(5): 1617–1660.

Machado, H., F. Santos, and S. Silva, Prisoners' expectations of the national forensic DNA database: Surveillance and reconfiguration of individual rights. *Forensic Sci Int*, 2011, **210**(1–3): 139–143.

Maguire, C.N., et al., Familial searching: A specialist forensic DNA profiling service utilising the national DNA database to identify unknown offenders via their relatives—The UK experience. *Forensic Sci Int Genet*, 2014, **8**(1): 1–9.

Mansel, C. and S. Davies, Minors or suspects? A discussion of the legal and ethical issues surrounding the indefinite storage of DNA collected from children aged 10–18 years on the national DNA database in England and Wales. *Med Sci Law*, 2012, **52**(4): 187–192.

Marjanovic, D., et al., Forensic DNA databases in Western Balkan region: Retrospectives, perspectives, and initiatives. *Croat Med J*, 2011, **52**(3): 235–244.

Martin, P.D., H. Schmitter, and P.M. Schneider, A brief history of the formation of DNA databases in forensic science within Europe. *Forensic Sci Int*, 2001, **119**(2): 225–231.

McEwen, J.E. and P.R. Reilly, A review of state legislation on DNA forensic data banking. *Am J Hum Genet*, 1994, **54**(6): 941–958.

McEwen, J.E., Forensic DNA data banking by state crime laboratories. *Am J Hum Genet*, 1995, **56**(6): 1487–1492.

McNamara, J.J. and R.J. Morton, Frequency of serial sexual homicide victimization in Virginia for a 10-year period. *J Forensic Sci*, 2004, **49**(3): 529–533.

Meester, R. and M. Sjerps, The evidential value in the DNA database search controversy and the two-stain problem. *Biometrics*, 2003, **59**(3): 727–732.

Melson, K.E., President's editorial—The journey to justice. *J Forensic Sci*, 2003, **48**(4): 705–707.

Mohanty, N.K. and S. Haldar, DNA test—A forensic boon. *J Indian Med Assoc*, 2006, **104**(2): 90–92, 94.

Murphy, E., The government wants your DNA. *Sci Am*, 2013, **308**(3): 72–77.

Myers, S.P., et al., Searching for first-degree familial relationships in California's offender DNA database: Validation of a likelihood ratio-based approach. *Forensic Sci Int Genet*, 2011, **5**(5): 493–500.

Nerko, C.J., Assessing Fourth Amendment challenges to DNA extraction statutes after Samson v. California. *Fordham Law Rev*, 2008, **77**(2): 917–949.

Nishimi, R.Y., et al., Genetic witness: Forensic uses of DNA tests. *J Int Bioethique*, 1991, **2**(1): 29–32.

Oz, C., M. Amiel, and A. Zamir, The Israel police DNA database: Recognizing the capability of databases in providing investigative leads. *Forensic Sci Int Genet*, 2012, **6**(1): e8–e10.

Parson, W. and M. Steinlechner, Efficient DNA database laboratory strategy for high through-put STR typing of reference samples. *Forensic Sci Int*, 2001, **122**(1): 1–6.

Pascali, V.L., G. Lago, and M. Dobosz, The dark side of the UK National DNA Database. *Lancet*, 2003, **362**(9386): 834.

Patyn, A. and K. Dierickx, Forensic DNA databases: Genetic testing as a societal choice. *J Med Ethics*, 2010, **36**(5): 319–320.

Peerenboom, E., Central criminal DNA database created in Germany. *Nat Biotechnol*, 1998, **16**(6): 510–511.

Pham-Hoai, E., F. Crispino, and G. Hampikian, The first successful use of a low stringency familial match in a French criminal investigation. *J Forensic Sci*, 2014, **59**(3): 816–819.

Raab, S., Cuomo seeks genetic data of offenders. *N Y Times*, May 10, 1992.

Reid, T.M., et al., Use of sibling pairs to determine the familial searching efficiency of forensic databases. *Forensic Sci Int Genet*, 2008, **2**(4): 340–342.

Reilly, P., Rights, privacy, and genetic screening. *Yale J Biol Med*, 1991, **64**(1): 43–45.

Rohlfs, R.V., S.M. Fullerton, and B.S. Weir, Familial identification: Population structure and relationship distinguishability. *PLoS Genet*, 2012, **8**(2): e1002469.

Rohlfs, R.V., et al., The influence of relatives on the efficiency and error rate of familial searching. *PLoS One*, 2013, **8**(8): e70495.

Rossy, Q., et al., Integrating forensic information in a crime intelligence database. *Forensic Sci Int*, 2013, **230**(1–3): 137–146.

Sankar, P., The proliferation and risks of government DNA databases. *Am J Public Health*, 1997, **87**(3): 336–337.

Scheck, B., DNA data banking: A cautionary tale. *Am J Hum Genet*, 1994, **54**(6): 931–933.

Schneider, P.M. and P.D. Martin, Criminal DNA databases: The European situation. *Forensic Sci Int*, 2001, **119**(2): 232–238.

Scott, G., M.F. Marshall, and C. Tindall, Of doctors and detectives. *N Y Times Web*, November 20, 2001, F7.

Shapiro, E.D. and M.L. Weinberg, DNA data banking: The dangerous erosion of privacy. *Clevel State Law Rev*, 1990, **38**(3): 455–486.

Shriver, M., T. Frudakis, and B. Budowle, Getting the science and the ethics right in forensic genetics. *Nat Genet*, 2005, **37**(5): 449–450; author reply 450–451.

Simoncelli, T., Dangerous excursions: The case against expanding forensic DNA databases to innocent persons. *J Law Med Ethics*, 2006, **34**(2): 390–397.

Sjerps, M. and R. Meester, Selection effects and database screening in forensic science. *Forensic Sci Int*, 2009, **192**(1–3): 56–61.

Smith, J.C., The precarious implications of DNA profiling. *Univ Pittsbg Law Rev*, 1994, **55**(3): 865–888.

Song, Y.S., et al., Average probability that a "cold hit" in a DNA database search results in an erroneous attribution. *J Forensic Sci*, 2009, **54**(1): 22–27.

Struse, H.M. and I.D. Montoya, Health services implications of DNA testing. *Clin Lab Sci*, 2001, **14**(4): 247–251.

Sullivan, K., et al., New developments and challenges in the use of the UK DNA database: Addressing the issue of contaminated consumables. *Forensic Sci Int*, 2004, **146**(Suppl): S175–S176.

Swinbanks, D., UK plans library of DNA profiles. *Nature*, 1993, **365**(6447): 596.

Taborda, J.G. and J. Arboleda-Florez, Forensic medicine in the next century: Some ethical challenges. *Int J Offender Ther Comp Criminol*, 1999, **43**(2): 188–201.

Taylor, D., et al., An investigation of admixture in an Australian Aboriginal Y-chromosome STR database. *Forensic Sci Int Genet*, 2012, **6**(5): 532–538.

Terry, S.F. and P.F. Terry, A consumer perspective on forensic DNA banking. *J Law Med Ethics*, 2006, **34**(2): 408–414.

Tvedebrink, T., et al., Analysis of matches and partial-matches in a Danish STR data set. *Forensic Sci Int Genet*, 2012, **6**(3): 387–392.

Van Camp, N. and K. Dierickx, The retention of forensic DNA samples: A socio-ethical evaluation of current practices in the EU. *J Med Ethics*, 2008, **34**(8): 606–610.

Voultsos, P., et al., Launching the Greek forensic DNA database. The legal framework and arising ethical issues. *Forensic Sci Int Genet*, 2011, **5**(5): 407–410.

Walsh, S.J., J.M. Curran, and J.S. Buckleton, Modeling forensic DNA database performance. *J Forensic Sci*, 2010, **55**(5): 1174–1183.

Walsh, S.J., et al., The collation of forensic DNA case data into a multi-dimensional intelligence database. *Sci Justice*, 2002, **42**(4): 205–214.

Werrett, D.J., The national DNA database. *Forensic Sci Int*, 1997, **88**(1): 33–42.

Williams, R. and P. Johnson, "Wonderment and dread": Representations of DNA in ethical disputes about forensic DNA databases. *New Genet Soc*, 2004, **23**(2): 205–223.

Williams, R. and P. Johnson, Inclusiveness, effectiveness and intrusiveness: Issues in the developing uses of DNA profiling in support of criminal investigations. *J Law Med Ethics*, 2005, **33**(3): 545–558.

Zamir, A., et al., The Israel DNA database—The establishment of a rapid, semi-automated analysis system. *Forensic Sci Int Genet*, 2012, **6**(2): 286–289.

Zimmermann, B., et al., Application of a west Eurasian-specific filter for quasi-median network analysis: Sharpening the blade for mtDNA error detection. *Forensic Sci Int Genet*, 2011, **5**(2): 133–137.

Evaluation of the Strength of Forensic DNA Profiling Results

25.1 A Review of Basic Principles of Genetics

25.1.1 Mendelian Genetics

Mendel's first law is the principle of segregation of alleles. Each pair of alleles segregates from others in the formation of gametes (mature reproductive cells such as spermatozoa and oocytes). One-half of the gamete carries one allele, and the other half carries the other allele.

Mendel's second law is the principle of independent assortment of alleles. The segregation of each pair of alleles is independent of the segregation of other pairs during the formation of gametes.

Gametes are formed during a process known as *meiosis*, in which cells with haploid chromosome numbers (23 in humans) are produced by the division of cells with diploid chromosome numbers (46 in humans). A fertilized human egg thus contains a diploid number of chromosomes. A diploid is composed of 22 pairs of autosomes and 2 sex chromosomes (XX in females and XY in males).

Based on Mendelian principles, genes on different chromosomes assort independently of one another in gamete production. Genes residing very closely together on the same chromosome are usually inherited together. Thus, they do not assort independently and are called *linked genes*. Genes distant from each other on the same chromosome are usually inherited separately. This results from an exchange of segments between a pair of *homologous* chromosomes when the chromosomes are paired during the early phases of meiotic division. These types of gene-exchange events on homologous chromosomes are collectively called *crossing over*, which results in the *recombination* of genes in a pair of chromosomes (Figure 25.1).

The Mendelian inheritances of genes can often be measured using *probabilities*. A probability is the ratio of the number of actual occurrences of an event to the number of possible occurrences. Additionally, the probability of two independent events occurring simultaneously is the product of each of their individual probabilities; this is known as the *product rule of probability*.

Mendelian principles apply to the inheritance of loci of the autosomal nuclear DNA genome commonly used for forensic testing. The Y-chromosomal genome is inherited paternally, which does not obey the rules for Mendelian principles. Mitochondria contain their own mitochondrial genomes and are inherited maternally. This maternal inheritance

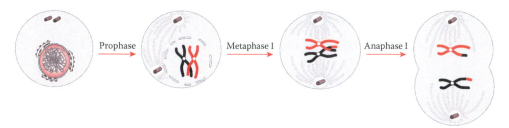

Figure 25.1 Crossing over hypothesis. Chromosomes are replicated prior to the first meiotic division. Each duplicate is called a chromatid, which forms into tetrads in the prophase. A crossover event occurs between the maternal and the paternal chromatids. Each chromatid breaks at the point of the cross and fuses with a portion of its counterpart. Chromosomes are associated with the microtubules during metaphase I and are pulled toward the spindle poles during anaphase I. The maternal and paternal homologs of one chromosome are shown. (© Richard C. Li.)

of mitochondrial genes also does not obey the rules for Mendelian principles. The inheritance of Y-chromosomal and mitochondrial genomes is referred to as non-Mendelian inheritance.

25.1.2 Population Genetics

Population genetics studies the causes of patterns of genetic variations within and among populations.

25.1.2.1 Allele Frequency

Allele frequency (p) can be calculated directly by counting the number of alleles of one type at a given locus and dividing it by the total number of alleles at that locus in a sampled population. This is called the *gene counting* method.

25.1.2.2 Genotype Frequency

Genotype frequency (P) observed at a given locus can be calculated by dividing the number of individuals with one genotype by the total number of individuals in a sampled population. Each genotype at the locus can be calculated separately. The summation of all genotype frequencies at that locus should equal 1.

25.1.2.3 Heterozygosity

Heterozygosity is the proportion of alleles, at a given locus, that are heterozygous and is calculated as

$$h = 1 - \sum_i p^2 \tag{25.1}$$

where:
   h = heterozygosity
   p = allelic frequency of the locus for homozygotes

The amount of heterozygosity at a locus in a sampled population is a measure of genetic variation. The higher the heterozygosity, the more variation there is at a given locus.

25.1.2.4 Hardy–Weinberg Principle

The Hardy–Weinberg (HW) principle, independently discovered by two scholars in the early 1900s, allows predictions of genotype frequencies to be made based on allelic frequencies. However, certain conditions must be met. The population must be large; mate randomly; and lack mutation, migration, and natural selection. If these conditions of the HW principle are met, the population will be in *equilibrium*, and the following results are expected:

1. The frequencies of the alleles will not change from one generation to the next.

2. Genotype frequencies can be predicted by the allelic frequencies (p^2 and q^2 for genotype frequencies of homozygotes and $2pq$ for genotype frequencies of heterozygotes). The sum of the genotype frequencies should equal 1.

$$p^2 + 2pq + q^2 = 1 \qquad (25.2)$$

If the observed genotype proportions are different from those expected, one or more of the HW assumptions have been violated.

25.1.2.5 Testing for HW Proportions of Population Databases

To determine whether the genotypes of a population in question obey the HW principle, a population database can be constructed. Samples (usually 100–200 samples for STR loci) are collected and analyzed at the loci of interest. Allelic frequencies are obtained by using the gene counting method. Table 25.1 shows the allelic frequencies of CODIS loci from a population database. The *observed genotype frequencies* at a given locus, as described earlier in Section 25.1.2.2, are calculated by dividing the number of individuals with one genotype by the total number of individuals in the population sampled. The *expected genotype frequencies* are calculated using p^2, $2pq$, and q^2.

The observed and expected genotype frequencies are then compared using a chi-square test. The significance of the differences between observed and expected genotype frequencies can then be determined. The chi-square is calculated using the following formula:

$$\chi^2 = \sum_{i=1}^{n} \frac{(O_i - E_i)^2}{E_i} \qquad (25.3)$$

where:
$O_i = i$th observed genotype frequency
$E_i = i$th expected genotype frequency
n = total number of genotypes

Chi-square (χ^2) is calculated as the sum of all genotypes of a given locus.

The chi-square value and the degrees of freedom (the number of genotypes minus the numbers of alleles) are then used to obtain a p value (not to be confused with the allelic frequency designated p) from a table of p values, and such tables can be found at the backs of most statistics textbooks. If the p value exceeds 0.05 (5% significance level), the deviation of the expected genotype frequencies from the observed genotype frequencies is not considered statistically significant. Thus, the null hypothesis that the observed genotype frequencies fit the expected genotype frequencies predicted by the HW principle is not rejected if the p value is greater than 0.05.

Table 25.1	Allelic Frequencies of 13 CODIS STR Loci		
Allele	Allelic Frequency (%)		
	African American	Caucasian	Hispanic
D3S1358	(N=210)	(N=203)	(N=209)
<12	0.476	0.000	0.000
12	0.238	0.000	0.000
13	1.190	0.246	0.239
14	12.143	14.039	7.895
15	29.048	24.631	42.584
15.2	0.000	0.000	0.000
16	30.714	23.153	26.555
17	20.000	21.182	12.679
18	5.476	16.256	8.373
19	0.476	0.493	1.435
>19	0.238	0.000	0.239
VWA	(N=180)	(N=196)	(N=203)
11	0.278	0.000	0.246
13	0.556	0.510	0.000
14	6.667	10.204	6.158
15	23.611	11.224	7.635
16	26.944	20.153	35.961
17	18.333	26.276	22.167
18	13.611	22.194	19.458
19	7.222	8.418	7.143
20	2.778	1.020	1.232
21	0.000	0.000	0.000
FGA	(N=180)	(N=196)	(N=203)
<18	0.278	0.000	0.000
18	0.833	3.061	0.246
18.2	0.833	0.000	0.000
19	5.278	5.612	7.882
19.2	0.278	0.000	0.000
20	7.222	14.541	7.143
20.2	0.000	0.255	0.246
21	12.500	17.347	13.054

Allele	Allelic Frequency (%)		
	African American	Caucasian	Hispanic
21.2	0.000	0.000	0.246
22	22.500	18.878	17.734
22.2	0.556	1.020	0.493
22.3	0.000	0.000	0.000
23	12.500	15.816	14.039
23.2	0.000	0.000	0.739
24	18.611	13.776	12.562
24.2	0.000	0.000	0.000
24.3	0.000	0.000	0.000
25	10.000	6.888	13.793
26	3.611	1.786	8.374
27	2.222	1.020	3.202
28	1.667	0.000	0.246
29	0.556	0.000	0.000
30	0.278	0.000	0.000
>30	0.278	0.000	0.000
D8S1179	($N = 180$)	($N = 196$)	($N = 203$)
<9	0.278	1.786	0.246
9	0.556	1.020	0.246
10	2.500	10.204	9.360
11	3.611	5.867	6.158
12	10.833	14.541	12.069
13	22.222	33.929	32.512
14	33.333	20.153	24.631
15	21.389	10.969	11.576
16	4.444	1.276	2.463
17	0.833	0.255	0.739
18	0.000	0.000	0.000
D21S11	($N = 179$)	($N = 196$)	($N = 203$)
24.2	0.279	0.510	0.246
24.3	0.000	0.000	0.000

Table 25.1   (Continued) Allelic Frequencies of 13 CODIS STR Loci

(continued)

Table 25.1 (Continued) Allelic Frequencies of 13 CODIS STR Loci			
Allele	Allelic Frequency (%)		
	African American	Caucasian	Hispanic
26	0.279	0.000	0.000
27	6.145	4.592	0.985
28	21.508	16.582	6.897
29	18.994	18.112	20.443
29.2	0.279	0.000	0.246
30	17.877	23.214	33.005
30.2	0.838	3.827	3.202
30.3	0.000	0.000	0.000
31	9.218	7.143	6.897
31.2	7.542	9.949	8.621
32	0.838	1.531	1.232
32.1	0.000	0.000	0.000
32.2	6.983	11.224	13.547
33	0.838	0.000	0.000
33.2	3.352	3.061	4.187
34	0.838	0.000	0.000
34.2	0.279	0.000	0.493
35	2.793	0.000	0.000
35.2	0.000	0.255	0.000
36	0.559	0.000	0.000
>36	0.559	0.000	0.000
D18S51	($N = 180$)	($N = 196$)	($N = 203$)
<11	0.556	1.276	0.493
11	0.556	1.276	1.232
12	5.833	12.755	10.591
13	5.556	12.245	16.995
13.2	0.556	0.000	0.000
14	6.389	17.347	16.995
14.2	0.000	0.000	0.000
15	16.667	12.755	13.793
15.2	0.000	0.000	0.000
16	18.889	10.714	11.576

Table 25.1 (Continued) Allelic Frequencies of 13 CODIS STR Loci			
Allele	**Allelic Frequency (%)**		
	African American	**Caucasian**	**Hispanic**
17	16.389	15.561	13.793
18	13.056	9.184	5.172
19	7.778	3.571	3.695
20	5.556	2.551	1.724
21	1.111	0.510	1.970
21.2	0.000	0.000	0.000
22	0.556	0.255	0.739
>22	0.556	0.000	1.232
D5S818	($N = 180$)	($N = 195$)	($N = 203$)
7	0.278	0.000	6.158
8	5.000	0.000	0.246
9	1.389	3.077	5.419
10	6.389	4.872	6.650
11	26.111	41.026	42.118
12	35.556	35.385	29.064
13	24.444	14.615	9.606
14	0.556	0.769	0.493
15	0.000	0.256	0.246
>15	0.278	0.000	0.000
D13S317	($N = 179$)	($N = 196$)	($N = 203$)
7	0.000	0.000	0.000
8	3.631	9.949	6.650
9	2.793	7.653	21.921
10	5.028	5.102	10.099
11	23.743	31.888	20.197
12	48.324	30.867	21.675
13	12.570	10.969	13.793
14	3.631	3.571	5.665
15	0.279	0.000	0.000
D7S820	($N = 210$)	($N = 203$)	($N = 209$)
6	0.000	0.246	0.239

(*continued*)

Table 25.1 (Continued) Allelic Frequencies of 13 CODIS STR Loci			
Allele	Allelic Frequency (%)		
	African American	Caucasian	Hispanic
7	0.714	1.724	2.153
8	17.381	16.256	9.809
9	15.714	14.778	4.785
10	32.381	29.064	30.622
10.1	0.000	0.000	0.000
11	22.381	20.197	28.947
11.3	0.000	0.000	0.000
12	9.048	14.039	19.139
13	1.905	2.956	3.828
14	0.476	0.739	0.478
CSF1PO	($N = 210$)	($N = 203$)	($N = 209$)
6	0.000	0.000	0.000
7	4.286	0.246	0.239
8	8.571	0.493	0.000
9	3.333	1.970	0.718
10	27.143	25.369	25.359
10.3	0.000	0.246	0.000
11	20.476	30.049	26.555
12	30.000	32.512	39.234
13	5.476	7.143	6.459
14	0.714	1.478	0.957
15	0.000	0.493	0.478
TPOX	($N = 209$)	($N = 203$)	($N = 209$)
6	8.612	0.000	0.478
7	2.153	0.246	0.478
8	36.842	54.433	55.502
9	18.182	12.315	3.349
10	9.330	3.695	3.349
11	22.488	25.369	27.273
12	2.392	3.941	9.330
13	0.000	0.000	0.239
THO1	($N = 210$)	($N = 203$)	($N = 209$)

Table 25.1 (Continued) Allelic Frequencies of 13 CODIS STR Loci			
Allele	Allelic Frequency (%)		
	African American	Caucasian	Hispanic
5	0.000	0.000	0.239
6	10.952	22.660	23.206
7	44.048	17.241	33.732
8	18.571	12.562	8.134
8.3	0.000	0.246	0.000
9	14.524	16.502	10.287
9.3	10.476	30.542	24.163
10	1.429	0.246	0.239
D16S539	(N = 209)	(N = 202)	(N = 208)
8	3.589	1.980	1.683
9	19.856	10.396	7.933
10	11.005	6.683	17.308
11	29.426	27.228	31.490
12	18.660	33.911	28.606
13	16.507	16.337	10.337
14	0.957	3.218	2.404
15	0.000	0.248	0.240

Source: Budowle, B., et al., *J Forensic Sci*, 44, 1277–1286, 1999. With permission.

25.1.2.6 Probability of Match

The discriminating power of genetic loci used above can be measured by *population match probability* (P_m). P_m is defined as the probability of having a matching genotype between two randomly chosen individuals. The lower the P_m, the less likely a match between two randomly chosen individuals will occur. This is calculated as follows:

$$P_m = \sum_i \left(p^2 \right)^2 + \sum_j \left(2pq \right)^2 \tag{25.4}$$

where:
  p and q = the frequencies of two different alleles
  P_m can also be used to compare the discriminating powers of different loci

Tables 25.2 through 25.4 show P_m values of loci commonly used for forensic applications, including SNP, VNTR, and STR.

Table 25.2 Heterozygosity and P_m Values of Six SNP Loci			
Locus	**Allele**	**Heterozygosity**	P_m
DQA1	7	0.828	0.053
LDLR	2	0.493	0.379
GYPA	2	0.498	0.376
HBGG	3	0.508	0.360
D7S8	2	0.476	0.388
GC	3	0.592	0.235
Average		0.566	
Product			2.5×10^{-4} (1 in 4000)

Source: Office of Justice Programs, National Institute of Justice, US Department of Justice, *The Future of Forensic DNA Testing: Predictions of the Research and Development Working Group*, 2000.

Table 25.3 Heterozygosity and P_m Values of Six VNTR Loci			
Locus	**Bins**	**Heterozygosity**	P_m
D1S7	28	0.945	0.0058
D2S44	26	0.926	0.0103
D4S139	19	0.899	0.0184
D10S28	24	0.943	0.0063
D14S13	30	0.899	0.0172
D17S79	19	0.799	0.0700
Average		0.902	
Product			8.26×10^{-12} (1 in 1.2×10^{11})

Source: Office of Justice Programs, National Institute of Justice, US Department of Justice, *The Future of Forensic DNA Testing: Predictions of the Research and Development Working Group*, 2000.

25.2 Statistical Analysis of DNA Profiling Results

It is desirable to evaluate the strength of DNA profiling results, particularly if two DNA profiles match. A DNA profile from crime scene evidence and a profile from a suspect may be the same for two reasons: (1) the crime scene sample may have come from the suspect or (2) the suspect happens to have the same profile as the individual who left the evidence found at the crime scene. The significance of a match between DNA profiles can be evaluated by using statistical calculations that determine the rarity of a specific DNA profile in a relevant population. The statistical evaluation of the significance can be included in a case report

Table 25.4 Heterozygosity and P_m Values for CODIS Loci

| Locus | Allele | Caucasian American | | African American | |
		Heterozygosity	P_m	Heterozygosity	P_m
CSF1PO	11	0.734	0.112	0.781	0.081
TPOX	7	0.621	0.195	0.763	0.090
TH01	7	0.783	0.081	0.727	0.109
VWA	10	0.811	0.062	0.809	0.063
D16S539	8	0.767	0.089	0.798	0.070
D7S820	11	0.806	0.065	0.782	0.080
D13S317	8	0.771	0.085	0.688	0.136
D5S818	10	0.682	0.158	0.739	0.112
FGA	19	0.860	0.036	0.863	0.033
D3S1358	10	0.795	0.075	0.763	0.094
D8S1179	10	0.780	0.067	0.778	0.082
D18S51	15	0.876	0.028	0.873	0.029
D2S11	20	0.853	0.039	0.861	0.034
Average		0.7812		0.7866	
Product			1.738×10^{-15} (1 in 5.753×10^{14})		1.092×10^{-15} (1 in 9.161 $\times 10^{14}$)

Source: Office of Justice Programs, National Institute of Justice, US Department of Justice, *The Future of Forensic DNA Testing: Predictions of the Research and Development Working Group,* 2000.

(Figure 25.2). Guidelines for DNA profile interpretation such as those issued by the National Research Council, the DNA Advisory Board, and the European DNA Profiling Group can be consulted.

25.2.1 Genotypes
The approaches to performing statistical analysis are (1) calculation of profile probability and (2) use of the likelihood ratio method. Although profile probability is the most commonly used method because of its simplicity, both approaches lead to the same conclusion.

25.2.1.1 Profile Probability
Profile probability can be calculated based on the following steps:

1. The locus genotype frequency can be calculated as follows; p and q are the allelic frequencies observed in the database for a given allele:

   Locus genotype frequency for homozygotes:

$$P_i = p^2 \tag{25.5}$$

Locus genotype frequency for heterozygotes:

$$P_j = 2pq \qquad (25.6)$$

2. Profile probability can then be calculated based on the product rule by multiplying all the locus genotype frequencies calculated as above.

The lower the profile probability is, the less likely that an individual chosen at random will have a coincident match with the DNA profile of the evidence sample. Table 25.5 shows calculations of profile probability from a DNA profile.

25.2.1.1.1 Structured Populations

The above calculation of profile probability is based on the assumption that a randomly selected individual is unrelated to a perpetrator. However, it is likely that the individual and the perpetrator are from the same subpopulation (groups within a population) and are thus not completely independent. Mating is more likely to occur within subpopulations than between subpopulations. As a result, the proportion of homozygotes increases and the proportion of heterozygotes decreases in a subpopulation because individuals in a subpopulation appear to be related.

DEPARTMENT OF FORENSIC BIOLOGY

LABORATORY REPORT

LAB NO: IQAS 2003, Lab # 259

SUSPECTS: Suspect 1, 2032
Suspect 2, 2033
Suspect 3, 2034

SUMMARY OF RESULTS:

Semen* was found on the crime scene sample 2031b, based on the presence of P30 antigen and sperm.

PCR DNA typing was done on crime scene sample 2031b. Results indicate the semen could have come from the suspect 1. This combination of DNA alleles would be expected to be found in approximately:

1 in greater than 1 trillion Blacks**
1 in greater than 1 trillion Caucasians
1 in greater than 1 trillion Hispanics
1 in greater than 1 trillion Asians

The DNA from crime scene sample 2031b could not have come from suspect 2 or suspect 3.

Amylase, a component of saliva, was not found on crime scene sample 2031b.

No semen was found on the control sample 2035.

Crime scene sample 2031a was not analyzed.

The DNA results in this case do not match any previous PCR (STR) DNA cases to date.

*Semen has two components: the seminal plasma (which contains the P30 antigen) and spermatozoa. Semen can be identified by detecting either P30 antigen and/or sperm.

** OCME STR database, National Research Council (1996) The Evaluation of Forensic DNA Evidence, Natl. Acad. Press, Washington DC.

Figure 25.2 An example of a laboratory report. (© Richard C. Li.)

Table 25.5		Calculation of Profile Probability of STR Profile with 13 CODIS Loci		
Locus	**Profile**	**Allelic Frequency[a]**	**Formula**	**Locus Genotype Frequency**
CSF1PO	10	0.25369	$2pq$	0.165
	12	0.32512		
D3S1358	14	0.14039	$2pq$	0.0691
	15	0.24631		
D5S818	11	0.41026	p^2	0.168
	11	0.41026		
D7S820	10	0.29064	$2pq$	0.117
	11	0.20197		
D8S1179	13	0.33929	p^2	0.115
	13	0.33929		
D13S317	11	0.31888	p^2	0.102
	11	0.31888		
D16S539	11	0.27228	$2pq$	0.185
	12	0.33911		
D18S51	15	0.12755	$2pq$	0.00911
	19	0.035710		
D21S11	30	0.23214	p^2	0.0539
	30	0.23214		
FGA	23	0.15816	$2pq$	0.0436
	24	0.13776		
TH01	8	0.12562	$2pq$	0.0767
	9.3	0.30542		
TPOX	8	0.54433	p^2	0.296
	8	0.54433		
VWA	17	0.26276	$2pq$	0.117
	18	0.22194		
			Profile probability $= 2.76 \times 10^{-14}$	

[a] See Table 25.1.

The effect of population structure should be considered and an appropriate correction should be made to estimate profile probabilities. The correction can be made by using the factor θ. Thus, the locus genotype frequency can be calculated as follows:

Locus genotype frequency for homozygotes:

$$P_i = \frac{\left[2\theta + (1-\theta)p\right]\left[3\theta + (1-\theta)p\right]}{(1+\theta)(1+2\theta)} \tag{25.7}$$

Locus genotype frequency for heterozygotes:

$$P_j = \frac{2\left[\theta + (1-\theta)p\right]\left[\theta + (1-\theta)q\right]}{(1+\theta)(1+2\theta)} \tag{25.8}$$

The θ value is 0.01 for the majority US population and 0.03 for the Native American population. Table 25.6 shows the calculation of profile probability with subpopulation correction using θ = 0.01. The profile probability is approximately three times higher than the value without the correction (Table 25.5). Additional corrections can be calculated for relatives, mixed stains, or database searches using formulas provided by the National Research Council's guidelines.

25.2.1.2 Likelihood Ratio

The *likelihood ratio* (LR) method is an alternative for evaluating the strength of a match. The method allows the calculation of the probability of the DNA profile under two hypothesis:

Hypothesis 1 (H₁) —The evidence and suspect profiles originated from the same source.

Hypothesis 2 (H₂) —The evidence and suspect profiles did not originate from the same source (i.e., the suspect happens to have the same profile as that of the individual who left the evidence).

The LR is the probability of hypothesis H₁ divided by the probability of hypothesis H₂. Where Pr₁ is the probability under hypothesis H₁, and Pr₂ is the probability under hypothesis H₂, this can be expressed as:

$$LR = \frac{Pr_1}{Pr_2} \tag{25.9}$$

The greater the numerator (Pr₁), the greater the likelihood ratio becomes. The result favors hypothesis H₁ (the evidence and suspect profile originated from the same source). Pr₁ is equal to 1 (100%) when a match occurs, and Pr₂ is equal to the profile probability. A LR of 1000 indicates that the evidence is 1000 times as probable if the evidence and suspect profiles originated from the same source.

25.2.2 Haplotypes

The term *haplotype* was first used to describe very closely linked polymorphic loci. During meiosis, alleles at neighboring loci cosegregate (both alleles segregate as a single allele) because of the close linkage of loci. The term also applies to genetic regions within which recombination events are very rare, that is, within mitochondrial and Y-chromosomal DNA. The entire mitochondrial DNA (mtDNA) sequence can be considered to be a single locus or haplotype because of the absence of recombination. Likewise, Y chromosome loci can also be considered haplotypes.

Where recombination is very rare, certain allelic combinations occur in populations much more frequently than would be expected. This phenomenon is called *linkage disequilibrium*. As a result, the HW principle cannot be applied. The two methods for evaluating the strength of

Locus	Profile	Allelic Frequency[a]	Formula	Locus Genotype Frequency
CSF1PO	10	0.25369	$\dfrac{2\big[\theta+(1-\theta)p\big]\big[\theta+(1-\theta)q\big]}{(1+\theta)(1+2\theta)}$	0.168
	12	0.32512		
D3S1358	14	0.14039	$\dfrac{2\big[\theta+(1-\theta)p\big]\big[\theta+(1-\theta)q\big]}{(1+\theta)(1+2\theta)}$	0.0734
	15	0.24631		
D5S818	11	0.41026	$\dfrac{\big[2\theta+(1-\theta)p\big]\big[3\theta+(1-\theta)p\big]}{(1+\theta)(1+2\theta)}$	0.180
	11	0.41026		
D7S820	10	0.29064	$\dfrac{2\big[\theta+(1-\theta)p\big]\big[\theta+(1-\theta)q\big]}{(1+\theta)(1+2\theta)}$	0.121
	11	0.20197		
D8S1179	13	0.33929	$\dfrac{\big[2\theta+(1-\theta)p\big]\big[3\theta+(1-\theta)p\big]}{(1+\theta)(1+2\theta)}$	0.126
	13	0.33929		
D13S317	11	0.31888	$\dfrac{\big[2\theta+(1-\theta)p\big]\big[3\theta+(1-\theta)p\big]}{(1+\theta)(1+2\theta)}$	0.113
	11	0.31888		
D16S539	11	0.27228	$\dfrac{2\big[\theta+(1-\theta)p\big]\big[\theta+(1-\theta)q\big]}{(1+\theta)(1+2\theta)}$	0.187
	12	0.33911		
D18S51	15	0.12755	$\dfrac{2\big[\theta+(1-\theta)p\big]\big[\theta+(1-\theta)q\big]}{(1+\theta)(1+2\theta)}$	0.012
	19	0.03571		
D21S11	30	0.23214	$\dfrac{\big[2\theta+(1-\theta)p\big]\big[3\theta+(1-\theta)p\big]}{(1+\theta)(1+2\theta)}$	0.0630
	30	0.23214		
FGA	23	0.15816	$\dfrac{2\big[\theta+(1-\theta)p\big]\big[\theta+(1-\theta)q\big]}{(1+\theta)(1+2\theta)}$	0.0473
	24	0.13776		
TH01	8	0.12562	$\dfrac{2\big[\theta+(1-\theta)p\big]\big[\theta+(1-\theta)q\big]}{(1+\theta)(1+2\theta)}$	0.0815
	9.3	0.30542		
TPOX	8	0.54433	$\dfrac{\big[2\theta+(1-\theta)p\big]\big[3\theta+(1-\theta)p\big]}{(1+\theta)(1+2\theta)}$	0.309
	8	0.54433		
VWA	17	0.26276	$\dfrac{2\big[\theta+(1-\theta)p\big]\big[\theta+(1-\theta)q\big]}{(1+\theta)(1+2\theta)}$	0.120
	18	0.22194		
				Profile probability $=7.8\times10^{-14}$

Table 25.6 Calculation of Profile Probability of STR with 13 CODIS Loci and Correction Factor ($\theta=0.01$)

[a] See Table 23.1.

a match between haplotypes are (1) mitotype frequency and (2) likelihood ratios. The current most common approach for interpreting mtDNA profiles is mitotype frequency carried out with the gene counting method, that is, the calculation of the number of occurrences of a particular sequence or haplotype. The interpretation of Y chromosome profiles is similar to the interpretation of mtDNA profiles. The estimation of the frequency of a mitotype can be calculated as shown below.

25.2.2.1 Mitotypes Observed in Database

If a mitotype is observed at least once in a database, Equation 25.10 can be used. P_{mt} is the mitotype frequency, x is the number of observations of the haplotype, and n is the size of the database (the number of mitotype entries):

$$P_{mt} = \frac{x+2}{n+2} \tag{25.10}$$

Any sampling error may be addressed by a confidence interval:

$$P_{mt} \pm 1.96 \sqrt{\frac{P_{mt}(1-P_{mt})}{n}} \tag{25.11}$$

In this case, P_{mt} is the mitotype frequency, and n is the size of the database. The conservative upper bound of the frequency is usually quoted.

25.2.2.2 Mitotype Not Observed in Database

If a mitotype has not been observed in a database, Equation 25.12 can be used to calculate the mitotype frequency; α is 0.05 for a 95% confidence interval and n is the size of the database.

$$P_{mt} = 1 - \alpha^{1/n} \tag{25.12}$$

Bibliography

Aitken, C.G.G., *Statistics and the Evaluation of Evidence for Forensic Scientists*, p. 260, 1995. New York: Wiley.

Aktulga, H.M., et al., Identifying statistical dependence in genomic sequences via mutual information estimates. *EURASIP J Bioinform Syst Biol*, 2007, **2007**(1): 14741.

Allen, R.W., et al., Considerations for the interpretation of STR results in cases of questioned half-sibship. *Transfusion*, 2007, **47**(3): 515–519.

Amorim, A., A cautionary note on the evaluation of genetic evidence from uniparentally transmitted markers. *Forensic Sci Int Genet*, 2008, **2**(4): 376–378.

Andersen, M.M., P.S. Eriksen, and N. Morling, The discrete Laplace exponential family and estimation of Y-STR haplotype frequencies. *J Theor Biol*, 2013, **329**: 39–51.

Andersen, M.M., et al., Estimating trace-suspect match probabilities for singleton Y-STR haplotypes using coalescent theory. *Forensic Sci Int Genet*, 2013, **7**(2): 264–271.

Balding, D.J., The prosecutor's fallacy and DNA evidence. *Crim Law Rev*, 1994, 711–721.

Balding, D.J., When can a DNA profile be regarded as unique? *Sci Justice*, 1999, **39**(4): 257–260.

Balding, D.J. and P. Donnelly, Inferring identify from DNA profile evidence. *Proc Natl Acad Sci U S A*, 1995, **92**(25): 11741–11745.

Balding, D.J. and P. Donnelly, Evaluating DNA profile evidence when the suspect is identified through a database search. *J Forensic Sci*, 1996, **41**(4): 603–607.

Balding, D.J. and R.A. Nichols, DNA profile match probability calculation: How to allow for population stratification, relatedness, database selection and single bands. *Forensic Sci Int*, 1994, **64**(2–3): 125–140.

Beecham, G.W. and B.S. Weir, Confidence interval of the likelihood ratio associated with mixed stain DNA evidence. *J Forensic Sci*, 2011, **56**(Suppl 1): S166–S171.

Behar, D.M., et al., A "Copernican" reassessment of the human mitochondrial DNA tree from its root. *Am J Hum Genet*, 2012, **90**(4): 675–684.

Bodner, M., et al., Inspecting close maternal relatedness: Towards better mtDNA population samples in forensic databases. *Forensic Sci Int Genet*, 2011, **5**(2): 138–141.

Botstein, D., et al., Construction of a genetic linkage map in man using restriction fragment length polymorphisms. *Am J Hum Genet*, 1980, **32**(3): 314–331.

Brenner, C.H., Difficulties in the estimation of ethnic affiliation. *Am J Hum Genet*, 1998, **62**(6): 1558–1560; author reply 1560–1561.

Brenner, C.H., Fundamental problem of forensic mathematics—The evidential value of a rare haplotype. *Forensic Sci Int Genet*, 2010, **4**(5): 281–291.

Bright, J.A., J.M. Curran, and J.S. Buckleton, Relatedness calculations for linked loci incorporating subpopulation effects. *Forensic Sci Int Genet*, 2013, **7**(3): 380–383.

Buckleton, J., C. Triggs, and S. Walsh, *Forensic DNA Evidence Interpretation*, p. 534, 2005. New York: CRC Press.

Buckleton, J.S., M. Krawczak, and B.S. Weir, The interpretation of lineage markers in forensic DNA testing. *Forensic Sci Int Genet*, 2011, **5**(2): 78–83.

Buckleton, J.S., S. Walsh, and S.A. Harbison, The fallacy of independence testing and the use of the product rule. *Sci Justice*, 2001, **41**(2): 81–84.

Budowle, B., The effects of inbreeding on DNA profile frequency estimates using PCR-based loci. *Genetica*, 1995, **96**(1–2): 21–25.

Budowle, B. and T.R. Moretti, Genotype profiles for six population groups at the 13 CODIS short tandem repeat core loci and other PCR based loci. *Forensic Sci Commun*, 1999, **1**(2). Retrieved from http://www.fbi.gov/about-us/lab/forensic-science-communications/fsc/july1999/index.htm/budowle.htm.

Budowle, B., F.S. Baechtel, and R. Chakraborty, Partial matches in heterogeneous offender databases do not call into question the validity of random match probability calculations. *Int J Legal Med*, 2009, **123**(1): 59–63.

Budowle, B., et al., Population data on the thirteen CODIS core short tandem repeat loci in African Americans, U.S. Caucasians, Hispanics, Bahamians, Jamaicans, and Trinidadians. *J Forensic Sci*, 1999, **44**(6): 1277–1286.

Budowle, B., et al., Validation and population studies of the loci LDLR, GYPA, HBGG, D7S8, and Gc (PM loci), and HLA-DQ alpha using a multiplex amplification and typing procedure. *J Forensic Sci*, 1995, **40**(1): 45–54.

Budowle, B., et al., Source attribution of a forensic DNA profile. *Forensic Sci Commun*, 2000, **2**(3). Retrieved from http://www.fbi.gov/about-us/lab/forensic-science-communications/fsc/july2000/index.htm/source.htm.

Budowle, B., et al., CODIS STR loci data from 41 sample populations. *J Forensic Sci*, 2001, **46**(3): 453–489.

Budowle, B., et al., Use of prior odds for missing persons identifications. *Investig Genet*, 2011, **2**(1): 15.

Butler, J.M., et al., Allele frequencies for 15 autosomal STR loci on U.S. Caucasian, African American, and Hispanic populations. *J Forensic Sci*, 2003, **48**(4): 908–911.

Carracedo, A., et al., New guidelines for the publication of genetic population data. *Forensic Sci Int Genet*, 2013, **7**(2): 217–220.

Chakraborty, R., Detection of nonrandom association of alleles from the distribution of the number of heterozygous loci in a sample. *Genetics*, 1984, **108**(3): 719–731.

Chakraborty, R., Sample size requirements for addressing the population genetic issues of forensic use of DNA typing. *Hum Biol*, 1992, **64**(2): 141–159.

Chakraborty, R. and D.N. Stivers, Paternity exclusion by DNA markers: Effects of paternal mutations. *J Forensic Sci*, 1996, **41**(4): 671–677.

Chakraborty, R., et al., The utility of short tandem repeat loci beyond human identification: Implications for development of new DNA typing systems. *Electrophoresis*, 1999, **20**(8): 1682–1696.

Clayton, D., On inferring presence of an individual in a mixture: A Bayesian approach. *Biostatistics*, 2010, **11**(4): 661–673.

Cockerton, S., K. McManus, and J. Buckleton, Interpreting lineage markers in view of subpopulation effects. *Forensic Sci Int Genet*, 2012, **6**(3): 393–397.

Collins, A. and N.E. Morton, Likelihood ratios for DNA identification. *Proc Natl Acad Sci U S A*, 1994, **91**(13): 5.

Condel, K. and M. Al Salih, A simple method for establishing concordance between short-tandem-repeat allele frequency databases. *Transfusion*, 2011, **51**(5): 986–992.

Crow, J.F., Hardy, Weinberg and language impediments. *Genetics*, 1999, **152**(3): 821–825.

Curran, J.M. and J. Buckleton, The appropriate use of subpopulation corrections for differences in endogamous communities. *Forensic Sci Int*, 2007, **168**(2–3): 106–111.

Curran, J.M. and J. Buckleton, Inclusion probabilities and dropout. *J Forensic Sci*, 2010, **55**(5): 1171–1173.

Curran, J.M. and J.S. Buckleton, An investigation into the performance of methods for adjusting for sampling uncertainty in DNA likelihood ratio calculations. *Forensic Sci Int Genet*, 2011, **5**(5): 512–516.

Curran, J.M., S.J. Walsh, and J. Buckleton, Empirical testing of estimated DNA frequencies. *Forensic Sci Int Genet*, 2007, **1**(3–4): 267–272.

Curran, J.M., et al., Interpreting DNA mixtures in structured populations. *J Forensic Sci*, 1999, **44**(5): 987–995.

Edwards, A., et al., Genetic variation at five trimeric and tetrameric tandem repeat loci in four human population groups. *Genomics*, 1992, **12**(2): 241–253.

Egeland, T. and A. Salas, Estimating haplotype frequency and coverage of databases. *PLoS One*, 2008, **3**(12): e3988.

Egeland, T. and A. Salas, A statistical framework for the interpretation of mtDNA mixtures: Forensic and medical applications. *PLoS One*, 2011, **6**(10): e26723.

Egeland, T. and N. Sheehan, On identification problems requiring linked autosomal markers. *Forensic Sci Int Genet*, 2008, **2**(3): 219–225.

Evett, I. and B. Weir, *Interpreting DNA Evidence: Statistical Genetics for Forensic Scientists*, p. 278, 1992. Sunderland, MA: Sinauer Associates.

Evett, I.W. and P. Gill, A discussion of the robustness of methods for assessing the evidential value of DNA single locus profiles in crime investigations. *Electrophoresis*, 1991, **12**(2–3): 226–230.

Evett, I.W., R. Pinchin, and C. Buffery, An investigation of the feasibility of inferring ethnic origin from DNA profiles. *J Forensic Sci Soc*, 1992, **32**(4): 301–306.

Fisher, R.A., Standard calculations for evaluating a blood-group system. *Heredity*, 1951, **5**(1): 95–102.

Foreman, L.A., A.F.M. Smith, and I. Evett, Bayesian validation of a quaduplex STR profiling system for identification purposes. *J Forensic Sci*, 1999, **44**: 378–486.

Foreman, L.A. and I.W. Evett, Statistical analyses to support forensic interpretation for a new ten-locus STR profiling system. *Int J Legal Med*, 2001, **114**(3): 147–155.

Foreman, L.A., et al., Interpreting DNA evidence: A review. *Int Stat Rev*, 2003, **71**(3): 13.

Gaensslen, R.E., Journal policy on the publication of DNA population genetic data. *J Forensic Sci*, 1999, **44**(4): 4.

Ge, J. and B. Budowle, Kinship index variations among populations and thresholds for familial searching. *PLoS One*, 2012, **7**(5): e37474.

Ge, J., B. Budowle, and R. Chakraborty, Interpreting Y chromosome STR haplotype mixture. *Leg Med*, 2010, **12**(3): 137–143.

Ge, J., et al., Haplotype block: A new type of forensic DNA markers. *Int J Legal Med*, 2010, **124**(5): 353–361.

Ge, J., et al., Pedigree likelihood ratio for lineage markers. *Int J Legal Med*, 2011, **125**(4): 519–525.

Gill, P. and H. Haned, A new methodological framework to interpret complex DNA profiles using likelihood ratios. *Forensic Sci Int Genet*, 2013, **7**(2): 251–263.

Gill, P., et al., A comparison of adjustment methods to test the robustness of an STR DNA database comprised of 24 European populations. *Forensic Sci Int*, 2003, **131**(2–3): 184–196.

Gjertson, D.W., et al., ISFG: Recommendations on biostatistics in paternity testing. *Forensic Sci Int Genet*, 2007, **1**(3–4): 223–231.

Guo, S.W. and E.A. Thompson, Performing the exact test of Hardy–Weinberg proportion for multiple alleles. *Biometrics*, 1992, **48**(2): 361–372.

Haned, H., Forensim: An open-source initiative for the evaluation of statistical methods in forensic genetics. *Forensic Sci Int Genet*, 2011, **5**(4): 265–268.

Hardy, G.H., Mendelian proportions in a mixed population. *Science*, 1908, **17**: 49–50.

Hartman, D., et al., Examples of kinship analysis where Profiler Plus was not discriminatory enough for the identification of victims using DNA identification. *Forensic Sci Int*, 2011, **205**(1–3): 64–68.

Henderson, J.P., The use of DNA statistics in criminal trials. *Forensic Sci Int*, 2002, **128**(3): 183–186.

Hill, C.R., et al., Concordance and population studies along with stutter and peak height ratio analysis for the PowerPlex ESX 17 and ESI 17 Systems. *Forensic Sci Int Genet*, 2011, **5**(4): 269–275.

Holt, C.L., et al., Practical applications of genotypic surveys for forensic STR testing. *Forensic Sci Int*, 2000, **112**(2–3): 91–109.

Hong-Sheng, G., et al., HGD-Chn: The database of genome diversity and variation for Chinese populations. *Leg Med*, 2009, **11**(Suppl 1): S201–S202.

Hopwood, A.J., et al., Consideration of the probative value of single donor 15-plex STR profiles in UK populations and its presentation in UK courts. *Sci Justice*, 2012, **52**(3): 185–190.

Hosking, L., et al., Detection of genotyping errors by Hardy–Weinberg equilibrium testing. *Eur J Hum Genet*, 2004, **12**(5): 395–399.

Janica, J., et al., Database of genetic profiles at the Department of Forensic Medicine, Medical University of Bialystok. *Adv Med Sci*, 2008, **53**(1): 64–68.

Kling, D., T. Egeland, and A.O. Tillmar, FamLink—A user friendly software for linkage calculations in family genetics. *Forensic Sci Int Genet*, 2012, **6**(5): 616–620.

Lareu, M.V., et al., Analysis of a claimed distant relationship in a deficient pedigree using high density SNP data. *Forensic Sci Int Genet*, 2012, **6**(3): 350–353.

Lauc, G., et al., Empirical support for the reliability of DNA interpretation in Croatia. *Forensic Sci Int Genet*, 2008, **3**(1): 50–53.

Lee, J.C., et al., A novel strategy for sibship determination in trio sibling model. *Croat Med J*, 2012, **53**(4): 336–342.

Li, L., et al., Maternity exclusion with a very high autosomal STRs kinship index. *Int J Legal Med*, 2012, **126**(4): 645–648.

Lincoln, P. and A. Carracedo, Publication of population data of human polymorphisms. *Forensic Sci Int*, 2000, **110**(1): 3–5.

Lins, A.M., et al., Development and population study of an eight-locus short tandem repeat (STR) multiplex system. *J Forensic Sci*, 1998, **43**(6): 1168–1180.

Matsumura, S., et al., Kinship analysis using DNA typing from five skeletal remains with an unusual post-mortem course. *Med Sci Law*, 2011, **51**(4): 240–243.

Montinaro, F., et al., Using forensic microsatellites to decipher the genetic structure of linguistic and geographic isolates: A survey in the eastern Italian Alps. *Forensic Sci Int Genet*, 2012, **6**(6): 827–833.

Morton, N.E., Genetic structure of forensic populations. *Proc Natl Acad Sci U S A*, 1992, **89**(7): 2556–2560.

Mueller, L.D., Can simple population genetic models reconcile partial match frequencies observed in large forensic databases? *J Genet*, 2008, **87**(2): 101–108.

Nei, M., Estimation of average heterozygosity and genetic distance from a small number of individuals. *Genetics*, 1978, **89**(3): 583–590.

Noris, G., et al., Mexican mestizo population sub-structure: Effects on genetic and forensic statistical parameters. *Mol Biol Rep*, 2012, **39**(12): 10139–10156.

Nothnagel, M., J. Schmidtke, and M. Krawczak, Potentials and limits of pairwise kinship analysis using autosomal short tandem repeat loci. *Int J Legal Med*, 2010, **124**(3): 205–215.

NRC (National Research Council), *The Evaluation of Forensic DNA Evidence*, 1996. Washington, DC: National Research Council Committee, National Academies Press.

Nunez, C., et al., Reconstructing the population history of Nicaragua by means of mtDNA, Y-chromosome STRs, and autosomal STR markers. *Am J Phys Anthropol*, 2010, **143**(4): 591–600.

O'Connor, K.L. and A.O. Tillmar, Effect of linkage between vWA and D12S391 in kinship analysis. *Forensic Sci Int Genet*, 2012, **6**(6): 840–844.

O'Connor, K.L., et al., Linkage disequilibrium analysis of D12S391 and vWA in U.S. population and paternity samples. *Forensic Sci Int Genet*, 2011, **5**(5): 538–540.

Office of Justice Programs, National Institute of Justice, US Department of Justice, *The Future of Forensic DNA Testing: Predictions of the Research and Development Working Group,* 2000.

Palo, J.U., et al., Genetic markers and population history: Finland revisited. *Eur J Hum Genet*, 2009, **17**(10): 1336–1346.

Parson, W. and H.J. Bandelt, Extended guidelines for mtDNA typing of population data in forensic science. *Forensic Sci Int Genet*, 2007, **1**(1): 13–19.

Parson, W. and L. Roewer, Publication of population data of linearly inherited DNA markers in the *International Journal of Legal Medicine*. *Int J Legal Med*, 2010, **124**(5): 505–509.

Pereira, L., et al., PopAffiliator: Online calculator for individual affiliation to a major population group based on 17 autosomal short tandem repeat genotype profile. *Int J Legal Med*, 2011, **125**(5): 629–636.

Pereira, R., et al., Straightforward inference of ancestry and admixture proportions through ancestry-informative insertion deletion multiplexing. *PLoS One*, 2012, **7**(1): e29684.

Phillips, C., et al., D9S1120, a simple STR with a common Native American-specific allele: Forensic optimization, locus characterization and allele frequency studies. *Forensic Sci Int Genet*, 2008, **3**(1): 7–13.

Phillips, C., et al., The recombination landscape around forensic STRs: Accurate measurement of genetic distances between syntenic STR pairs using HapMap high density SNP data. *Forensic Sci Int Genet*, 2012, **6**(3): 354–365.

Prieto, L., et al., The GHEP-EMPOP collaboration on mtDNA population data—A new resource for forensic casework. *Forensic Sci Int Genet*, 2011, **5**(2): 146–151.

Pu, C.E. and A. Linacre, Increasing the confidence in half-sibship determination based upon 15 STR loci. *J Forensic Leg Med*, 2008, **15**(6): 373–377.

Pu, C.E. and A. Linacre, Systematic evaluation of sensitivity and specificity of sibship determination by using 15 STR loci. *J Forensic Leg Med*, 2008, **15**(5): 329–334.

Puch-Solis, R., S. Pope, and I. Evett, Calculating likelihood ratios for a mixed DNA profile when a contribution from a genetic relative of a suspect is proposed. *Sci Justice*, 2010, **50**(4): 205–209.

Rajeevan, H., et al., Introducing the forensic research/reference on genetics knowledge base, FROG-kb. *Investig Genet*, 2012, **3**(1): 18.

Ravid-Amir, O. and S. Rosset, Maximum likelihood estimation of locus-specific mutation rates in Y-chromosome short tandem repeats. *Bioinformatics*, 2010, **26**(18): i440–i445.

Raymond, M. and F. Rousset, Population genetics software for exact tests and ecumenicism. *J Hered*, 1995, **86**: 12.

Robertson, B. and G.A. Vigneaux, Expert evidence: Law, practice, and probability. *Oxford J Legal Stud*, 1992, **12**(1): 392–403.

Rodig, H., et al., Evaluation of haplotype discrimination capacity of 35 Y-chromosomal short tandem repeat loci. *Forensic Sci Int*, 2008, **174**(2–3): 182–188.

Roeder, K., DNA fingerprinting: A review of the controversy. *Stat Sci*, 1994, **9**(2): 222–278.

Rowold, D.J. and R.J. Herrera, On human STR sub-population structure. *Forensic Sci Int*, 2005, **151**(1): 59–69.

Royall, R., *Statistical Evidence: A Likelihood Paradigm*, 1997. New York: Chapman & Hall.

Sanchez, J.J., et al., Forensic typing of autosomal SNPs with a 29 SNP-multiplex—Results of a collaborative EDNAP exercise. *Forensic Sci Int Genet*, 2008, **2**(3): 176–183.

Sanchez-Diz, P., et al., Population and segregation data on 17 Y-STRs: Results of a GEP-ISFG collaborative study. *Int J Legal Med*, 2008, **122**(6): 529–533.

Santos, C., et al., Frequency and pattern of heteroplasmy in the control region of human mitochondrial DNA. *J Mol Evol*, 2008, **67**(2): 191–200.

Silva, N.M., et al., Human neutral genetic variation and forensic STR data. *PLoS One*, 2012, **7**(11): e49666.

Stockmarr, A., Likelihood ratios for evaluating DNA evidence when the suspect is found through a database search. *Biometrics*, 1999, **55**(3): 671–677.

Taroni, F., et al., Evaluation and presentation of forensic DNA evidence in European laboratories. *Sci Justice*, 2002, **42**(1): 21–28.

Taylor, C. and P. Colman, Forensics: Experts disagree on statistics from DNA trawls. *Nature*, 2010, **464**(7293): 1266–1267.

Thompson, W.C. and E.L. Schumann, Interpretation of statistical evidence in criminal trials: The prosecutor's fallacy and the defense attorney's fallacy. *Law Hum Behav*, 1987, **11**(3): 11.

Toscanini, U., et al., Evaluating methods to correct for population stratification when estimating paternity indexes. *PLoS One*, 2012, **7**(11): e49832.

Tvedebrink, T., et al., Estimating the probability of allelic drop-out of STR alleles in forensic genetics. *Forensic Sci Int Genet*, 2009, **3**(4): 222–226.

van Oven, M., et al., Unexpected island effects at an extreme: Reduced Y chromosome and mitochondrial DNA diversity in Nias. *Mol Biol Evol*, 2011, **28**(4): 1349–1361.

Walsh, S.J., et al., Evidence in support of self-declaration as a sampling method for the formation of sub-population DNA databases. *J Forensic Sci*, 2003, **48**(5): 1091–1093.

Walsh, S.J., et al., Use of subpopulation data in Australian forensic DNA casework. *Forensic Sci Int Genet*, 2007, **1**(3–4): 238–246.

Weir, B.S. and W.G. Hill, Estimating F-statistics. *Annu Rev Genet*, 2002, **36**: 721–750.

Wenk, R.E., Detection of genotype recycling fraud in U.S. immigrants. *J Forensic Sci*, 2011, **56**(Suppl 1): S243–S246.

Willuweit, S., et al., Y-STR frequency surveying method: A critical reappraisal. *Forensic Sci Int Genet*, 2011, **5**(2): 84–90.

Quality Assurance and Quality Control

Quality assurance (QA) for forensic services requires certain processes to ensure that a service will meet laboratory requirements for the integrity of testing. A QA program must include components that address:

- Continuing education, training, and certification of personnel
- Specification and calibration of equipment and reagents
- Documentation and validation of analytic methods
- Use of appropriate standards and controls
- Sample handling procedures
- Proficiency testing
- Data interpretation and reporting
- Audits (internal and external) and laboratory accreditation
- Corrective actions to address deficiencies and assessments for laboratory competence

Over the years, many guidelines for quality assurance in forensic DNA laboratories have been established. These guidelines will be introduced in this chapter.

Quality control (QC) for forensic services refers to the operational procedures necessary to meet quality requirements. QC procedures may include maintenance of calibration records for equipment and instruments as well as the testing of chemical reagents and supplies used in analysis to ensure reliable results.

26.1 US Quality Standards

DNA profiling methods were first used in criminal investigations in the 1980s. By the early 1990s, emerging forensic DNA techniques had undergone detailed reviews by the *National Research Council* (NRC) of the National Academy of Sciences. In 1992, the first published NRC report included recommendations in the areas of (1) technical considerations, (2) statistical interpretation, (3) laboratory standards, (4) data banks and privacy, (5) legal considerations, and (6) societal and ethical issues related to forensic DNA testing. The NRC report attempted to explain the basic scientific principles of forensic DNA technology and made suggestions for applications and improvements. However, the report received negative criticism from both the forensic and the legal communities.

In 1996, a second NRC committee was formed "to update and clarify discussion of the principles of population genetics and statistics as they apply to DNA evidence." Its report stated:

> The central question that the report addresses is this: What information can a forensic scientist, population geneticist, or statistician provide to assist a judge or jury in drawing inferences from the finding of a match? To answer this question, the committee reviewed the scientific literature and the legal cases and commentaries on DNA profiling, and it investigated the criticisms that have been voiced about population data, statistics, and laboratory error. Much has been learned since the last report. The technology for DNA profiling as well as the methods for estimating frequencies and related statistics have progressed to the point where the reliability and validity of properly collected and analyzed DNA data should not be in doubt. The new recommendations presented here should pave the way to more effective use of DNA evidence. (*The Evaluation of Forensic DNA Evidence* also known as the NRC II report)

The NRC II report consisted of (1) an introduction describing the 1992 report, changes made subsequent to that report, and the validity and application of DNA typing techniques; (2) assurance of high standards of laboratory performance; (3) population genetics issues; (4) statistical issues; and (5) DNA evidence in the legal system.

In 1995, The *DNA Advisory Board* (*DAB*) was formed as a result of the DNA Identification Act (1994) passed by Congress. The DAB served from 1995 to 2000 to develop guidelines for quality assurance in forensic laboratories. During that time, the DAB provided two sets of guidelines: Quality Assurance Standards for Forensic DNA Testing Laboratories (1998) and Quality Assurance Standards for Convicted Offender DNA Databasing Laboratories (1999). These standards describe the requirements to ensure quality and integrity of the data as well as competency of laboratories.

These standards were built on the previous standards set by the *Scientific Working Group on DNA Analysis Methods* (*SWGDAM*). After the DAB's assignment ended in 2000, the SWGDAM became responsible for providing guidelines to the US forensic community. The SWGDAM was established in 1988 by the FBI Laboratory. It was initially called the Technical Working Group on DNA Analysis Methods; the name changed in 1998. The SWGDAM comprises forensic scientists from DNA laboratories in the United States and Canada. Its purpose is to facilitate forensic DNA community discussions regarding necessary laboratory methods and to share protocols for forensic DNA testing. The FBI sponsors and hosts its meetings and plays an important role in its activities.

The SWGDAM established and revised several guidelines including the Guidelines for a Quality Assurance Program for DNA Analysis, published in 1989, 1991, and 1995 (the validation section was revised in 2003); Quality Assurance Standards for Forensic DNA Testing Laboratories (2011); Quality Assurance Standards for DNA Databasing Laboratories (2011); Validation Guidelines for DNA Analysis Methods (2012); and SWGDAM Training Guidelines, published in 2001 and revised in 2013.

The SWGDAM also formed subcommittees to provide guidelines in more specific areas of forensic DNA testing, for example, SWGDAM's Interpretation Guidelines for Autosomal STR Typing by Forensic DNA Testing Laboratories (2010), Y-chromosome Short Tandem Repeat (Y-STR) Interpretation Guidelines (2009), and Interpretation Guidelines for Mitochondrial DNA Analysis by Forensic DNA Testing Laboratories (2013). The SWGDAM also organized a number of interlaboratory and validation studies for new techniques.

26.2 International Quality Standards

In the early stages of forensic DNA testing, the *International Society for Forensic Genetics* (*ISFG*) recognized the potential of DNA testing for criminal investigations and made a number of recommendations related to the forensic application of DNA polymorphisms.

The ISFG provided various recommendations for forensic DNA testing of STR, mtDNA, and Y chromosome markers for the international community. It formed a working group called the *European DNA Profiling Group* (*EDNAP*) in 1991. The EDNAP has investigated systems for DNA profiling, has organized a number of collaborative exercises for the evaluation of new methods, and has published reports of its studies.

The *European Network of Forensic Science Institutes* (*ENFSI*) formed its DNA working group about a decade ago to address issues of quality and standards for forensic DNA testing. The *Interpol European Working Party on DNA Profiling* (*IEWPDP*) also makes recommendations for applying DNA evidence to criminal investigations in Europe. Based on the EDNAP exercises and recommendations by the ENFSI and IEWPDP, the *European Standard Set* (*ESS*) for autosomal STR core loci was established. The Standardization of DNA Profiling Techniques in the European Union (STADNAP) group has been working on the selection of forensic DNA profiling systems, methods for use among European countries, and the maintenance of European population databases.

26.3 Laboratory Accreditation

Accreditation is the process used to assess the qualification of a laboratory to meet established standards. During an accreditation process, the services and performance of a laboratory are evaluated, particularly in the areas of management, operations, personnel, procedures, equipment, physical plants, security, and personnel safety procedures. The accreditation process generally involves several components such as self-evaluation, the preparation of supporting documents, on-site inspection and reports, and accreditation review reports.

Accreditation in the United States is offered by the *Laboratory Accreditation Board* of the *American Society of Crime Laboratory Directors* (*ASCLD/LAB*) for forensic laboratories performing casework. The *American Association of Blood Banks* (*AABB*) provides accreditation for laboratories performing DNA parentage testing according to the AABB's standards.

The accreditation of a forensic laboratory is granted for 5 years. To remain in compliance, a laboratory must undergo audits to evaluate its operation according to established guidelines. The FBI has published the Quality Assurance Standards Audit for Forensic DNA Testing Laboratories (2011) and the Quality Assurance Standards Audit for DNA Databasing Laboratories (2011). The areas of operating protocols, instruments and equipment, and personnel training are evaluated based on guidelines. Problems identified during an audit must be documented and actions to resolve the problems must be addressed. Annual internal audits and external audits during alternate years are required under the guidelines.

26.4 Laboratory Validation

Validation is the process of confirming that a laboratory procedure is sufficiently robust, reliable, and reproducible. A robust method maintains successful performance and can cope with errors. A reliable method produces accurate results. A reproducible method achieves the same or very similar results each time a sample is tested.

Two types of laboratory validations, developmental and internal validations, are used for modifying methods for forensic DNA analysis. An *internal validation* is required when adopting a procedure for forensic applications. Based on the SWGDAM's Validation Guidelines for DNA Analysis Methods, the internal validation is "an accumulation of test data within the laboratory to demonstrate that established methods and procedures perform as expected in the laboratory." The *developmental validation* "is the acquisition of test data and determination of conditions and limitations of a new or novel DNA methodology for use on forensic, database, known or casework reference samples." For example, the characterization of a new genetic marker requires studies in the inheritance, the genomic location, the detection method, and the polymorphism of the genetic marker. However, developmental validation studies in the same subject but from fields other than forensic DNA analysis may be acceptable for forensic applications.

Additionally, the precision and the accuracy of the test are important criteria for developmental validation. Based on the SWGDAM, *precision* "characterizes the degree of mutual agreement among a series of individual measurements, values and/or results. Precision depends only on the distribution of random errors and does not relate to the true value or specified value." The measure of precision can be expressed as a standard deviation of test results. *Accuracy* is referred to as "the ability of a measuring instrument to give responses close to a true value," which can be assessed by a *performance check*, a quality assurance procedure monitoring the performance of instruments and equipment affecting the accuracy of forensic DNA testing.

26.5 Proficiency Testing

Proficiency testing is an important component of quality control and quality assurance. It evaluates a laboratory's performance of DNA analyses according to the laboratory's standard protocols. Proficiency testing also evaluates the quality of performance by individual analysts following laboratory protocols.

Proficiency tests of DNA analysts must be conducted every 6 months based on DNA Advisory Board Standards. The testing usually involves mock forensic case samples including questioned bodily fluid stains and reference samples. The test is assigned to an analyst for processing according to the laboratory procedures. A report must be prepared and is then reviewed. The proficiency test can be administered as either an open or a blind test. In the blind test, the analyst is not aware that he or she is being tested. Blind testing is considered a more effective means of evaluating performance.

The tests may be administered internally or by any of a number of external testing organizations. For example, Orchid Cellmark provides the *International Quality Assessment Scheme (IQAS)* DNA proficiency test, and the Collaborative Testing Services (CTS) Forensics Testing Program offers similar proficiency tests. These proficiency tests may include a selection of sample types (neat or mixture) for serological tests and/or DNA analysis (autosomal STR, Y-STR, and mitochondrial DNA). The College of American Pathologists (CAP) has a paternity testing proficiency program offering external proficiency tests. In Europe, the *German DNA Profiling Group (GEDNAP)* provides proficiency testing for participating European laboratories.

26.6 Certification

Certification is a voluntary process that recognizes the attainment of professional qualifications needed for practice in forensic services. Certification is not required, but is desired by some laboratories. In 2009, a report entitled Strengthening Forensic Science in the United States: A Path Forward was published by the Committee on Identifying the Needs of the Forensic Science Community at the National Academy of Sciences. The report recommends that certification should be mandatory for forensic science professionals, which includes written examinations, supervised practice, proficiency testing, and compliance to a code of ethics (Section 26.8). In the United States, the *American Board of Criminalistics (ABC)* offers three types of certification for forensic scientists. A diplomate must pass a general knowledge examination. ABC also requires a bachelor's degree in a natural science and 2 years of experience in a forensic laboratory. To obtain fellow status (higher than diplomate status), an applicant must have 2 years of experience in his or her specialty and must have met the diplomate requirements in addition to passing a written specialty examination and a proficiency test.

ABC has added a third certification that is completely separate from the other certification programs. The technical specialist certification for molecular biology was created to recognize the qualifications required for the analysis of biological materials through DNA profiling. An applicant must take a specialist examination containing questions from the general knowledge examination and a subset of questions from the forensic biology fellow examination. This certification also requires a bachelor's degree in a natural science and 2 years of experience along with successful completion of a proficiency examination within 12 months of taking the technical specialist certification examination.

26.7 Forensic DNA Analyst Qualifications

DAB's Quality Assurance Standards for Forensic DNA Testing Laboratories require that an examiner or analyst have "at a minimum a BA/BS degree or its equivalent degree in a biology, chemistry or forensic science-related area and must have successfully completed college course work (graduate or undergraduate level) covering the subject areas of biochemistry, genetics and molecular biology," as well as "course work and/or training in statistics and population genetics as it applies to forensic DNA analysis."

Additionally, SWGDAM Training Guidelines discuss the course work requirements in detail. At least nine cumulative semester hours are required for the course work covering the required subject areas. Particularly, the section in Fundamental Scientific Knowledge of the SWGDAM Training Guidelines defines required key elements to be covered in course work. The key elements also aid in evaluating the contents of the course work.

Biochemistry "refers to the nature of biologically important molecules in living systems, DNA replication and protein synthesis, and the quantitative and qualitative aspects of cellular metabolism," and may include but is not limited to:

- Structure and function of cellular components such as proteins, carbohydrates, lipids, nucleic acids, and other biomolecules
- Chemistry of enzyme-catalyzed reactions
- Metabolism
- DNA and RNA
- Protein synthesis
- Cell membrane transport
- Signal transduction

Genetics "refers to the study of inherited traits, genotype/phenotype relationships, and population/species differences in allele and genotype frequencies," and may include but is not limited to:

- Heredity
- Function of genes
- Gene expression
- Recombinant DNA
- Mitosis/meiosis

Molecular biology "covers theories, methods, and techniques used in the study and analysis of gene structure, organization, and function," and may include, but is not limited to:

- Interrelationship of DNA, RNA, and protein synthesis
- Central dogma

- Transcription, translation, replication
- Recombinant DNA techniques
- PCR
- Cloning

Course work that is similar in content and scope to those described above may be qualified as required course work. Compliance with the required course contents can be evaluated through transcripts, syllabi, and letters from the instructors.

DAB's Quality Assurance Standards for Forensic DNA Testing Laboratories also describes requirements for an analyst to meet prior to initiating independent casework analysis using DNA technology. First, an analyst must have at least 6 months of forensic DNA laboratory experience "including the successful analysis of a range of samples typically encountered in forensic casework." Second, an analyst must have "successfully completed a qualifying test before beginning independent casework responsibilities."

26.8 Code of Ethics of Forensic Scientists

Ethical codes for forensic scientists are used as guides for individuals making their decisions in distinguishing the difference between correctness and incorrectness. These codes are usually adopted by forensic organizations to regulate the profession. Failure to comply with a code of forensic practice can raise doubt in an individual's fitness for providing forensic services. Additionally, it may result in the expulsion of an individual from a forensic organization. A typical document of ethical codes for forensic scientists generally contains five sections: (1) ethics relating to the scientific method, (2) ethics relating to opinions and conclusions, (3) ethical aspects of court presentation, (4) ethics relating to the general practice of forensics, and (5) ethical responsibilities to the profession.

For the first section, the ethical application of scientific methods is discussed in detail. True scientific methods should be utilized to adequately analyse all the evidence. Such analyses should not be conducted by secret processes. Conclusions should be drawn from the analyses of evidence that appears representative, typical, or reliable.

Moreover, ethical standards relating to opinions and conclusions are very important to the forensic profession. Conclusions should always be drawn from the application of proven scientific methods. The purpose of experimental design and the interpretation of results is to reveal facts. During an analysis, experimental controls should always be utilized. If necessary, the results of analyses should be verified by repeating the analysis or using additional techniques. Explanations should be provided where inconclusive results are obtained. The opinion provided by a forensic scientist should be unbiased and should not be influenced by matters unrelated to the specific evidence in question. A forensic scientist should not choose the interpretation that is in favor of the side of his or her employer.

Pertaining to the courtroom, ethical standards relating to expert witness testimonies is a crucial component for a forensic scientist. An expert opinion is defined as a formal consideration of a subject within an individual's knowledge, skill, education, training, and experience. A forensic scientist should not extend an opinion beyond his or her competence. Appropriate terms should be used to represent the degrees of certainty of an expert opinion. A forensic scientist should not only present the evidence that supports the view of the side of his or her employer. When testifying as an expert witness, language that can be understood by lay jurors should be used to avoid misinterpretation.

Furthermore, in terms of the general practice of forensic science, a few things should be mentioned here. A forensic scientist in private practice should set a reasonable fee for the services provided, which should not be rendered on a contingency fee basis. When a different opinion is presented by another analyst in the reexamination of evidence, both analysts should resolve their contradiction before the trial. A forensic scientist may serve, in an advisory capacity, either the prosecutor or defense in the cross-examination of another expert.

Lastly, ethical responsibilities to the profession are an important component of ethical codes. Information regarding new developments or techniques of forensic analysis should be made available to other forensic scientists. Likewise, any information regarding methods in use that may appear unreliable should be brought to the attention of others. Individual forensic scientists should not seek publicity for the association of their name with specific cases, developments, publications, or organizations solely for the purpose of gaining personal prestige.

Bibliography

Andersen, J., et al., Report on the third EDNAP collaborative STR exercise. *Forensic Sci Int*, 1996, **78**(2): 83–93.

Balazic, J. and I. Zupanic, Quality control and quality assurance in DNA laboratories: Legal, civil and ethical aspects. *Forensic Sci Int*, 1999, **103**: S1–S5.

Bar, W., et al., DNA recommendations. Further report of the DNA commission of the ISFH regarding the use of short tandem repeat systems. International Society for Forensic Haemogenetics. *Int J Legal Med*, 1997, **110**(4): 175–176.

Buckleton, J., Validation issues around DNA typing of low level DNA. *Forensic Sci Int Genet*, 2009, **3**(4): 255–260.

Budowle, B., et al., A perspective on errors, bias, and interpretation in the forensic sciences and direction for continuing advancement. *J Forensic Sci*, 2009, **54**(4): 798–809.

Carracedo, A., et al., DNA commission of the International Society for Forensic Genetics: Guidelines for mitochondrial DNA typing. *Forensic Sci Int*, 2000, **110**(2): 79–85.

Carracedo, A., et al., Results of a collaborative study of the EDNAP group regarding the reproducibility and robustness of the Y-chromosome STRs DYS19, DYS389 I and II, DYS390 and DYS393 in a PCR pentaplex format. *Forensic Sci Int*, 2001, **119**(1): 28–41.

Cormier, K., L. Calandro, and D.J. Reeder, Evolution of the quality assurance documents for DNA laboratories. *Forensic Mag*, 2005, 2(1): 16–19.

Council, N.R., *NRC II: The Evaluation of Forensic DNA Evidence*, 1996. Washington, D.C.: National Academy Press.

Coyle, H.M., The importance of scientific evaluation of biological evidence—Data from eight years of case review. *Sci Justice*, 2012, **52**(4): 268–270.

DNA Advisory Board, Federal Bureau of Investigation, US Department of Justice, Quality assurance standards for forensic DNA testing laboratories. *Forensic Sci Commun*, 2000, **2**(3). Retrieved from http://www.fbi.gov/about-us/lab/forensic-science-communications/fsc/july2000/codis2a.htm.

Duewer, D.L., K.L. Richie, and D.J. Reeder, RFLP band size standards: NIST standard reference material 2390. *J Forensic Sci*, 2000, **45**(5): 1093–1105.

Duewer, D.L., et al., NIST mixed stain studies #1 and #2: Interlaboratory comparison of DNA quantification practice and short tandem repeat multiplex performance with multiple-source samples. *J Forensic Sci*, 2001, **46**(5): 1199–1210.

Frank, W.E., et al., Validation of the AmpFlSTR Profiler Plus PCR amplification kit for use in forensic casework. *J Forensic Sci*, 2001, **46**(3): 642–646.

Gill, P., et al., Report of the European DNA profiling group (EDNAP)—Towards standardisation of short tandem repeat (STR) loci. *Forensic Sci Int*, 1994, **65**(1): 51–59.

Gill, P., et al., Report of the European DNA profiling group (EDNAP): An investigation of the complex STR loci D21S11 and HUMFIBRA (FGA). *Forensic Sci Int*, 1997, **86**(1–2): 25–33.

Gill, P., et al., Report of the European DNA profiling group (EDNAP)—An investigation of the hypervariable STR loci ACTBP2, APOAI1 and D11S554 and the compound loci D12S391 and D1S1656. *Forensic Sci Int*, 1998, **98**(3): 193–200.

Gill, P., et al., DNA commission of the International Society of Forensic Genetics: Recommendations on forensic analysis using Y-chromosome STRs. *Int J Legal Med*, 2001, **114**(6): 305–309.

Gill, P., et al., DNA commission of the International Society of Forensic Genetics: Recommendations on the interpretation of mixtures. *Forensic Sci Int*, 2006, **160**(2–3): 90–101.

Gill, P., et al., DNA commission of the International Iociety of Forensic Genetics: Recommendations on the evaluation of STR typing results that may include drop-out and/or drop-in using probabilistic methods. *Forensic Sci Int Genet*, 2012, **6**(6): 679–688.

Gusmao, L., et al., DNA Commission of the International Society of Forensic Genetics (ISFG): An update of the recommendations on the use of Y-STRs in forensic analysis. *Forensic Sci Int*, 2006, **157**(2–3): 187–197.

Holt, C.L., et al., TWGDAM validation of AmpFlSTR PCR amplification kits for forensic DNA casework. *J Forensic Sci*, 2002, **47**(1): 66–96.

Houck, M., et al., The balanced scorecard: Sustainable performance assessment for forensic laboratories. *Sci Justice*, 2012, **52**(4): 209–216.

Jarman, P.G., S.L. Fentress, and D.E. Katz, Mitochondrial DNA validation in a state laboratory. *J Forensic Sci*, 2009, **54**(1): 95–102.

Kimpton, C., et al., Report on the second EDNAP collaborative STR exercise. European DNA profiling group. *Forensic Sci Int*, 1995, **71**(2): 137–152.

Kline, M.C., et al., Interlaboratory evaluation of short tandem repeat triplex CTT. *J Forensic Sci*, 1997, **42**(5): 897–906.

Kline, M.C., et al., NIST mixed stain study 3: DNA quantitation accuracy and its influence on short tandem repeat multiplex signal intensity. *Anal Chem*, 2003, **75**(10): 2463–2469.

Kline, M.C., et al., Production and certification of NIST standard reference material 2372 human DNA quantitation standard. *Anal Bioanal Chem*, 2009, **394**(4): 1183–1192.

LaFountain, M.J., et al., TWGDAM validation of the AmpFlSTR Profiler Plus and AmpFlSTR COfiler STR multiplex systems using capillary electrophoresis. *J Forensic Sci*, 2001, **46**(5): 1191–1198.

Lanning, K.A., et al., Scientific working group on materials analysis position on hair evidence. *J Forensic Sci*, 2009, **54**(5): 1198–1202.

Laux, D.L., Development of biological standards for the quality assurance of presumptive testing reagents. *Sci Justice*, 2011, **51**(3): 143–145.

Levin, B.C., et al., Comparison of the complete mtDNA genome sequences of human cell lines—HL-60 and GM10742A—From individuals with pro-myelocytic leukemia and leber hereditary optic neuropathy, respectively, and the inclusion of HL-60 in the NIST human mitochondrial DNA standard reference material—SRM 2392-I. *Mitochondrion*, 2003, **2**(6): 387–400.

Lincoln, P.J., DNA recommendations—Further report of the DNA commission of the ISFH regarding the use of short tandem repeat systems. *Forensic Sci Int*, 1997, **87**(3): 181–184.

Marquez, M.C., Interpretation guidelines of mtDNA control region sequence electropherograms in forensic genetics. *Methods Mol Biol*, 2012, **830**: 301–319.

Micka, K.A., et al., TWGDAM validation of a nine-locus and a four-locus fluorescent STR multiplex system. *J Forensic Sci*, 1999, **44**(6): 1243–1257.

Morling, N., et al., Paternity testing commission of the International Society of Forensic Genetics: Recommendations on genetic investigations in paternity cases. *Forensic Sci Int*, 2002, **129**(3): 148–157.

Ong, B.B. and N. Milne, Quality assurance in forensic pathology. *Malays J Pathol*, 2009, **31**(1): 17–22.

Patterson, D. and R. Campbell, The problem of untested sexual assault kits: Why are some kits never submitted to a crime laboratory? *J Interpers Violence*, 2012, **27**(11): 2259–2275.

Peterson, J.L., et al., The feasibility of external blind DNA proficiency testing. I. Background and findings. *J Forensic Sci*, 2003, **48**(1): 21–31.

Peterson, J.L., et al., The feasibility of external blind DNA proficiency testing. II. Experience with actual blind tests. *J Forensic Sci*, 2003, **48**(1): 32–40.

Presley, L.A., The evolution of quality standards for forensic DNA analyses in the United States. *Profiles DNA*, 1999, 3(2): 10–11.

Prieto, L., et al., 2006 GEP-ISFG collaborative exercise on mtDNA: Reflections about interpretation, artefacts, and DNA mixtures. *Forensic Sci Int Genet*, 2008, **2**(2): 126–133.

Prieto, L., et al., GHEP-ISFG proficiency test 2011: Paper challenge on evaluation of mitochondrial DNA results. *Forensic Sci Int Genet*, 2013, **7**(1): 10–15.

Prinz, M., et al., DNA commission of the International Society for Forensic Genetics (ISFG): Recommendations regarding the role of forensic genetics for disaster victim identification (DVI). *Forensic Sci Int Genet*, 2007, **1**(1): 3–12.

Rand, S., M. Schurenkamp, and B. Brinkmann, The GEDNAP (German DNA profiling group) blind trial concept. *Int J Legal Med*, 2002, **116**(4): 199–206.

Rand, S., et al., The GEDNAP blind trial concept part II. Trends and developments. *Int J Legal Med*, 2004, **118**(2): 83–89.

Reardon, S., Faulty forensic science under fire. *Nature*, 2014, **506**(7486): 13–14.

Reeder, D.J., Impact of DNA typing on standards and practice in the forensic community. *Arch Pathol Lab Med*, 1999, **123**(11): 1063–1065.

Roewer, L. and M. Geppert, Interpretation guidelines of a standard Y-chromosome STR 17-plex PCR-CE assay for crime casework. *Methods Mol Biol*, 2012, **830**: 43–56.

Schneider, P.M., Scientific standards for studies in forensic genetics. *Forensic Sci Int*, 2007, **165**(2–3): 238–243.

Schneider, P.M., et al., Results of collaborative study regarding the standardization of the Y-linked STR system DYS385 by the European DNA Profiling (EDNAP) group. *Forensic Sci Int*, 1999, **102**(2–3): 159–165.

Scott, C.L., Believing doesn't make it so: Forensic education and the search for truth. *J Am Acad Psychiatry Law*, 2013, **41**(1): 18–32.

SWGDAM (Scientific Working Group on DNA Analysis Methods), Revised validation guidelines. *Forensic Sci Commun*, 2004, **6**(3). Retrieved from http://www.fbi.gov/about-us/lab/forensic-science-communications/fsc/july2004/index.htm/standards/2004_03_standards02.htm.

Szibor, R., et al., Cell line DNA typing in forensic genetics—The necessity of reliable standards. *Forensic Sci Int*, 2003, **138**(1–3): 37–43.

Tully, G., et al., Considerations by the European DNA profiling (EDNAP) group on the working practices, nomenclature and interpretation of mitochondrial DNA profiles. *Forensic Sci Int*, 2001, **124**(1): 83–91.

Tully, G., et al., Results of a collaborative study of the EDNAP group regarding mitochondrial DNA heteroplasmy and segregation in hair shafts. *Forensic Sci Int*, 2004, **140**(1): 1–11.

Wallin, J.M., et al., TWGDAM validation of the AmpFISTR blue PCR amplification kit for forensic casework analysis. *J Forensic Sci*, 1998, **43**(4): 854–870.

Welch, L.A., et al., European Network of Forensic Science Institutes (ENFSI): Evaluation of new commercial STR multiplexes that include the European Standard Set (ESS) of markers. *Forensic Sci Int Genet*, 2012, **6**(6): 819–826.

Index

A

ABO blood group system, 331–332
  antigens, biosynthesis, 333
  A and B antigens, inheritance, 336
  molecular basis, 333–335
  secretors, 335–336
Absorption–elution assay, 338, 339
Acid phosphatase (AP)
  colorimetric assays, 261–264
    chemical structures, 262
    MUP assay, 262, 264
    α-naphthylphosphate, 262–264
    removal of phosphate, 261
  fluorometric assays, 264, 265
  in semen, 259
  vaginal, identification, 293–294
Aerosol tuberculosis pathogens, 3
Agarose, 160–161; see also DNA electrophoresis
  chains, 161
  chemical components, 161
  gel electrophoresis, 360
Agglutination, 197–198
  based assays, 208–210
    direct, 208, 210
    inhibition assays, 208
    passive, 208–210
  initial binding, 198
  lattice formation, 198
Albumin, 246
Alkaline phosphatase, 180
Allele-specific oligonucleotide (ASO)
  assay, 466
  hybridization technique, 440–442
    biotinylated primers, DNA amplification, 440
    DNA sample, PCR-amplified fragments, 440
    HLA-DQA1 locus, DNA typing, 441, 442
    PCR product, 440–441
    reverse blot format, 440, 441
Alternate light source (ALS), 8, 279
Alu elements, 68
Amelogenin locus, 416–418
  AMELY gene, structure, 416, 417–418
  human X chromosome and sex-typing loci, 416, 417
  human Y chromosome and sex-typing loci, 416, 417
  makers and homologous genes, 416
  sex typing, 416, 418
American Board of Criminalistics (ABC), 526
American Society of Crime Laboratory Directors (ASCLD/LAB), 525
Amplification, 143–150
  cycle parameters, 147–149
    annealing temperature, 147
    denaturation, 147
    first cycle of PCR, 148
    PCR cycling protocols, 147

    temperature parameters, 147
    thermal cycler, 148–149
  denaturation and renaturation of DNA, 143
  PCR (see Polymerase chain reaction (PCR))
  RT PCR for RNA-based assays, 150–153
Amplified fragment length polymorphism (AFLP), 363–364
  DNA polymorphism, 70
  D1S80 loci, 363–364
  PCR amplification, 363
Amylase(s), 277–285
  α-amylase, 281–285
    ELISA, 284–285
    human pancreatic α-amylase (HPA), 277, 279
    human salivary α-amylase (HSA), 277, 279
    immunochromatographic assays, 281–284
    RNA-based assays, 285
  β-amylases, 277
  colorimetric assays, 281, 282
  starch, 277–281
    glucose polymers (amylose and amylopectin), 277, 278
    iodine (I_2) test, 279
    radial diffusion assay, 280
    starch-iodine assay, 279–281
    Y-chromosome-specific quantitative PCR assays, 281
Anthropology, forensic, 58–60
  comparing the striations, 59
  scanning electron microscope, 59
  skeletal remains, 58
  trace evidence, 60
Antibodies, 245–247
  antigen-antibody binding, 247
  antihuman hemoglobin (Hb), 246
  glycophorin A (GPA), 246
  specificity, 246–247
  titration of, 246
  types, 245–246
Antigen–antibody binding reactions, 193–198
  affinity, 193, 194
  avidity, 193, 194
  primary reactions, 194, 195
  secondary reactions, 194–198
    agglutination, 197–198
    postzone, 197
    prozone, 196
    zone of equivalence, 196–197
Antigen polymorphism, 65–66
Antiglobulins, 192
Avian myeloblastosis virus (AMV), 150, 151
Azoospermia, 257

B

Benzidine, 234–236
Biological evidence, 3–7, 18–20, 53–57

collection, 18–20
    bloodstain pattern evidence, 18
    control (known or blank) samples, 20
    handling sharp objects, 19
    labels with the chain of custody, 19
    multiple analysis of evidence, 20
    size of stain, 20
    trace evidence, 20
    wet evidence, 20
  laboratory analysis, 53–57
    comparison of individual characteristics, 54–57
    crime scene investigation, 54
    digital imaging system, 57
    forensic biology laboratory, 56
    forensic DNA profiling, 54
    gas chromatograph, 56
    individual characteristics, comparison, 54
    latent fingerprints by dusting methods, 55
    reporting results and expert testimony, 55–57
    striation marks on fired bullets, 57
    tissue samples, 57
  recognition, 3–7
    case-to-case linkage, 7
    corpus delicti (body of crime), 3
    Locard exchange principle, 6
    log sheet, documenting authorized personnel, 7
    modus operandi (MO), 6
    victim-to-perpetrator linkage, 6
  sources, 77–102
    bodily fluids, 77–81
    cells, 81–86
    tissues, 87–102
Biotinylation, DNA, 180–182
  biotin, 180
  colorimetric and chemiluminescent reactions, 182
  forensic DNA testing, 181
  nonradioactive tags, labeling and detecting nucleic
    acids, 181
  reporter enzyme assay, 181–182
Blood, identification, 231–241
  biological properties, 231–232
    blood cells, 231, 232
    clots, 35
    composition, 231, 232
    plasma, 35
    platelets, 231
    red blood cells, 231
    white blood cells, 231
  confirmatory assays, 239–241
    chromatographic and electrophoretic
      methods, 240
    immunological methods, 241
    microcrystal assays, 239–240
    RNA-based assays, 241
    spectrophotometric methods, 240–241
  presumptive assays, 232–239
    chemiluminescence and fluorescence assays,
      236–238
    colorimetric assays, 234–236
    mechanisms of presumptive assays, 232–234
    oxidants, 238–239
    oxidation–reduction reactions, 234

    plant peroxidases, 239
    reductants, 239
Blood group typing, 331–338
  ABO blood group system, 331–332
    biosynthesis of antigens, 333, 334
    inheritance of A and B antigens, 336
    molecular basis of ABO system, 333, 334–335
    secretors, 335–336
  absorption–elution assay, 338
  blood groups, 331, 332
  forensic applications, 336
  Lattes crust assay, 336–338
Bloodstains, 44–51
  evidence, chemical enhancement and
    documentation, 37–38
    documenting bloodstain patterns, 37
    photographic documentation, 38
    photographs, 38
    videotaping, 38
  formation, 35–37
    of a blood drop from a source, 36
    morphologies of falling blood drops, 37
    surface tension and stains, 37
    viscosity of blood, 35
  passive, 44–45
    drip pattern, 44
    flow pattern, 45
    pool and bubble ring patterns, 45
    splash pattern, 44
  projected, 46–48
    bloody impressions, 46–47
    cast-off pattern, 47
    expiration pattern, 47
    forward spatter, 47–48
  transfer, 46–51
    cast-off patterns, 50
    expiration pattern, 50, 51
    impact bloodstain pattern, 50
    perimeter stains, 49
    swipe pattern, 46
    wipe pattern, 47–48
Bodily fluids, 77–81
  blood components, 77
  extracellular nucleic acids, 77–80
    apoptosis, 80
    apoptotic bodies, 79
    biosafety cabinet, 79
    cell-free nucleic acids, 79
    exosomes and microvesicles, healthy cells, 79, 80
    extracellular vesicles (EVs), 79
    multivesicular bodies (MVBs), 80
    nucleic acids, 77
    types of evidence samples, 77, 78
  transcellular fluids, 77
Boiling lysis and chelation, extraction, 118
  boiling, 117
  centrifugation, 117
  washing, 117
Bone, 92–98
  bone matrix, 96
  collagens, 93
  cortical bone, 97

DNA in cortical bone, 95
epiphysis, 93
human rib bone fragments, 98
marrow cavity, 92
osteon, 97
skeletal remains, 98
source of DNA evidence, 94–98
structure, 96
Bovine serum albumin (BSA), 149
Bowman's capsule, 307
Buccal epithelial cells, 279, 280

C

Cambridge reference sequence (CRS), 461
Capillary electrophoresis, 164–165
    ABI PRISM 310® Genetic Analyzer, 166
    capillary array systems, 165
    components, 166
    electrophoretic separation models, 164
    instruments with detection systems, 165
    linear polydimethylacrylamide, 164
    multicapillary electrophoresis instrument, 166
    vertical slab polyacrylamide gel, 165
Cells, 81–86
    carbohydrate-containing glycoproteins and
            glycolipids, 81
    cell surface markers, 81
    cytosol, 84–85
        messenger RNAs, 84–85
        microRNAs (miRNAs), 85–86
    membrane, 81
    mitochondria and other organelles, 83–84
    nucleated cells, 81–83
    and tissue disruption, 111–114
        automated mechanical disruption, 113
        barocycler, 113
        cryogenic grinding, 114
        DNA polynucleotide chain, 112
        structure of a nucleotide, 112
Certification, 526–527
    American Board of Criminalistics (ABC), 526
    technical specialist certification for molecular
            biology, 527
Chain-termination or Sanger method, 468–470
Chemiluminescence and fluorescence assays, 236–238
    fluorescin, 238
    luminol (3-aminophthalhydrazide), 237–238
Christmas tree stain, 265, 266
Chromatographic and electrophoretic methods, 240
Code of ethics, forensic scientists, 528–529
    five sections, 528
    forensic scientist, 528–529
Colorimetric assays, 234–236
    benzidine and derivatives, 234–236
    leucomalachite green (LMG) assay, 234
    phenolphthalin assay, 234
Combined DNA Index System (CODIS), 70, 485
    database, growth, 490
    indexes, 488–489
        CODIS-compatible DNA profile, 489
        example, 488

forensic profiles, 487
National Missing Person DNA Database
        (NMPDD), 487
rapid DNA instruments, 488–489
Rapid DNA Program Office, 488
infrastructure, 485–487
    Local DNA Index System (LDIS), 485
    National DNA Index System (NDIS), 485
    State DNA Index System (SDIS), 485
Complement fixation, 194
Counterimmunoelectrophoresis (CIE), *see* Crossed-over
        immunoelectrophoresis
Crime scene bloodstain pattern analysis, 35–51
    analyzing spatter stains, 39–44
    chemical enhancement and documentation, 37–38
    formation of bloodstains, 35–37
    human blood, biological properties, 35
    types of bloodstain patterns, 44–51
Crime scene investigation, biological evidence, 3–30
    barricades, 6
    barrier tape, 4
    chain of custody, 17–18
    collection of biological evidence, 18–20
    crime scene unit vehicle, 4
    documentation, 16–18
    final survey and release, 25–29
        alleged diluted blood, 27
        blood cards, 26
        dried bodily fluid stains, 27
        druggist's fold, example, 28
        evidence pouch, 28
        FTA filter paper, application, 26
        knife, photographic documentation, 27
        proper marking of sealed evidence, 29
    investigation team, 4
    marking evidence, 20
    packaging and transportation, 21–25
    police officer, 5
    privacy screen and tent, 5
    protection of crime scene, 3
        biosafety procedures, 3
        contamination, prevention of, 3
    recognition of biological evidence, 3–7
    reconstruction, 29–30
        based on the information, 29
        examples, 29–30
        scientific process, 30
    searches, 8–15
Crossed immunoelectrophoresis (CRIE), 207
Crossed-over electrophoresis procedure, 251, 253–255
Crossed-over immunoelectrophoresis, 208
C stretches, 465
Cycle sequencing technique, 470–471

D

DAB's Quality Assurance Standards for Forensic DNA
        Testing Laboratories, 528
Dane's staining method, 293
D-dimer assay, 299–300
    agglutination process, 299
    ELISA, 299

Index

fibrin polymer, formation and degradation, 299
immunochromatographic assays, 299, 300
menstrual blood samples, identification, 299
Denaturation and renaturation, DNA, 143–144, 163
  base pairing, 144
  DNA double helix, 143
  hydrogen bonding, 143
  melting curve, 144
  melting temperature (Tm), 143
  nucleotide content, 143
  single strands in solution, 143
Deoxynucleoside triphosphates (dNTPs), 147
Detection methods, 175–186
  direct detection of DNA in gels, 175–177
  DNA probes in hybridization-based assays, 177–182
  for PCR-based assays, 182–186
Differential extraction, 121–122
  differential extraction process, 121
  DTT reaction, 121
  sperm plasma membrane, 121
Dimethylaminocinnamaldehyde (DMAC) assay, 308, 312
Displacement loop (D loop), 462, 464
Dithiothreitol (DTT), 114
DNA Advisory Board (DAB), 524
DNA electrophoresis, 159–168
  apparatus and forensic applications, 163–168
    capillary electrophoresis, 164–165
    microfluidic devices, 165–168
    slab gel electrophoresis, 163–164
  principles, 159
    electrical potential, 159
    electrophoretic mobility of macromolecules, 159
    short tandem repeat (STR) analysis, 159
  size, estimation, 168–171
    of DNA fragment, 169
    of electromobility, 169
    local Southern method, 170–171
    relative mobility, 168–169
    of unknown DNA fragment sizes, 170
  supporting matrices, 160–163
    agarose, 160–161
    matrix, 160
    nucleic acid separation mechanisms, 160
    polyacrylamide, 161–163
DNA evidence, 89–102
  hair, 91–94
    dendritic processes, 92
    hair root of a pulled hair, 94
    melanocytes, 92
    mtDNA sequence, 91
    nuclear DNA analysis, 91
    scalp hair cycle, 92
    scalp hair follicle, 93
  skin, 87
    cigarette butt, 89
    worn glove, 89
  teeth, 100–102
    dental pulp tissue, 101
    dissecting teeth for DNA isolation, 102
    extraction of pulp tissue, 102
    odontoblast processes, 101
    tools for dissecting teeth, 101

DNA extraction, 111–115
  cell and tissue disruption, 111–114
  contamination, 115
  differential extraction, 121–122
  ethylenediaminetetraacetic acid (EDTA), 114
  extraction by boiling lysis and chelation, 117–118
  extraction reagent blanks, 115
  lysis of cellular and organelle membranes, 114
  mercaptoethanol or dithiothreitol (DTT), 114
  with phenol–chloroform, 115–117
  removal of proteins and cytoplasmic
    constituents, 114
  silica-based extraction, 118–120
  storage of DNA solutions, 115
  TE buffer, 115
DNA fingerprinting, 357
DNA Identification Act, 524
DNA in gels, direct detection, 175–177
  fluorescent intercalating dye staining, 175–176
    destaining to reduce nonspecific staining, 175
    DNA-dye complex emitting fluorescence on
      UV-light exposure, 176
    ethidium bromide, 175
    intercalating agents, nucleic acid stains, 176
    staining of DNA in agarose gels, 175
    SYBRr stains, 175
  silver staining, 175–177
    of DNA, 177
    electrophoretically separated DNA fragments, 175
    process, 176
    variable number tandem repeat (VNTR),
      method, 175–176
DNA methylation assays, bodily fluid identification,
    210–212
  bisulfite sequencing, 211–212
  MeDIP, 211, 212
  methylation-sensitive restriction enzyme, 211
  MSRE-PCR, 210–211
  TDMRs, 210–212
DNA polymorphism, 66–70
  AFLP, 70
  Alu elements, 68
  cis-regulatory sequences, 66
  CODIS, 70
  exons, 66
  genes and related sequences, 66–67
  homozygous, 69
  human DNA polymorphic markers, 68–69
  intergenic noncoding sequences, 66–68
  introns, 66
  minisatellites, 67
  mitochondrial DNA (mtDNA) profiling, 70
  mobile elements classification, 67
  PCR technique, 69
  profiling, 69–70
  promoter sequences, 66
  retrotransposition, 67
  retrotransposons, 67
  tandem repeats, 66
DNA probes, hybridization-based assays, 177–182
  biotinylation, colorimetric reporting systems,
    180–182

enzyme-conjugated probe, chemiluminescence
  reporting system, 180
radioisotope labeled probes, 177–179
DNA profiles, 491–492
DNA quantitation, 133–140
  fluorescent intercalating dye assay, 134–137
  quantitative PCR assay, 137–140
  slot blot assay, 133–134, 135
Documentation, 16–17
  photographic, bloodstain patterns, 17
  photograph log sheet, 17, 18
  sketches, photographs and videographs, 16
Double immunodiffusion assays, 251
  Ouchterlony assay, 251
  ring assay, 251
Down's syndrome, 380–381

E

Entomology, forensic, 62
  multiwavelength viewing and imaging device, 62
  photographic documentation, 62
Enzyme-conjugated probe, chemiluminescence
  reporting system, 180
  alkaline phosphatase, 180
  detection system using AP-conjugated probe, 180
  Lumi-Phos Plus, 180
Enzyme-linked immunosorbent assay (ELISA), 201–202,
  265
  α-amylase, 284–285
  antibody-antigen-antibody sandwich, 201, 284
  enzyme-labeled antiglobulin, 202
  prostate-specific antigen, identification, 268
  semenogelin (Sg) in semen, identification, 270
Epitope, 189
Erythrocyte protein polymorphisms, 342–344
  hemoglobin (Hb), 342–344
    fetal hemoglobin, 342–343
    hemoglobin S, 343–344
    isoelectric focusing electrophoresis, 342, 343
    normal and sickle-cell hemoglobin β chains,
      342, 343
  isoenzymes, 342
Erythrocytes, *see* Red blood cells
Ethylenediaminetetraacetic acid (EDTA), 114
European DNA Profiling Group (EDNAP), 525
European minimal haplotype locus set, 410
European Standard Set (ESS), 525
Extracellular matrix (ECM), 302–303

F

Fecal matter, identification, 318–324
  in criminal investigations, 318–319
  digestive system, 319, 320
  fecal bacterial identification, 324
  fecal formation, 319, 320
  macroscopic and microscopic examination,
    320–321
  urobilinoids tests, 322–324
    bilirubin glucuronides in liver, formation,
      322, 323

formation, 322
Schlesinger and Edelman tests, 323
urobilin and stercobilin, 323
Ferroprotoporphyrin, 231, 233
Flinders Technology Associates (FTA) filter paper,
  application, 26
Fluorescence assays, *see* Chemiluminescence and
  fluorescence assays
Fluorescent intercalating dye assay, 134–137
  DNA quantitation using intercalating dye, 136
  intercalating dyes, 134
  Quant-iTTM PicoGreen® dsDNA reagent
    (Invitrogen), 134
Fluorescent resonance energy transfers (FRET), 138
Fluorescin, 238
Forensic biology, 53–70
  development, 65–70
    antigen polymorphism, 65–66
    DNA polymorphism, 66–70
    forensic serology, 66
    protein polymorphism, 66
  laboratory analysis, 53–57
  science services, 57–65
    anthropology, 58–60
    comparison of two items, 58
    dental cast, 64
    entomology, 64–65
    fired hollowpoint bullets, 58
    odontology, 64
    pathology, 58
Forensic DNA analyst qualifications, 527–528
  biochemistry, 527
  DAB's Quality Assurance Standards for Forensic
    DNA Testing Laboratories, 528
  genetics, 527
  molecular biology, 527
  SWGDAM Training Guidelines, 527
Forensic DNA databases, 485–498
  characteristics, 486
  CODIS, 485–489
  database expansion, 490, 491
  DNA profiles, 489, 491–492
  familial searches, 495–498
  history of, 485
  SDIS database, 489
  searches for investigations, 492–495
  SWGDAM, 486
Forensic DNA polymorphism profiling, 69–70
Forensic protein profiling, 338–345
  erythrocyte protein polymorphisms, 342–344
    erythrocyte isoenzymes, 342
    hemoglobin (Hb), 342–344
  matrices, protein electrophoresis, 340
  protein polymorphic markers, 338–339
  separation by isoelectric point, 341
  separation by molecular weight, 340–341
    detergents, 340–341
    reducing agents, 340
  serum protein polymorphisms, 344–345
Forensic scientist, 528–429
Forensic STR analysis, 375–376; *see also* Short tandem
  repeat (STR) profiling

capillary electrophoresis separation, 375
DNA profiling, 371–374
    CODIS, 371
    cytogenetic map, 374
    DNA profile, 371, 372–373
    European Standard Set (ESS), 371
    population match probability (P_m), 371
    second-generation multiplex (SGM), 371
electropherogram, 375
genotypes of STR fragments, 376
    AmpFlSTR® COfiler® PCR Amplification Kit, 376, 377
    AmpFlSTR® Identifiler® Plus Kit, 376, 378
    individual DNA profile (COfiler), 377, 379
    individual DNA profile (Identifiler), 378, 380
profiling results, 376
    exclusion, 376
    inclusion (match), 376
    inconclusive result, 376
relative fluorescence units (RFU), 376

G

Genetics, 527
Genotyping, STR loci, 376–383; *see also* Short tandem repeat (STR) profiling
    amplification artifacts, 382–383
        allelic dropout, 383
        heterozygote imbalance, 383, 384
        nontemplate adenylation, 383
        stuttering, 382
    challenging forensic samples, 384–389
        degraded DNA, 384, 386, 387
        low copy number (LCN) DNA testing, 386
        mixtures, 386–389
    electrophoretic artifacts, 384
        pull-up peaks, 384, 385
        spikes, 384, 385
    mutations, 377–381
        chromosomal and gene duplications, 380–381
        point mutations, 381
        at STR core repeat regions, 377–380
German DNA Profiling Group (GEDNAP), 526
Ground-penetrating radar (GPR), 11

H

Hair, 88–95
    biology, 89–91
        anagen-phase hair follicle, 91
        catagen phase, 90
        dermal papilla, 89
        hair follicles, 88
        hair shaft, 90
        scalp hair follicle, 90
        telogen phase, 90
    evidence collection, swabbing, 88
    melanosome transportation, 95
    scalp and pubic hairs, 88
    source of DNA evidence, 91–94
    telogen hair roots, 95
Hapten, 189

Haptoglobin (Hp), 344
    Hardy–Weinberg principle, 504
Hemagglutination, 198
Hematin crystal assay, 240
Heme, chemical structures, 231, 233
Hemochromagen crystal assay, 239
Hemoglobin (Hb), 231, 233, 342–344
    fetal hemoglobin, 342–343
    hemoglobin S, 343–344
    isoelectric focusing electrophoresis, 342, 343
    normal and sickle-cell hemoglobin β chains, 342, 343
Hepatitis B virus (HBV), 3
Hepatitis C virus (HCV), 3
Heteroplasmy, 463–465
    description, 463–464
    length, 465
    sequence, 464–465
Hexagon OBTI, 247
High-intensity light-emitting diode (LED) device, 10, 13
HLA-DQA1 locus, 438–442
    allele-specific oligonucleotide hybridization, 440–442
    chromosomal locations of SNP, 438, 439
    DQ_α AmpliType and polymarker assays, 438–439
    polymarker system, 438, 439
Human immunodeficiency virus (HIV), 3
Human leukocyte antigen (HLA) family, 438
Human pancreatic α-amylase (HPA), 277, 279
Human salivary α-amylase (HSA), 277, 279
Hybridoma cells, 191, 192

I

Identity-by-state (IBS) method, 496
Immunochromatographic assays, 202–204, 247–250
    α-amylase, 281–284
        anti-HSA antibody, 283–284
        identification of saliva, 281, 283
        RSID®-Saliva kit, 284
    antibody-antigen-antibody sandwich, 202
    high-dose hook effect, 203–204
    human glycophorin A protein, identification, 248–250
        anti-GPA Ab, 248, 250
        RSID™-Blood, 248–249, 250
        structure, 250
    human hemoglobin protein, identification, 247–248
        ABAcard HemaTrace®, 247–248, 249
        antibody-antiglobulin complex, 247, 248
        Hexagon OBTI, 247, 249
    immunochromatographic membrane device, 202–203
    prostate-specific antigen, identification, 265–268
    seminal vesicle-specific antigen, identification, 269–270
Immunodiffusion, 204–206
    double, 205–206
    single, 204–205
Immunoelectrophoresis (IEP), 206–207
Immunoglobulins, 189
    amino acid sequences, 191
    classes, 189
    isotypes, 189
    structures, 190

Immunological methods, 241
Intergenic noncoding sequences, 66
Internal positive control (IPC), 139
International Quality Assessment Scheme (IQAS) DNA
        proficiency test, 526
International quality standards, 525
    European DNA Profiling Group (EDNAP), 525
    European Standard Set (ESS), 525
    Interpol European Working Party on DNA Profiling
        (IEWPDP), 525
Interpol European Working Party on DNA Profiling
        (IEWPDP), 525
Iodine (I_2) test, 279
Isoelectric focusing (IEF) technique, 341, 342

K

Kastle–Meyer assay, 234, 248
17-Ketosteroid, identification, 310–311, 315–316, 317
Kinship Index (KI) method, 496

L

Laboratory accreditation, 525
    American Society of Crime Laboratory Directors
        (ASCLD/LAB), 525
    Quality Assurance Standards Audit for DNA
        Databasing Laboratories, 525
Laboratory validation, 525–526
    accuracy, 526
    developmental validation, 525
    performance check, 526
Lactate dehydrogenase (LDH) assay, 301–302
    colorimetric assay, 301–302
    electrophoretic separation, 301
    isozymes, 301
    Meldola's Blue, reduction and oxidation, 301, 302
Lactobacillus bacteria, 294
Laser capture microdissection (LCM), 265
Lattes crust assay, 336–338
    agglutination patterns, 336–337
    blood group typing, 337
    Landsteiner's experiments, principles, 337
    procedure, 337
    results, 337–338
Leucocytes, *see* White blood cells
Leucomalachite green (LMG) assay, 234, 235
Local DNA Index System (LDIS), 485
Low copy number (LCN) of DNA, 149
Lugol's iodine staining, 290–292
    epithelial cells, 292
    glycogen, structure, 290–291
    glycogen-iodine complex, 292
Luminol (3-aminophthalhydrazide), 237–238

M

Male-specific Y (MSY) region, 407–408
Marking evidence, 20
    identification, 20
    information, log or tag, 20
Maryland's DNA Collection Act, 491

Matrix metalloproteinase (MMP) genes, 302–303
Mendelian genetics, 503–504
    crossing over hypothesis, 503–504
    first law, 503
    probabilities, 503
    product rule of probability, 503
    second law, 503
Menstruation, 295–298
    uterine cycle, 296–298
        changes in functionalis, uterine mucosa, 297–298
        endometrium, phases, 296–297
    uterine endometrial hemostasis, 298
Mercaptoethanol, 114
Messenger RNA (mRNA), 84–85, 212–214
    markers, 212, 213–214
    reference genes, 212
    ribosomal complex, 85
    tissue-specific genes, 84, 212
Methylation-sensitive restriction enzyme (MSRE), 211
Methylation-sensitive restriction enzyme digestion
        polymerase chain reaction (MSRE-PCR),
        210–211
Methyl-DNA immunoprecipitation (MeDIP), 211, 212
4-Methylumbelliferone phosphate (MUP) assay,
        262, 264
Microbial DNA analysis, bodily fluid identification,
        217–220
    concept of microbiome, 218
    DNA markers, 219
    Escherichia coli rRNA operon, 218
    human microbiota, 217–218
    intergenic spacer region (ISR), 219
    metagenomics, 218
    ribosomal RNA (rRNA) genes, 219
Microcrystal assays, 239–240
    hematin crystal assay, 240
    hemochromagen crystal assay, 239
Microfluidic devices, 165–168
    integrated, 168
        data analysis and genotyping, 168
        rapid DNA instrument, 168
    modular, 167–168
        electrophoretic assay, 167
        for gel electrophoresis, 167
        polymer matrix, 168
MicroRNAs (miRNAs), 85–86
    based assays, 214–215
        markers, 214–215
        miRNA responsive element (MRE), 214
        RNA-induced silencing complex (RISC), 214
    biogenesis, 86
    definition, 85
    genomic location and structure, 85
    intergenic genes, 85
    intronic, 85
    monocistronic, 85
    polycistronic, 85
Minimal haplotype loci, *see* European minimal
        haplotype locus set
Minisatellites, 67, 353, 354
Mitochondrial DNA (mtDNA) profiling, 70, 461–474
    DNA sequencing reactions, 468

chain-termination or Sanger method, 468–470
cycle sequencing, 470–471
electrophoresis and mitotype designations, 471–473
deletions, 472
heteroplasmic sites, 472
insertions, 472
reporting format, 471–472
forensic mtDNA testing, 465–474
general considerations, 465–466
human mitochondrial genome, 461–463
genetic contents, 461–463
maternal inheritance, 463, 464
PCR amplification, 467–468
polymorphic regions, 463–465
heteroplasmy, 463–465
hypervariable regions, 463
profiling results, 472–474
screen assay, 466, 467
Mitotype, 463
Mixture interpretation, 386, 388–389
Molecular biology, 527
Moloney strain of the murine leukemia virus
(MMLV), 151
Monoclonal antibody, 191, 192
Multilocal Y-STR loci (MLL), 411, 415
Multilocus probe (MLP) technique, 357–358
Mutations, STR loci, 377–381
chromosomal and gene duplications, 380–381
triallelic or three-peaks pattern, 380–381
trisomy, 380
point mutations, 381
nucleotide substitution mutations, 381
null allele or silent allele, 381
at STR core repeat regions, 377–380
germ-line mutation, 377
somatic mutations, 379–380

N

National DNA Index System (NDIS), 485
National Missing Person DNA Database (NMPDD), 487
National Research Council (NRC), recommendations,
523
Next-generation sequencing (NGS) technologies, 444,
446
chemistry, 447–450
cyclic reversible termination, 447, 450
pyrosequencing technology, 447, 449
sequencing by synthesis and sequencing by
ligation, 447
coverage, 447–451
disadvantage, 450
or sequencing depth, 447, 450
de novo sequencing, 444
enrichment method, 444
PCR-based approach, 444, 446
resequencing, 444
Nondestructive assays, bodily fluids identification, 220
fluorescence spectroscopy, 220
Raman spectroscopy, 220
Non-Mendelian inheritance, 463
Nonrecombining Y (NRY) region, 407

Nonsynonymous SNPs (nsSNPs), 444
Nontemplate addition, 383
Nuclear Fast Red (NFR), 265
Nucleic acid extraction, 111–126
DNA extraction, 115–122
DNA extraction, basic principles, 111–115
essential features of RNA, 122–124
RNA extraction, 124–126
Null allele or silent allele, 381

O

Odontology, forensic, 64
Oligodeoxynucleotide priming, 151
gene-specific primers, 151
oligo (dT) primer, 151
random hexamer primers, 151
Oligospermia, 257
Orthotolidine, 236
Ouchterlony assay, 206, 251–253
identity, 251
nonidentity, 251
partial identity, 251
results, 251, 253
Oxidants, 234, 238–239
Oxidation–reduction reactions, 234

P

Packaging and transportation, crime scene
investigation, 21–25
evidence collection, kits/methods, 21–22, 24
evidence from different sources, 24
fingernail swabbing, 24
folding of evidence, 23
hand bags, 25
liquid evidence, 24
packaged evidence, 25
packing materials, 23
swabs, 23
trace evidence, 25
Pathology, forensic, 63–64
buried human skeletons, 63
forensic pathology facility, 63
human skull, 64
insects found on a dead animal, 64
photographic documentation prior to autopsy, 63
PCR-based assays, detection methods, 182–185
fluorescent dyes, 182–185
applications in STR kits, examples, 183
chemical structures of, 184
direct labeling of nucleic acids, 185
emission spectra of common fluorescent dyes, 183
fluorescent dye-labeled primer, 184
labeled dideoxynucleotides (ddNTPs), 185
fluorophore detection, 185
absorption and emission, 185
charge-coupled device (CCD), 185
definition, 185
excitation spectrum, 185
lasers, excitation sources, 185
matrix, 185

labeling methods, 182–185
Periodic acid–Schiff method, 292
Phadebas reagent, 281
Phenol–chloroform extraction, 115–117
    cell lysis and protein digestion, 115
    concentrating DNA, 115–117
    concentrating DNA solutions, 116
    DNA extraction using organic solvent, 116
    extraction with organic solvents, 115
Phenolphthalin assay, 234, 235
Phosphoglucomutase (PGM), 342
Photographic documentation, 38
Plant peroxidases, 239
Plasmin, 298
Platelets, 231
Polyacrylamide; *see also* DNA electrophoresis
    capillary electrophoresis matrix, 162
    chemical structures, 162
    cross-linking reaction, 161
    electrophoresis, denaturing, 163
    formation of gel, 162
    gel electrophoresis, 363
    gel matrix, 161
    polymerization reaction, 161
Polyclonal antiserum, 191
Polymerase chain reaction (PCR), 143–150; *see also*
        Amplification
    essential components, 145–147
        deoxynucleoside triphosphates (dNTPs), 147
        divalent cations, Mg^{2+}, 147
        multiplex PCR, 147
        other components, 147
        PCR primers, 145–147
        positive control, 147
        Primer3, 146
        primer dimers, 147
        thermostable DNA polymerases, 145–146
    factors affecting, 149–150
        amplification-negative controls, 149
        bovine serum albumin (BSA), 149
        contamination, 149–150
        extraction reagent blanks, 149
        inhibitors, 149
        internal positive control, 149
        low copy number (LCN) of DNA, 149
        stochastic effect, 149
        template quality, 149
    principles, 143–145
        amplicons, 143
        concept of synthesizing DNA, 144
        PCR amplification curve, 145
        PCR amplification curve, S-shaped, 145
    technique, 69
Population genetics, 504–512
    allele frequency, 504
    genotype frequency, 504
    Hardy–Weinberg principle, 504
    heterozygosity, 504
    probability of match, 511–412
    testing for HW proportions of population
        databases, 505
Precipitation-based assays, 204–208

immunodiffusion, 204–206
    double, 205–206
    single, 204–205
    immunoelectrophoretic methods, 206–208
        crossed immunoelectrophoresis (CRIE), 207
        crossed-over immunoelectrophoresis, 208
        immunoelectrophoresis (IEP), 206–207
        rocket immunoelectrophoresis, 207–208
Precipitins, 195
Primary binding assays, 201–204
    enzyme-linked immunosorbent assay (ELISA),
        201–202
    immunochromatographic assays, 202–204
Primer dimers, 147
Proficiency testing, 526
    German DNA Profiling Group (GEDNAP), 526
    International Quality Assessment Scheme (IQAS)
        DNA proficiency test, 526
Prostate-specific antigen (PSA), 259, 267
    identification, 265–268
        ELISA method, 268
        immunochromatographic assays, 265–268
Protein polymorphism, 66
Proteomic approaches, mass spectrometry, 214–215
    analysis strategies for protein identification,
        216–217
        break-then-sort (shotgun approach), 217
        liquid chromatography-mass spectrometry (LC-
        MS) system, 216
        sort-then-break approach, 216–217
    mass spectrometric instrumentation for protein
        analysis, 215–216
        ionization, 215
        mass analyzer, 215
        peptide sequencing, 216
        precursor ions, 216
        product ions, 216
        tandem mass spectrometry, 215–216
Pseudoautosomal regions (PARs), 407

Q

Quality assurance (QA)
    certification, 526–527
    code of ethics of forensic scientists, 528–529
    components, 523
    DNA Advisory Board (DAB), 524
    DNA Identification Act, 524
    forensic DNA analyst qualifications, 527–528
    international quality standards, 525
    laboratory accreditation, 525
    laboratory validation, 525, 526
    National Research Council (NRC),
        recommendations, 523
    NRC II report, 524
    proficiency testing, 526
    SWGDAM, 524
    US quality standards, 523–524
Quality Assurance Standards Audit for DNA Databasing
    Laboratories, 525
Quality Assurance Standards for DNA Databasing
    Laboratories, 524

Index

Quality control (QC), *see* Quality assurance (QA)
Quantitative PCR (qPCR) assay, 137–140
    commercial qPCR kits, 138
    end-point PCR using SYBR green detection, 137
    real-time quantitative PCR, advantages, 137, 140
    TaqMan method, 138–140

R

Radial immunodiffusion assay, 205
Radioisotope labeled probes, 177–179
    autoradiography, 179
    denatured probes, 177
    dNTPs, 178–179
    labeling, 177
Rapidly mutating Y-STR (RM Y-STR) loci, 411–412
Recombination fraction (or recombination frequency),
    411, 415
Red blood cells, 231
Reductants, 239
Reflected ultraviolet imaging system (RUVIS) imager, 15
Restriction fragment length polymorphism (RFLP),
    353–363
    detection, 359–360
    DNA degradation, 360
    electrophoresis and blotting artifacts, 362–363
        bands running off gel, 363
        partial stripping, 362
        separation resolution limits and band shifting,
        362–363
    hybridization with probes, 357–359
    restriction digestion-related artifacts, 361–362
        partial restriction digestion, 361
        point mutations, 362
        star activity, 361–362
    restriction endonuclease digestion, 354–356
        bacterial DNA, 355–356
        forensic DNA testing, 356
        HaeIII, 355
        HinfI and *PstI* sites, 356
        restriction sites, 354, 356
        type I and type III restriction endonucleases, 355
        Type II restriction endonucleases, 354–355
    Southern transfer, 356–357
Retrotransposition, 67
Reverse blot assay, 466
Reverse transcriptase PCR (RT-PCR), 151–153, 241, 295;
    see also Amplification
    amplified products, analyzing, 153
    cDNA synthesis, 151
    end-point PCR, 153
    multiplex assays, 153
    for RNA-based assays, 150–151
        central dogma, 150
        oligodeoxynucleotide priming, 151
        pathway for the flow of genetic information, 150
        RNA synthesis or transcription, 150
        translation, 150
    strategies, 152
    two-step PCR, 153
Reverse transcription, 150–151
    avian myeloblastosis virus (AMV), 150, 151

Moloney strain of the murine leukemia virus
    (MMLV), 151
Revised Cambridge reference sequence (rCRS), 461
Ring assay procedure, 205, 251, 252
RNA-based assays, 241, 270–271
    and RNA profiling, 212–215
        messenger RNA (mRNA)-based assays, 212–214
        microRNA-based assays, 214–215
RNA extraction, 122–126
    features, 122–124
        chemical structure, 122
        mRNA, structure, 123
        ribonucleotide, structure, 122
        ribose and deoxyribose, comparison, 123
        uracil and thymine, comparison, 123
    methods, 124–126
        miRNA extraction, 124–126, 126
        RNA–DNA coextraction, 124
        silica-based method, 125
Rocket immunoelectrophoresis, 207–208

S

Saliva, identification of, 277–285
    α-amylase, 281–285
        ELISA, 284–285
        immunochromatographic assays, 281–284
        RNA-based assays, 285
    amylase activity, determination, 279–281
        colorimetric assays, 281
        starch-iodine assay, 279–281
    amylases, 277–279
    human salivary glands, 277, 278
    visual examination, 279
Sanger sequencing, 467–468, 472
Scientific Working Group on Bloodstain Pattern
    Analysis (SWGSTAIN), 39
Scientific Working Group on DNA Analysis Methods
    (SWGDAM), 376, 486, 524, 527
Searches
    crime scene investigation, 7–15
        alternate light source (ALS), 8
        biological stains, 8
        blood on clothing, 10
        cadaver-sniffing dogs, 12
        compact alternate light source devices, 13
        compact Rapid DNA device, 14
        electrostatic dust print, 9
        enhancement reagents, 15
        fingerprints, 15
        glove, disposable and extended cuff, 8
        grid search pattern, outdoor scene, 11
        ground-penetrating radar (GPR), 11
        high-intensity light-emitting diode (LED) device,
        10, 13
        items of evidence, 7
        LED light sources, 13
        line search pattern, outdoor scene, 11
        personal protection wear and devices, 8
        phenolphthalin and leucomalachite green
        tests, 14
        photographic documentation, 9

portable and field-deployable instruments, 15
reflected ultraviolet imaging system (RUVIS)
imager, 15
scales, photographic documentation, 15
sketch documentation, 14
tracking dog, 12
familial, forensic DNA databases, 495–498
excluding candidates through Y-STR
screening, 498
focusing on rare alleles, 496–497
"Grim Sleeper" case, 498
identity-by-state (IBS) method, 496
Kinship Index (KI) method, 496
legal and ethical issues, 495–496
workflow, example, 497
for investigations, 492–495
case-to-case searches, 493
case-to-offender searches, 492–493
hits and investigations, 493
match stringencies, 494
search stringency and partial matches, 493–495
weekly routine CODIS search, 492
Secondary binding assays, 204–210
agglutination-based assays, 208–210
precipitation-based assays, 204–208
immunodiffusion, 204–206
immunoelectrophoretic methods, 206–208
Semen, identification, 257–268
alternate light sources (ALSs), 260–261
biological characteristics, 257–260
acid phosphatase (AP), 259
prostate-specific antigen (PSA), 259
seminal vesicle-specific antigen (SVSA), 259–260
spermatozoa, 257–259
confirmatory assays, 264–271
microscopic examination of spermatozoa,
264–265
prostate-specific antigen, identification, 265–268
RNA-based assays, 270–271
seminal vesicle-specific antigen, identification,
269–270
presumptive assays, 260–264
acid phosphatase techniques, 261–264
lighting techniques for visual examination,
semen stains, 260–261
Seminal vesicle-specific antigen (SVSA), 259–260
ELISA, 270
immunochromatographic assays, 269–270
Serology, forensic, 66, 199–201
class and individual characteristics, biological
evidence, 199–200
presumptive and confirmatory assays, 200
primary and secondary binding assays, 200–201
scope of, 199
Serology concepts, 189–198
antigen-antibody binding reactions, 194–198
primary reactions, 194
secondary reactions, 194–198
strength of, 193
reagents, 189–192
antibodies, 189–191
antiglobulins, 192

immunogens and antigens, 189, 190
monoclonal antibodies, 191–192
polyclonal antibodies, 191
Serum component, blood, 191
Sex chromosome haplotyping and gender identification,
407–419
sex typing, 416–419
amelogenin locus, 416–418
DXYS156 locus, 419
sex-determining region Y (SRY) gene, 418
steroid sulfatase (STS) gene, 419
TSPY-like (TSPYL) gene, 419
TSPY locus, 418–419
X chromosome haplotyping, 412–416
Y chromosome haplotyping, 407–412
human Y chromosome genome, 407–408
Y-STR, 409–412
Sex-determining region Y (SRY) gene, 418
Short tandem repeat (STR) profiling, 159, 369–389
advantages, 369
core repeat and flanking regions, 369–370
factors affecting genotyping results, 376–383
amplification artifacts, 382–383
electrophoretic artifacts, 384
mutations, 377–381
forensic STR analysis, 375–376
capillary electrophoresis separation, 375
determining the genotypes of STR
fragments, 376
electropherogram, 375
interpretation of STR profiling results, 376
relative fluorescence units (RFU), 376
genotyping of challenging forensic samples, 384–389
degraded DNA, 384–386
low copy number DNA testing, 386
mixtures, 386–389
microsatellites or simple sequence repeats, 369
repeat unit length, 370
repeat unit sequences, 370–371
complex repeats, 370
compound repeats, 370
nonconsensus alleles, 371
simple repeats, 370
STR loci commonly used for forensic DNA profiling,
371–374
Sickle-cell anemia, 344
Silica-based extraction, 118–120
automated DNA sample preparation, 120
cell lysis and protein digestion, 118
DNA adsorption onto silica, 118
DNA extractions, 119
elution of DNA, 119
silica-membrane spin column, 118
washing, 119
Single-locus probe (SLP) technique, 358–359
Single nucleotide polymorphism (SNP) profiling, 437–451
allele-specific hybridization, 444
allele-specific oligonucleotide ligation, 444
assays, SNP typing, 444, 445
characteristics, 437–438
advantages, 437
biallelic SNP, 437

classes, 437, 438
    forensic DNA analysis, 437–438
DNA samples and template preparation, 446–447
    autosomal genome sequencing, 446
    emulsion PCR, 447
    solid-phase PCR, 447, 448
forensic applications, 438–444
HLA-DQA1 locus, 438–442
invasive cleavage method, 444
mitochondrial DNA (mtDNA) profiling, 443
next-generation sequencing (NGS) technologies,
    444–451
potential applications of SNPs for phenotyping, 444
primer extension methods, 444
Skin, 87–88
    biology, 87
    sectional view, 87
    source of DNA evidence, 87, 89
    transfer DNA, 87
Slab gel electrophoresis, 163–164
    agarose gel electrophoresis, 163–164
        horizontal apparatus, 163
        separated DNA samples, bands, 164
    polyacrylamide gel electrophoresis, 164
        cross-contamination, 164
        detection of DNA bands, 164
Slot blot assay, 133–134, 135
    biotinylated probe, 133–134
    colorimetric detection, 134
    D17Z1 probe, 133
    human DNA quantitation, 134
    human genomic DNA, detecting, 133
    slot blot device, 134
Sodium dodecylsulfate (SDS), 339–340
Southern blotting, *see* Southern transfer
Southern transfer, 356–357
Spatter stains, analysis, 39–44
    angles of impact, 40–41
    area of origin, 41–44
    directionality of stains, 39
    impact angle, 41
    string method, 42
    tangent method, 43
    travelling velocities, categories, 39
Species identification, 245–255
    antibody specificity, 246–247
    antigen-antibody binding, 247
    crossed-over electrophoresis, 251–255
    DNA analysis, 245
    double immunodiffusion assays, 251
        Ouchterlony assay, 251
        ring assay, 251
    immunochromatographic assays, 247–250
        human glycophorin A protein, 248–250
        human hemoglobin protein, 247–248
    precipitation-based assays, 245
    titration of antibodies, 246
    types of antibodies, 245–246
Spectrophotometric methods, 240–241
Spermatogenesis, 257, 258
Spermatozoa, 257–265
    description, 257

male reproductive system, anatomy, 257, 258
microscopic examination, 264–265
seminal fluid, 257
spermatogenesis, 257, 258
structures, 257–259
vasectomy, 257
State DNA Index System (SDIS), 485, 489
Statistical analysis, DNA profiling results, 513–518
    genotypes, 513–516
        likelihood ratio, 516
        profile probability, 513–514
        structured populations, 514–516
    haplotypes, 516, 518
        linkage disequilibrium, 516
        mitotype not observed in database, 518
        mitotypes observed in database, 518
Stochastic effect, 149
Stutter ratio, 383
Sweat, identification, 316–318
    biology of perspiration, 316–317
        apocrine sweat glands, 316, 318
        eccrine sweat glands, 316, 318
    dermcidin assays, 318, 319
    presumptive assays, 310–311, 317–318

T

Takayama crystal assay, 239
Tamm–Horsfall protein (THP), 310–311, 315–316
    ELISA, 315
    immunochromatographic assays, 315–316
Tandem repeats, 66
TaqMan method, 138–139
    cycle threshold, 139
    DNA double helix, 138
    fluorescent resonance energy transfers (FRET), 138
    internal positive control (IPC), 139
    real-time PCR, 139
    tetramethyl-rhodamine (TAMRA), 139
Technical specialist certification for molecular biology, 527
Teeth, 98–102
    biology, 98–100
        adult tooth, 100
        cementoblasts, 100
        deciduous or primary teeth, 98
        odontoblast process, 99
        secondary or permanent dentition, 98
        tools for cutting bone samples, 99
    source of DNA evidence, 100–102
Teichmann crystal assay, 240
Tetramethylbenzidine (TMB), 236
Tetramethylrhodamine (TAMRA), 139
Thrombin, 298
Thrombocytes, *see* Platelets
Tissues, 87–102
    bone, 92–98
    hair, 88–92
    skin, 87–88
    teeth, 98–102
Tissue-specific differentially methylated regions
    (TDMRs), 210–212
Trisomy, 380

U

Urine, identification, 307–316
    confirmative assays, 315–316
        17-ketosteroids, 315–316
        Tamm–Horsfall protein, 315
    creatinine, 314
        formation, 314
        Jaffe's assay, 314
    formation, 307–308
        nephron, 307, 309
        reabsorption process, 307
        secretion, 307
        urea, 307–308
        urinary system, 307, 308
    human urinary system, 308
    urea, 308–313
        DMAC assay, 308, 310–311, 312
        fluorometric method, 312, 313
        urease, biochemical reaction, 310–311, 312, 313
Uromodulin, *see* Tamm–Horsfall protein (THP)

V

Vaginal secretions and menstrual blood, identification, 289–303
    acid phosphatase, 293–294
    confirmatory assays, 295–296
        markers of mRNA-based assays, 295, 296
        miRNA Markers, 295, 296
        RT-PCR technique, 295
    D-dimer assay, 299–300
    lactate dehydrogenase assay, 301
    menstruation, 295–298
        uterine cycle, 296–298
        uterine endometrial hemostasis, 298
    RNA-based assays, 302–303
    vaginal bacteria, 294–295
    vaginal stratified squamous epithelial cells, 289–293
        Dane's staining method, 293
        layers of cells, 289–290, 291
        Lugol's iodine staining, 290–292
        Periodic acid–Schiff method, 292
        squamous mucosa and submucosa, 289, 290
Variable number tandem repeat (VNTRs), 175–176, 353–364

W

Wet evidence, 20
White blood cells, 231

X

X-chromosomal STR (X-STR) profiling, 412
X chromosome haplotyping, 412–416

Y

Y chromosome haplotyping, 407–412
    human Y chromosome genome, 407–408
        male-specific Y (MSY) region, 407–408
        patrilineage, 407, 408
        polymorphic sequences, 408
        pseudoautosomal regions (PARs), 407
        structure, 407, 408
    Y-STR, 409–412
        core Y-STR loci, 410–411, 412
        disadvantage, 409
        forensic DNA testing, 409
        genotype of DNA profile, 410, 411
        haplotypes, 410
        human cytogenetic map, 409
        multilocal Y-STR loci, 411
        rapidly mutating, 411–412
        repeat unit length, 410
Yield gel, *see* Agarose gel electrophoresis

AFLP, 363–364
minisatellites, 353, 354
population match probability (P_m), 353
RFLP, 353–363
    detection, 359–360
    factors affecting RFLP results, 360–363
    hybridization with probes, 357–359
    restriction endonuclease digestion, 354–356
    Southern transfer, 356–357
Videotaping, 38
Vomitus, identification, 324–326
    assays, 325–326
    biology of gastric fluid, 324–325
    pepsin-proteolytic assay, 326
    vomiting, act of, 325–326